System
signal

System analysis and signal processing

with emphasis on the use of MATLAB

PHILIP DENBIGH

University of Sussex

Harlow, England • Reading, Massachusetts • Menlow Park, California
New York • Don Mills, Ontario • Amsterdam • Bonn • Sydney • Singapore
Tokyo • Madrid • San Juan • Milan • Mexico City • Seoul • Taipei

Addison Wesley Longman Limited
Edinburgh Gate
Harlow
Essex CM20 2JE
England

and Associated Companies throughout the World.

Typset in 10/12.5 Times by 25
Printed and bound in the United States of America

First printed 1998-Reprinted 1998

ISBN 0-201-17860-5

British Library Cataloguing-in-Publication Data
A catalogue record for this book is available from the British Library

Library of Congress Cataloging-in-Publication Data
Denbigh, Philip.
System analysis and signal processing : with emphasis on the use of MATLAB / Philip Denbigh.
p. cm.
Includes bibliographical references (p. 504) and index.
ISBN 0-201-17860-5 (alk. paper)
1. Signal processing—Computer simulation. 2. MATLAB. I. Title.
TK5102.9.D46 1998
621.382′2—dc21 97-40611
CIP

Contents

Preface

This is intended to be very much a *teaching* book, and the effort in writing it has been directed more at developing clear explanations of difficult concepts than at producing a definitive text. It covers a very broad syllabus on system analysis and continuous and discrete-time signal processing that is aimed at being suited to all years of an undergraduate programme, but also includes a small amount of more advanced material.

The book commences at a much more elementary level than would normally be associated with the later content of the book. The motive behind this is to ensure that the more difficult concepts are based upon a solid foundation. It is intended that the book should be substantially self-contained, showing how all the concepts build on each other – how the pieces of the jigsaw fit together. This means that a wide range of topics is covered and, in order to keep the book of reasonable length, the treatment is constrained to the essentials that are thought necessary for a good understanding. What has been left out is considered almost as important as what has been included. Details that some other texts include are deliberately omitted. This is for brevity, so as not to detract from those aspects which are thought to be more important, and so that the reader should not find the book heavy going or daunting. It is accepted that it may sometimes be useful to supplement the material with that from other texts. The bibliography contains a short list of books that have been selected for their readability and/or extended coverage of relevant material.

One special feature of the book is that an attempt is made to throw some physical insight into the Laplace transform, somewhat similar to that which is commonly and much more easily done with the Fourier transform. It is felt that the common practice of introducing the Laplace transform by a *definition*, without any *interpretation* of that definition, is unconvincing and too mathematical for most readers. It is hoped that the treatment given will be more acceptable.

A second major feature of the book is the extensive use in some chapters of the scientific software MATLAB. This is done, firstly because it produces impressive displays that help to explain the text, and secondly because it has become an invaluable and widely accepted software tool whose existence has affected signal processing design procedures. As such, an aquaintance with its power is considered to be an essential part of a signal processing syllabus. A particularly impressive feature of MATLAB is that student versions of it are available for the price of an average hardback textbook. These student versions do have limitations compared with the professional version of MATLAB, but are nevertheless very powerful and totally adequate for many applications. For use with Microsoft Windows 3.1 the relevant student edition is Version 4, which is available as a diskette plus manual (MATLAB 1995). This can also be used on Microsoft Windows 95 and Windows NT, but the updated student edition for these is Version 5, and this is available as a CD-ROM plus manual (MATLAB 1997). Versions 4 and 5 are also available for the appropriate Macintosh computers. All the MATLAB commands in this book are restricted to commands that are valid for both of these student editions, as well as being fully compatible with the professional versions of MATLAB which may be available to students through a copy on their university network.

Although primarily concerned with system analysis and signal processing, a secondary function of this book is to provide an introductory tutorial on the use of MATLAB. For this reason the code used is listed in the text rather than on an attached diskette. Sometimes MATLAB is used is solely for producing the figures, and it is felt inappropriate to include code in the main text. Since, however, this coding may be of interest to the reader learning MATLAB programming, a selected number of these figures are marked by an asterisk (for example Fig. 3.3*) and the code is listed in Appendix A.

The problems at the end of each chapter are few in number but have been carefully designed to make a significant contribution to the learning process. If the material is understood most of them can be solved quite quickly. A solutions manual is available to instructors.

It is important to end with some acknowledgements. I should like to express my gratitude to the anonymous reviewers of the original manuscript for making a variety of constructive suggestions, many of which were subsequently incorporated. Also, I should like to thank Ahmad Shamsoddini for passing on some of his knowledge on the less well-documented aspects of MATLAB and for his meticulous reading of my corrected proofs. As well as finding a number of remaining typos he pointed out and helped remedy a few unclear, and even erroneous, statements. However it is almost inevitable that some mistakes will remain and if readers find any, or can suggest improvements or areas that need clarification, I would be grateful to hear of them. My e-mail address is p.n.denbigh@sussex.ac.uk.

Philip Denbigh
December 1997

Chapter one

Getting started in MATLAB and an introduction to systems and signal processing

1.1 ● Preview

This introductory chapter begins by presenting some of the main commands and capabilities of the powerful and widely used scientific software sold as MATLAB®. It is a very condensed version of what is described more fully in the manuals that accompany the student editions of MATLAB, constraining itself to the very minimum that is needed for getting started with its use in this book.

The chapter continues by classifying various types of signal and defining the linear time-invariant systems that form the subject of the book. Some simple examples of analogue and digital processing are given and reasons are presented for why digital processing is increasingly the preferred option. Finally, the special importance of sinusoidal, impulse function and step function signals is justified.

1.2 Getting started in MATLAB

MATLAB has become an invaluable and widely accepted software tool whose existence has affected signal processing design procedures; as such, an acquaintance with its power has become an important part of a signal processing syllabus. This section is intended as a very brief tutorial on some of its most important features. A particularly useful feature of MATLAB is that student editions are available from scientific bookshops for no more than the price of a typical hardback textbook, and yet they retain many of the most important capabilities of the full version. The relevant student edition for use with Microsoft Windows 3.1 is Version 4 and is available as a diskette plus manual (MATLAB 1995). This can also be used on Microsoft Windows 95 and Windows NT, but the updated student edition for these is Version 5, and this is available as a CD-ROM plus manual (MATLAB 1997). Versions 4 and 5 are also available for the appropriate Macintosh computers. Although fully compatible with the professional versions of MATLAB which may well be available to students through a copy on their university network, *all the MATLAB commands in this book are restricted to commands that are included in both of these student editions.*

Many sections of this book use MATLAB to demonstrate signal processing, and at these times the code will often be included in the main text. This is partly to demonstrate the power of the software and partly to provide further guidance in its use. In other sections MATLAB is used solely for producing figures, and on these occasions the code is usually omitted. Sometimes however, when there is something new to be learnt about the use of MATLAB for visualization, the figure is marked with an asterisk and the code is included in Appendix A.

The purpose of this present section is to describe some of the most basic features of MATLAB code so that the listings included later in the book will not need too much explanation to be understood.

1.2.1 Variables

When a number or a sequence of numbers is given a name it is termed a variable. A variable name can contain up to 19 characters, excluding punctuation characters. The first character must be a letter, but the remainder can be letters, digits or underscores and MATLAB distinguishes between lower case and upper case letters. An example of allocating a specific value to a variable is given by the command

```
b = 7
```

This causes the variable b to have the value 7. After entering this command the following appears on the screen

```
b =

      7
```

If it is wished to suppress such a message one follows the command with a semicolon. For example

```
b=7;
```

results in no such message on the screen.

A calculation can be involved when defining a variable. For example, the command

```
my_fingers=2*5;
```

introduces a variable *my_fingers* that has the value 10.

A few variables are built into MATLAB, namely *pi*, *i*, *j*, *ans*, *eps*, *inf*, *NaN*, *realmin* and *realmax*. Only *pi*, *j* and *ans* are used in this text. The variable *pi* is π. The variable j (which is the same as i) is $\sqrt{-1}$. The variable *ans* is a default variable name used for results. If, for example, we use MATLAB as a simple calculator by entering

```
4*5
```

we see

```
ans =
        20
```

1.2.2 Sequences, arrays and vectors

The previous subsection considered the situation where the variable had a single numerical value (it can be termed a *scalar* variable). Very often the variable is a *sequence* of numbers and one example of a four-element sequence might be noon temperatures over a four-day period. A set of one or more sequences can be termed an *array* and an example would be the daily noon temperatures over a four-day period at three locations, say London, New York and Tokyo. A hypothetical set of such temperatures could be entered as a MATLAB variable by encasing the values in square brackets separated by spaces and with the different number sequences separated by semicolons. Thus the command

```
x = [12.3 13.6 14.5 16.2; 18.1 18.8 19.2 18.4; 20.9 21.7 19.0 19.4]
```

results in the display

```
x =

  12.3000   13.6000   14.5000   16.2000
  18.1000   18.8000   19.2000   18.4000
  20.9000   21.7000   19.0000   19.4000
```

where x is an array in which the top row represents the London data, the second row the New York data, and the third row the Tokyo data. This array can also be thought of as a *matrix* of numbers. In this example it is a 3×4 matrix (3 rows by 4 columns). Generalizing this, an $m \times n$ matrix A is an array of numbers given by

$$A = \begin{bmatrix} a_{11} & a_{12} & \cdots & a_{1n} \\ a_{21} & a_{22} & \cdots & a_{2n} \\ \vdots & \vdots & \ddots & \vdots \\ a_{m1} & a_{m2} & \cdots & a_{mn} \end{bmatrix}$$

This matrix A contains m rows each of n elements. Alternatively it can be considered as containing n columns each of m elements.

When the sequence is a one-dimensional array of numbers it is commonly termed a *vector*. In the example of temperatures given by the x array variable above, each set of numbers representing the temperatures at a single location is a *row vector*. If on the other hand we consider the temperatures at all three locations, but on one specific day, we would have a *column vector*. This book will deal mainly with vectors, and particularly with row vectors.

1.2.3 Generating vectors

Generating a row vector is a limited case of generating an array, and specific values can be assigned to it by encasing these values in square brackets separated by spaces. For example, entering

x=[1 3 5 7]

results in the display

```
x =
        1   3   5   7
```

and signifies the creation of a variable x that is this sequence of numbers arranged as a row vector.

Column vectors are sometimes used in this book and may be regarded as another limited case of generating an array. This time, assignment is done by separating the specific values of the sequence by semicolons. For example

x=[1; 3; 5; 7]

results in column vector indicated by the display

```
x =

        1
        3
        5
        7
```

We can include calculations as part of the assignation of values. For example

x=[.1*pi .2*pi .3*pi .4*pi .5*pi]

results in the display

```
x =
        0   0.3142   0.6283   0.9425   1.2566   1.5708
```

A quick and easy way of generating a vector of *successive integers* is to separate the first and final values by a colon and to enclose them within parentheses. For example

x=(3:6)

results in the display

```
x =
        3   4   5   6
```

This is fine for integer values, one apart, but the concept can be extended to any set of uniformly spaced numbers by inserting *an increment* between the first and final values and separating all values by colons. If wished, the command can also include a mathematical operation. For example

x=(0: 0.1: 0.5)*pi

results in the display

```
x =
        0  0.3142  0.6283  0.9425  1.2566  1.5708
```

These are the same values as were obtained before when they were entered individually as a sequence. When long sequences are involved this last method of entry is clearly much easier. Another efficient method of doing the same involves the 'linspace' command, which generates a linear ramp of numbers. Here the first number within parentheses is the first number of the sequence. The second, separated by a comma, is the last value and the third is the total number of terms. Thus the instruction

```
x=linspace(0, 0.5*pi, 6)
```

results in the display

```
x =
        0  0.3142  0.6283  0.9425  1.2566  1.5708
```

where we have six terms commencing at zero and ending at 0.5π. Thus we have a different way of generating the same variable as in the previous example. It should be noted that the increment in element values equals the last value divided by the number of terms less one.

It is very common to generate one vector from another based upon a mathematical relationship. For example we might have

```
t=linspace(0, 50*pi, 501);
y=sin(t);
```

This generates a variable t that is a vector of 501 numbers 0.1π apart between 0 and 50π. It then calculates the sine of these 501 values of t and assigns them to a variable y. It will be noted that each of the two commands has been ended with a semicolon. This suppresses the display and avoids the computer screen being filled with a space-consuming list of 501 values for each of the two variables.

Vectors can be concatenated to make a single longer vector. For example, the commands

```
a=[3 4 6];
b=[1 7 9 11];
c=[a b a]
```

result in the generation of the vector variable c whose values are shown up on the computer display as

```
c =
        3    4    6    1    7    9    11    3    4    6
```

A vector which is often useful is one consisting solely of zeros. This can readily be generated by placing a constraint on a more general command for generating a *two-dimensional* array of zeros. The command

```
a=zeros(3,5)
```

defines a variable a that is a matrix of three rows and five columns all of which are zeros. Thus the display resulting from this command is

```
a =

     0  0  0  0  0
     0  0  0  0  0
     0  0  0  0  0
```

In order to obtain just a row vector of zeros we constrain the array to one row using

b=zeros(1,5)

This produces

```
b =

     0  0  0  0  0
```

There is a comparable function for producing a vector of ones. The command

c=ones(1,8)

produces

```
c =

     1  1  1  1  1  1  1  1
```

If a variable *d* describing some other array of the correct dimensions is already in existence there is a shortcut. The function **ones(size(d))** generates an array of identical dimensions containing all ones. Similarly, **zeros(size(d))** generates an array of identical dimensions containing all zeros. If *d* happens to be a $1 \times n$ array (an n-element vector) these instructions generate n-element vectors of ones or zeros.

If elements *l* to *k* of a new variable are assigned to specified values a vector of length *k* is created in which all unassigned elements are given zero values. Thus the command

g(4:6)=[8 2 9]

produces a six-element vector variable in which the elements below number 4 are filled with zeros. This is indicated by the display

```
g =

     0  0  0  8  2  9
```

If we wish to extend this vector by augmenting it with additional zeros we can assign some larger numbered element to be zero. Thus, if we continue with the command

g(10)=0

the variable *g* becomes a 10-element vector that accords with the display

```
g =

     0  0  0  8  2  9  0  0  0  0
```

It may be noted that this leads to another very neat way of creating a vector that is all zeros. If the variable *f* has not previously been assigned values, the command

f(8) = 0

results in the vector *f* for which the values are shown by the display to be

```
f =

        0  0  0  0  0  0  0  0
```

1.2.4 Addressing vectors

The command

x = (3:10);

has been shown to generate a vector of the eight numbers 3, 4, 5, ..., 10. Following this by the command

x(3)

causes the third element in the vector defined by the variable *x* to be addressed and results in the display

```
ans =
        5
```

An alternative would be the command

y = x(3)

in which case we obtain the display

```
y =
      5
```

Here we have defined a new scalar variable *y* and allocated the value 5 to it.

We can address a sub-vector within the vector. Following the assignment command **x = (3:10);** the instruction

y = x(2:6)

creates a variable *y* that is a vector of the 2nd, 3rd, 4th, 5th and 6th values of the *x* vector. Thus this instruction results in the display

```
y =
      4  5  6  7  8
```

One of the options in *creating* a vector was to specify the increment. We can do the same in *addressing* a vector. For example

y = x(6:-2:2)

causes the 6th, 4th and 2nd elements of the *x* variable to be addressed and results in the display

```
y =
      8  6  4
```

1.2.5 Array mathematics

The mathematical operations that follow will be directed primarily at one-dimensional arrays, or vectors. However, they are also fully applicable to *multi*-dimensional arrays in MATLAB 5 (though limited to two dimensional arrays in MATLAB 4). As our first example we might have:

```
y=[8 6 4];
z=y*2-3^2
```

This multiplies each element in the y vector by 2 and subtracts 3^2 from it. The variable z that results from this is shown on the display as

```
z =
      7     3     -1
```

It should be noted that the order of precedence in mathematical operations is

1. power operation
2. multiplication and division (equal precedence)
3. addition and subtraction (equal precedence)

In the case of equal precedence, operations are done in order of occurrence; for example, 4/3*2 = (4/3)*2 and not 4/(3*2).

Although it makes no difference to its execution, the command may be easier for the programmer to interpret if written with gaps. For example we could just as well have the command **z=y*2 - 3^2**.

Complete arrays can be added, subtracted, multiplied, divided, or raised to powers on an element by element basis. Note that array operations are quite different to the matrix operations described in Section 1.2.9. An obvious constraint of array additions, subtractions, multiplications or divisions is that the array involved must have the same dimensions. We might generate two vectors of six elements using

```
x=(0:2:10)
```

to produce

```
x =
      0    2    4    6    8    10
```

followed by

```
y=(0:.2:1)
```

to produce

```
y =
      0    0.2000    0.4000    0.6000    0.8000    1.0000
```

The command

```
z=x+y
```

then results in an array variable z whose values are given on the computer display as

```
z =
      0    2.2000    4.4000    6.6000    8.8000    11.0000
```

Any operation between two arrays precedes the usual arithmetic symbol with a dot. For example, multiplication of arrays uses the dot multiplication symbol **.*** such that the command

```
z=x.*y
```

results in an element by element multiplication of x and y to produce the display

```
z =
      0    0.4000    1.6000    3.6000    6.4000    10.0000
```

Division of arrays uses the dot division symbol **./** such that the command

```
z=x./y
```

results in an element by element division of x by y. The command

```
z=x.^y
```

causes each element of x to be raised to the power y, while

```
z=x.^2
```

causes each element of x to be raised to the power 2. Unlike the operation **x*2** this last command requires the dot since it involves an operation between two *vectors*; that is, between x and itself.

1.2.6 Multiple commands on one line and comments

For economy of space in stored programs it is quite common to place multiple commands on one line. This is permissible so long as they are separated by commas or semicolons. Commas cause the outcome of each command to be displayed, whereas semicolons suppress the display. Also, any text after a percentage sign (%) is taken as a comment and is neither acted on nor displayed. For example

```
a=(3:6); b=a.^2, c=a.^3;      % test on the use of commas and semicolons
```

results in the display

```
b =
      9    16    25    36
```

It should be noted that the % sign only applies to the line which it is on. Any additional line of comments must commence with another % sign.

1.2.7 Element indexing and time indexing

As already discussed, an element in a vector is addressed by its numerical position in that vector. These positions will be referred to as the *element* index, and they commence at 1. Frequently, however, it may be wished to attach a *time* index to each element which is different to this element index. For example it frequently happens that the first element of a vector corresponds to time $t = 0$ such that it is useful to have a timing index 0 associated with it. We can achieve the necessary indexing by involving *two* vectors, one corresponding to the main variable and the other, of equal length, providing the timing indices of the elements. Thus the commands

x=[3 2 1], n=(0:2)

result in the creation of two vector variables whose values are described by the display

```
x =

     3   2   1

n =

     0   1   2
```

If we address the first elements of both of these two vectors with the command

x(1),n(1)

we obtain the display

```
ans =

     3

ans =

     0
```

In a similar way, the command

x=[17 22 19], n=[1995 1996 1997]

produces two variables described by the display

```
x =

     17   22   19

n =

     1995        1996        1997
```

If we choose, the second of these variables could be a timing index associated with the first variable. For example, x might correspond to the price of some commodity in the three years 1995, 1996 and 1997.

A quite commonly used alternative term to *time* indexing is *sample* indexing. This has the major advantage that it is applicable to signals that are not functions of time (such as amplitude as a function of distance). However, the term *time* indexing will be adopted since most signals in this book are functions of time and the terminology makes a clearer distinction from *element* indexing.

1.2.8 Augmenting a vector with zeros to correspond with an extended timing index vector

Quite often in digital signal processing one needs to extend a sequence to have zeros added at either end while maintaining the time indexing of the existing elements. For example, starting with the vector variable x arising from the command **x = [8 2 7]** and the time indexing vector associated with it arising from the command **n = (3:5)** we might wish to create a new 'zero-padded' vector variable xx that is associated with a new time indexing vector whose values extend from -2 to $+8$. Clearly the most straightforward way of doing this is to enter the two new variables manually as

```
xx = [0 0 0 0 0 8 2 7 0 0 0], nn = (-2:8)
```

This produces two vectors described by the display

```
xx =

     0  0  0  0  0  8  2  7  0  0  0

nn =

    -2 -1  0  1  2  3  4  5  6  7  8
```

However, it can be very useful to find an alternative way of producing xx, particularly when the vector lengths involved are very much longer than in this example. Having defined the new timing index vector nn by the command **nn = (-2:8);** the first step is to recognize that the number of zeros preceding the original sequence can be obtained by subtracting $nn(1)$ from $n(1)$. In the example just given $n(1) - nn(1) = 3 - (-2) = 5$ and it is noted that this is the number of zeros that is added to the beginning of the x vector to create xx. Hence the first step in an alternative way of deriving xx begins by concatenating a vector of 5 zeros with the original x vector. The complete set of commands is

```
x = [8 2 7]; n = (3:5);
nn = (-2:8);
xx = [zeros(1,n(1)-nn(1)) x]
```

This produces

```
xx =

     0  0  0  0  0  8  2  7
```

We can next determine the total number of elements in vector nn and then assign the nnth element of xx to have value zero. This causes xx to be of the same length as nn, with any elements as yet unassigned to be given zero values. Thus we continue with the command

```
xx(length(nn))=0
```

where **length(nn)** is a function that determines the number of elements in *nn*. The final result is a vector variable of the same length as *nn* whose values are shown by the display to be

```
xx =

      0   0   0   0   0   8   2   7   0   0   0
```

1.2.9 Linear algebra and matrix operations

The 'MAT' of MATLAB is not an abbreviation of 'maths' as is sometimes supposed, but rather of 'matrix'. However, although matrix operations are the foundation and one of the most powerful attributes of MATLAB they are not used significantly in this book. The treatment will therefore be very cursory, with the only examples presented being those of matrix multiplication and of converting a row vector into a column vector.

If we wish to perform the matrix operation $\mathbf{C}=\mathbf{AB}$, where

$$\mathbf{A}=\begin{bmatrix}1 & 2 & 1\\ 3 & 3 & 4\end{bmatrix} \quad \text{and} \quad \mathbf{B}=\begin{bmatrix}6 & 2\\ 2 & 0\\ 1 & 3\end{bmatrix}$$

we can enter

```
A=[1 2 1; 3 3 4];
B=[6 2; 2 0; 1 3]
C=A*B
```

and obtain

```
C =

      11     5
      28    18
```

Matrix multiplication should not be confused with array multiplication. For array multiplication the two arrays must have identical dimensions and the elements of one are directly multiplied by the corresponding elements of the other using dot multiplication, as indicated by the symbol **.***. This contrasts with the totally different mathematical operation of matrix multiplication, where the multiplication symbol is an asterisk without the preceding dot. Similarly, the slash and power symbols *on their own* are used for *matrix* division and power operations.

One matrix application that will be shown in Chapter 4 to be very valuable in network analysis is the application for solving simultaneous equations. If, for example, we have the three equations

$$32.7x_1 + 12.4x_2 - 17.4x_3 = 13.7$$
$$12.9x_1 - 48.8x_2 - 39.8x_3 = 52.9$$
$$51.2x_1 + 66.4x_2 - 23.7x_3 = 63.2$$

we can write this in matrix form as

$$\begin{bmatrix} 32.7 & 12.4 & -17.4 \\ 12.9 & -48.8 & -39.8 \\ 51.2 & 66.4 & -23.7 \end{bmatrix} \times \begin{bmatrix} x_1 \\ x_2 \\ x_3 \end{bmatrix} = \begin{bmatrix} 13.7 \\ 52.9 \\ 63.2 \end{bmatrix}$$

or

$$\mathbf{AX}=\mathbf{B}$$

Analytically the solution is $\mathbf{X} = \mathbf{A}^{-1}\mathbf{B}$.

Thus, to solve the column vector **X** using MATLAB we enter the **A** matrix and the **B** column vector with

```
A = [32.7 12.4 -17.4; 12.9 -48.8 -39.8; 51.2 66.4 -23.7];
B = [13.7; 52.9; 63.2];
```

The command **X = inv(A)*B** then solves the equation, producing the output

```
X =
 -1.7660
  1.1372
 -3.2959
```

and thus signifying $x_1 = -1.7660$, $x_2 = 1.1372$, $x_3 = -3.2959$.

An alternative to the command **X = inv(A)*B** that is faster and more accurate is **X = A\B** (note the direction of the slash).

Matrices can be transposed using the dot apostrophe. A special case of this matrix transposition is the conversion of a row vector into a column vector, and vice versa. For example, the commands **x=[1; 3; 5; 7]; y=x.';** create a variable x that is a column vector and a variable y that is the corresponding row vector. (Note: an apostrophe on its own causes a complex conjugate transpose; this is identical to the dot apostrophe for vectors or arrays of *real* numbers and could have been used in the commands above.)

1.2.10 Plotting data values

Although commands are available in MATLAB for producing sophisticated three-dimensional colour plots, the plots in this book are limited to two-dimensional black and white displays. In those cases where the plots are associated with some new signal processing procedure the code for producing the plots is often included in the main text. In cases where the plots are included purely to help visualization, but where there is something new to be learnt about the use of MATLAB for producing plots, the code is usually excluded from the main text but may instead be listed in Appendix A. When this is done the corresponding figure is marked with an asterisk (for example, Figure 1.14*). In all circumstances where code is given some explanatory comments are included.

The purpose of this section is to establish some of the most basic commands, and this is done by means of some examples. The following set of commands in bold type results in Figure 1.1.

```
t=(0:500);
```

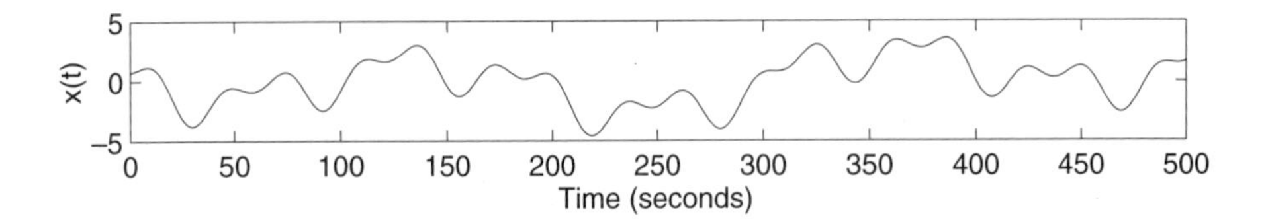

Fig. 1.1 ● Example of plot using MATLAB

This causes the array variable t to have the values 0, 1, 2, ..., 500.

```
x=1.5*cos(.1*t)+2*cos(.03*t+2)+1.7*cos(.01*t+1)-cos(.2*t+.5)+.9*sin(.006*t);
```

This generates a variable x that is the sum of four sinusoids.

```
plot(t,x)
```

This plots x versus t, where t is the parameter plotted on the horizontal axis, and x is the parameter plotted on the vertical axis (note that the symbol x is commonly used in systems work to represent a signal amplitude, and is rarely used as the symbol for the parameter plotted on the x-axis). The scales of the axes are selected automatically to fit the data values. In this example we could have had the simpler command **plot(x)**. This plots each element of x against the value of the element index. In this particular case, but not in general, the t values are identical to these index values. If the first statement had been **t=(0:500)*2** the two plots would have been different.

```
axis([0 500 -5 5])
```

This controls the horizontal axis to extend from 0 to 500 and the vertical axis to extend from −5 to +5. (In this particular example this statement is unnecessary since the automatic scaling gives the same axes.) It should be noted that in cases where the axes are changed by the axis command the data is *replotted* in accordance with the specified new axes.

```
xlabel('Time (seconds)'),ylabel('x(t)')
```

This adds the labels within the quote marks to the x and y axes respectively.

Very often it is wished to make some changes to the plot. In Figure 1.1, for example, it may well be decided that the font size is too big. We can change the font size of the axes of the current plot as follows

```
set(gca,'fontsize',10)
```

gca causes the remaining commands within the brackets to be applied to the axes of the current plot. In this example there is a single operation, and this is that specified within quotes, namely **fontsize**. The **10** changes the size of the font from the default value of 12 to 10. The result of this command causes the display of Figure 1.1 to change to that of Figure 1.2. It will be noted that the modification of font size is applied only to the labelling of the tick marks that result automatically from the plot command, and does not apply to the x and y labels, which were added as separate commands. If we wish to change the font size of the labels we need to *repeat* the labelling commands (although this would not be necessary if the font-setting command had been included in the original set of commands *prior* to the labelling command).

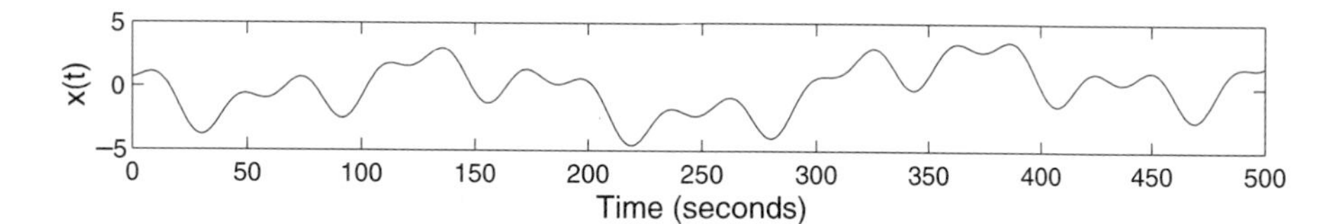

Fig. 1.2 Modification of font size on the axes

It is often useful to add a title to the plot. Doing this and relabelling the axes in the smaller font we could have

```
xlabel('Time (seconds)'), ylabel('x(t)'), title('Example of MATLAB plot')
```

This now changes Figure 1.2 to Figure 1.3. It is useful to note that either commas or semicolons may be used to separate the commands in this line of code. This is because there is no text to be displayed or suppressed other than that which is involved with the plot.

The command **plot(t,x)** interpolates between data points to produce the continuous curves of Figures 1.1, 1.2 and 1.3. When arrays contain a *small* number of elements it is often preferable to display the individual points without any interpolation. As seen in the next example this is done using the **stem(n,x)** command.

```
n = (0:50);
x = 1.5*cos(n) + 2*cos(.3*n + 2) + 1.7*cos(.1*n + 1)-cos(2*n + .5) + .9*sin(.06*n);
% the array variable x contains only 51 values
stem(n,x)                % this is the command for plotting data points individually
axis([0 50 -5 5]); xlabel('n'); ylabel('x[n]')
```

This results in Figure 1.4. The **stem** command causes each sample amplitude to be shown by a circle at the end of a stem.

Very often it is useful to display several different plots in the same figure window. To do this we can precede a plot command by the 'subplot' command. For example, **subplot(4,1,1)** divides the figure window into a matrix of 4 × 1 rectangular panes (or 4 'rows' and 1 'column') and, counting through these rectangular panes as one would read a book, selects the first of these for the forthcoming plot. Subsequent use of **subplot(4,1,2)**, **subplot(4,1,3)** and **subplot(4,1,4)** would make panes available for further plots in the same figure window.

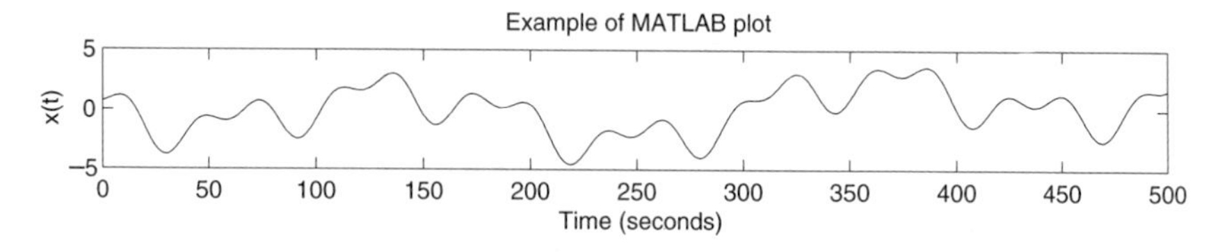

Fig. 1.3 Labelling of plots

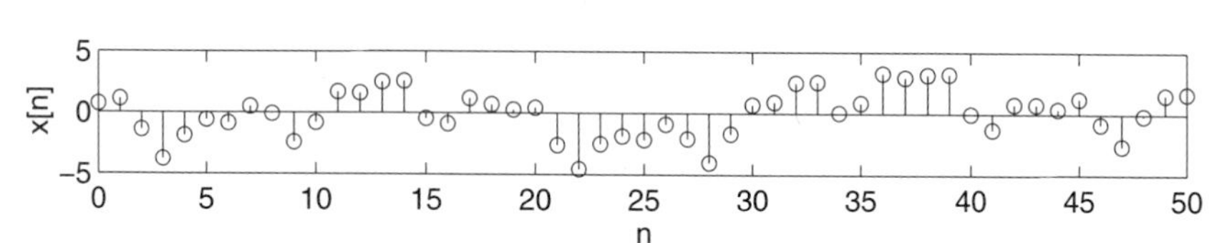

Fig. 1.4 Use of the stem command for displaying discrete data values

1.2.11 Hardcopy

It is common to require a hardcopy of a figure, and for this one clicks on the File command in the menu above the MATLAB figure window and follows this with the Print command.

If a hardcopy of reduced dimensions is wanted (the same reduction in both x and y dimensions), this can be achieved by preceding the print command with the Print setup facility and using it to apply print scaling. An alternative and more flexible procedure in which the x and y dimensions can be changed independently, and where the position on the printed page can be controlled, uses the command

set(gcf,'Paperposition',[a b c d])

This does not change the image on the computer screen but specifies the location and size of the MATLAB figure window as it will appear in the hardcopy from the printer. It should be noted that the figure window itself is larger than the plots contained within it. Hence the plots themselves will occupy less space in the hardcopy than indicated by this command. The default units are inches. Here the dimensions of the figure window in the hard copy are c inches by d inches, and the bottom left-hand corner of it is a inches from the left and b inches up. Care should be taken with any subsequent hardcopies since this instruction will remain in force unless modified by a new **set(gcf,'Paperposition',[a b c d])** command or else deleted by the command **reset(gcf)** to return to default values.

If it is wished to print out a 'letter-box' shaped display, such as Figure 1.4, a lazy alternative to the above is to use part of the figure window by preceding a plot command with a subplot command such as **subplot(3,1,1)**. The print command then generates a complete figure window in hardcopy, but with only the top third containing a plot.

1.2.12 Transfer of a figure to a word processor document

To insert a figure into a word processor document one can use the copy command within the edit menu of MATLAB and then paste it in the word processor document. For black and white figures the best results are obtained if any colour is first removed. In Version 4 this requires the command **blt**. In Version 5 it requires **nodither ('save')** (with colour being returned using **nodither ('restore')**). If desired a reversal of black and white can be achieved by preceding the copy command with **whitebg**. Changes to the sizes of figures can be done by dragging the borders or corners of the window. Note that the size of the labelling is maintained if the dragging is applied to the original MATLAB window but that everything is scaled if this dragging is applied to the figure window in the document.

1.2.13 M-files

Rather than writing and executing commands one line at a time, it is very common to prefer to write a complete program of commands which is stored as a file, and then run this as an entity. To do this in MATLAB one clicks on the File menu in the command window, followed by New and then M-file. A default editor then appears on the screen on which one enters the program. On completion one clicks on the editor's File menu, followed by Save As. One then enters a suitable filename, which must end with the extension 'm'. An example would be Fig1_4.m. After okaying this entry the file is stored within the current directory. This is 'stdntmat' when using the student edition of MATLAB. In order to run the M-file one then simply types the filename, without the extension 'm' and enters this command (**Fig1_4** if the M-file is Fig1_4.m). The program then runs. If it is wished to modify the program one clicks on the File menu in the MATLAB command window, followed by Open M-file. This takes one back to the editor. After editing the program one should click on Save, close down the editor, and then finally return to the MATLAB command window. To run the edited program one does as before. One types the filename and then enters it.

1.3 Some signals and systems terminology

Although some of the terms used to describe signals are widely used it is important to ensure that they are clearly defined.

A *discrete-time* signal is defined only at particular instants in time. If, for example, the horizontal axis of Figure 1.4 denotes time, then the plot is of a discrete-time signal.

An *analogue* time-varying signal is one that has a continuous range of amplitudes and which is defined continuously over time. An example is the fluctuating signal $x(t)$ shown in Figure 1.1. (Some caution is needed here, however, because Figure 1.1 was actually created in MATLAB by generating 501 *discrete* values of an analogue signal, and the *continuous* plot of Figure 1.1 arose because the plotting routine caused an interpolation between the discrete values. In other words, the plot portrays as an analogue signal what is really a discrete-time signal.)

A *continuous-time* signal is defined at all instants in time but, as in the quantized version of $x(t)$ shown in Figure 1.5, does not necessarily contain a continuous range of amplitudes. Although this separates its definition from that of an *analogue* signal the distinction is rarely relevant and the two terms tend to be used synonomously. (The MATLAB code used to produce Figure 1.5 involves the MATLAB function **round(x)**, which rounds off data values to the nearest integer value. The code was the same as that used to produce Figure 1.1 except that command **plot(t,x)** was replaced by **plot(t,round(x))** and **ylabel('x(t)')** was replaced by **ylabel('xq(t)')**.)

Many discrete-time signals arise from the sampling of an analogue signal. Thus a suitable notation is $x(nt_s)$, where t_s is the time between samples and n is an integer. An alternative notation takes the sample period as understood and known, and writes the signal as $x[n]$. Here *square brackets are used to indicate that* $x[n]$ *is a discrete-time signal.* This convention is very widespread, although it should be noted that many authors use parentheses (for example $x(n)$). Some of the arguments in

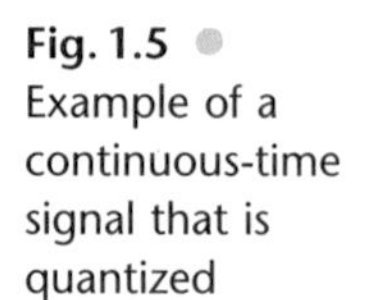
Fig. 1.5 Example of a continuous-time signal that is quantized

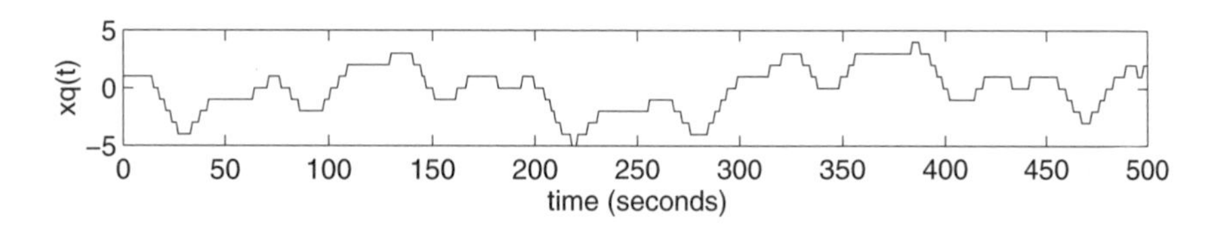

favour of square brackets are as follows:

- The use of parentheses for continuous-time signals and square brackets for discrete-time signals *clarifies* which type of signal we are dealing with.
- Two very important functions defined in Chapter 2 are the continuous-time impulse function $\delta(t)$ and the discrete-time impulse function $\delta[n]$. In spite of the general adoption of the same δ symbol for both, these two functions are substantially different. Usually the appropriate definition is obvious through the use of the parameter t or n, but a much greater rigour is achieved by the use of a different form of bracket.
- In MATLAB an element of a signal vector is addressed using parentheses (for example **x(1)**). The index within the parentheses is the position of the element within the vector and is not the timing index. It is useful to have a separate notation in which the index within square brackets is an unambiguous specification of the time of a sample. If, for example, we have the commands **x = [3 5 2]; n = [-1 0 1];** we are saying that we have a signal vector for which $x[-1] = 3$, $x[0] = 5$, $x[1] = 2$. However, the MATLAB commands **x(1)**, **x(2)**, **x(3)** produce the results 3, 5 and 2. We see that $x[1]$ is different from the $x(1)$ of MATLAB.

To summarize, the discrete-time signal arising from sampling the analogue signal $x(t)$ can be written either as $x(nt_s)$ or $x[n]$. Use of the notation $x(n)$ will be disallowed within this text.

If the samples of a discrete-time signal have a *continuous* range of amplitudes, a discrete-time signal can be further classified as being a *sampled-data* signal. In contrast, if the samples are *quantized* such that they have a finite set of values, the signal is a *digital* signal. Usually the quantized amplitudes of a digital signal would be described by a codeword of binary numbers. A very simple example of a coding system would be to use the codeword 0001 to describe the value 1, 0010 to describe 2, 0011 to describe 3, and so on. In practice, the terms *discrete-time* and *digital* tend to be interchanged quite loosely without adhering to these strict differences regarding quantization.

Many signals of interest do not involve time. For example, a signal might represent the height of a river as a function of distance from the sea, or the blackness of an image as a function of pixel position within an image. These would be referred to simply as continuous or discrete signals (as opposed to continuous-time or discrete-time signals).

Signals that can be given a *functional* description are said to be deterministic. A simple example of a deterministic analogue signal would be $x(t) = 9\sin(200t)$. A comparable example for a signal that is sampled at regular intervals would be $x(nt_s) = 9\sin(200nt_s)$, where n is an integer and t_s is the time between discrete values. The time t_s is likely to be known and if, for example, $t_s = 0.1$ we could also write this as $x[n] = 9\sin(20n)$, where it is noted that square brackets are used for the signal $x[n]$ but that any type of bracket could be used for $9\sin(20n)$; parentheses are chosen for simplicity.

For discrete signals, however, the possibility also exists for a *sequence* description that lists specific values of the deterministic signal. For example a sequence description of the discrete samples of a rectangular pulse might be

$$x[n] = \{0, 0, 5, 4, \underset{\uparrow}{3}, 2, 1, 0, 0\}$$

Here the marker arrow indicates the time origin, such that $x[-4] = 0$, $x[-3] = 0$, $x[-2] = 5$, $x[-1] = 4$, $x[0] = 3$, $x[1] = 2$, $x[2] = 1$, $x[3] = 0$, and $x[4] = 0$.

Data values not included in the sequence, $x[-5]$ for example, are assumed to be zero. Indeed, an alternative and adequate description of the same signal would be

$$x(n) = \{5, 4, \underset{\uparrow}{3}, 2, 1\}$$

If the sequence starts at 0 the marker arrow may be omitted. For example $\{3, 2, 4, 1\}$ can be assumed to be the same as $\{\underset{\uparrow}{3}, 2, 4, 1\}$.

In order to emphasize that $x[n]$ is a listed sequence it is quite common to write $x[n]$ inside curly brackets, that is

$$\{x[n]\} = \{5, 4, \underset{\uparrow}{3}, 2, 1\}$$

However, this convention will not be used in this book.

Many signals cannot be given any explicit mathematical description, even though much is known about them. Such signals can only be described by their statistical properties. If, for example, the very small signal generated across a resistance R, due to the thermal agitation of electrons within it, is amplified by an ideal noiseless amplifier of voltage gain A and bandwidth B, the output signal at an arbitrary instant cannot be predicted. In the absence of any prior or subsequent knowledge of the signal all that can be said about the signal at a particular instant is that it has a certain probability of lying in a certain voltage band. For example it can be shown that the probability that the voltage at that instant lies between voltage levels v and $v + \mathrm{d}v$ is given by $p(v)\mathrm{d}v$, where

$$p(v) = \frac{1}{\sqrt{2\pi\sigma^2}} \exp\left(-\frac{v^2}{2\sigma^2}\right)$$

and where $\sigma = A\sqrt{4kTBR}$, k is Boltzmann's constant and T is the temperature of the resistance. Signals such as this are known as analogue *random* or *stochastic* signals. This expression above for $p(v)$ is an example of a partial description of a random signal by its *first-order statistics*. For a more comprehensive description of the random signal we require to know about the second and higher order statistics of the signal. These indicate how the probability of the signal lying in a particular band is affected by the previous history of the signal, but this subject will not be examined any further in this book except to say that the second-order statistics of a random signal are related to the spectral properties of that signal.

We can also have *discrete* random signals: an example would be the sequence of numbers arising from the tossing of a die, for example $x[n] = \{5, 2, 6, 1, 1, 3, 5, 4\}$. Assuming the die is not loaded, the outcome of any throw will not be affected by what has gone before and the description of the outcome cannot extend beyond the first-order statistics, which in this case simply say that the probability of any of the six numbers occurring is one sixth.

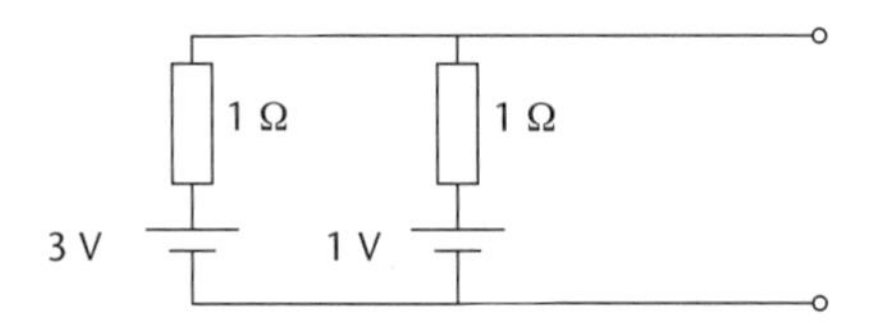

Fig. 1.6 Example of a simple linear time-invariant system

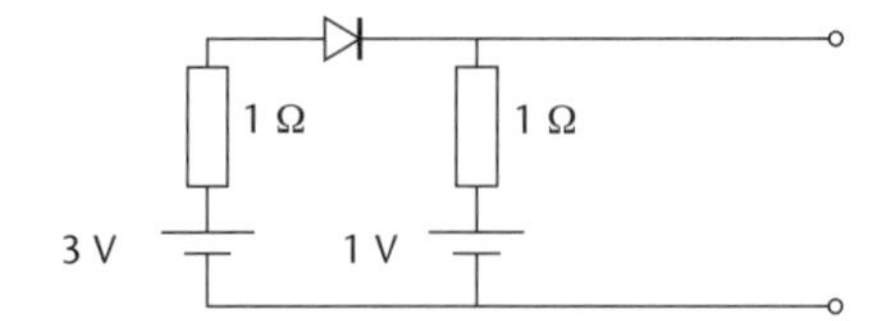

Fig. 1.7 Example of a simple non-linear time-invariant system

Most of this book will be concerned with *linear time-invariant* (or LTI) systems. A time-invariant system is one whose properties do not change with time. A linear system is one for which the output arising from multiple inputs is the sum of the outputs due to the individual inputs (that is, *superposition* applies). Figure 1.6 shows an example of a linear system. Due to the 3 V source on its own (that is, with the 1 V source short-circuited) the output voltage is 1.5 V. Due to the 1 V source on its own the output voltage is 0.5 V. The sum is 2 V. It is easy to show by circuit analysis that this same result of 2 V is obtained when both sources are present simultaneously.

In contrast, the network of Figure 1.7 containing an ideal diode is a *non*-linear time-invariant system. With the 3 V source on its own the diode is turned on and the output voltage is 1.5 V. With the 1 V source on its own the diode is turned off, such that the output voltage is 1 V. By superposition the output voltage would be 2.5 V. However, with both sources present simultaneously the diode is turned on, and the output voltage is therefore the same 2 V as for Figure 1.6. The output due to the two voltage sources together is not the superposition of the outputs due to two voltage sources separately, and this signifies that the circuit is non-linear. Another example of a linear time-invariant system is an ideal amplifier.

An example of a linear time-variant system is an amplifier with a time-varying gain, such as might be used in a radar to enhance weak echoes from distant targets relative to strong echoes from nearby targets. Here the receiver gain would be made to increase with time, in accordance with an appropriate law, following the transmission of each radar pulse.

1.4 Some examples of analogue systems and analogue signal processing

Most signals that require processing are analogue signals. Examples are signals from biomedical sensors, echoes from radars or sonars, images from television cameras and speech signals. Usually these signals require some processing, of which one simple example is the filtering needed to remove unwanted high-frequency noise from a signal of interest. An example of an appropriate lowpass filter is shown in Figure 1.8. One objective of this book is to develop the tools to *analyse* such a network, and in particular to learn how to determine its frequency response. A second, and perhaps even more important, objective is to learn how to *synthesize* such a network to achieve a *required* frequency response.

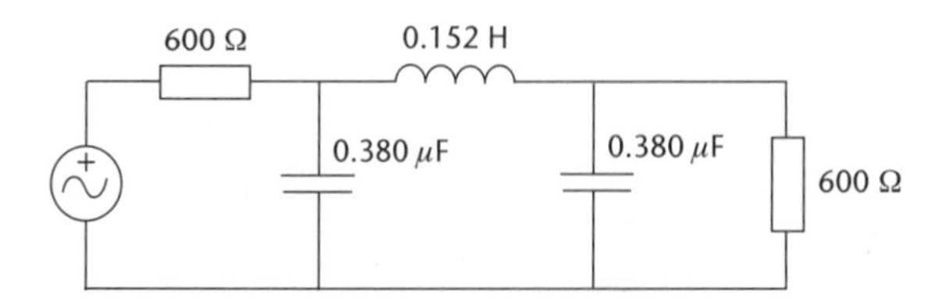

Fig. 1.8 The circuit of a simple lowpass filter

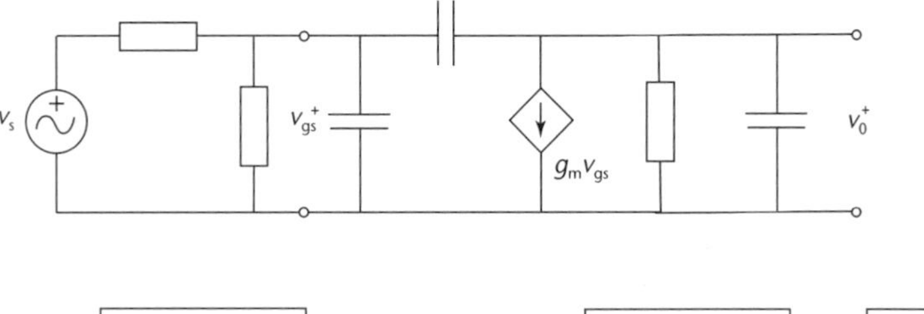

Fig. 1.9 High-frequency equivalent circuit of an FET amplifier

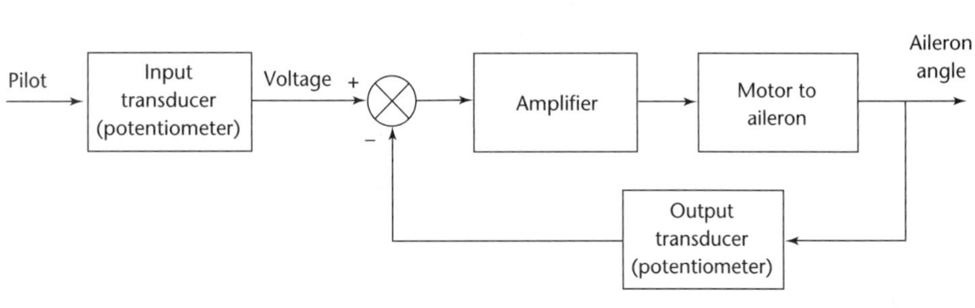

Fig. 1.10 Block diagram of the control system for an aircraft aileron

Being a filter, the network of Figure 1.8 is intended to have some special and useful frequency response. However, it will often be required to find the frequency responses of systems other than filters. An example is the network of Figure 1.9, which is the high-frequency equivalent circuit of a field-effect transistor amplifier. Here the diamond-shaped symbol is used to denote a source that is controlled by some other voltage or current. The arrow inside the diamond indicates that it is a current source. Besides determining the frequency response, it may also be useful to determine the response of the circuit to impulsive or step excitations, that is, its impulse response or step function response.

Other important examples of analogue systems include control systems. These occur widely and require analysis. An example is shown in Figure 1.10 and is for adjusting the angle of an aircraft aileron. The objective is that the aileron should follow rapidly an electrical signal actuated by the pilot. A notable feature of the system is the use of feedback. A sensor on the aileron drive shaft generates an electrical signal proportional to its angle, and the difference (or error) between this and the actuating signal is amplified and causes the drive motor to rotate. Only when the aileron is in the desired position is there is no error and hence no further drive from the motor.

At first sight the concept of feedback appears a perfect method for achieving the correct angle on the aileron. However, there is much more to the problem than so far indicated. We would like the aileron to respond very quickly to the electrical input, preferably without overshoot or other undesirable features. To achieve this the two networks must have suitable gains and phase shifts as a function of frequency. Once again there is the need for understanding how to analyse and synthesize electrical systems.

1.5 Some examples of digital systems and digital signal processing

Frequently we are limited to a discrete-time description of a signal that is actually continuous. For example, the temporal variation of rainfall may be considered a continuous or analogue signal, but the measurement equipment may be of a type that only provides discrete values *once a day* of the total rainfall throughout the previous day. The result would often be referred to as a 'sampled data' signal. A simple example of a system operating on sampled data is the digital filter shown in Figure 1.11. The rectangular boxes represent delays of one sample period, and it is apparent therefore that the output $y[n]$ equals the sum of the present input and the four previous inputs, all divided by five. The network is a 'five-point averager' and the effect would be to smooth the data. Worded differently, it acts as a lowpass filter, very much like the analogue network of Figure 1.8. Mathematically, the effect of the filter can be expressed by the 'difference equation' relating output to input, which, in this case, is

$$y[n] = \tfrac{1}{5}(x[n] + x[n-1] + x[n-2] + x[n-3] + x[n-4]) \qquad (1.1)$$

Here the filter output relies solely on *input* values (that is, on feedforward terms being applied to the adder), and such a filter is known as a *non-recursive* digital filter.

It is interesting to note that the output one sample earlier is given by

$$y[n-1] = \tfrac{1}{5}(x[n-1] + x[n-2] + x[n-3] + x[n-4] + x[n-5]) \qquad (1.2)$$

This means that the output $y[n]$ could be written

$$y[n] = y[n-1] - \tfrac{1}{5}x[n-5] + \tfrac{1}{5}x[n] \qquad (1.3)$$

and this leads to the alternative configuration in Figure 1.12. This has a feedback term as well as 'feedforward' terms and is therefore known as a *recursive* digital filter. Recursive filters are often preferable because of the fewer mathematical calculations required.

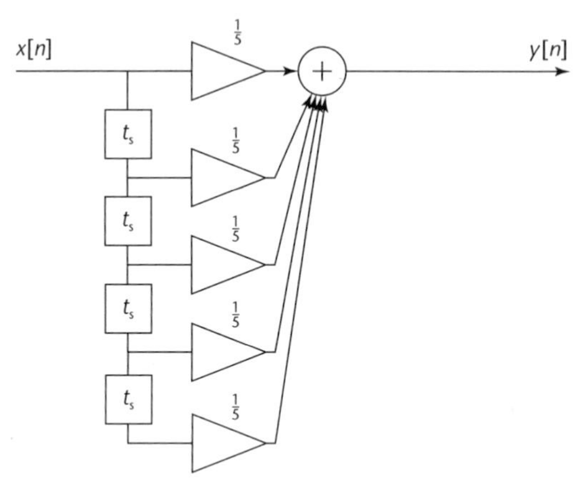

Fig. 1.11 A simple non-recursive digital filter

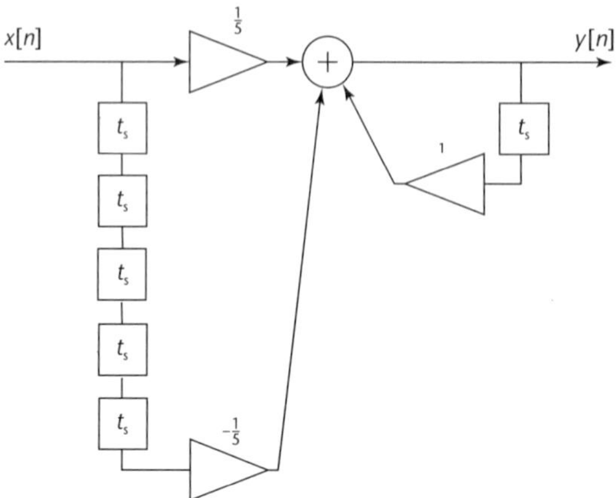

Fig. 1.12 A recursive filter that can replace the non-recursive filter of Figure 1.11

It should be noted that the configurations of Figures 1.11 and 1.12 are somewhat symbolic, in that it is not normal to realize such a filter with analogue delays and analogue amplifiers. In the rainfall example it is likely to be achieved in software, using data files of past measurements. To apply the non-recursive difference equation of Equation (1.1) to 15 days' worth of measurements of rainfall contained within a vector x, a MATLAB program stored within an M-file called 'rain.m' might be

```
x=[0 0 0 0 2.4 3.3 5.9 1.1 9.2 3.4 6.1 2.7 3.3 5.2 1.9 3.8 5.3 1.2 1.4];
                              % this is the rainfall data preceded by four zeros
y=zeros(size(x));                                        % initialized output
for n=5:19
y(n)=0.2*(x(n)+x(n-1)+x(n-2)+x(n-3)+x(n-4));             % final output
end
y
```

This program introduces a pair of MATLAB instructions that control the flow of command execution. The **for** instruction terminated by the **end** instruction causes the intermediate commands to be repeated with the value of n incremented each time. Since in this example four *previous* values of the input data are needed, the loop is not started until the fifth element of the data set. Because the input data elements are indexed from one upwards these intermediate commands therefore commence with a value of n equal to 5 and are then repeated with values of n rising to 19. The output values are initially set at zero. They stay that way until changed by the command that implements the difference equation for the appropriate index value.

On execution of this M-file the following display is produced:

```
y =

Columns 1 through 7

      0      0      0      0    0.4800    1.1400    2.3200

Columns 8 through 14

      2.5400    4.3800    4.5800    5.1400    4.5000    4.9400    4.1400

Column 15 through 19

      3.8400    3.3800    3.9000    3.4800    2.7200
```

It should be noted that, since valid rainfall data is needed from four previous days, the required smoothed data is not obtained until the ninth day. There is a 'start-up' transient.

The alternative to the non-recursive difference equation is the recursive difference equation of Equation (1.3), and for this the program might be

```
% rain2.m
x=[0 0 0 0 0 2.4 3.3 5.9 1.1 9.2 3.4 6.1 2.7 3.3 5.2 1.9 3.8 5.3 1.2 1.4]
% this is the rainfall data supplemented by five zeros
y=zeros(size(x));                                        % initialized output
for n=6:20
y(n)=y(n-1)+0.2*(x(n)-x(n-5));                           % final output
end
y
```

This produces the output

```
y =

Columns 1 through 7

     0      0      0      0      0    0.4800    1.1400

Columns 8 through 14

     2.3200  2.5400  4.3800  4.5800  5.1400  4.5000  4.9400

Columns 15 through 20

     4.1400  3.8400  3.3800  3.9000  3.4800  2.7200
```

Several points are worthy of note when comparing the programs and results of the recursive and non-recursive filters. Firstly, since five *previous* values of the input data are needed in the recursive difference equation, the loop is not started in this last program until the sixth element of the data set. Secondly, if the recursive difference equation is to give the correct result, it must commence with the correct value of $y[n-1]$. Otherwise $y[n]$ will be in error by the error in $y[n-1]$, and $y[n+1]$, $y[n+2]$, $y[n+3]$ will maintain this error. Having preceded the input data with five zeros this correct initial value of $y[n-1]$ is achieved by initializing the output values to zero, since this signifies $y[5]=0$ when the loop cycle commences at $n=6$. Thirdly, the start-up transients periods are different in the two cases. For the non-recursive system the difference equation is not applied until $n=5$, such that the correct result is not obtained until $n=9$. In the case of the recursive filter the supplementation of the data set by five initial zeros means that the correct answer is not obtained until $n=10$.

In the above example the rainfall data has been entered into the program itself. For large data sets it would be more usual to extract the measurements from a previously recorded data file.

Another very common scenario, particularly in situations where the data samples are more closely spaced in time, is where *real-time* processing is required. This can mean that the processing needed to produce the current output sample must be completed before the next input sample arrives. Such processing may be needed:

- because the data set is too large to be stored in a data file;
- because there is an urgent need for the processed signal. As an example it might be wished to display a noisy cardiac signal live on a monitor, having first smoothed it using the five-point averager.

It will be shown in Chapter 7 that no information is lost by sampling a signal so long as the samples are close enough. Hence, in the example of smoothing a cardiac signal, the cardiac signal would first be sampled and converted into a digital signal every t_s seconds by means of an analogue to digital converter (also known as an A/D converter, or ADC). These digital samples would then enter a microprocessor or special digital signal processing chip where a machine code program, equivalent to one of the MATLAB programs above, would perform the processing in some time less than t_s, such that the processing is completed before the next input sample is entered. In other words, if t_m is the machine cycle time and n is the number of machine cycles needed to achieve the required multiplications and additions, it is necessary that $nt_{\mathrm{m}} < t_{\mathrm{s}}$. Operating at the same rate as the input A/D conversion, the output digital number would be converted back to an analogue output sample using a digital to analogue converter (also known as a D/A converter, or DAC). These output samples would then be smoothed to give a continuous analogue signal to be applied to the monitor. The block diagram of a complete real-time digital system is shown in Figure 1.13. It includes an analogue input filter whose function will be considered further in Chapter 7.

As another example of where signals gain from real-time processing, consider a sonar transmitting a 'Chirp' pulse whose frequency changes with time, such as shown in Figure 1.14(a). The weak echo from some distant target might be as shown in Figure 1.14(b). In the likely event that the echo is masked by background noise, such as that of Figure 1.14(c), the actual return signal would be the superposition of Figures 1.14(b) and 1.14(c), which is that shown in Figure 1.14(d). The objective would be to process this return signal to reveal the existence and position of the echo. The optimum technique for achieving this is to determine if and when the waveshape of the receive signal shows a good degree of match to the waveshape of the transmit pulse. One way of doing this is to have a receiver that produces replicas of the transmit pulse with a very large number of different delays, and tests the 'match' or 'correlation' between each of these and the receive signal. Noting that a multiplication of the transmit pulse by itself produces a signal that is always positive and therefore has a large integral, the criterion for judging a good match between the receive signal and a delayed transmit pulse by correlation is that the product of the two should have a large integral. For example, Figure 1.14(e) shows the transmit pulse with an identical delay to the echo. The result of multiplying this with the clean echo of Figure 1.14(b) is shown in Figure 1.14(f), and this waveform is seen to be always positive and hence to have a positive integral. In contrast, multiplying the delayed transmit pulse with the noise of Figure 1.14(c) produces the waveform shown in Figure 1.14(g), and this is seen to be bipolar, such that its integral is less. The result when the signal and noise are present together may be predicted by superposition, and Figure 1.14(h) shows the product of the delayed transmit pulse and the noisy received signal of Figure 1.14(d). Its integral is clearly positive. In this way, the presence of the echo is revealed when the replica of the transmit pulse has the correct delay. Indeed, Figure 1.14(i) shows the correlation between the noisy receive signal of Figure 1.14(d) and a replica of the transmit pulse as a function of the delay of the replica. The presence of the weak echo is clearly revealed.

Parts of the MATLAB program for producing Figure 1.14 use functions and procedures that have not yet been introduced. For this reason the listing of the program has been assigned to Appendix A. In accordance with the procedure mentioned earlier, the convention is adopted where the figure is marked with an asterisk to indicate that this is what is done (for example Figure 1.14*).

Fig. 1.13 Block diagram of a real-time system for processing cardiac signals

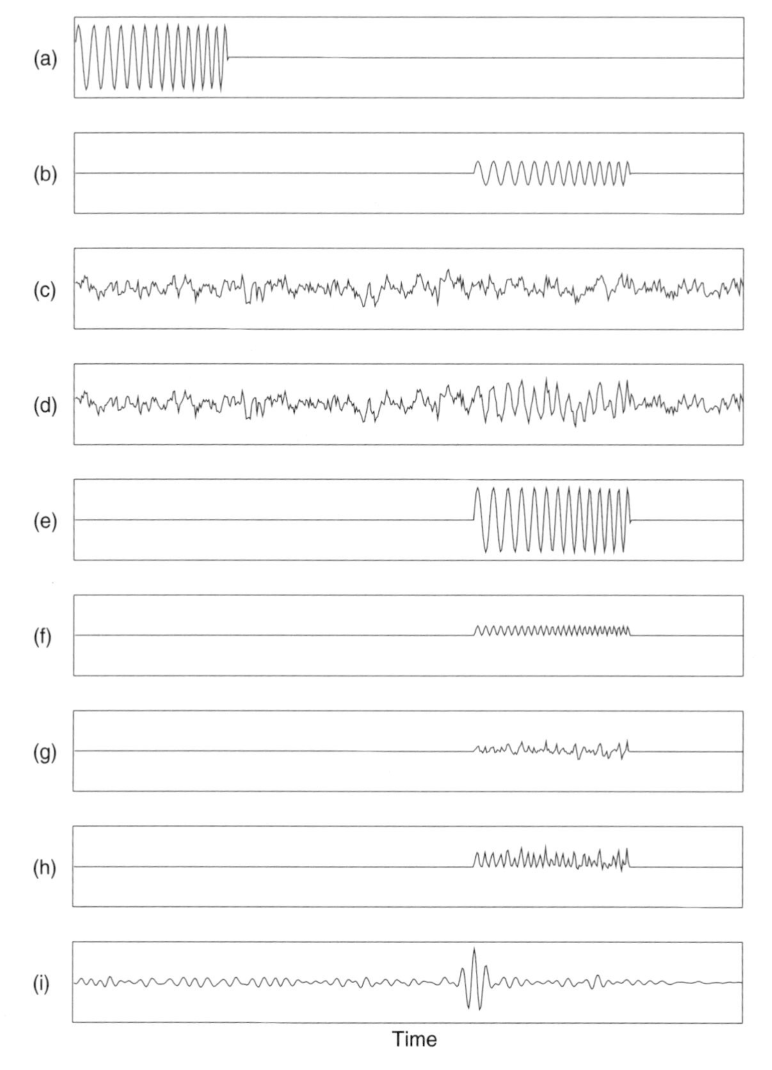

Fig. 1.14 Example of digital processing for enhancing weak sonar echoes by correlation processing: (a) transmit chirp pulse; (b) echo in the absence of noise; (c) noise (d) echo plus noise; (e) replica of transmit pulse with same delay as echo; (f) product of replica and clean echo; (g) product of replica and noise; (h) product of replica with echo plus noise; (i) correlation as a function of the delay of the replica

Correlation detection is discussed in much more detail in Chapter 16. For the moment, it should be noted that, by using a superior transmit pulse to that of Figure 1.14(a), an even greater enhancement of weak echoes can be obtained than is indicated by the difference between Figure 1.14(d) and Figure 1.14(i).

Enormous improvements in target detection are possible by using an appropriate transmit pulse and correlation processing. Digital implementation is feasible, and correlation is one of very many examples of the usefulness and importance of digital signal processing.

An even more complex example of processing is that of computerized axial tomography, where a weak X-ray source is rotated around the human body and the strength of the signal transmitted through the body is measured by a detector coupled to the source, but on the opposite side of the body. The transmissivity through the body is measured for a very large number of rotational positions and a complex mathematical operation can be applied to these measurements to determine the transmissivities of very small local regions within a cross-sectional slice of the body. These CAT (or CT) scans can thus be used to create a highly meaningful image that is useful for diagnosis. The mathematical operation required is too complex to be performed by analogue circuits, but is ideally suited to digital processing on a computer. There are many other examples where the processing to be done is too complex for analogue circuits.

More and more, analogue systems are being replaced by digital systems. Taking the control system of Figure 1.10 as an example, everything can be digital with the exception of the motor. The network functions can be replaced by computer or microcomputer algorithms. Even the angular sensor can be a device (a shaft encoder) that directly provides a digital output.

1.6 Justification for the digital processing of signals

The majority of signals that require processing are analogue signals, and the main disadvantage of digital processing is the need to convert them into digital form by means of an analogue to digital converter and subsequently back to analogue form by means of a digital to analogue converter. The cost and limited speed of these converters is sometimes a problem. In return, however, digital processing has the following major advantages over analogue processing.

- *Accuracy and reproducibility.* For example, the digital multiplication of two numbers, as by a pocket calculator, gives a precise and accurate result that is not affected by temperature, ageing, or the processor used. An analogue multiplier would produce a far less accurate result.
- *Capability for very much more complex operations than are possible with analogue circuits.* The example of computerized tomography has been mentioned. It is also very easy to introduce (though much less easy to analyse) useful non-linear operations such as conditional statements. If, for example, it is wished to smooth some data we might begin by discarding values that are obviously erroneous, based upon unrealistic deviations from neighbouring values. A suitable rejection algorithm could be implemented very simply with a digital (or computer) processor.
- *Flexibility.* The processing can be changed without any change to the hardware. The change is achieved merely by reprogramming.
- *Reliability.* Computers, microprocessors and special-purpose signal processing chips manage complex operations using few devices and few interconnections.

As a result of these advantages digital processing is rapidly increasing in its importance relative to analogue processing.

1.7 Some signals of special importance

A signal $x(t)$ that varies sinusoidally with time at a frequency f may be written

$$x(t) = A\sin(2\pi ft + \theta)$$

where A is the peak amplitude and θ is the phase of the sinusoid at $t = 0$. The waveform repeats itself with a period $1/f$, and is as illustrated in Figure 1.15(a).

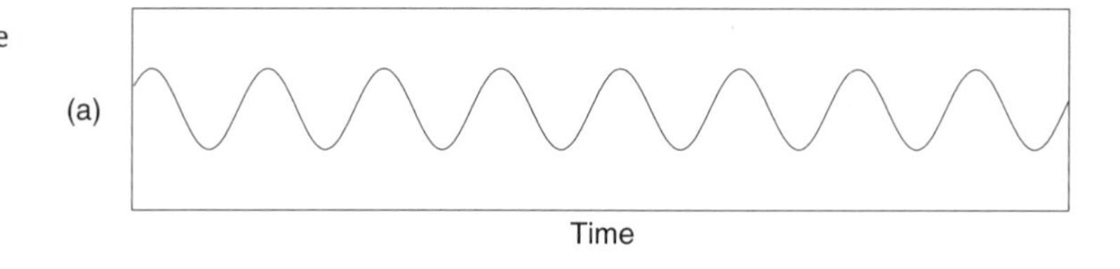

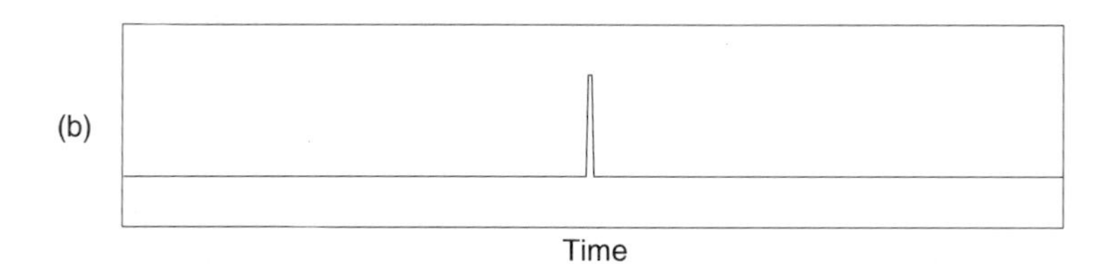

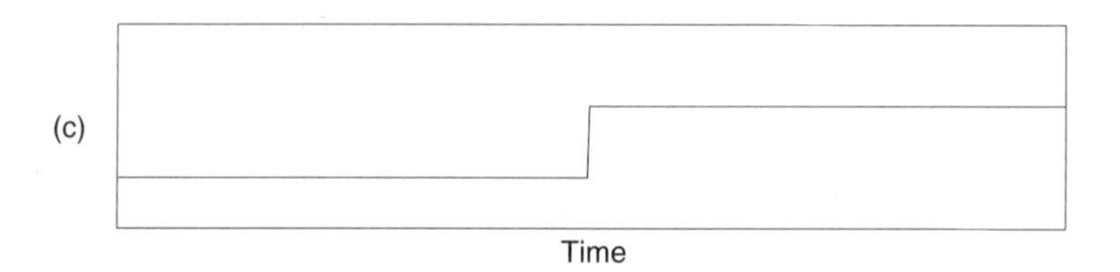

Fig. 1.15 Some signals of special importance: (a) sinusoidal signal; (b) short pulse approximation of impulse function; (c) step function

An alternative expression for the sinusoidal signal is $x(t) = A\sin(\omega t + \theta)$, where $\omega = 2\pi f$ and is known as a radian frequency, with units of rad/s. Sometimes f is termed a cyclic frequency to distinguish it clearly from a radian frequency. The units of f are hertz (abbreviated Hz), which is the modern-day accepted standard that replaces the older but more descriptive term of cycles per second.

There are many reasons why sinusoids are important.

- They are *common in nature*, being associated with *resonance*.
- Any *periodic* signal can be considered as comprised of the *sum of sinusoids* of different frequencies.
- Sinusoids change in amplitude and phase when they pass through linear networks but nevertheless *stay sinusoids*.
- The effect of a linear network can be *totally characterized* by the gain and phase shift as a function of frequency, $A(\omega)$ and $\phi(\omega)$, respectively. For example, a signal $V_{\mathrm{m}}\sin(\omega_0 t + \theta)$ passing through such a linear network would emerge as $V_{\mathrm{m}}A(\omega_0)\sin\{\omega_0 t + \theta + \phi(\omega_0)\}$.

For these reasons the sinusoid is perhaps the most important waveform in physics and engineering.

Probably the only competitor to the sinusoid in terms of its importance is the *impulse function*. Here we need to distinguish between the impulse function that is relevant to continuous-time signals and the impulse function that is relevant to discrete-time signals. In the latter case the term *sample function* can also be used, and this has the advantage of distinguishing between the two types of function. The impulse function that is relevant to continuous-time signals can be considered to be the limiting case of a rectangular pulse that is made very, very short (see Figure 1.15(b) for an appropriate rectangular pulse), but this is examined more thoroughly in the next chapter. For analogue systems the importance of this function arises because any *continuous* signal may be considered as comprised of an infinite number of adjacent impulse functions that are spaced infinitely closely. If the system response to a single impulse function is known, it is then possible to extrapolate from this to determine the response to the complete signal. In a similar way, any discrete-time signal may be considered as comprised of samples, one for each data value, and the response of a digital system to one such sample can be used to predict the response to the complete discrete-time signal. Impulse functions and impulse responses are considered further in the next chapter.

After the sinusoid and impulse function the step function shown in Figure 1.15(c) is probably next in importance, particularly in automatic control systems. One might for example have a system in which the pointing direction of a closed-circuit TV camera is proportional to a voltage generated by a security guard sitting at a console. If the guard sees something suspicious he may wish to redirect the camera by rapidly applying a voltage appropriate to the new direction. It is important to know how the camera moves in response to this step change in voltage. The generally recognized symbol for a continuous-time unit step function is $u(t)$. This is defined as

$$\begin{aligned} u(t) &= 1 \qquad t \geqslant 0 \\ &= 0 \qquad t < 0 \end{aligned} \tag{1.4}$$

Besides its applications as an input signal, the unit step function also enables a convenient mathematical notation for a signal that is zero up to some moment in time and is finite thereafter. For example, the signal

$$\begin{aligned} x(t) &= \sin \omega t \qquad t \geqslant 0 \\ &= 0 \qquad t < 0 \end{aligned}$$

can be written more elegantly and concisely as

$$x(t) = u(t) \sin \omega t$$

Figure 1.15 is a very simple plot compared with what follows later in this book, but, since a major part of this chapter is an introductory tutorial on the use of MATLAB, it provides a useful opportunity to demonstrate some more of its plotting commands. The commands not yet met are followed by explanatory comments.

```
% fig1_15.m
t=(1:500);
x=.4*sin(.1*t+.5);
subplot(4,1,1)
plot(x)
set(gca,'xtick',[],'ytick',[])          % this turns the tick marks off
axis([0 500 -1 1])
text(-35, 0,'(a)')
% this places the text message (a) at a position (-35,0), where these units are in
% accordance with the current axes as specified by the axis command above
```

```
xlabel('time')
y = zeros(size(t));
y(249:251) = [1 1 1];
% note that when assigning these three values of y to unity we have to equate them
% to a vector of three ones. In MATLAB 5 we can simply write y(249:251) = 1.
subplot(4,1,2)
plot(y)
set(gca,'xtick',[],'ytick',[])
axis([0 500 -.5 1.5])
text(-35, .5,'(b)')
% this places the text message (b) at a position (-35,0.5), where these units are in
% accordance with the current axes as specified by the axis command above
xlabel('time')
y = zeros(size(t));
y(250:500) = .7*ones(1,251);
% this assigns the subvector of y to a vector of ones that is of the same size.
subplot(4,1,3)
plot(y)
set(gca,'xtick',[],'ytick',[])
axis([0 500 -.5 1.5])
text(-35, .5,'(c)')
% this places the text message (c) at a position (-35,0.5), where these units are in
% accordance with the current axes as specified by the axis command above
xlabel('time')
```

1.8 ● Summary

After a brief tutorial in the use of MATLAB some important classifications of signal are defined, namely analogue, continuous-time, discrete-time, sampled data, digital, deterministic and random. It is emphasized that most signals of interest in nature are analogue, but that the tendency is increasingly to sample them and then process them digitally. This is done because of the capacity of digital signal processing for sophisticated operations and for its accuracy, reliability and programming flexibility. As a further justification, many analogue signals are in any case only *known* at discrete moments in time (one such example would be rainfall measurements at 24 hour intervals).

1.9 ● Problems

1. Give the MATLAB instructions to derive a vector variable containing 100 '1's, followed by 100 '2's, followed by 50 zeros, followed by 80 '3's. Display on the screen elements 245 to 255.

2. A vector variable x has been created in MATLAB using the command **x=[2 -1 4 3 -5]**; Give the commands needed to create a new variable y of 500 elements in which elements 300 to 304 have the same values as x, and in which the other elements have zero values.

3. A pulse is defined by

$$x(t) = \sin(2000\pi t) \qquad 0 < t < 10 \text{ ms}$$
$$= 0 \qquad \text{elsewhere}$$

Give the MATLAB instructions needed to derive and plot a sequence of these signal values 0.1 ms apart occupying the timespan $0 < t < 30$ ms.

4. A pulse is defined by

$$x(t) = \sin(2000\pi t) \qquad -5 < t < 5 \text{ ms}$$
$$= 0 \qquad \text{elsewhere}$$

Give the MATLAB instructions needed to derive and plot a sequence of these signal values 0.1 ms apart occupying the timespan $-15 < t < 15$ ms.

5. Give MATLAB instructions needed to produce on the computer screen three separate plots, one of $\cos(2000\pi t)$, one of $\cos(10000\pi t)$, and one of $\cos(2000\pi t)\cos(10000\pi t)$. The plots should be based upon 401 points occupying the timespan $0 < t < 4$ ms.

Chapter two

Impulse functions, impulse responses and convolution

2.1 ● Preview

Any discrete-time signal may be considered as comprised of samples. If the response of a digital system to a *single* sample of unit magnitude is known it is possible to extrapolate the time-domain response to the *complete* discrete-time signal.

This concept may be extended such that any *continuous* signal may be considered as being approximately comprised of a very large number of very short adjacent pulses. The concept of an impulse function is introduced. This is the limiting case of a short pulse whose duration becomes infinitely small but whose area is finite. The effect of a continuous signal on a linear system may then be considered the same as if the continuous signal is considered to be comprised of an *infinite* number of impulse functions that are spaced *infinitely closely.* If the system response to a *single* impulse is known it is possible to extrapolate the time-domain response to the *complete* signal. It is primarily for this reason that impulses are of great importance.

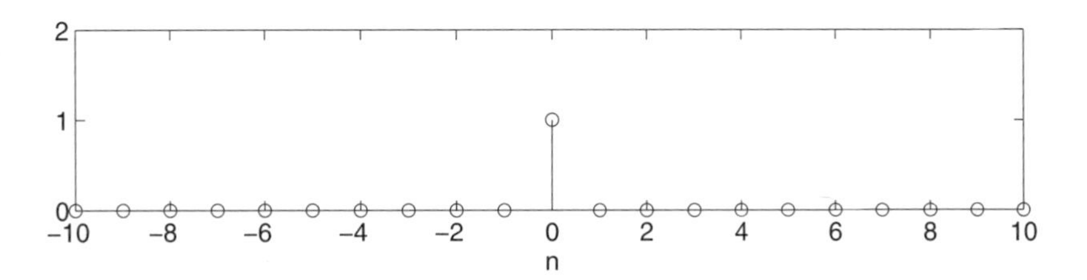

Fig. 2.1 Graphical representation of the unit sample sequence $\delta[n]$

2.2 The unit sample function (or discrete-time unit impulse function)

The unit sample function $\delta[n]$ is a very special discrete-time signal whose samples have unit amplitude when $n = 0$, but are zero for all other values of n, that is:

$$\begin{aligned} \delta[n] &= 1 \qquad n = 0 \\ \delta[n] &= 0 \qquad n \neq 0 \end{aligned} \tag{2.1}$$

A plot of $\delta[n]$ is shown in Figure 2.1, where, following the usual convention, samples of discrete signals are shown as 'stems' capped with a small circle on top. In this case the amplitude of the single stem is unity. Alternative terms to amplitude are 'strength', 'height' and 'weight'.

Samples can have any amplitude or position. For example, $1.6\delta[n-2]$ represents a sample of amplitude 1.6 at $n = 2$, as shown in Figure 2.2.

Any discrete signal may be expressed as the sum of samples. For example, the sequence

$$\{2, \underset{\uparrow}{3}, 4, 1\}$$

may be written as $2\delta[n+1] + 3\delta[n] + 4\delta[n-1] + \delta[n-2]$. Generalizing this, an alternative way of expressing any discrete signal $x[n]$ is as

$$x[n] = \sum_{k=-\infty}^{\infty} x[k]\delta[n-k] \tag{2.2}$$

The term *discrete-time unit impulse function* is used quite commonly as an alternative to the term *unit sample function.*

2.3 Discrete-time impulse responses and the convolution sum

It may well be asked why it is useful to replace $x[n]$ by the more complicated expression $\sum_{k=-\infty}^{\infty} x[k]\delta[n-k]$. The reason concerns the simple manner in which

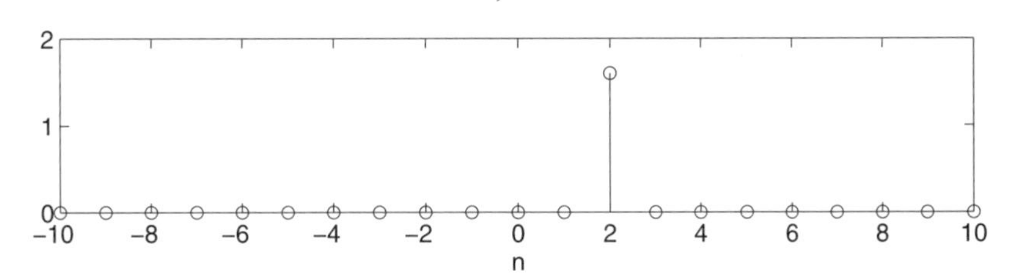

Fig. 2.2 Graphical representation of $1.6\delta[n-2]$

these expressions can be modified to provide the *output* of a system to an input $x[n]$ when the system impulse response is known.

Let the response of a discrete LTI system to a unit sample function $\delta[n]$ be $h[n]$, where $h[n]$ is termed the impulse response. The response to a sample function $x[k]\delta[n-k]$ will then be $x[k]h[n-k]$. Replacing $x[k]\delta[n-k]$ by $x[k]h[n-k]$ within the equation $x[n] = \sum_{k=-\infty}^{\infty} x[k]\delta[n-k]$ we find that the response of the system to an input $x[n]$ is

$$y[n] = \sum_{k=-\infty}^{\infty} x[k]h[n-k] \tag{2.3}$$

This is known as the convolution sum. A shorthand notation for this operation of convolution is $x[n] * h[n]$, where the asterisk is used to denote convolution, that is

$$x[n] * h[n] = \sum_{k=-\infty}^{\infty} x[k]h[n-k] \tag{2.4}$$

The asterisk used for convolution must not be confused with the asterisk commonly used in computer languages to denote multiplication.

Example 2.1

Determine the impulse response of the non-recursive digital network of Figure 2.3. Hence determine the output sequence when the input sequence is

$$x[n] = \{2, \underset{\uparrow}{1}, 1.5, .5\}$$

Solution If a discrete-time unit impulse (a unit sample function) is applied to the network of Figure 2.3 it is influenced by each of the three input coefficients in turn. Thus the impulse response is given by the sequence

$$h[n] = \{\underset{\uparrow}{.5}, 1.5, 1\}$$

as shown in Figure 2.4(a).

The input sequence $x[n]$ is shown in Figure 2.4(b) and one can consider the response to each of these input samples separately. For example the first sample of the input sequence $x[-1]$ corresponds to $k = -1$ and, having a weight of 2, gives rise to the sequence $2h[n+1]$ shown in Figure 2.4(c). The second sample of the sequence corresponds to $k = 0$, has a weight of one, and gives rise to the sequence $h[n]$ as shown in Figure 2.4(d). Similarly the third and fourth samples give rise to

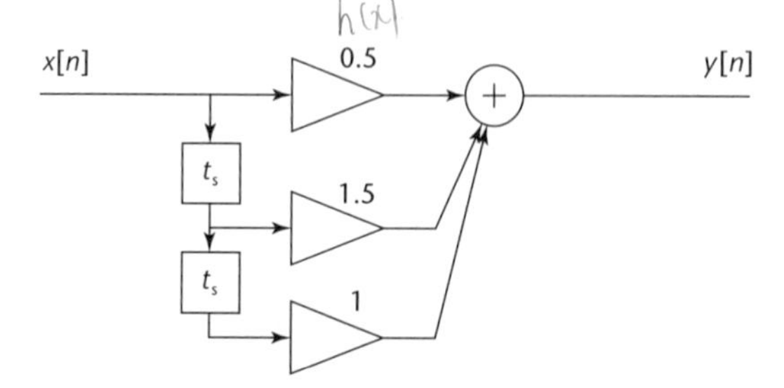

Fig. 2.3 ● Simple non-recursive digital network

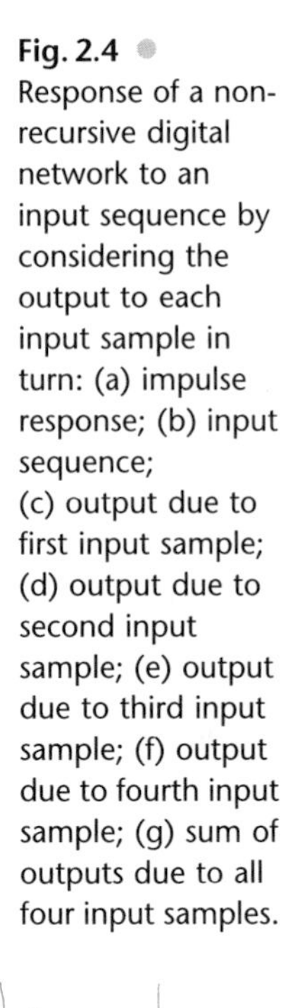

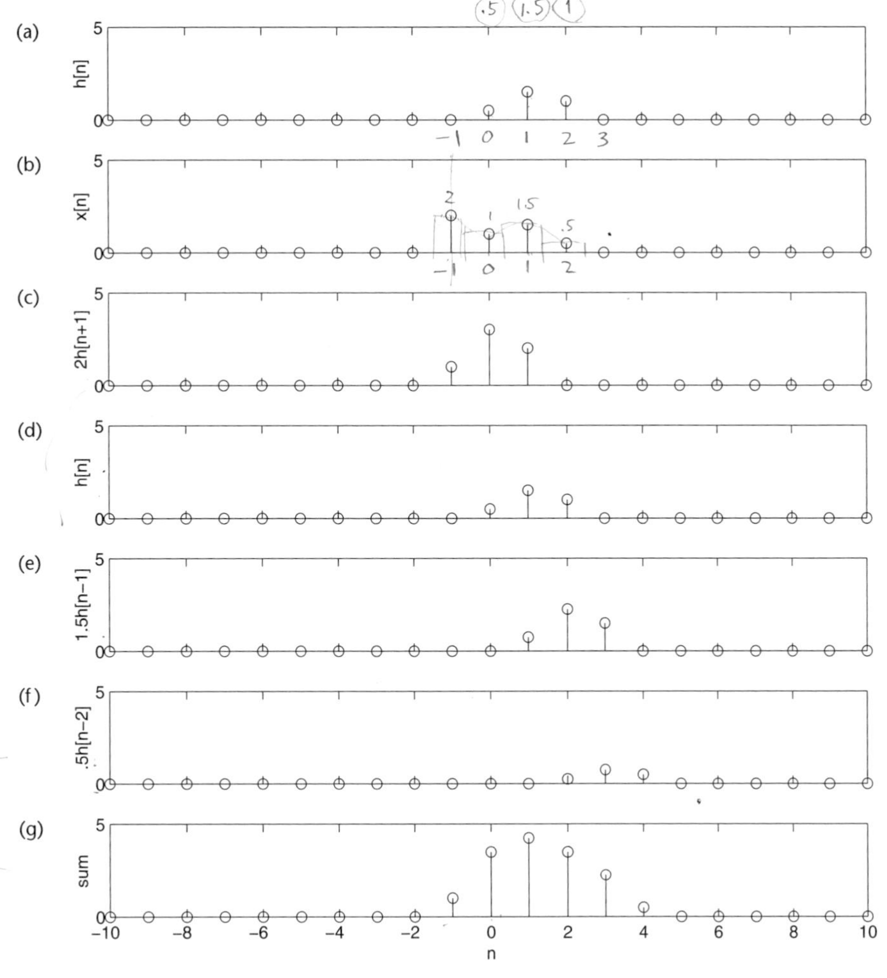

Fig. 2.4 ● Response of a non-recursive digital network to an input sequence by considering the output to each input sample in turn: (a) impulse response; (b) input sequence; (c) output due to first input sample; (d) output due to second input sample; (e) output due to third input sample; (f) output due to fourth input sample; (g) sum of outputs due to all four input samples.

$1.5h[n-1]$ and $0.5h[n-2]$, or $x[1]h[n-1]$ and $x[2]h[n-2]$, as shown in Figures 2.4(e) and (f).

The response to all four input samples is the sum of Figures 2.4(c), (d), (e) and (f) and is the convolution sum $\sum_{k=-\infty}^{\infty} x[k]h[n-k]$. As shown in Figure 2.4(g), it is

$$\{1, 3.5, 4.25, 3.5, 2.25, .5\}$$

It is important to recognize that the convolution sum can be evaluated without resort to a diagram. It can be done by a systematic tabulation and addition of the individual responses to each input sample.

Example 2.2

Repeat the convolution of Example 2.1 by tabulating and adding the individual responses to each input sample.

Solution The first input sample of weight 2 at $n = -1$ gives an output which is the impulse response multiplied by 2 and advanced by one sample, thus producing the sequence

$$\{1, \underset{\uparrow}{3}, 2\}$$

The second input sample of weight 1 at $n = 0$ gives an output which is simply the impulse response multiplied by unity, thus producing the sequence

$$\{\underset{\uparrow}{.5}, 1.5, 1\}$$

In a similar way, the third sample produces

$$\{\underset{\uparrow}{0}, .75, 2.25, 1.5\}$$

and the fourth produces

$$\{\underset{\uparrow}{0}, 0, .25, .75, .5\}$$

The net resultant is is determined by aligning these sequences and adding them, as shown below

1	$\underset{\uparrow}{3.0}$	2.0			
	$\underset{\uparrow}{0.5}$	1.5	1.0		
	$\underset{\uparrow}{0.0}$	0.75	2.25	1.5	
	$\underset{\uparrow}{0.0}$	0.0	0.25	0.75	0.5
1.0	$\underset{\uparrow}{3.5}$	4.25	3.5	2.25	0.5

As in Example 2.1 the result is the sequence

$$\{1, \underset{\uparrow}{3.5}, 4.25, 3.5, 2.25, .5\}$$

There is an interesting 'trick' for achieving the convolution sum (one which turns out to have additional significance later in the book). This is to replace each sequence by a polynomial of a variable z in which the power of each term conveys the position of that term relative to the origin. The two polynomials are multiplied and the inverse operation is then applied to the resulting product.

Example 2.3

Repeat the convolution of Example 2.1 by replacing the sequences by appropriate polynomials in z, multiplying these polynomials, and then replacing the resulting polynomial by a sequence.

Solution Following this procedure let the input sequence

$$\{2, \underset{\uparrow}{1}, 1.5, .5\}$$

be 'transformed' into the polynomial $2z + 1 + 1.5z^{-1} \times 0.5z^{-2}$, where multiplication of a sample amplitude by z indicates that the sample is advanced by one sample interval relative to the time origin. Similarly multiplications of sample amplitudes by z^{-1} and z^{-2} indicate delays of one and two sample intervals respectively.

In a similar way the impulse response

$$\{\underset{\uparrow}{.5}, 1.5, 1\}$$

is transformed into the polynomial $0.5 + 1.5z^{-1} + z^{-2}$.

The product of the two polynomials can then be achieved by long multiplication as follows:

$$
\begin{array}{llllll}
0.5 + 1.5z^{-1} + z^{-2} & & & & & \\
2z + 1 & + 1.5z^{-1} & + 0.5z^{-2} & & & \\
\hline
z \quad + 3 & + 2z^{-1} & & & & \\
\quad .5 & + 1.5z^{-1} & + z^{-2} & & & \\
 & .75z^{-1} & + 2.25z^{-2} & + 1.5z^{-3} & & \\
 & & 0.25z^{-2} & + 0.75z^{-3} & + 0.5z^{-4} & \\
\hline
z \quad + 3.5 & + 4.25z^{-1} & + 3.5z^{-2} & + 2.25z^{-3} & + 0.5z^{-4} &
\end{array}
$$

Performing the inverse procedure (or transform) to regain a sequence from this resulting polynomial, we obtain

$$\{1, \underset{\uparrow}{3.5}, 4.25, 3.5, 2.25, .5\}$$

which is the same result as before.

The justification for the procedure of Example 2.3 is clear by noting the similarity of long multiplication with the procedure of summing the impulse responses from the individual input terms that was shown previously in Example 2.2. What happens is that the powers of the z terms establish the timings of the individual impulse responses so that they are forced to add up correctly.

The transform of a sequence into a polynomial in the way shown is known as the z transform of the sequence, and will be examined from a different viewpoint in Chapter 14. It is a *transform* in that an inverse procedure can regain the original sequence, that is

$$\{\underset{\uparrow}{0.5}, 1.5, 1\} \xrightarrow{z \text{ transform}} 0.5 + 1.5z^{-1} + z^{-2} \xrightarrow{\text{inverse } z \text{ transform}} \{\underset{\uparrow}{0.5}, 1.5, 1\}$$

Transforms only become established procedures if they are useful. In this case the justification is that the somewhat complex operation of convolution can be achieved by an alternative and simpler operation, namely that of taking the z transforms of the input and of the impulse response, multiplying them, and taking the inverse z transform. Chapter 14 will find further and more substantial justifications for the z transform.

This section concerns impulse responses and convolution. It has been shown that the output of a discrete-time system in response to an input may be deduced by means of the convolution sum if the impulse response is known, and it is for this reason that the impulse response is such an important means of characterizing a system.

2.4 Alternative forms and interpretations of the convolution sum

Equation (2.3) gives the convolution between an input $x[n]$ and an impulse response $h[n]$ as $y[n] = \sum_{k=-\infty}^{\infty} x[k]h[n-k]$. Physically this has been interpreted as the sum of a set of impulse responses, where each of the impulse responses has arisen from one of the input samples. In the equation k is a 'dummy' variable which disappears after summing over all values of k.

An interchange of the variables n and k gives the alternative equation

$$y[k] = \sum_{n=-\infty}^{\infty} x[n]h[k-n] \tag{2.5}$$

This time the outcome is in terms of the variable k and n is the dummy variable which disappears after summing over all values of n. Equations (2.3) and (2.5) each have their merits. Equation (2.3) has the advantage that the answer is in terms of the original variable n, but the disadvantage that a change in variable occurs between the original sequences and those inserted into the convolution sum. Equation (2.5) has the advantage that the the original variable is retained in the sequences that are inserted into the convolution sum, but the disadvantage that the answer is in terms of a new variable k. In many respects the operation of Equation (2.5) is easier to visualize since, by retaining the original variable within the summation, it indicates very clearly that the impulse response $h[n]$ is reversed in time to produce $h[-n]$, that this is shifted by different values of k, that for each value of k this is then multiplied by $x[n]$, and that the resultant terms are then summed. This procedure leads to the concept of graphical convolution, which is best understood by an example.

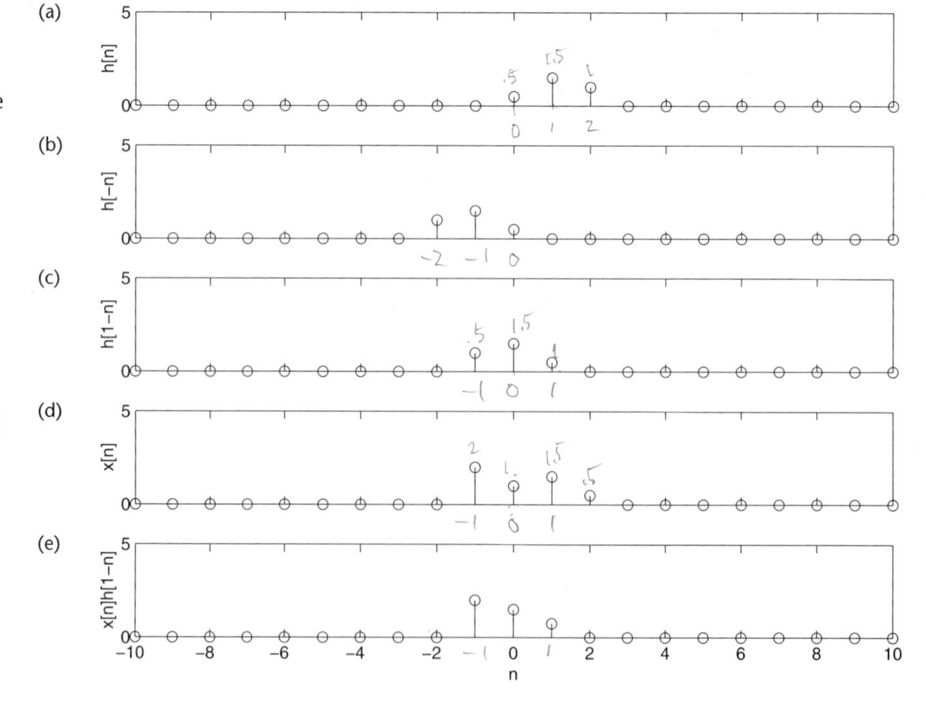

Fig. 2.5 Procedure of graphical convolution for the case of $k=1$: (a) impulse response; (b) time-reversed impulse response; (c) time-reversed impulse response delayed by one sample; (d) input sequence; (e) product of $x[n]$ and $h[1-n]$

Example 2.4

Perform the graphical convolution of the sequences used in the previous examples, namely

$$\{2, \underset{\uparrow}{1}, 1.5, 5\} \qquad \text{and} \qquad \{.5, 1.5, 1\}$$

Solution The procedure for evaluating $y[k] = \sum_{k=-\infty}^{\infty} x[n]h[k-n]$ is illustrated by Figure 2.5 for the specific case of $k = 1$. Figure 2.5(a) shows $h[n]$ and Figure 2.5(b) shows $h[-n]$. Figure 2.5(c) shows $h[-n]$ advanced by one sample period such that we have $h[1-n]$, corresponding to $k = 1$. Figure 2.5(d) shows $x[n]$ and Figure 2.5(e) shows the product $x[n]h[1-n]$, which in this case is the sequence

$$\{2, \underset{\uparrow}{1.5}, .75\}$$

The sum of these three terms is $(2 + 1.5 + 0.75)$, giving the result

$$y[1] = \sum_{n=-\infty}^{\infty} x[n]h[1-n] = 4.25$$

The same procedure can be adopted for other values of k to give the overall result that

$$y[k] = \sum_{n=-\infty}^{\infty} x[n]h[k-n] = \{1, \underset{\uparrow}{3.5}, 4.25, 3.5, 2.25, .5\}$$

Returning to the relative merits of writing the convolution sum as $y[n] = \sum_{k=-\infty}^{\infty} x[k]h[n-k]$ as compared to $y[k] = \sum_{n=-\infty}^{\infty} x[n]h[k-n]$ an advantage of the former is that it allows the result to be written $x[n] * h[n] = y[n]$, since the same variable is retained for the outcome as is used for the functions convolved. In contrast, one cannot write $y[k] = \sum_{n=-\infty}^{\infty} x[n]h[k-n] = x[n] * h[n]$ since the final variable changes from n to k. A way around this would be to write:

$x[n] * h[n] = y[n]$ where, making a change of variable from n to k,

$$y[k] = \sum_{n=-\infty}^{\infty} x[n]h[k-n]$$

However, this is a fairly minor argument and it should be emphasized that it is important to be familiar with both forms of the convolution sum, as both are widely used and the differences in their form do sometimes cause confusion.

So far the applications of convolution have concerned the effect of the impulse response of a discrete LTI system on an input sequence. There are many situations, however, where neither sequence convolved would normally be considered an impulse response. For example, a digital communication signal could be 'scrambled' for security purposes by convolving it with a codeword known only to legitimate users. The codeword could be also used for unscrambling the message. Hence it is useful to recognize that equivalent formulae to Equations (2.3) and (2.5) are available for convolving two sequences $x_1[n]$ and $x_2[n]$. These are

$$y[n] = \sum_{k=-\infty}^{\infty} x_1[k]x_2[n-k] \tag{2.6}$$

and

$$y[k] = \sum_{n=-\infty}^{\infty} x_1[n]x_2[k-n] \tag{2.7}$$

2.5 Convolution using MATLAB

The treatment of how to evaluate the convolution sum has so far been directed at providing a physical interpretation of convolution. However, the simple way of performing a convolution, and one which is particularly advantageous with long sequences, is to use MATLAB. A specific function exists in MATLAB for convolution, and this makes the task trivial. Considering for example the convolution of the sequences $x_1[n] = \{2.4, 3.6, 0.2, 5.3, 1.4, 4.4, 3.6\}$ and $x_2[n] = \{3.1, 3.8, 5.2, 7.7, 2.5, 4.6\}$, a suitable program is

```
x1=[2.4 3.6 0.2 5.3 1.4 4.4 3.6];
x2=[3.1 3.8 5.2 7.7 2.5 4.6];
y=conv(x1,x2)                    % this convolves sequences x1 and x2
```

The outcome is

```
y =

Columns 1 through 7

 7.4400  20.2800  26.7800  54.3900  59.2400  68.1000  93.0300

Columns 8 through 12

 61.5100  80.4800  45.1600  29.2400  16.5600
```

In other words

$$x_1[n] * x_2[n] = \{7.44, 20.28, 26.78, 54.39, 59.24, 68.10, 93.03, 61.51, 80.48, 45.16, 29.24, 16.56\}$$

It will be noted that there are no marker arrows in the original expressions for $x_1[n]$ and $x_2[n]$, meaning, by default, that these sequences commence at $n = 0$. Consideration of the graphical interpretation of convolution shows that the first term of $x_1[n] * x_2[n]$ also occurs at $n = 0$, thus confirming that it is also appropriate to omit the marker from the result for $x_1[n] * x_2[n]$.

The problem is somewhat more difficult when the sequences do not commence at $n = 0$, because the outcome of the **conv** function does not reveal timing information. A specific example will be examined and by means of this a general set of commands will be derived. What follows adopts the use of n and k that is in accordance with Equations (2.5) and (2.7), namely that $y[k] = \sum_{n=-\infty}^{\infty} x_1[n]x_2[k-n]$.

Consider the convolution of the same two sequences but with different origins, that is

$$x_1[n] * x_2[n] \quad \text{where } x_1[n] = \{2.4, 3.6, \underset{\uparrow}{0.2}, 5.3, 1.4, 4.4, 3.6\}$$

and

$$x_2[n] = \{3.1, 3.8, 5.2, 7.7, \underset{\uparrow}{2.5}, 4.6\}$$

Whereas the element indices of a vector in MATLAB always commence at 1 it will be noted that the timing (or sample) indices can be quite different and in our case begin at -2 for $x_1[n]$ and at -3 for $x_2[n]$. It follows that we must declare the timing indices together with the data sequences. Thus we now begin with the modified commands

```
x1=[4 2 6 3 8 1 5]; n1=(-2:4);
x2=[3 8 6 9 6 7]; n2=(-4:1);
```

A consideration of graphical convolution readily shows that the timing indices of the convolution sum will be from $(-2-4)$ to $(4+1)$, or from -6 to $+5$. Generalizing this, the lowest timing index of the convolution sum is the sum of the lowest timing indices of $x_1[n]$ and $x_2[n]$, while the highest timing index of the convolution sum is the sum of the highest timing indices of $x_1[n]$ and $x_2[n]$. We can find these automatically in MATLAB with the commands

```
kmin=n1(1)+n2(1);
kmax=max(n1)+max(n2);
```

We can next evaluate and display the convolution sum while simultaneously generating and displaying an array of its samples indices. This is done using the command

```
y=conv(x1,x2), k=(kmin:kmax)
```

The result obtained is

```
y =

    12   38   58   105   126   179   148   178   108   107   37   35

k =

    -6   -5   -4   -3   -2   -1   0   1   2   3   4   5
```

We can easily plot this last result using the command **stem(k,y)**, and for many purposes the outcome is totally adequate. However, it can often be very useful to accompany a plot of the convolution sum with plots of the sequences themselves. One possibility for obtaining these three plots would be the sequence of instructions given by

```
subplot(3, 1, 1)
stem(n1, x1)
subplot(3, 1, 2)
stem(n2, x2)
subplot(3, 1, 3)
stem(k,y)
```

Unfortunately, since $x_1[n]$, $x_2[n]$ and $y[n]$ are all of different lengths, the scales of the horizontal axes would be different for all three plots, and this is far from ideal. One way of overcoming this difficulty is to determine the lower and upper indices of the two data sets and of the convolution sum, to select the lowest and highest of all three, and then to augment all three sequences with zeros such that they are all defined between the lower and upper indices. Since the complete set of instructions involves several commands and may be required many times this is a good opportunity to show how users can generate *their own functions* in MATLAB. Such functions can be used subsequently in just the same way as **sin(x)**, **stem(k,y)**, or any other command. The objective in what follows is to generate a new MATLAB function **convstem(x1, n1, x2, n2)** which, when *preceded by the specification of variables* in the order $x1$, $n1$, $x2$, $n2$, performs the complete process of producing plots of $x_1[n]$, $x_2[n]$ and $x_1[n]*x_2[n]$, all on the same horizontal scale.

In order to create a new MATLAB function it is necessary to develop an M-file of which the first line specifies that a new function is being generated, and specifies the name of that function. The filename must be the same as the function name. The following is a suitable M-file.

```
% convstem.m is for plotting two sequences and their convolution sum such that
% all three plots have the same scale of indices and the index scale extends from
% the lowest index amongst the three to the highest index amongst the three.
function convstem(x1, n1, x2, n2)
y=conv(x1,x2);
% We now need to find the lowest index kmin of the convolution sum.
kmin=n1(1)+n2(1);
% We next make the lowest index of the plot equal to the lowest index amongst
% x1, x2 and the convolution sum.
kmin_plot=min([n1(1) n2(1) kmin]);
% We need to find the the highest index kmax of the convolution sum
kmax=max(n1)+max(n2);
% We next make the highest index of the plot equal to one plus the highest index
% amongst x1, x2 and the convolution sum. The one is needed since we shall later
% assign the last element of each vector to zero and we must not destroy any data.
kmax_plot=max([max(n1) max(n2) kmax])+1;
% We next generate a vector of indices that is appropriate to plots of x1, x2 and
% the convolution sum.
k_plot=(kmin_plot : kmax_plot);
% Next come the plots themselves. Each one must be padded with zeros outside the
% range of the original data. The technique follows that presented in Sec.1.2.8
xx1=[zeros(1,n1(1)-kmin_plot) x1];
xx1(length(k_plot))=0;      % This extenson of the xx1 vector by means of this
% assignment command is why we extended kmax_plot by one.
xx2=[zeros(1,n2(1)-kmin_plot) x2];
xx2(length(k_plot))=0;
yy=[zeros(1,kmin-kmin_plot) y];
yy(length(k_plot))=0;
% We now plot xx1
subplot(3, 1, 1)
stem(k_plot, xx1)
% We now set the limits of the horixontal axis of the plots avoiding stems on the
% left or right boundaries by extending the boundaries beyond the lowest and
```

```
% highest timing index. We also reduce the font size and specify a tick mark for
% every lag defined in the k_plot vector.
a = length(k_plot)/50;          % The horizontal axis will be extended by this much.
set(gca,'XLim',[kmin_plot - a  kmax_plot + a],'fontsize',10,'XTick',[k_plot])
xlabel('n'),ylabel('x1[n]')
% We now plot xx2
subplot(3, 1, 2)
stem(k_plot,xx2)
set(gca,'XLim',[kmin_plot - a  kmax_plot + a],'fontsize',10,'XTick',[k_plot])
xlabel('n'),ylabel('x2[n]')
% Finally we plot the convolution sum yy.
subplot(3, 1, 3)
stem(k_plot,yy)
set(gca,'XLim',[kmin_plot - a  kmax_plot + a],'fontsize',10,'XTick',[k_plot])
xlabel('Lag k of x2[-n] relative to x1[n], or of x1[-n] relative to x2[n]')
ylabel('x1[n]*x2[n]')
```

Example 2.5

Using the **convstem(x1,n1,x2,n2)** function introduced into your computer by the preceding M-file, plot the sequences

$$x_1[n] = \{\underset{\uparrow}{0}, 0, .8, .8, .8, .8, .8, .8, \},$$

$$x_2[n] = \{0.7, 0.7, 0.7, 0.7, 0.7, \underset{\uparrow}{0.7}, 0.7, 0.7, 0.7, 0.7\}$$

and their convolution sum.

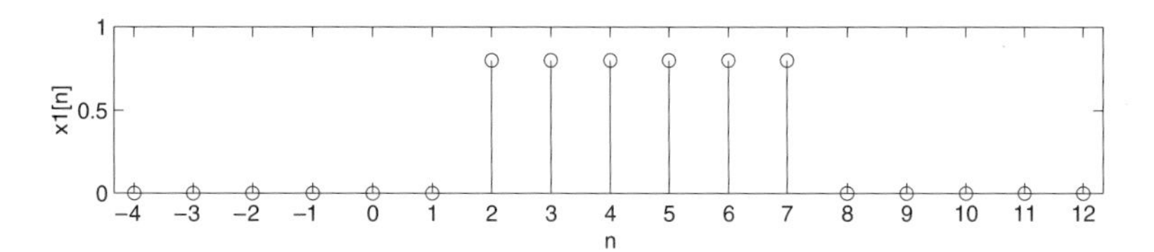

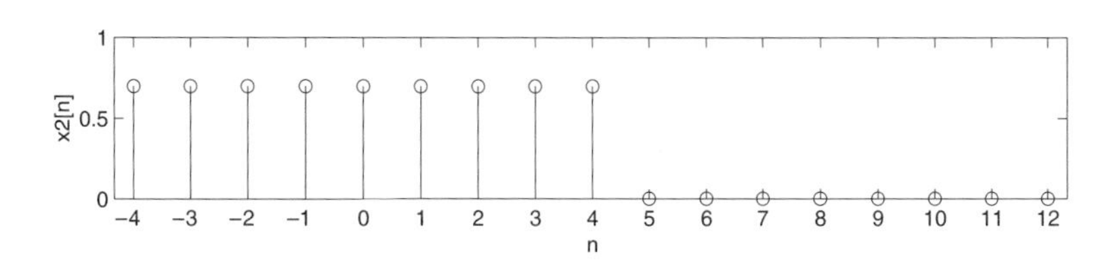

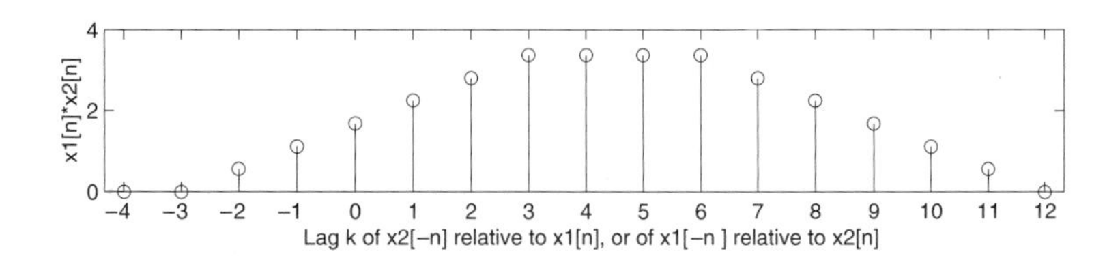

Fig. 2.6 Use of a specially created MATLAB function **convstem** for plotting two sequences and their convolution sum

Solution We require only

```
x1=[.8 .8 .8 .8 .8 .8]; n1=(2:7);
x2=[.7 .7 .7 .7 .7 .7 .7 .7 .7]; n2=(-4:4);
convstem(x1,n1,x2,n2)
```

The result is shown in Figure 2.6

2.6 Continuous-time impulse functions

For *continuous* signals the nearest equivalent to the unit sample function (or discrete-time unit impulse function) is the continuous-time unit impulse function or delta function $\delta(t)$. It may be thought of as a pulse centred at zero time that is infinitely narrow but which is of unit area. Being infinitely narrow its exact shape is not important. Conceptually, however, a rectangular shape is probably the easiest to consider. In this event the delta function can be considered to be the pulse of Figure 2.7(a), where $T \rightarrow 0$ and $V \rightarrow \infty$, but where the values of V and T are coupled by the relationship $VT = 1$.

In contrast to the 'stem' notation for a discrete impulse function, a continuous impulse function is represented by a vertical arrow whose height denotes its area (or strength). The case of the unit impulse function is shown in Figure 2.7(b).

It should be noted that the continuous-time unit impulse function $\delta(t)$ is substantially different from the discrete-time unit impulse function $\delta[n]$, the first of these having an *infinite* amplitude and the second having a *unity* amplitude. This difference is one of several reasons for making a clear distinction between the two notations, using parentheses for $\delta(t)$ and square brackets for $\delta[n]$.

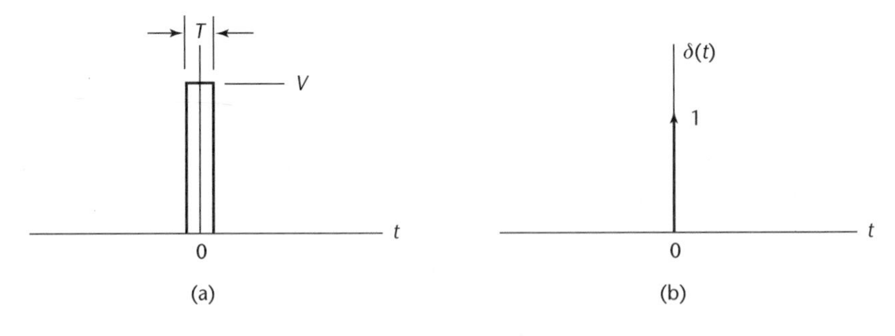

Fig. 2.7 The unit impulse function. (a) A rectangular pulse that gives the continuous unit impulse function in the limiting case of $T \rightarrow 0$ but $VT = 1$; (b) representation of the unit impulse function

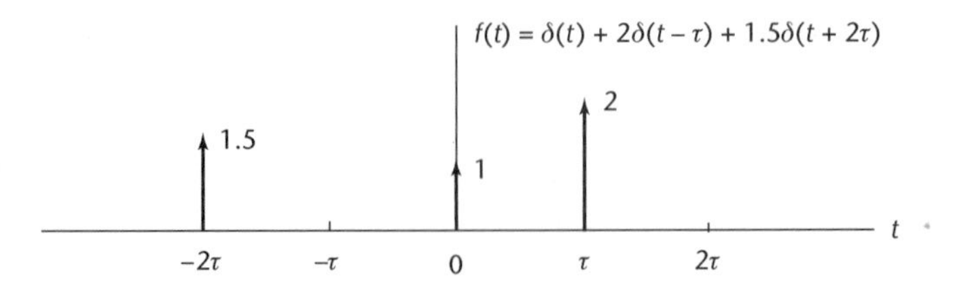

Fig. 2.8 Representation of
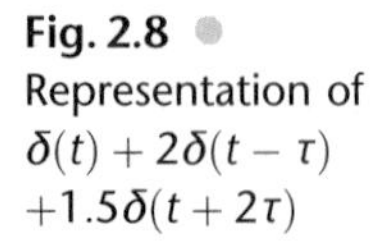
$\delta(t) + 2\delta(t - \tau) + 1.5\delta(t + 2\tau)$

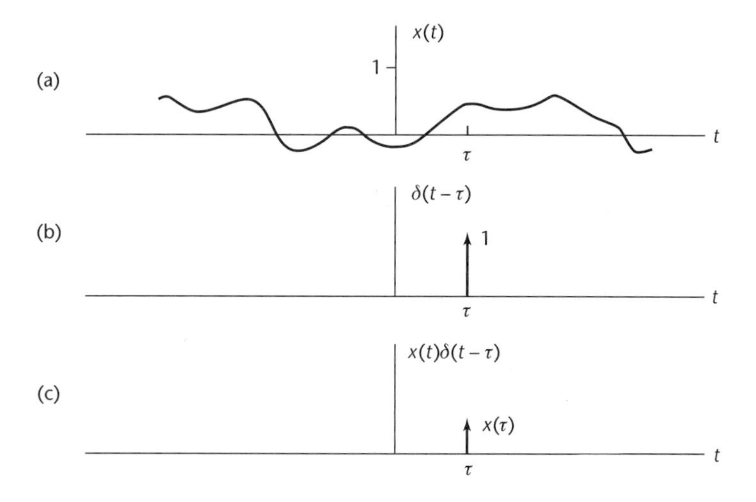

Fig. 2.9 The sifting property of the impulse

Continuous impulses may be centred at times other than zero and can have strengths other than unity. For example $\delta(t) + 2\delta(t - \tau) + 1.5\delta(t + 2\tau)$ represents one impulse of strength 1 centred at time zero (a unit impulse function), plus one impulse of strength 2 centred at time τ, plus one impulse of strength 2.5 centred at time -2τ. This is illustrated by Figure 2.8.

As another example, $x(\tau)\delta(t - \tau)$ represents a shifted impulse of strength $x(\tau)$ centred at time τ.

An important property of the impulse is its ability to *sift* a continuous signal at a particular instant. For example, as shown in Figure 2.9, a continuous signal $x(t)$ multiplied by the unit impulse $\delta(t - \tau)$ gives an impulse of weight $x(\tau)$ centred at time τ, that is:

$$x(t)\delta(t - \tau) = x(\tau)\delta(t - \tau) \tag{2.8}$$

It is important to note that the sifted signal is different from a single sample of $x(t)$. Although it has a finite weight $x(\tau)$, the impulse of Figure 2.9(c) has an infinite amplitude ($V \to \infty$), whereas a sample at time τ would have an amplitude $x(\tau)$.

It is also important to note that impulses such as $\delta(t)$ and $x(\tau)\delta(t - \tau)$ are never met in practical systems, since they have infinite amplitudes. They do, however, have considerable theoretical importance, since a network is usefully characterized by its response to an impulse function. In practice, a unit impulse function can be adequately approximated by a pulse of finite duration and finite amplitude so long as the duration is small enough and the area of the pulse is unity.

2.7 Continuous-time impulse responses and the convolution integral

A continuous signal $x(t)$ can be approximated by a 'staircase' waveform with steps $\Delta\tau$ apart, as shown in Figure 2.10. Thus it can be thought of as the linear sum of an infinite number of adjacent rectangular pulses each of width $\Delta\tau$. The times of the leading edges of these rectangular pulses increase progressively, and are $k\Delta\tau$, where k represents the integers between $-\infty$ and $+\infty$. The amplitude of the rectangular pulse at time $k\Delta\tau$ is $x(k\Delta\tau)$, and its area is therefore $x(k\Delta\tau)\Delta\tau$.

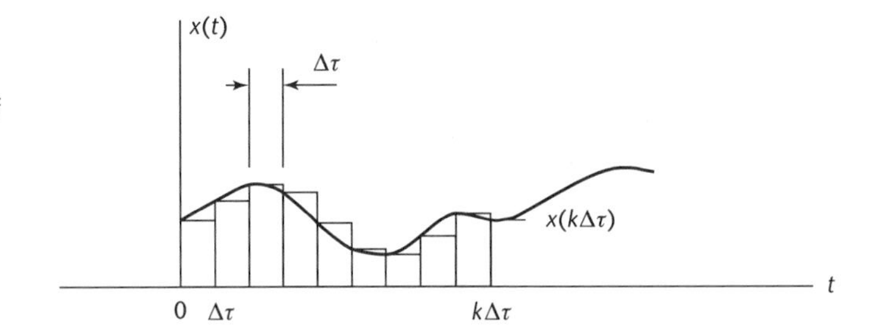

Fig. 2.10 Staircase approximation of an analogue waveform

If each of the rectangular pulses is replaced by one of half the duration but twice the amplitude, such as shown in Figure 2.11, the waveform is clearly changed. If applied to a system such as an amplifier with a limited dynamic range it could cause operation to change from the system's linear region to its non-linear region. If, however, this new waveform is applied to a *linear* system with some finite response time that smooths the waveform, the output of that system will be the same as for the original waveform so long as $\Delta\tau$ is small compared with the response time.

Taking this a stage further, we can replace the rectangular pulses of Figure 2.10 by impulses of appropriate weights and yet expect the response of a linear system to be the same. Thus, if the rectangular pulse of area $x(k\Delta\tau)\Delta\tau$ at time $k\Delta\tau$ is replaced by an impulse $x(k\Delta\tau)\Delta\tau.\delta(t - k\Delta\tau)$, we can sum all such impulses to obtain a new waveform $x'(t)$ that has a similar effect on a linear system to the original waveform $x(t)$. The full expression for $x'(t)$ is

$$x'(t) = \sum_{k=-\infty}^{\infty} x(k\Delta\tau)\delta(t - k\Delta\tau)\Delta\tau \qquad (2.9)$$

This waveform $x'(t)$ is the set of impulse functions shown in Figure 2.12.

Consider next what happens if the signal $x'(t)$ given by the summation $\sum_{k=-\infty}^{\infty} x(k\Delta\tau)\delta(t - k\Delta\tau)\Delta\tau$ enters a system with an impulse response $h(t)$. A

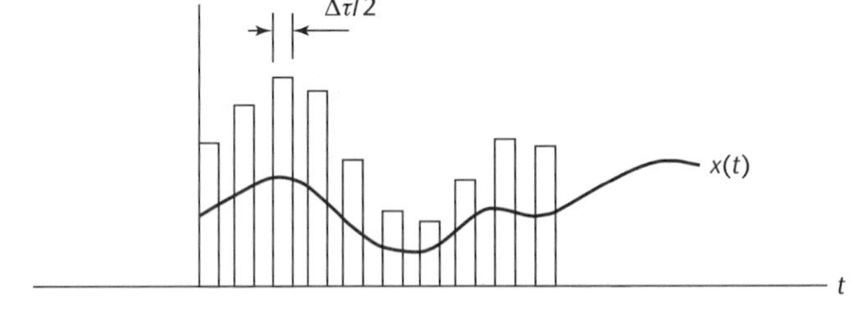

Fig. 2.11 Input to a linear system that gives the same output as the staircase waveform of Figure 2.10

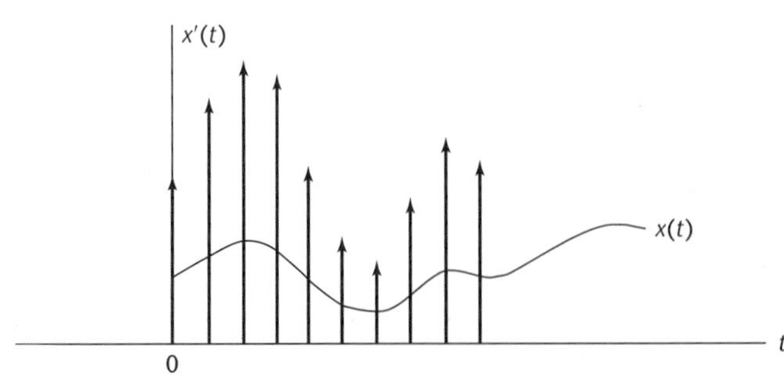

Fig. 2.12 Replacement of an analogue waveform by a set of impulses

delayed impulse $\delta(t - k\Delta\tau)$ would leave the system as $h(t - k\Delta\tau)$ and thus each impulse within the summation is modified by the system to leave as $x(k\Delta\tau)h(t - k\Delta\tau)\Delta\tau$. This causes the output $y(t)$ to be related to the input $x'(t)$ by the expression

$$y(t) = \sum_{k=-\infty}^{\infty} x(k\Delta\tau)h(t - k\Delta\tau)\Delta\tau \tag{2.10}$$

Taking the limiting condition of $\Delta\tau \to 0$ and $k\Delta\tau \to \tau$ this becomes

$$y(t) = \int_{-\infty}^{\infty} x(\tau)h(t - \tau)\ \mathrm{d}\tau \tag{2.11}$$

Since this is the output due to the impulsive waveform $x'(t)$ *it is also the output due to the original analogue waveform* $x(t)$.

The integral of Equation (2.11) is the convolution integral. It uses the dummy variable τ. *It enables the output to be calculated for any input and thus emphasizes the importance of the impulse response as a means of characterizing a system.*

As with the convolution sum a shorthand notation for this operation of convolution is the asterisk, that is

$$y(t) = x(t) * h(t) = \int_{-\infty}^{\infty} x(\tau)h(t - \tau)\ \mathrm{d}\tau \tag{2.12}$$

2.8 A graphical interpretation of the convolution integral

In a similar manner to what was done in Section 2.4, we can interchange the two variables in Equation (2.11). Interchanging t with τ produces the equation

$$y(\tau) = \int_{-\infty}^{\infty} x(t)h(\tau - t)\ \mathrm{d}t \tag{2.13}$$

This equation is well suited to a graphical interpretation. The graphical procedure for convolving a continuous input signal with an impulse response is similar to that applied to the convolution sum in Section 2.4, except that it now involves integration rather than summation. It provides considerable physical insight into the convolution operation.

The integral $\int_{-\infty}^{\infty} x(t)h(\tau - t)\mathrm{d}t$ suggests that the input signal, $x(t)$, should be multiplied by a time-reversed version of $h(t)$ that is displaced by τ, and that the product should then be integrated. This should be done for all values of τ.

Example 2.6

Use graphical convolution to determine the output of an *RC* network when the input is a rectangular pulse.

Solution It can be shown that the impulse response of the simple *RC* circuit of Figure 2.13 is $(1/RC)\mathrm{e}^{-t/RC}$. The procedure of graphical convolution is shown in Figure 2.14.

Figure 2.14(a) shows the impulse response as a function of a time variable t, while Figure 2.14(b) shows it time-reversed and delayed by a specific time τ_0. Figure 2.14(c) shows the input signal as a function of the time variable t and Figure 2.14(d)

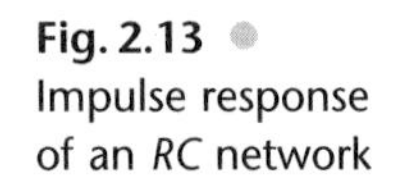

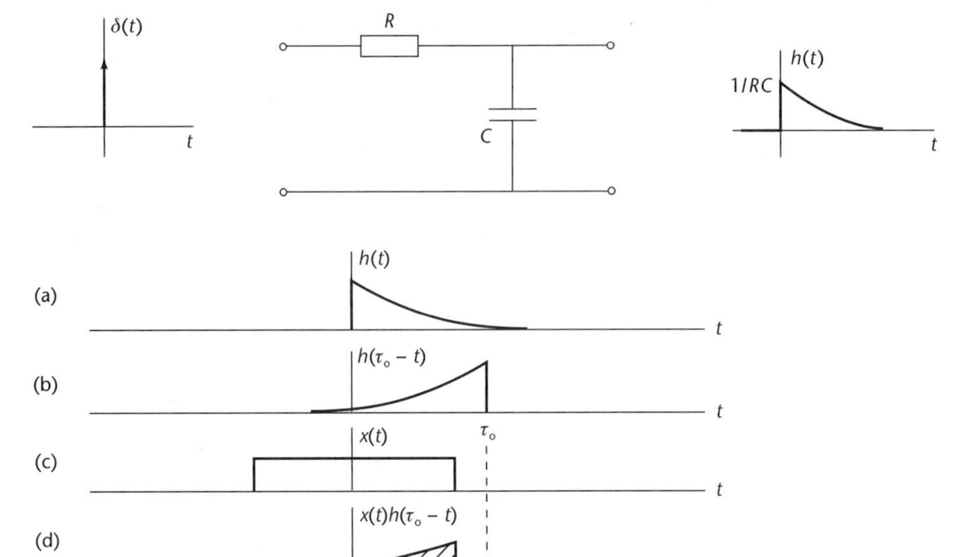

Fig. 2.13 Impulse response of an *RC* network

Fig. 2.14 Graphical convolution of rectangular pulse with impulse response

shows the product of $x(t)$ and $h(\tau_0 - t)$. Figure 2.14(e) denotes the area of the product by a cross at $\tau = \tau_0$. Figure 2.14(f) shows the result of evaluating $\int_{-\infty}^{\infty} x(t)h(\tau - t)\mathrm{d}t$ for all values of τ. It represents the output of the *RC* network to the rectangular pulse. It is the same result that is achieved by the alternative, more physical, argument that the capacitor charges up exponentially until the end of the pulse, and then discharges exponentially.

In a similar way to that discussed in Section 2.4 regarding the convolution sum there are merits in using t as the dummy variable in the convolution integral as in Equation (2.13), but there are other merits in using τ as the dummy variable in the convolution integral as in Equation (2.12). One advantage of the latter is that it permits the equation $x(t) * h(t) = \int_{-\infty}^{\infty} x(\tau)h(t - \tau)\mathrm{d}\tau$, whereas the equation $x(t) * h(t) = \int_{-\infty}^{\infty} x(t)h(\tau - t)\mathrm{d}t$ is not strictly correct since the variable remaining after the integration is τ and this conflicts with the variable used on the left-hand side of the equation. A more correct application of Equation (2.13) in this way would be to write:

'$x(t) * h(t) = y(t)$ where

$y(\tau) = \int_{-\infty}^{\infty} x(t)h(\tau - t)\mathrm{d}t$ and we then require a change of variable from τ to t.'

However, as with the convolution sum, this is again a fairly minor argument and it is important to be familiar with both forms of the convolution integral, as both are widely used.

2.9 Convolution of analogue signals using MATLAB

Section 2.5 has shown how sequences of samples can be convolved using MATLAB. It has been argued that a continuous signal is akin to a set of closely spaced impulses rather than to a set of closely spaced samples, and it might therefore be concluded that such discrete-time convolution would be inappropriate to continuous signals. However, the numerical way of convolving analogue signals is in fact to convolve sampled versions of those signals and then to interpolate the result between samples. The justification for this is that the difference between a set of impulse functions and a set of samples applies equally well to the convolved signal as to the original signals.

Example 2.7

Example 2.6 used graphical convolution to estimate the response of an *RC* network to a rectangular pulse. Repeat using MATLAB. Do this for the case where the rectangular pulse is of 1 ms duration and where $RC = 0.5$ ms.

Solution As in Example 2.5 the impulse response is $h(t) = (1/RC)\mathrm{e}^{-t/RC}$. If $RC = 0.5$ ms we have $h(t) = 2000\mathrm{e}^{-2000t}$. A sampled version of this is $h(nt_s) = 2000\mathrm{e}^{-2000nt_s}$. Let the sampling interval t_s be very much less than the time constant and equal to 0.01 ms. If we consider $h(t)$ to be significant only between 0 and 5 ms we can generate the corresponding samples in MATLAB using

```
n=(0:500);
ts=0.00001;
h=2000*exp(-2000*n*ts);
```

We can approximate a 1 ms pulse by 100 samples of ones using

```
x=zeros(size(n));
x(1:100)=ones(1,100);
```

We can now perform the convolution using

```
y=conv(x,h);
```

This contains a number of samples equal to the number of samples of $h[n]$ plus the number of samples of $x[n]$ less one. We can derive a time axis appropriate to a lesser number of points 500 using

```
t=(0:499)*0.00001;
```

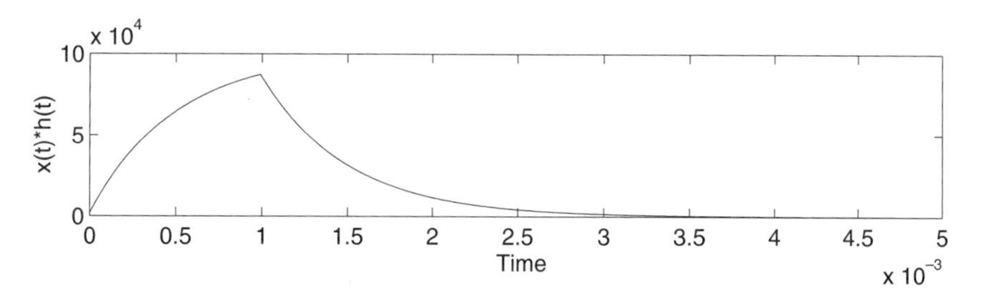

Fig. 2.15 Convolution of rectangular pulse with the impulse response of an *RC* network using MATLAB

Finally we can plot an interpolated set of 500 points of the convolution sum using

```
plot(t,y(1:500))
xlabel('Time');ylabel('x(t)*h(t)')
```

The result of this is Figure 2.15.

One point that can easily cause confusion and needs great care concerns the fact that the numbering of elements in MATLAB vectors starts at one rather than zero. In the preceding example this means that the first point in the convolution vector y corresponds to time $t = 0$ and not to time $t = t_s$. A plot of y versus t takes this into account.

It should be noted that, in contrast to the **stem** command, the **plot** instruction automatically interpolates between the discrete points evaluated by the MATLAB **conv** command. It is thus appropriate for displaying analogue signals.

2.10 Some examples of convolution not related to networks

The convolution integral given by Equation (2.11) applies to time waveforms, but it should not be supposed that the applications of convolution are constrained to the time domain. There are many other situations where convolution is important and relevant.

Example 2.8

Show how convolution can explain the blurring of a moving object in a photograph.

Solution Consider the horizontal section AA of the image of a car shown in Figure 2.16. In the absence of movement a plot of blackness versus distance might appear as in Figure 2.17(a), where the main body of the car is hypothetically assumed to have a set of discrete reflectivity levels and where the thin car aerial gives rise to a very narrow rectangular pulse, as shown in Figure 2.17(a). If the aerial were infinitely narrow while reflecting the same amount of light the plot of blackness versus distance would be an impulse function, as shown in Figure 2.17(b). If however the car moves a significant distance in the exposure time of the photograph, the image becomes blurred and the plot of blackness versus distance might appear as in Figure 2.17(c).

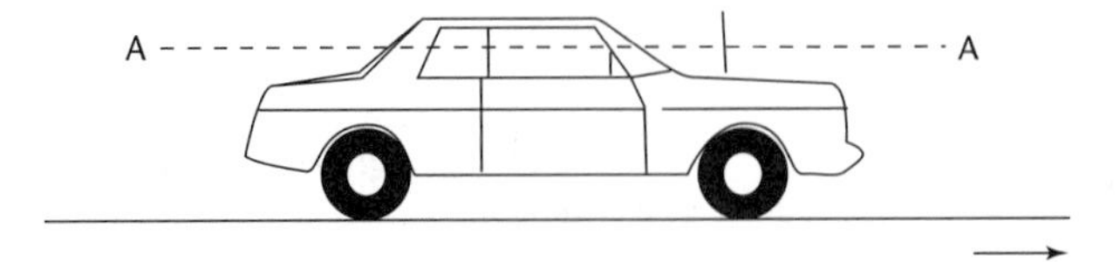

Fig. 2.16 Simplified image of a car

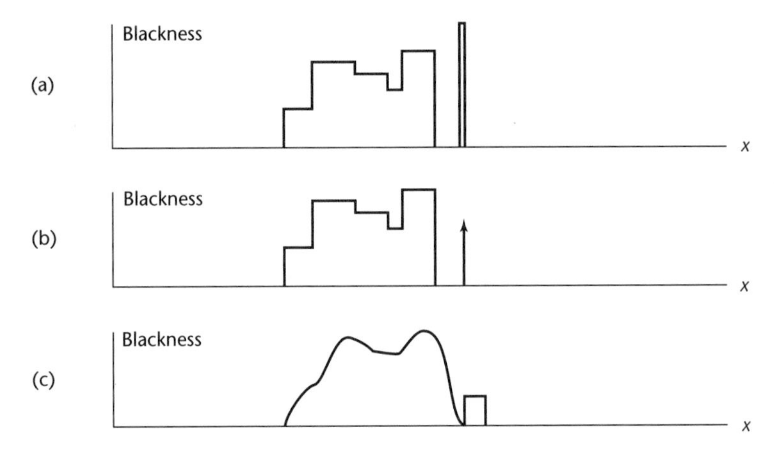

Fig. 2.17 Blackness of photographic image along section *AA*: (a) for stationary vehicle; (b) for stationary vehicle with impulsive response from aerial; (c) for moving vehicle

For the case of a car moving at constant speed the impulse function that corresponded to the aerial now becomes a rectangle. Thus the impulse response of the system may be considered to be a rectangular function of blackness versus distance (note that the concept of impulse functions and impulse responses does not need to be constrained to functions of time). Again the operation of convolution is applicable. Away from the aerial, the actual image is the true image convolved with a rectangular function. The effect is that the image is smeared.

Example 2.9

Determine how the radar signal received from the various scatterering points on an aircraft is affected by the nature and duration of the transmitted pulse.

Solution If the radar were able to transmit an impulse the return signal might be as shown in Figure 2.18.

Fig. 2.18 Echoes from a target due to an impulse function radar transmission

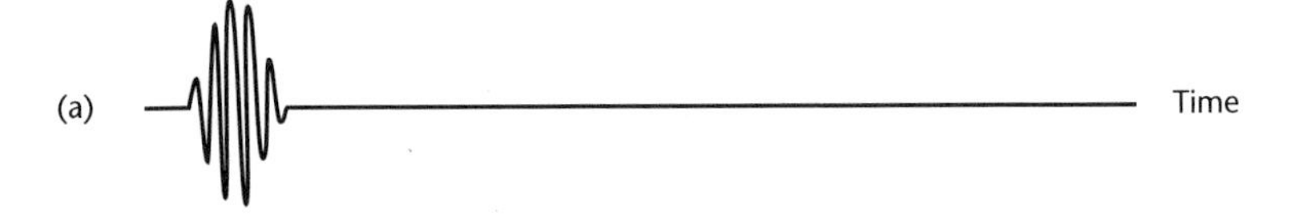

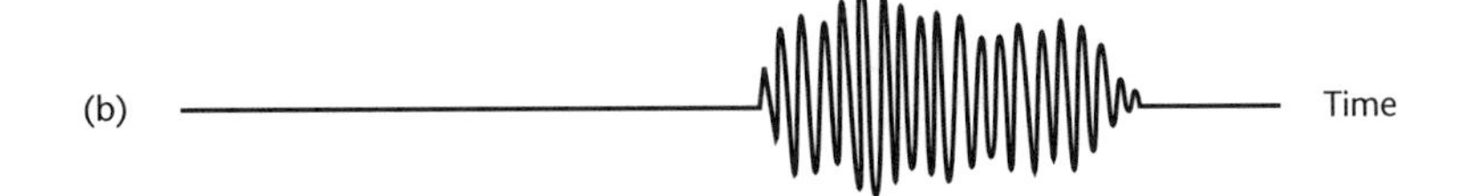

Fig. 2.19 Radar waveforms: (a) transmit pulse; (b) signal retured from aircraft

A typical radar transmits a short tone burst such as shown in Figure 2.19(a). The signal actually returned is the result of convolving this with the impulse response of Figure 2.18 and is as shown in Figure 2.19(b). It is noted that much of the detail of scattering centres is now absent.

Example 2.10

Show how the recording of the sound made by a cap gun in a room can be used to predict the frequency response of the room.

Solution When a person listens to a hi-fi set within a room, the received signal comes not only by a direct path between loudspeaker and listener, but also via echoes from the floor, walls, ceiling and other surfaces. These multipath signals constitute what is known as reverberation and are clearly shown up by the response of a microphone to a very abrupt sound from the position of the loudspeaker, such as that from a cap gun. The received sound would typically be as shown in Figure 2.20 and can be considered a good approximation to the impulse response of the room for those specific positions of source and receiver.

Therefore, in order to determine how a signal $\cos \omega t$ from the loudspeaker is modified at the listener it is necessary only to convolve $\cos \omega t$ with the impulse response. One *practical* way of doing this would be to approximate $\cos \omega t$ and the impulse response by many closely spaced samples and then perform a discrete signal convolution on a computer.

If this is done for many different values of ω it is found that reverberation causes the level of sound heard by the listener to change *rapidly and substantially* with the frequency of the sound source. Since the quality of the sound can be good in spite of this the implication is that any designers of audio equipment who put their efforts solely into achieving a very flat frequency response are missing some of the important attributes of true high fidelity!

An easier (but related) way of determining the frequency response is to evaluate the Fourier transform of the impulse response, but this will be discussed later in the book.

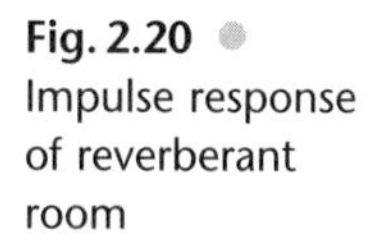

Fig. 2.20 ● Impulse response of reverberant room

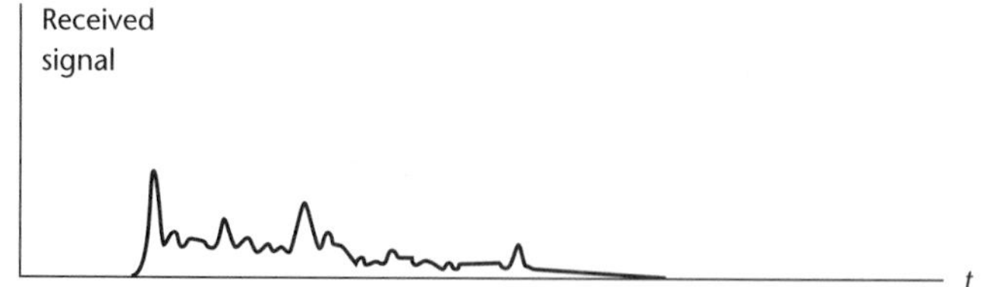

The applications of convolution are not restricted to convolving an input signal with an impulse response in order to determine the output. If two signals $x_1(t)$ and $x_2(t)$ are convolved, where neither is an impulse response, the result is of the same form as in Equation (2.11) or (2.13) and is

$$y(\tau) = \int_{-\infty}^{\infty} x_1(t)x_2(\tau - t)\, \mathrm{d}t \tag{2.14}$$

or

$$y(t) = \int_{-\infty}^{\infty} x_1(\tau)x_2(t - \tau)\, \mathrm{d}\tau \tag{2.15}$$

It will be shown in Chapter 6 that convolution is not restricted to time domain signals and that it can be useful to convolve two spectra.

2.11 Some properties of convolution

Signals are commonly convolved by other signals and the concept of convolution is of great importance in signal processing. Three important properties of convolution are the commutative, distributive, and associative laws, which are, respectively

$$x_1(t) * x_2(t) = x_2(t) * x_1(t) \tag{2.16}$$

$$x_1(t) * [x_2(t) + x_3(t)] = x_1(t) * x_2(t) + x_1(t) * x_3(t) \tag{2.17}$$

$$x_1(t) * [x_2(t) * x_3(t)] = [x_1(t) * x_2(t)] * x_3(t) \tag{2.18}$$

All of these can be proved from the definition of convolution provided by the convolution integral. For example, in terms of the variable τ

$$x_1(t) * x_2(t) = \int_{-\infty}^{\infty} x_1(t)x_2(\tau - t)\, \mathrm{d}t$$

Putting $\tau - t = a$, such that $\mathrm{d}a = -\mathrm{d}t$, this becomes $-\int_{-\infty}^{\infty} x_1(\tau - a)x_2(a)\, \mathrm{d}a$. Replacing a by t and rearranging gives $\int_{-\infty}^{\infty} x_2(t)x_1(\tau - t)\, \mathrm{d}t$, or $x_2(t) * x_1(t)$, thus proving the commutative law. It is important to note that the order in which the operations of convolution and multiplication are undertaken can *not* be interchanged, that is:

$$x_1(t)[x_2(t) * x_3(t)] \neq [x_1(t)x_2(t)] * x_3(t)$$

2.12 Deconvolution

It will have been noticed in Examples 2.7, 2.8 and 2.9 that the convolution of a wanted signal by some other signal results in the elongation or *smearing* of the wanted signal. The insight that this smearing is the mathematical process of convolution leads to the exciting possibility that an *inverse* operation might be applied, whereby the smearing could be *removed* by *deconvolution*. For example, blurred photographs of moving cars have been successfully deblurred to render readable images of number plates that were previously unrecognizable, even though this tends to be contrary to intuition. Other applications of deconvolution include the sharpening of out-of-focus or abberated images, of which the most famous example is that of the Hubble Space Telescope before the defects in its optics were corrected by astronauts. Unfortunately this topic is outside the scope of this book, but it should be noted that deconvolution is fraught with difficulties, usually relating to noise and measurement inaccuracies, and that the improvements are frequently disappointing. Problem 9 gives some insight into one technique of deconvolution and the errors resulting from it.

2.13 ● Summary

A common requirement is the need to determine the response of a system to an input. The mathematical operation for doing this requires that the input be convolved with the system impulse response. For the case of a discrete-time signal $x[n]$ and a discrete-time impulse response $h[n]$ the output is given by the convolution sum, and this is defined as $y[k] = \sum_{n=-\infty}^{\infty} x[n]h[k-n]$. For the case of an analogue signal $x(t)$ and an analogue impulse response $h(t)$ the output is given by the convolution integral and this is defined as $y(\tau) = \int_{-\infty}^{\infty} x(t)h(\tau - t)\, dt$.

This chapter has presented graphical interpretations of the convolution sum and convolution integral. It has also shown how a function exists within MATLAB for the numerical evaluation of convolved signals. Some applications of convolution have been described and it has been shown that they are not constrained solely to functions of time. For example, they can apply to functions of distance.

It will be shown later in the book that numerical methods exist for performing convolution that are computationally more efficient than indicated by the operations implied by the definition of the convolution sum.

2.14 ● Problems

1. Convolve the two sequences

$x[n] = \{2, 5, \underset{\uparrow}{3}, -1, 0, 1\}$ and

$y[n] = \{-2, \underset{\uparrow}{1}, -3, 4, 2\}$

(a) manually, following the procedures indicated by the definition of the convolution sum

(b) by transforming the sequences into polynomials in z using the z transform, multiplying the polynomials, and applying the inverse z transform to the outcome

(c) using MATLAB

2. Evaluate $x[n] * y[n]$ and $y[n] * x[n]$ for the two sequences

$x[n] = \{4, -5, \underset{\uparrow}{2}, -2, -1\}$ and

$y[n] = \{-3, -2, \underset{\uparrow}{3}, -2, 1, 1\}$

and demonstrate that the two results are the same.

3. By considering the graphical interpretation of convolution determine:

(a) the convolution of a 1 ms rectangular pulse commencing at $t = 0$ with itself

(b) the convolution of the 1 ms rectangular pulse with a 2 ms rectangular pulse that also commences at $t = 0$

4. A resonant circuit has an impulse response given by

$$h(t) = e^{-1000t}\cos(7200t + \tan^{-1}(1000/7200)) \qquad t \geqslant 0$$
$$= 0 \qquad \text{elsewhere}$$

By using the graphical interpretation of the convolution integral, sketch the response of the circuit to a rectangular pulse 5 ms long and commencing at $t = 0$. Obtain a more accurate plot using MATLAB.

5. For the three sequences

$$x_1[n] = \{-2, 1, \underset{\uparrow}{1}, 3\}, \quad x_2[n] = \{2, -1; 3, \underset{\uparrow}{1}, 1\}$$
$$\text{and} \qquad x_3[n] = \{-4, 1, \underset{\uparrow}{2}, 2\}$$

evaluate $x_1[n] * (x_2[n] + x_3[n])$ and $x_1[n] * x_2[n] + x_1[n] * x_3[n]$ and confirm that they give the same result.

6. A digital filter has the impulse response $h[n] = \{1, 1, 1, 1, -1, -1, -1, -1\}$. What will its effect be

(a) on an input that is a sampled d.c. signal (an infinite sequence of 1s)?
(b) on the input signal $\sin(\pi n/4)$?
(c) on the input signal $\sin(\pi n/2)$?

Make some deductions about the properties of the digital filter.

7. In a non-return-to zero (NRZ) digital communication signal the data rate is 4 Mbit/s and each bit period of 0.25 µs is fully occupied by a 1 or a 0. By approximating each bit by 10 discrete-time unit samples or by 10 discrete-time zero samples derive a MATLAB signal that is an approximation to the signal 01110111101. This signal is transmitted down a fibre optic cable which causes some dispersion (spreading) of the signal. Neglecting the overall delay down the cable, the impulse response is approximately given by

$$h(t) = \frac{t}{\tau^2} e^{-t^2/2\tau^2} \qquad t \geqslant 0$$
$$= 0 \qquad t < 0$$

where $\tau = 0.3$ µs. Use MATLAB to determine the output over the period of the input signal. If the receiver samples the incoming waveform at the end of each bit period and then imposes a threshold halfway between the minimum and maximum received amplitudes, above which a 1 is deemed to have occurred and below which a 0 is deemed to have occurred, deduce the signal received.

8. Two analogue filters are cascaded using a buffer between them such that the impulse response of each is not influenced by the other. The impulse responses of the two filters are

$$h_1(t) = \frac{1}{\tau_1} e^{-t/\tau_1} \qquad t \geqslant 0$$
$$= 0 \qquad \text{elsewhere}$$
$$h_2(t) = \frac{1}{\tau_2} e^{-t/\tau_2} \qquad t \geqslant 0$$
$$= 0 \qquad \text{elsewhere}$$

where $t_1 = 2.2$ ms and $t_2 = 1.5$ ms. Determine the overall impulse response:

(a) analytically
(b) using MATLAB (derive the necessary code and plot the result)

9. A signal $x[n]$ is applied to a digital network whose impulse response is $h[n]$ and results in an output $y[n]$. Use the z transform method of multiplying polynomials in order to perform the convolution $y[n] = x[n] * h[n]$ where

$$x[n] = \{\underset{\uparrow}{1}, 3, 3, 2\} \qquad \text{and} \qquad h[n] = \{\underset{\uparrow}{4}, 3, 2, 1\}$$

Confirm the result by assigning appropriate values to the variables x and h in MATLAB followed by the command **conv(x,h)**. Consider next the possibility that the output $y[n]$ and the impulse response $h[n]$ are known, but not the input $x[n]$. Apply the *reverse* procedure of dividing the z transform of $y[n]$ by the z transform $h[n]$ in order to determine $x[n]$. Demonstrate that the same answer is obtained in MATLAB using the command **deconv(y,h)**.

Add some noise to y by using the command **z=y+.3*randn(size(y))** and determine now the outcome of deconvolving z with h.

Chapter three

The steady state response of analogue networks to cosinusoids and to the complex exponential $e^{j\omega t}$

3.1 ● Preview

A common requirement is to find the response of a network to a cosinusoidal excitation. This task is greatly simplified if the cosinusoid is replaced by a complex exponential excitation.

3.2 The properties of network elements

There are three important linear elements that make up networks.

(a) Resistance R ohms (abbreviated Ω). In accordance with Ohm's law the voltage v volts across a resistance is proportional to the current i amperes through it, such that

$$v = iR \tag{3.1}$$

(b) Inductance L henrys (abbreviated H). The voltage across an inductance is proportional to the rate of change of current through it in accordance with the formula

$$v = L\frac{\mathrm{d}i}{\mathrm{d}t} \tag{3.2}$$

(c) Capacitance C farads (abbreviated F). The voltage across a capacitor is proportional to the electrical charge q on it in accordance with the formula

$$v = \frac{q}{C} \tag{3.3}$$

Because current is the rate of change of charge, the charge on a capacitance at a time t may be determined by integrating the current through the capacitance over all time up to time t, that is $q = \int_{-\infty}^{t} i \, dt$. Therefore we have the alternative formula

$$v = \frac{1}{C} \int_{-\infty}^{t} i \, dt \tag{3.4}$$

It is important to note that *practical* resistors, inductors, and capacitors may not behave as the ideal network elements of resistance, inductance and capacitance. Wirewound resistors, used when large amounts of power must be dissipated, are constructed as coils and can be inductive at quite low frequencies. Carbon and metal oxide resistors are much closer to the ideal network element of resistance, although they can become inductive at high frequencies. Inductors usually have a significant series resistance associated with them due to finite resistance of the winding. They also have some capacitance between neighbouring turns of the winding, and this self-capacitance can be significant at high frequencies. Capacitors are perhaps the closest of the three components to ideal network elements and, with the exception of electrolytic capacitors, are usually close to ideal capacitances.

3.3 The difficulty of solving network equations

By making use of Equations (3.1), (3.2) and (3.4), the voltage across a series combination of a resistance, inductance and capacitance would be given by the differential equation

$$v = iR + L\frac{di}{dt} + \frac{1}{C}\int i \, dt \tag{3.5}$$

In network analysis we often wish to solve equations of this type when v is an applied voltage of the form $V_m \cos \omega t$.

The first point to realize is that there will generally be a *transient* immediately after the voltage is first applied and that this will depend upon the exact point in the cosinusoidal cycle that the turn-on occurs. However, this transient soon decays, and the following discussion concerns the settled or 'steady state' response.

An elegant approach for finding the steady state response involves complex exponentials and will be given in the next section. But to demonstrate that this is justified, a more obvious technique will be considered first. It begins by arguing that frequency does not change in a linear system and that the current is therefore also cosinusiodal with frequency ω, but has a different amplitude and phase to that of the voltage, such that it is given by $i = I_m \cos(\omega t + \beta)$. The technique will be to insert this current into Equation (3.5) *and then to apply the equality to enable the unknown amplitude and phase to be determined.*

As the purpose is to argue that this technique is an *inefficient* one which should not therefore be repeated, the reader may wish to bypass the remainder of this section.

Example 3.1

Considering the very simple circuit of just a resistance and inductance, as shown in Figure 3.1, determine the current resulting from an applied cosinusoidal voltage $V_m \cos \omega t$. Do this by assuming a current equal to $I_m \cos(\omega t + \beta)$ and substituting this into the differential equation of the network.

Solution We have

$$V_m \cos \omega t = iR + L\frac{di}{dt} \tag{3.6}$$

Differentiating the current $I_m \cos(\omega t + \beta)$ to obtain di/dt we can substitute for i and di/dt in the differential equation to obtain

$$V_m \cos \omega t = I_m R \cos(\omega t + \beta) - I_m \omega L \sin(\omega t + \beta) \tag{3.7}$$

Expanding this using trigonometric identities we obtain

$$V_m \cos \omega t = I_m R \cos \omega t \cos \beta - I_m R \sin \omega t \sin \beta - I_m \omega L \sin \omega t \cos \beta - I_m \omega L \cos \omega t \sin \beta$$

From this we can derive two equations, one demanding the equality of $\cos \omega t$ terms, and the other of $\sin \omega t$ terms

$$V_m = I_m R \cos \beta - I_m \omega L \sin \beta \tag{3.8}$$

$$0 = -I_m R \sin \beta - I_m \omega L \cos \beta \tag{3.9}$$

From Equation (3.9) we can solve for β:

$$\tan \beta = -\frac{\omega L}{R} \quad \text{or} \quad \beta = -\tan^{-1}\left(\frac{\omega L}{R}\right) \tag{3.10}$$

Obtaining I_m is more difficult but can be done by eliminating β from Equations (3.8) and (3.9). This can be done by squaring these two equations and adding them. From Equation (3.8):

$$V_m^2 = I_m^2 R^2 \cos^2 \beta + I_m^2 (\omega L)^2 \sin^2 \beta - 2I_m^2 R \omega L \sin \beta \cos \beta \tag{3.11}$$

From Equation (3.9):

$$0 = I_m^2 R^2 \sin^2 \beta + I_m^2 (\omega L)^2 \cos^2 \beta + 2I_m^2 R \omega L \sin \beta \cos \beta \tag{3.12}$$

Adding these gives

$$V_m^2 = I_m^2 R^2 + I_m^2 (\omega L)^2$$

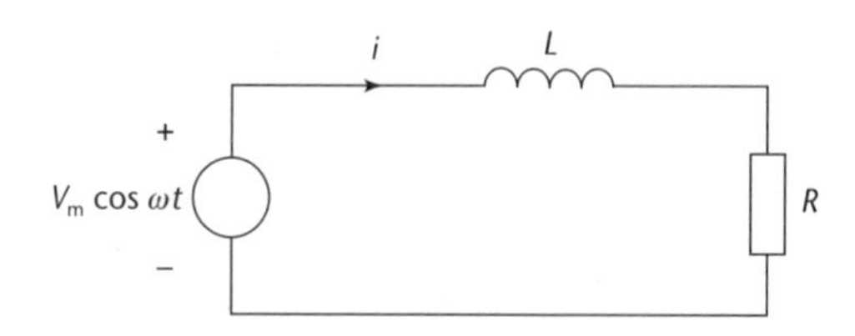

Fig. 3.1 Series *RL* circuit

Hence

$$I_m = \frac{V_m}{\sqrt{R^2 + (\omega L)^2}} \tag{3.13}$$

Therefore, since $i = I_m \cos(\omega t + \beta)$, we obtain

$$i = \frac{V_m}{\sqrt{R^2 + (\omega L)^2}} \cos\left[\omega t - \tan^{-1}\left(\frac{\omega L}{R}\right)\right] \tag{3.14}$$

This is the required solution.

In the above example the technique of solving the network equation is to assume the form of the solution, insert it into the equation, and then apply the equality to enable the unknown amplitude and phase to be determined. It is important to appreciate why this technique for solving such a simple differential equation proves so difficult. It arises because differentiation of a cosine function produces a *different function*, namely a sine function. After the insertion of the assumed solution into the network equation, the result contains an awkward mixture of sinusoidal and cosinusoidal functions that makes it difficult to determine the unknown constants. Everything would be very much simpler if we could find some way of circumventing this change of function caused by differentiation. This is achieved by involving complex exponentials.

3.4 The use of the complex exponential waveform exp(jωt)

The square of a real number is positive, and hence the square root of a negative number cannot physically exist; it is considered to be an 'imaginary' number. The imaginary number corresponding to the square root of minus one is given the symbol 'j'.

Multiplication of a real voltage waveform by the imaginary number j makes that voltage waveform become a mathematical abstraction that is not physically realizable, a waveform that is considered to be imaginary. If such an imaginary voltage is applied to a network the resulting current will also be imaginary. The advantage of considering such a possibility arises because the current resulting from an imaginary voltage will be totally separable from any current caused by a real voltage. The easiest way to solve an equation such as Equation (3.5) is to replace the cosinusoidal excitation by a *hypothetical* excitation that has an imaginary sinusoidal component added to it, namely the complex voltage $(V_m \cos\omega t + jV_m \sin\omega t)$, or simply $V_m e^{j\omega t}$, and to recognize that the real part of the resulting current is solely due to the real part of that applied voltage.

Suppose the current resulting from this complex exponential time waveform, $V_m e^{j\omega t}$, is another complex exponential time waveform, $I_m e^{j(\omega t + \beta)}$. Then the real part of this complex current must be caused by the real part of the complex voltage, while the imaginary part of this complex current must be caused by the imaginary part of the complex voltage.

The reason why it is so useful to consider the complex exponential time waveform $e^{j\omega t}$ is that $e^{j\omega t}$ stays of the same form after differentiation or integration. Insertion into a network differential equation of the excitation voltage, $V_m e^{j\omega t}$, and of the assumed resulting current $I_m e^{j(\omega t+\beta)}$, results in the common factor $e^{j\omega t}$ which can be divided out. This is in total contrast to the substitution into the differential equation of the excitation voltage $V_m \cos \omega t$ and an assumed resulting current $I_m \cos(\omega t+\beta)$. This term $I_m \cos(\omega t+\beta)$ changes from a cosinusoid to a sinusoid when differentiated, thus making the differential equation an awkward mixture of cosinusoids and sinusoids that is difficult to solve.

Example 3.2

Solve the differential equation of Equation (3.6) by replacing the cosinusoidal voltage with a complex exponential voltage, assuming that a complex exponential current results from it, and then taking the real part of this current.

Solution We can replace the excitation function $V_m \cos \omega t$ by $\tilde{v} = V_m e^{j\omega t}$, where a tilde has been used over the v to emphasize that it is a complex waveform. Equation (3.6) then becomes

$$V_m e^{j\omega t} = iR + L\frac{di}{dt} \tag{3.15}$$

Because the voltage waveform has real and imaginary components, the current must also have real and imaginary components and it is assumed that the current is of the form

$$\tilde{i} = I_m e^{j(\omega t+\beta)} \tag{3.16}$$

where, once again, a tilde has been used to emphasize that i is a complex waveform. This equation can be rewritten

$$\tilde{i} = I_m e^{j\beta} e^{j\omega t} = \tilde{I}_m e^{j\omega t} \tag{3.17}$$

where, for mathematical elegance, $I_m e^{j\beta}$ has been replaced by the complex number $\tilde{I}_m$.

Inserting this current into Equation (3.15) and differentiating as appropriate gives

$$V_m e^{j\omega t} = R\tilde{I}_m e^{j\omega t} + j\omega L\tilde{I}_m e^{j\omega t} \tag{3.18}$$

The $e^{j\omega t}$ terms divide out. *This is why it is so advantageous to use complex exponentials.* The result of doing this is

$$\tilde{I}_m = \frac{V_m}{R + j\omega L} \tag{3.19}$$

But $R + j\omega L$ may be written in polar form as $\sqrt{R^2 + (\omega L)^2}\, e^{j\tan^{-1}(\omega L/R)}$;

$$\therefore\ \tilde{I}_m = \frac{V_m e^{j\beta}}{\sqrt{R^2 + (\omega L)^2}} \tag{3.20}$$

where

$$\beta = -\tan^{-1}(\omega L/R) \tag{3.21}$$

But $\tilde{i} = \tilde{I}_{\mathrm{m}} \mathrm{e}^{\mathrm{j}\omega t}$,

$$\therefore \tilde{i} = \frac{V_{\mathrm{m}}}{\sqrt{R^2 + (\omega L)^2}} \mathrm{e}^{\mathrm{j}(\omega t + \beta)} \tag{3.22}$$

Taking the real part of this as being the component caused by the real part of the applied voltage, $V_{\mathrm{m}} \mathrm{e}^{\mathrm{j}\omega t}$, we obtain

$$i = \frac{V_{\mathrm{m}}}{\sqrt{R^2 + (\omega L)^2}} \cos(\omega t + \beta) \tag{3.23}$$

where $\beta = -\tan^{-1}(\omega L/R)$.

The method used in Example 3.2 for solving the differential equation of Equation (3.6) is much easier than that used in Example 3.1. With practice, and by invoking the impedance functions introduced in the following sections and the phasors introduced in the next chapter, a solution by means of complex exponentials becomes even easier. The contrast between the two methods increases with increasing network complexity. *The use of complex exponentials is the generally adopted technique for determining the steady state response of a network to a sinusoid.*

3.5 Impedance functions

The impedance function of a network element is defined as the ratio of voltage to current *when the current through the element is a complex exponential waveform*; that is, when it is of the form $\mathrm{e}^{\mathrm{j}\omega t}$. The units of an impedance function are ohms.

3.5.1 Inductance

Consider an inductance for which $v = L\, \mathrm{d}i/\mathrm{d}t$, and let the current through it be the complex exponential current $\tilde{i} = \tilde{I}_{\mathrm{m}} \mathrm{e}^{\mathrm{j}\omega t}$. Differentiating this current enables us to determine $L\, \mathrm{d}i/\mathrm{d}t$ and hence the complex exponential voltage $\tilde{v}$ across the inductance:

$$\tilde{v} = \mathrm{j}\omega L \tilde{I}_{\mathrm{m}} \mathrm{e}^{\mathrm{j}\omega t} = \mathrm{j}\omega L \tilde{i}$$

$$\therefore \tilde{v}/\tilde{i} = \mathrm{j}\omega L \tag{3.24}$$

This is the impedance function of the inductance. It is sometimes denoted by $\tilde{Z}(\omega)$, or by $\tilde{Z}(\mathrm{j}\omega)$, where the tilde and j are used to emphasize that this impedance is a complex quantity. An alternative to the tilde is to use boldface: $\mathbf{Z}(\omega)$ or $\mathbf{Z}(\mathrm{j}\omega)$. Very often, some or all of these clarifying modifications are omitted for economy, and the impedance function written simply as $\mathbf{Z}$, $\mathbf{Z}(\omega)$, or even as Z. This text will henceforth use the boldface notation on its own; that is:

$$\mathbf{Z} = \mathrm{j}\omega L \tag{3.25}$$

The quantity $\mathrm{j}\omega L$ is the ratio between voltage and current for an inductance *when, and only when*, the excitation is *complex exponential* in form. As such it has little *physical* meaning, since we do not have complex exponential waveforms in real life. It is, however, a very useful function in that it provides a *stepping stone* for finding the ratio between voltage and current when dealing with real signals.

Suppose, for example, that we require to find the current that results from the application of a real voltage $v = V_m \cos \omega t$ across the inductance. Instead of $V_m \cos \omega t$ we suppose the complex voltage $\tilde{v} = V_m e^{j\omega t}$ to be applied, comprised of a real part $V_m \cos \omega t$ and an imaginary part $V_m \sin \omega t$. The resulting current $\tilde{i}$ is given by

$$\tilde{i} = \frac{\tilde{v}}{j\omega L} = \frac{V_m e^{j\omega t}}{j\omega L} = \frac{V_m e^{j\omega t}}{\omega L e^{j\pi/2}} = \frac{V_m}{\omega L} e^{j(\omega t - \pi/2)} \tag{3.26}$$

But this can be expanded as

$$\tilde{i} = \frac{V_m}{\omega L} \cos(\omega t - \pi/2) + j \frac{V_m}{\omega L} \sin(\omega t - \pi/2) \tag{3.27}$$

The real (or physical) part of this current is that which is caused by the real (or physical) part of the voltage. Hence

$$i = \frac{V_m}{\omega L} \cos\left(\omega t - \frac{\pi}{2}\right) = \frac{V_m}{\omega L} \sin \omega t \tag{3.28}$$

The real voltage and resulting real current are plotted in Figure 3.2, and it is observed that the current lags the voltage by 90°.

Whereas the ratio between complex voltage and complex current is simply $j\omega L$ for an inductance, it will be noted that the ratio of *real* voltage to *real* current is given by

$$\frac{v}{i} = \frac{V_m \cos \omega t}{(V_m/\omega L) \cos[\omega t - (\pi/2)]} = \omega L \cot \omega t \tag{3.29}$$

This is a *time-varying* quantity with a value varying from $+\infty$ to $-\infty$. This is illustrated in Figure.3.3.

The ratio v/i could be thought of as an *instantaneous* impedance. It has more *physical* meaning than the impedance function so far introduced, which is the very much simpler ratio between complex voltage and complex current, but is *far more awkward* to use. The advantage of the impedance function over an 'instantaneous impedance' is even more pronounced when other network elements are added in

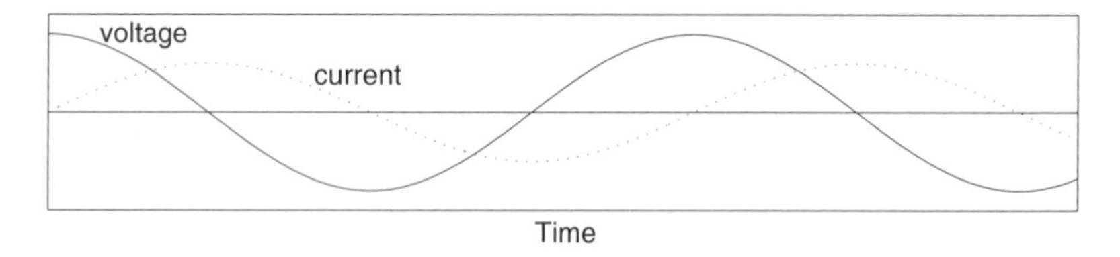

Fig. 3.2* ● Voltage and current waveforms for an inductance

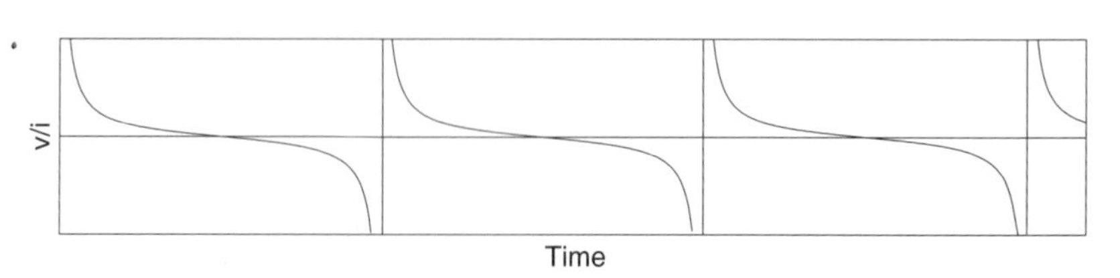

Fig. 3.3* ● Ratio of voltage to current for an inductance as a function of time when $v = V_m \cos \omega t$

series or in parallel. Indeed it is *very unusual* ever to invoke the concept of an *instantaneous* impedance.

The result of Equation (3.25), that $\mathbf{Z} = \mathrm{j}\omega L$, may also be written as

$$\mathbf{Z} = \mathrm{j}X_L \tag{3.30}$$

where X_L is termed the *reactance* of the inductance. Its units are ohms. For an inductance we have

$$X_L = \omega L \tag{3.31}$$

3.5.2 Resistance

Consider next a resistance R for which $v = iR$. If the current is the complex current $\tilde{i} = \tilde{I}_\mathrm{m}\mathrm{e}^{\mathrm{j}\omega t}$, substitution gives us that the resulting complex voltage $\tilde{v} = R\tilde{I}_\mathrm{m}\mathrm{e}^{\mathrm{j}\omega t} = R\tilde{i}$. Hence

$$\tilde{v}/\tilde{i} = R \tag{3.32}$$

This time the result is no different to the ratio of real voltage to real current. We have

$$\mathbf{Z} = R \tag{3.33}$$

3.5.3 Capacitance

The final element type is the capacitance, for which $v = (1/C)\int_{-\infty}^{t} i\,\mathrm{d}t$. Let the current be a complex current $\tilde{i} = \tilde{I}_\mathrm{m}\mathrm{e}^{\mathrm{j}\omega t}$ such that there is a complex voltage $\tilde{v}$ across the capacitance. Integrating $\tilde{i}$ in the equation for v gives

$$\tilde{v} = \frac{\tilde{I}_\mathrm{m}}{C\mathrm{j}\omega}\mathrm{e}^{\mathrm{j}\omega t} = \frac{\tilde{i}}{\mathrm{j}\omega C}$$

$$\therefore \frac{\tilde{v}}{\tilde{i}} = \frac{1}{\mathrm{j}\omega C} \tag{3.34}$$

or

$$\mathbf{Z} = 1/\mathrm{j}\omega C \tag{3.35}$$

As with the inductance there may be a conceptual difficulty in having an impedance which is complex. What matters, though, is that it is easy to derive from $\mathbf{Z}$ the relationship between the real voltage across the capacitance and the real current through it when one of them is a real cosinusoid. Suppose $v = V_\mathrm{m}\cos\omega t$. The corresponding complex voltage is $\tilde{v} = V_\mathrm{m}\cos\omega t + \mathrm{j}V_\mathrm{m}\sin\omega t$ or $V_\mathrm{m}\mathrm{e}^{\mathrm{j}\omega t}$. Using $\tilde{v}/\tilde{i} = \mathbf{Z}$ the resulting complex current is given by

$$\tilde{i} = \tilde{v}/\mathbf{Z} = \mathrm{j}\omega C\tilde{v} = \mathrm{j}\omega CV_\mathrm{m}\mathrm{e}^{\mathrm{j}\omega t} = \omega CV_\mathrm{m}\mathrm{e}^{\mathrm{j}(\omega t + \pi/2)} \tag{3.36}$$

The physical current resulting from V_m is then the real part of this, $\omega CV_\mathrm{m}\cos(\omega t + \pi/2)$ or $-\omega CV_\mathrm{m}\sin\omega t$. The real voltage and real current are plotted in Figure 3.4, and this time it is observed that the current leads the voltage by 90°. (A useful mnemomic that may be helpful for remembering whether current leads voltage or vice versa is the word 'CIVIL'. The 'IV' is associated with 'C' and indicates that current leads voltage for a capacitance. The 'VI' is associated with 'L' and thus indicates that voltage leads current for an inductance.)

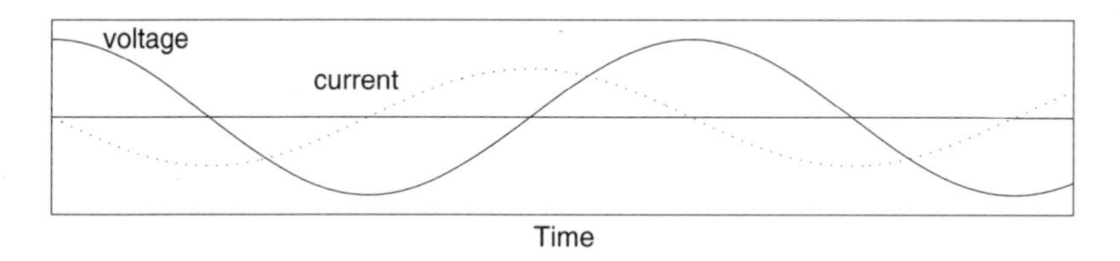

Fig. 3.4 Voltage and current waveforms for a capacitance

As with inductance, the ratio of *real* voltage to *real* current for a capacitance *varies with time*, that is

$$\frac{V_{\mathrm{m}} \cos \omega t}{\omega C V_{\mathrm{m}} \cos(\omega t + \pi/2)} = -\frac{1}{\omega C} \cot \omega t \tag{3.37}$$

This ratio of real voltage to real current (that is, of actual instantaneous voltage to actual instantaneous current) is again difficult to deal with, in a way that the ratio of complex exponential voltage to complex exponential current is not. This reaffirms the major reason for preferring impedance functions; the impedance function for a capacitance is simply $1/\mathrm{j}\omega C$.

The result of Equation (3.35) is that $\mathbf{Z} = 1/\mathrm{j}\omega C$ may also be written as

$$\mathbf{Z} = \mathrm{j}X_{\mathrm{C}} \tag{3.38}$$

where X_{C} is the *reactance* of the capacitance in ohms. For a capacitance we have

$$X_{\mathrm{C}} = -1/\omega C \tag{3.39}$$

3.6 Admittance functions

The admittance function of a network element is simply the reciprocal of the impedance function $\mathbf{Z}$ and is therefore defined as the ratio of current to voltage *when the voltage across the element is a complex exponential waveform* (that is, when it is of the form $e^{j\omega t}$). It is the ratio $\tilde{i}/\tilde{v}$ and its symbol is $\mathbf{Y}$. Its units are siemens (abbreviated S). Thus for a resistance R the admittance in siemens is

$$\mathbf{Y} = 1/R \tag{3.40}$$

It may also be written as

$$\mathbf{Y} = G \tag{3.41}$$

where G is the *conductance* of the element in siemens and we have

$$G = 1/R \tag{3.42}$$

For an inductance L the admittance is

$$\mathbf{Y} = 1/\mathrm{j}\omega L \tag{3.43}$$

It may also be written as

$$\mathbf{Y} = \mathrm{j}B_L \tag{3.44}$$

where B_L is the *susceptance* of the inductance in siemens, and we have

$$B_L = -1/\omega L \tag{3.45}$$

For a capacitance the admittance is

$$\mathbf{Y} = \mathrm{j}\omega C \tag{3.46}$$

It may also be written as

$$\mathbf{Y} = \mathrm{j}B_{\mathrm{C}} \tag{3.47}$$

where B_{C} is the *susceptance* of the capacitance in siemens, and we have

$$B_{\mathrm{C}} = \omega C \tag{3.48}$$

3.7 Elements in series and in parallel

The concept of complex impedance functions can be extended to elements in series, and the mathematical advantage of using the impedance functions becomes even greater when this happens. Consider the series *RLC* circuit shown in Figure 3.5 and suppose we wish to relate the complex voltage across it to the complex current through it. The complex voltage is the sum of the voltages across the three elements. Since there is a common complex current $\tilde{i}$ through all three elements we obtain

$$\tilde{v} = R\tilde{i} + \mathrm{j}\omega L\tilde{i} + \frac{1}{\mathrm{j}\omega C}\tilde{i} = \tilde{i}\mathbf{Z} \tag{3.49}$$

where

$$\mathbf{Z} = R + \mathrm{j}\omega L + 1/\mathrm{j}\omega C \tag{3.50}$$

Here we have a simple addition of impedances. The simple addition arises because elements in series carry the same current. In general the impedance of elements in series is given by

$$\mathbf{Z} = \mathbf{Z}_1 + \mathbf{Z}_2 + \mathbf{Z}_3 + \cdots \tag{3.51}$$

Assuming a mixture of resistances, inductances and capacitances in series the resultant impedance may be expressed as

$$\mathbf{Z} = R + \mathrm{j}X \tag{3.52}$$

where R is the sum of the individual resistances, and X is the sum of the individual reactances. (It will be noted that the reactances of inductances and capacitances are of opposite sign and therefore reactances can annul one other.)

Elements in parallel have different currents but have the same voltage across them. In this situation the admittances add, that is

$$\mathbf{Y} = \mathbf{Y}_1 + \mathbf{Y}_2 + \mathbf{Y}_3 + \cdots \tag{3.53}$$

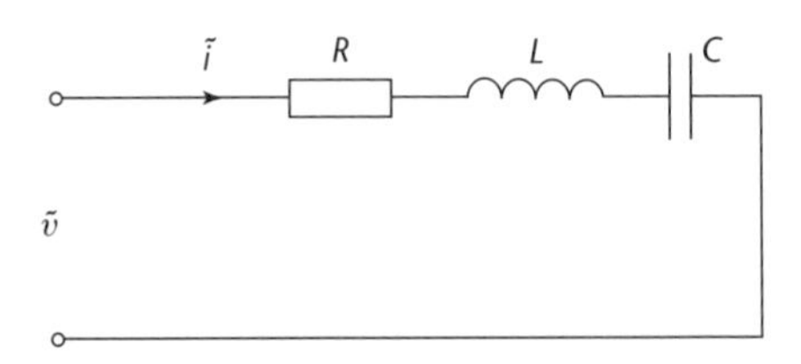

Fig. 3.5 Series *RLC* circuit

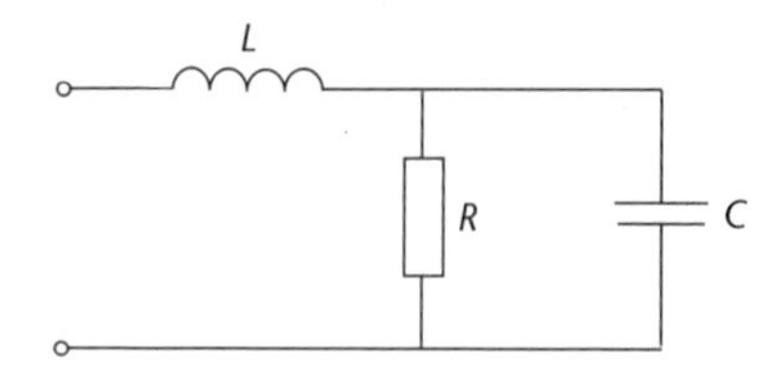

Fig. 3.6 Impedance analysis

Assuming a mixture of resistances, inductances and capacitances in parallel the resultant admittance may be expressed as

$$\mathbf{Y} = G + \mathrm{j}B \tag{3.54}$$

where G is the the sum of the individual conductances and B is the sum of the individual susceptances.

Example 3.3

As a simple example that combines the use of impedance and admittance, determine the impedance of the network in Figure 3.6.

Solution The resistance and capacitance are in parallel such that we can add their admittances to obtain a combined admittance of $(1/R) + \mathrm{j}\omega C$. The corresponding impedance is the reciprocal of this, or $R/(1 + \mathrm{j}\omega CR)$. We now note that this is in series with the inductance, so that, adding impedances, the total impedance is therefore

$$\mathbf{Z} = \mathrm{j}\omega L + \frac{R}{1 + \mathrm{j}\omega CR} \tag{3.55}$$

Impedance functions are useful because they enable us to find the relationship between an applied voltage and the resulting current when one of them is a cosinusoid. To reiterate the procedure, one first finds the complex current that arises when the applied voltage is a complex exponential time waveform. One then takes the real part of this as being the contribution caused by the real part of the applied voltage.

Example 3.4

Taking the configuration of Figure 3.6 as an example, determine the total input current arising from an applied voltage $v = V_{\mathrm{m}} \cos(\omega t + \theta)$.

Solution As previously suggested, we add an imaginary component to the voltage to make it become the complex exponential $\tilde{v}$, of which $V_{\mathrm{m}} \cos(\omega t + \theta)$ is the real part. In other words we replace v by $\tilde{v}$, where

$$\tilde{v} = V_{\mathrm{m}} \mathrm{e}^{\mathrm{j}(\omega t + \theta)}$$

Then, using the impedance of Equation (3.55):

$$\tilde{i} = \frac{\tilde{v}}{\mathbf{Z}} = \frac{\tilde{v}}{\mathrm{j}\omega L + R/(1 + \mathrm{j}\omega CR)} = V_{\mathrm{m}} \mathrm{e}^{\mathrm{j}\omega t} \mathrm{e}^{\mathrm{j}\theta} \frac{1 + \mathrm{j}\omega CR}{R - \omega^2 LCR + \mathrm{j}\omega L} \tag{3.56}$$

The best way to simplify this is to express the various components of this equation in polar coordinates. This is because it is very much easier to multiply and divide complex quantities expressed in polar coordinates than complex quantities expressed in rectangular coordinates. Proceeding in this way:

$$\tilde{i} = V_{\mathrm{m}} \mathrm{e}^{\mathrm{j}\theta} \mathrm{e}^{\mathrm{j}\omega t} \frac{\sqrt{1 + (\omega CR)^2} \angle \tan^{-1} \omega CR}{\sqrt{(R - \omega^2 LCR)^2 + \omega^2 L^2} \angle \tan^{-1}[\omega L/(R - \omega^2 LCR)]}$$

$$\therefore \tilde{i} = V_{\mathrm{m}} \mathrm{e}^{\mathrm{j}\omega t} \frac{\sqrt{1 + (\omega CR)^2}}{\sqrt{(R - \omega^2 LCR)^2 + \omega^2 L^2}} \times \exp\{\mathrm{j}(\theta + \tan^{-1} \omega CR - \tan^{-1}[\omega L/(R - \omega^2 LCR)])\} \quad (3.57)$$

The wanted input current is the real part of this complex current:

$$i = V_{\mathrm{m}} \frac{\sqrt{1 + (\omega CR)^2}}{\sqrt{(R - \omega^2 LCR)^2 + \omega^2 L^2}} \times \cos\{\omega t + \theta + \tan^{-1} \omega CR - \tan^{-1}[\omega L/(R - \omega^2 LCR)]\} \quad (3.58)$$

3.8 Frequency transfer functions and Bode plots

A frequency transfer function is the ratio of the complex exponential time waveform at one point of a network to the complex exponential time waveform causing it at some other point in the network. It may be a ratio of two voltages, or of two currents, or of a voltage to a current, or of a current to a voltage. Its most common use is as the ratio of an output voltage to an input voltage and, if not stated otherwise, this will be assumed to be the case. In this event it is denoted by the symbol $H(\omega)$ or often, in order to emphasize that is a complex quantity, by $H(\mathrm{j}\omega)$. A tilde could be used over the H to further emphasize that $H(\omega)$ is a complex quantity, but this is uncommon. This book will use the $H(\omega)$ notation, or $H(f)$ when expressed in terms of cyclic frequency rather than radian frequency.

Example 3.5

Determine the frequency transfer function of the network of Figure 3.7.

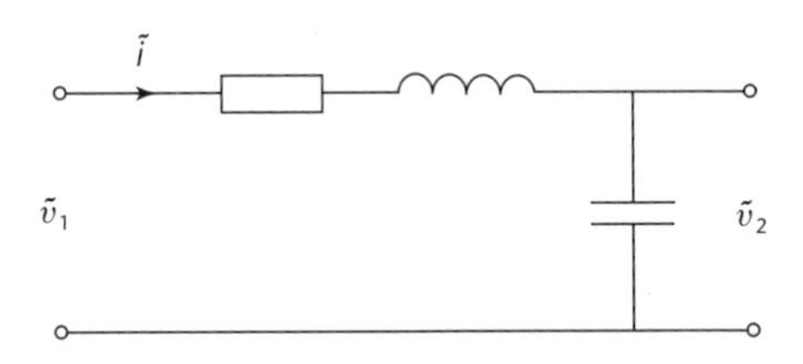

Fig. 3.7 Transfer function analysis

Solution If we apply a complex exponential voltage $\tilde{v}_1$ the current is given by the equation $\tilde{v}_1 = R\tilde{i} + j\omega L\tilde{i} + (1/j\omega C)\tilde{i}$. The resulting output voltage is given by $\tilde{v}_2 = (1/j\omega C)\tilde{i}$. It follows that

$$H(\omega) = \frac{\tilde{v}_2}{\tilde{v}_1} = \frac{1/j\omega C}{R + j\omega L + 1/j\omega C} = \frac{1}{1 - \omega^2 LC + j\omega CR} \tag{3.59}$$

The frequency transfer function can be written in polar coordinates as $H(\omega) = A(\omega)e^{j\phi(\omega)}$ where $A(\omega)$ indicates the amplitude response and $\phi(\omega)$ the phase response.

Example 3.6

Determine the amplitude and phase response of the network of Figure 3.7.

Solution Equation (3.59) can be expressed in polar coordinates as

$$\mathrm{H}(\omega) = \frac{1}{\sqrt{(1 - \omega^2 LC)^2 + \omega^2 C^2 R^2}\ \exp[j\tan^{-1}(\omega CR/(1 - \omega^2 LC))]} \tag{3.60}$$

Therefore

$$A(\omega) = \frac{1}{\sqrt{(1 - \omega^2 LC)^2 + \omega^2 C^2 R^2}} \tag{3.61}$$

and

$$\phi(\omega) = -\tan^{-1}(\omega CR/(1 - \omega^2 LC)) \tag{3.62}$$

The frequency transfer function of a system, defined as $A(\omega)e^{j\phi(\omega)}$, is one of the most commonly used ways of describing the properties of a system. As indicated in Figure (3.8) it is the *multiplier* that acts on an input $e^{j\omega t}$ to give the output $A(\omega)e^{j\phi(\omega)}e^{j\omega t}$, or $A(\omega)e^{j\{\omega t + \phi(\omega)\}}$.

If we consider the real part of the output as being caused by the real part of the input the practical significance of this is that the output is $A(\omega)\cos[\omega t + \phi(\omega)]$ when the input is $\cos\omega t$. The cosinusoid has its amplitude *multiplied* by $A(\omega)$ but its *phase* shifted by $\phi(\omega)$. This is emphasized in Figure 3.9.

The frequency transfer function can be measured experimentally by applying an oscillator to the input of the system, examining the output, and determining the gain $A(\omega)$ and phase shift $\phi(\omega)$ as the frequency of the oscillator is varied. It is an important *alternative* to the impulse response for characterizing the system. However, as will be shown in Chapter 6, the two are related by the Fourier transform.

The information of the frequency transfer function is commonly conveyed by two separate plots, one of $20\log_{10} A(\omega)$ versus the logarithm of frequency , and the other of $\phi(\omega)$ versus the logarithm of frequency. The quantity $20\log_{10} A(\omega)$ is the amplitude ratio $A(\omega)$ expressed in decibels (abbreviated as dB) and is also referred to as the gain. The use of logarithmic scales for $A(\omega)$ and ω leads to compact plots showing performance over a wide range of frequencies. An example of such plots for the simple RC network of Figure 3.10 is shown in Figure 3.11.

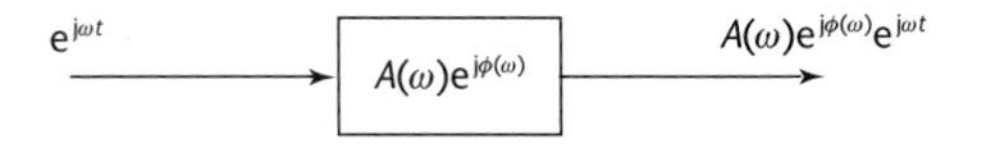

Fig. 3.8 ● The transfer function acting as a multiplier on a complex exponential input

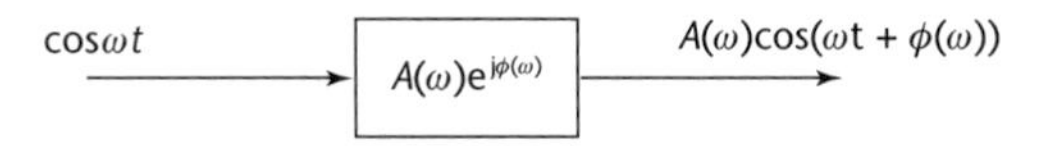

Fig. 3.9 ● Effect of a transfer function on a cosinusoidal input

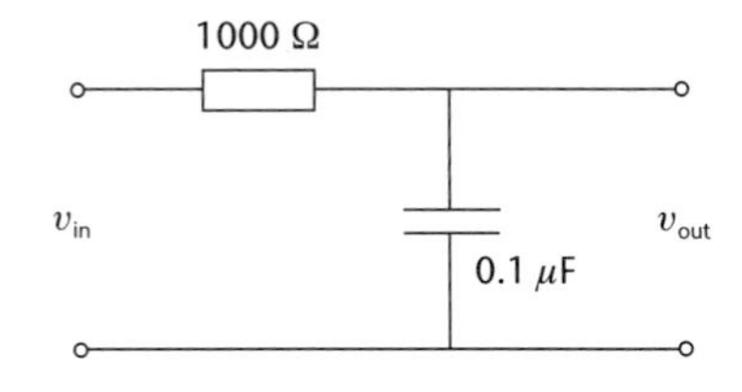

Fig. 3.10 ● *RC* network

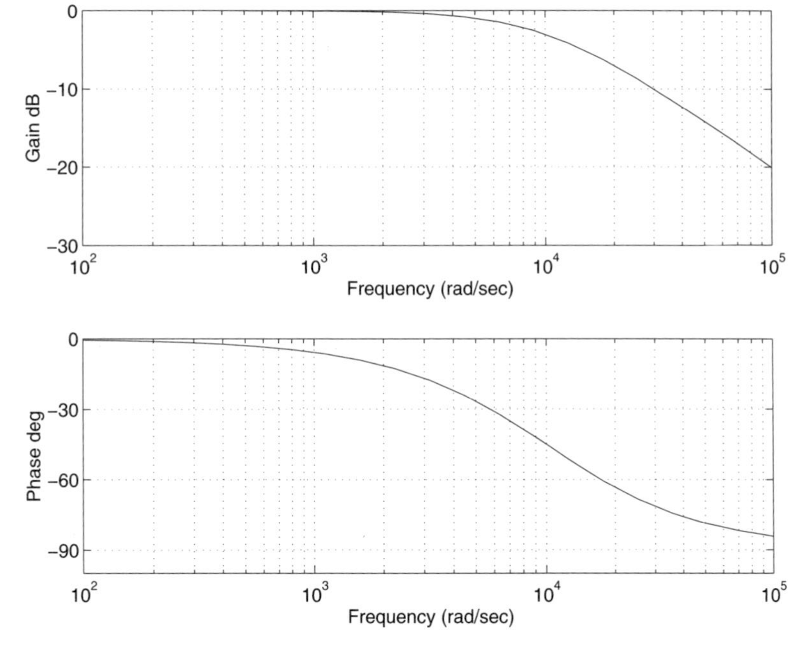

Fig. 3.11 ● Example of frequency response plots for *RC* network

These were obtained using a simple MATLAB command that will be described in Chapter 13.

One important virtue of frequency response plots that have magnitude on a logarithmic scale is that the response curves for the product of two frequency transfer functions are obtained by simple addition. The same applies to the phase plots when phase is on a linear scale.

The importance in feedback system analysis of frequency response plots using scales of log-amplitude versus log-frequency and linear phase versus log-frequency was recognized by H. W. Bode (1945). For this reason these plots are often called Bode plots. Sometimes asymptotic approximations to these Bode plots are used, in which case they are termed asymptotic Bode plots. Bode plots are particularly useful in the analysis of systems with feedback, but for this the reader is recommended to refer to books on control theory.

3.9 Summary

The response of an electrical network composed of linear resistances, inductances and capacitances is determined by a linear differential equation. It is common to have a cosinusoidal excitation, but it is mathematically very much easier to replace this by a complex exponential excitation and then to take the real part of the resulting response. Because of this the impedance functions used to characterize network elements are the ratio of complex exponential voltages to complex exponential currents. The reciprocals of these are admittance functions.

For linear networks the ratio of an output complex exponential voltage to an input complex exponential voltage is known as the frequency transfer function $H(\omega)$. This can be broken down into an amplitude response $A(\omega)$ and a phase response $\phi(\omega)$. Plots of $20\log_{10} A(\omega)$ versus $\log_{10}\omega$ together with $\phi(\omega)$ versus $\log_{10}\omega$ are often known as Bode magnitude and Bode phase plots; they provide a complete characterization of the network.

3.10 Problems

1. A series *RLC* circuit is driven by a cosinusoidal voltage source $V_m \cos \omega t$ such that the current i is given by the network equation

$$V_m \cos \omega t = iR + L\frac{di}{dt} + \frac{1}{C}\int i\,dt$$

Determine the current

(a) by replacing the driving voltage by a complex exponential voltage $\tilde{v} = V_m e^{j\omega t}$, assuming a solution of the form $\tilde{i} = I_m e^{j(\omega t+\beta)}$, and substituting these into the differential equation to determine $I_m e^{j(\omega t+\beta)}$ and thence i;

(b) without using complex exponentials but by assuming a current $i = I_m \cos(\omega t + \beta)$, substituting this into the differential equation and solving for I_m and β.

2. Determine the impedance of the circuit in Figure 3.12.

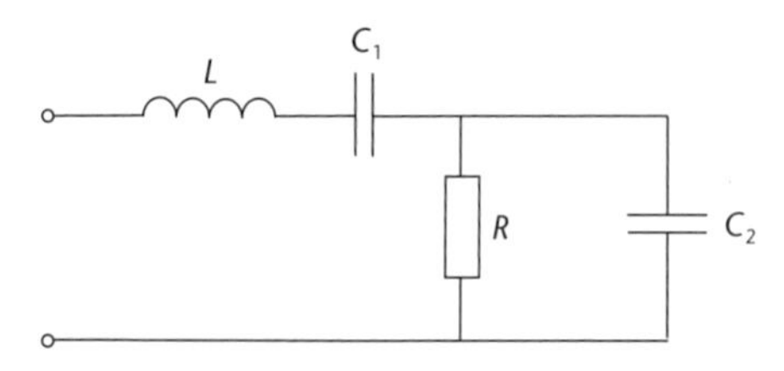

Fig. 3.12

3. A resistance of 5 Ω is in series with a reactance of 4 Ω. Determine the conductance and susceptance of the circuit.

4. Determine the frequency transfer function of the circuit shown in Figure 3.13.

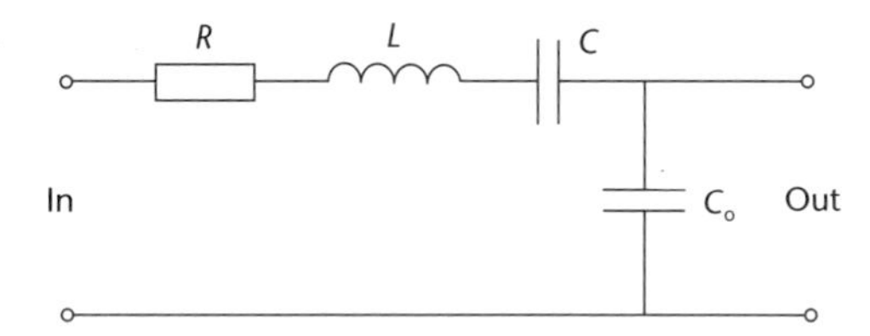

Fig. 3.13

5. An *RC* network has the transfer function $H(\omega) = 1/(1 + j\omega CR)$. Find simplified expressions for $H(\omega)$ that are close approximations to the true transfer function for the two cases $\omega \ll 1/CR$ and $\omega \gg 1/CR$. Hence sketch an asymptotic Bode magnitude plot. Determine the 'break frequency' B at which the two asymptotes intersect. Sketch next an asymptotic Bode phase plot in which a third asymptote is added to intersect the low-frequency and high-frequency asymptotes at $0.1B$ and $10B$.

Chapter four

Phasors

4.1 Preview

It has been established in Chapter 3 that it is useful in network analysis to replace cosinusoidal voltages and currents by complex exponential voltages and currents. This chapter shows how these complex exponentials can be represented as phasors. These phasors provide a simplified notation for the purposes of analysis, but also have the advantage that they can very easily be portrayed pictorially. Phasor diagrams often provide considerable physical insight into network behaviour. Phasor diagrams can also be useful for representing signals in communications and related systems.

4.2 Vector representation of cosinusoids

A complex quantity $Ae^{j\theta}$ can be portrayed in the complex plane by a vector of length A at an angle of θ to the real axis. In a similar manner the complex exponential waveform $Ae^{j\omega t}$, found useful in the previous chapter for determining the response of networks to cosinusoids, can be portrayed in the complex plane by a vector of length A at an angle of ωt to the real axis (Figure 4.1). This can be thought of as a vector that is of constant amplitude but which rotates counterclockwise such that its angle ωt increases with time.

The significance of this is that $Ae^{j\omega t}$ can be expanded as $A\cos\omega t + jA\sin\omega t$, such that the real part of the vector is given by

$$\mathrm{Re}\{Ae^{j\omega t}\} = A\cos\omega t. \tag{4.1}$$

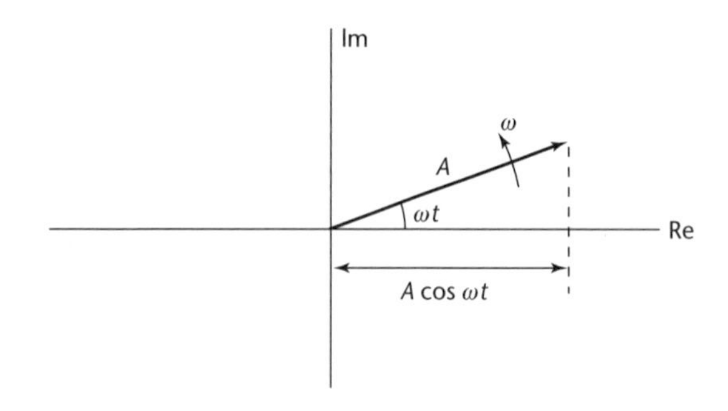

Fig. 4.1 ● Portrayal of $Ae^{j\omega t}$ by a vector in the complex plane

It follows that a projection of the rotating vector onto the real axis generates the cosinusoidal oscillation $A \cos \omega t$. This is illustrated by Figure 4.2.

If this same cosinusoidal signal is now shifted in time such that it has a phase shift of θ at $t = 0$, it becomes $A \cos(\omega t + \theta)$. However,

$$A \cos(\omega t + \theta) = \text{Re}\{Ae^{j(\omega t + \theta)}\} \tag{4.2}$$

The vector is now $Ae^{j(\omega t + \theta)}$ and is as shown in Figure 4.3.

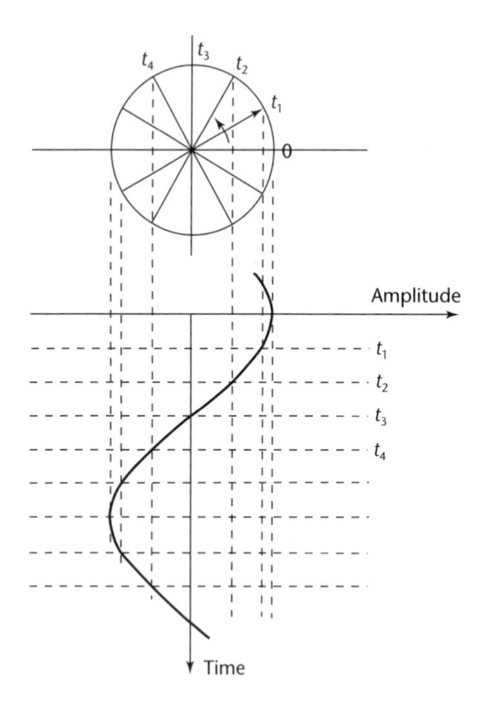

Fig. 4.2 ● Cosinusoidal oscillation by the projection of a rotating vector on to the real axis

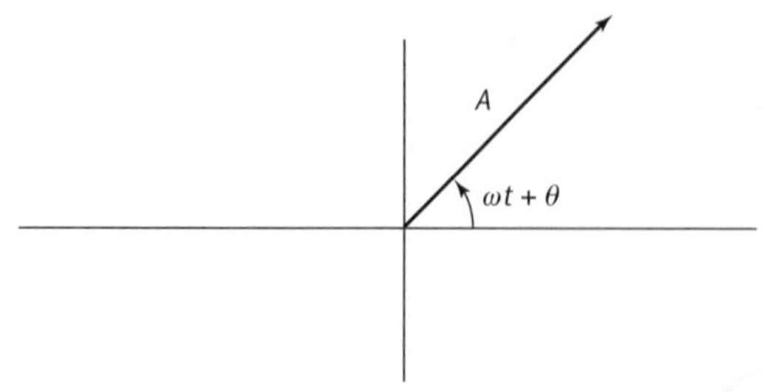

Fig. 4.3 ● Portrayal of $Ae^{j(\omega t+\theta)}$ by a vector in the complex plane

A diagram of the rotating vector $Ae^{j(\omega t+\theta)}$ is thus shown to convey all the information of the signal $A\cos(\omega t+\theta)$, but in a different form to that of a plot of amplitude versus time. It is certainly easier to draw than the waveform itself. Its main advantages, though, arise when we consider several different signals, or when we consider a signal of several components.

Example 4.1

Give a pictorial vector representation of the two signals $v_1(t) = 10\cos(\omega t + 30°)$ and $v_2(t) = 6\cos(\omega t + 75°)$.

Solution The vectors representing the two signals are shown in Figure 4.4. This representation contains all of the information of the two time waveforms drawn in Figures 4.5(a) and 4.5(b), but in a much simpler form.

Example 4.2

Give a pictorial vector representation of the signal $v(t) = 10\cos(\omega t + 300°) + 6\cos(\omega t + 75°)$.

Solution The two separate components have already been shown in Figure 4.4. Their vector sum is shown in Figure 4.6, where both components are assumed to rotate at ω rad/s. The signal $v(t)$ is then the sum of their two projections on the real axis.

It will be noted in this example that, as the two vectors rotate, they maintain their phase difference relative to one another. Thus the resultant is of constant length and also rotates at ω rad/s.

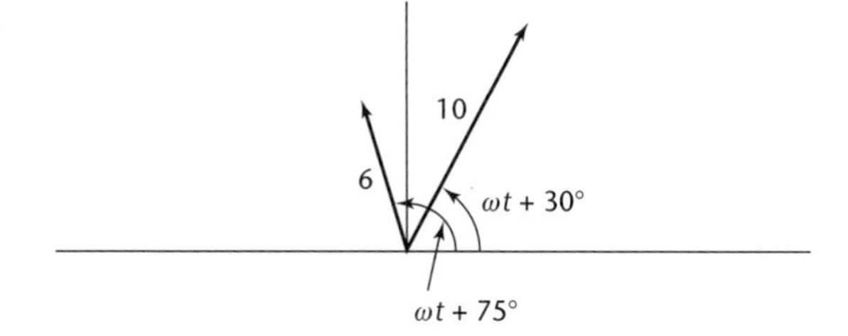

Fig. 4.4 Vector representation of $v_1(t) = 10\cos(\omega t + 30°)$ and $v_2(t) = 6\cos(\omega t + 75°)$

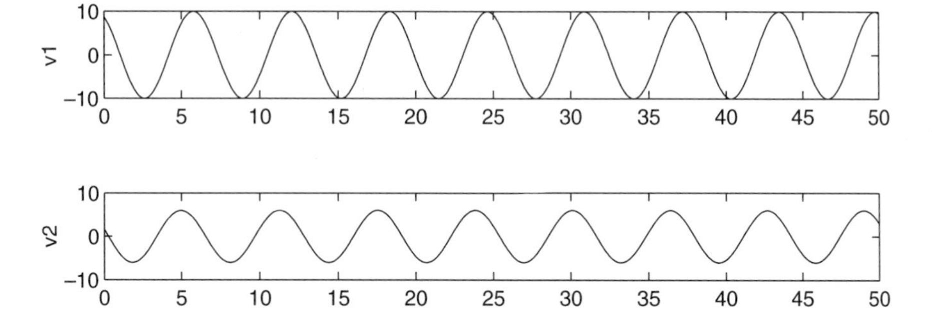

Fig. 4.5 The waveforms $v_1(t) = 10\cos(\omega t + 30°)$ and $v_2(t) = 6\cos(\omega t + 75°)$

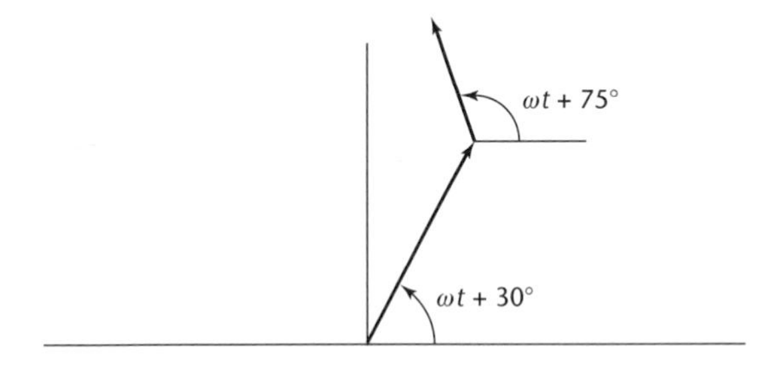

Fig. 4.6 Vector representation of the signal $v(t) = 10\cos(\omega t + 30°) + 6\cos(\omega t + 75°)$

The projection of the resultant onto the real axis is a *single* cosinusoid. Since trigonometric identities can be used to reduce the expression $10\cos(\omega t + 30°) + 6\cos(\omega t + 75°)$ into a single cosinusoid, this is to be expected.

In both these examples it will be noted that the two vectors maintain their phase difference relative to one another as they rotate. This leads to the possibility of an even simpler representation. So long as all components have *the same frequency*, the rotation at ω rad/s can be taken as *understood*, and can be omitted from the diagram. In effect the vectors are then drawn for a single instant in time, namely $t = 0$.

Example 4.3

Repeat Example 4.1 but omitting the rotation.

Solution This is done in Figure 4.7 where the two components are drawn of the correct magnitudes and phases corresponding to time $t = 0$. Apart from the fact that rotation is ignored, such that the complete angle of the cosinusoid $(\omega t + \theta)$ is replaced by θ alone, the diagram is identical to Figure 4.4.

A major advantage of the vector portrayal of cosinusoidal signals is the way it readily enables them to be added without resort to trigonometric identities.

Example 4.4

Use the vector representation of the signal $v(t) = 10\cos(\omega t + 30°) + 6\cos(\omega t + 75°)$ to determine a single cosinusoid that can replace it.

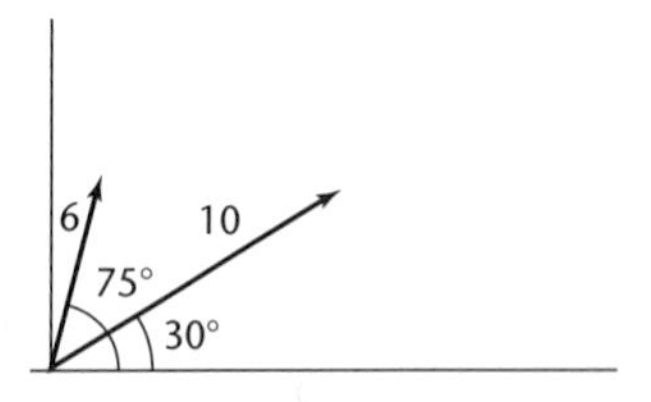

Fig. 4.7 Vector representation of $v_1(t) = 10\cos(\omega t + 30°)$ and $v_2(t) = 6\cos(\omega t + 75°)$ with the rotation omitted

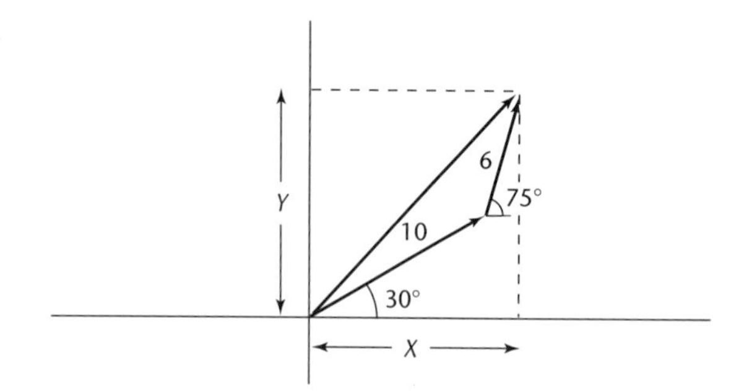

Fig. 4.8 ● Vector representation of the signal $v(t) = 10\cos(\omega t + 30°) + 6\cos(\omega t + 75°)$ corresponding to time $t = 0$

Solution Figure 4.8 is a repeat of Figure 4.6 except that

(a) it corresponds to $t = 0$ such that the angle ωt is zero. The vector rotation is not portrayed but is taken as understood;

(b) the projections of the resultant onto the real and imaginary axes are labelled as X and Y respectively.

It is seen that the resultant at $t = 0$ has an amplitude equal to $\sqrt{X^2 + Y^2}$ and a phase angle equal to $\tan^{-1}(Y/X)$. If the sinusoidal fluctuation with time is now imposed we see that that resulting signal is

$$v(t) = \sqrt{X^2 + Y^2}\cos[\omega t + \tan^{-1}(Y/X)]$$

We can readily evaluate X and Y by summing the projections of the two vectors separately on the real and imaginary axes:

$$X = 10\cos 30° + 6\cos 75° = 10.21$$
$$Y = 10\sin 30° + 6\sin 75° = 10.80$$

Hence

$$v(t) = \sqrt{10.21^2 + 10.8^2}\cos[\omega t + \tan^{-1}(10.8/10.21)]$$
$$= 14.86\cos(\omega t + 46.6°)$$

It will be noted that, besides providing a useful visual portrayal of the two signals, the use of vectors in this way provides an elegant mathematical means of combining two cosinusoids of different amplitudes and phases into a single cosinusoid.

4.3 ● Phasors and phasor diagrams

A diagram such as Figure 4.8 is known as a *phasor diagram*, where the terms 'phasor' and 'vector' are used almost synonomously. The main distinguishing feature is that the term 'phasor' emphasizes that the vectors portrayed refer to *signals* rather than to something quite different such as a force or a velocity.

Note: In communication systems and disciplines reliant on signal processing, such as radar and sonar, the usual convention is to designate the amplitude of a phasor as the *peak* amplitude of the signal it represents. This is what has been done in Figure 4.8. *In network analysis, however*, the usual convention is to designate the amplitude of a phasor as the *r.m.s.* amplitude of the signal it represents. This is done *because a.c. voltages and currents are usually described by their r.m.s. values* (such as 230 V for the European mains supply or 110 V for the North American mains supply).

In network analysis the symbols for voltage and current phasors are **V** and **I** respectively, where upper-case bold symbols are used. If the voltage and current waveforms are $V_m\cos(\omega t+\theta_1)$ and $I_m\cos(\omega t+\theta_2)$, such that the corresponding complex exponential waveforms are $V_m e^{j(\omega t+\theta_1)}$ and $I_m e^{j(\omega t+\theta_2)}$, these phasors are given by

$$\mathbf{V}=\frac{V_m}{\sqrt{2}}e^{j\theta_1} \tag{4.3}$$

and

$$\mathbf{I}=\frac{I_m}{\sqrt{2}}e^{j\theta_2} \tag{4.4}$$

Alternatively, using 'Steinmetz' notation

$$\mathbf{V}=\frac{V_m}{\sqrt{2}}\angle\theta_1 \tag{4.5}$$

and

$$\mathbf{I}=\frac{I_m}{\sqrt{2}}\angle\theta_2 \tag{4.6}$$

If normal italic symbols V and I are used to denote the r.m.s. amplitudes of the voltage and current these can be rewritten as

$$\mathbf{V}=Ve^{j\theta_1}=V\angle\theta_1 \tag{4.7}$$

and

$$\mathbf{I}=Ie^{j\theta_2}=I\angle\theta_2 \tag{4.8}$$

V and **I** are sometimes known as the phasor *representations* of the signals $V_m\cos(\omega t+\theta_1)$ and $I_m\cos(\omega t+\theta_2)$. They are also often known as the phasor *transforms* of these signals. They are *transforms*:

(a) because they are no longer time functions;

(b) because, apart from the frequency, which is assumed known and understood, they do contain *all* the information of the time waveforms; they are just in a different form.

Most often the term *transform* is omitted for shorthand convenience, and **V** and **I** are simply known as phasors.

Example 4.5

Give a phasor portrayal of the UK mains supply, $230\sqrt{2}\cos\omega t$ volts and of a resulting current, $5\sqrt{2}\cos(\omega t-20°)$amps.

Solution We have the two phasors, $\mathbf{V}=230\angle 0°$ and $\mathbf{I}=5\angle-20°$, where the magnitudes correspond to the r.m.s. values. These are shown in Figure 4.9.

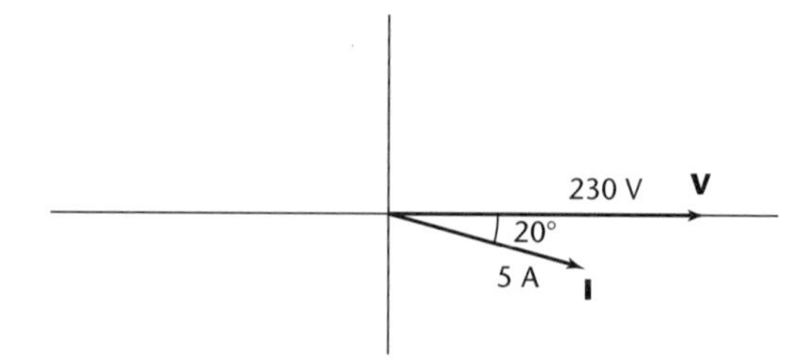

Fig. 4.9 Phasor portrayal of the UK mains voltage and of a resulting current

4.4 Phasors in network analysis

Suppose we wish to determine the current resulting from the application of a voltage $V_m \cos(\omega t + \theta_1)$ to a network whose impedance is **Z**. The procedure adopted in Chapter 3 was to replace the real voltage waveform by the complex voltage waveform, $\tilde{v} = V_m e^{j(\omega t + \theta_1)}$, and then to determine the complex current waveform using the equation

$$\tilde{i} = \tilde{v}/\mathbf{Z} \tag{4.9}$$

The actual current i would then be determined by extracting the real part of $\tilde{i}$.

However, writing $\tilde{i}$ as $I_m e^{j(\omega t+\theta_2)}$, Equation (4.9) becomes

$$I_m e^{j(\omega t+\theta_2)} = \frac{V_m e^{j(\omega t+\theta_1)}}{\mathbf{Z}}$$

Dividing through by $e^{j\omega t}$, this becomes

$$I_m e^{j\theta_2} = \frac{V_m e^{j\theta_1}}{\mathbf{Z}}$$

Dividing both sides by $\sqrt{2}$ gives

$$I e^{j\theta_2} = \frac{V e^{j\theta_1}}{\mathbf{Z}}$$

where V and I are now the r.m.s. amplitudes.

Using Equations (4.7) and (4.8), this may be expressed more elegantly using phasors, simply as

$$\mathbf{I} = \mathbf{V}/\mathbf{Z} \tag{4.10}$$

For cosinusoidal excitations it is in general much more convenient, and hence much more common, to analyse networks using phasors than using complex time waveforms.

Example 4.6

A voltage $v = 325.3 \cos(314t - 25°)$ volts is applied to a network consisting of $R = 10$ ohms and $C = 360\ \mu\text{F}$ in series. Determine the current i.

Solution The network is shown in Figure 4.10. We have $\mathbf{Z} = R + 1/j\omega C = 10 - j/(314 \times 360 \times 10^{-6}) = 10 - j8.846 = 13.35\angle -41.5°$. Using the phasor representation of the voltage waveform

$$\mathbf{V} = (325.3/\sqrt{2})\angle -25° = 240\angle -25°\ \text{V}$$

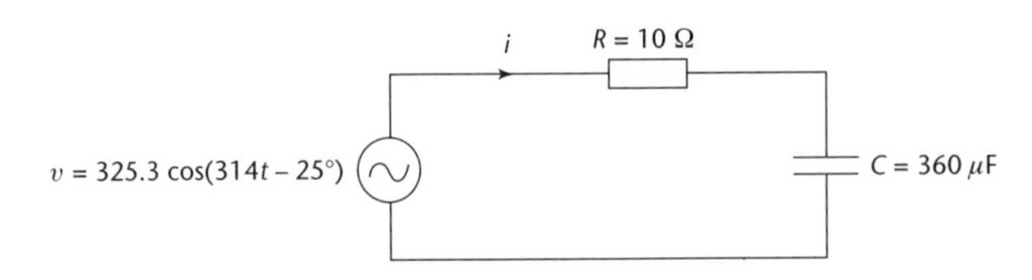

Fig. 4.10 A series *RC* network

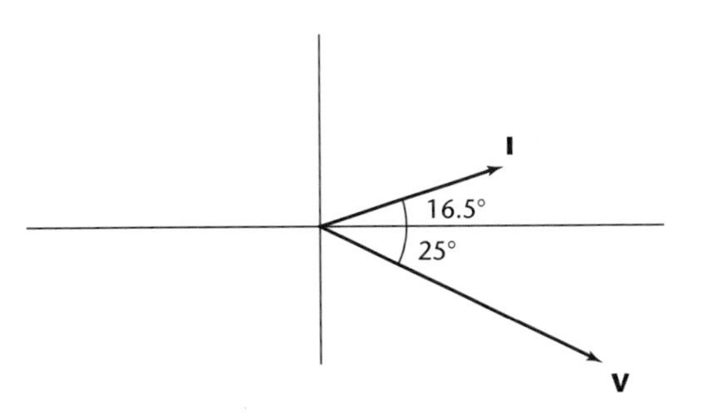

Fig. 4.11 Phasor diagram of the voltage and current in *RC* network

Then

$$\mathbf{I} = \frac{\mathbf{V}}{\mathbf{Z}} = \frac{230\angle -25^\circ}{13.35\angle -41.5^\circ} = 17.23\angle 16.5^\circ \text{ A}$$

Hence

$$i = 17.23\sqrt{2}\cos(314t + 16.5^\circ) \text{ amperes}$$

The phasor diagram in Figure 4.11 shows the voltage and current phasors. The current leads the voltage by the angle of the impedance, namely 41.5°.

Phasors can often provide considerable physical insight into system behaviour, and a good example is that of resonance.

Example 4.7

A cosinusoidal current generator is applied to the parallel *RLC* circuit shown in Figure 4.12. Draw a phasor diagram for the three currents $\mathbf{I}_R$, $\mathbf{I}_L$, $\mathbf{I}_C$, the combined current **I**, and the common voltage **V**. Use this phasor diagram to explain how the voltage across the circuit can attain a maximum at one frequency, thus demonstrating *resonance*.

Solution The first point to note is that, in line with common practice, the labelled voltage and currents are actually phasor representations of the voltage and currents, rather than the true voltage and currents. Ideally they are shown in bold, as they are in

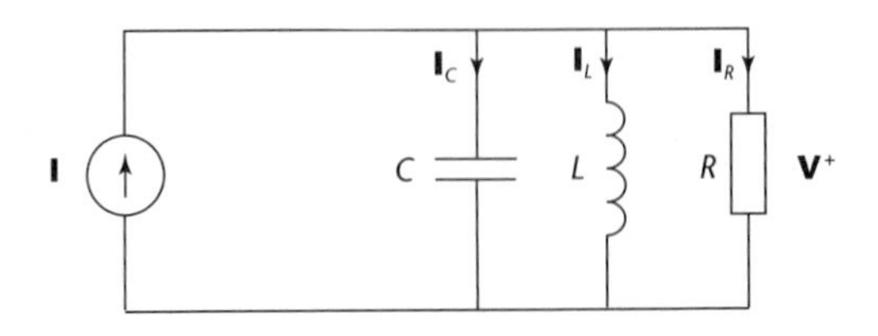

Fig. 4.12 Parallel *RLC* network

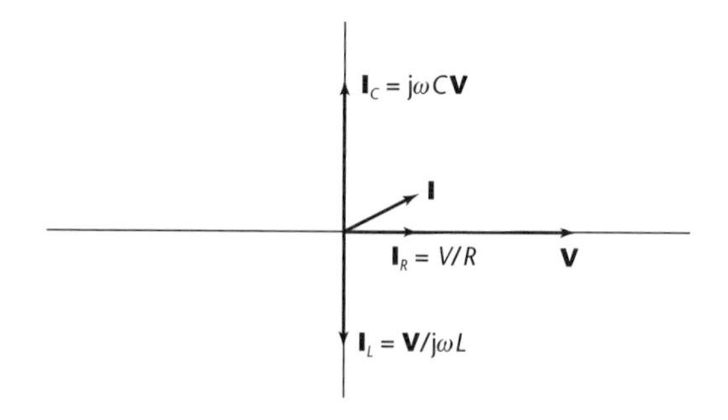

Fig. 4.13 Phasor diagram of voltage and currents in a parallel *RLC* network

Figure 4.12, but this is obviously difficult with hand-drawn sketches and is often not done. It will also be noted that the plus sign above the **V** denotes the sign convention for the voltage, namely that a voltage which is greater at the top end of the resistance than at the bottom end is considered to be positive.

When drawing the phasor diagram it is easiest to start with that which is *common* to all the elements, which is the voltage across them, rather with that which is known, which is the generator current. Thus, in Figure 4.13, the phasor **V** is drawn first. It is drawn of arbitrary length, since the amplitude V of this voltage is not yet known. The remaining phasors are then drawn *relative* to this reference. The phasor $\mathbf{I}_R$ is drawn in phase with **V** and of length V/R. The phasor $\mathbf{I}_L$ is drawn lagging **V** by 90° and of length $V/\omega L$. The phasor $\mathbf{I}_C$ is drawn leading **V** by 90° and of length ωCV. The total current is then given by

$$\mathbf{I} = \mathbf{I}_R + \mathbf{I}_L + \mathbf{I}_C$$

and is the vector sum of the three component currents, as shown in Figure 4.13. However, **I** must equal the generator current and, since this is set, this now establishes the *scale* of the current phasors.

It will be noted that $|\mathbf{I}_R|$ is less than the generator current $|\mathbf{I}|$; also that $\mathbf{I}_L$ and $\mathbf{I}_C$ have magnitudes $|\mathbf{I}_L|$ and $|\mathbf{I}_C|$ which have a dependence on frequency but which are always in antiphase. This signifies that there is some frequency at which they cancel each other. When this happens $\mathbf{I}_R$ equals the generator current and therefore acquires its maximum value. Since $V = I_R R$ this also means that the *voltage* then has *its* maximum value. This is the resonant condition.

The frequency at which the cancellation of inductance and capacitance currents occurs is that which causes the voltage to be a maximum, and is the resonant frequency.

4.5 The use of phasors in power calculations

Before considering the use of phasors for power calculations a preliminary discussion of dissipative and reactive power in networks is given. Consider a load across which there is a voltage $v(t) = \sqrt{2}\,V\cos(\omega t + \alpha)$ and through which there is a current $i(t) = \sqrt{2}\,I\cos(\omega t + \alpha + \theta)$. The phasor diagram for this voltage and current is given in Figure 4.14.

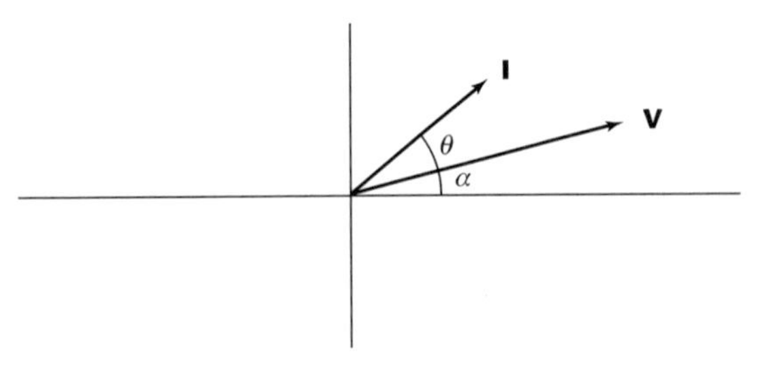

Fig. 4.14 Phasor diagram of voltage across and current through a load

The instantaneous power in the load is given by

$$p(t) = v(t)i(t) \tag{4.11}$$

Therefore

$$\begin{aligned} p(t) &= 2VI\cos(\omega t + \alpha)\cos(\omega t + \alpha + \theta) \\ &= VI[\cos(2\omega t + 2\alpha + \theta) + \cos\theta] \end{aligned} \tag{4.12}$$

This instantaneous power has an alternating component, signifying energy going in and out of the load, plus a d.c. component. The *mean* power corresponds to the d.c. component and, from Equation (4.12), is given by

$$P = VI\cos\theta \text{ watts} \tag{4.13}$$

This is the power actually *dissipated* by the load.

The product of r.m.s. voltage and r.m.s. current, VI, is called the apparent power and has the units of volt-amperes (VA). It is often given the symbol S.

The ratio of mean power to apparent power is defined as the power factor (p.f.), that is:

$$\text{p.f.} = \frac{\text{mean power}}{VI} \tag{4.14}$$

such that

$$\text{mean power} = VI \times \text{p.f.} \tag{4.15}$$

For the sinusoidal waveforms under consideration the power factor is given by

$$\text{p.f.} = \frac{VI\cos\theta}{VI} = \cos\theta \tag{4.16}$$

A positive value of θ gives the same value of $\cos\theta$ as an equal but negative value of θ. To overcome this ambiguity the power factor should be specified as lagging or leading, depending, respectively, on whether the current lags or leads the voltage.

Example 4.8

Determine the power factor of the load in Figure 4.15 when it is driven by a 50 Hz source.

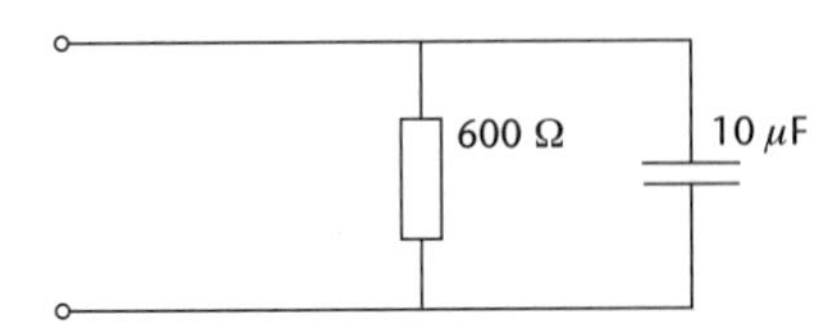

Fig. 4.15 Example of a reactive load

Solution

$$\mathbf{Z}_C = \frac{-j}{100\pi \times 10^{-5}} = -j318.3\ \Omega$$

$$\therefore\ \mathbf{Y} = \frac{1}{600} - \frac{1}{j318.3} = \frac{j318.3 - 600}{j318.3 \times 600} = 0.34 + j0.653 = 0.737\angle 62.5^\circ$$

This signifies that the current leads the voltage by 62.5°.

$$\therefore\ \text{p.f.} = \cos 62.5^\circ = 0.46 \text{ leading}$$

The mean, or dissipated, power is due to the component of current that is in phase with the voltage. The current that is 90° out of phase with the voltage causes what is termed 'reactive power', and which is given by

$$Q = VI\sin\theta \tag{4.17}$$

The units are volt-amperes reactive (var).

Based upon this definition, Q can be positive or negative depending on whether the current leads or lags the voltage, but a more usual convention is to describe it by its modulus and to add the specification of whether it is leading or lagging depending, respectively, on whether the current leads or lags the voltage. For example we might specify a reactive power of 1400 var, lagging.

Another term of importance is the complex power **S**, and this is where the phasor representations of voltages and currents are involved. The complex power is defined by

$$\mathbf{S} = \mathbf{VI}^* \tag{4.18}$$

where the asterisk is used to denote a complex conjugate (thus if $\mathbf{I} = Ie^{j\theta}$ then $\mathbf{I}^* = Ie^{-j\theta}$).

To see the significance of this, consider the voltage $\sqrt{2}\,V\cos(\omega t + \alpha)$ and the current $\sqrt{2}\,I\cos(\omega t + \alpha + \theta)$. We have

$$\mathbf{V} = Ve^{j\alpha}$$

$$\mathbf{I} = Ie^{j(\alpha+\theta)}$$

Hence

$$\mathbf{S} = \mathbf{VI}^* = Ve^{j\alpha}Ie^{-j(\alpha+\theta)} = VIe^{-j\theta} = VI\cos\theta - jVI\sin\theta \tag{4.19}$$

We see that

(a) the magnitude of this complex power is the apparent power VI;

(b) the real part of this complex power is the average power;

(c) the modulus of the imaginary part of this complex power is the reactive power. (Contrary to what one might expect, this reactive power is leading if the imaginary component of **S** is negative, and lagging if the imaginary component of **S** is positive.)

The use of phasors for network analysis and the method of calculating power are brought together in the following example.

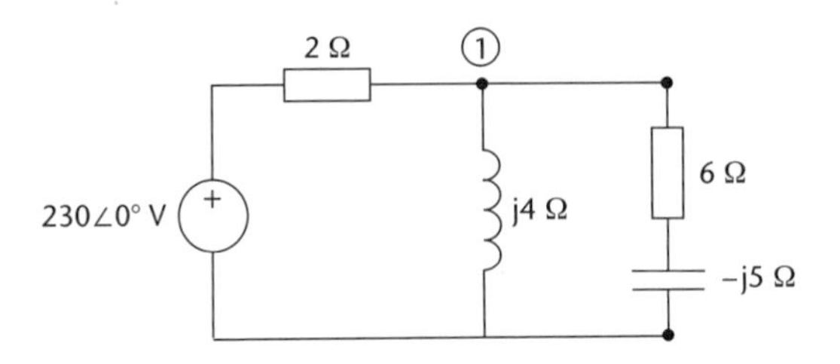

Fig. 4.16 Network example for power calculation

Example 4.9

In the circuit of Figure 4.16 determine the power dissipated in the load consisting of a 6 Ω resistance in series with a capacitance of −j5 Ω.

Solution Assume that the unknown voltage at node 1 is $\mathbf{V}_1$. There are three branch currents flowing out of this node and their sum must be zero.

$$\therefore\ (\mathbf{V}_1 - 230\angle 0°)/2 + \mathbf{V}_1/\text{j}4 + \mathbf{V}_1/(6 - \text{j}5) = 0$$

Hence

$$0.5\mathbf{V}_1 - 115\angle 0° - \text{j}0.25\mathbf{V}_1 + 0.0984\mathbf{V}_1 + \text{j}0.0820\mathbf{V}_1 = 0$$

$$\therefore\ \mathbf{V}_1 = \frac{115\angle 0°}{0.5984 - \text{j}0.1680} = 185.0\angle 15.68°$$

The current in the load is given by

$$\mathbf{I} = \frac{\mathbf{V}_1}{6 - \text{j}5} = \frac{185.0\angle 15.68°}{7.81\angle -39.8°} = 23.69\angle 55.48°$$

The complex power in the load is given by

$$\mathbf{S} = \mathbf{V}_1\mathbf{I}^* = 185.0\angle 15.68° \times 23.69\angle -55.48° = 4382.8\angle -39.8°$$

$$\therefore\ P = \text{Re}(\mathbf{S}) = 4382.8\ \cos(-39.8°) = 3367 \text{ watts}$$

4.6 The application of MATLAB to network analysis

One important technique of network analysis is mesh current analysis where an appropriate number of closed loops within the network are assigned mesh currents and the voltage drop around each of these loops is then equated to zero. A second and probably even more important technique is nodal voltage analysis where independent nodes are assigned nodal voltages and the sum of all currents out of (or into) each one is equated to zero. This was done in Example 4.9 for a single unknown node voltage. Considering the network of Figure 4.17 as a two node example, the

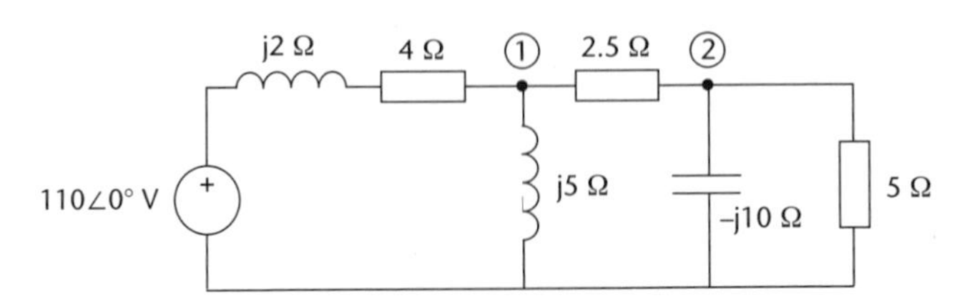

Fig. 4.17 Network example solved using MATLAB

nodal voltage equations expressed in matrix form are

$$\begin{bmatrix} \left(\frac{1}{4+j2}+\frac{1}{j5}+\frac{1}{2.5}\right) & -\left(\frac{1}{2.5}\right) \\ -\left(\frac{1}{2.5}\right) & \left(\frac{1}{2.5}-\frac{1}{j10}+\frac{1}{5}\right) \end{bmatrix} \begin{bmatrix} \mathbf{V}_1 \\ \mathbf{V}_2 \end{bmatrix} = \begin{bmatrix} 110/(4+j2) \\ 0 \end{bmatrix} \tag{4.20}$$

Using abbreviated matrix notation this may be written

$$\mathbf{YV} = \mathbf{I} \tag{4.21}$$

where **Y** is an admittance matrix. To solve this we simply require the MATLAB commands

```
Y = [1/(4 + 2i) + 1/5i + 1/2.5)  - 1/2.5; -1/2.5 (1/2.5 - 1/10i + 1/5)];
I = [110/4 + 2i); 0];
V = Y\I                % alternatively, although slower, we could use V=inv(Y)*I
```

This produces

```
V =
56.7162 + 10.4606i
37.9198 + 0.6538i
```

In other words $\mathbf{V}_1 = 56.7162 + j10.4606$ and $\mathbf{V}_2 = 37.9198 + j0.6538$

4.7 Phasors not related to network analysis

As stated in Section 4.2, the usual convention when phasors relate to cosinusoids in network analysis is that they have lengths equal to the *r.m.s.* amplitudes of the cosinusoids, but in other applications the usual convention is that phasors have lengths equal to the *peak* amplitudes of the cosinusoids. For example, the phasor representation of a modulated communication signal involves peak amplitudes. The uses of phasors outside of network analysis, and the physical insights provided by them, are best demonstrated by examples.

Example 4.10

Use phasors to show that two sinusoids of slightly different frequencies result in an amplitude-modulated waveform whose envelope *beats* at the difference frequency.

Solution Consider the signal $v(t) = a_1 \cos(\omega_1 t) + a_2 \cos(\omega_2 t + \theta)$, where $a_2 < a_1$. This may be rewritten

$$v(t) = a_1 \cos(\omega_1 t) + a_2 \cos[\omega_1 t + (\omega_2 - \omega_1)t + \theta]$$

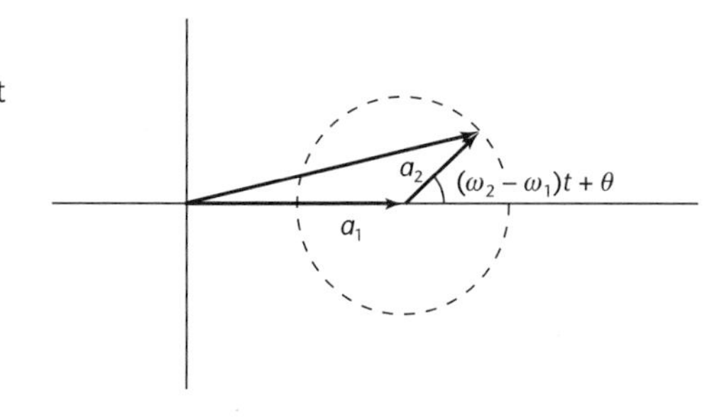

Fig. 4.18 Two phasors of different frequencies, showing how the resultant envelope has a beat frequency

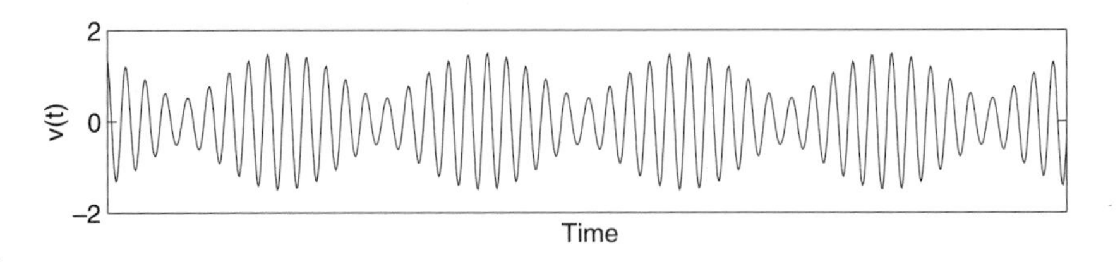

Fig. 4.19 The waveform arising from the interference between two sinusoids of different frequency

Thus the second term may be considered as having a time-varying phase angle $(\omega_2 - \omega_1)t + \theta$ relative to the first term. Using a stationary phasor to represent the first term the second term is then represented by a *rotating* phasor. The phasor diagram is then as shown in Figure 4.18, where it is noted that the phasor lengths correspond to the *peak* amplitudes. (In many respects this is more like a vector diagram, but the term phasor is used to emphasize that the arrows relate to *signals*.)

As the shorter phasor rotates relative to the longer phasor there is a resultant which is amplitude modulated at a frequency $(\omega_2 - \omega_1)$. This is the beating effect. With $a_2 = 0.5a_1$, $\omega_2 = 1.1\omega_1$ and $\theta = \pi/4$ the waveform will appear as shown in Figure 4.19. There will also be some phase modulation at the same frequency, but this is less apparent to the eye.

The same result could easily have been obtained using trigonometric identities to manipulate the expression for $v(t)$. A phasor diagram, however, provides some very useful physical insight into the mechanism of the effect.

Example 4.11

A simple array used for sonar consists of three omnidirectional elements whose outputs are added. Use a phasor diagram to help visualize how the sensitivity of the array varies with the angle of arrival of a sonar echo. Assume that the incoming signal is of a single frequency and that the elements are five wavelengths (5λ) apart.

Solution To understand this problem consider a signal from a distant source to strike the array such that the angle of incidence is α. Figure 4.20 shows the ray paths from the distant source to the left-hand and centre receivers.

If the source is sinusoidal of radian frequency ω, the signal received by the left-most element will depend upon the delay between source and element but can arbitrarily be considered to be $\cos \omega t$. The path length from the source to the next

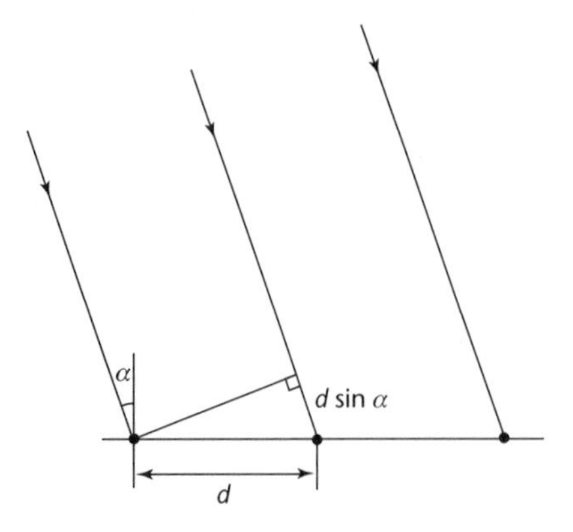

Fig. 4.20 Ray paths of signal incident on an array of three receivers

element is greater by $d \sin \alpha$, where d is the element spacing, and hence the signal on this element is the same except delayed by $(d \sin \alpha)/c$, where c is the velocity at which acoustic waves propagate in water (which is approximately 1500 m/s). The delay on the third element is $(2d \sin \alpha)/c$. Therefore the signal resulting from the addition of all three receiver outputs is $v(t)$, where

$$v(t) = \cos \omega t + \cos\left[\omega\left(t - \frac{d \sin \alpha}{c}\right)\right] + \cos\left[\omega\left(t - \frac{2d \sin \alpha}{c}\right)\right] \tag{4.22}$$

Defining a term k as

$$k = \omega/c \tag{4.23}$$

this becomes

$$v(t) = \cos \omega t + \cos(\omega t - kd \sin \alpha) + \cos(\omega t - 2kd \sin \alpha) \tag{4.24}$$

The differences between the various elements have now been expressed as phase shifts rather than as time delays.

Using the relationship $\lambda = c/f$, Equation (4.23) may also be expressed as

$$k = 2\pi/\lambda \tag{4.25}$$

and, since $1/\lambda$ is the number of wavelengths in one metre, $2\pi/\lambda$ is the number of radians in one metre. For this reason k is often known as the wavenumber.

In our case $d = 5\lambda$, such that $kd = 10\pi$ radians. This gives

$$v(t) = \cos \omega t + \cos(\omega t - 10\pi \sin \alpha) + \cos(\omega t - 20\pi \sin \alpha)$$

Figures 4.21(a), (b), (c) and (d) show phasor diagrams for angles of incidence of 0°, 1.91°, 3.82° and 5.74°, corresponding to $\sin \alpha$ values of 0, $\frac{1}{30}$, $\frac{1}{15}$, $\frac{1}{10}$ respectively. The amplitude of the resultant represents the sensitivity of the array. It is seen that there is a peak in the sensitivity at 0°, as is to be expected, but a complete null when $\alpha = 3.82°$, corresponding to the case of Figure 4.21(c).

A complete description of the sensitivity as a function of angle, $D(\alpha)$, is obtained from a mathematical evaluation of the length of the resultant phasor. This is given by

$$D(\alpha) = |\,1 + e^{-j10\pi \sin \alpha} + e^{-j20\pi \sin \alpha}\,| \tag{4.26}$$

This directional response is shown in Figure 4.22 for all angles between −20° and +20°, with the four values treated by the phasor diagrams of Figure 4.21 denoted by asterisks.

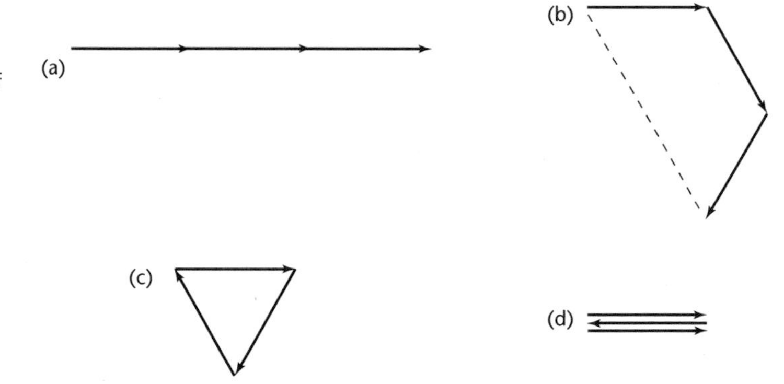

Fig. 4.21 Phasor diagrams for the resultant of the signals on three receivers 5λ apart for different angles of incidence:
(a) $\alpha = 0°$;
(b) $\alpha = 1.91°$;
(c) $\alpha = 3.82°$;
(d) $\alpha = 5.74°$

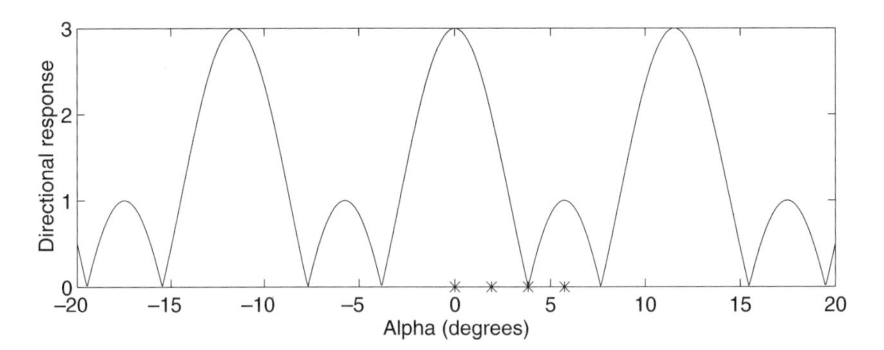

Fig. 4.22 Directional response of three-element array with elements 5λ apart

It will be noted that phasor diagrams can give considerable insight into the *physical mechanism* by which this directional response arises. They show how the signals on the various array elements can interfere constructively or destructively depending on direction. As the direction changes from a zero angle of incidence the phasors begin to curl in on each other. This causes the resultant to get smaller and demonstrates a reduced sensitivity. With a further increase in the angle of incidence the phasor diagram unwinds and indicates an increasing sensitivity. A still further increase in the angle of incidence causes the phasor diagram to again curl in on itself such that the sensitivity again decreases.

Example 4.12

A bandpass filter in the intermediate frequency (IF) stage of an FM receiver passes the wanted FM channel while rejecting neighbouring channels. Unfortunately, some of the noise generated in front-end electronics of the receiver also emerges from the same filter. This noise and the wanted signal then enter an FM demodulator. Use phasors to provide a physical explanation of why the noise out of the demodulator is usually small but increases very sharply and becomes very impulsive as the signal weakens, such that the signal to noise ratio (SNR) drops below a threshold value of around 12 dB. (Note: the 'snow' appearing on a TV picture when the signal is very weak is a manifestation of this phenomenon.)

Solution If the bandpass filter is centred at a frequency f_0, the noise passing through it will be similar to a cosine wave of frequency f_0 except that it will be amplitude and phase modulated in a noiselike manner. It would appear on an oscilloscope much as shown in Figure 4.23. The noise passing through the IF bandpass filter and entering the FM demodulator can thus be written as

$$n(t) = r(t)\cos[\omega_0 t + \phi(t)] \tag{4.27}$$

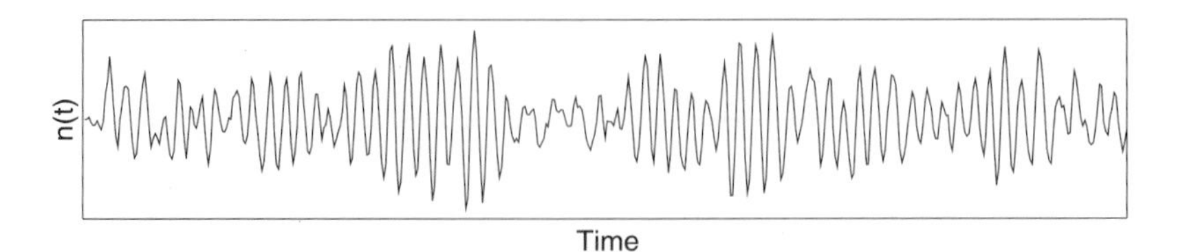

Fig. 4.23 Example of narrowband noise

where $r(t)$ represents some random fluctuation in the *envelope* of this noise signal, and $\phi(t)$ represents a random fluctuation in its *phase*.

Relative to the cosinusoid $\cos \omega_0 t$, a phasor representation of the noise is thus as shown in Figure 4.24.

For simplicity let us now consider a wanted signal that is *unmodulated*, namely $A \cos(\omega_0 t)$, and determine the effect of the noise on this. The resultant signal is given by

$$v(t) = A \cos(\omega_0 t) + r(t) \cos[\omega_0 t + \phi(t)] \tag{4.28}$$

A phasor diagram of the two components of $v(t)$ is shown in Figure 4.25.

The noise phasor $n(t)$ changes its length and phase as a function of time and, if it is weak compared with the wanted signal, the locus of its tip might perhaps be as shown in Figure 4.26. The wanted signal is unmodulated, but the addition of noise produces a resultant that is amplitude and phase modulated by the noise. The phase angle of the resultant is $\theta(t)$ and the r.m.s. value of $\theta(t)$ in this particular example is seen to be quite small, perhaps 0.25 radians.

Consider next what happens if the average noise level is increased. The phasor diagram and the locus of the tip of the noise phasor might now be as shown in Figure 4.27.

It will be noted that the tip of the noise phasor sometimes encircles the origin. When this happens there is a rapid change of 2π radians in the phase $\theta(t)$ of the resultant. The phase of the resultant as a function of time might be as shown in Figure 4.28,

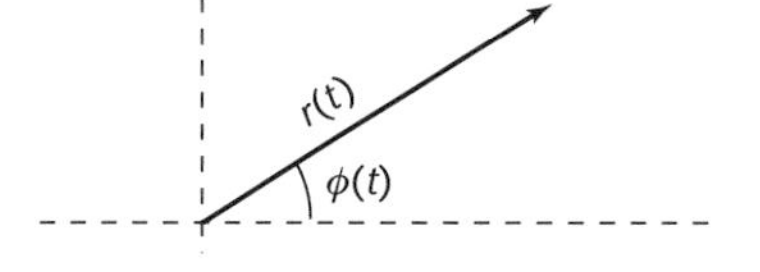

Fig. 4.24 ● Phasor representation of narrowband noise

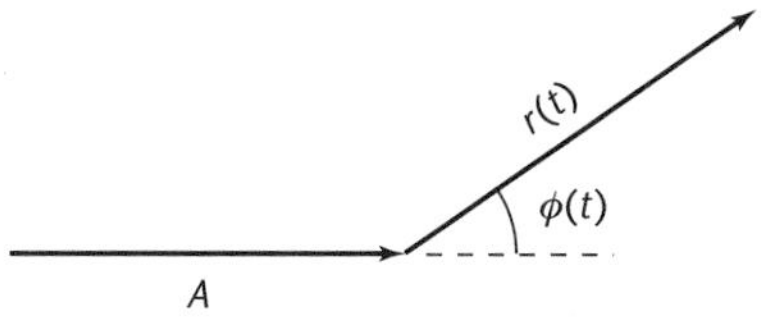

Fig. 4.25 ● Phasor diagram of an unmodulated carrier signal plus noise

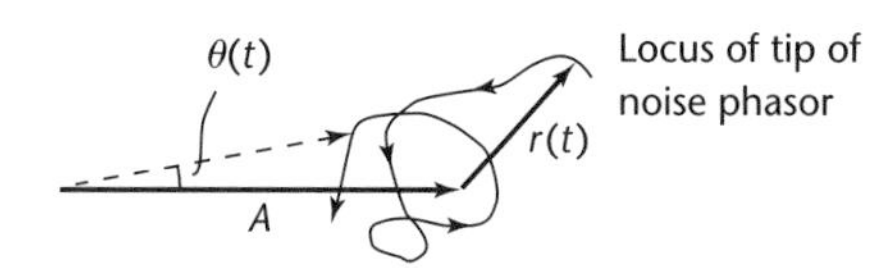

Fig. 4.26 ● Phasor diagram of $A \cos(\omega_0 t) + n(t)$ for the case of high SNR

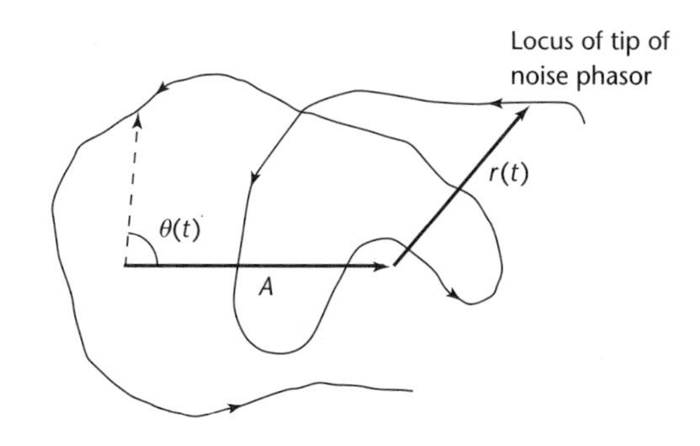

Fig. 4.27 ● Phasor diagram of $A \cos(\omega_0 t) + n(t)$ for the case of low SNR

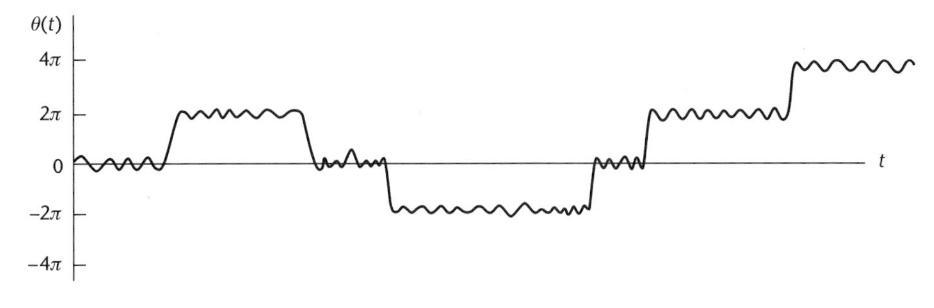

Fig. 4.28 Variation with time in the phase of $A \cos \omega_0 t + n(t)$ for the case of low SNR

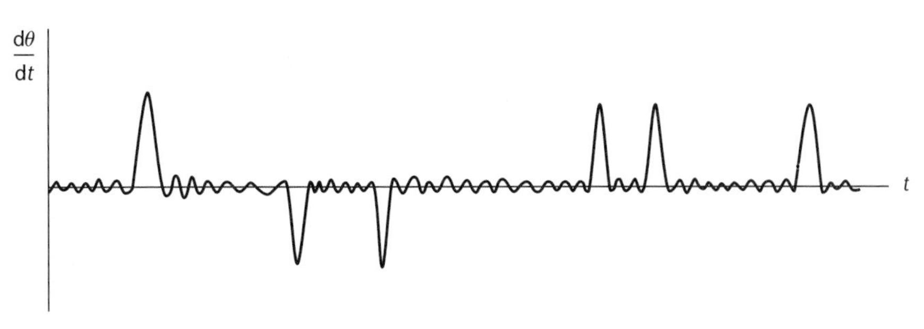

Fig. 4.29 Variation with time in the instantaneous frequency of $A \cos \omega_0 t + n(t)$ for the case of low SNR

where some 2π jumps in phase are shown. Depending on the direction in which the noise phasor rotates when encircling the origin the jump can be either $+2\pi$ or -2π.

By definition the instantaneous frequency of a signal is the rate of change of phase. Thus the output of the FM demodulator is proportional to $\mathrm{d}\theta/\mathrm{d}t$ and is as shown in Figure 4.29. It is seen that there are output noise spikes arising whenever there is a change in phase of 2π radians.

A more detailed analysis shows that the likelihood of these noise spikes is negligible when the ratio of signal power to mean noise power is greater than about 16 dB. However, as this input signal to noise ratio decreases to around 12 dB the likelihood of these noise spikes becomes significant and the output noise power increases very greatly. It should be noted that this increase in output noise power is likely to occur in practice because of a fall in input signal power rather than because of a rise in input noise power. The sudden degradation of output signal to noise ratio as the input signal to noise ratio falls below about 12–13 dB is known as the threshold phenomenon in frequency demodulation. The treatment given in this example is a good example of how phasors can provide a simple physical explanation of a complex phenomenon in a communication system.

4.8 Contra-rotating phasors

So far phasors have been considered either as rotating, in which case the projection of the phasor on the real axis is a time varying quantity, or as stationary, in which case the projection of the phasor on the real axis is the value of the signal at zero time. In this latter case it is taken as understood that the signal is fluctuating cosinusoidally at some known angular frequency ω.

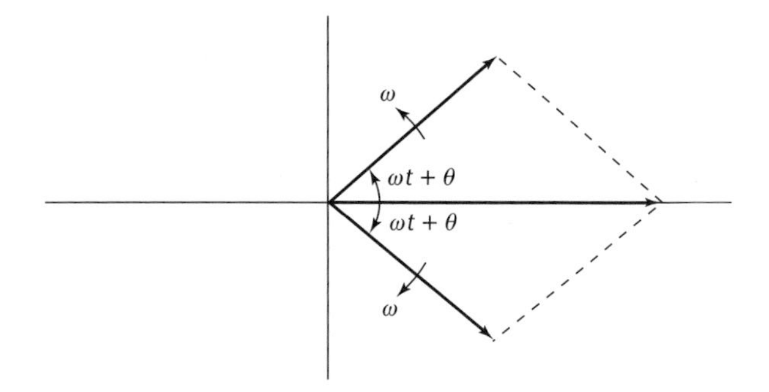

Fig. 4.30 Portrayal of $A\cos(\omega t + \theta)$ by two contra-rotating phasors

An alternative way of representing a signal $\cos(\omega t + \theta)$ makes use of the result that $\cos x = \frac{1}{2}e^{jx} + \frac{1}{2}e^{-jx}$, such that

$$A\cos(\omega t + \theta) = \tfrac{1}{2}Ae^{j(\omega t + \theta)} + \tfrac{1}{2}Ae^{-j(\omega t + \theta)} = \tfrac{1}{2}Ae^{j(\omega t + \theta)} + \tfrac{1}{2}Ae^{j(-\omega t - \theta)} \tag{4.29}$$

This shows that the signal $A\cos(\omega t + \theta)$ may be portrayed as two contra-rotating phasors (or vectors), as shown in Figure 4.30. The resultant of the two phasors always lies on the real axis and has an amplitude that fluctuates cosinusoidally with time. This representation is rather clumsy compared with that of projecting a single phasor onto the real axis and is rarely used.

4.9 ● Summary

Phasors are used to express sinusoidal signals solely in terms of their magnitude and phase. They are a simplified form of the complex exponential representation of signals in which the frequency and temporal variation are omitted, these being taken as known and understood. Network analysis is usually done using phasors. Phasors have the additional advantage that they are easy to portray visually and phasor diagrams of voltages and currents can often provide significant physical insight into the behaviour of a network. Phasors are also useful for portraying communication signals, but the conventions are different than for network analysis. In network analysis the phasor magnitudes are generally taken to be the r.m.s. amplitudes, whereas they are taken to be the peak amplitudes in communication applications.

4.10 ● Problems

1. A voltage 155.6 cos(377*t*) is applied to a network consisting of $R = 15\ \Omega$ in series with $L = 0.1$ H. Determine the current using phasors.
2. A circuit consists of a 10 Ω resistor, a 20 mH inductance and a 1 μF capacitance in series and is driven by an alternating voltage source of 1 V r.m.s. Draw a phasor diagram for the circuit (a) at the resonant frequency and (b) at half the resonant frequency.
3. The circuit shown in Figure 4.31 has unknown voltages $\mathbf{V}_1$ and $\mathbf{V}_2$ at nodes 1 and 2. Derive two equations on the basis that the sum of the currents into each node is zero, and solve them to determine $\mathbf{V}_2$ and hence the current in the 4 Ω resistance.

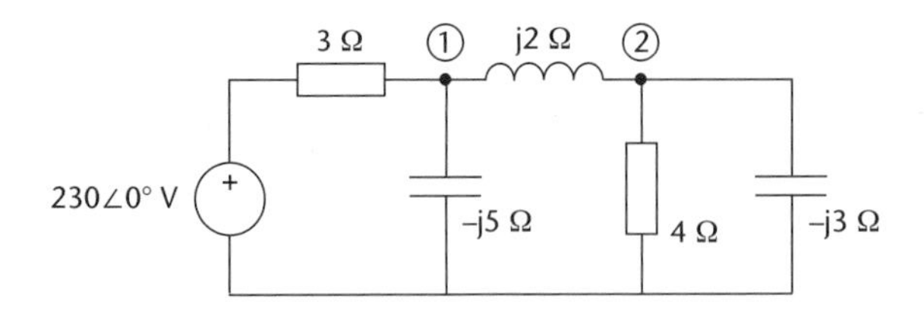

Fig. 4.31

4. The amplitude modulated signal $A(1 + m \cos pt) \cos(\omega_c t)$ can be expanded in terms of a carrier and two sidebands as $A \cos(\omega_c t)$ $+(Am/2) \cos[(\omega_c - p)t] + (Am/2) \cos[(\omega_c + p)t]$. Draw a phasor diagram of the three components and demonstrate that the resultant phasor is solely amplitude modulated. Using a second phasor diagram show next that the signal $A \cos(\omega_c t) - (Am/2) \cos(\omega_c - p)t$ $+(Am/2) \cos(\omega_c + p)t$ contains little amplitude modulation but significant phase modulation.

5. An unmodulated carrier wave $A \cos \omega_c t$ and an unwanted unmodulated interference signal $B \cos(\omega_c + \omega_1)t$ enter an FM demodulator whose output is $(k/2\pi)(d\theta/dt)$ volts, where θ is the phase perturbation due to the interference. Use a phasor diagram of the signal $A \cos \omega_c t + B \cos(\omega_c + \omega_1)t$ to estimate the maximum and minimum voltage from the demodulator if $A = 1$, $B = 0.2$, $\omega_1 = 2000$ rad/s, and $k = 1$mv/Hz.

6. A carrier wave of frequency ω_c that is frequency modulated by a sinusoid of frequency p can be written $\cos(\omega_c t + \beta \sin pt)$. This can be expanded mathematically as

$$\begin{aligned}\cos(\omega_c t + \beta \sin pt) &= \mathrm{J}_0(\beta) \cos \omega_c t \\ &+ \mathrm{J}_1(\beta)\,[\cos(\omega_c + p)t - \cos(\omega_c - p)t] \\ &+ \mathrm{J}_2(\beta)\,[\cos(\omega_c + 2p)t + \cos(\omega_c - 2p)t] \\ &+ \mathrm{J}_3(\beta)\,[\cos(\omega_c + 3p)t - \cos(\omega_c - 3p)t] + \cdots\end{aligned}$$

where $\mathrm{J}_n(\beta)$ is a Bessel function of the first kind and order n. When $\beta=1$, $\mathrm{J}_0(\beta) = 0.77$, $\mathrm{J}_1(\beta) = 0.44$, $\mathrm{J}_2(\beta) = 0.11$, $\mathrm{J}_3(\beta) = 0.02$ and higher order Bessel functions are negligible. Draw a phasor diagram of the components for the cases of: (a) $\beta = 1$, $t = 0$; (b) $\beta = 1$, $pt = \pi/2$. Confirm from the phasor diagram that the resultant amplitude is unchanged between the two cases but that the phase has changed by 1 radian. Draw a phasor diagram for the case of $\beta = 1$, $pt = \pi/4$.

7. A sonar array consists of six equispaced hydrophones in a straight line. Their outputs are summed to give a composite output that is dependent on the direction of a single-frequency incoming signal. If the separation of the hydrophones is two wavelengths at the signal frequency, draw the phasor diagram of the six hydrophone outputs and their sum for the cases:

(a) when the signal arrives from a direction normal to the line of the hydrophones (sometimes termed the boresight direction);

(b) when the signal arrives from a direction offset from the boresight direction by 5°.

8. The radar scattering properties of a low-flying aircraft can be approximated as being caused by just three scattering points in the horizontal plane, each of equal scattering strength. These are at the vertices of an equilateral triangle ABC. The sides of the triangle are 10 m, and this corresponds to 250 wavelengths at the radar frequency. Draw a phasor diagram for the three echo components (a) when the distant radar is equidistant from B and C, and (b) when it is offset by 0.08° from this direction. Draw some conclusions about the detectability of an aircraft using a single radar pulse.

Chapter five

Line spectra and the Fourier series

5.1 Preview

The last two chapters have shown the importance of cosinusoids. This chapter shows how any periodic waveform may be considered to be comprised of a set of harmonically related cosinusoids. A closely related alternative is that the periodic waveform is considered to be comprised of a set of harmonically related complex exponential waveforms.

5.2 One-sided frequency domain descriptions of sinusoids and cosinusoids

The importance of cosinusoids has already been emphasized. The expression $x(t) = A_1 \cos(\omega_1 t + \theta_1) + A_2 \cos(\omega_2 t + \theta_2)$, giving the variation of signal amplitude with time, is the *time domain* description of a signal containing two such cosinusoids. A *frequency domain* description of the same signal would take it as *understood* that the signal is considered to consist of cosinusoids, and would describe this particular signal solely by the magnitudes and phases of these cosinusoids as a function of their frequency. (Note: it is probably more common in conversation to talk about *sinusoidally* varying signals than about *cosinusoidally* varying signals, and $\sin \omega t$ clearly has a different phase to $\cos \omega t$. In fact, for reasons discussed in Section 2.3, which concerned the fact that $\cos \omega t$ is the real part of $e^{j\omega t}$, *it is preferable to specify phase relative to a cosinusoidal reference*.)

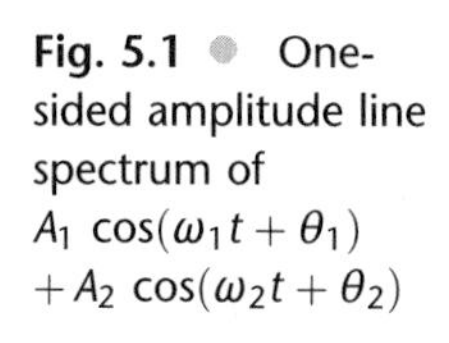

Fig. 5.1 One-sided amplitude line spectrum of $A_1\cos(\omega_1 t+\theta_1)+A_2\cos(\omega_2 t+\theta_2)$

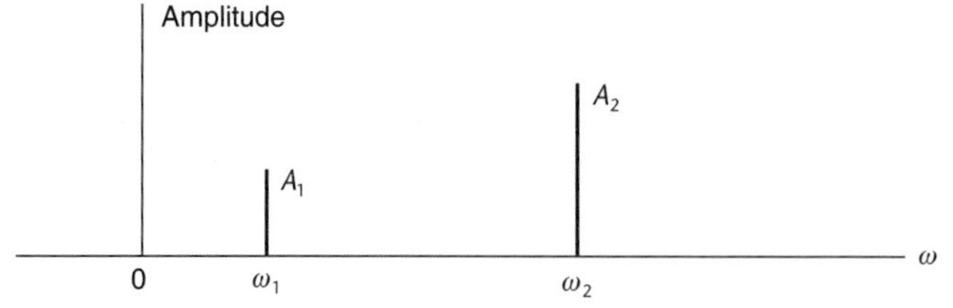

Fig. 5.2 One-sided phase spectrum of $A_1\cos(\omega_1 t+\theta_1)+A_2\cos(\omega_2 t+\theta_2)$

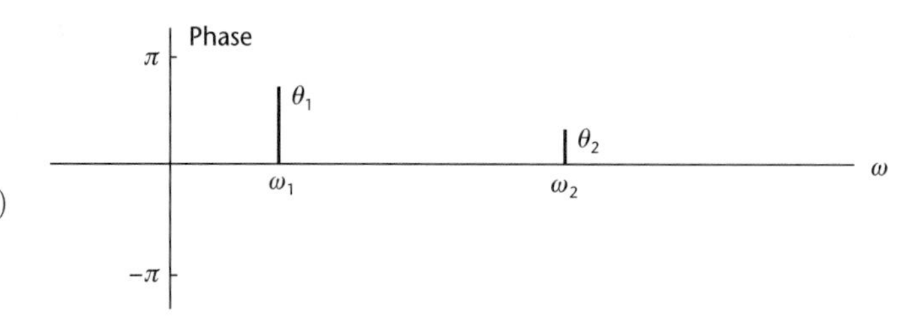

Fig. 5.3 One-sided phase spectrum of $A_1\sin(\omega_1 t+\theta_1)+A_2\sin(\omega_2 t+\theta_2)$

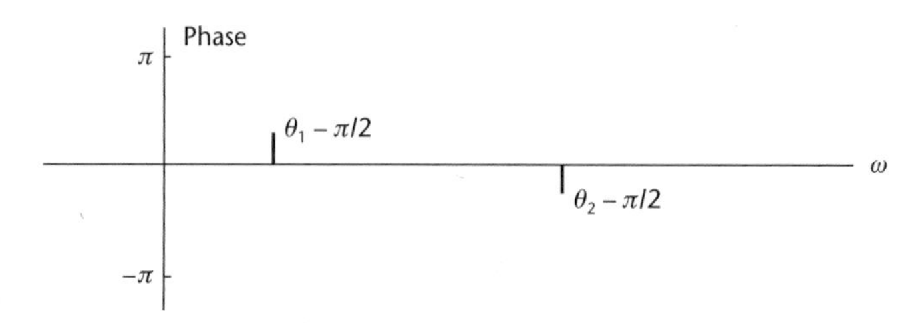

As an example, the frequency domain description of the signal $A_1\cos(\omega_1 t+\theta_1)+A_2\cos(\omega_2+\theta_2)$ is that there is one term at radian frequency ω_1 of amplitude A_1 and phase θ_1, and a second term at radian frequency ω_2 of amplitude A_2 and phase θ_2. A convenient diagramatic description of this is by means of a *line spectrum* of amplitude such as in Figure 5.1, plus a *line spectrum* of phase such as in Figure 5.2. These plots are known as one-sided line spectra, where the term 'one-sided' signifies that, in accordance with physical intuition, frequencies are never negative, but vary from zero to infinity. This contrasts with the two-sided spectra considered in Section 5.3, where the concept of *negative* frequencies is explained and found to be useful.

Returning to the convention of using cosinusoids as the reference for phase we would rewrite the signal $x(t)=A_1\sin(\omega_1 t+\theta_1)+A_2\sin(\omega_2 t+\theta_2)$ as $x(t)=A_1\cos(\omega_1 t+\theta_1-90°)+A_2\cos(\omega_2 t+\theta_2-90°)$. The amplitude spectrum of this is the same as that of $A_1\cos(\omega_1 t+\theta_1)+A_2\cos(\omega_2 t+\theta_2)$, but the phase spectrum terms are 90° less. Thus we have Figure 5.3 in place of Figure 5.2.

5.3 Two-sided frequency domain descriptions of cosinusoids

The single cosinusoid, $A\cos(\omega t+\theta)$, can be written as the sum of two complex exponentials of the same amplitude, but having exponents of opposite sign; that is

$$A\cos(\omega t+\theta)=\frac{A}{2}\,e^{j(\omega t+\theta)}+\frac{A}{2}\,e^{-j(\omega t+\theta)} \tag{5.1}$$

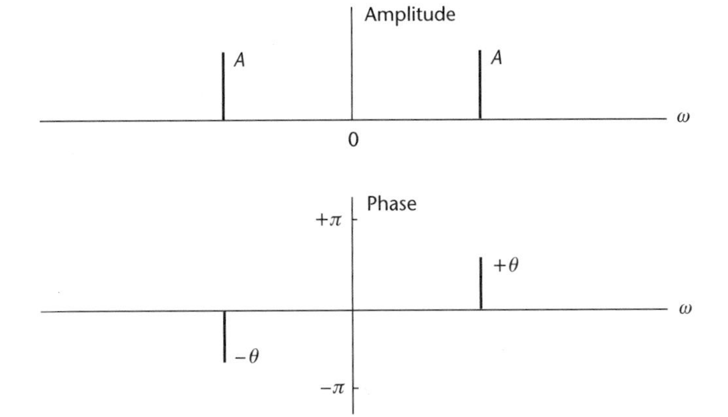

Fig. 5.4 Two-sided amplitude line spectrum of $A\ \cos(\omega t + \theta)$

Fig. 5.5 Two-sided phase spectrum of $A\ \cos(\omega t + \theta)$

Rearranging the exponent of the last term this may be written alternatively as

$$A\ \cos(\omega t + \theta) = \frac{A}{2}\ \mathrm{e}^{\mathrm{j}(\omega t + \theta)} + \frac{A}{2}\ \mathrm{e}^{\mathrm{j}(-\omega t - \theta)} \tag{5.2}$$

The first of these terms, $(A/2)\mathrm{e}^{\mathrm{j}(\omega t + \theta)}$, may be considered to be a complex exponential that has a frequency ω, an amplitude $A/2$, and a phase angle θ. The second term, $(A/2)\mathrm{e}^{\mathrm{j}(-\omega t - \theta)}$, may be considered to be a complex exponential that has a frequency $-\omega$, an amplitude $A/2$, and a phase angle $-\theta$. Thus we have complex exponential terms of both positive *and negative* frequencies. By considering the signal to be composed of complex exponentials rather than of cosinusoids we have emerged with positive and negative frequencies and hence with a *two-sided* line spectrum. The amplitude and phase components of this two-sided line spectrum are given in Figures 5.4 and 5.5. It will be noted that the amplitude spectrum has even symmetry and that the phase spectrum has odd symmetry. It follows from Equation (5.2) that this is always the case with the spectra of real signals.

If we next consider the signal $A_1 \cos(\omega_1 t + \theta_1) + A_2 \cos(\omega_2 t + \theta_2)$, treated in Section 5.1, we now have four complex exponential terms: one that has a frequency ω_1, an amplitude $A_1/2$, and a phase angle θ_1; one that has a frequency $-\omega_1$, an

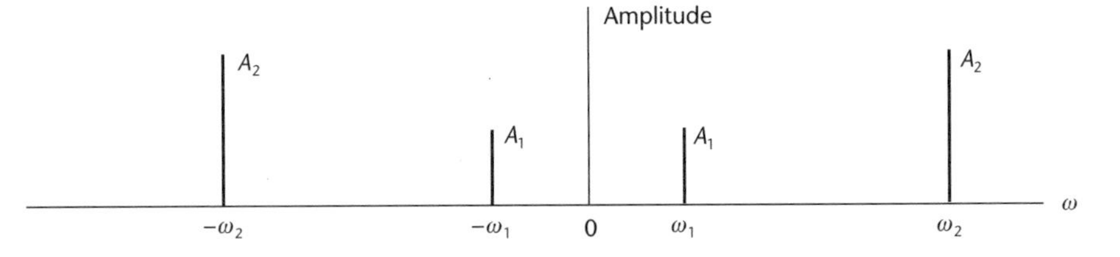

Fig. 5.6 Two-sided amplitude line spectrum of $A_1\ \cos(\omega_1 t + \theta_1) + A_2\ \cos(\omega_2 t + \theta_2)$

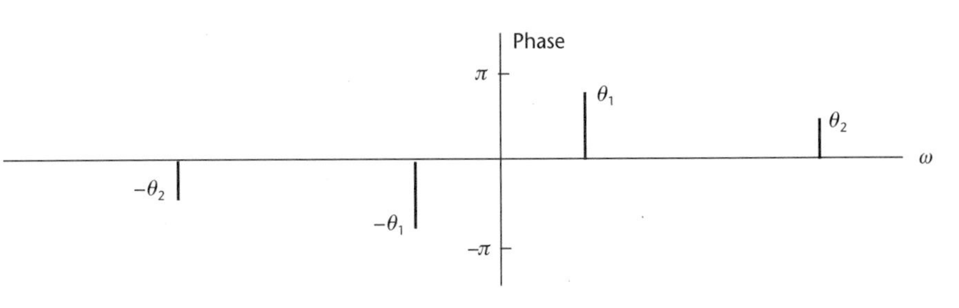

Fig. 5.7 Two-sided phase spectrum of $A_1\ \cos(\omega_1 t + \theta_1) + A_2\ \cos(\omega_2 t + \theta_2)$

amplitude $A_1/2$, and a phase angle $-\theta_1$; one that has a frequency ω_2, an amplitude $A_2/2$, and a phase angle θ_2; and one that has a frequency $-\omega_2$, an amplitude $A_2/2$, and a phase angle $-\theta_2$. We thus emerge with the two-sided amplitude line spectrum shown in Figure 5.6, and the two-sided phase spectrum shown in Figure 5.7.

It should be noted that the terms *amplitude spectrum* and *phase spectrum* used on their own generally refer to *two-sided* spectra.

5.4 Negative frequencies

Negative frequencies often cause a conceptual difficulty and may be easier to accept if it is realized:

(a) that negative frequencies arise only in the context of representing signals as complex exponentials of the form $e^{j\omega t}$, and not in the representation of signals as sinusoids or cosinusoids;

(b) that, for real (that is, physical) waveforms, a complex exponential such as $A_1 e^{j(-\omega_1 t-\theta_1)}$, having the negative frequency $\omega = -\omega_1$, is always accompanied by a complex exponential $A_1 e^{j(+\omega_1 t+\theta_1)}$, having the same amplitude, opposite phase, and the positive frequency $\omega = +\omega_1$. *The two components add together to constitute the real waveform and this has a positive frequency.*

Complex exponentials, and hence negative frequencies, are used extensively because of their mathematical elegance for solving the differential equations associated with system analysis.

5.5 Periodic signals and the trigonometric Fourier series

Let $f_p(t)$ be a periodic signal with period T, such as the train of pulses shown in Figure 5.8. Fourier showed that such a signal can be considered to be composed of a d.c. term, a fundamental cosinusoid of frequency $1/T$ (that is, of period T), plus an infinite set of harmonics of radian frequencies $2/T$, $3/T$, $4/T$ etc. The sum of these terms is known as a trigonometric Fourier series.

The harmonics are not generally of the same amplitude or phase. Hence the Fourier series is

$$f_p(t) = c_0 + \sum_{n=1}^{\infty} c_n \cos(n\omega_0 t + \theta_n) \tag{5.3}$$

where c_0 is a d.c. term, $c_1 \cos(\omega_0 t + \theta_1)$ is a component of fundamental frequency given by $\omega_0 = 2\pi/T$ rad/s, $c_2 \cos(2\omega_0 t + \theta_2)$ is the second harmonic, $c_3 \cos(3\omega_0 t + \theta_3)$ is the third harmonic, and so on.

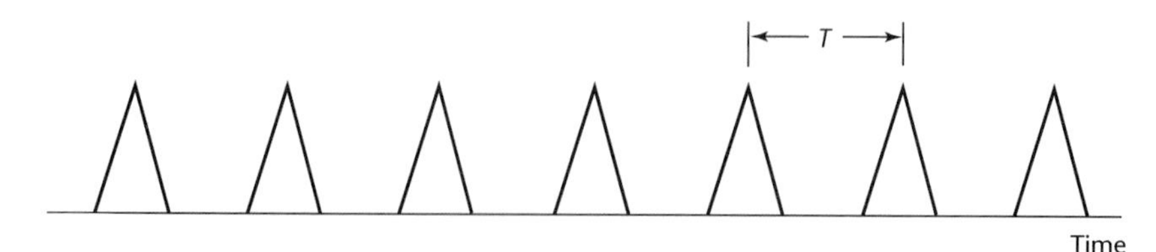

Fig. 5.8 Example of a periodic signal

In order to determine the c and θ coefficients it helps to first expand the cosine terms in Equation (5.3) using the trigonometric identity

$$\cos(A+B) = \cos A \cos B - \sin A \sin B \tag{5.4}$$

This gives

$$f_p(t) = c_0 + \sum_{n=1}^{\infty} \{c_n \cos(\theta_n)\cos(n\omega_0 t) - c_n \sin(\theta_n)\sin(n\omega_0 t)\}$$

or

$$f_p(t) = a_0 + \sum_{n=1}^{\infty} \{a_n \cos(n\omega_0 t) + b_n \sin(n\omega_0 t)\} \tag{5.5}$$

where

$$a_0 \equiv c_0, \qquad a_n \equiv c_n \cos(\theta_n), \qquad b_n \equiv -c_n \sin(\theta_n). \tag{5.6}$$

Equations (5.3) and (5.5) are alternative forms of the trigonometric Fourier series. If we know $f_p(t)$ and wish to find a particular one of the a_n coefficients, namely a_k, we multiply both sides of Equation 5.5 by $\cos(k\omega_0 t)$ and integrate over a complete period T. This gives

$$\int_{-T/2}^{T/2} f_p(t)\cos(k\omega_0 t)\mathrm{d}t = a_0 \int_{-T/2}^{T/2} \cos(k\omega_0 t)\mathrm{d}t + \sum_{n=1}^{\infty}\int_{-T/2}^{T/2} a_n \cos(n\omega_0 t)\cos(k\omega_0 t)\mathrm{d}t$$
$$+ \sum_{n=1}^{\infty}\int_{-T/2}^{T/2} b_n \sin(n\omega_0 t)\cos(k\omega_0 t)\mathrm{d}t$$

Using the trigonometric identities

$$\cos A \cos B = \tfrac{1}{2}\cos(A+B) + \tfrac{1}{2}\cos(A-B) \tag{5.7}$$

and

$$\sin A \cos B = \tfrac{1}{2}\sin(A+B) + \tfrac{1}{2}\sin(A-B) \tag{5.8}$$

this equation becomes

$$\int_{-T/2}^{T/2} f_p(t)\cos(k\omega_0 t)\mathrm{d}t = a_0 \int_{-T/2}^{T/2} \cos(k\omega_0 t)\mathrm{d}t$$
$$+ \sum_{n=1}^{\infty}\int_{-T/2}^{T/2} \frac{a_n}{2}\{\cos(n\omega_0 + k\omega_0)t + \cos(n\omega_0 - k\omega_0)t\}\mathrm{d}t$$
$$+ \sum_{n=1}^{\infty}\int_{-T/2}^{T/2} \frac{b_n}{2}\{\sin(n\omega_0 + k\omega_0)t + \sin(n\omega_0 - k\omega_0)\mathrm{d}t \tag{5.9}$$

Because of the odd symmetry of sine functions, all terms within the last summation are zero and the equation simplifies to

$$\int_{-T/2}^{T/2} f_p(t)\cos(k\omega_0 t)\mathrm{d}t = a_0 \int_{-T/2}^{T/2} \cos(k\omega_0 t)\mathrm{d}t$$
$$+ \sum_{n=1}^{\infty}\int_{-T/2}^{T/2} \frac{a_n}{2}\{\cos(n\omega_0 + k\omega_0)t + \cos(n\omega_0 - k\omega_0)t\}\mathrm{d}t \tag{5.10}$$

If $k = 0$, each of the terms within the remaining summation have a non-zero frequency $n\omega_0$ and their integrals equal zero since the limits of integration encompass a whole number of complete cycles. Therefore Equation (5.10) then reduces to

$$\int_{-T/2}^{T/2} f_{\mathrm{p}}(t)\cos(0 \cdot t)\mathrm{d}t = a_0 \int_{-T/2}^{T/2} \cos(0 \cdot t)\mathrm{d}t = a_0 T$$

that is

$$a_0 = \frac{1}{T}\int_{-T/2}^{T/2} f_{\mathrm{p}}(t)\mathrm{d}t$$

If $k \neq 0$ each of the terms within the summation, with one exception, contains a whole number of complete cycles of a cosinusoid, and hence their integrals also equal zero. The one exception is the kth term of the summation (that is, $n = k$). The summation then reduces from

$$\sum_{n=1}^{\infty} \int_{-T/2}^{T/2} \frac{a_n}{2} \{\cos(n\omega_0 + k\omega_0)t + \cos(n\omega_0 - k\omega_0)t\}\mathrm{d}t$$

to

$$\int_{-T/2}^{T/2} \frac{a_k}{2} \cos(0 \cdot t)\mathrm{d}t$$

But

$$\int_{-T/2}^{T/2} \frac{a_k}{2} \cos(0 \cdot t)\mathrm{d}t = \frac{a_k T}{2} \tag{5.11}$$

Equation (5.10) then becomes

$$\int_{-T/2}^{T/2} f_{\mathrm{p}}(t)\cos(k\omega_0 t)\mathrm{d}t = \frac{a_k T}{2}$$

that is

$$a_k = \frac{2}{T}\int_{-T/2}^{T/2} f_{\mathrm{p}}(t)\cos(k\omega_0 t)\mathrm{d}t \tag{5.12}$$

In a similar manner, if both sides of Equation (5.4) are multiplied by $\sin(k\omega_0 t)$ and integrated over a complete period T, we obtain

$$b_k = \frac{2}{T}\int_{-T/2}^{T/2} f_{\mathrm{p}}(t)\sin(k\omega_0 t)\mathrm{d}t \tag{5.13}$$

Therefore, summarizing so far

$$f_{\mathrm{p}}(t) = a_0 + \sum_{n=1}^{\infty} \{a_n \cos(n\omega_0 t) + b_n \sin(n\omega_0 t)\} \tag{5.14}$$

where

$$a_0 = \frac{1}{T}\int_{-T/2}^{T/2} f_{\mathrm{p}}(t)\mathrm{d}t \tag{5.15}$$

and, considering the nth coefficient rather than involving a kth coefficient, Equations (5.12) and (5.13) become

$$a_n = \frac{2}{T}\int_{-T/2}^{T/2} f_{\text{p}}(t)\cos(n\omega_0 t)\text{d}t \tag{5.16}$$

and

$$b_n = \frac{2}{T}\int_{-T/2}^{T/2} f_{\text{p}}(t)\sin(n\omega_0 t)\text{d}t \tag{5.17}$$

In general, the Fourier series representation of Equation (5.3) is preferred to that of Equation (5.5) (repeated in Equation (5.14)) since it is easier to make a physical interpretation of a single cosinusoid specified by its magnitude and phase than to make a physical interpretation of a sinusoid/cosinusoid combination where the specification is of the two magnitudes.

Using the relations $a_0 \equiv c_0$, $a_n \equiv c_n \cos(\theta_n)$, $b_n \equiv -c_n \sin(\theta_n)$ from Equation (5.6) we can obtain

$$a_n^2 + b_n^2 = c_n^2(\cos^2(\theta_n) + \sin^2(\theta_n)) = c_n^2$$

and

$$\frac{b_n}{a_n} = -\tan\theta_n$$

Thus

$$f_{\text{p}}(t) = c_0 + \sum_{n=1}^{\infty} c_n \cos(n\omega_0 + \theta_n) \tag{5.18}$$

where

$$c_0 = a_0 \tag{5.19}$$

$$c_n = \sqrt{a_n^2 + b_n^2} \tag{5.20}$$

$$\theta_n = \tan^{-1}(-b_n/a_n) \tag{5.21}$$

and where a_0, a_n and b_n are given by Equations (5.15)–(5.17).

A plot of the c coefficients versus frequency gives the *one-sided* amplitude spectrum of the periodic signal. A plot of the θ coefficients versus frequency gives the *one-sided* phase spectrum of the periodic signal. Both spectra are *line* spectra.

5.6 The exponential Fourier series

A much more compact form of the Fourier series than the trigonometric Fourier series of the previous section consists of having a sum of *complex exponential* time waveforms. This *exponential Fourier series* is composed of terms having both positive and negative frequencies which, when taken together, comprise cosinusoids of different amplitudes and phases. We have

$$f_{\text{p}}(t) = \sum_{n=-\infty}^{\infty} F_n e^{jn\omega_0 t} \tag{5.22}$$

We could obtain the F_n coefficients from Equation (5.18) by splitting each cosinusoid within this summation into two complex exponentials, that is

$$c_n \cos(n\omega_0 t + \theta_n) = \frac{c_n}{2}\, e^{j(n\omega_0 t + \theta_n)} + \frac{c_n}{2}\, e^{-j(n\omega_0 t + \theta_n)}$$

$$= \frac{c_n}{2}\, e^{j\theta_n} e^{jn\omega_0 t} + \frac{c_n}{2}\, e^{-j\theta_n} e^{-jn\omega_0 t}$$

The two exponentials can be considered to have positive and negative frequencies $n\omega_0$ and $-n\omega_0$. They have identical amplitudes but equal and opposite phases, that is

$$F_n = \frac{c_n}{2}\, e^{j\theta_n}, \qquad F_{-n} = \frac{c_n}{2}\, e^{-j\theta_n}$$

However, the elegance of the exponential Fourier series is more apparent if the coefficients are derived from first principles, rather than as an extension of the trigonometric Fourier series. To determine the F_n coefficients directly from

$$f_p(t) = \sum_{n=-\infty}^{\infty} F_n e^{jn\omega_0 t}$$

we multiply both sides of this equation by $e^{-jk\omega_0 t}$ and integrate over one cycle; that is

$$\int_{-T/2}^{T/2} f_p(t) e^{-jk\omega_0 t}\, dt = \int_{-T/2}^{T/2} \left\{ e^{-jk\omega_0 t} \sum_{n=-\infty}^{\infty} F_n e^{jn\omega_0 t} \right\} dt$$

This can be rearranged to give

$$\int_{-T/2}^{T/2} f_p(t) e^{-jk\omega_0 t}\, dt = \int_{-T/2}^{T/2} \sum_{n=-\infty}^{\infty} F_n e^{j(n-k)\omega_0 t}\, dt$$

$$= \sum_{n=-\infty}^{\infty} F_n \int_{-T/2}^{T/2} \{\cos[(n-k)\omega_0 t] + j\,\sin[(n-k)\omega_0 t]\} dt$$

The integrals of all the sine terms are zero because of their odd symmetry. Also, with one important exception, the integrals of the cosine terms are zero because their limits of integration encompass one or more complete periods. The one exception is when $n = k$, and this produces $\int_{-T/2}^{T/2} F_k \cos(0)\, dt$. Since $\cos(0) = 1$ this simplifies to $F_k T$. Hence

$$\int_{-T/2}^{T/2} f_p(t) e^{-jk\omega_0 t}\, dt = F_k T$$

Thus the exponential Fourier series is given by Equation (5.22), where, considering the nth coefficient rather than the kth coefficient

$$F_n = \frac{1}{T} \int_{-T/2}^{T/2} f_p(t) e^{-jn\omega_0 t}\, dt \tag{5.23}$$

An alternative result with different limits of integration is

$$F_n = \frac{1}{T} \int_0^T f_p(t) e^{-jn\omega_0 t}\, dt \tag{5.24}$$

It will be noted that these F_n coefficients are complex, since the integral for evaluating them in Equation (5.23) can be expanded as

$$F_n = \frac{1}{T}\int_{-T/2}^{T/2} f_p(t)\left[\cos(n\omega_0 t) - j\sin(n\omega_0 t)\right]dt$$

Since cosines have even symmetry, F_{-n} has a real part that is the same as the real part of F_n. In contrast, since sines have odd symmetry, F_{-n} has an imaginary part that is the negative of the imaginary part of F_n. It follows that F_n and F_{-n} are complex conjugates; that is

$$F_{-n} = F_n^* \tag{5.25}$$

This means that, if $F_n = |F_n|e^{j\theta_n}$, then $F_{-n} = |F_n|e^{-j\theta_n}$

A plot of $|F_n|$ versus the associated frequency $n\omega_0$ gives the *two-sided* amplitude spectrum of $f_p(t)$. It follows from Equation (5.25) that $|F_n|$ has *even* symmetry.

A plot of θ_n versus $n\omega_0$ gives the *two-sided* phase spectrum and it follows from Equation (5.25) that this has *odd* symmetry.

Both these spectra are *line* spectra.

Example 5.1

Derive the exponential Fourier series of the rectangular pulse train shown in Figure 5.9 where, for convenience, the mid-point of one of the pulses is taken to be centred at the origin.

Solution Over the period $-T/2 < t < T/2$ we have

$$f_p(t) = A \qquad -\tau/2 < t < \tau/2$$
$$= 0 \qquad \text{elsewhere}$$

Therefore Equation (5.23) gives

$$F_n = \frac{1}{T}\int_{-\tau/2}^{\tau/2} Ae^{-jn\omega_0 t}\,dt$$

where $\omega_0 = 2\pi/T$.

$$\therefore \quad F_n = \frac{-A}{jn\omega_0 T}e^{-jn\omega_0 t}\Big|_{-\tau/2}^{\tau/2} = A\,\frac{e^{jn\omega_0\tau/2} - e^{-jn\omega_0\tau/2}}{jn\omega_0 T} = \frac{2A}{n\omega_0 T}\sin\left(\frac{n\omega_0\tau}{2}\right)$$

This may be expressed in terms of cyclic frequency and rearranged to be written as

$$F_n = \frac{A\tau}{T}\,\frac{\sin(n\pi f_0\tau)}{n\pi f_0\tau} \tag{5.26}$$

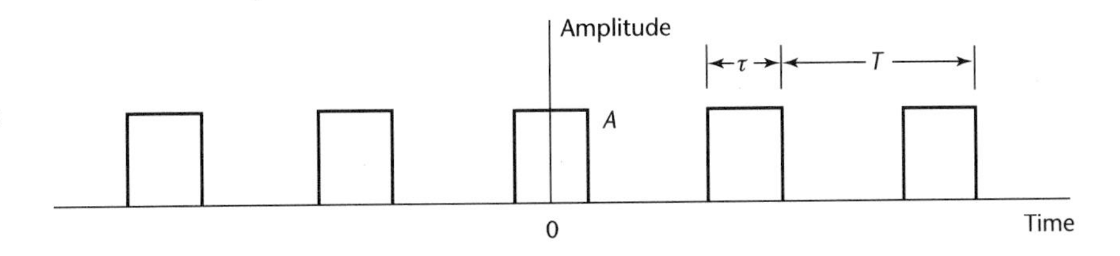

Fig. 5.9 Periodic train of rectangular pulses

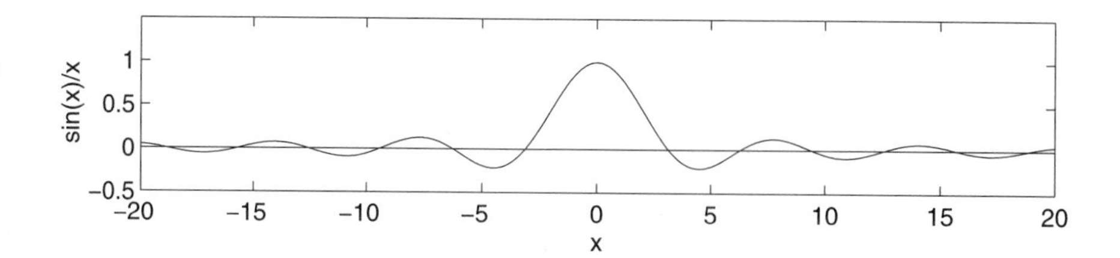

Fig. 5.10 The function sin $(x)/x$

Thus, inserting this into Equation (5.22), the exponential Fourier series for the rectangular pulse train may be written

$$f_{\mathrm{p}}(t) = \frac{A\tau}{T} \sum_{n=-\infty}^{\infty} \frac{\sin(n\pi f_0 \tau)}{n\pi f_0 \tau} \mathrm{e}^{\mathrm{j}2\pi n f_0 t} \tag{5.27}$$

The $\sin(x)/x$ function that appears in this expression for $f_{\mathrm{p}}(t)$ is frequently encountered in signal processing applications and is as shown in Figure 5.10. It is closely related to the sinc function, defined as

$$\operatorname{sinc}(x) = \frac{\sin(\pi x)}{\pi x} \tag{5.28}$$

This sinc function is identical in shape to the $\sin(x)/x$ function, the only difference being that its nulls occur at integer values of x rather than at integer multiples of π.

Using Equation (5.26), Figure 5.11 shows the F_n coefficients for various values of τ/T, where T is fixed. It shows how the spectrum widens as the duty cycle τ/T decreases. In contrast, Figure 5.12 is for various values of τ/T, where τ is fixed. This shows how the spectral envelope remains constant but the harmonics become closer together as T increases.

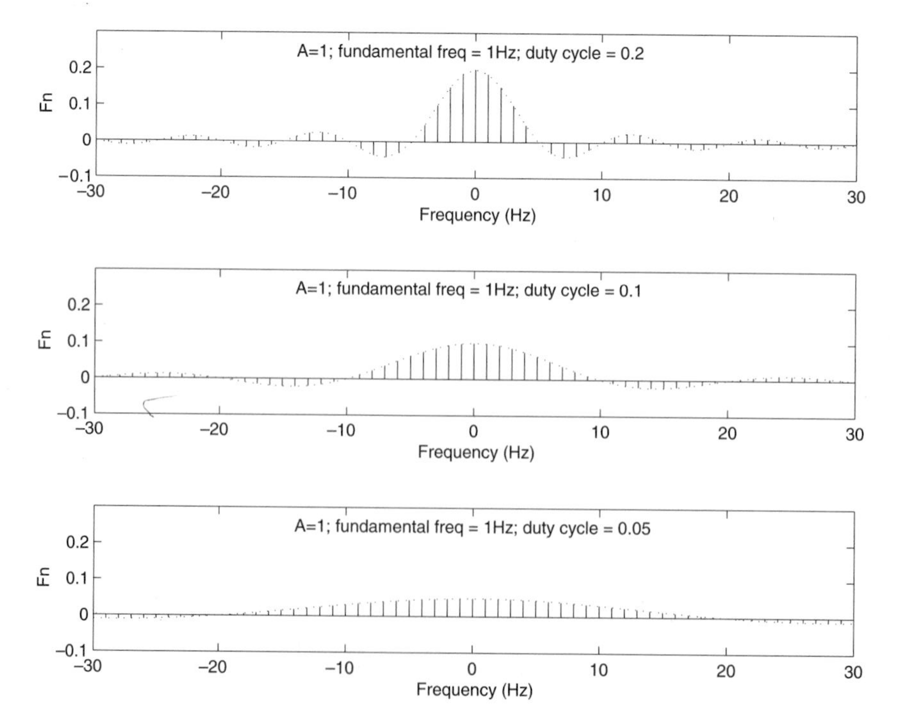

Fig. 5.11 Line spectrum of a rectangular pulse train for various values of τ/T, T fixed

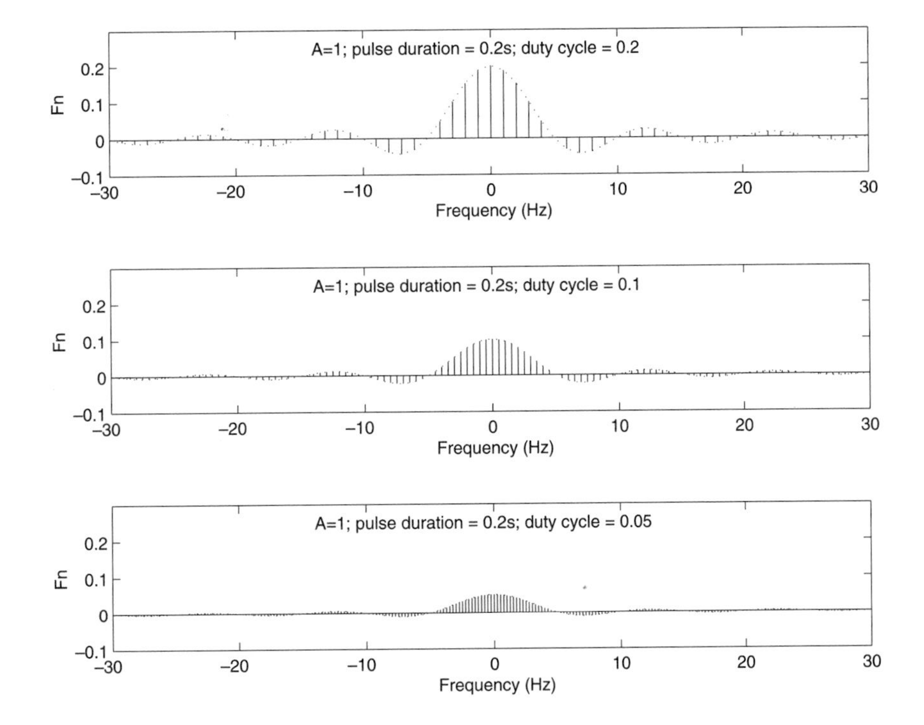

Fig. 5.12 Line spectrum of a rectangular pulse train for various values of τ/T, τ fixed

Figures 5.11 and 5.12 both represent *two-sided* amplitude spectra. They are *line* spectra, and the discrete lines are $1/T$ Hz apart in cyclic frequency (or $2\pi/T$ rad/s apart in radian frequency). Their envelopes are of the form $\sin(x)/x$ and have their first nulls when $\omega\tau/2 = \pi$, that is when $f = 1/\tau$, or $2\pi/\tau$. Because of the even symmetry of the pulse train about the time origin, the F_n coefficients in this example are all real. Thus the phase spectra are zero everywhere.

5.7 Plotting line spectra using MATLAB

The 'stem' plot is a standard function in MATLAB and produces plots similar to Figures 5.11 and 5.12, except that it causes the stems to be capped with small circles. In order to produce stems that are not capped we can make a minor modification to the stem function and store it as a new function in our personal MATLAB library.

In order to find the M-file that gives the stem function we do the following. In the MATLAB command window we:

1. click on the File menu
2. click on the Open M-file menu item
3. select the toolbox directory
4. select the matlab directory
5. select the plotxy directory
6. select the stem.m file

At this point we see the M-file that generates the function **stem(x,y,linetype)** that is used to plot **y** versus **x** using a specified type of line or symbol. An important part of it generates the stems. The line which needs modifying comes after this and is

```
h = plot(x,y,'o',xx(:),yy(:),linetype);
```

The first part of this, **h = plot(x,y,'o')**, performs a plot of **y** versus **x** using circular symbols at each data point. The second part plots the stems using a specified linestyle. In order to produce lines without the circles we merely need to delete the first part. Thus we change the line to

```
h = plot(xx(:),yy(:),linetype);
```

We do not wish to lose the stem function but rather to supplement our library with a new function, for which an appropriate name might be 'stalk'. Thus we also change the first line of the program from **function stem(x,y,linetype)** to **function stalk(x,y,linetype)** and then save this modified M-file as c:\stdntmat\stalk.m.

It may be useful to generate yet another function – one which is identical to the stalk function except that the envelope of the line spectrum is shown by joining the data points with a dotted line. To do this we can do as before except that the plot command now becomes

```
h = plot(x,y,':',xx(:),yy(:),linetype);
```

where ':' is the specification for a dotted linestyle. A suitable name for this alternative function might be 'stalk2', in which case we can change the first line of the program to

```
function stalk2(x,y,linetype)
```

and then save this second modified M-file as c:\stdntmat\stalk2.m.

Example 5.2

Develop a MATLAB program to produce Figure 5.11.

Solution We require to determine values of the coefficients in Equation (5.27) for various values of τ/T, where T is fixed. The following M-file produced Figure 5.11 and uses the adapted 'stem' function 'stalk2' that plots stems without circles, while also showing the envelope of the data points by joining them with a dotted line.

```
% fig5_11.m
A=1;
f0=1;                          % the fundamental frequency is 1Hz
T=1/f0;                        % the fundamental period is 1 second
n=(-30:30);                    % the harmonic numbers
freq=n*f0;                     % the frequencies of the harmonics
%
tau=0.2;                       % the pulse duration
Fn=A*tau/T*sinc(freq*tau);     % the Fourier coefficients
subplot(3,1,1)
stalk2(freq,Fn)                % plot the Fourier coefficients versus
```

```
% frequency using a specially developed function for plotting stems without
% heads, together with the envelope of the data points
set(gca,'fontsize',10)
ylabel('Fn'),xlabel('Frequency')
text(-14, 0.25, 'A=1; fundamental freq=1Hz; duty cycle=0.2','fontsize',10)
axis([-30 30 -0.1 0.3])
%
tau=0.1;                        % repeat for a new pulse duration
Fn=A*tau/T*sinc(freq*tau);
subplot(3,1,2)
stalk2(freq,Fn)
set(gca,'fontsize',10)
ylabel('Fn'),xlabel('Frequency')
text(-14, 0.25, 'A=1; fundamental freq=1Hz; duty cycle=0.1','fontsize',10)
axis([-30 30 -0.1 0.3])
%
tau=0.05;                       % repeat for a new pulse duration
Fn=A*tau/T*sinc(freq*tau);
subplot(3,1,3)
stalk2(freq,Fn)
set(gca,'fontsize',10)
ylabel('Fn'),xlabel('Frequency')
text(-14, 0.25, 'A=1; fundamental freq=1Hz; duty cycle=0.05','fontsize',10)
axis([-30 30 -0.1 0.3])
```

The result of running this M-file is Figure 5.11.

5.8 Further properties and examples of Fourier series

Special cases of the rectangular pulse train already examined are a train of impulses and a square wave. These are examined in the following examples.

Example 5.3

Derive the exponential Fourier series of a train of unity amplitude *impulses*.

Solution This is an important special case of the train of rectangular pulses treated in Example 5.1, where the duration τ of the pulses tends to zero but their area $A\tau = 1$ such that

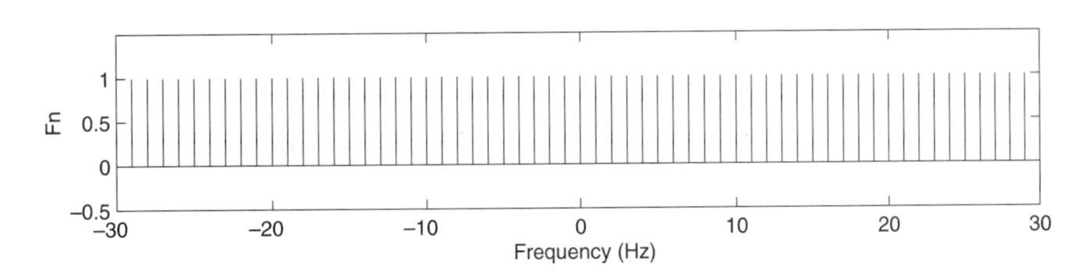

Fig. 5.13 The line spectrum of a train of unit impulses

their amplitude A increases to infinity. Therefore, using Equation (5.26),

$$F_n = \lim_{\substack{\tau \to 0 \\ A\tau \to 1}} \frac{A\tau}{T} \frac{\sin(n\pi f_0 \tau)}{n\pi f_0 \tau} = \frac{1}{T} \tag{5.29}$$

Thus the amplitude spectrum becomes a set of equal amplitude lines of magnitude $1/T$. This is shown in Figure 5.13.

The phase spectrum of the rectangular pulse train in Example 5.1 was zero because of its even symmetry about the time origin. The phase spectrum would not be zero for other choices of time origin. This is illustrated by the next example.

Example 5.4

Derive the exponential Fourier series of the displaced rectangular pulse train shown in Figure 5.14.

Solution This time we can use Equation (5.24) to obtain

$$F_n = \frac{1}{T}\int_0^{\tau} f_{\mathrm{p}}(t)\mathrm{e}^{-\mathrm{j}n\omega_0 t}\,\mathrm{d}t = \frac{1}{T}\int_0^{\tau} A\mathrm{e}^{-\mathrm{j}n\omega_0 t}\,\mathrm{d}t = \frac{-A}{\mathrm{j}n\omega_0 T}\,\mathrm{e}^{-\mathrm{j}n\omega_0 t}\Big|_0^{\tau} = A\,\frac{1-\mathrm{e}^{-\mathrm{j}n\omega_0 \tau}}{\mathrm{j}n\omega_0 T}$$

Taking out the factor $\mathrm{e}^{-\mathrm{j}n\omega_0\tau/2}$ this becomes

$$F_n = A\mathrm{e}^{-\mathrm{j}n\omega_0\tau/2}\,\frac{\mathrm{e}^{\mathrm{j}n\omega_0\tau/2} - \mathrm{e}^{-\mathrm{j}n\omega_0\tau/2}}{\mathrm{j}n\omega_0 T}$$

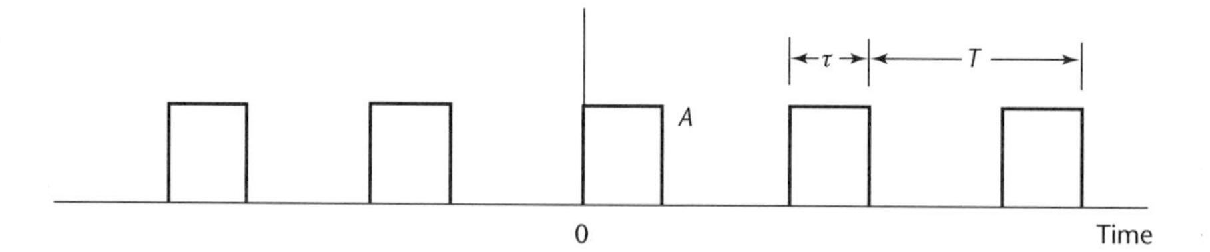

Fig. 5.14 Train of rectangular pulses that is asymmetrical about the origin

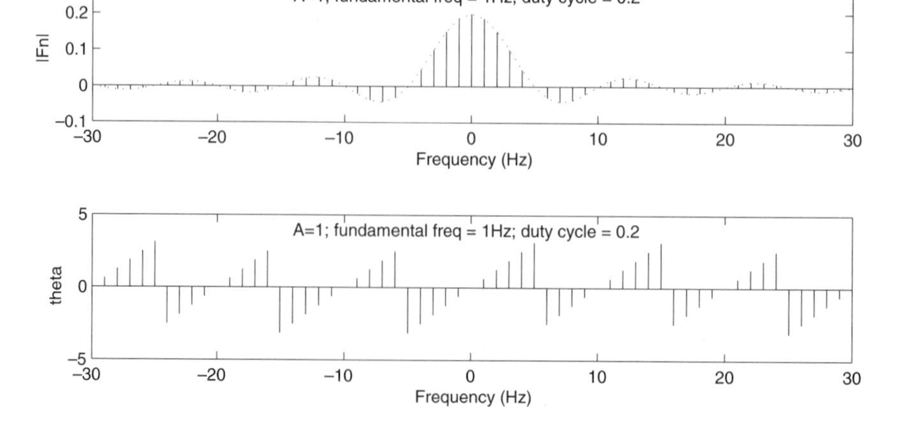

Fig. 5.15 Amplitude and phase spectrum of the asymmetrical train of rectangular pulses shown in Figure 5.14

This can be simplified to give

$$F_n = \frac{A\tau}{T}\frac{\sin(n\omega_0\tau/2)}{n\omega_0\tau/2}\,\mathrm{e}^{-\mathrm{j}n\omega_0\tau/2}$$
$$= |F_n|\mathrm{e}^{\mathrm{j}\theta_n}$$

where

$$|F_n| = \frac{A\tau}{T}\frac{\sin(n\pi f_0\tau)}{n\pi f_0\tau} \tag{5.30}$$

and

$$\theta_n = -\pi n f_0\tau \tag{5.31}$$

A comparison with Equation (5.26) for the symmetrical pulse train of Example 5.1 shows that the amplitude spectrum has been unaffected by the change in time origin, but the phase spectrum is now non-zero and changes linearly with frequency. Figure 5.15 shows the amplitude and phase spectra for the displaced rectangular pulse train of Figure 5.14, for the case of $\tau/T = 0.2$. Since a cosinusoid of phase θ is indistinguishable from a cosinusoid of the same frequency of phase $\theta \pm 2n\pi$ the phase is shown constrained to the limits $\pm\pi$.

It will be noted in Example 5.4 that a time delay of $\tau/2$ between the waveforms of Figure 5.14 and Figure 5.9 causes a phase shift of $-\pi n\omega_0\tau$, or hence of $-\pi f\tau$. This result can be easily generalized. A delay of t_0 always introduces the frequency-dependent phase shift of $-2\pi f t_0$, or $-\omega t_0$.

Section 5.5 gave a self-contained derivation of the coefficients of the trigonometric Fourier series. However, if wished, each pair of positive and negative frequency terms in the exponential Fourier series can be joined, using $\frac{1}{2}(\mathrm{e}^{\mathrm{j}n\omega_0 t} + \mathrm{e}^{-\mathrm{j}n\omega_0 t}) = \cos(n\omega_0 t)$, to give an alternative derivation of the *trigonometric* Fourier series. This is illustrated by the next example.

Example 5.5

Determine the trigonometric Fourier series of the rectangular pulse train of Example 5.1.

Solution The coefficients of the trigonometric Fourier series can be obtained using Equations (5.15)–(5.17) and Equations (5.19)–(5.21). Alternatively, they can be obtained by combining the negative frequency components of the exponential Fourier series with the positive frequency components using $\frac{1}{2}(\mathrm{e}^{\mathrm{j}n\omega_0 t} + \mathrm{e}^{-\mathrm{j}n\omega_0 t}) = \cos(n\omega_0 t)$. Applying this latter procedure to Equation (5.27) we obtain

$$f_{\mathrm{p}}(t) = \frac{A\tau}{T}\left(1 + 2\sum_{n=1}^{\infty}\frac{\sin(n\pi f_0\tau)}{n\pi f_0\tau}\cos(2\pi n f_0 t)\right) \tag{5.32}$$

The coefficients of the trigonometric Fourier series correspond to a one-sided

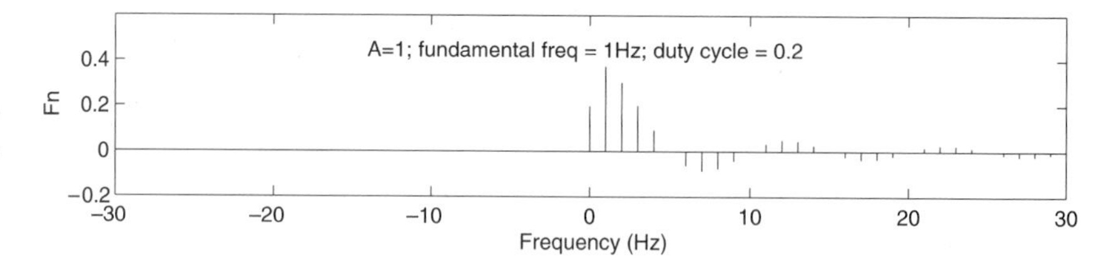

Fig. 5.16 One-sided spectrum of a rectangular pulse train having a 20% duty cycle

spectrum and the result of Equation (5.32) for the case of $\tau/T = 0.2$ is shown in Figure 5.16. It is to be compared with the two-sided spectrum for the case of $\tau/T = 0.2$ in Figure 5.11. The one-sided spectrum may be thought of as arising from folding the negative frequency components to lie over and to be added to the positive frequency components. It should be noted that the vertical scale of Figure 5.16 is twice that of Figure 5.11.

Example 5.6

Determine the trigonometric Fourier series of the square wave signal shown in Figure 5.17, having a peak amplitude of unity and a zero mean.

Solution The square wave can be considered a special case of the rectangular pulse train where $\tau/T = 0.5$, $A = 2$ and the d.c. component is absent. From Equation (5.32), putting $f_0 = 1/T$, $\tau/T = 0.5$, $A = 2$, and taking away the d.c. component,

$$f_p(t) = 2\sum_{n=1}^{\infty} \frac{\sin(n\pi/2)}{n\pi/2} \cos(2\pi n f_0 t)$$

$$= \frac{4}{\pi}\left[\cos(\omega_0 t) - \frac{1}{3}\cos(3\omega_0 t) + \frac{1}{5}\cos(5\omega_0 t) - \frac{1}{7}\cos(7\omega_0 t) + \cdots\right] \quad (5.33)$$

This is an infinite series consisting of a fundamental and odd harmonics.

A plot of the corresponding two-sided spectrum is shown in Figure 5.18 together with the $\sin(x)/x$ shaped envelope. It is noted that the fundamental frequency lies halfway between zero frequency and the first null of the spectral envelope, while the harmonics lie midway between subsequent nulls.

To demonstrate that Equation (5.33) does indeed represent a square wave, the first five terms and their sum are shown in Figure (5.19). It is seen that the general form of the square wave is beginning to emerge, even with just five terms.

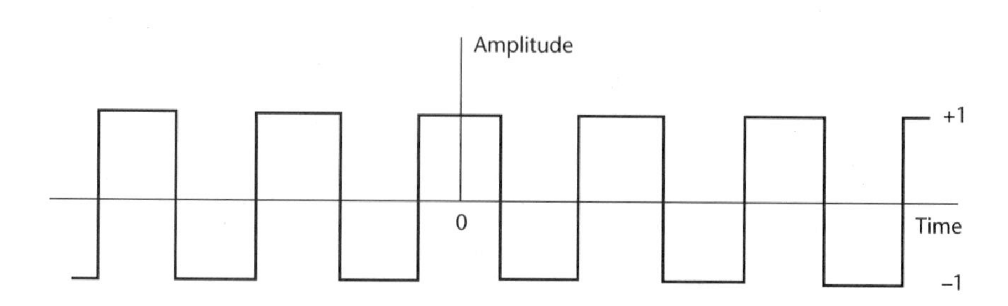

Fig. 5.17 Square wave of zero mean

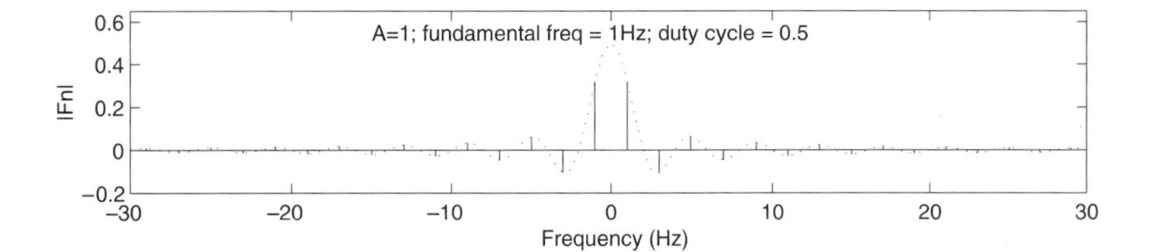

Fig. 5.18 Line spectrum of a square wave

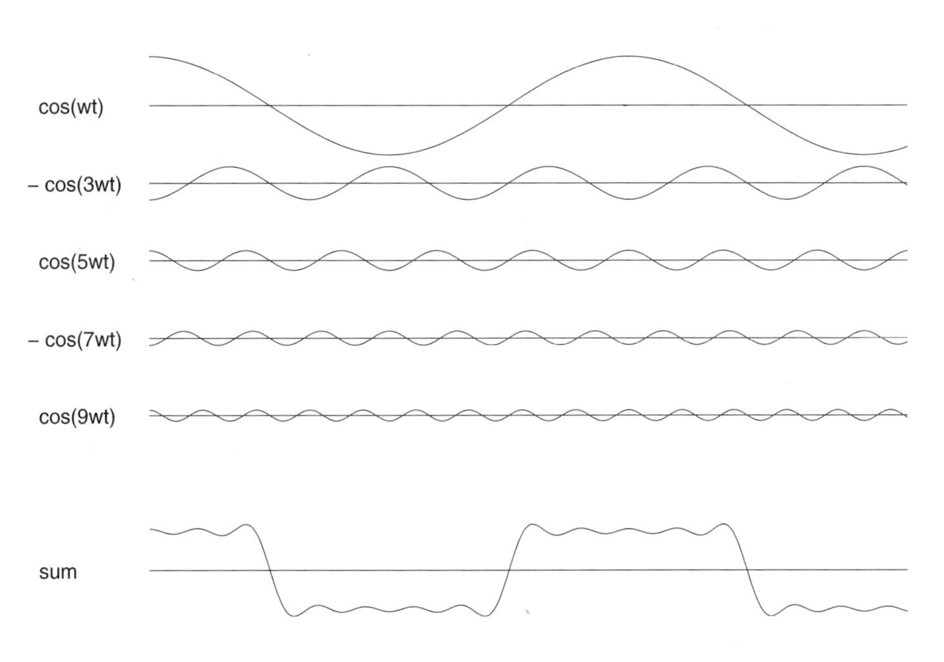

Fig. 5.19* The first five terms of the Fourier series of a square wave, and their sum

5.9 Summary

The essential concept of Fourier techniques is that any waveform may be considered to be composed of the sum of cosinusoids of different frequencies.

In the case of a *periodic* signal the frequencies of the cosinusoids are constrained to *discrete* values, namely the fundamental frequency of the signal and its harmonics. Fourier series analysis determines the appropriate amplitude and phase angle of each cosinusoid. These amplitudes and phases can be portrayed as line spectra.

Making use of the identity $\cos x = \frac{1}{2}e^{jx} + \frac{1}{2}e^{-jx}$, and hence $\cos(\omega t + \theta) = \frac{1}{2}e^{j(\omega t + \theta)} + \frac{1}{2}e^{-j(\omega t + \theta)}$ it is common practice to replace each cosinusoid by two complex exponentials, one of a positive frequency and one of a negative frequency. The amplitudes and phases of these can also be portrayed by line spectra.

5.10 Problems

1. Derive the exponential Fourier series of the sawtooth waveform shown in Figure 5.20. Use the result to plot a two-sided line spectrum and a one-sided line spectrum of the waveform.

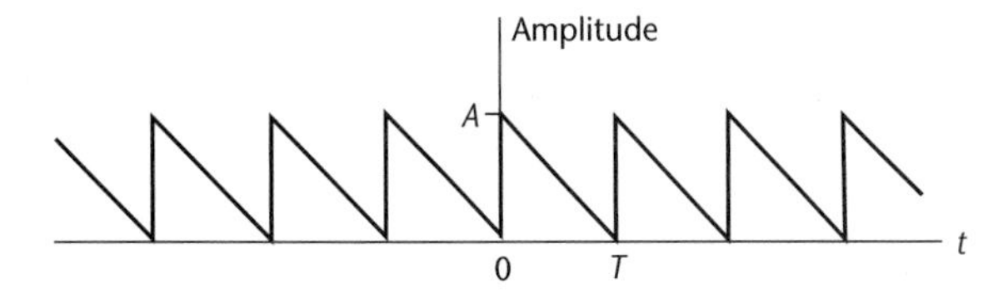

Fig. 5.20

2. Using the trigonometric identity $\cos A \cos B = \frac{1}{2}\cos(A - B) + \frac{1}{2}\cos(A + B)$ express the amplitude modulated signal $A(1 + m \cos pt)\cos(\omega_c t)$ as the sum of cosinusoids. Hence determine and plot the one-sided spectrum. By using the result $\cos x = \frac{1}{2}(e^{jx} + e^{-jx})$ to expand the signal further, determine and plot the two-sided spectrum.

3. A power supply has a step-down transformer producing the voltage $v_1(t) = 40\cos(377t)$. This is followed by a full-wave rectifier whose output is $v_2(t) = |v_1(t)|$. Determine the spectrum of $v_2(t)$. If this signal passes through a smoothing *RC* network determine the value of *RC* that causes the largest harmonic in the output to have a peak value equal to 1% of the d.c. voltage.

4. As in Problem 5.3, the full-wave rectified signal arising from the voltage $v_1(t) = 40\cos(377t)$ passes through a smoothing *RC* network whose output is a d.c. voltage plus some ripple. Using the result of the Fourier analysis obtained in Problem 5.3 write a MATLAB program that plots the output waveform as a function of *RC*. Confirm that the correct output waveform is obtained if $RC = 0$. Change the value of *RC* to determine empirically the condition for the peak ripple to be 1% of the d.c. voltage. Check that this agrees with the result of Problem 5.3.

5. A MATLAB program is included in Appendix A for generating and summing the first five terms of the Fourier series of a square wave, thus leading to Figure 5.19. Extend this program to plot the sum of the first five terms when these terms retain their amplitudes but each undergoes a phase shift of $-90°$.

6. A square wave passes through an *RC* network for which the frequency transfer function is $H(\omega) = 1/(1 + j\omega CR)$. Determine the first five terms of the Fourier series of the output signal if the period of the square wave equals $2CR$. Develop a MATLAB program that generates and displays the sum of these five Fourier components.

CHAPTER 6

Spectral density functions and the Fourier transform

6.1 ● Preview

For the case of periodic signals, the last chapter showed that the frequencies of Fourier series components are constrained to *discrete* values that are multiples of the periodic frequency. The component at each of these frequencies is described by an amplitude and a phase. An aperiodic (or non-periodic) signal may be considered as the limiting case of a periodic signal whose period tends to infinity. Thus the frequency spacing of the components is infinitesimally small, such that a Fourier description must involve amplitude and phase spectra that are *continuous* rather than discrete. The complex function $F(f)$ that simultaneously conveys this amplitude and phase information is the Fourier transform of the signal. For a signal whose units are volts the amplitude information of the trigonometric or exponential Fourier series is conveyed by a *line* spectrum, measured in volts. In contrast the information of a Fourier transform is conveyed by the real and imaginary parts of a voltage *density* spectrum, measured in V/Hz. These real and imaginary components are commonly combined to produce an amplitude (or magnitude) spectrum measured in V/Hz and a phase spectrum measured in radians. The outcome of a Fourier transform is often referred to as a *spectral density function*.

With periodic signals the spectral density becomes infinite at those harmonic frequencies where there is power present. The Fourier transform copes with this by means of impulse functions.

6.2 Energy signals and power signals

If a voltage $f(t)$ is present across a resistance R, the *instantaneous* power dissipated in the resistance is $f^2(t)/R$. The energy dissipated is $\int_{-\infty}^{\infty} f^2(t)/R \, dt$. If normalized to a resistance of 1 Ω the instantaneous power is $f^2(t)$ and the dissipated energy equals $\int_{-\infty}^{\infty} f^2(t) \, dt$. The normalized *mean* power is

$$\lim_{T\to\infty} \frac{1}{T} \int_{-\infty}^{\infty} f^2(t) \, dt$$

A signal $f(t)$ is termed an *energy signal* if $\int_{-\infty}^{\infty} f^2(t) \, dt$ is finite. It is termed a *power signal* if the energy $\int_{-\infty}^{\infty} f^2(t) \, dt$ is infinite but the mean power

$$\lim_{T\to\infty} \frac{1}{T} \int_{-\infty}^{\infty} f^2(t) \, dt$$

is finite.

Examples of energy signals are the single rectangular pulse shown in Figure 6.1(a) and the tone burst shown in Figure 6.1(b).

Examples of power signals are the sinewave shown in Figure 6.2(a), the noise signal shown in Figure 6.2(b) and the step signal shown in Figure 6.2(c). They all

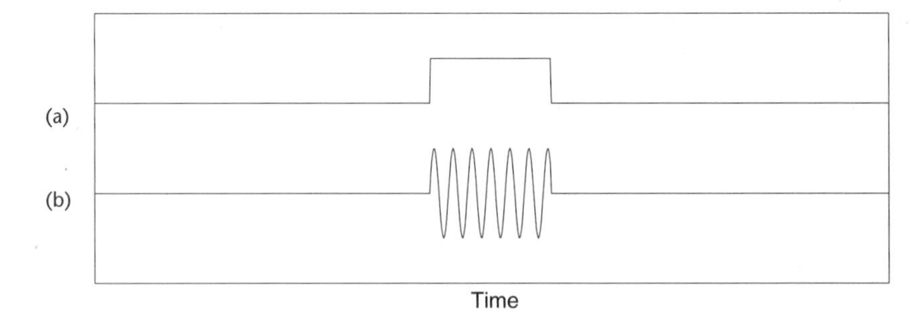

Fig. 6.1* Examples of energy signals: (a) rectangular pulse; (b) tone burst

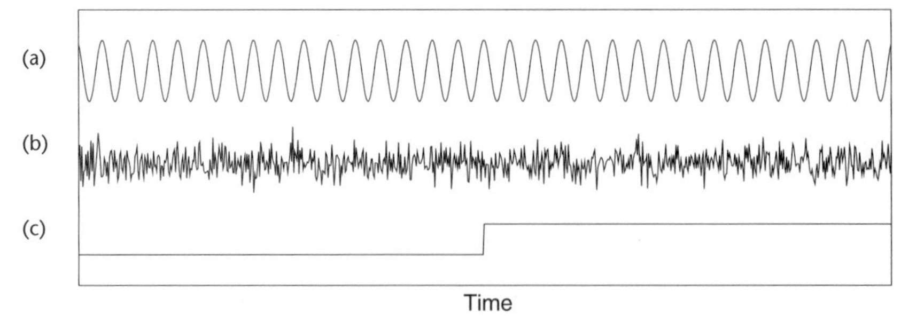

Fig. 6.2 Examples of power signals : (a) sinewave; (b) noise; (c) step function

have infinite energy because they are non-zero for an infinite period of time.

An interesting special situation is that of an impulse function. An impulse of strength A has infinite energy because, considering it to be the limiting case of a rectangular pulse of duration T and amplitude V as $T \rightarrow 0$, $V \rightarrow \infty$ and $VT = A$, its energy is V^2T. This equals VA, is infinite and hence, unlike the rectangular pulse, the impulse cannot be termed an energy signal. It is also difficult to term it a power signal since its mean power (the power averaged over all time) is almost meaningless.

6.3 The Fourier transform of an aperiodic energy signal

An aperiodic (or non-periodic) energy signal may be considered as the limiting case of a periodic signal whose period tends to infinity. Thus the single rectangular pulse of Figure 6.1(a) may be considered to be a periodic train of rectangular pulses where the period is infinite. Thus, if we consider the periodic signal $f_p(t)$ shown in Figure 6.3 obtained by making the rectangular pulse $f(t)$ repeat itself with a *finite* period T, we may consider $f(t)$ to be the limiting case of $f_p(t)$ as $T \longrightarrow \infty$; that is

$$f(t) = \lim_{T \longrightarrow \infty} f_p(t) \tag{6.1}$$

However, the exponential Fourier series given by Equation 5.22 is appropriate for a periodic signal $f_p(t)$. This is a summation of terms given by

$$f_p(t) = \sum_{n=-\infty}^{\infty} F_n e^{jn\omega_0 t} = \sum_{n=-\infty}^{\infty} F_n e^{j2\pi n f_0 t} \tag{6.2}$$

where

$$f_0 = 1/T \tag{6.3}$$

and where, from Equation (5.23)

$$F_n = \frac{1}{T}\int_{-T/2}^{T/2} f_p(t) e^{-j2\pi n f_0 t}\, dt \tag{6.4}$$

Because of the $1/T$ term outside the integral the coefficients become smaller in magnitude as $T \longrightarrow \infty$. Simultaneously, because the terms are $1/T$ Hz apart, they become more and more closely packed. In the limit as $T \longrightarrow \infty$, the spectrum becomes continuous. It is useful to introduce a continuous spectrum $F(f)$ such that the strength of this in a frequency band $1/T$ wide, namely $(1/T)F(f)$, is the same as the strength of the discrete component F_n that it replaces This means that

$$F(f) = \lim_{T \longrightarrow \infty} F_n T \tag{6.5}$$

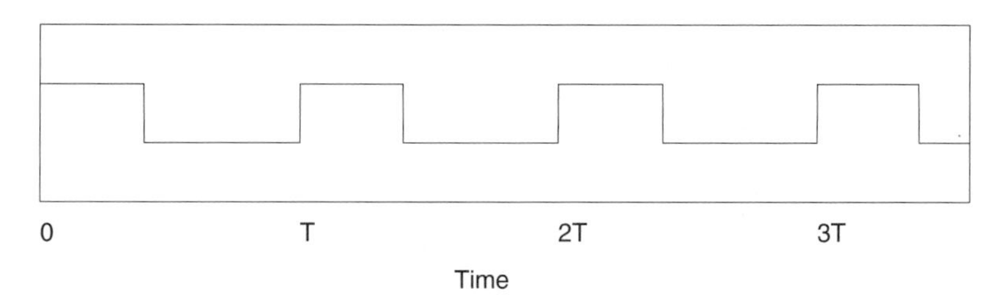

Fig. 6.3 Periodic train of rectangular pulses

To find $F(f)$ we can now substitute the value of F_n from Equation (6.4) into Equation (6.5), and replace nf_0 by f, to obtain

$$F(f) = \lim_{T\to\infty} \int_{-T/2}^{T/2} f_{\mathrm{p}}(t)\mathrm{e}^{-\mathrm{j}2\pi ft}\,\mathrm{d}t \tag{6.6}$$

However, because the signals $f(t)$ and $f_{\mathrm{p}}(t)$ are identical over the period T, a valid alternative to this expression is

$$F(f) = \int_{-\infty}^{\infty} f(t)\mathrm{e}^{-\mathrm{j}2\pi ft}\,\mathrm{d}t \tag{6.7}$$

This is known as the Fourier transform of the energy signal $f(t)$; it is also sometimes known as the Fourier integral.

In general, $F(f)$ is a *complex* function of f and has real and imaginary components, both of which are spectral density functions with units of volts/Hz. $F(f)$ may also be written as $|F(f)|\,\mathrm{e}^{\mathrm{j}\theta(f)}$, or $A(f)\,\mathrm{e}^{\mathrm{j}\theta(f)}$, where $A(f)$ is the amplitude spectrum and $\theta(f)$ is the phase spectrum. The units of $A(f)$ are V/Hz and those of $\theta(f)$ are radians.

There are some signals for which a meaningful Fourier transform does not exist. The conditions for a unique and meaningful Fourier transform are known as the Dirichlet conditions, but the reader is referred elsewhere for these since they are rarely infringed for signals met in practical situations.

6.4 The Fourier transform of a rectangular pulse

Because of its importance the shape of a rectangular pulse is given its own name and symbol. A rectangle function of duration τ is defined as

$$\begin{aligned} \mathrm{rect}\left(\frac{t}{\tau}\right) &= 1 \qquad -\frac{\tau}{2} < t < \frac{\tau}{2} \\ &= 0 \qquad \text{elsewhere} \end{aligned} \tag{6.8}$$

This is shown in Figure 6.4. The rectangular pulse is frequently encountered as a gating signal used to turn another signal on and off, and hence a rectangle function is also known as the unit gate function.

A rectangular pulse of amplitude A is described mathematically as $A\,\mathrm{rect}(t/\tau)$. Therefore, applying Equation (6.7)

$$F(f) = \int_{-\infty}^{\infty} A\,\mathrm{rect}(t/\tau)\mathrm{e}^{-\mathrm{j}2\pi ft}\,\mathrm{d}t = \int_{-\tau/2}^{\tau/2} A\mathrm{e}^{-\mathrm{j}2\pi ft}\,\mathrm{d}t$$

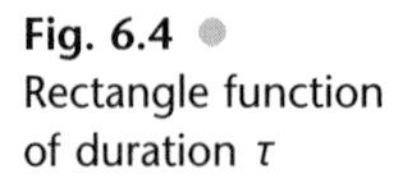
Fig. 6.4 Rectangle function of duration τ

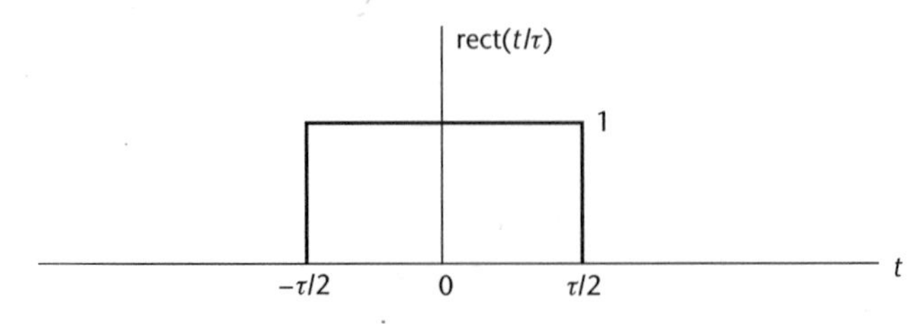

The integration produces

$$F(f) = \frac{A}{-\mathrm{j}2\pi f}\,\mathrm{e}^{-\mathrm{j}2\pi ft}\Big|_{-\tau/2}^{\tau/2} = \frac{A}{-\mathrm{j}2\pi f}\,[\mathrm{e}^{-\mathrm{j}2\pi f\tau/2} - \mathrm{e}^{\mathrm{j}2\pi f\tau/2}] = \frac{A}{\pi f}\,\sin(\pi f\tau)$$

This can be rearranged as

$$F(f) = A\tau\,\frac{\sin(\pi f\tau)}{\pi f\tau} \tag{6.9}$$

This is the $\sin(x)/x$ shape shown in Figure 6.5 and has nulls when $f = \pm 1/\tau$, $\pm 2/\tau$, $\pm 3/\tau$ and so on. It is a continuous version of the discrete line spectrum given in Equation (5.26) for a rectangular pulse train.

If radian frequency is preferred to cyclic frequency we can substitute ω for $2\pi f$ and the Fourier transform of the rectangular pulse can be expressed as

$$F(\omega) = A\tau\,\frac{\sin(\omega\tau/2)}{\omega\tau/2} \tag{6.10}$$

The spectral densities $F(f)$ and $F(\omega)$ given by Equations (6.9) and (6.10) have the same numerical values at all frequencies (for example $A\tau$ when f and ω are zero) and the amplitude spectra $|F(f)|$ and $|F(\omega)|$ both therefore have the same units of V/Hz. Thus an amplitude spectrum plotted using Equation (6.10) is identical to one plotted using Equation (6.9) except that, in terms of radian frequencies, the nulls occur at $\omega = \pm 2\pi/\tau$, $\pm 4\pi/\tau$, $\pm 6\pi/\tau$, $\pm 6\pi/\tau$ and so on.

It will be noted that $F(f)$ is entirely real in this particular example and can hence be represented in Figure 6.5 by a single plot. Usually Fourier transforms are complex and, because of this, it is common to express Fourier transforms in terms of their amplitude and phase spectra, $|F(f)|$ and $\angle F(f)$. The amplitude and phase spectra for the rectangular pulse are shown in Figure 6.6. It will be noted that since

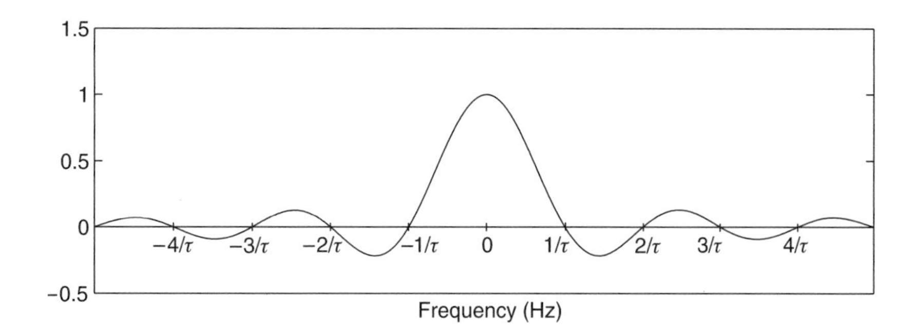

Fig. 6.5 Fourier transform of a rectangular pulse of duration τ and amplitude $1/\tau$

(a)

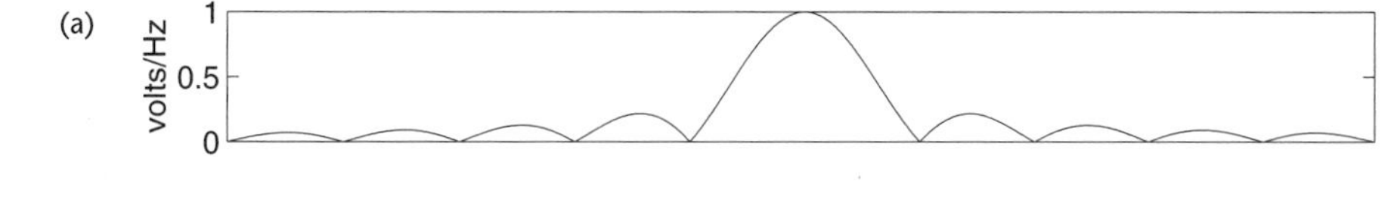

(b)

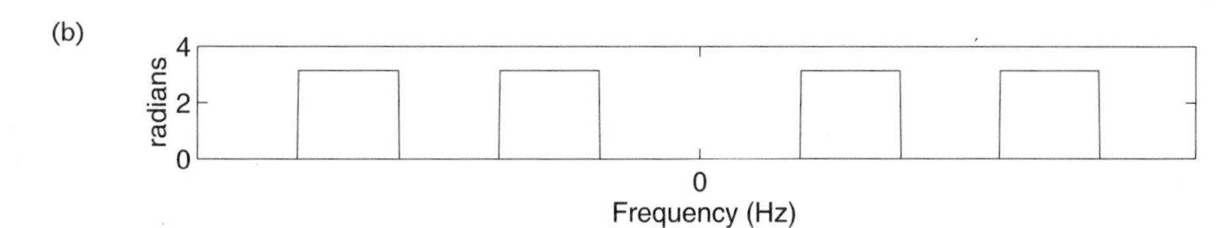

Fig. 6.6 Spectrum of a rectangular pulse: (a) amplitude; (b) phase

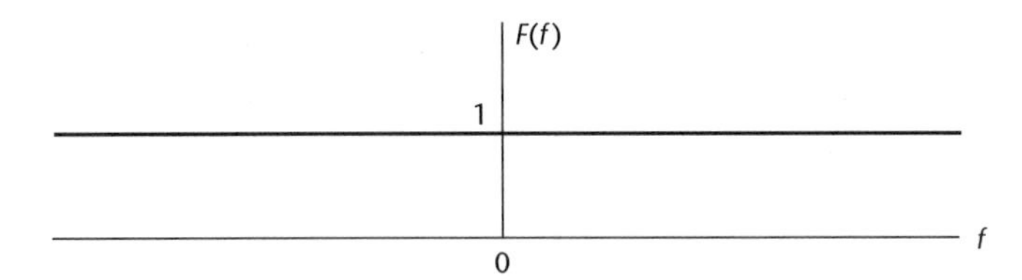

Fig. 6.7 ● Fourier transform of a unit impulse

$-\cos \omega t = \cos(\omega t - \pi) = \cos(\omega t + \pi)$ it is equally valid to consider the phase shifts of $+\pi$ in Figure 6.6 as being phase shifts of $-\pi$.

6.5 ● The Fourier transform of an impulse

As pointed out earlier, the continuous-time unit impulse function is not a finite energy signal. It may however be considered a limiting case of the rectangular pulse considered in Section 6.3 for which $\tau \to 0$ while $A\tau = 1$. Therefore, using Equation (6.9):

$$F(f) = \lim_{\tau \to 0} \frac{\sin(\pi f \tau)}{\pi f \tau} = 1 \tag{6.11}$$

The result is a continuous spectrum of unit magnitude, as shown in Figure 6.7. In this example $F(f)$ is entirely real, which means that the amplitude spectrum is of unit magnitude and the phase spectrum is zero.

It is important to realize that the phase spectrum is just as significant as the amplitude spectrum. The fact that it is zero for the impulse is extremely informative. To illustrate this, consider a long but *finite segment* of the wideband noise signal shown in Figure 6.2(b). Such a signal may be expected intuitively, and correctly, to have an almost flat amplitude spectrum, very similar to that for the impulse. What makes the Fourier transform of this noise signal different to that of the impulse is the phase spectrum. The phase spectrum of the segment of the noiselike signal will not be *zero* but will instead show random fluctuations across the frequency band.

6.6 ● Symmetry in Fourier transforms

Using the relationship $e^{-jx} = \cos x - j \sin x$ the Fourier transform given by Equation (6.7) can be rewritten as

$$F(f) = \int_{-\infty}^{\infty} f(t)\cos(2\pi f t)\mathrm{d}t - \mathrm{j}\int_{-\infty}^{\infty} f(t)\sin(2\pi f t)\,\mathrm{d}t \tag{6.12}$$

The even symmetry of cosine functions means that $\cos(2\pi ft)$ equals $\cos(-2\pi ft)$, and this causes the first of these two terms to give the same results for positive and negative values of frequency. In contrast, the odd symmetry of sine functions causes the second of these two terms to give results that are equal but of opposite polarity for positive and negative frequencies. It follows that Fourier transforms of real signals have amplitude spectra with even symmetry, but phase spectra with odd symmetry. The even symmetry in the phase spectrum of Figure 6.6(b) is deceptive since a phase shift of $+\pi$ at some of the negative frequencies could just as well be denoted by a phase shift of $-\pi$, thus giving the phase spectrum odd symmetry.

Note: Although it seems reasonable at this stage to expect that all waveforms are real, it will be shown in Chapter 17 that complex signals are useful concepts and can even have a physical meaning. The Fourier transforms of such signals do not have the constraints of even symmetry in the amplitude spectrum and odd symmetry in the phase spectrum.

6.7 Some physical insight into the evaluation of Fourier transforms

Section 6.3 gave an example of how the Fourier integral could be evaluated analytically in the case of a rectangular pulse. For many waveforms an analytical solution is not possible and numerical techniques are necessary. The discrete Fourier transform (DFT) and fast Fourier transform (FFT) are used extensively for this purpose and are examined in Chapters 8 and 9. In these the waveform is approximated by a finite number of discrete samples and the Fourier transform is evaluated at a finite number of discrete frequencies. Before this, however, some useful physical insight into the continuous Fourier transform can be obtained by considering the application of numerical techniques for evaluating the Fourier transform of a continuous waveform $f(t)$. The rectangular pulse has already been studied and one of unit amplitude that is 12 ms long is a suitable example. Such a pulse is shown in Figure 6.8(a).

Since the Fourier transform is a continuous function, an answer is clearly required for all frequencies. All that will be done here, however, is to consider how we might evaluate the Fourier transform at a single frequency, say 100 Hz.

The expression for the Fourier transform, namely $F(f) = \int_{-\infty}^{\infty} f(t)\mathrm{e}^{-\mathrm{j}\pi ft}\,\mathrm{d}t$, may be expanded into $F(f) = \int_{-\infty}^{\infty} f(t)\,[\cos(2\pi ft) - \mathrm{j}\sin(2\pi ft)]\,\mathrm{d}t$. It is seen to have a real component $A(f)$ resulting from integrating $f(t)$ multiplied by a cosine wave, and an imaginary component $B(f)$ resulting from integrating $f(t)$ multiplied by a sine wave. Figure 6.8(b) shows a 100 Hz cosine wave and Figure 6.8(c) shows the product $f(t)\cos(200\pi t)$. The real component $A(100)$ of the Fourier transform is therefore the area of this product, and this can be determined graphically. It turns out that the answer is -0.156 V s, or -0.156 V/Hz.

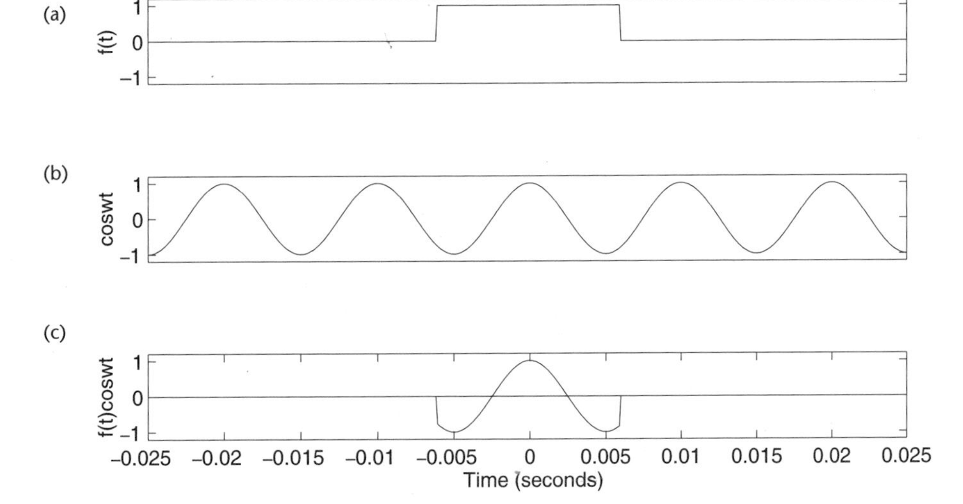

Fig. 6.8 Numerical evaluation of the real part of a Fourier transform at 100 Hz: (a) the waveform; (b) a 100 Hz cosine wave; (c) the product

Figure 6.9(a) repeats the waveform $f(t)$ and, in a similar manner to Figure 6.8, Figures 6.9(b) and 6.9(c) show a 100 Hz sine wave and the product $f(t)\sin(200\pi t)$. The imaginary component $B(100)$ of the Fourier transform is the area of this product. The area is zero in this case because of the symmetry of the rectangular pulse about zero time.

The magnitude of the Fourier transform at 100 Hz is therefore $\sqrt{A^2(100) + B^2(100)}$. In this example of an even function this reduces to $A(100)$, and a numerical value is given by the measured area of 0.156 V/Hz. The phase of the Fourier transform at 100 Hz is the four-quadrant arctangent of $B(100)/A(100)$ and is π radians. Thus $F(f) = 0.156\angle\pi = -0.156$. We can cross-check this result with Figure 6.5. Since $\tau = 12$ ms, the 100 Hz frequency corresponds to $f = 1.2/\tau$. We see that Figure 6.5 also gives $F(f) = -0.156$ when $f = 1.2/\tau$.

A very important observation that results from Figure 6.8 is that the real part of the Fourier integral $A(f)$ depends upon *how well matched* the signal is to $\cos(2\pi ft)$. In our example the small net negative area of Figure 6.8(c) shows that there is a small negative match (the fact that the match is negative merely signifies that the pulse is better matched to $-\cos(2\pi ft)$ than to $\cos(2\pi ft)$). Similarly, the imaginary part of the Fourier integral $B(f)$ depends upon *how well matched* the signal is to $\sin(\pi ft)$. In our case there is a zero match, but this is only because the signal in this example has even symmetry.

By taking the magnitude $\sqrt{A^2(f) + B^2(f)}$, the dependence of the Fourier integral on the position of the signal disappears. Hence the *magnitude* of the Fourier transform at frequency f depends upon how well matched the signal is to an *ideally* phased sinusoid of frequency f. A detailed discussion of correlation is postponed to Chapter 16, but it is interesting to note that these statements are equally valid if the word *matched* is replaced by the word *correlated*. For example, the magnitude of the Fourier transform at frequency f depends upon how well *correlated* the signal is to an ideally phased sinusoid of frequency f.

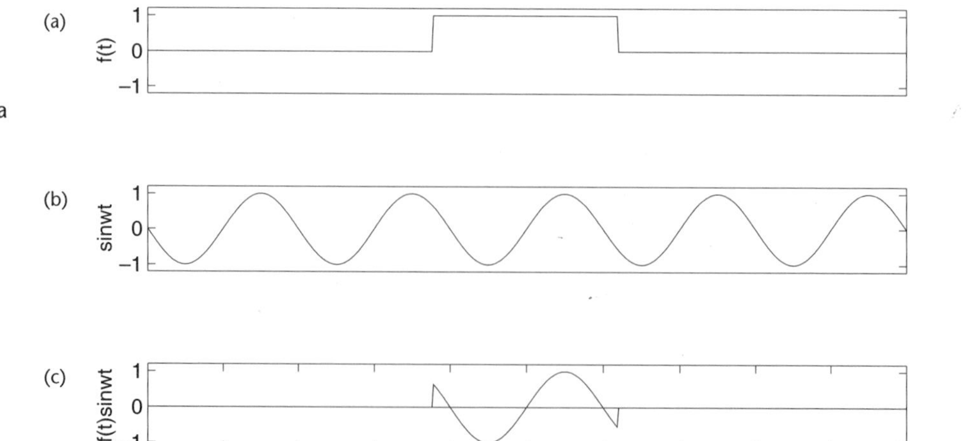

Fig. 6.9 Numerical evaluation of the imaginary part of a Fourier transform at 100 Hz: (a) the waveform; (b) a 100 Hz sine wave; (c) the product

6.8 The inverse Fourier transform

Equation (6.7) gives the integral that expresses $F(f)$ as a function of $f(t)$. It is useful also to have an inverse relationship, that gives $f(t)$ as a function of $F(f)$. Such a relationship is the *inverse* Fourier transform.

In Section 6.3 it was argued that an aperiodic energy signal can be thought of as a periodic signal $f_p(t)$ with a period T that tends to infinity, and hence with a fundamental frequency $f_0 = 1/T$ that tends to zero. The nth harmonic of this periodic signal has a frequency nf_0 and has a complex amplitude given by the coefficient F_n in the exponential Fourier series. This coefficient has the same strength as the area of the spectral density function $F(f)$ over the same bandwidth f_0, that is

$$F_n = \lim_{f_0 \to 0} f_0 F(nf_0) \tag{6.13}$$

Equation (6.1) gave $f(t) = \lim_{T\to\infty} f_p(t)$ and can be rewritten as $f(t) = \lim_{f_0\to 0} f_p(t)$. Combining this with the result of Equation (6.2) that $f_p(t) = \sum_{n=-\infty}^{\infty} F_n e^{j2\pi n f_0 t}$, and substituting for F_n in this from Equation (6.13) gives

$$f(t) = \lim_{f_0 \to 0} \sum_{n=-\infty}^{\infty} f_0 F(nf_0) e^{j2\pi n f_0 t} \tag{6.14}$$

In the limit the summation becomes integration. Hence, replacing the term nf_0 by the continuous variable f, and considering the separation of the discrete spectral lines which is now vanishingly small to be df, where $df = f_0$, we obtain

$$f(t) = \int_{-\infty}^{\infty} F(f) e^{j2\pi f t}\, df \tag{6.15}$$

This is known as the *inverse* Fourier transform.

Equations (6.7) and (6.15) are repeated below as Equations (6.16) and (6.17) and form what is known as a Fourier transform pair.

$$F(f) = \int_{-\infty}^{\infty} f(t) e^{-j2\pi f t}\, dt \tag{6.16}$$

$$f(t) = \int_{-\infty}^{\infty} F(f) e^{j2\pi f t}\, df \tag{6.17}$$

Very often these equations are expressed in terms of radian frequency by making the substitution $\omega = 2\pi f$, such that $d\omega = 2\pi\, df$. This leads to a notation that is more compact, but one which is somewhat less easy to interpret and which contains less symmetry between the two equations, namely

$$F(\omega) = \int_{-\infty}^{\infty} f(t) e^{-j\omega t}\, dt \tag{6.18}$$

and

$$f(t) = \frac{1}{2\pi} \int_{-\infty}^{\infty} F(\omega) e^{j\omega t}\, d\omega \tag{6.19}$$

It is sometimes useful to convey the operations of taking a Fourier transform and an inverse Fourier transform by using the symbols $\mathscr{F}$ and $\mathscr{F}^{-1}$, for example

$$\mathscr{F}[f(t)] = F(f), \qquad \mathscr{F}^{-1}[F(f)] = f(t) \tag{6.20}$$

6.9 Reconstruction of a waveform from its Fourier transform

The Fourier transform provides a description of a signal that is a complete one but which is in the frequency domain rather than in the time domain. Some useful physical understanding of the Fourier transform is gained by demonstrating that a time waveform can indeed be reconstructed from a Fourier transform by applying the inverse Fourier transform to it.

Consider the result for the Fourier transform of a rectangular pulse given by Equation (6.9) and the approximation to the inverse transform given by Equation (6.14). Combining the two and recognizing that nf_0 is equivalent to the continuous variable f we obtain

$$f(t) = \lim_{f_0 \to 0} \sum_{n=-\infty}^{\infty} f_0 A\tau \frac{\sin(\pi n f_0 \tau)}{\pi n f_0 \tau} e^{j2\pi n f_0 t}$$

Using the result that $e^{j2\pi n f_0 t} + e^{-j2\pi n f_0 t} = 2\cos(2\pi n f_0 t)$, this can be rewritten as

$$f(t) = f_0 A\tau + \lim_{f_0 \to 0} \sum_{n=1}^{\infty} 2 f_0 A\tau \,\mathrm{sinc}(n f_0 \tau)\cos(2\pi n f_0 t) \tag{6.21}$$

The objective now is to find what this equation produces and to show that the result is indeed similar to the rectangular pulse.

Let $A = 1$ and $\tau = 1$s. A reasonable value of f_0 to approximate the condition that $f_0 \to 0$ is that $f_0 = 0.01$ Hz, since this signifies a periodic waveform with a period of 100 times the pulse duration. Figure 6.10(a) shows six terms taken from the summation, for n values of 20, 40, 60, 80, 100 and 120 respectively. If we add these

(a)

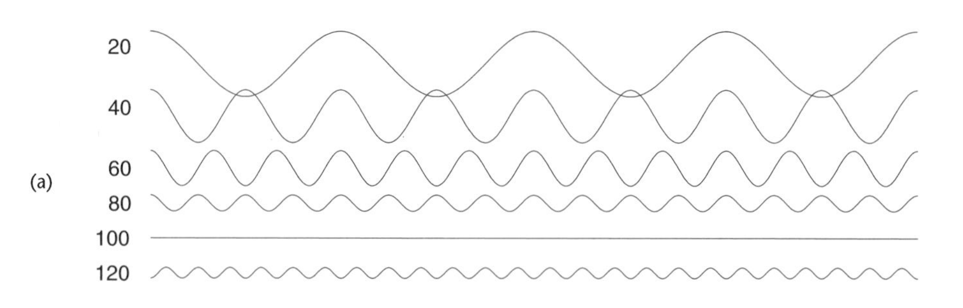

(b)

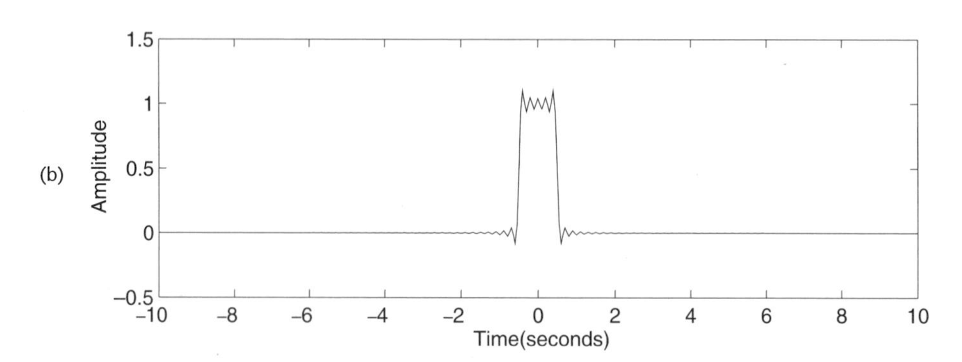

Fig. 6.10 Reconstruction of a waveform from its Fourier transform: (a) contributions from six harmonically related frequencies; (b) reconstruction from 501 terms

plus all the other 495 terms between $n = 0$ and $n = 500$ we obtain the result shown in Figure 6.10(b). It is seen that this sum of 501 terms is a reasonable approximation to the rectangular pulse. The shape becomes more and more like that of the rectangular pulse as even more terms are included. Of course it should be realized that the finite value of f_0 means that the result obtained is actually periodic and that this periodicity would become apparent if Figure 6.10 had a timespan exceeding ± 100s.

6.10 Fourier transforms of power signals

The energy $\int_{-\infty}^{\infty} f^2(t)\, \mathrm{d}t$ of power signals is infinite. One example of a power signal is continuous noise, such as shown in Figure 6.2(b) and consideration of such a signal shows that the integrals $\int_{-\infty}^{\infty} f(t)\cos(2\pi ft)\, \mathrm{d}t$ and $\int_{-\infty}^{\infty} f(t)\sin(2\pi ft)\, \mathrm{d}t$ do not converge, and hence the Fourier transform cannot be evaluated. However, there are other power signals which do have meaningful Fourier transforms. These usually have Fourier transforms which are infinite at certain frequencies and which can be represented by impulse functions in the frequency domain.

6.11 Fourier transforms of periodic signals

Consider first the simplest periodic signal, a cosine wave $A\cos(2\pi f_0 t)$. Section 6.6 has shown that, since this is an even function, the imaginary component of $F(f)$ must equal zero. Thus the equation $F(f) = \int_{-\infty}^{\infty} f(t)\mathrm{e}^{-\mathrm{j}2\pi ft}\, \mathrm{d}t$ can be simplified to become

$$F(f) = \int_{-\infty}^{\infty} A\cos(2\pi f_0 t)\cos(2\pi ft)\, \mathrm{d}t$$

$$= \frac{A}{2}\int_{-\infty}^{\infty} \{\cos[2\pi(f_0 - f)t] + \cos[2\pi(f_0 + f)t]\}\, \mathrm{d}t$$

If $f = \pm f_0$ this contains $(A/2)\int_{-\infty}^{\infty} \cos(0)\, \mathrm{d}t$, *which is infinite*. If $f \neq \pm f_0$ the two integrals are finite *but indeterminate*. It is clear that there is a problem with the Fourier transform of a periodic signal.

The easiest way of determining the Fourier transform of a cosine wave is to adopt a physical approach to the problem. The signal $A\cos(2\pi f_0 t)$ can be expressed as $(A/2)\mathrm{e}^{\mathrm{j}2\pi f_0 t} + (A/2)\mathrm{e}^{-\mathrm{j}2\pi f_0 t}$ and the exponential Fourier series therefore has two spectral lines, one of magnitude $A/2$ at frequency $+f_0$ and one of magnitude $A/2$ at frequency $-f_0$, as shown in the line spectrum of Figure 6.11(a). Remembering that the result of the Fourier transform is a spectral *density* it is clear that the two spectral lines of the Fourier series must be replaced by a spectral density function whose *area* equals the magnitude of these lines. Since, by definition, an impulse function of strength $A/2$ has an area of $A/2$ it follows immediately that

$$F(f) = \frac{A}{2}\,\delta(f - f_0) + \frac{A}{2}\,\delta(f + f_0) \tag{6.22}$$

This result is illustrated in Figure 6.11(b) where, because $F(f)$ is entirely real, a single plot conveys all the information. It should be noted that the arrows have a very different meaning to the lines of Figure 6.11(a).

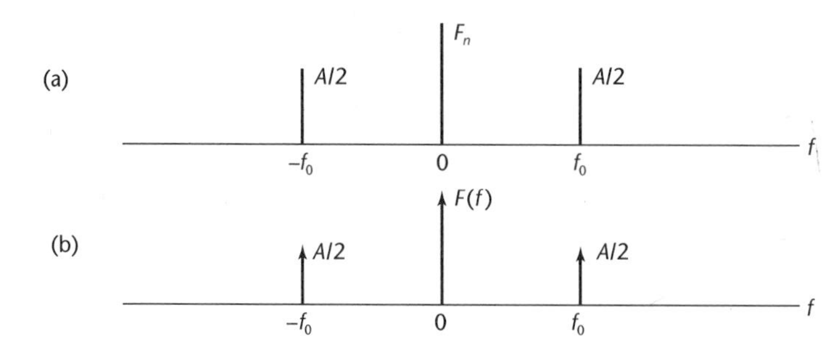

Fig. 6.11 Frequency representation of a cosine wave: (a) line spectrum; (b) Fourier transform

Plots of $|F_n|$ and $|F(f)|$ versus frequency are different and yet both would commonly be referred to as amplitude (or magnitude) spectra. Any incorrect interpretation of these terms is unlikely, however, except in the case of periodic signals. If the possibility of an ambiguity is considered a problem it may be preferred to expand the terms, referring to an amplitude *line* spectrum for a plot of $|F_n|$ versus frequency and to an amplitude *density* spectrum for a plot of $|F(f)|$ versus frequency.

In a similar way to what has been done with the signal $A\cos(2\pi f_0 t)$ the signal $A\sin(2\pi f_0 t)$ can be expressed as $(A/2\mathrm{j})\mathrm{e}^{\mathrm{j}2\pi f_0 t} - (A/2\mathrm{j})\mathrm{e}^{-\mathrm{j}2\pi f_0 t}$ and it follows therefore that

$$F(f) = \frac{A}{2\mathrm{j}}\,\delta(f-f_0) - \frac{A}{2\mathrm{j}}\,\delta(f+f_0) \tag{6.23}$$

This time both terms are imaginary and the Fourier transform can be conveyed by a single portrayal of $\mathrm{j}F(f)$, shown in Figure 6.12. A portrayal that is less compact but more generally applicable, and is in line with previous spectral plots, is an amplitude spectrum $A(f)$ that is the same as Figure 6.11(b), accompanied by a phase spectrum $\theta(f)$ that is $-\pi/2$ at $f=f_0$ and $\pi/2$ at $f=-f_0$.

It is reassuring to confirm these results mathematically and the easiest way is to work backwards, by showing that the *inverse* transform of the suggested result is indeed the original time waveform. For example, using the definition of Equation 6.15, the inverse Fourier transform of $(A/2)\,\delta(f+f_0)+(A/2)\,\delta(f-f_0)$ is given by

$$f(t) = \int_{-\infty}^{\infty} F(f)\mathrm{e}^{\mathrm{j}2\pi ft}\,\mathrm{d}f = \int_{-\infty}^{\infty}\left(\frac{A}{2}\,\delta(f+f_0) + \frac{A}{2}\,\delta(f-f_0)\right)\mathrm{e}^{\mathrm{j}2\pi ft}\,\mathrm{d}f$$

Remembering that the unit impulse has unit area at the position at which it is centred this simplifies to

$$f(t) = \frac{A}{2}\,\mathrm{e}^{-\mathrm{j}2\pi f_0 t} + \frac{A}{2}\,\mathrm{e}^{\mathrm{j}2\pi f_0 t} = A\cos(2\pi f_0 t)$$

This confirms the result of Equation (6.22).

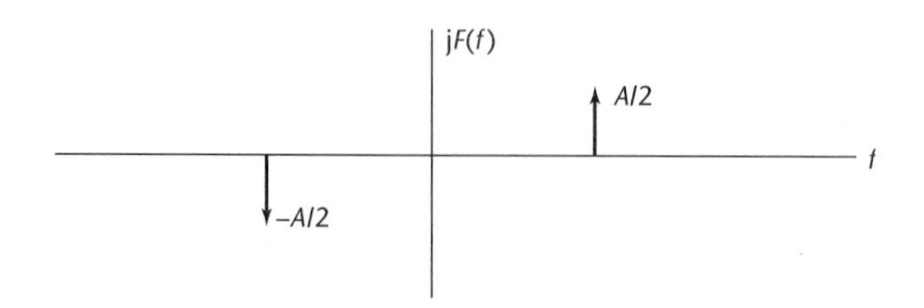

Fig. 6.12 Portrayal of the Fourier transform of a sine wave

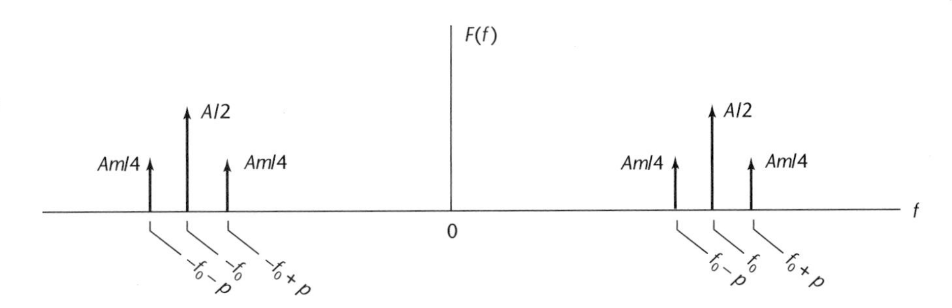

Fig. 6.13 Fourier transform of a cosinusoidally modulated AM signal

Example 6.1

Determine the Fourier transform of the amplitude modulated signal $f(t) = A(1 + m\cos pt)\cos(\omega_0 t)$.

Solution Expanding the signal by use of trigonometric identities we obtain

$$f(t) = \cos(\omega_0 t) + \frac{Am}{2}\cos(\omega_0 - p)t + \frac{Am}{2}\cos(\omega_0 + p)t$$

This consists of a carrier at frequency f_0, accompanied by upper and lower sidebands. It follows from Equation (6.22) that

$$F(f) = \frac{A}{2}\delta(f - f_0) + \frac{A}{2}\delta(f + f_0) + \frac{Am}{4}\delta(f - f_0 - p) + \frac{Am}{4}\delta(f - f_0 + p) + \frac{Am}{4}\delta(f + f_0 - p) + \frac{Am}{4}\delta(f + f_0 + p)$$

Since the phase spectrum is zero the Fourier transform can thus be represented by the single plot of Figure (6.13).

For other periodic signals the same arguments may be repeated to show that the Fourier transform is identical to the spectrum resulting from the exponential Fourier series, except that each spectral line is replaced by an impulse function having the same strength as the amplitude of that line. Thus, if a periodic signal $f(t)$ is expressed in terms of its exponential Fourier series as

$$f(t) = \sum_{n=-\infty}^{\infty} F_n e^{j2\pi n f_0 t}$$

where f_0 is the fundamental frequency, then

$$F(f) = \sum_{n=-\infty}^{\infty} F_n \delta(f - nf_0) \tag{6.24}$$

Example 6.2

Determine the Fourier transform $P(f)$ of a periodic sequence of unit impulse functions.

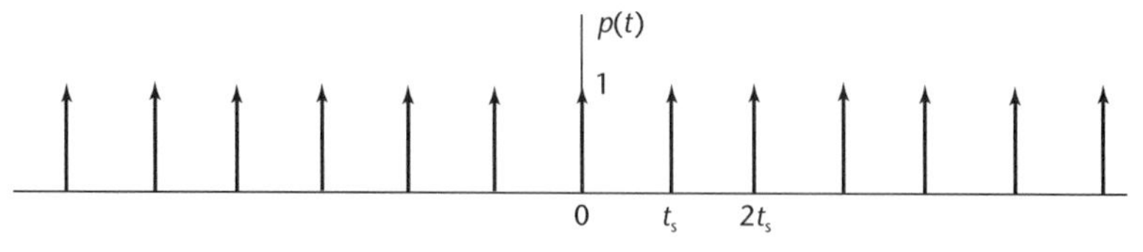

Fig. 6.14 Periodic sequence of unit impulses

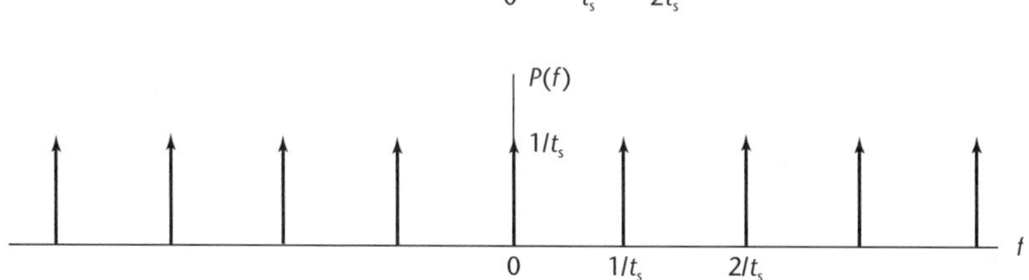

Fig. 6.15 Fourier transform of periodic sequence of unit impulses

Solution It should be noted that the periodic sequence of unit impulse functions is of great importance since, when multiplied by some other signal, it produces samples of that signal. It will therefore be termed a *sampling function* and will be denoted within this book by the symbol $p(t)$. The sequence is also sometimes known as a *comb* function. In keeping with the sampling application the spacing of the impulses is given the symbol t_s. Thus the waveform is as shown in Figure (6.14).

Equation (5.29) showed that the coefficients of the exponential Fourier series of a train of unity amplitude impulses were given by $F_n = 1/T$, where T is the period. The components of the line spectrum are $f_0 = 1/T$ apart. Thus, for impulses t_s apart, the Fourier transform is given by

$$P(f) = \frac{1}{t_s} \sum_{n=-\infty}^{\infty} \delta(f - nf_s) \tag{6.25}$$

where the periodic frequency is the sampling frequency given by $f_s = 1/t_s$. The Fourier transform $P(f)$ is illustrated in Figure 6.15.

6.12 Cyclic frequency versus radian frequency

Fourier transforms can be expressed in terms of cyclic frequency to give a function $F(f)$, or in terms of radian frequency to give a function $F(\omega)$. It is easy to become confused about the relationship between the two because of the different way in which these functions must be handled, depending on whether they consist of continuous functions or impulse functions. In the case of a continuous spectral density function it is necessary simply to make the substitution $\omega = 2\pi f$. For example, the Fourier transform of a rectangular pulse is given in Equation 6.9 by

$$F(f) = A\tau \frac{\sin(\pi f \tau)}{\pi f \tau} \quad \text{V/Hz}$$

and in Equation (6.10) by

$$F(\omega) = A\tau \frac{\sin(\omega\tau/2)}{\omega\tau/2} \quad \text{V/Hz}$$

However, for the impulse functions associated with a periodic signal, this is not the case. For example Equation (6.22) gave the Fourier transform of $A\cos(2\pi f_0 t)$ as

$$F(f) = \frac{A}{2}\,\delta(f - f_0) + \frac{A}{2}\,\delta(f + f_0)$$

but, in terms of the variable ω, the correct result for the Fourier transform of the signal $A\cos(\omega_0 t)$ is given by

$$F(\omega) = A\pi\delta(\omega - \omega_0) + A\pi\delta(\omega + \omega_0) \tag{6.26}$$

The change in the weights of the impulse functions between these two equations can be very confusing and arises because, just as for $|F(f)|$, the units of $|F(\omega)|$ are V/Hz and not V/rad/s. In Equation (6.26) the area of the first impulse function in volts is not $\int_{-\infty}^{\infty} A\pi\delta(\omega - \omega_0)\,\mathrm{d}\omega$, or $A\pi$ volts. Rather it is

$$\int_{-\infty}^{\infty} A\pi\delta(\omega - \omega_0)\,\mathrm{d}f = \int_{-\infty}^{\infty} \frac{A}{2}\,\delta(\omega - \omega_0)\,\mathrm{d}\omega = \frac{A}{2}\text{ volts}$$

just as is obtained from Equation (6.22)

As a final check that Equation (6.26) is correct, we can undertake the inverse Fourier transform of the function $F(\omega)$ in Equation (6.26) using the definition provided by Equation (6.19). This gives

$$\begin{aligned} f(t) &= \frac{1}{2\pi}\int_{-\infty}^{\infty} F(\omega)\mathrm{e}^{\mathrm{j}\omega t}\,\mathrm{d}\omega = \frac{1}{2\pi}\int_{-\infty}^{\infty} \{A\pi\delta(\omega - \omega_0) + A\pi\delta(\omega + \omega_0)\}\mathrm{e}^{\mathrm{j}\omega t}\,\mathrm{d}\omega \\ &= \frac{A\pi}{2\pi}\,\mathrm{e}^{\mathrm{j}\omega_0 t} + \frac{A\pi}{2\pi}\,\mathrm{e}^{-\mathrm{j}\omega_0 t} = A\cos(\omega_0 t) \end{aligned}$$

as required.

Partly because cyclic frequencies are in more common everyday use (how many people know the mains frequency in rad/s?), partly because of the greater symmetry of the Fourier transform pair when cyclic frequency is used, and partly because of the confusion that can sometimes arise from using a Fourier transform whose parameter is ω but whose amplitude spectrum has the units of V/Hz, this book generally prefers to express Fourier transforms in terms of cyclic frequency. However the ω representation is used on occasion and the argument in its favour is that it is more concise and efficient to write ω than $2\pi f$, particularly when such terms occur many times.

6.13 Signal transmission through linear networks

Equation (2.11) showed that the output of a network, $g(t)$, is obtained by convolving the input with the network impulse response, that is

$$g(t) = \int_{-\infty}^{\infty} f(\tau)h(t - \tau)\,\mathrm{d}\tau$$

where, in terms of the time variable τ, the input waveform is $f(\tau)$ and the impulse response is $h(\tau)$. This is the time domain approach. The alternative approach is to work in the frequency domain using the Fourier transform, and this is often advantageous.

In Chapter 3 the frequency transfer function of a network, $H(\omega)$, was shown to be the ratio of output signal to input signal when the input signal was the complex

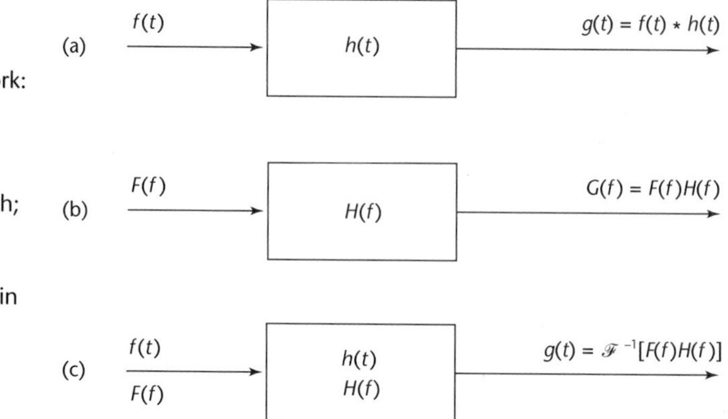

Fig. 6.16 Transmission through a network: (a) time domain approach; (b) Frequency domain approach; (c) time domain result using a frequency domain approach

exponential time waveform $e^{j\omega t}$. Thus, if a periodic input signal is made up of harmonics whose amplitudes are the F_n's given by the Fourier series, $f(t) = \sum_{n=-\infty}^{\infty} F_n e^{jn\omega_0 t}$, passage through the network will cause each coefficient to be multiplied by the value of $H(\omega)$ at the corresponding frequency.

The Fourier transform is similar to the Fourier series except that the complex exponential content is described by a spectral density, rather than by discrete amplitudes. If therefore the Fourier transform of an input signal is $F(\omega)$ and the frequency transfer function of a network is $H(\omega)$, the Fourier transform of the output signal, $G(\omega)$, is given by

$$G(\omega) = F(\omega)H(\omega) \tag{6.27}$$

We can then determine the output time waveform by taking the inverse Fourier transform of $G(\omega)$:

$$g(t) = \mathscr{F}^{-1}[F(\omega)H(\omega)] = \frac{1}{2\pi}\int_{-\infty}^{\infty} F(\omega)H(\omega)e^{j\omega t}\, d\omega \tag{6.28}$$

Alternatively, working in terms of preferred mode of cyclic frequency,

$$G(f) = F(f)H(f) \tag{6.29}$$

and

$$g(t) = \mathscr{F}^{-1}[F(f)H(f)] = \int_{-\infty}^{\infty} F(f)H(f)e^{j2\pi ft}\, df \tag{6.30}$$

These results are summarized in Figure 6.16. Here Figure 6.16(a) shows a time domain approach to transmission through a network, Figure 6.16(b) shows a frequency domain approach, and Figure 6.16(c) shows how a time domain output can be obtained using a frequency domain approach.

Example 6.3

Determine the impulse response of a lowpass filter having an ideal brickwall response.

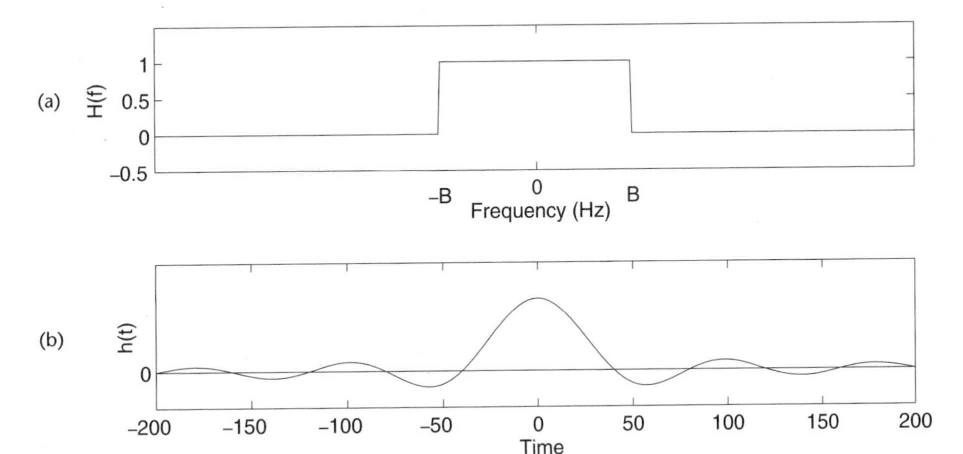

Fig. 6.17 Response of ideal brickwall filter: (a) transfer function of filter; (b) impulse response

Solution From Equation (6.11) the Fourier transform of a unit impulse input is $F(f) = 1$. Hence the Fourier tranform of the output is given by $F(f)H(f) = H(f)$ and the impulse response is $\mathscr{F}^{-1}[H(f)]$. However, by brickwall response we mean a transfer function whose amplitude response is as shown in Figure 6.17(a). Assuming a zero phase response this means

$$H(f) = 1 \qquad -B<f<B$$
$$= 0 \qquad \text{elsewhere}$$

Hence

$$h(t) = \int_{-\infty}^{\infty} H(f)e^{j2\pi ft}\,df = \int_{-B}^{B} e^{j2\pi ft}\,df = \frac{1}{j2\pi t}\,e^{j2\pi ft}\bigg|_{-B}^{B} = \frac{1}{j2\pi t}\,[e^{j2\pi Bt} - e^{-j2\pi Bt}]$$
$$= 2B\,\text{sinc}(2Bt)$$

This is shown in Figure 6.17(b).

The impulse response of an ideal brickwall filter has been found to be a sinc function centred about zero and the interesting observation that there is a response for $t<0$, before the impulse is ever applied to the filter, proves that such a filter is not physically realizable.

Any system that responds to an excitation before the excitation occurs is termed to be *non-causal*. All physically realizable systems are causal.

6.14 The importance of the Fourier transform in network analysis

The application of Fourier techniques to linear networks described in the previous section suggests the possibility of working in the frequency domain rather than in the time domain, and it is natural to wonder why we might wish to do this. The following are some of the reasons.

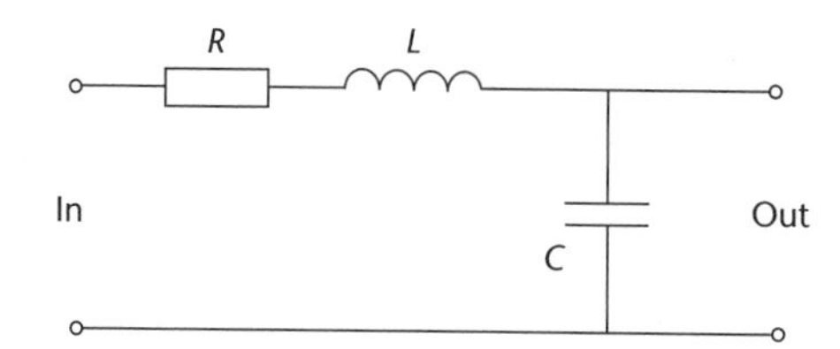

Fig. 6.18 Example of a network whose transfer function is easy to evaluate

(1) Although the input signal to a network is often described in the time domain (for example, a cosine wave or rectangular pulse) the properties of the network are generally much easier to obtain and describe in the frequency domain. For example, a bandpass filter would normally be thought of in terms of its centre frequency and bandwidth, rather than by its impulse response. As another example the frequency transfer function of the network in Figure 6.18 is easily shown by network analysis to be given by

$$H(\omega) = \frac{1/\mathrm{j}\omega C}{R + \mathrm{j}\omega L + 1/\mathrm{j}\omega C}$$

whereas its impulse response $h(t)$ is rather less easy to obtain.

(2) Many input signals (such as a cosine wave or rectangular pulse) have a simple Fourier transform $F(\omega)$. The operation for evaluating the frequency domain output is then the simple multiplication, $F(\omega)H(\omega)$. This is much easier than the convolution operation, $f(t) * h(t)$, needed for producing the time domain output.

(3) Although the evaluation of $F(\omega)H(\omega)$ ends up with a frequency domain output rather than with a time domain output, this frequency domain description is often sufficient; indeed it is sometimes preferable.

Example 6.4

A bandpass filter is centred at 455 kHz and has a bandwidth of 6 kHz. Determine whether it affects the amplitude modulated signal given by $f(t) = [1 + 0.8\cos(15000t)]\cos(2.87 \times 10^6 t)$

Solution Trigonometric expansion of the expression for $f(t)$ gives a carrier at a radian frequency of 2.87×10^6 rad/s, or 456.77 kHz, plus sidebands displaced above and below this by 15 000 rad/s, or 2387Hz. The signal therefore occupies the band 454.39 to 459.16 kHz. The filter has its passband extending from 452 to 458 kHz and it is seen that the upper sideband will be attenuated, thus causing distortion.

This very simple example shows the importance of a frequency domain approach to a system.

6.15 The energy density spectrum

As discussed in Section 6.2 the energy dissipated when a voltage $f(t)$ is present across a resistance R is given by $\int_{-\infty}^{\infty} [f^2(t)/R]\mathrm{d}t$ such that, after normalizing it to a resistance of 1Ω, the energy is given by

$$E = \int_{-\infty}^{\infty} f^2(t)\, \mathrm{d}t \tag{6.31}$$

It is useful to have a function $E(f)$ which describes the *frequency distribution* of this energy. Since the total energy is the sum of contributions from all frequencies, we have

$$E = \int_{-\infty}^{\infty} E(f)\, \mathrm{d}f \tag{6.32}$$

To find $E(f)$ we re-express Equation (6.31) in terms of $F(f)$ and then compare the result with Equation (6.32) to determine $E(f)$ in terms of $F(f)$.

Replacing one $f(t)$ in Equation (6.31) by the inverse Fourier transform of $F(f)$ we obtain

$$E = \int_{-\infty}^{\infty} f^2(t)\, \mathrm{d}t = \int_{-\infty}^{\infty} f(t) \left[\int_{-\infty}^{\infty} F(f) \mathrm{e}^{\mathrm{j}2\pi ft}\, \mathrm{d}f \right] \mathrm{d}t$$

The order of integration can be changed for this to become

$$E = \int_{-\infty}^{\infty} F(f)\, \mathrm{d}f \int_{-\infty}^{\infty} f(t)^{\mathrm{j}2\pi ft}\, \mathrm{d}t$$

Hence

$$E = \int_{-\infty}^{\infty} F(f)F(-f)\, \mathrm{d}f \tag{6.33}$$

But, for real signals, $F(f)$ has an amplitude spectrum with even symmetry and a phase symmetry with odd symmetry, as shown in Figure 6.19.

The transform $F(-f)$ is the same as $F(f)$ except reversed in frequency, and is therefore as shown in Figure 6.20.

It follows from Figures 6.19(b) and 6.20(b) that $F(f)F(-f)$ has zero phase and therefore that Equation (6.33) becomes

$$E = \int_{-\infty}^{\infty} f^2(t)\, \mathrm{d}t = \int_{-\infty}^{\infty} |F(f)|^2\, \mathrm{d}f \tag{6.34}$$

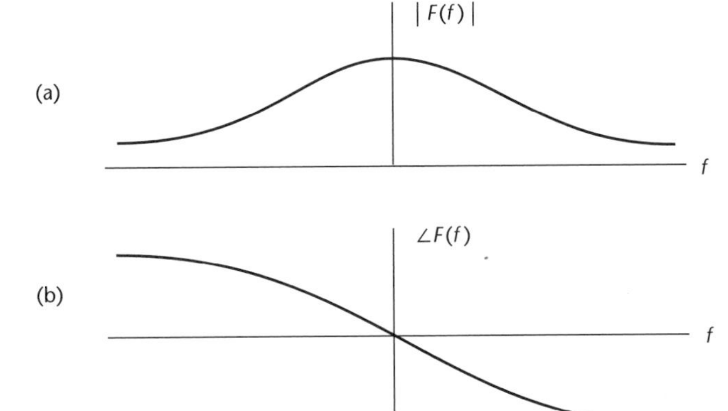

Fig. 6.19 Symmetry properties of a transform $F(f)$: (a) amplitude spectrum; (b) phase spectrum

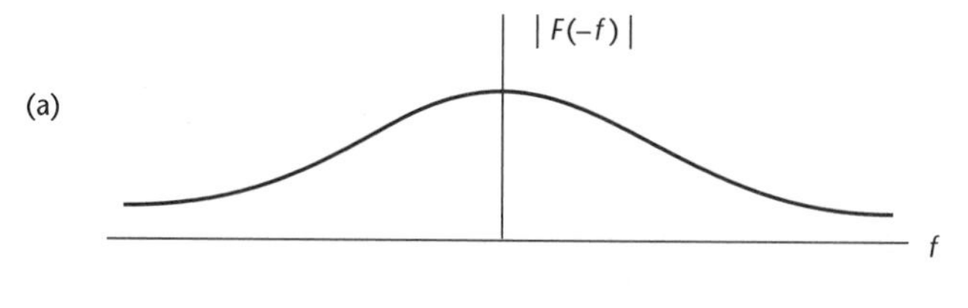

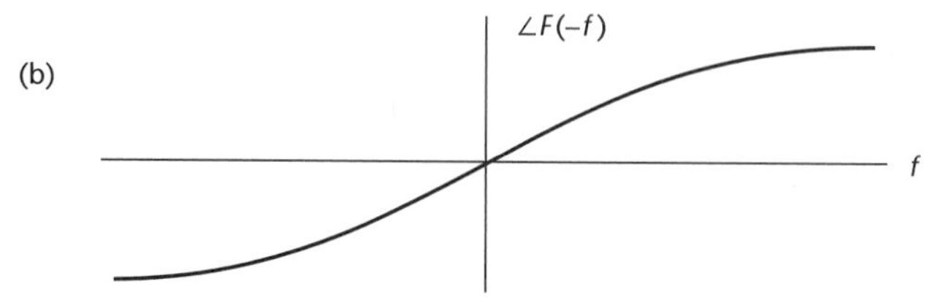

Fig. 6.20 Symmetry properties of the transform $F(-f)$: (a) amplitude spectrum; (b) phase spectrum

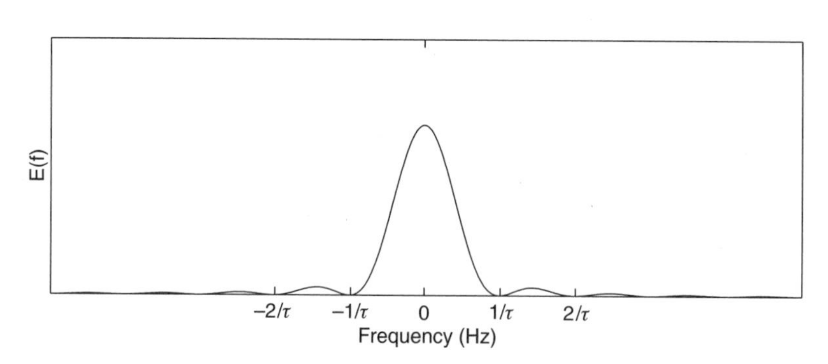

Fig. 6.21 Energy spectrum of a rectangular pulse

The equality between $\int_{-\infty}^{\infty} f^2(t)\,dt$ and $\int_{-\infty}^{\infty} |F(f)|^2\,df$ is often known as Parseval's theorem for energy signals.

Comparing Equation (6.32) with Equation (6.34) the result that we want is that the energy spectral density is given by

$$E(f) = |F(f)|^2 \tag{6.35}$$

The units of energy spectral density are J/Hz.

Example 6.5

Section 6.4 showed that the Fourier transform of a rectangular pulse of amplitude A and width τ was $A\tau[\sin(\pi f\tau)/\pi f\tau]$. What is its energy spectrum?

Solution From Equation (6.35) the energy spectrum of the rectangular pulse is $[A\tau(\sin(\pi f\tau)/\pi f\tau)]^2$. This is shown in Figure 6.21.

It is useful to extend Equation (6.35) to the case where the signal $f(t)$ is modified by a network with transfer function $H(f)$. Here the transform of the output is $F(f)H(f)$ and the energy spectral density of the output is given by

$$E(f) = |F(f)H(f)|^2 = |F(f)|^2\,|H(f)|^2 \tag{6.36}$$

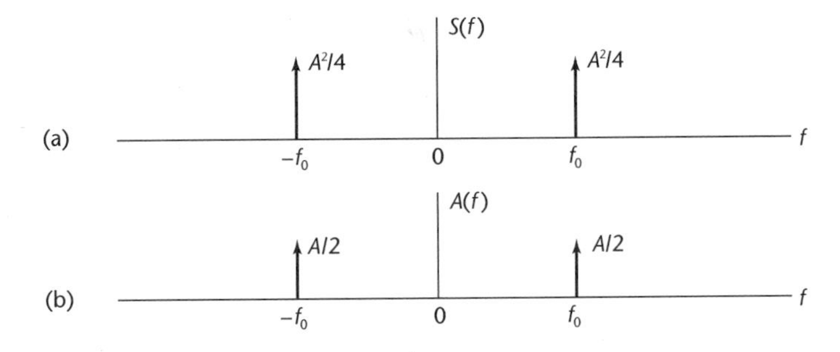

Fig. 6.22 Spectrum of a cosinusoid: (a) power spectral density; (b) amplitude spectrum

6.16 Power spectral density (PSD)

In the previous section it was found useful with finite energy signals to invoke an energy spectral density function $E(f)$ for which $\int_{-\infty}^{\infty} E(f)\, df$ was the total energy. For similar reasons it is convenient with power (infinite energy) signals to invoke a power spectral density $S(f)$ for which $\int_{-\infty}^{\infty} S(f)\, df$ is the total power.

The simplest example of a periodic signal is the cosine wave, $A\cos(\omega_0 t)$, and for this the mean power dissipated in a 1Ω resistance is $A^2/2$. In order that a two-sided power spectral density $S(f)$ should give the result that $\int_{-\infty}^{\infty} S(f)\, df = A^2/2$, it follows that $S(f)$ should consist of two impulse functions, one of strength $A^2/4$ at frequency $-f_0$ and one of strength $A^2/4$ at frequency $+f_0$, as shown in Figure 6.22(a)

For comparison, the amplitude spectrum of the consinusoid is shown in Figure 6.22(b) and it will be seen that the strengths of the impulse functions in the power spectrum are the square of the strengths of the impulse functions in the amplitude spectrum. Note that this is different from the relationship between the energy density spectrum and the Fourier transform for a finite energy signal as given by Equation (6.35) in that, because the height of an impulse function is infinite, $S(f)$ does *not* equal $|F(f)|^2$.

It will be remembered that the Fourier transform of a periodic signal may be obtained by converting the line spectrum of the exponential Fourier series into a set of impulse functions whose strengths equal the amplitudes of the lines. It follows that the power spectrum of a periodic signal can be obtained by replacing the line spectrum of the exponential Fourier series by a set of impulse functions whose strengths equal the square of the amplitudes of the lines; that is

$$S(f) = \sum_{n=-\infty}^{\infty} |F_n|^2 \delta(f - nf_0) \qquad (6.37)$$

Example 6.6

Example 6.1 considered the Fourier transform of the amplitude modulated signal $f(t) = A(1 + m\cos pt)\cos(\omega_0 t)$. Determine its power spectral density.

Solution As in Example 6.1 the signal can be expanded by means of trigonometric identities to give

$$f(t) = A\cos(\omega_0 t) + \frac{Am}{2}\cos(\omega_0 - p)t + \frac{Am}{2}\cos(\omega_0 + p)t$$

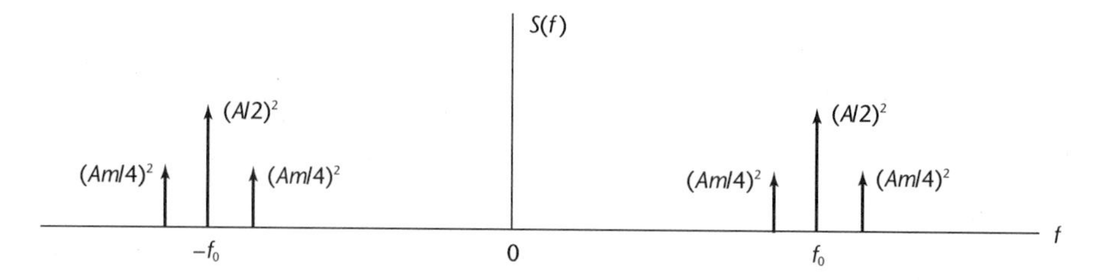

Fig. 6.23 The power spectral density of a cosinusoidally modulated AM signal

It follows from Equation (6.37) that

$$S(f) = \left(\frac{A}{2}\right)^2 \delta(f - f_0) + \left(\frac{A}{2}\right)^2 \delta(f + f_0) + \left(\frac{Am}{4}\right)^2 \delta(f - f_0 - p)$$
$$+ \left(\frac{Am}{4}\right)^2 \delta(f - f_0 + p) + \left(\frac{Am}{4}\right)^2 \delta(f + f_0 - p) + \left(\frac{Am}{4}\right)^2 \delta(f + f_0 + p)$$

The power spectral density is thus as shown in Figure 6.23.

Let us next consider a power signal that is finite over an infinite period. An example of such a signal is continuous noise. As discussed in Section 6.9 this cannot be Fourier transformed. The way we determine the spectral properties of the signal is to restrict our consideration to a long but *finite segment* of the waveform that then constitutes an *energy* signal. Let the segment be of duration T and let the waveform within this segment be called $f_T(t)$, with Fourier transform $F_T(f)$. Using Equation 6.35 the energy of this segment is $\int_{-\infty}^{\infty} |F_T(f)|^2 \, df$. The power averaged over the duration T of the segment is this energy divided by T. As $T \to \infty$ this becomes the mean power of the aperiodic power signal. Hence, if $S(f)$ is the power spectral density of the noiselike power signal,

$$\int_{-\infty}^{\infty} S(f) \, df = \lim_{T \to \infty} \frac{1}{T} \int_{-\infty}^{\infty} |F_T(f)|^2 \, df$$

or

$$S(f) = \lim_{T \to \infty} \frac{1}{T} |F_T(f)|^2 \tag{6.38}$$

If the digital techniques of Chapter 9 are used for estimating the Fourier transform of a waveform segment, this equation provides the basis for a practical way of estimating the power spectrum of a power signal.

If the signal passes through a network with transfer function $H(f)$ the output power spectrum $S_o(f)$ is related to the input power spectrum $S_i(f)$ by

$$S_o(f) = S_i(f) |H(f)|^2 \tag{6.39}$$

Example 6.7

Given a recording of automobile noise how might one determine its spectral properties?

Solution One can take a finite segment of the sound of duration T and determine the Fourier transform of this segment, $F_T(f)$, using the digital techniques to be discussed in

Chapter 9. From this $(1/T)\int_{-\infty}^{\infty}|F_T(f)|^2\,df$ can be derived and plotted, and this will indicate the spectral properties of the signal. Such a plot can reveal much about the speeds and imbalances of engine components.

If the properties of the signal are *stationary*, which means that they do not change with time, and if T is long enough to make insignificant the effect of the discontinuities at the beginning and end of the segment, then *any* segment of duration T should give a similar result. In order to smooth out any spurious effects and to thereby achieve a more reliable result it is advantageous to average several such spectra.

If the segment of a stationary signal is long enough to indicate its true spectral properties, the derived spectrum should be unchanged by any increase in the length of the segment. In practice, many signals of interest are *non-stationary*, and a typical example would be a speech signal. The technique of displaying the power spectrum here is to divide the waveform up into segments over which the spectral properties are approximately stationary and to determine the spectrum of each. The complete set may be put side by side to constitute what is known as a time-frequency plot or *spectrogram*. A particularly compact way of doing this uses blackness to denote power level, and an example of a typical result is shown in Figure 6.24.

The main problem in producing spectrograms concerns the choice of segment duration. If the segment is too long the spectral properties of the signal may change significantly during that period. If the segment is too short the segment may not be truly representative of the signal at that time.

6.17 The equivalence of convolution in one domain to multiplication in the other domain

It has already been shown in Section 6.13 that, working in the frequency domain, the Fourier transform of a system output is the input Fourier transform *multiplied* by the Fourier transform of the impulse response; that is $G(f) = F(f)H(f)$ This contrasts with the time domain result that the output of a network is the input *convolved* with

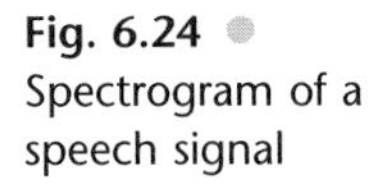
Fig. 6.24 Spectrogram of a speech signal

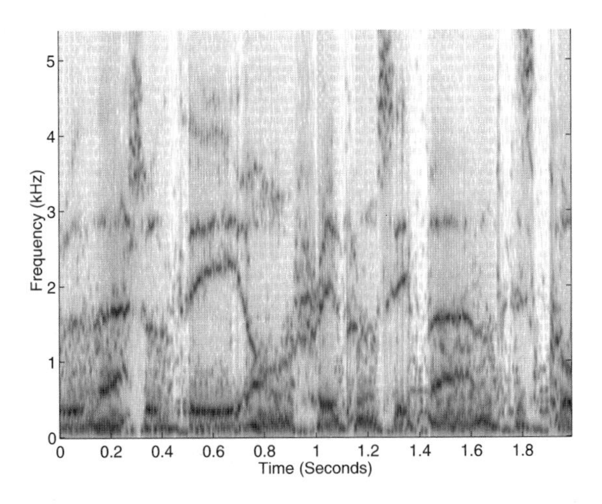

the network impulse response: $g(t) = f(t) * h(t)$, The two results can be combined to give

$$\mathscr{F}[f(t) * h(t)] = F(f)H(f) \tag{6.40}$$

and

$$\mathscr{F}^{-1}[F(f)H(f)] = f(t) * h(t) \tag{6.41}$$

These results are not confined to networks. If two waveforms $f_1(t)$ and $f_2(t)$ are convolved in some quite different application, then the analogous equations are

$$\mathscr{F}[f_1(t) * f_2(t)] = F_1(f)F_2(f) \tag{6.42}$$

and

$$\mathscr{F}^{-1}[F_1(f)F_2(f)] = f_1(t) * f_2(t) \tag{6.43}$$

These equations are sometimes described as the time-convolution theorem.

Example 6.8

Use the time convolution theorem to determine the Fourier transform and the energy density spectrum of the triangular pulse τ wide shown in Figure 6.25.

Solution A very simple way of doing this is to recognize that this triangular pulse can be obtained by convolving the two rectangular functions, each $\tau/2$ wide, that are shown in Figures 6.26.

It follows that the Fourier transform of the triangular pulse is obtained by multiplying the Fourier transforms of the two rectangle functions, which is the same as squaring the Fourier transform of one. The Fourier transform of a rectangle function $\tau/2$ wide is shown in Figure 6.27(a) and the square of it shown in Figure 6.27(b). This then is the Fourier transform of the triangular pulse.

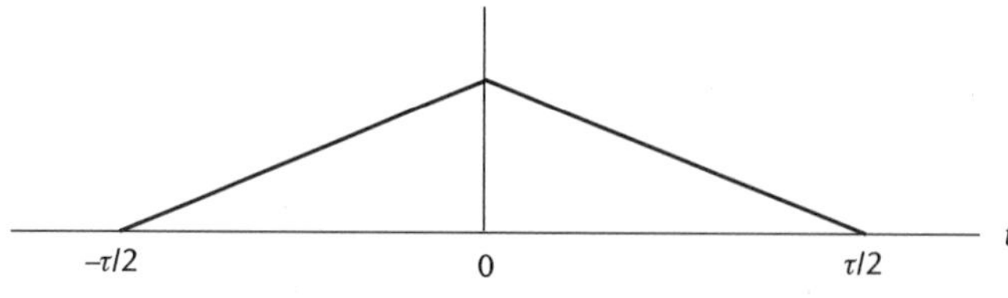

Fig. 6.25 ● Triangular pulse

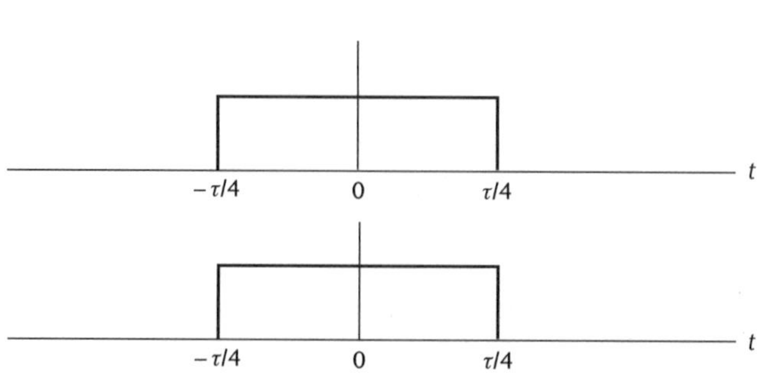

Fig. 6.26 ● Two rectangle functions that can be convolved to realize a triangular pulse

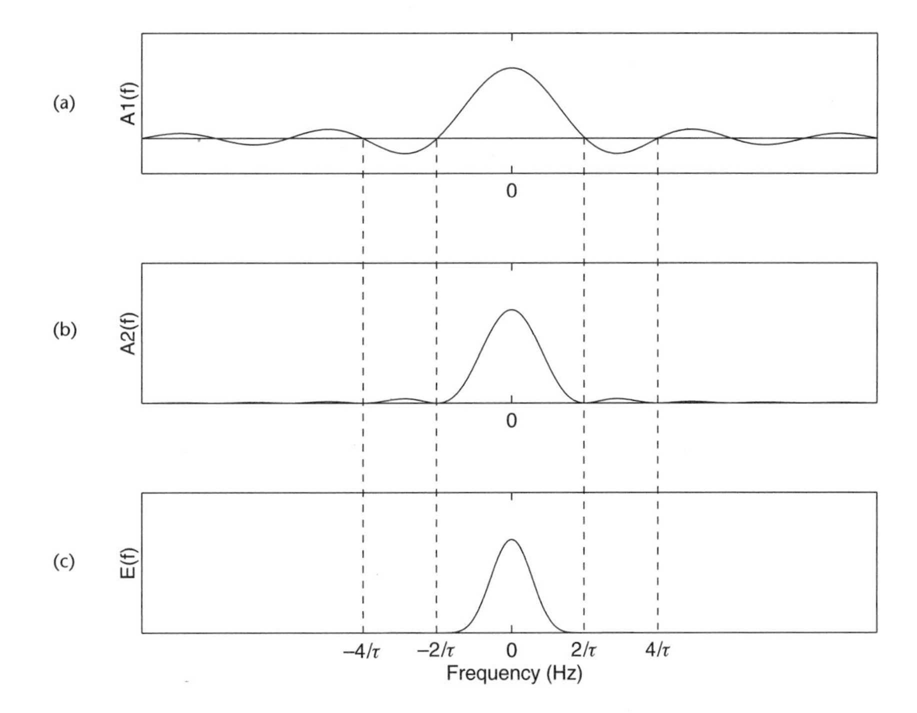

Fig. 6.27 Spectra: (a) amplitude spectrum of rectangular pulse of duration $\tau/2$; (b) amplitude spectrum of triangular pulse; (c) energy spectrum of triangular pulse

Using Equation (6.35) the energy spectrum of the triangular function is obtained by squaring the Fourier transform, and this is shown in Figure 6.27(c). It is seen to have much lower sidelobes than in the energy spectrum of the rectangular pulse derived in Example 6.5.

The Fourier transform pair given by Equations (6.16) and (6.17) show that the mathematical operation of performing a Fourier transform is almost identical to the mathematical operation of performing an inverse Fourier transform. Because of this similarity there is a dual to the result of Equation (6.42) that $\mathscr{F}[f_1(t) * f_2(t)] = F_1(f)F_2(f)$. This is that

$$\mathscr{F}^{-1}[F_1(f) * F_2(f)] = f_1(t)f_2(t) \tag{6.44}$$

This may be written alternatively as

$$\mathscr{F}[f_1(t)f_2(t)] = F_1(f) * F_2(f) \tag{6.45}$$

Equations (6.44) and (6.45) tell us that *convolution* in the *frequency* domain signifies *multiplication* in the *time* domain. This statement is sometimes described as the frequency-convolution theorem.

Example 6.9

Determine the Fourier transform, and hence the energy density spectrum, of the radar transmit pulse shown in Figure 6.28.

Solution The simplest way to do this is to recognize that this 'tone burst' is the product of a continuous cosine wave and a rectangular pulse: $r(t)\cos(2\pi f_0 t)$. Thus, using

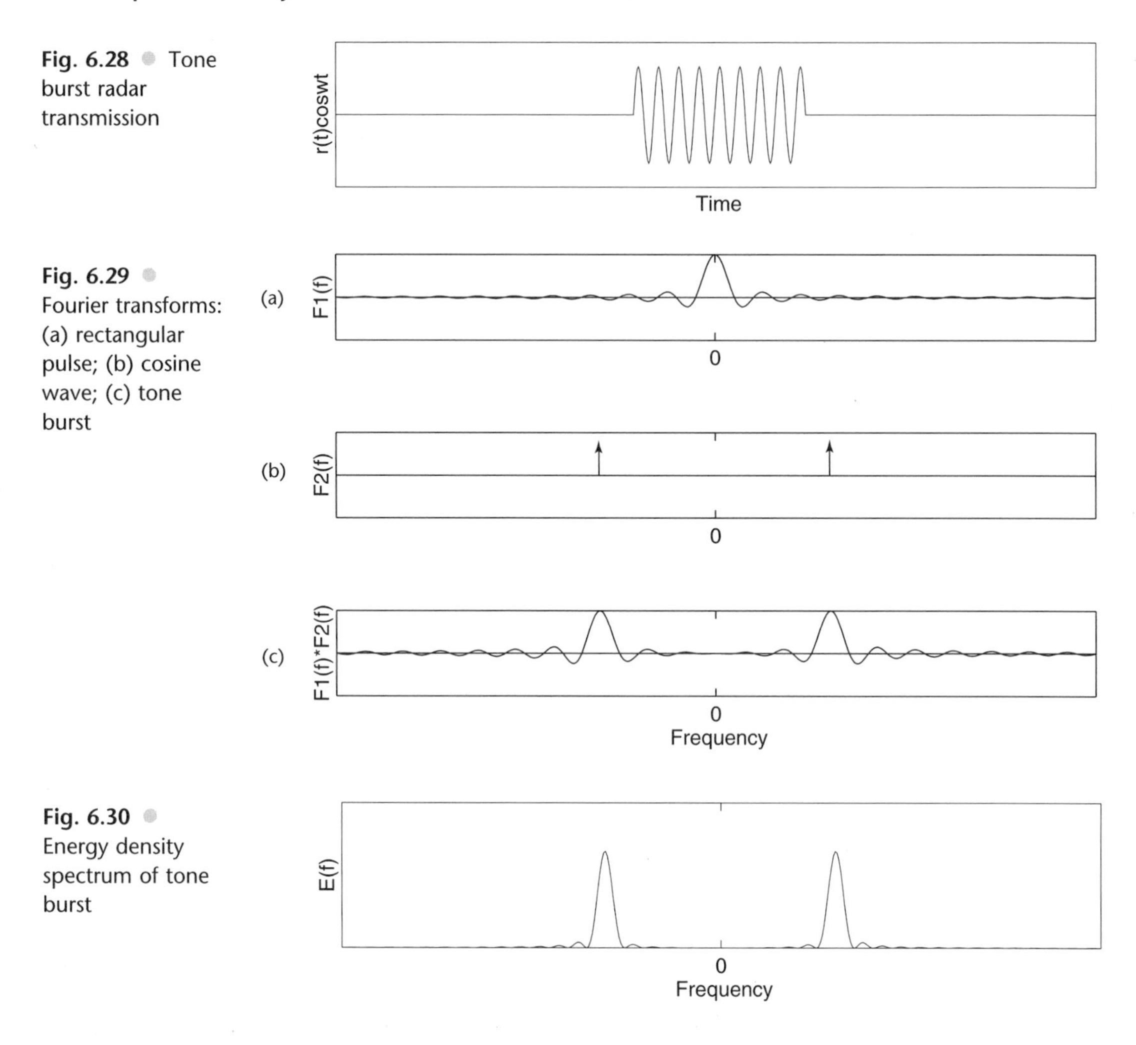

Fig. 6.28 Tone burst radar transmission

Fig. 6.29 Fourier transforms: (a) rectangular pulse; (b) cosine wave; (c) tone burst

Fig. 6.30 Energy density spectrum of tone burst

Equation (6.45), we can obtain the Fourier transform of $r(t)\cos(\pi f_0 t)$ by convolving the Fourier transform of $r(t)$ with the Fourier transform of $\cos(2\pi f_0 t)$. This is done in Figures 6.29(a), (b) and (c). The energy spectrum of $r(t)\cos(2\pi f_0 t)$ is the magnitude of the Fourier transform squared and is shown in Figure 6.30.

Some words of caution are needed regarding convolution in the frequency domain. In Chapter 2 the principles of graphical convolution were outlined and these are what are used to produce Figure 6.29(c). However, in contrast to physical time waveforms which are entirely real, Fourier transforms generally have real and imaginary components. The graphical interpretation of convolution that was portrayed in Section 2.7 is not applicable when complex quantities are involved. In general the amplitude spectrum resulting from the convolution of the two Fourier transforms is not obtainable by convolving the corresponding two *amplitude* spectra. Graphical convolution of Figures 6.29(a) and 6.29(b) has produced the correct answer in Figure 6.29(c) because one of the Fourier transforms is a set of impulses. In other situations graphical convolution can produce the wrong result. Great care is needed.

6.18 Some other properties of Fourier transforms

6.18.1 Superposition

The Fourier transform of the sum of two waveforms $f_1(t)$ and $f_2(t)$ is the sum of their separate transforms; that is

$$\mathscr{F}\{a_1 f_1(t) + a_2 f_2(t)\} = a_1 F_1(f) + a_2 F_2(f) \tag{6.46}$$

This follows directly from the integral definition of the Fourier transform. Some caution is needed, however, because it must be remembered that $F_1(f)$ and $F_2(f)$ have real and imaginary components, and that complex addition is thus required. In general the sum of the individual amplitude spectra does not give the amplitude spectrum of the sum.

6.18.2 Duality

Suppose, knowing the Fourier transform $X(f)$ of $x(t)$, we wish to determine the Fourier transform of $X(t)$. (Note that the notation used for a waveform has been changed from $f(t)$ to $x(t)$ since what follows involves a function $x(f)$ and this is less confusing than $f(f)$. In any case $x(t)$ and $X(f)$ are very commonly used notations for a signal and its Fourier transform.)

The definition of an inverse Fourier transform gives us $\int_{-\infty}^{\infty} X(f)e^{j2\pi ft}\,df = x(t)$. Changing the variable f to t such that $df = dt$, and the variable t to $-f$, this last equation becomes $\int_{-\infty}^{\infty} X(t)e^{-j2\pi ft}\,dt = x(-f)$. However the left-hand half of this equation is seen to equal the Fourier transform of $X(t)$. Therefore, if

$$\mathscr{F}[x(t)] = X(f) \tag{6.47}$$

then

$$\mathscr{F}[X(t)] = x(-f) \tag{6.48}$$

These two equations demonstrate the duality of Fourier transforms.

Example 6.10

It was shown in Section 6.3 that the Fourier transform of the rectangle time function $\text{rect}(t/\tau)$ is the sinc shaped frequency function $\tau[\sin(\pi f\tau)/\pi f\tau]$. Determine the Fourier transform of the sinc shaped function of time given by $f_a[\sin(\pi f_a \tau)/\pi f_a \tau]$.

Solution Using the notation of Equation (6.47) we have $x(t) = \text{rect}(t/\tau)$ with transform $X(f) = \tau[\sin(\pi f\tau)/\pi f\tau]$. The notation appropriate to Equation 6.48 is that $X(t) = f_a[\sin(\pi f_a t)/\pi f_a t]$. Hence, using the duality result of Equation (6.48) the Fourier transform of $X(t)$ is given by $x(f) = \text{rect}(-f/f_a)$. However, because of the symmetry of the rectangle function we can also write

$$x(f) = \text{rect}(f/f_a) \tag{6.49}$$

This duality is illustrated in Figure 6.31.

It will be noted that this example is closely related to Example 6.3, where a brickwall filter of bandwidth B was shown to have a sinc function impulse response.

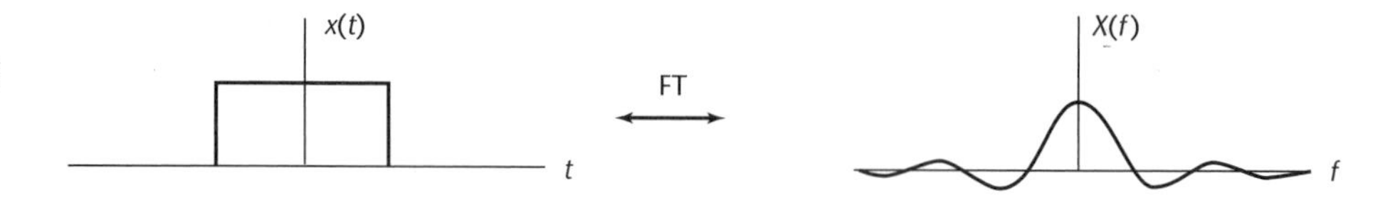

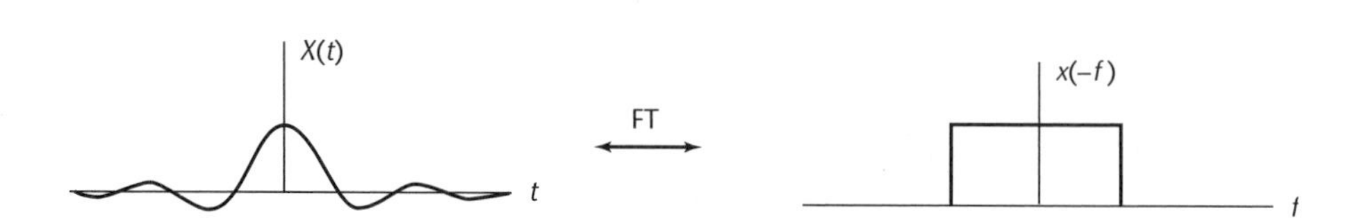

Fig. 6.31 Demonstration of duality in Fourier transforms

Here a sinc function signal (not necessarily an impulse response) is shown to have a rectangle spectral density function. A close comparison shows that B in Example 6.3 is equivalent to $f_a/2$ in this present example.

It should be noted that a particular energy spectrum is not unique to a specific pulse. There are many waveforms with the same amplitude spectrum (and hence the same energy spectrum) but having different phase spectra.

6.18.3 Compression and expansion

If the Fourier transform of $f(t)$ is $F(f)$ then the Fourier transform of a function that is the same as $f(t)$ except compressed in time by a factor a is a function that is similar to $F(f)$ except expanded in frequency by a and divided in amplitude by a. This is proved as follows. We have

$$\mathscr{F}\,[f(at)] = \int_{-\infty}^{\infty} f(at)\mathrm{e}^{-\mathrm{j}2\pi ft}\,\mathrm{d}t$$

Making the change of variable $x = at$, this becomes

$$\mathscr{F}\,[f(at)] = \frac{1}{a}\int_{-\infty}^{\infty} f(x)\mathrm{e}^{-\mathrm{j}2\pi fx/a}\,\mathrm{d}x = \frac{1}{a}\,F(f/a) \tag{6.50}$$

Example 6.11

If a pulse is 'stretched' by a factor of two while retaining its shape, what happens to its bandwidth?

Solution From Equation (6.50) the bandwidth of the pulse is reduced by a factor of two.

6.18.4 The effect of a time shift on the Fourier transform

If the Fourier transform of $f(t)$ is $F(f)$ we now examine the Fourier transform of

$f(t - t_0)$. Using the definition of the Fourier transform we have

$$\mathscr{F}[f(t - t_0)] = \int_{-\infty}^{\infty} f(t - t_0)e^{-j2\pi ft}\, dt$$

Putting $t - t_0 = x$ such that $dt = dx$

$$\mathscr{F}[f(t - t_0)] = \int_{-\infty}^{\infty} f(x)e^{-j2\pi f(x + t_o)}\, dx = e^{-j2\pi ft_0}F(f)$$

In other words, if the Fourier transform of $f(t)$ is $A(f)e^{j\theta(f)}$, the Fourier transform of $f(t - t_0)$ is

$$\mathscr{F}[f(t - t_0)] = A(f)e^{j\{\theta(f) - 2\pi ft_0\}} \quad (6.51)$$

Thus the effect of delaying a waveform by t_0 is to introduce into the phase spectrum a modification of $-2\pi ft_0$ (or $-\omega t_0$) that decreases linearly with frequency.

Example 6.12

The Fourier transform of a rectangular pulse of amplitude A and width τ has been shown to equal $A\tau[\sin(\pi f\tau)/\pi f\tau]$ when it is centred at $t = 0$. What is its Fourier transform when its leading edge is positioned at $t = 0$, as shown in Figure 6.32?

Solution Compared with the rectangular pulse centred at zero there is a delay of $\tau/2$. Therefore, from Equation (6.51), the new Fourier transform is $A\tau[\sin(\pi f\tau)/\pi f\tau]\, e^{-j\pi f\tau}$. The amplitude spectrum is unchanged but the phase spectrum is no longer zero. The amplitude and phase spectra are shown in Figure 6.33.

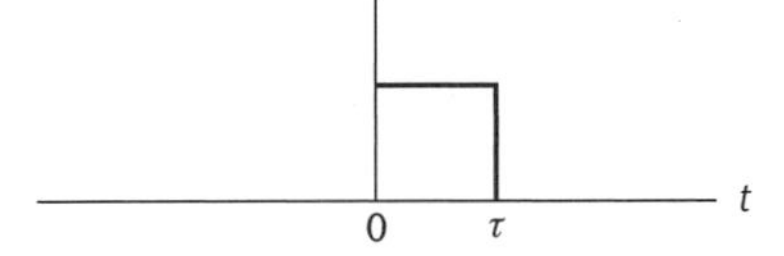

Fig. 6.32 Asymmetrical delayed rectangular pulse

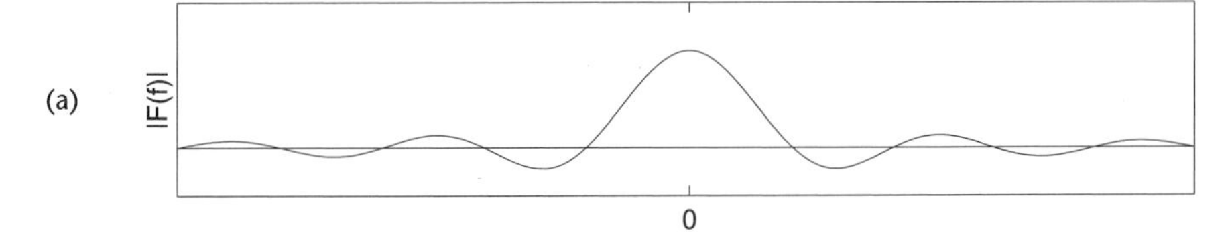

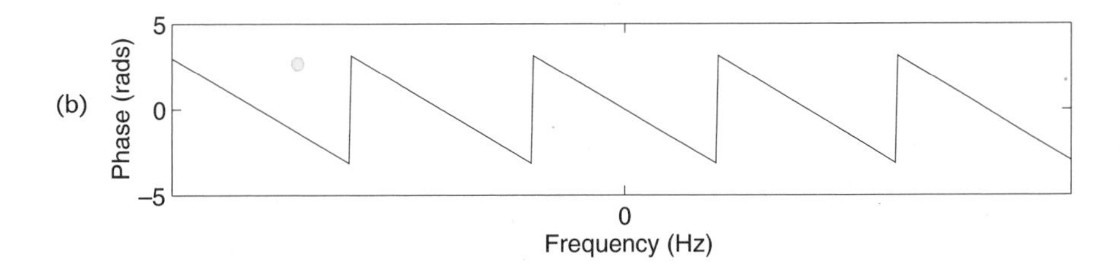

Fig. 6.33 Fourier transform of an offset rectangular pulse: (a) amplitude spectrum; (b) phase spectrum

6.18.5 The effect of a frequency shift on the inverse Fourier transform

The dual to the time shift relationship is that, if $F(f)$ is shifted up in frequency by a constant f_0, its inverse transform is multiplied by $e^{j2\pi f_0 t}$. The converse is that, if a time function is multiplied by $e^{j2\pi f_0 t}$, its transform is shifted by f_0. The proof of this converse is easier and is as follows. From the definition of the Fourier transform

$$\begin{aligned}\mathscr{F}\left[f(t)e^{j2\pi f_0 t}\right] &= \int_{-\infty}^{\infty} f(t)e^{j2\pi f_0 t}e^{-j2\pi ft}\,dt \\ &= \int_{-\infty}^{\infty} f(t)e^{-j2\pi(f-f_0)t}\,dt\end{aligned}$$

It follows directly that

$$\mathscr{F}\left[f(t)e^{j2\pi f_0 t}\right] = F(f-f_0) \tag{6.52}$$

Example 6.13

Determine the Fourier transform of $f(t)\cos(\omega_0 t)$ in terms of the Fourier transform of $f(t)$.

Solution The signal $f(t)\cos\omega_0 t$ may be re-expressed as $f(t)\left[\frac{1}{2}e^{j\omega_0 t} + \frac{1}{2}e^{-j\omega_0 t}\right]$. If the Fourier transform of $f(t)$ is $F(f)$, it follows from Equation (6.52) that the Fourier transform of $f(t)\cos(\omega_0 t)$ is $\frac{1}{2}F(f+f_0) + \frac{1}{2}F(f-f_0)$.

This provides a method for determining the Fourier transform of a radar tone burst that is an alternative to that used in Example 6.9. The multiplication of $f(t)$ by $\cos(\omega_0 t)$ causes the Fourier transform of $f(t)$ to be divided by two and shifted up and down in frequency by f_0.

6.18.6 Differentiation and integration

The Fourier transform of a differentiated finite-energy time waveform is the Fourier transform of the time waveform before differentiation multiplied by $j2\pi f$. For example, if $f(t) \leftrightarrow F(f)$, then

$$\mathscr{F}\left[\frac{d}{dt}f(t)\right] = j2\pi fF(f) \tag{6.53}$$

The easiest way to prove that Equation (6.53) achieves the differentiation of $f(t)$ is by differentiating the inverse Fourier transform of $F(f)$. We have

$$f(t) = \int_{-\infty}^{\infty} F(f)e^{j2\pi ft}\,df$$

and therefore

$$\begin{aligned}\frac{d}{dt}f(t) &= \frac{d}{dt}\int_{-\infty}^{\infty} F(f)e^{j2\pi ft}\,df \\ &= \int_{-\infty}^{\infty} \frac{\partial}{\partial t}\{F(f)e^{j2\pi ft}\}\,df \\ &= \int_{-\infty}^{\infty} j2\pi fF(f)e^{j2\pi ft}\,df\end{aligned}$$

But the RHS of this equation is the inverse Fourier transform of $j2\pi fF(f)$. Equation (6.53) follows directly. The consequence of the multiplying factor of $j2\pi f$ in Equation (6.53) is that the effect of differentiating a signal is to enhance its high-frequency content.

The corresponding integration property is

$$\mathscr{F}\left\{\int_{-\infty}^{T} f(\tau)\,\mathrm{d}\tau\right\} = \frac{1}{j2\pi f}F(f) + \frac{F(0)\delta(f)}{2} \tag{6.54}$$

where

$$F(0) = \int_{-\infty}^{\infty} f(t)\,\mathrm{d}t \tag{6.55}$$

The significance of the $1/f$ factor in Equation (6.54) is that the effect of integrating a signal is to decrease its high frequency content.

6.19 A selection of some important Fourier transforms

In what follows an important new function is introduced. This is the signum function for which the standard symbol is $\mathrm{sgn}(t)$. This is defined as

$$\mathrm{sgn}(t) = \begin{cases} -1 & t<0 \\ 1 & t>0 \end{cases} \tag{6.56}$$

The transforms of this and other important functions are given in Table 6.1. The last of these is an interesting one because it signifies the same shape in both the time and frequency domains. This shape is known as the normal or Gaussian distribution and an important parameter of such a distribution is its standard deviation, defined as the displacement for which the function is reduced to $e^{-\frac{1}{2}}$ of its maximum value. For $\exp(-a^2t^2)$ this occurs when $t = 1/a\sqrt{2}$ and for $(\sqrt{\pi}/a)\,e^{-(\pi f/a)^2}$ it occurs when $f = a/\pi\sqrt{2}$.

Table 6.1 Some important Fourier transforms

$f(t)$	$F(f)$
$\delta(t)$	1
1	$\delta(f)$
$\cos(2\pi f_0 t + \theta)$	$\frac{1}{2}e^{j\theta}\,\delta(f - f_0) + \frac{1}{2}e^{-j\theta}\,\delta(f + f_0)$
$\mathrm{rect}(t/\tau)$	$\tau\,\mathrm{sinc}(\tau f)$
$\mathrm{sinc}(Bt)$	$\frac{1}{\lvert B\rvert}\,\mathrm{rect}(f/B)$
$\mathrm{sgn}(t)$	$1/j\pi f$
$u(t)$	$\frac{1}{2}\delta(f) + 1/j2\pi f$
$e^{-at}u(t) \quad a>0$	$1/(a + j2\pi f)$
$\sum_{-\infty}^{\infty}\delta(t - nT)$	$\frac{1}{T}\sum_{-\infty}^{\infty}\delta(f - n/T)$
$\exp(-a^2t^2)$	$\frac{\sqrt{\pi}}{a}\,e^{-(\pi f/a)^2}$

6.20 ● Summary

The Fourier transform of a time domain signal conveys the complete information of the signal in the frequency domain. It is termed a transform because an inverse operation enables the original time domain signal to be regained. The frequency domain representation consists of an amplitude spectrum and a phase spectrum. If the time domain signal is in volts the units of the amplitude spectrum are V/Hz and and those of the phase spectrum are radians.

Useful but incomplete measures of the spectrum are the energy density spectrum (applicable in the case of signals with finite energy) or the power spectral density (applicable in the case of signals with infinite energy; that is, power signals). These measures are useful because they convey the frequency content of the signal, and these frequently provide insight into the nature of the signal. They are incomplete measures because they have discarded phase information and the original time domain signal cannot be regained from them.

6.21 ● Problems

1. Sketch the one-sided line spectrum, the two-sided line spectrum, and the density spectrum of the signal $A \cos \omega_0 t$.

2. The signal $\cos(10t) + \cos(11t)$ passes through a square law device. Determine the Fourier transform of the output:

(a) by applying trigonometric identities to $[\cos(10t) + \cos(11t)]^2$;

(b) by applying the frequency convolution theorem.

3. Fourier series analysis can be used to find the spectral content of the full-wave rectified cosinusoid $|\cos \omega_0 t|$ (see for example Problem 5.3). Apply the alternative (and probably easier) technique of considering $|\cos \omega_0 t|$ to be the product of $\cos \omega_0 t$ and a square wave of the same frequency in order to produce an approximate sketch of the spectrum.

4. Sketch the Fourier transform of a signal $\cos(pt)$. Sketch the Fourier transform of a signal $\cos(\omega_c t)$ where $\omega_c = 10p$. Using the frequency convolution theorem, sketch the spectrum of $\cos(pt)\cos(\omega_c t)$.

5. A sonar pulse is obtained by gating on the output of a 10 kHz free running oscillator for two cycles such that the waveform is given by

$$f(t) = \text{rect}(t/T)\cos(2\pi f_0 t) = \text{rect}(f_0 t/2)\cos(2\pi f_0 t)$$

Sketch $f(t)$ Determine the spectrum of $\text{rect}(f_0 t/2)$. Apply the frequency convolution theorem to sketch the amplitude spectrum of the sonar pulse. Repeat for the case where the transmit pulse is given by $\text{rect}(f_0 t/2)\sin(2\pi f_0 t)$. Highlight the main differences between the two amplitude spectra.

6. An on–off keying signal used for a digital communication link is obtained by using the digital sequence to turn a 1 MHz carrier wave on and off. If the spectral content of the digital sequence is contained within the bandwidth 0 to 10 kHz determine the bandwidth of the on–off keying signal.

The same digital sequence is used to generate a phase shift keying signal in which $\cos \omega t$ is transmitted when there is a 1 and $-\cos \omega t$ when there is a 0. Determine the bandwidth of the phase shift keying signal.

7. A frequency shift keying signal used for a digital communication link is obtained by transmitting a frequency f_1 when there is a 1 and a frequency f_2 when there is a 0. The detection scheme separates the two frequencies by passing the receive signal through two bandpass filters, one centred at f_1 and the other at f_2, and decides whether a 1 or a 0 has been transmitted by determining which of the filter outputs is the largest. By considering a data stream that is alternate 1s and 0s, where the resulting frequency shift keying signal is the superposition of two on–off keying signals, determine the approximate ratio of $(f_2 - f_1)$ to bit rate that is necessary for the spectrum to be separable in this way.

8. A comb filter has the frequency transfer function $H(f) = 1 + \cos(2\pi f/f_0)$. Determine its impulse response and show that it is not physically realizable. Find the transfer function of a filter having the same amplitude response that is physically realizable and show how the filter may be constructed, assuming ideal delay elements are available.

Another comb filter has the transfer function

$$H(f) = 1 + \frac{1}{\pi}\left[\cos(2\pi f/f_0) - \tfrac{1}{3}\cos(6\pi f/f_0) + \tfrac{1}{5}\cos(10\pi f/f_0)\right]$$

Sketch this amplitude response and determine the impulse response of a physically realizable filter that achieves it. Show how the filter may be constructed using ideal delay elements.

9. A filter has an amplitude response given by $|H(f)| = \exp(-0.001f^2)$. Determine the impulse response if the phase response is zero. Determine the impulse response if the phase response is given by $\theta(f) = kf$. Determine the value of k such that 99% of the area of the impulse response occurs for $t < 0$. Compare this with the initial gradient of the phase response of an RC network having the same 3 dB bandwidth.

10. For an even function the expression for a Fourier transform simplifies to $F(f) = \int_{-\infty}^{\infty} f(t)\cos(2\pi ft)\,dt$. If $f(t)$ is a rectangular pulse of duration T sketch the waveform $f(t)\cos(2\pi ft)$ for different values of f and hence determine the values of f for which the Fourier transform is zero. Confirm that the result agrees with the well-known analytical result for the Fourier transform of $\mathrm{rect}(t/T)$.

Sketch the waveform $\mathrm{rect}(t/T)\cos(2\pi ft)$ for $f = 0,\ f = 3/(2T),\ f = 5/(2T)$ and $f = 7/(2T)$. Based upon these sketches deduce the values of the Fourier transform at these frequencies.

11. A Hilbert transform causes the phases of all frequency components of a signal to be retarded by 90° without altering their amplitudes. Thus its transfer function is $H(f) = -\mathrm{j}\,\mathrm{sgn}(f)$. Determine the corresponding impulse response (Hint: duality may be useful).

Hilbert transformers have applications in the hardware generation of single sideband communication signals. Comment on an obvious fundamental difficulty in realizing one and suggest how it might be largely overcome.

12. Fourier transforms are relevant to signals other than those which are functions of time. For example it can be shown that in sonar or radar the directional response of a line array is given by

$$D(\theta) = \int_{-\infty}^{\infty} T(x)\mathrm{e}^{-\mathrm{j}kx\sin\theta}\,dx$$

where $k = 2\pi/\lambda$, λ is the wavelength of the incoming signal, whose angle of incidence on the array is θ, and $T(x)$ is the sensitivity of the array as a function of position x on the array. For an array of length L with uniform sensitivity we have $T(x) = \mathrm{rect}(x/L)$. The equation for $D(\theta)$ has a strong similarity with the equation for the Fourier transform of a time waveform, namely $F(\omega) = \int_{-\infty}^{\infty} f(t)\mathrm{e}^{-\mathrm{j}\omega t}\,dt$. Use the result for the Fourier transform of a rectangular pulse to determine the directional response of a uniform array of length L.

Suggest how one might obtain a directional response that is uniform over a sector θ_0 and zero outside: $D(\theta) = \mathrm{rect}(\theta/\theta_0)$.

Chapter seven

The sampling and digitization of signals

7.1 ● Preview

It is common to process an analogue signal digitally, and this involves sampling and coding the signal with an analogue to digital converter (ADC) and then finally, after processing, decoding it using a digital to analogue converter (DAC). A brief review is given of coding types and of some of the most important types of converters.

The chapter then examines what sampling rates are needed to avoid any loss of information from the signal and discusses why anti-aliasing filters are commonly included to truncate the spectrum of the original signal. The quantization effects of coding are also discussed and the degradation due to this quantization is estimated.

7.2 Overview of sampling and digitization

Most signals that require processing are analogue in nature. However, the power and flexibility of computers, microprocessors and special purpose chips have made the *digital* processing of these signals become increasingly attractive. A typical real-time digital processing system is shown in Figure 7.1. The initial analogue to digital conversion, and the return from a digital signal back to an analogue signal after

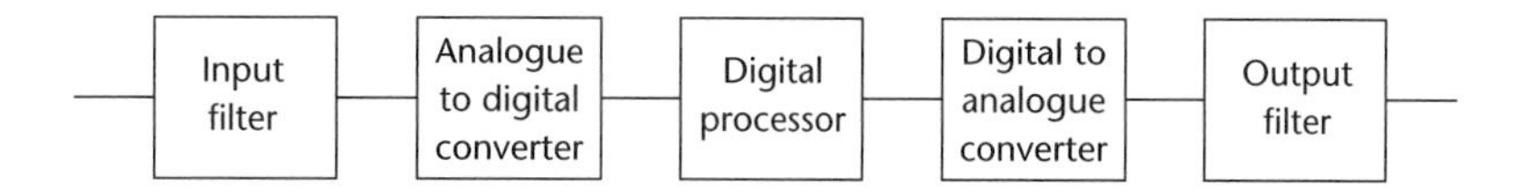

Fig. 7.1 Block diagram of a digital processing system

processing, are essential ingredients of the system. The purpose of the input filter is to avoid aliasing effects, as will be discussed in Section 7.7.

The analogue to digital conversion, although usually achieved within a single device, may be thought of as comprised of several processes:

1. The sampling of the signal at regular intervals.
2. The use of a *sample and hold* circuit prior to the encoder. This is done because a finite time is needed to convert a sample amplitude into an appropriate code and it is better to hold sample amplitudes at a constant level between samples while conversion to the code takes place (see Figure 7.3), rather than having an input to the encoder that is changing. The sample and hold circuit is frequently an integral part of the ADC, but may be separate.
3. The replacement of each sample by a binary code that conveys its amplitude. For example a 16 bit analogue to digital converter (ADC) would generate 2^{16}, or 65 536, binary codes starting with 0000000000000000, 0000000000000001, 0000000000000010 and extending up to 1111111111111111. A more modest 4 bit ADC would generate 2^4, or 16, binary codes, starting with 0000, 0001, 0010, 0011 and extending up to 1111. These cover the decimal equivalents of 0 to 15. A simple use of such a set of codes would be to make them correspond to 16 bands, q apart, starting with the code 0000 for Band 0 extending from 0 to q, and ending with the code 1111 for Band 15 extending from $15q$ to $16q$. If the signal level lies in a particular band the corresponding code would be generated. An alternative way of looking at this is to regard the binary codes as corresponding to the *quantization levels* $0.5q$, $1.5q$, $2.5q$, up to $15.5q$. An example of these concepts is presented in Figure 7.2, where three successive samples lying in Bands 3, 8 and 11

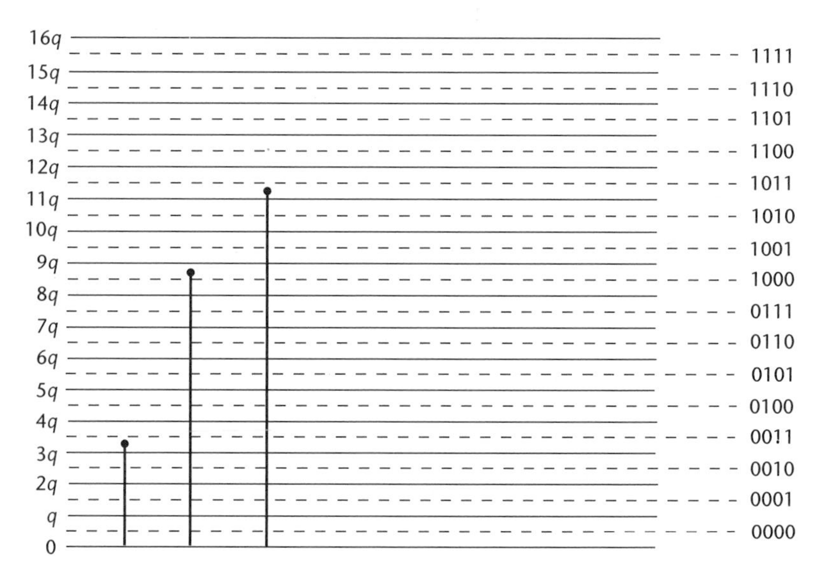

Fig. 7.2 The concept of quantization bands and quantization levels

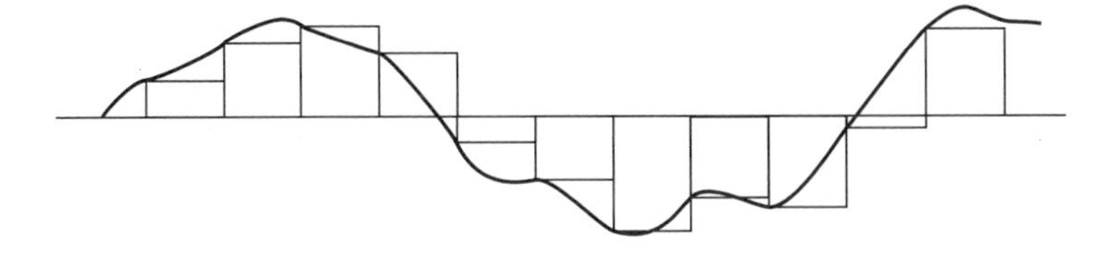

Fig. 7.3 ● The effect of a sample and hold

are coded as 0011, 1000 and 1011 respectively. (It will be noted that, since the numbering system begins with a Band 0, Bands 3, 8 and 11 actually correspond to the 4th, 8th and 12th bands.)

The digital to analogue converter (DAC) converts digital codes back into analogue sample levels. In many DACs each output analogue sample level is *held* at a constant level by means of a capacitance until the conversion of the next input word is complete, and this produces a staircase-type output waveform similar to that shown in Figure 7.3, which was for the output of the sample and hold circuit preceding the ADC. The only difference between the two is that the amplitudes of the steps are now *quantized*, corresponding, in the case of a 4 bit system, to the levels $0.5q$, $1.5q$, $2.5q$, up to $15.5q$. Finally, this quantized staircase waveform is smoothed using a lowpass reconstruction filter to produce an analogue output waveform. The process of holding amplitudes between samples is termed that of a *zero-order hold* (ZOH). It is essentially synonymous with the term *sample and hold*, the only difference being that *sample and hold* implies that a sampling operation is involved whereas *zero-order hold* implies an operation on existing samples. Two important reasons for holding the decoder output levels between samples are:

- Short samples contain little energy and would mean that the output voltage would be very small after smoothing by the reconstruction filter. Held samples have much more energy.
- The holding of the samples helps the smoothing process.

With many DACs a practical alternative to holding the output samples is to hold the *input* digital word until the next digital word arrives, using an internal or external *latch*. This approach is very common, but, because of the finite time taken by the decoding process, does have the problem that glitches can be generated when the input word changes. An important figure of merit of a DAC is that this conversion time, or *settling* time, should be very short.

The precision with which an analogue sample can be conveyed by a digital code depends upon the number of possible codes available. An 8 bit coder has 2^8 (or 256) possible codes, a 16 bit coder has 2^{16} (or 65 536) possible codes, and an n bit coder has 2^n possible codes. Each code corresponds to a specific quantization level, and each sample has to be approximated by the nearest of these quantization levels. The difference between the analogue sample level and the quantization level by which it is approximated is known as the quantization error. Commonly available ADCs deliver between 8 and 16 bits, thus providing between 256 and 65 536 different quantization levels, and thus the opportunity of small quantization errors. For specialist applications 32 or 64 bits might be used. In general it is advisable, on grounds of cost and efficiency, to use the fewest bits and the lowest sampling rate that are needed for a particular application. For this reason it is important to

Fig. 7.4 Use of a parallel to serial conversion for a digital communication link

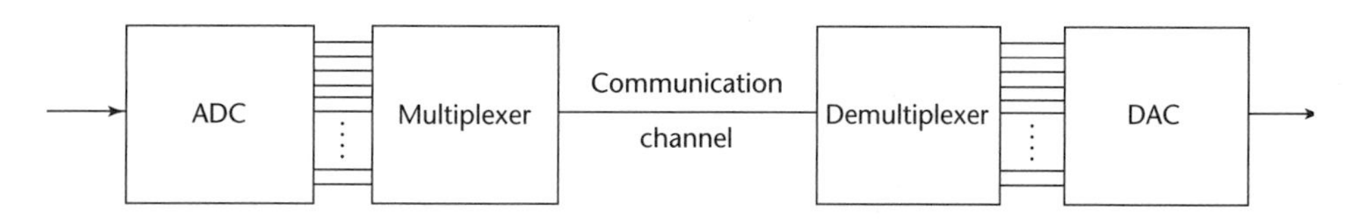

understand the degradation introduced into the signal by quantization and sampling. This will be dealt with in subsequent sections.

Most ADC chips have separate pins for each bit and produce a *parallel* output. The digital signal processor (DSP) works directly on these and, in turn, produces a parallel bit pattern to be decoded by the digital to analogue converter (DAC). For some applications, however, it is necessary to perform a parallel to serial conversion after the ADC by means of a multiplexer, and conversely a serial to parallel conversion before the DAC. An example of where this approach would be needed is in a two-line or radio communication link, such as shown in Figure 7.4.

The great strength of digital communication links compared with analogue communication links is that the code can be 'cleaned up' at each of a large number of repeaters. Suppose, for example, that one particular sample of a speech waveform is encoded as 01011100 and is transmitted as the sequence of voltage levels −5, 5, −5, 5, 5, 5, −5 and −5 volts, where −5 volts is used to convey a digital zero and +5 volts to convey a digital one. Due to attenuation and noise the sequence entering a repeater might be the voltage levels −0.52, 0.64, −0.43, 0.71, 0.48, 0.55, −0.66, −0.49. The repeater can then impose a threshold of zero volts and designate any voltage larger than zero as signifying a digital 1, and any voltage less than zero as signifying a digital 0. In this way a new sequence of voltage levels −5, 5, −5, 5, 5, 5, −5 and −5 volts can be generated and sent on its way to the next repeater. In spite of the changes to the received sequence due to attenuation and noise this regenerated sequence is exactly the same the original sequence. So long as the repeaters are close enough that errors are extremely rare (perhaps 1 in 10^9) the degradation due to attenuation and noise after a large number of repeaters is negligible. This is in contrast to analogue systems where the noise entering a repeater is amplified along with the wanted signal, and thus accumulates over a long link.

7.3 Binary codes

For unipolar signals the natural binary code is generally used, whereby 0000 corresponds to decimal zero, 0001 corresponds to decimal one, 0010 corresponds to decimal 2, 0011 to decimal 3 and so on. Generalizing this, the n bit code $b_{n-1}, b_{n-2}, \ldots, b_2, b_1, b_0$ corresponds to the decimal number

$$2^{n-1}b_{n-1} + 2^{n-2}b_{n-2} + \cdots + 2^2b_2 + 2^1b_1 + 2^0b_0$$

Because it has the greatest effect on the decimal number, the first term of the code, b_{n-1}, is known as the most significant bit (MSB) and the last term of the code, b_0, is known as the least significant bit (LSB).

For bipolar signals we have the problem of how to convey the polarity as well as the magnitude, and the three most common solutions are:

1. To apply a known positive bias to the signal such that it becomes always positive, and then to use natural binary code. This is known as offset binary code. Considering as an example a signal having quantized values that correspond to the integers between −8 and +7, the number −8 could be offset by 8 to become 0 and be represented by the four bit code 0000, while 0 becomes 8 to be represented by 1000, and 7 becomes 15 to be represented by the code 1111. Following processing this d.c. offset can subsequently be removed in the DAC.
2. To adopt *sign magnitude* code, in which the left-hand bit is used to convey polarity and the rest to convey magnitude. Thus, the most significant bit to represent the magnitude is now the second bit. The usual convention for the sign bit is a 0 for positive and a 1 for negative. Thus the four bit code 0 101 would correspond to decimal 5, whereas 1 101 would correspond to −5. These codes have been written here with a small gap between the first two bits purely to emphasize that the first bit conveys polarity. One major problem with sign-magnitude code is that there are two different representations for the number zero.
3. To adopt *two's complement* code. For positive numbers this is identical to sign magnitude code. Considering a negative number, one takes the code for the corresponding positive number, complements each digit, and then adds 1. Thus +5 is represented by 0 101, whereas −5 is given by (1 010 + 1), or 1 011. As with sign magnitude notation a small gap has been inserted between the first two bits to emphasize that the first bit conveys polarity, but it should be noted that the remaining bits are different from when sign-magnitude notation is used. The advantage of two's complement notation is that the addition of two numbers of opposite polarity is achieved by direct addition of their two codes. Considering (+5 − 5), for example, we have (0 101 + 1 011) or 10 000. Discarding the extra left-hand bit gives 0 000 and hence zero as required. As another example the code 0 110 represents +6. Its complement is 1 001. Thus, after adding one to obtain 1 010, we have 1 010 as the two's complement representation of −6. Then (+5 − 6) leads to (0 101 + 1 010) = 1 111. Since the first digit is a 1 it is apparent that we are concerned with a negative number, and this means that the inverse operation of two's complementing must be applied: subtract one and then complement. Subtracting one from 1 111 gives 1 110, and complementing each digit gives 0 001. Thus, remembering that we are dealing with a negative number, this corresponds to the decimal number −1, as required.

7.4 Analogue to digital and digital to analogue converters

7.4.1 The 'flash' ADC

The parallel-comparator, or 'flash', analogue to digital converter is capable of achieving higher sampling rates than other types of ADC and yet is one of the simplest techniques conceptually. The operation of a unipolar 3-bit flash ADC is illustrated by Figure 7.5. In Figure 7.5 a reference voltage, V, is applied to a resistor chain to produce the seven voltages $7V/8$, $6V/8$, $5V/8$, $4V/8$, $3V/8$, $2V/8$ and $V/8$.

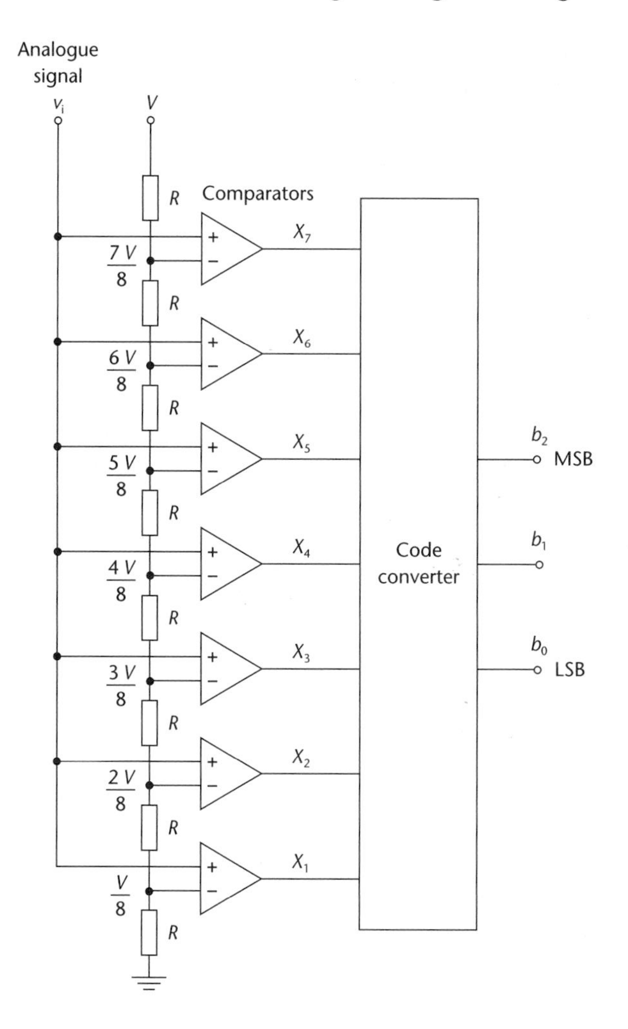

Fig. 7.5 Block diagram of a unipolar 3 bit flash ADC

Each of these is compared with the input signal voltage by means of a comparator, thus producing 1s or 0s at each of the comparator outputs, X_7, X_6, X_5, X_4, X_3, X_2 and X_1. These outputs vary from all 0s if the input voltage is 0 volts, up to all 1s if the input voltage is V volts. The seven bit parallel sequence then enters a combinational digital circuit that converts them into three output bits b_2, b_1 and b_0, in accordance with the truth table given in Table 7.1. The successive rows correspond to increasing input levels.

It is seen that all eight possible 3 bit codes are produced and correspond correctly with the eight input quantization bands. The code converter changes the so-called *temperature code* at the output leaving the comparators into the required binary code which, in this case, is natural binary code.

For bipolar signals it is merely necessary to extend the resistor chain down to a negative reference voltage of $-V$, as shown in Figure 7.6. This time the code converter would have a different truth table, one that accords with the required bipolar code; probably offset binary code, although sign magnitude or two's complement code are equally possible.

Table 7.1 ● Truth table for converting codes

Input	X_7	X_6	X_5	X_4	X_3	X_2	X_1	b_2	b_1	b_0
$0 \leqslant v_i < V/8$	0	0	0	0	0	0	0	0	0	0
$V/8 \leqslant v_i < 2V/8$	0	0	0	0	0	0	1	0	0	1
$2V/8 \leqslant v_i < 3V/8$	0	0	0	0	0	1	1	0	1	0
$3V/8 \leqslant v_i < 4V/8$	0	0	0	0	1	1	1	0	1	1
$4V/8 \leqslant v_i < 5V/8$	0	0	0	1	1	1	1	1	0	0
$5V/8 \leqslant v_i < 6V/8$	0	0	1	1	1	1	1	1	0	1
$6V/8 \leqslant v_i < 7V/8$	0	1	1	1	1	1	1	1	1	0
$7V/8 \leqslant v_i < V$	1	1	1	1	1	1	1	1	1	1

Flash ADCs have the advantage of a very short conversion time, typically 10 ns or less, corresponding to sampling frequencies of 100 MHz or more. Their greatest disadvantage is they require $2^n - 1$ comparators for an n-bit device, and this number becomes excessive when the number of bits is large. Even for an 8 bit device the number of comparators needed is 255. For this reason, few flash ADCs exceed 8 bits.

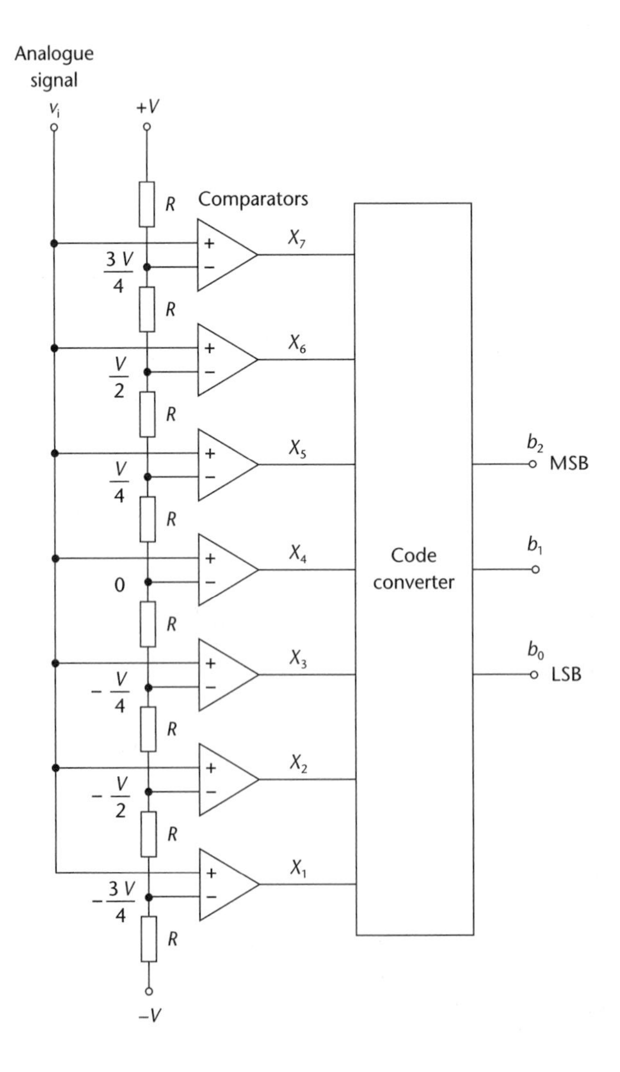

Fig. 7.6 ● Modification of a 3 bit flash ADC for bipolar signals

7.4.2 The weighted-resistor-network DAC

Many ADCs incorporate DACs internally as an essential part of their operation and, before considering any other form of ADC, it is useful therefore to consider DACs. Conceptually the weighted-resistor DAC illustrated in Figure 7.7 is one of the simplest forms of DAC. The switches shown are fast-acting electronic switches operated by the incoming digital word. A zero causes them to be connected to ground, and a one causes them to be connected to the reference voltage V. With the switches as shown in this 4 bit example the incoming code $b_3b_2b_1b_0$ is 1110 and the current entering the virtual ground of the inverting input of the op-amp is therefore given by

$$V\left[\frac{1}{R/8}+\frac{1}{R/4}+\frac{1}{R/2}+\frac{0}{R}\right]$$

Equating this to the forward current in the feedback resistance, $-V_0/R_f$, we obtain

$$V_0 = -V\frac{R_f}{R}(8+4+2+0)$$

Generalizing this to an n bit code in which the most significant bit (MSB) is denoted by b_{n-1} and the least significant bit (LSB) by b_0, the output would be given by

$$V_0 = -V\frac{R_f}{R}[2^{n-1}b_{n-1} + 2^{n-2}b_{n-2} + \cdots + 4b_2 + 2b_1 + b_0]$$

or

$$V_0 = q[2^{n-1}b_{n-1} + 2^{n-2}b_{n-2} + \cdots + 4b_2 + 2b_1 + b_0] \tag{7.1}$$

where the quantization step size q is given by

$$q = -V\frac{R_f}{R} \tag{7.2}$$

and it is seen that a negative value of V is needed to obtain a positive value of q.

The DAC of Figure 7.7 gives a unipolar output, but bipolar outputs can be obtained from offset binary codes by having the bits cause switching between a negative and positive reference voltage rather than between a negative reference voltage and ground. Similarly, bipolar outputs can be obtained from sign-magnitude codes by using the sign bit to reverse the polarity of the reference voltage. Other bipolar codes can be converted into offset binary or sign-magnitude code prior to the DAC.

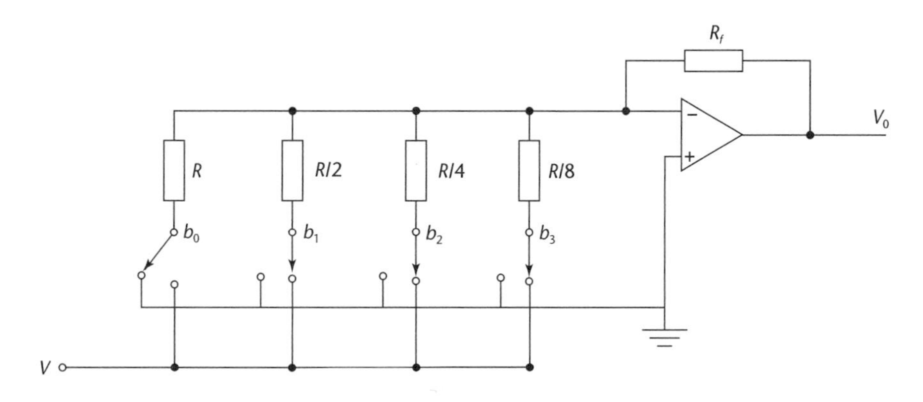

Fig. 7.7 Schematic of a weighted-resistor DAC

Unfortunately, the simple concept of Figure 7.7 encounters practical problems with the precision needed of resistance values, and with speed limitations imposed by stray capacitances across the larger resistances. A much more satisfactory and common design is the R–$2R$ style DAC. This and its advantages over the weighted-resistor DAC are considered next.

7.4.3 The *R*–2*R* DAC

A unipolar 4 bit R–$2R$ style DAC is shown in Figure 7.8. To understand its operation let us initially assume all the inputs are grounded as shown (that is, at logic 0) and begin by determining the resistances between node 2 and ground, looking separately to the left and right of node 2.

Looking left from node 1 we see R in series with the parallel combination of $2R$ and $2R$, thus giving $2R$. Hence looking left from node 2 we also see R in series with the parallel combination of $2R$ and $2R$, again giving $2R$.

The inverting input of the op-amp is a virtual ground and hence, when looking right from node 2, we see R in series with the parallel combination of $2R$ and $2R$, yet again giving $2R$.

Generalizing these arguments it is easily seen that the resistance looking right or left from any node is always $2R$ so long as the digital input bits are all zero. For example, the resistance looking right and left of node 1 is $2R$ in the configuration of Figure 7.8.

Consider next the effect of changing b_1 from a logic 0 to a logic 1 (from 0 volts to V volts). The equivalent circuit at node 1 is shown in Figure 7.9. The resistance seen by the b_1 input is $2R$ in series with the parallel combination of $2R$ and $2R$, which is $3R$. Hence the current I is given by

$$I = \frac{V}{3R} \tag{7.3}$$

This current divides equally at node 1 with $I/2$ flowing to the right. At node 2, however, this current encounters an onward path plus a downward path, both with resistance $2R$. Hence this current divides equally, with $I/4$ flowing onwards to node 3. In a similar way this current divides at node 3, with $I/8$ flowing through the onward path to the virtual ground. To summarize, a b_1 value of 1 causes a current into the virtual ground of $I/2^3$.

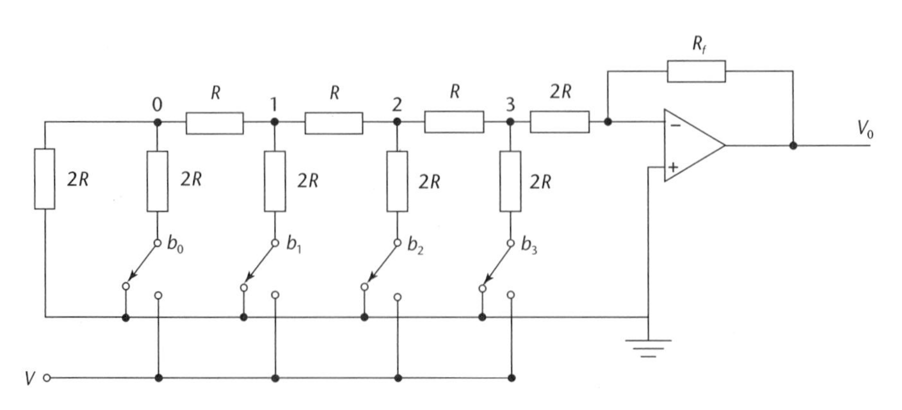

Fig. 7.8 Schematic of a 4 bit $R-2R$ style DAC

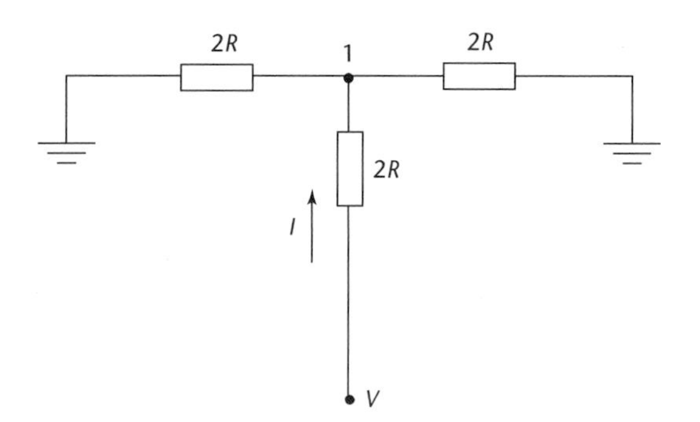

Fig. 7.9 Equivalent circuit for node 1 when input sequence is 0010

Following a similar argument, the condition $b_0 = 1$, $b_1 = 0$, $b_2 = 0$, $b_3 = 0$ would cause a current into the virtual ground of $I/2^4$, the condition $b_0 = 0$, $b_1 = 0$, $b_2 = 1$, $b_3 = 0$ would cause a current into the virtual ground of $I/2^2$, and the condition $b_0 = 0$, $b_1 = 0$, $b_2 = 0$, $b_3 = 1$ would cause a current into the virtual ground of $I/2^1$. If we have a combination of inputs we can find the total current into the virtual ground by superpostion. It is given by

$$I_{\text{total}} = I\left[\frac{b_3}{2} + \frac{b_2}{2^2} + \frac{b_1}{2^3} + \frac{b_0}{2^4}\right]$$

But, in an op-amp circuit, this current must equal the current in the feedback resistor, and this is given in Figure 7.8 by $(0 - V_0)/R_f$. Hence, equating the two currents and multiplying both sides of the equation by R_f, we obtain

$$V_0 = \frac{-R_f I}{2^4}\left[2^3 b_3 + 2^2 b_2 + 2^1 b_1 + b_0\right]$$

It is seen that the output voltage is proportional to the decimal equivalent of the binary code, thus demonstrating the action of the DAC.

The configuration of Figure 7.8 can be extended to produce an n bit DAC and, as was done with the weighted-resistor DAC, the result just obtained for the output voltage V can be generalized to give

$$V_0 = q\left[2^{n-1} b_{n-1} + 2^{n-2} b_{n-2} + \cdots + 4b_2 + 2b_1 + b_0\right] \tag{7.4}$$

where, using Equation (7.3)

$$q = -\frac{R_f}{2^n}\frac{V}{3R} \tag{7.5}$$

and a negative value of V is again needed to obtain a positive value of q.

The R–$2R$ style DAC has many advantages compared with the weighted-resistor DAC of Figure 7.7. If the $2R$ resistors are made up from two resistors in series, each of value R, only one value of resistance is needed. This makes it easy to obtain very precise values with similar temperature coefficients. Furthermore, the small 2 : 1 spread in resistance values means that R can be a low value that will be relatively immune to stray capacitance. This enables short settling times and high conversion rates.

As discussed in Section 7.1, the output of the DAC would normally be held at a constant level by means of a capacitance until the next digital input has caused the converted output to settle at its final value. Steps and glitches are then removed by a lowpass filter.

7.4.4 The successive-approximation ADC

The basic concept of an n bit successive-approximation ADC is that the n bits are determined in sequence, starting with the MSB and ending with the LSB. Beginning with the MSB, this is obtained by using a comparator to determine whether the input voltage is above or below the mid-point of the permitted voltage range. This allocates the input voltage to one half of the voltage range. The next most significant bit is then obtained by determining whether the voltage is above or below the mid-point of this selected half. This procedure continues with bits of progressively less and less significance, and hence leads to the term 'successive-approximation'. A system block diagram for an 8 bit ADC is shown in Figure 7.10, where the successive approximation register (SAR) is a digital device whose properties are best explained by an example. The system converts input voltages lying in the range 0 to $256q$ volts, and it will be supposed that, in this particular example, the input voltage level happens to be $125q$ volts.

The following are the events at successive clock pulses:

1. The SAR generates a sequence which, after passing through the DAC, produces half the maximum allowable input signal (that is, the SAR generates the sequence 10000000 such that the output of the DAC is $128q$ volts).
2. The input voltage of $125q$ volts is compared with the DAC output of $128q$ volts. Since the DAC output is too high it is reduced by a quarter of its range (by $64q$ volts, from $128q$ volts to $64q$ volts) by changing the output of the SAR to 01000000.
3. The input voltage of $125q$ volts is compared with the DAC output of $64q$ volts. Since the DAC output is too low it is increased by an eighth of its range (by $32q$ volts, from $64q$ volts to $96q$ volts) by changing the output of the SAR to 01100000.
4. The input voltage of $125q$ volts is compared with the DAC output of $96q$ volts. Since the DAC output is too low it is increased by $\frac{1}{16}$ of its range (by $16q$ volts, from $96q$ volts to $112q$ volts) by changing the output of the SAR to 01110000.

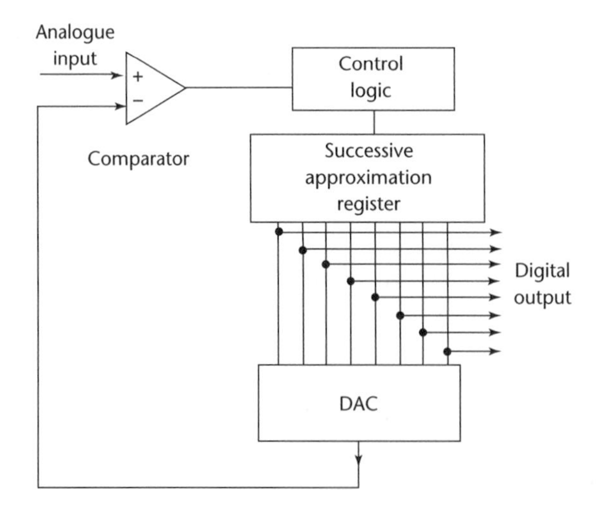

Fig. 7.10 ● Block diagram for 8 bit successive-approximation ADC

5. The input voltage of $125q$ volts is compared with the DAC output of $112q$ volts. Since the DAC output is too low it is increased by $\frac{1}{32}$ of its range (by $8q$ volts, from $112q$ volts to $120q$ volts) by changing the output of the SAR to 01111000.
6. The input voltage of $125q$ volts is compared with the DAC output of $120q$ volts. Since the DAC output is too low it is increased by $\frac{1}{64}$ of its range (by $4q$ volts, from $120q$ volts to $124q$ volts) by changing the output of the SAR to 01111100.
7. The input voltage of $125q$ volts is compared with the DAC output of $124q$ volts. Since the DAC output is too low it is increased by $\frac{1}{128}$ of its range (by $2q$ volts, from $124q$ volts to $126q$ volts) by changing the output of the SAR to 01111110.
8. The input voltage of $125q$ volts is compared with the DAC output of $126q$ volts. Since the DAC output is too high it is decreased by $\frac{1}{256}$ of its range (by q volts, from $126q$ volts to $125q$ volts) by changing the output of the SAR to 01111101. The conversion is now complete.

Generalizing from the above example, the conversion of an n bit successive-approximation ADC is complete after n clock cycles. In 1997 a medium-cost successive-approximation 12 bit ADC operates with a typical maximum clock frequency of 6 MHz, corresponding to a minimum conversion time of 2 μs and a sampling frequency of 500 kHz. Compared with a flash converter the maximum sampling rate is much less, but the number of bits, and hence accuracy, can be much greater.

7.5 Distortion due to sampling, and the sampling theorem

Neglecting any processing between the ADC and DAC, the actions shown in Figure 7.1 of digitizing an analogue signal and subsequently reconstructing an analogue signal can be broken down into the operations shown in Figure 7.11.

Consideration of the anti-aliasing filter and quantizer will be postponed until Sections 7.7 and 7.9 respectively. Neglecting practical problems of finite conversion times and so on, the output of the decoder is identical to the input to the encoder and thus, without the quantizer, the effect of the ADC and DAC reduces to that of a sample and hold followed by a zero-order hold. The two holding actions may be taken together and are considered in Section 7.8. The purpose of this section is to consider only the effect of sampling the signal and then filtering these samples with the final lowpass filter. The act of replacing an analogue signal by discrete samples has itself the potential to cause a loss of information about that signal and it is essential to determine the conditions for this loss not to occur.

We need first to distinguish between the practical or 'natural' sampling of a signal in which the signal is effectively multiplied by a periodic train $q(t)$ of finite amplitude short rectangular pulses, and the much more theoretical 'ideal' sampling of a signal

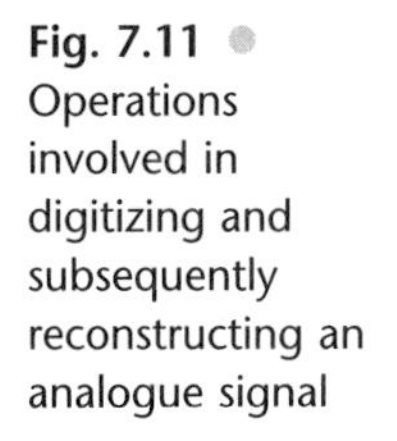

Fig. 7.11 Operations involved in digitizing and subsequently reconstructing an analogue signal

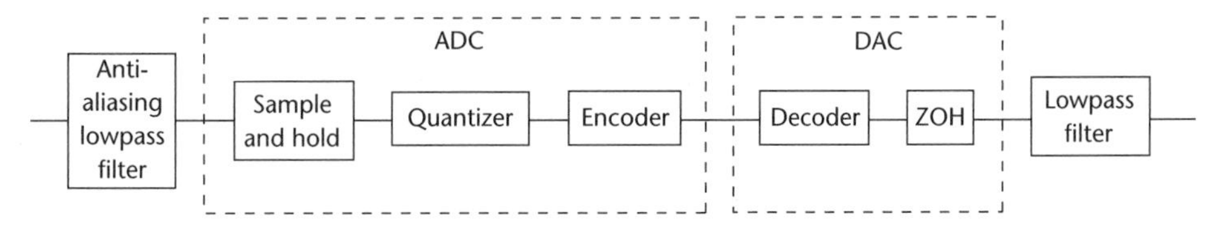

in which the signal is multiplied by a periodic train $p(t)$ of *impulses*. Ideal sampling is discussed in Section 7.6. This section deals with natural sampling.

Figure 7.12(a) shows a signal waveform and Figure 7.12(b) shows a train of rectangular pulses having a small duty cycle. Figure 7.12(c) shows the sampled waveform obtained by taking the product of the two. Such a signal is sometimes referred to as a sampled-data signal.

It is instructive to compare the spectrum of the sampled signal with that of the signal before sampling and to do this one can take a Fourier series approach or a Fourier transform approach. Because it provides such an easy physical interpretation, this section adopts the trigonometric Fourier series approach. Equation (5.32) gave the result that a train of rectangular pulses τ wide, T apart, and of amplitude A, could be written as

$$f(t) = \frac{A\tau}{T}\left(1 + 2\sum_{n=1}^{\infty} \frac{\sin(n\pi f_0\tau)}{n\pi f_0\tau} \cos n\omega_0 t\right)$$

where $\omega_0 = 2\pi/T$.

If τ/T is small the value of the sinc function is close to unity for the first few values of n. Hence, putting $A = 1$ and substituting t_s for T to signify that we are now dealing specifically with a sampling signal, we obtain

$$q(t) \approx \frac{\tau}{t_s}\,[1 + 2\cos(\omega_s t) + 2\cos(2\omega_s t) + 2\cos(3\omega_s t) + \cdots]$$

where $q(t)$ denotes a periodic train of finite duration sampling pulses and the subscript of ω is changed to s to emphasize that we are dealing with a sampling frequency (that is, $\omega_s = 2\pi f_s$, where $f_s = 1/t_s$).

The simplest example of a signal to be sampled is a cosinusoid $\cos p_1 t$ of radian frequency p_1. The product of this signal and the pulse train is then $s(t)$, where

$$s(t) = \frac{\tau}{t_s}\cos(p_1 t)\,[1 + 2\cos(\omega_s t) + 2\cos(2\omega_s t) + 2\cos(3\omega_s t) + \cdots] \tag{7.6}$$

Using the trigonometric identity $\cos A \cos B = \frac{1}{2}\cos(A - B) + \frac{1}{2}\cos(A + B)$,

$$s(t) = \frac{\tau}{t_s}\,[\cos(p_1 t) + \cos(\omega_s - p_1)t + \cos(\omega_s + p_1)t + \cos(2\omega_s - p_1)t + \cos(2\omega_s + p_1)t + \cos(3\omega_s - p_1)t + \cos(3\omega_s + p_1)t + \cdots] \tag{7.7}$$

(a)

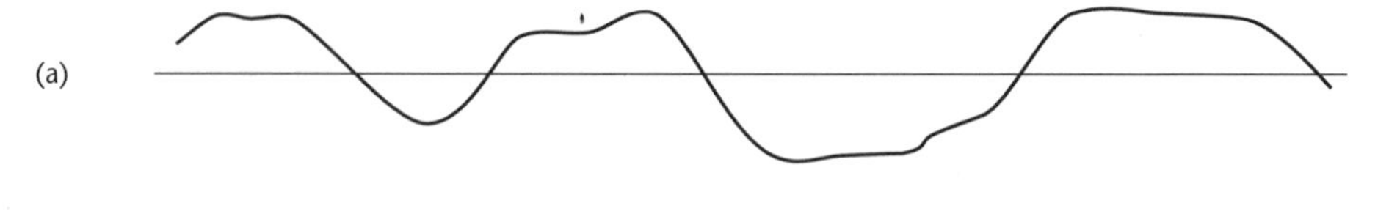

(b)

(c)

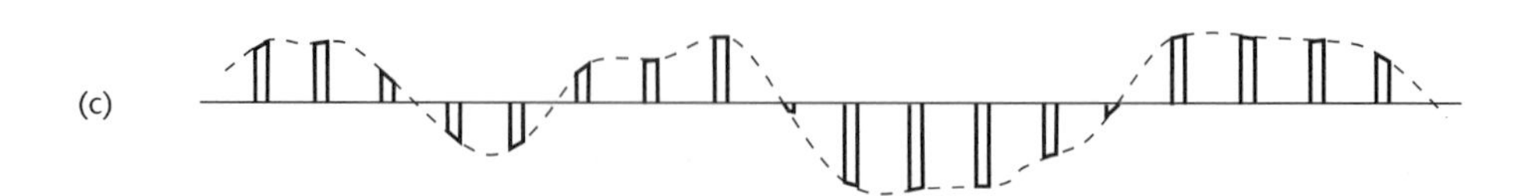

Fig. 7.12 The sampling process: (a) signal; (b) periodic train of pulses; (c) naturally sampled signal

This consists of a 'baseband' term, $\cos p_1 t$, and then lower and upper 'sidebands' centred about each harmonic of the sampling frequency. The one-sided amplitude line spectra of $\cos(p_1 t)$ and $s(t)$ (Fourier series coefficients) are shown in Figures 7.13(a) and 7.13(b). It is seen that the original signal $\cos p_1 t$ is significantly lower in frequency than the other terms in the spectrum, and can thus be regained by passing $s(t)$ through a lowpass filter whose cutoff frequency lies between p_1 and $(\omega_s - p_1)$. The fact that this is possible demonstrates, perhaps surprisingly, that no information has been lost by sampling the waveform.

If an ideal brickwall lowpass filter with a cutoff frequency $\omega_s/2$ were feasible the frequency p_1 could be increased to $\omega_s/2$ and yet be separable from the component at $(\omega_s - p_1)$. If $p_1 > \omega_s/2$ however, then, as demonstrated by the spectrum of Figure 7.14, the term of frequency $(\omega_s - p_1)$ will pass through this ideal lowpass filter. The interaction of terms in this situation is known as aliasing. It leads to distortion in the reconstructed signal.

A physical understanding of aliasing may be helped by also examining the phenomenon in the time domain. In Figure 7.15 the infrequent samples of the high-frequency sinusoid are seen also to correspond to the samples expected from a low frequency sinusoid. Depending on the cutoff frequency of the lowpass filter used to regain an analogue signal from this sampled data signal the output could be a single sinusoid of a different frequency from that of the original, thus providing an extreme example of distortion due to aliasing.

The need for sampling a signal adequately often is perhaps clearer when the signal has a continuous spectrum. Figure 7.16 is the symbolic representation of the amplitude spectrum of a telephony channel and shows components ranging between 300 Hz and 3.4 kHz. The reason for using a triangle is not to suggest that speech has such a distribution of frequency components. It does not. Rather it is because the

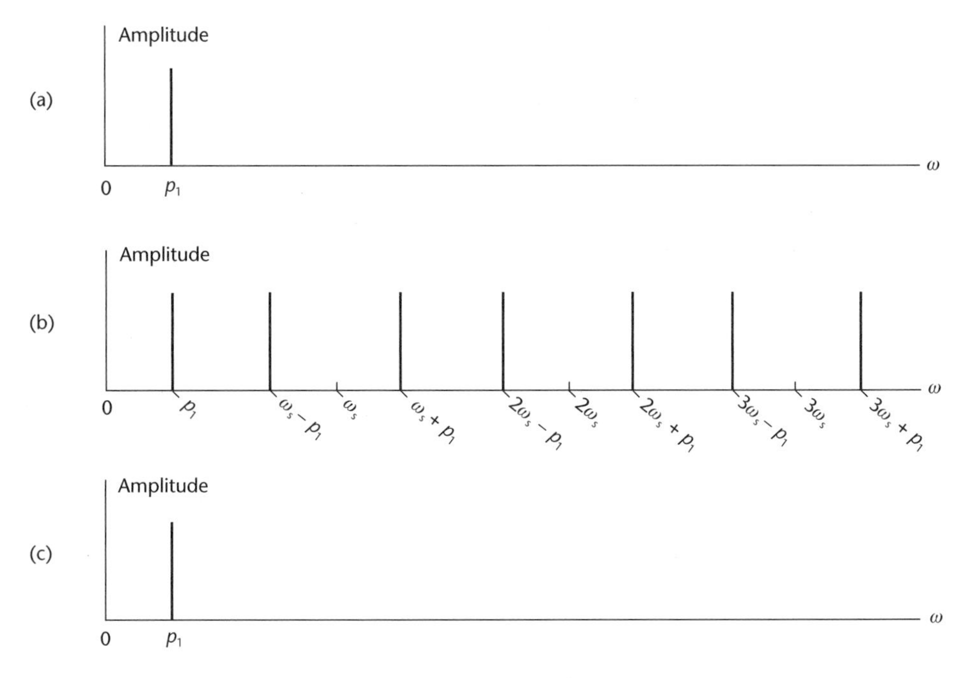

Fig. 7.13 One-sided amplitude line spectra: (a) of input signal $\cos(p_1 t)$; (b) of input signal after natural sampling; (c) of output signal after sampled signal is passed through a lowpass filter whose cutoff frequency is $\omega_s/2$

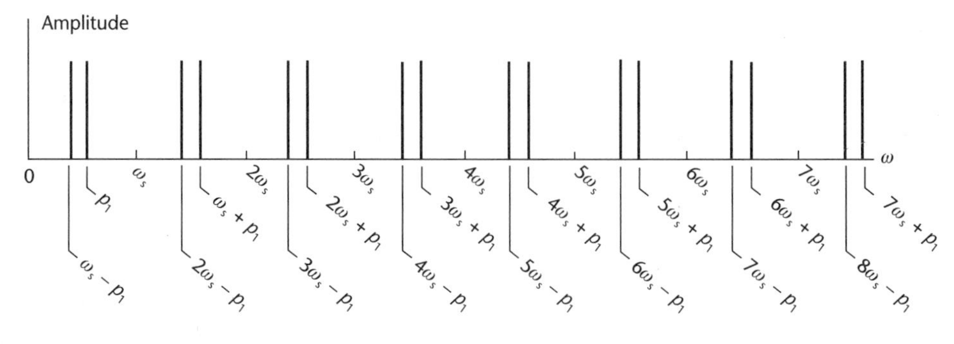

Fig. 7.14 Aliasing when the sampling frequency is too low

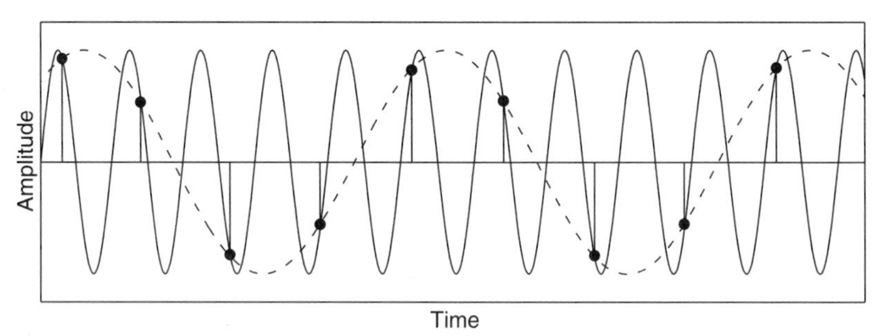

Fig. 7.15 A time domain demonstration of aliasing

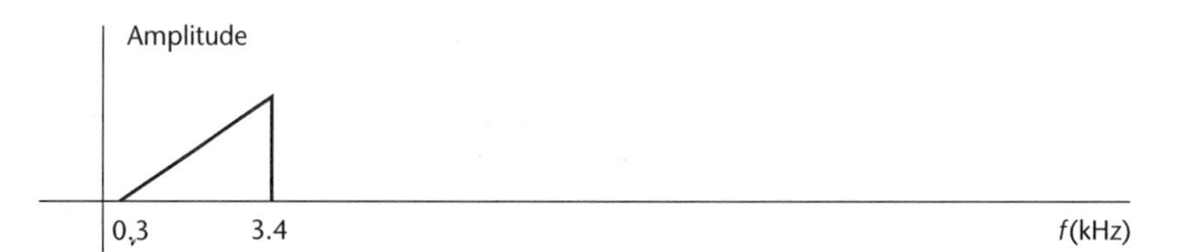

Fig. 7.16 Symbolic representation of the one-sided spectrum of a telephone channel

triangle is simple to draw and readily reveals any spectral inversions when part of a modulation scheme. For example, an extension of Figure 7.13(b) to an infinite number of adjacent sinusoids occupying the band 300 Hz and 3.4 kHz reveals that if such a signal is sampled at 8 kHz, the resulting signal has the amplitude spectrum shown in Figure 7.17. It is seen once again that a lowpass filter can separate out the spectrum of the original signal, thus indicating that no information has been lost by sampling. The limiting condition for adequate sampling is that the sampling rate should be more than twice the highest signal frequency. Such a sampling rate is known as the Nyquist rate. In practice, because perfect filters are impossible, the sampling rate must exceed the Nyquist rate by some margin, typically about 15%. This is why 3.4 kHz telephone channels are sampled at 8 kHz rather than at the Nyquist rate of 6.8 kHz. Even with such a relaxation in sampling rate, high specification filters are required. For example, Problem 7.2 shows that two *RC* filters in cascade are inadequate to reconstruct an undistorted sinusoid in spite of a sampling rate that is *four times* the Nyquist rate!

The Nyquist *rate* should not be confused with the Nyquist *frequency*. This latter term is widely used to denote *half the sampling frequency*, and this is irrespective of whatever that sampling frequency might be. The Nyquist frequency is also known as the folding frequency.

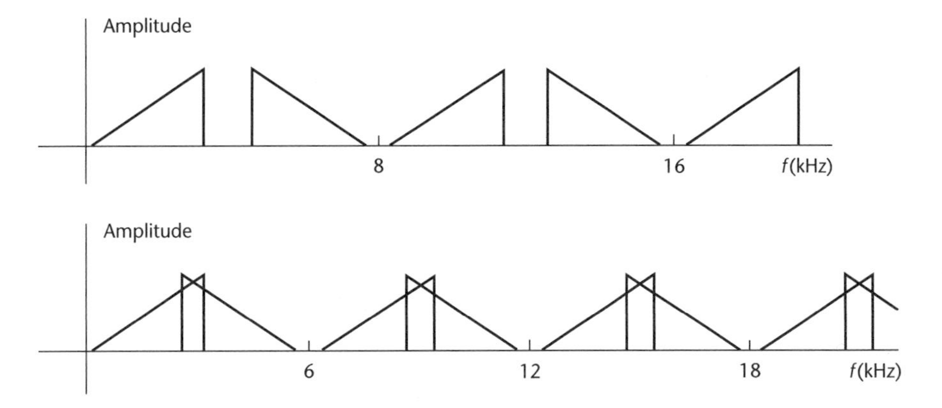

Fig. 7.17 One-sided spectrum of a telephone channel sampled at 8 kHz

Fig. 7.18 One-sided spectrum of a sampled telephone channel showing aliasing

An example of the spectrum arising when a telephone is sampled at an inadequate rate is shown in Figure 7.18 and shows aliasing. It is not possible to regain the original telephone channel from such a sampled signal.

It should be noted that the aliased components will not have similar phases and cannot therefore be simply added arithmetically.

If a two-sided spectrum is preferred then, as demonstrated in Section 5.6, this can be obtained from a one-sided spectral description by expressing the cosinusoidal components of the signal in terms of complex exponential components. Thus the signal $\cos(p_1 t)$ may be rewritten as $\frac{1}{2}[e^{jp_1 t} + e^{-jp_1 t}]$, and the sampled signal of Equation (7.7) expressed as

$$s(t) = \frac{\tau}{2t_s}\,[e^{jp_1 t} + e^{-jp_1 t} + e^{j(\omega_s - p_1)t} + e^{-j(\omega_s - p_1)t} + e^{j(\omega_s + p_1)t} + e^{-j(\omega_s + p_1)t} + \cdots] \quad (7.8)$$

This expression leads to a two-sided spectrum that is similar to the one-sided spectrum of Figure 7.13 except that it includes a 'mirror image' of the one-sided spectrum and the terms are of half the amplitude. In a similar way, the two-sided equivalents to Figures 7.17(b) and 7.18 can be readily obtained.

7.6 Ideal sampling and the Fourier transform approach to the spectra of sampled signals

The term 'ideal sampling' refers to the hypothetical situation where a signal waveform $f(t)$ is multiplied by a periodic train of unit impulses, $p(t)$, to produce a sampled waveform $s(t) = f(t)p(t)$. It is hypothetical, since impulses have infinite amplitudes and it is therefore not possible to generate them in practice. An analysis of $f(t)p(t)$ can be useful, however, since, at low frequencies, the spectrum of a periodic train of unit impulses is very similar to the spectrum of a periodic train of very short rectangular pulses of unit area. It should be noted that the sampled waveform $s(t)$ is an analogue signal and is different from the discrete-time waveform that is a sequence of values given by $f[n] = f(nt_s)$. By definition the amplitudes of the impulses within $s(t)$ are infinite, whereas the amplitude values within the discrete-time signal are finite. It is the *weights* of the impulses within $s(t)$ are the same as the amplitudes of $f[n]$.

The Fourier transform approach to the problem of determining the two-sided spectrum of an ideally sampled signal is mathematically elegant and uses the result of

Equation (6.45) that

$$\mathscr{F}\,[f_1(t)f_2(t)] = F_1(f) * F_2(f)$$

Applying this to the multiplication of a signal $f(t)$ by a periodic train of unit impulses $p(t)$ to produce a sampled waveform $s(t) = f(t)p(t)$, we obtain

$$S(f) = F(f) * P(f) \tag{7.9}$$

The Fourier transform of the original signal, $F(f)$, is a two-sided complex spectrum whose amplitude spectrum can again be portrayed symbolically using triangles, as in Figure 7.19.

From Equation (6.25) $P(f)$ is a series of impulse functions and, so long as $p(t)$ is symmetrical about the time origin, these are all of zero phase such that it is fully represented by the amplitude spectrum shown in Figure 7.20.

The graphical interpretation of convolving waveforms given in Section 2.8 was that of reversing one, shifting it by different amounts and, for each shift, multiplying it with the other waveform and integrating the result. Such an interpretation is generally unsuitable for convolving Fourier transforms, because each is a complex function comprised of an amplitude spectrum and a phase spectrum. However, since $P(f)$ in our case consists of widely spaced impulse functions that are fully represented by the amplitude spectrum of Figure 7.20, the product of $F(f)$ with a shifted and reversed $P(f)$ gives a set of impulse functions in the frequency domain that are of equal phase and whose integration is therefore straightforward, such that graphical interpretation is permissible. This leads to the amplitude spectrum of $F(f) * P(f)$ shown in Figure 7.21.

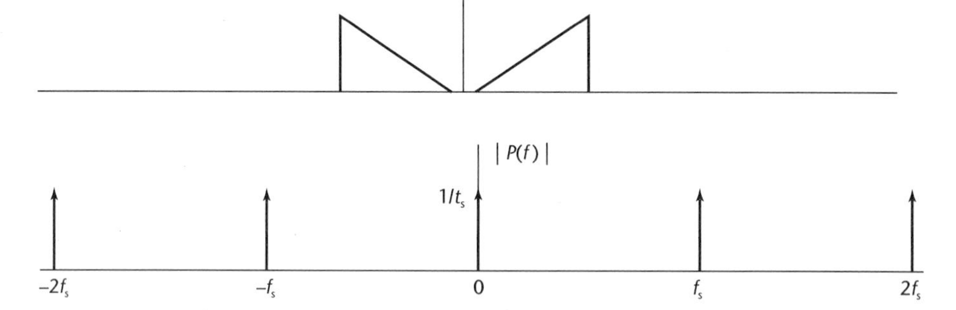

Fig. 7.19 Amplitude spectrum $|F(f)|$ of an input signal

Fig. 7.20 The Fourier transform of a periodic train of impuses

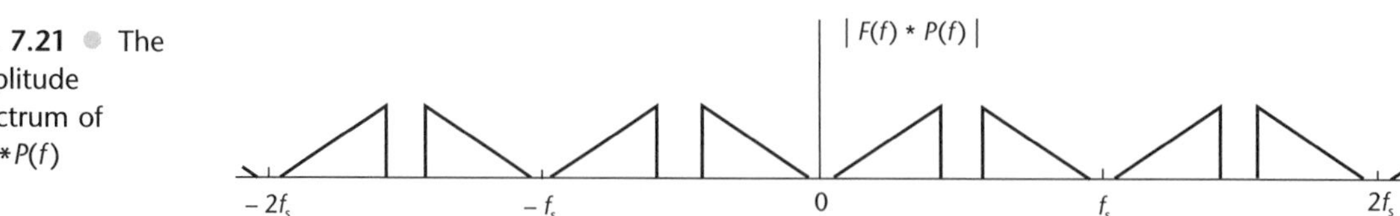

Fig. 7.21 The amplitude spectrum of $F(f)*P(f)$

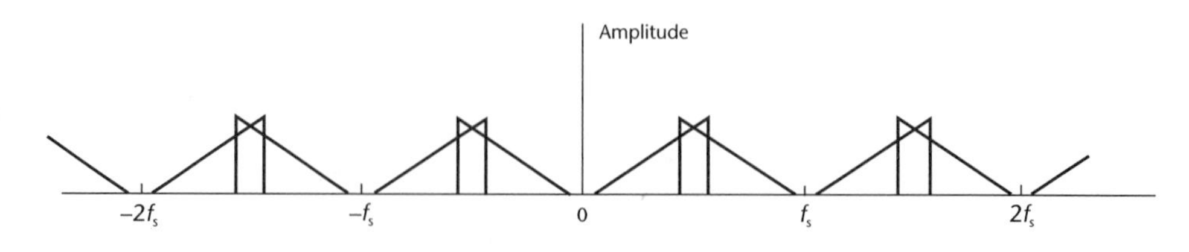

Fig. 7.22 Amplitude spectrum showing aliasing

If the sampling frequency is reduced the sidebands can overlap to give the situation shown symbolically in Figure 7.22. It should be noted that overlapping sidebands will generally contain components of different phase, and therefore cannot be simply added. However, as in Section 7.5, based upon the Fourier series, the conclusion remains that aliasing occurs if the sampling frequency is less than twice the highest signal frequency, but can otherwise, in theory at least, be avoided.

7.7 Anti-aliasing filters

The true one-sided spectrum of a speech signal is totally unlike the symbolic triangle of Figure 7.16 and is much more as shown in Figure 7.23, where there is a peak power density at around 800 Hz and a small amount of high-frequency power extending above the 4 kHz that is the theoretical maximum for a sampling frequency of 8 kHz. The result of this is that a small amount of aliasing can occur, as shown in the two-sided spectrum of the sampled signal shown in Figure 7.24.

To avoid this aliasing when digitizing speech it is necessary to eliminate speech components above 3.4 kHz using an analogue 'anti-aliasing' filter that precedes the ADC, as shown in Figure 7.11. The distortion introduced by this truncation of the speech spectrum is naturally undesirable. However, it is small and is in any case more acceptable than the interference that would otherwise be introduced into a regenerated signal due to aliasing. The same arguments apply to other signals that require digitizing.

Another justification for an anti-aliasing filter arises when a wanted signal is contaminated by an *unwanted* signal of greater bandwidth, perhaps noise. If the combination is sampled at a rate appropriate to the wanted signal, but below that appropriate to the noise, aliasing will cause the noise to increase. In Figure 7.25, for

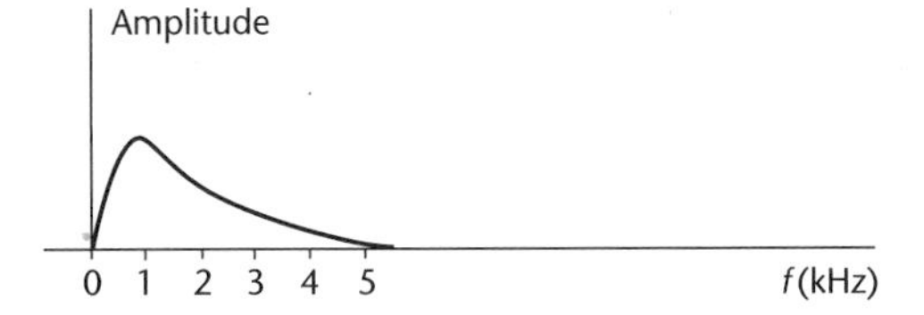

Fig. 7.23 Time-averaged one-sided spectrum of a typical speech signal

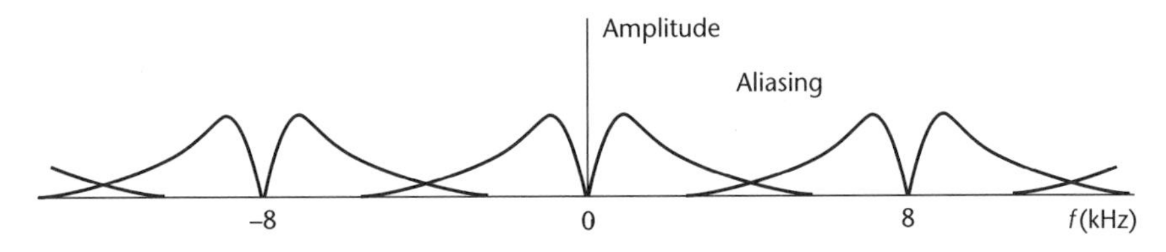

Fig. 7.24 Aliasing arising from the 8 kHz sampling of speech

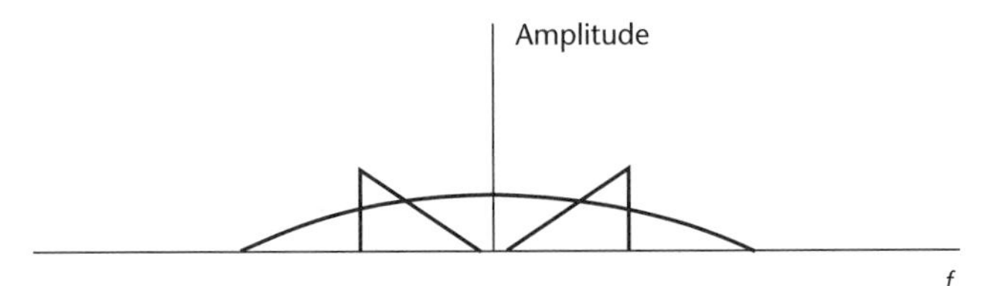

Fig. 7.25 Symbolic representation of the spectra of speech and noise

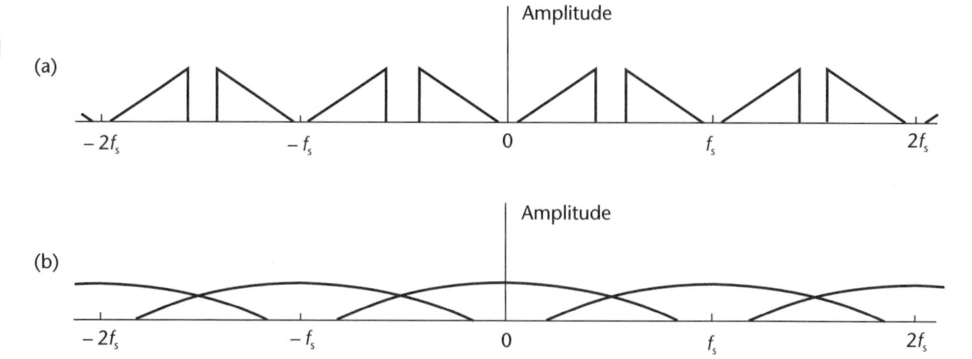

Fig. 7.26 Spectra of sampled signals: (a) of the speech alone; (b) of the noise alone

example, a speech signal denoted by a triangle is accompanied by noise, denoted symbolically by the segment of a circle, that is of greater bandwidth.

Figure 7.26(a) shows the spectrum of the speech signal after being sampled at a frequency that satisfies the Nyquist rate for the highest speech frequency. Figure 7.26(b) shows the spectrum of the noise after being sampled at the same rate. The spectrum of the total sampled signal is the superposition of *complex* spectra associated with the simplified *amphitude* representations of Figures 7.26(a) and 7.26(b). If the speech signal were to be restored by passing this total sampled signal through a lowpass filter, it can be seen that the ratio of speech to noise will be worse than before sampling. This is due to the aliasing of the noise. An anti-aliasing filter will prevent this increase in noise.

7.8 The effect of holding sample amplitudes

Sections 7.6 and 7.7 have shown that, whether natural or ideal sampling is used, undistorted signals can be regenerated from a sampled signal by means of a lowpass filter so long as the sampling frequency is high enough to avoid aliasing. Figure 7.11 gave a block diagram of the operations involved in converting an analogue signal into a digital signal, and back again into an analogue signal. Neglecting any effects of quantization for the moment, and remembering that the decoder provides an exact inverse operation to the encoder, it will be apparent that the signal leaving the DAC is like a sampled version of the input signal except that the sample values are *held* between samples by the zero-order hold. Effectively the signal leaving the DAC is what is termed a 'flat-topped' sampled signal. It is relevant to examine the effect on the spectrum of flat-topped sampling.

Figure 7.27(a) shows a sampled waveform with samples t_s apart, while Figure 7.27(b) shows a rectangular pulse of duration t_s. The convolution of these two produces the sampled and held waveform of Figure 7.27(c).

Thus, if the sampled waveform is given by $s(t) = f(t)p(t)$, the sampled-and-held waveform is given by

$$s'(t) = [f(t)p(t)] * \text{rect}(t/t_s) \qquad (7.10)$$

In accordance with Equation (6.42) convolution in the time domain corresponds to multiplication in the frequency domain. Thus

$$S'(f) = [F(f) * P(f)]\,\text{sinc}(ft_s) \qquad (7.11)$$

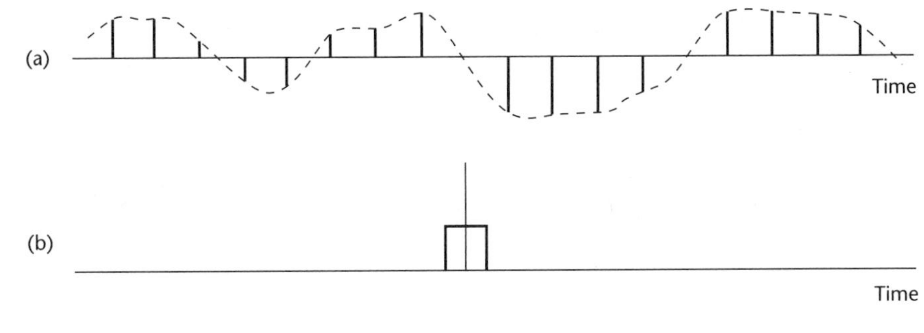

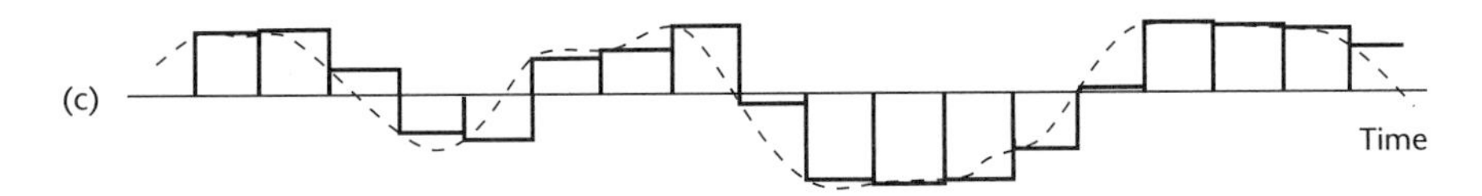

Fig. 7.27 Mathematical explanation of a sample and hold: (a) a signal sampled with period t_s; (b) a rectangular pulse of duration t_s; (c) the convolution of the two

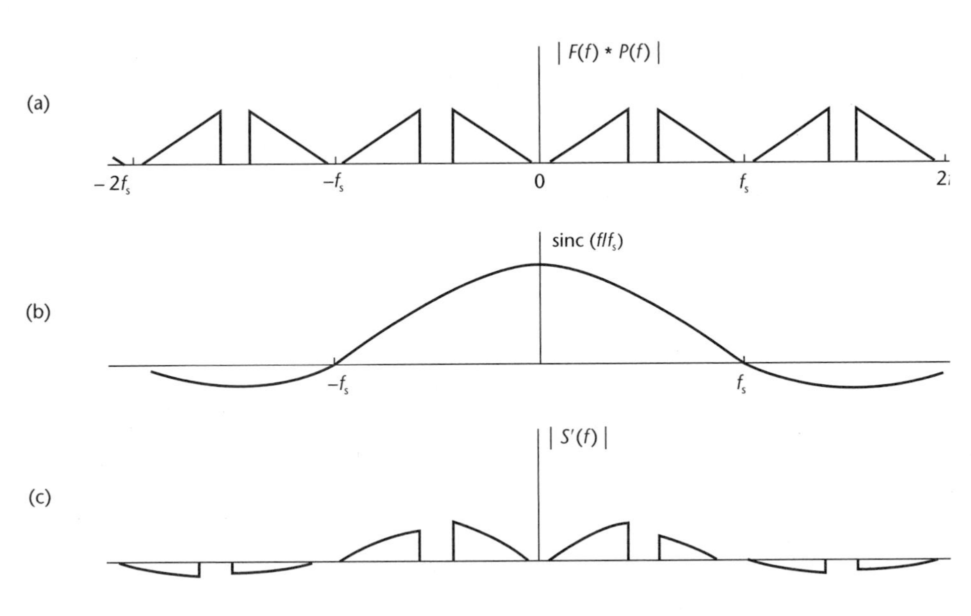

Fig. 7.28 Effect of a sample and hold: (a) the spectrum of a signal after ideal sampling; (b) the spectrum of a rectangular pulse; (c) the spectrum of the signal after a sample and hold

Figure 7.28(a) is a repeat of Figure 7.21 and shows the spectrum of $F(f) * P(f)$. Figure 7.28(b) shows the spectrum of $\text{sinc}(ft_s)$, while Figure 7.28(c) shows the product of Figures 7.28(a) and 7.28(b).

It is noted that the holding operation goes some way to achieving the function of the final lowpass reconstruction filter in that it reduces frequency components above half the sampling frequency. Within the wanted passband there is some slight frequency distortion which, in special circumstances, may need to be corrected. In most cases the advantages gained by the holding of samples greatly outweigh the disadvantages. Prior to encoding, the main advantage of holding the amplitude is to avoid a changing signal during the finite conversion time of the ADC. After decoding, the main advantage of the ZOH is to increase the energy of the pulses entering the reconstruction filter.

7.9 Dynamic range and quantization errors

If the 2^n quantization bands are q wide the largest sine wave that can enter an ADC without saturating it is one whose peak-to-peak amplitude is $2^n q$. The smallest sine wave that can can have any meaningful encoding is one whose peak-to-peak amplitude is just greater than q, since any smaller sine wave could be contained entirely within one quantization cell, such that all samples would be given the same code. If this were to happen the sine wave would emerge from a subsequent DAC as a d.c. level. On the basis that a sine wave whose peak-to-peak amplitude is q is the smallest suitable for encoding, the dynamic range would be $2^n q/q$ or 2^n in amplitude. Expressed in dB this is $20 \log_{10} (2^n)$ or $6.02n$ dB. However, this figure is somewhat pessimistic, since

- it assumes that the weak sine wave of peak-to-peak amplitude q is *centred* in a quantization band such that the ADC generates the same code at all sampling instant, and
- it is more usual to be interested in the ratio of largest to weakest signal when *both are present simultaneously*.

For these reasons the dynamic range is normally taken to be a ratio involving quantization noise, where the meaning of quantization noise and its relevance to dynamic range are now examined.

Figure 7.29 shows an analogue signal and the output samples resulting from sampling and quantizing. Since the decoder in Figure 7.11 provides a precise inverse of the encoder operation, it follows that these quantized samples also represent the output of the decoder (neglecting any zero-order-hold). It has been shown that, so long as the samples are close enough, the original signal could be exactly constructed from the samples if there were no quantization. What we now wish to determine is the degradation arising from the quantization. The easiest way to do this begins by recognizing that the output is unchanged if the sampler and quantizer are put in reverse order such that the quantizer comes first. The staircase waveform of Figure 7.30(a) shows the outcome when the analogue waveform of Figure 7.29 enters a quantizer directly. If this quantized signal is now sampled, it is easy to see that the result is the same as given by the samples of Figure 7.29.

The difference between the quantizer input and its staircase output is shown in Figure 7.30(b) and is termed the quantization noise. The quantized staircase waveform that is sampled can be thought of as the sum of the original analogue

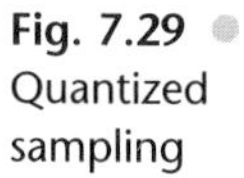
Fig. 7.29 Quantized sampling

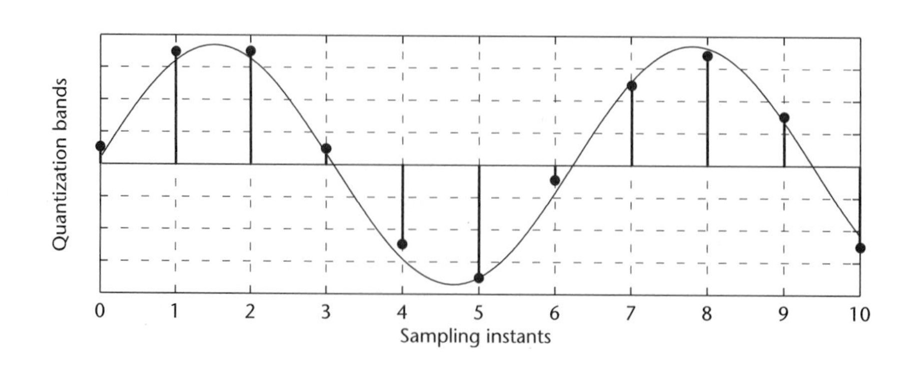

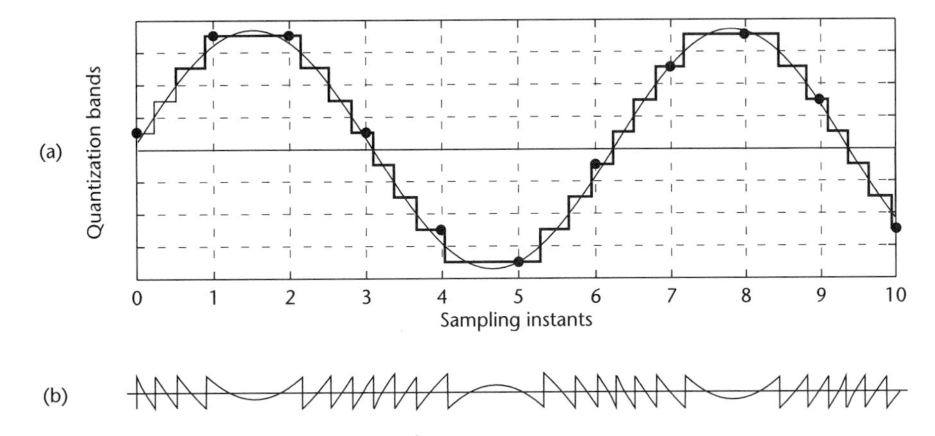

Fig. 7.30 Effects of quantization: (a) quantized waveform; (b) quantization noise

waveform and this quantization noise. It follows that the error in the output samples is that which would occur if the quantization noise alone were sampled. The error in the output signal is the outcome of passing these samples of quantization noise through the reconstruction filter.

It will be apparent from Figure 7.30(b) that the quantization noise is closely approximated by a sawtooth waveform of varying period and with a peak amplitude of $q/2$. The mean square voltage may be determined by averaging the squared voltage over any one of the periods. If a period of the sawtooth waveform extends from $-T/2$ to $T/2$ its waveform over that period is given by

$$\epsilon = \frac{q}{T}t \tag{7.12}$$

and the mean square voltage is given by

$$\overline{\epsilon^2} = \frac{1}{T}\int_{-T/2}^{T/2} \epsilon^2 \mathrm{d}t = \frac{1}{T}\left(\frac{q}{T}\right)^2 \frac{t^3}{3}\Bigg]_{-T/2}^{T/2} = \frac{q^2}{12} \tag{7.13}$$

An alternative approach often adopted (for which the necessary background in probability is not included in this book) is to recognize that the quantization noise is a random variable that is uniformly distributed between $-q/2$ and $q/2$ such that its probability density function $p(\epsilon)$ is a rectangular function of height $1/q$ and the variance is hence given by

$$\overline{\epsilon^2} = \int_{-\infty}^{\infty} \epsilon^2 p(\epsilon)\mathrm{d}\epsilon = \int_{-q/2}^{q/2} \frac{\epsilon^2}{q}\mathrm{d}\epsilon = \frac{\epsilon^3}{3q}\Bigg]_{-q/2}^{q/2} = \frac{q^2}{12} \tag{7.14}$$

This gives the same result.

The sawtooth quantization noise of Figure 7.30(b) contains sharp transitions and can be expected to have a bandwidth very much greater than that of the signal being quantized. It might be supposed that much of this noise would therefore be removed by the reconstruction filter, thus rendering inaccurate the estimates of its mean square voltage in Equations (7.13) or (7.14). In fact this is not the case, as the following frequency domain argument now shows.

Figure 7.31(a) shows a hypothetical two-sided spectrum of the quantization noise with its peak power centred at the mean frequency of the sawtooth-like waveform (this

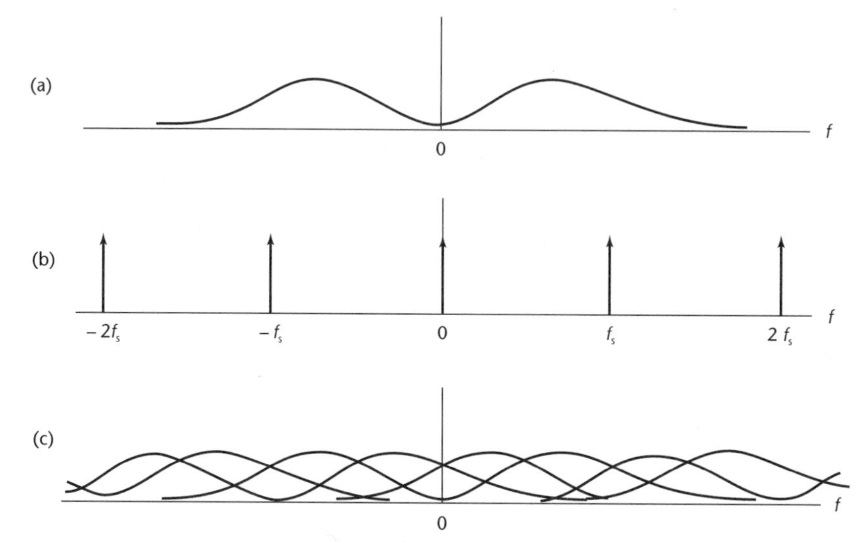

Fig. 7.31 Effect of sampling quantization noise: (a) spectrum of quantization noise; (b) spectrum of sampling waveform; (c) spectrum of sampled quantization noise

depends of the average time that the signal takes to pass through one quantization level and hence on the signal frequency). It is noted that much of this noise spectrum extends far above half the sampling frequency and it is easy to imagine that most of it might be removed by the reconstruction filter. However, this is not the case. If we consider the quantization noise to be sampled, the spectrum of this noise becomes convolved with the spectrum of the sampling waveform shown in Figure 7.31(b) and the resultant spectrum shown in Figure 7.31(c) suffers from severe aliasing. This aliasing causes the high frequency noise to be folded into the low frequency region such that the noise below half the sampling frequency has the same power as the total noise before sampling. It follows that the mean square quantization noise voltage of $q^2/12$ is a relevant and meaningful measure of the degradation of the output due to quantization.

The largest sine wave that can be accommodated by an n bit ADC has a peak-to-peak voltage of $2^n q$ and therefore

$$\frac{\text{mean square voltage of the largest sine wave not to cause saturation}}{\text{mean square voltage of the quantization noise}}$$

$$= \frac{(2^n q/2\sqrt{2})^2}{q^2/12} = \frac{3}{2}2^{2n} \quad (7.15)$$

Expressed in dB we have

$$\left[\frac{\text{mean square voltage of the largest sine wave not to cause saturation}}{\text{mean square voltage of the quantization noise}}\right]_{\text{dB}}$$

$$= 10\ \log_{10}(\tfrac{3}{2}\,2^{2n}) = 6.02n + 1.76\ \text{dB} \quad (7.16)$$

Very often this ratio is known as the 'coding dynamic range' but this term can be misleading, since it implies that the system is totally unsuited to signals outside of this range. In fact it will sometimes be possible for a signal to be weaker than the quantization noise and yet be detectable if some additional processing is involved. Spectral analysis is one example of such processing. Spectral analysis effectively partitions the frequency spectrum into narrow bands such that the noise in any one of these is much less than the total noise. This will cause a weak signal lying within one of these bands to be enhanced relative to the total quantization noise.

In practice, the advantages gained by subsequent processing are unreliable, such that the coding dynamic range of Equation (7.16) is in fact relevant. For example, consider the signal to be made up of a strong sinusoid and a weak sinusoid. The quantization noise due to the strong sinusoid will be periodic and will therefore exist as harmonics or aliased harmonics of the sinusoid. These harmonics will be confined to a few narrow bands within the spectrum and may be mistaken for a weak input sinusoid, thus effectively inhibiting the detection of the weak input sinusoid. If one considers the extreme (though unrealistic) situation where all the quantization noise is a single harmonic of a peak-to-peak amplitude equal to q it would be difficult to distinguish this from a weak input signal whose peak-to-peak amplitude is q. On this basis the amplitude ratio of the largest sinusoid to the weakest sinusoid would be $(2^n q)/q$, or $6.02n$ dB. These arguments suggest that it is unlikely to achieve in practice a dynamic range that is very much greater than given by Equation (7.16).

Another argument supporting Equation (7.16) as a useful measure of dynamic range is that it closely agrees with the figure obtained in the first paragraph of this section where $6.02n$ dB represented the ratio of the strongest sinusoid that could be quantized without saturation to the weakest sinusoid that could exist on its own without the possibility of being quantized into a single level. Unfortunately, a detailed analysis of dynamic range is outside the scope of this text.

7.10 An introduction to the sampling of bandpass signals

A common requirement is to process digitally a signal whose bandwidth B is small compared with its centre frequency f_0. Such a signal will be referred to as a bandpass signal and its spectrum might be as shown in Figure 7.32. The sampling theorem, as derived so far, demands that the sampling rate be greater than twice the highest frequency; that is greater than $2(f_0 + B/2)$. It is in fact possible to avoid any loss of information about the signal using a density of samples that is very much less than implied by such a high sampling frequency. The matter will be treated in some detail in Chapter 17, but an indication that it is feasible comes from recognizing the possibility that the bandpass signal could be downconverted to some lower centre frequency before being sampled. Since such downconversion can be reversed (by upconversion) it is clear that no information about the signal need be lost. Indeed, it will be shown in Chapter 17 that, by having two channels of downconversion, in both of which the signal is downconverted to become centred at zero frequency, it is possible with ideal restoration filters to require no more than $2B$ samples/s, where B is the signal bandwidth.

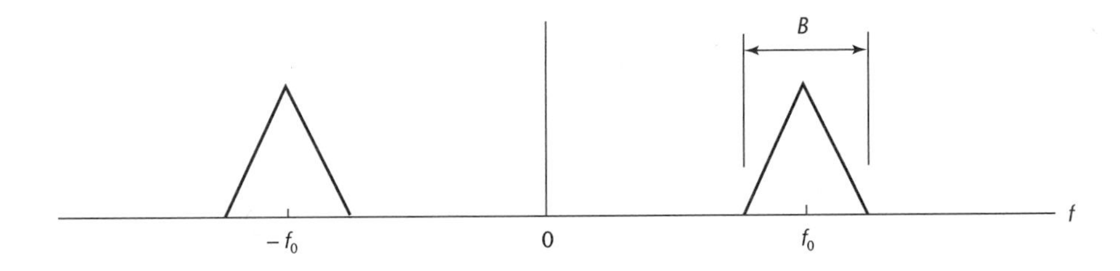

Fig. 7.32 Spectrum of a bandpass signal

7.11 Summary

It is shown that the minimum theoretical frequency at which a signal may be sampled to avoid any loss of information is twice the highest signal frequency, but that a somewhat higher sampling frequency is necessary to allow reconstruction of an undistorted analogue signal using physically realizable filters after the DAC. It is shown that it is beneficial to incorporate a zero-order hold at the output of the DAC. The use of anti-aliasing filters for constraining the highest signal frequency prior to the ADC is justified. The effect of coding signal samples is shown to be equivalent to a quantization of the signal, and the associated degradation is quantified. The permissible dynamic range of signal amplitudes is related to the number of bits in the ADC.

The chapter also describes some of the more important types of ADC and DAC and gives some indication of their complexities and conversion speeds.

7.12 Problems

1. Express the decimal numbers 48 and -72 in 8 bit sign magnitude code. Repeat using 8 bit two's complement code and show how the sum of these two numbers is achieved using it.
2. A 1 kHz cosinusoid is sampled at 8 kHz. Use MATLAB to plot the reconstructed analogue signal if the sampled waveform is passed through a simple *RC* lowpass network with a -3 dB cutoff frequency of 2 kHz. Repeat for the case where a second identical (but non-interacting) filter is cascaded with the first. Note that the reconstructed waveforms are very imperfect even through the signal frequency is only one eighth of the Nyquist rate!
3. A 4.5 kHz cosinusoid is sampled at 8 kHz. Sketch the spectrum of the sampled signal. An attempt is made to reconstruct the original signal by passing the samples through:
   (a) an ideal brickwall lowpass filter with a cutoff at 4 kHz (that is, at half the sampling frequency);
   (b) an ideal brickwall lowpass filter with a cutoff at 5 kHz.

   Give the waveform of the output signal in both cases.
4. The signal $\cos(2000t) + 0.5\cos(14000t)$ is digitized using an ADC operating at 8 kHz. Neglecting quantization effects, determine the reconstructed signal obtained by passing the coded signal through a DAC that incorporates a zero-order hold, and then through an ideal 4 kHz lowpass filter.
5. Generate in MATLAB 1000 samples 0.1 ms apart of the signal $7\cos(100t)$. Quantize it by using the **round(x)** command to round off numbers to the nearest integer. Plot within the same figure window, one above the other, the original signal and the difference between the original signal and this quantized signal. Repeat for the signal $3\cos(60t) + 2\cos(95t) + 3.5\sin(138t)$.
6. Generate in MATLAB 1000 samples 0.1 ms apart of the signal $8\cos(100t) + 0.4\cos(1030t)$. Quantize it by using the **round(x)** command to round off numbers to the nearest integer. Plot within the same figure window, one above the other, the original signal and the quantized signal. Note that due to quantization noise, the weaker signal component cannot be perceived in the display of the quantized signal.

Chapter eight

The discrete Fourier transform

8.1 ● Preview

The continuous Fourier transform has applications that include spectral analysis and the convolution of time domain signals, but there are few signals for which it can be evaluated analytically. In contrast, the discrete Fourier transform is amenable to numerical evaluation; hence its importance. The discrete Fourier transform is commonly abbreviated as the DFT.

The discrete Fourier transform is *not* simply a discrete or sampled version of the continuous Fourier transform. For this reason this chapter develops the DFT as an *entity in itself* rather than as a modification of the continuous Fourier transform. It is only later in the chapter that the connections between the DFT and the continuous Fourier transform are established.

It is felt that more physical insight is obtained by deriving the *inverse* DFT before the DFT itself and this is what is done. The chapter begins, however, with a discussion of basis functions.

8.2 An introduction to basis functions

It is helpful to begin with a short review of Fourier techniques, putting emphasis on the decomposition of a signal into a set of basis functions.

The trigonometric Fourier series of a periodic waveform $f_p(t)$, of period T, expresses that waveform as the summation of a d.c. term, a fundamental cosinusoid, a fundamental sinusoid and a set of harmonics, that is

$$f_p(t) = a_0 + \sum_{n=1}^{\infty} a_n \cos(n\omega_0 t) + b_n \sin(n\omega_0 t), \qquad \text{where } \omega_0 = 2\pi/T \tag{8.1}$$

The constant term and each of the sinusoids and cosinusoids can each be considered to be a *basis function* and it will be noted that the signal is the superposition of the complete set of basis functions. As was done in Section 5.5, the coefficient of any one of these is found by multiplying both sides of this equation by that basis function and then integrating over all time. On the right-hand side of the equation all of the resulting integrals are then zero except for that containing the wanted coefficient; this enables the coefficient to be determined.

The knowledge that sines and cosines are the basis functions being used, plus a specification of their amplitudes as a function of frequency $n\omega_0$, is enough to provide a complete description of the signal. Because this description of the signal is now in terms of the a and b coefficients, and these are a function of $n\omega_0$, it is a *frequency domain* description rather than a *time domain* description. Hence a *transform* has been made.

There are some variations on the Fourier series above. We can re-express the Fourier series as

$$f_p(t) = c_0 + \sum_{n=1}^{\infty} c_n \cos(n\omega_0 t + \theta_n) \tag{8.2}$$

This time we have only one type of basis function, namely a cosine waveform, but each one needs be described by its phase as well as by its amplitude; that is, by the coefficients c_n and θ_n. Two spectra are now needed; a discrete amplitude spectrum giving the coefficients c as a function of $n\omega_0$ and a discrete phase spectrum giving the coefficients θ as a function of $n\omega_0$.

Another representation of a periodic waveform uses the exponential Fourier series of Equation (5.22)

$$f_p(t) = \sum_{n=-\infty}^{\infty} F_n \, e^{jn\omega_0 t} \tag{8.3}$$

and thus expresses $f_p(t)$ as the sum of a set of complex exponential basis functions that have both positive and negative frequencies. The coefficients F_n are now complex. They are described by a complex spectrum; that is, by an amplitude spectrum and by a phase spectrum. It will be noted that the amplitude spectrum has even symmetry and that the phase spectrum has odd symmetry. An example of such spectra was given in Figure 5.15 for the case of an asymmetrical train of rectangular pulses.

The concepts may be extended to aperiodic signals by considering them to be periodic with an infinite period. Equation (6.14) expressed the aperiodic signal $f(t)$ by the formula

$$f(t) = \lim_{f_0 \to 0} \sum_{n=-\infty}^{\infty} f_0 F(nf_0) e^{j2\pi nf_0 t} \tag{8.4}$$

This can be considered to be a summation of an infinite number of basis functions that are the given by the complex exponentials $e^{j2\pi nf_0 t}$. Their magnitudes are $f_0 F(nf_0)$. An example of this was given in Section 6.9 where the summation of a large but finite number of basis functions of appropriate amplitude and phase was shown to give a good approximation to a rectangular pulse.

The infinite summation may be replaced by the integral that is the inverse Fourier transform, namely

$$f(t) = \int_{-\infty}^{\infty} F(f)e^{j2\pi ft}\, df \tag{8.5}$$

This may be regarded as being the limiting case of expressing $f(t)$ in terms of complex exponential basis functions. This time the basis functions are infinitely closely spaced in frequency as well as being infinite in number.

8.3 Alternative sets of basis functions and orthogonality

Generalizing the concepts above, we can express $f_p(t)$ as

$$f_p(t) = f_1\Phi_1(t) + f_2\Phi_2(t) + f_3\Phi_3(t) + \cdots = \sum_n f_n\Phi_n(t) \tag{8.6}$$

where $\Phi_n(t)$ are the basis functions and the f_n are a set of numbers. These numbers may be real or complex, but the important point is that they are just numbers; they are not functions of time.

Sinusoids, cosinusoids and complex exponentials are not the only possible basis functions. If some other type of basis function is chosen we can evaluate the consequent f_n values by extending the procedure adopted when the basis functions are complex exponentials, which is to multiply both sides of Equation (8.6) by one of the basis functions, $\Phi_k(t)$, and to integrate over all time. Multiplying Equation (8.6) by $\Phi_k(t)$ and integrating gives

$$\int_{-\infty}^{\infty} f_p(t)\Phi_k(t)\, dt = f_1 \int_{-\infty}^{\infty} \Phi_1(t)\Phi_k(t)\, dt + f_2 \int_{-\infty}^{\infty} \Phi_2(t)\Phi_k(t)\, dt + \cdots + f_k \int_{-\infty}^{\infty} \Phi_k^2(t)\, dt + \cdots \tag{8.7}$$

Suppose the following condition is imposed on the choice of basis functions

$$\int_{-\infty}^{\infty} \Phi_n(t)\Phi_k(t)\, dt = 0, \qquad \text{except when } n = k \tag{8.8}$$

This is the condition that the basis functions are *orthogonal*. Its effect is that all the terms except for one disappear from the right-hand side of Equation (8.7), leaving

$$\int_{-\infty}^{\infty} f_p(t)\Phi_k(t)\, dt = f_k \int_{-\infty}^{\infty} \Phi_k^2(t)\, dt \tag{8.9}$$

The condition of orthogonality has thus achieved two important results:

- it has led to a *means* of evaluating the f_n values;
- it has caused there to be a *unique* set of f_n values.

Sine waves, cosine waves and complex exponentials are all orthogonal functions. In contrast, a set of square waves of different frequencies are not. If, for example, the

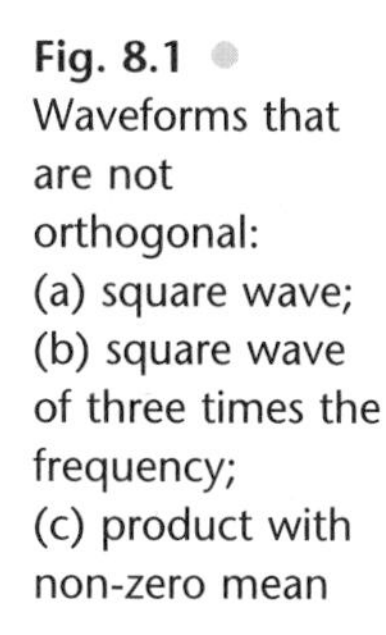
Fig. 8.1 Waveforms that are not orthogonal: (a) square wave; (b) square wave of three times the frequency; (c) product with non-zero mean

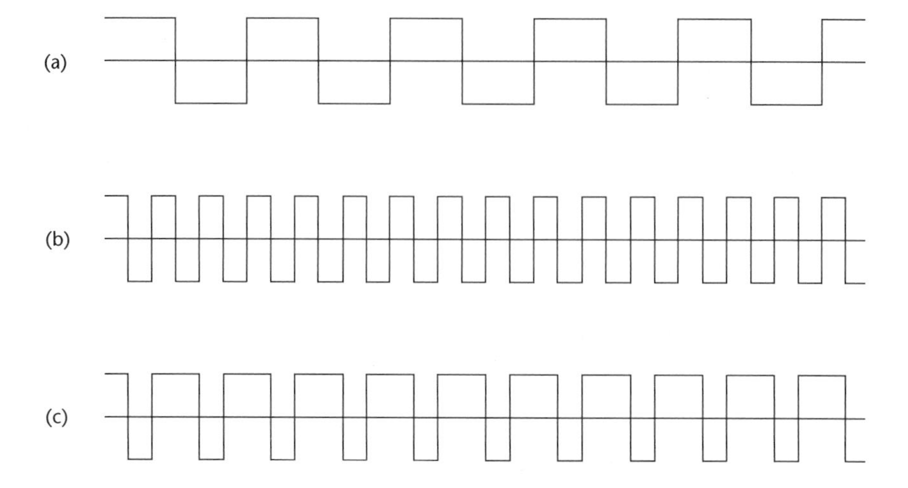

square wave of fundamental period T shown in Figure 8.1(a) is multiplied by the square wave of period $T/3$ shown in Figure 8.1(b) the result is the waveform shown in Figure 8.1(c), which does not have a zero mean. Thus, although it turns out to be possible to find a set of square waves that can be superimposed to realize an arbitrary signal $f_p(t)$, the set would not be unique. It makes such a set have no value.

Trigonometric and exponential functions are not the only basis functions that are orthogonal. Some examples of other sets are Walsh functions and Haar functions. The first eight Walsh functions are shown in Figure 8.2, and it is easy to verify orthogonality by multiplying any one of these by any of the others, and noting that the mean value of this product is zero. There is an infinite set of Walsh functions and their orthogonality means that any signal can be decomposed into a unique set of these, each one having an appropriate amplitude.

A description of a waveform by the coefficients of the Walsh functions would be referred to as the Walsh transform. Similarly, the description of a waveform by the coefficients of Haar functions would be the Haar transform. A description of a waveform by the coefficients of any set of orthogonal basis functions can be considered a transform. This is because the coefficients provide a complete description of the signal and can be used to restore the signal.

The next question to be asked is why one might wish to describe a signal as the summation of basis functions, or in a transformed form as the coefficients of these basis functions The reasons are several.

- Depending on what basis functions are chosen, the basis functions may have a strong *physical* significance. This is what happens in the case of Fourier analysis, because the presence of a sinusoid of a particular frequency is indicative of a specific physical mechanism. For example the presence of a strong spectral line in an acoustical signal indicates a vibration or a resonance at that frequency.
- The breakdown of a signal into certain classes of basis function can be helpful in determining the response of a linear system to that signal. The familiar example of this is when the basis functions are complex exponentials. It arises because the response of a linear system is described by a differential equation, and it has been shown that the solution of this differential equation is easy if the input happens to be one of these complex exponential basis functions. If the input is comprised of more than one

complex exponential the response to each can be determined, and the response to their totality found by simple summation. This is largely what Fourier analysis is all about. In it the input signal, $f(t)$, is replaced by a *frequency domain* description, $F(f)$, that just gives the amplitudes and phases of the complex exponentials as a function of their frequency. The frequency transfer function of the network, $H(f)$, gives its gain and

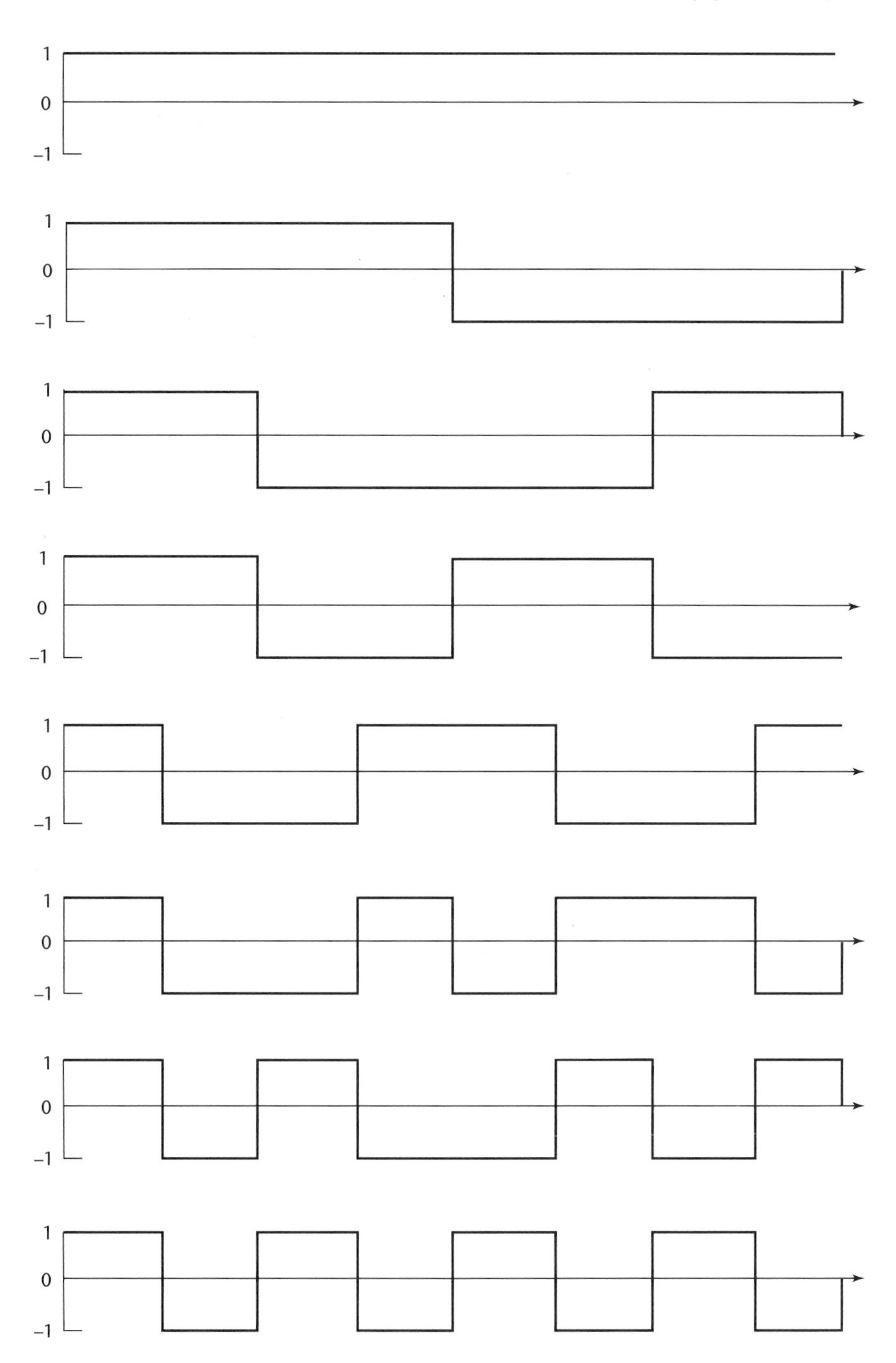

Fig. 8.2 The first eight Walsh functions as examples of functions that are orthogonal

phase shift as a function of frequency such that a frequency domain description of the output is $G(f) = F(f)H(f)$. The output time waveform, $g(t)$, is then determined by evaluating the integral that constitutes the inverse Fourier transform.

- Depending on the physical mechanism that generates the signal, the number of basis function coefficients needed to give an accurate representation of the signal may sometimes be less than the number of samples needed to convey the signal directly. If this is the case a compression of data has occurred and it would allow the signal to be transmitted using a reduced bandwidth. For example, many speech and image processing techniques involve a transform closely related to the Fourier transform known as the Discrete Cosine Transform. This has data reduction properties. Data reduction is also the main reason for studies into the Walsh transform. However, these other transforms do not have a *physical* significance that is comparable to that of the Fourier transform.

8.4 Trigonometric basis functions for discrete signals

Consider the 8 point discrete-time waveform $x[n]$ shown in Figure 8.3. As there are eight independent samples it follows that the amplitudes of eight orthogonal basis functions should be adequate to completely characterize the waveform.

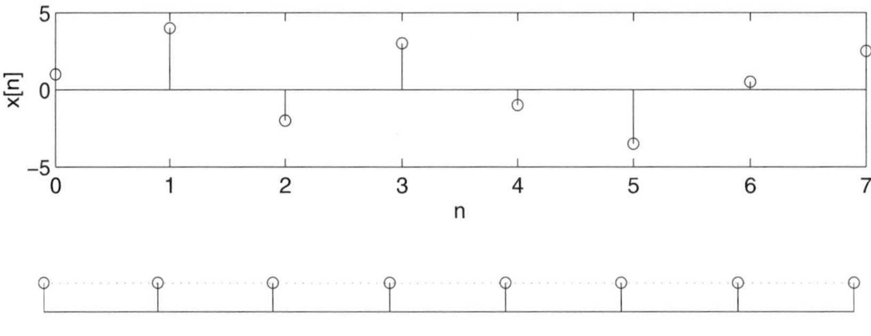

Fig. 8.3 Example of an 8 point discrete-time waveform

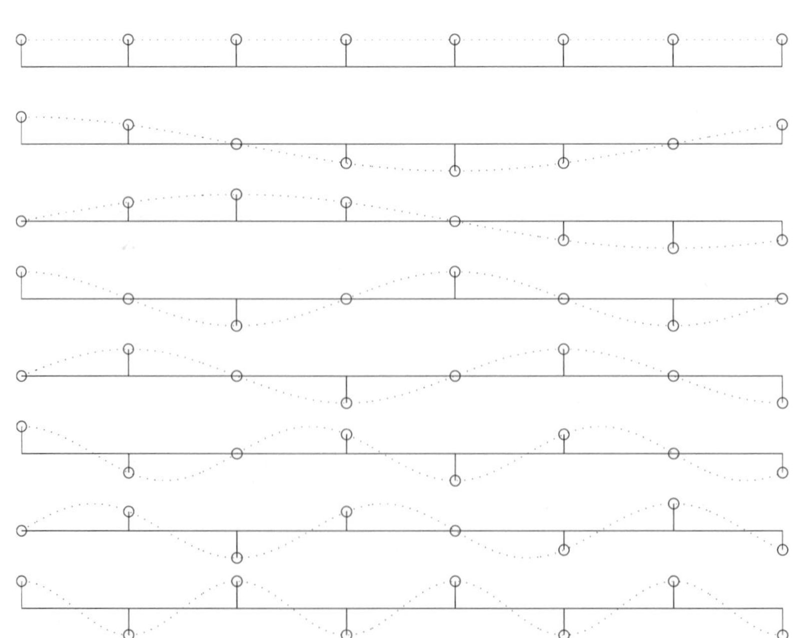

Fig. 8.4* A set of eight discrete-time basis functions that are orthogonal

A set of eight suitable discrete basis functions is shown in Figure 8.4. Envelopes have been added to these basis functions to show that they can be considered sampled versions of cosinusoids and sinusoids.

Thus it is proposed that

$$\begin{aligned} x[n] &= \sum \text{basic functions} \\ &= a_0 \cos 0 + a_1 \cos(2\pi n/8) + b_1 \sin(2\pi n/8) + a_2 \cos(4\pi n/8) \\ &\quad + b_2 \sin(4\pi n/8) + a_3 \cos(6\pi n/8) + b_3 \sin(6\pi n/8) + a_4 \cos(8\pi n/8) \end{aligned} \tag{8.10}$$

It is noted that this is very similar to a trigonometric Fourier series except that the terms are *discrete* sinusoidal and cosinusoidal signals.

The transform that is suggested by Equation (8.10) is that the information of $x[n]$ should be conveyed by the eight coefficients, $a_0, a_1, b_1, a_2, b_2, a_3, b_3$ and a_4, rather than by the eight discrete values of $x[n]$.

The orthogonality of the basis functions means that if we wish to find the coefficient a_k we multiply Equation (8.10) by $\cos(2\pi kn/8)$ and then sum the resulting samples on both sides of the equation. Similarly, if we wish to find coefficient b_k we multiply Equation (8.10) by $\sin(2\pi kn/8)$ and sum the resulting samples.

Example 8.1

If we have the 8 point sequence

$$x[n] = \{\underset{\uparrow}{1}, 1, 1, 1, 0, 0, 1, 0\}$$

determine b_1.

Solution Multiplying both sides of Equation (8.10) by $\sin(2\pi n/8)$ and summing the resulting samples on both sides of the equation we obtain

$$\begin{aligned} \sum_{n=0}^{7} x[n] \sin(2\pi n/8) &= \sum_{n=0}^{7} a_0 \sin(2\pi n/8) + \sum_{n=0}^{7} a_1 \cos(2\pi n/8) \sin(2\pi n/8) \\ &+ \sum_{n=0}^{7} b_1 \sin^2(2\pi n/8) + \sum_{n=0}^{7} a_2 \cos(4\pi n/8) \sin(2\pi n/8) \\ &+ \sum_{n=0}^{7} b_2 \sin(4\pi n/8) \sin(2\pi n/8) + \sum_{n=0}^{7} a_3 \cos(6\pi n/8) \sin(2\pi n/8) \\ &+ \sum_{n=0}^{7} b_3 \sin(6\pi n/8) \sin(2\pi n/8) + \sum_{n=0}^{7} a_4 \cos(8\pi n/8) \sin(2\pi n/8) \end{aligned} \tag{8.11}$$

On the right-hand side of Equation (8.11) the orthogonality causes all of the eight summations, except for one, to equal zero. For example $\cos(4\pi n/8)$, $\sin(2\pi n/8)$ and the product $\cos(4\pi n/8) \sin(2\pi n/8)$ are shown in Figures 8.5(a), (b) and (c). It is readily seen from Figure 8.5(c) that $\sum_{n=0}^{7} a_2 \cos(4\pi n/8) \sin(2\pi n/8) = 0$.

The one exception is $\sum_{n=0}^{7} b_1 \sin^2(2\pi n/8)$. Using a trigonometric expansion to rewrite this as $\sum_{n=0}^{7} 0.5b_1[1 - \cos(4\pi n/8)]$, it is seen to equal $4b_1$. Therefore

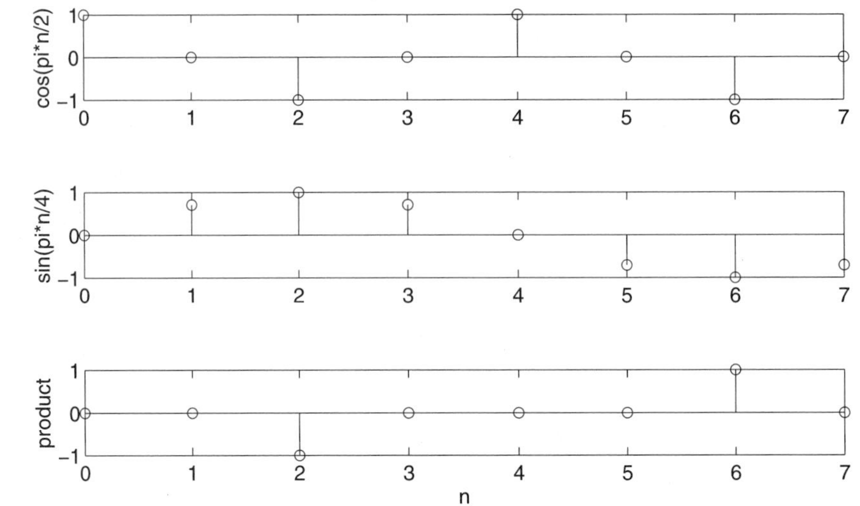

Fig. 8.5 Demonstration of orthogonality between two discrete-time trigonometric basis functions: (a) one basis function; (b) another basis function; (c) their product, showing that the summation of the samples is zero

Equation (8.11) becomes

$$\sum_{n=0}^{7} x[n]\sin(2\pi n/8) = 4b_1$$

or

$$b_1 = \frac{1}{4}\sum_{n=0}^{7} x[n]\sin(2\pi n/8) \tag{8.12}$$

Figures 8.6(a) and (b) show $x[n]$ and $\sin(2\pi n/8)$, while Figure 8.6(c) shows the product $x[n]\sin(2\pi n/8)$. It is seen that $\sum_{n=0}^{7} x[n]\sin(2\pi n/8) = 0 + 0.707 + 1 + 0.707 + 0 + 0 - 1 + 0 = 1.414$.

$\therefore \quad b_1 = 0.3535$

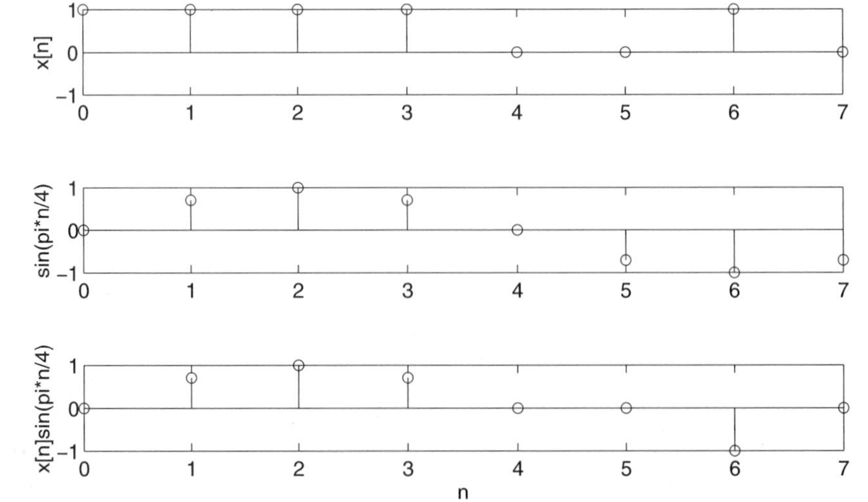

Fig. 8.6 Evaluation of the coefficient b_1: (a) the input signal; (b) the trigonometric basis function associated with the coefficient; (c) the product of the two

Example 8.2

If we have the 8 point sequence

$$x[n] = \{\underset{\uparrow}{1}, 1, 1, 1, 0, 0, 1, 0\}$$

for which b_1 was determined in Example 8.1, determine the coefficients of all eight cosinusoidal and sinusoidal basis functions. Confirm that the sum of the basis functions does indeed equal $x[n]$.

Solution The coefficient b_1 has already been determined in Example 8.1. In a similar manner we can multiply both sides of Equation (8.10) by other trigonometric basis functions and then sum the resulting samples on both sides of the equation to obtain seven more equations comparable to Equation (8.12). Repeating Equation (8.12), all eight equations are

$$a_0 = \frac{1}{8}\sum_{n=0}^{7} x[n] = 0.625$$

$$a_1 = \frac{1}{4}\sum_{n=0}^{7} x[n]\cos(2\pi n/8) = 0.25$$

$$b_1 = \frac{1}{4}\sum_{n=0}^{7} x[n]\sin(2\pi n/8) = 0.3535$$

$$a_2 = \frac{1}{4}\sum_{n=0}^{7} x[n]\cos(4\pi n/8) = -0.25$$

$$b_2 = \frac{1}{4}\sum_{n=0}^{7} x[n]\sin(4\pi n/8) = 0$$

$$a_3 = \frac{1}{4}\sum_{n=0}^{7} x[n]\cos(6\pi n/8) = 0.25$$

$$b_3 = \frac{1}{4}\sum_{n=0}^{7} x[n]\sin(6\pi n/8) = 0.3535$$

$$a_4 = \frac{1}{4}\sum_{n=0}^{7} x[n]\cos(8\pi n/8) = 0.125$$

Table 8.1 The values of the eight basis functions at eight values of n

	n							
Basis function	0	1	2	3	4	5	6	7
$a_0 \cos 0$	0.625	0.625	0.625	0.625	0.625	0.625	0.625	0.625
$a_1 \cos(2\pi n/8)$	0.25	0.177	0	−0.177	−0.25	−0.177	0	0.177
$b_1 \sin(2\pi n/8)$	0	0.25	0.354	0.25	0	−0.25	−0.354	−0.25
$a_2 \cos(4\pi n/8)$	−0.25	0	0.25	0	−0.25	0	0.25	0
$b_2 \sin(4\pi n/8)$	0	0	0	0	0	0	0	0
$a_3 \cos(6\pi n/8)$	0.25	−0.177	0	0.177	−0.25	0.177	0	−0.177
$b_3 \sin(6\pi n/8)$	0	0.25	−0.354	0.25	0	−025	0.354	−0.25
$a_4 \cos(8\pi n/8)$	0.125	−0.125	0.125	−0.125	0.125	−0.125	0.125	−0.125
Sum	1	1	1	1	0	0	1	0

Thus it is proposed that $x[n]$ is the sum of eight basis functions of amplitudes given by these coefficients, that is

$$x[n] = 0.625 + 0.25\cos(2\pi n/8) + 0.3535\sin(2\pi n/8) - 0.25\cos(4\pi n/8) + 0 \\ + 0.25\cos(6\pi n/8) + 0.3535\sin(6\pi n/8) + 0.125\cos(8\pi n/8) \qquad (8.13)$$

Table 8.1 gives the values of the eight basis functions at each value of n. It also gives the sum of the basis functions for each value of n, and this is seen to correspond to $x[n]$, as expected.

Example 8.2 has confirmed the validity of Equation (8.10) for representing an 8 point discrete signal as the sum of eight trigonometric basis functions, and Equation (8.10) can now be generalized to an N point discrete signal. We obtain

$$x[n] = \sum_{m=0}^{N/2} a_m \cos(2\pi mn/N) + \sum_{m=0}^{N/2} b_m \sin(2\pi mn/N) \quad 0 \leqslant n \leqslant (N-1) \qquad (8.14)$$

where each summation contains $(N/2 + 1)$ terms. It is noted that two of the terms in the second summation are zero since $\sin(2\pi mn/N) = \sin(0) = 0$ when $m = 0$, and $\sin(2\pi mn/N) = \sin(\pi n) = 0$ when $m = N/2$. This leaves a total of N non-zero coefficients. Thus, as would be expected, the coefficients of N basis functions are needed to convey the information of N independent samples within a discrete signal.

Just as was done with the Fourier series in Section 5.5, an alternative to a summation of sinusoids and cosinusoids is a summation of cosinusoids of differing phase. Thus an alternative to Equation (8.14) is

$$x[n] = \sum_{m=0}^{n/2} c_m \cos(2\pi mn/N + \theta_m) \qquad 0 \leqslant n \leqslant (N-1) \qquad (8.15)$$

8.5 Digital frequency

For discrete signals it is common and useful to involve the concept of *digital* frequency. Its symbol is F and it is conventional frequency *expressed as a fraction of the sampling frequency*, such that its units are *cycles per sample*:

$$F = \frac{f}{f_s} = ft_s \text{ cycles per sample} \qquad (8.16)$$

where f_s is the sampling frequency and t_s is the sampling period. Since there are generally several samples per cycle it is probably easier to visualize the number of samples per cycle than the number of cycles per sample, and the digital frequency is the reciprocal of this. Thus a frequency of 10 Hz sampled four times per cycle, corresponding to a sampling frequency of 40 Hz, signifies a digital frequency of 0.25 cycles/sample. A frequency of 100 Hz sampled at 400 Hz corresponds to the same digital frequency, demonstrating that digital frequency is a frequency which is normalized with respect to the sampling frequency. It should be noted, however, that plots of frequency response obtained with MATLAB use the term *normalized* frequency for a frequency which is normalized with respect to *half* the sampling frequency (for example Section 14.11). Although somewhat confusing, this convention will be adopted henceforth and means that a digital frequency of one corresponds to a normalized frequency of two.

Just as an 'analogue' frequency can be expressed in rad/s as well as in Hz, digital frequency can be expressed in *radians/sample* as well as in cycles per sample. Its symbol is Ω (an upper-case ω) and

$$\Omega = \frac{\omega}{f_s} = \omega t_s \text{ radians/sample} \tag{8.17}$$

As with F it is probably easier to visualize the reciprocal of this, which is the number of samples per radian.

Example 8.3

If a 5 Hz sinusoid is sampled at 25 Hz, what is its digital frequency?

Solution From Equation (8.16) the digital frequency can be expressed as

$$F = (5/25) = 0.2 \text{ cycles/sample}$$

or, using Equation (8.17), as

$$\Omega = (5/25)2\pi = 1.257 \text{ rad/sample}$$

It is useful to relate the concept of digital frequency to the set of basis functions shown in Figure 8.4. If we consider the two shown in Figures 8.4(a) and 8.4(b) whose envelopes have the lowest frequencies we see that there are 8 samples per cycle. Thus their digital frequency is $\frac{1}{8}$ cycle/sample. Generalizing to the case where we have N samples the basis functions whose envelopes have the lowest frequencies have N samples per cycle and hence the fundamental digital frequency is given by

$$F = \frac{1}{N} \text{ cycles/sample} \tag{8.18}$$

It follows that Equation (8.14) can be rewritten as

$$x[n] = \sum_{m=0}^{N/2} a_m \cos(2\pi mFn) + \sum_{m=0}^{N/2} b_m \sin(2\pi mFn) \qquad 0 \leqslant n \leqslant (N-1) \tag{8.19}$$

where the digital frequencies of the terms in the summation equal mF, and thus vary between 0 and $NF/2$, and hence between 0 and 0.5. It has already been established that the last term in the summation of sinusoids is zero, but substitution of $mF = 0.5$ into Equation (8.19) shows that the last term in the summation of sampled cosinusoids always contains just two samples/cycle.

An alternative to Equation (8.19) arises from Equation (8.15) which, substituting $F = 1/N$, can be rewritten as

$$x[n] = \sum_{m=0}^{N/2} c_m \cos(2\pi mFn + \theta_m) \qquad 0 \leqslant n \leqslant (N-1) \tag{8.20}$$

Example 8.4

Express an 8 point discrete-time waveform as the sum of eight sinusoidal and cosinusoidal basis functions that are written in terms of a fundamental digital frequency.

Solution Using Equation (8.18), the fundamental digital frequency is $\frac{1}{8}$cycle/sample. Thus, using Equation (8.19)

$$x[n] = a_0 \cos 0 + a_1 \cos(2\pi Fn) + b_1 \sin(2\pi Fn) + a_2 \cos(4\pi Fn) + b_2 \sin(4\pi Fn)$$
$$+ a_3 \cos(6\pi Fn) + b_3 \sin(6\pi Fn) + a_4 \cos(8\pi Fn)$$

Alternatively, if the fundamental digital frequency is expressed in radians/sample

$$x[n] = a_0 \cos(0) + a_1 \cos(\Omega n) + b_1 \sin(\Omega n) + a_2 \cos(2\Omega n) + b_2 \sin(2\Omega n)$$
$$+ a_3 \cos(3\Omega n) + b_3 \sin(3\Omega n) + a_4 \cos(4\Omega n)$$

8.6 The inverse discrete Fourier transform

Remembering that $\cos x = \frac{1}{2}(e^{jx} + e^{-jx})$, the summation of cosinusoids in Equation (8.15) can be expressed in terms of complex exponentials as

$$x[n] = c_0 + \frac{c_1}{2} e^{j(2\pi n/N + \theta_1)} + \frac{c_1}{2} e^{j(-2\pi n/N - \theta_1)}$$
$$+ \frac{c_2}{2} e^{j(4\pi n/N + \theta_2)} + \frac{c_2}{2} e^{j(-4\pi n/N - \theta_2)} + \cdots \tag{8.21}$$

or as

$$x[n] = \sum_{m=0}^{N/2} \frac{c_m}{2} \left(e^{j(2\pi mn/N + \theta_m)} + e^{j(-2\pi mn/N - \theta_m)}\right) \qquad 0 \leqslant n \leqslant (N-1) \tag{8.22}$$

The complex exponentials are seen to go in pairs and, apart from the case of $m = 0$, the first one in the pair can be considered to have a digital frequency of m times the fundamental digital frequency of $1/N$ cycles/sample, namely m/N cycles/sample, while the second one in the pair can be considered to have a digital frequency of $-m/N$ cycles/sample. Thus an alternative to Equation (8.22) is a summation of complex exponentials having *positive* and *negative* frequencies. Changing the coefficients used in the summation we obtain

$$x[n] = \frac{1}{N} \sum_{m=-N/2}^{N/2} X[m] e^{j2\pi mn/N} \qquad 0 \leqslant n \leqslant (N-1) \tag{8.23}$$

where

$$\frac{1}{N} X[0] = c_0; \qquad \frac{1}{N} X[1] = \frac{c_1}{2} e^{j\theta_1}; \qquad \frac{1}{N} X[-1] = \frac{c_1}{2} e^{-j\theta_1} \quad \text{and so on} \tag{8.24}$$

and square brackets are used to indicate that $X[m]$ represents a *discrete* set of coefficients.

Generalizing Equation (8.24) we see that the complex exponential terms with negative frequencies in Equation (8.23) have the same magnitude as those with positive frequencies, but have phases of opposite polarity, that is

$$X[-k] = X^*[k] \qquad \text{where } k \text{ is a positive integer} \tag{8.25}$$

There is yet another way of expressing $x[n]$ as the sum of basis functions and this, in fact, is the most important.

Since $e^{j2\pi n} = 1$, the complex exponential $e^{-j2\pi kn/N}$ in Equation (8.23), having the negative digital frequency of $-k/N$ cycles/sample, can be multiplied by $e^{j2\pi n}$ to be rewritten as

$$e^{-j2\pi kn/N} = e^{-j2\pi kn/N}\, e^{j2\pi n}$$

This can be rearranged as

$$e^{-j2\pi kn/N} = e^{j(2\pi/N)(N-k)n} \tag{8.26}$$

Equation (8.26) shows that a complex exponential in Equation (8.23) having a *negative* digital frequency of $-k/N$ cycles/sample can be replaced by a complex exponential having a *positive* digital frequency of $(N-k)/N$ cycles/sample. Thus the limits of the summation from $-N/2$ to $+N/2$ in Equation (8.23) can be changed to be from 0 to $(N-1)$ such that the equation becomes

$$x[n] = \frac{1}{N}\sum_{m=0}^{N-1} X[m]e^{j2\pi mn/N} \qquad 0 \leqslant n \leqslant (N-1) \tag{8.27}$$

Equation (8.24) is the *inverse* discrete Fourier transform (inverse DFT) whereby a signal can be resynthesized by adding a set of complex exponential basis functions, each having the appropriate complex amplitude. The basis functions in this inverse DFT are $1/N$ cycles/sample apart in digital frequency.

The coefficients in Equation (8.27) are not all independent. From Equation (8.25) the coefficient of a complex exponential term having the negative digital frequency of $-k/N$ cycles/sample is the conjugate of the coefficient of the complex exponential term having the positive digital frequency of $+k/N$ cycles/sample. In Equation (8.27) it follows that the coefficient of a complex exponential term having the positive digital frequency $(N-k)/N$ cycles/sample is the conjugate of the coefficient of the complex exponential term having the positive digital frequency $+k/N$ cycles/sample, that is

$$X[N-k] = X^*[k] \tag{8.28}$$

From Equations (8.27) and (8.28) it appears that only $N/2$ coefficients are needed to convey the information of N independent signal samples (that is, there are more independent samples than there are independent frequency coefficients). This may appear to contravene laws of nature. Physically the phenomenon arises because the N frequency coefficients have real and imaginary components such that $N/2$ coefficients have N degrees of freedom, just as there are N degrees of freedom in the original N samples of the signal.

Example 8.5

If $N = 16$ and $X(2) = 3 + j5$ in the summation of Equation (8.27), what is $X[14]$?

Solution Using Equation (8.28)

$$X[14] = X[16 - 2] = X^*[2] = 3 - j5$$

The discussion so far has related to the representation of *any* sampled signal as the sum of discrete basis functions and is not constrained to signals that are functions of time. When the sequence $x[n]$ is a *time* sequence we could emphasize this by writing the signal as $x(nt_s)$, where the signal samples are t_s seconds apart. In terms of 'analogue' frequency this would mean that the basis functions would be ($1/N$ cycles/sample) × ($1/t_s$ samples/second) apart. Using the symbol f_d to denote this frequency separation of the basis functions we have

$$f_d = 1/(Nt_s) \text{ Hz} \tag{8.29}$$

We can also write

$$f_d = 1/T \tag{8.30}$$

where $T = Nt_s$, and is the length of the sequence (including half a sample spacing at either end of the sequence of samples). Using Equation (8.29) to replace $1/N$ by $f_d t_s$, Equation (8.27) can be rewritten as

$$x(nt_s) = \frac{1}{N}\sum_{m=0}^{N-1} X(mf_d)e^{j2\pi mnf_d t_s}$$

$$= \frac{1}{N}\sum_{m=0}^{N-1} X(mf_d)e^{j2\pi mf_d nt_s} \qquad 0 \leqslant nt_s \leqslant (N-1)t_s \tag{8.31}$$

This equation has strong similarities with Equation (6.7) for a continuous inverse Fourier transform, as given in Equation 6.15, which was $x(t) = \int_{-\infty}^{\infty} X(f)e^{j2\pi ft}\,df$. Following a derivation of the (forward) DFT in Section 8.9 a detailed examination of the similarites and differences between it and the continuous Fourier transform will be included in Section 8.11.

8.7 Notation and symbols

There is a wide diversity of symbols used by different authors which can be very confusing. The purpose of this section is to emphasize those used in this book and to establish where the confusions can arise.

For analogue signals the lower-case symbols of f and ω are universally adopted for cyclic frequency and radian frequency. This text uses the upper-case symbols of F and Ω for digital frequency. The symbols f_s and t_s are used for sampling frequency and sampling period respectively. The symbol T is used for the duration of a waveform segment; it is also used for the period of a periodic waveform except when that periodic waveform is the sampling waveform $p(t)$, in which case the period is taken to be t_s. The symbol f_d is used for the frequency separation of the sampled complex exponentials that make up a sampled waveform in the inverse DFT of Equations (8.27) and (8.31). There is no particular reason for choosing the subscript d but, once chosen, it may be helpful to associate it with the word 'discrete'.

The terms which need special care are F and T. Many texts only express digital frequency in terms of Ω radians/sample and then have the symbol F available to use in place of f_{d} for the frequency separation of terms in the inverse DFT. This approach is rejected here because it is felt that it is generally preferable to use cyclic frequencies to radian frequencies and, if the upper-case Ω is the digital version of ω, then the upper-case F is clearly the logical choice for the digital version of f. This then leads to the need for an alternative symbol for the frequency separation of the sampled complex exponentials that make up a sampled waveform in the inverse DFT and is why f_{d} is used (it should be noted that the use of this symbol is unique to this book and arises because the symbol F is reserved for digital frequency).

Many texts give the symbol T multiple uses, of which one is the sampling period. This multiple usage can be confusing and hence it is preferred to let sampling period have the exclusive use of the symbol t_{s}.

8.8 Evaluating the coefficients of complex exponential basis functions

The basis functions in the inverse DFT of Equation (8.27) are orthogonal such that the coefficient of any one can be obtained by multiplying both sides of the equation by the conjugate of that basis function, and then summing the resulting terms. All of the terms on the right-hand side of the equation except for one will disappear, thus allowing that coefficient to be determined. Before obtaining a generalized result it is instructive to consider some specific examples.

Example 8.6

If

$$x[n] = \{\underset{\uparrow}{1}, 1, 1, 1, 0, 0, 1, 0\}$$

determine the complex amplitudes of the $X(2)$ and $X(6)$ terms in the summation that defines the inverse DFT, as given by Equation (8.27).

Solution We have $N = 8$ such that Equation (8.27) can be written out in full as

$$\begin{aligned} x[n] &= \tfrac{1}{8}X[0]\mathrm{e}^{\mathrm{j}0} + \tfrac{1}{8}X[1]\mathrm{e}^{\mathrm{j}2\pi n/8} + \tfrac{1}{8}X[2]\mathrm{e}^{\mathrm{j}4\pi n/8} + \tfrac{1}{8}X[3]\mathrm{e}^{\mathrm{j}6\pi n/8} \\ &\quad + \tfrac{1}{8}X[4]\mathrm{e}^{\mathrm{j}8\pi n/8} + \tfrac{1}{8}X[5]\mathrm{e}^{\mathrm{j}10\pi n/8} + \tfrac{1}{8}X[6]\mathrm{e}^{\mathrm{j}12\pi n/8} + \tfrac{1}{8}X[7]\mathrm{e}^{\mathrm{j}14\pi n/8} \end{aligned} \tag{8.32}$$

To determine $X[2]$ we multiply both sides of this by $\mathrm{e}^{-\mathrm{j}4\pi n/8}$ and sum the resulting samples on both sides of the equation to obtain

$$\begin{aligned} \sum_{n=0}^{7} x[n]\mathrm{e}^{-\mathrm{j}4\pi n/8} &= \tfrac{1}{8}X[0]\sum_{n=0}^{7}\mathrm{e}^{\mathrm{j}0}\mathrm{e}^{-\mathrm{j}4\pi n/8} + \tfrac{1}{8}X[1]\sum_{n=0}^{7}\mathrm{e}^{\mathrm{j}2\pi n/8}\mathrm{e}^{-\mathrm{j}4\pi n/8} \\ &\quad + \tfrac{1}{8}X[2]\sum_{n=0}^{7}\mathrm{e}^{\mathrm{j}4\pi n/8}\mathrm{e}^{-\mathrm{j}4\pi n/8} + \tfrac{1}{8}X[3]\sum_{n=0}^{7}\mathrm{e}^{\mathrm{j}6\pi n/8}\mathrm{e}^{-\mathrm{j}4\pi n/8} \\ &\quad + \tfrac{1}{8}X[4]\sum_{n=0}^{7}\mathrm{e}^{\mathrm{j}8\pi n/8}\mathrm{e}^{-\mathrm{j}4\pi n/8} + \tfrac{1}{8}X[5]\sum_{n=0}^{7}\mathrm{e}^{\mathrm{j}10\pi n/8}\mathrm{e}^{-\mathrm{j}4\pi n/8} \\ &\quad + \tfrac{1}{8}X[6]\sum_{n=0}^{7}\mathrm{e}^{\mathrm{j}12\pi n/8}\mathrm{e}^{-\mathrm{j}4\pi n/8} + \tfrac{1}{8}X[7]\sum_{n=0}^{7}\mathrm{e}^{\mathrm{j}14\pi n/8}\mathrm{e}^{-\mathrm{j}4\pi n/8} \end{aligned} \tag{8.33}$$

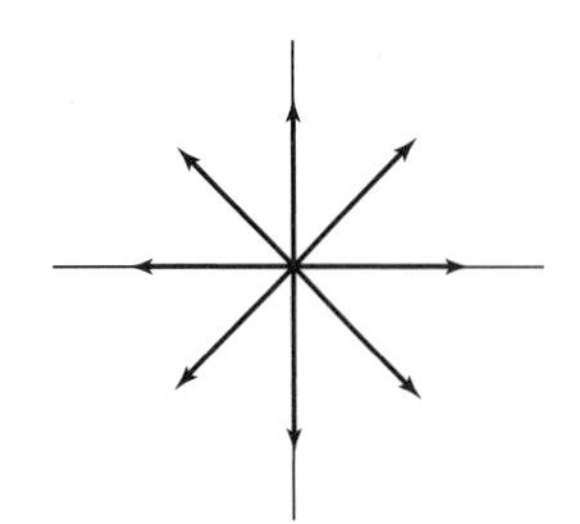

Fig. 8.7 Phasor diagram showing the summation of terms given by $\sum_{n=0}^{7} e^{-j2\pi n/8}$

Considering as an example the summation $\frac{1}{8}X[1]\sum_{n=0}^{7} e^{j2\pi n/8}e^{-j4\pi n/8}$ we can combine the two complex exponentials by adding their exponents and simplify the expression to become $\frac{1}{8}X[1]\sum_{n=0}^{7} e^{j2\pi n/8}$. The eight components within the summation of $\frac{1}{8}X[1]\sum_{n=0}^{7} e^{j2\pi n/8}$ can then be represented by the eight phasors of Figure 8.7. By inspection it is clear that their sum is zero. The same applies to six of the other seven summations on the right-hand side of the equation. The only exception is that involving $X[2]$. Thus Equation (8.33) becomes

$$\sum_{n=0}^{7} x[n]e^{-j4\pi n/8} = \frac{1}{8}X[2]\sum_{n=0}^{7} e^{j0} = \frac{1}{8}X[2] \times 8 = X[2]$$

Hence

$$X[2] = \sum_{n=0}^{7} x(n)e^{-j4\pi n/8} \tag{8.34}$$

Inserting the values $x[0]=1$, $x[1]=1$, $x[2]=1$, $x[3]=1$, $x[4]=0$, $x[5]=0$, $x[6]=1$ and $x[7]=0$, we obtain

$$X[2] = 1\cdot e^{-j0} + 1\cdot e^{-j\pi/2} + 1\cdot e^{-j\pi} + 1\cdot e^{-j3\pi/2} + 0\cdot e^{-j4\pi/2} + 0\cdot e^{-j5\pi/2} + 1\cdot e^{-j6\pi/2} + 0\cdot e^{-j7\pi/2}$$

Figure 8.8 shows the five non-zero terms in phasor form. Four of these phasors cancel one another, leaving

$$X[2] = -1$$

In a similar manner we can evaluate $X[6]$. Alternatively, and requiring much less effort, we can apply Equation (8.28) to obtain $X[6] = X^*[2] = -1$.

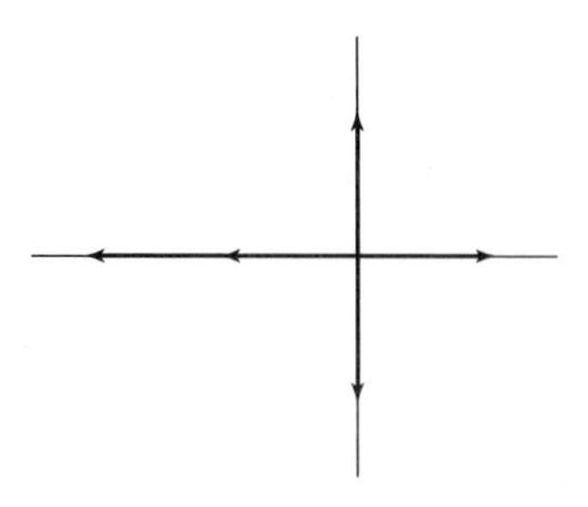

Fig. 8.8 Phasors corresponding to terms within the summation $\sum_{n=0}^{7} x[n]e^{-j4\pi n/8}$

Example 8.7

Example 8.2 took the sequence

$$x[n] = \{\underset{\uparrow}{1}, 1, 1, 1, 0, 0, 1, 0\}$$

and expressed it as the sum of trigonometric basis functions. For the same sequence Example 8.6 found $X[2] = -1$ and $X[6] = -1$. Reconcile the two sets of results.

Solution The inverse DFT definition of Equation (8.27) expresses an 8 coefficient sequence as the sum of complex exponential basis functions whose frequencies are harmonics of the digital frequency of $\frac{1}{8}$ cycles/sample. The one whose coefficient is $X[6]$ has a digital frequency of $\frac{6}{8}$ cycles/sample and can be written $\frac{1}{8}X[6]e^{6(2\pi/8)n}$. It can equally well be written $\frac{1}{8}X[6]e^{(8-2)(2\pi/8)n} = \frac{1}{8}X[6]e^{2\pi n}e^{-2(2\pi/8)n} = \frac{1}{8}X[6]e^{-2(2\pi/8)n}$ and be considered to have a negative digital frequency of $-\frac{2}{8}$ cycles/sample. The combination of this with the complex exponential basis function whose digital frequency is $+\frac{2}{8}$ cycles/sample, and which is therefore $\frac{1}{8}X[2]e^{2(2\pi/8)n}$, produces $\frac{1}{8}(X[2]e^{j4\pi n/8} + X[6]e^{-j4\pi n/8})$. Inserting $X[2] = X[6] = -1$, this reduces to $\frac{1}{8}(-e^{j4\pi n/8} - e^{-j4\pi n/8})$ and hence to $-0.25\cos(4\pi n/8)$. As expected, this corresponds to the term of digital frequency $+\frac{2}{8}$ cycles/sample present in the sum of trigonometric basis functions given by Equation (8.13) of Example 8.2.

8.9 The discrete Fourier transform

The technique of determining the X coefficients described in Example 8.6 of Section 8.8 can now be generalized and leads to the (forward) discrete Fourier transform. Equation (8.27) gave

$$x[n] = \frac{1}{N}\sum_{m=0}^{N-1} X[m]e^{j2\pi mn/N} \qquad 0 \leqslant n \leqslant (N-1)$$

In order to determine a specific coefficient $X[k]$ we multiply both sides of this equation by $e^{-j2\pi kn/N}$ to give

$$x[n]e^{-j2\pi kn/N} = \frac{1}{N}\sum_{m=0}^{N-1} X[m]e^{j2\pi(m-k)n/N} \qquad 0 \leqslant n \leqslant (N-1)$$

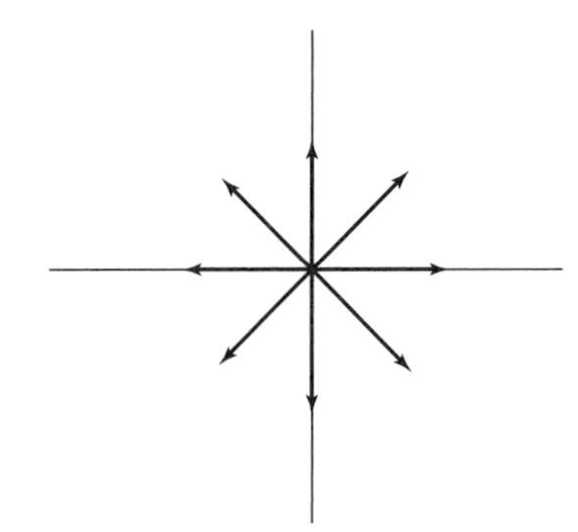

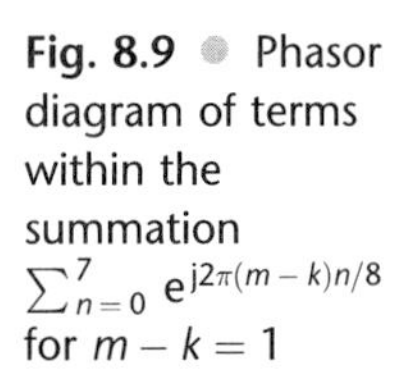

Fig. 8.9 Phasor diagram of terms within the summation $\sum_{n=0}^{7} e^{j2\pi(m-k)n/8}$ for $m - k = 1$

If we now sum over the N possible values of n we obtain

$$\sum_{n=0}^{N-1} x[n]\mathrm{e}^{-\mathrm{j}2\pi kn/N} = \frac{1}{N}\sum_{n=0}^{N-1}\left(\sum_{m=0}^{N-1} X[m]\ \mathrm{e}^{\mathrm{j}2\pi(m-k)n/N}\right)$$

Changing the order of the summation

$$\sum_{n=0}^{N-1} x[n]\mathrm{e}^{-\mathrm{j}2\pi kn/N} = \frac{1}{N}\sum_{m=0}^{N-1} X[m]\left(\sum_{n=0}^{N-1} \mathrm{e}^{\mathrm{j}2\pi(m-k)n/N}\right) \tag{8.35}$$

Considering as an example $m-k=1$ and $N=8$, the terms within the inner summation would be given by $(e^{j0}+e^{j2\pi/8}+e^{j4\pi/8}+e^{j6\pi/8}+e^{j\pi}+e^{j10\pi/8}+e^{j12\pi/8}+e^{j14\pi/8})$. A phasor diagram of these terms is as shown in Figure 8.9 and the result is seen to be zero.

In a similar manner the inner summation on the right-hand side of Equation (8.35) is seen to be zero for all values of m except for the case where $m=k$. For this special case all the terms in the summation equal unity and

$$\sum_{n=1}^{N-1} \mathrm{e}^{\mathrm{j}2\pi(m-k)n/N} = N$$

It follows that Equation (8.35) becomes

$$\sum_{n=0}^{N-1} x[n]\mathrm{e}^{-\mathrm{j}2\pi kn/N} = \frac{1}{N}\cdot NX[k] = X[k] \qquad 0 \leqslant n \leqslant (N-1)$$

Hence, reverting from the variable k in this expression to the variable m, we obtain

$$X[m] = \sum_{n=0}^{N-1} x[n]\mathrm{e}^{-\mathrm{j}2\pi mn/N} \qquad 0 \leqslant m \leqslant (N-1) \tag{8.36}$$

This expression is the discrete Fourier transform (or DFT).

If $x[n]$ refers to the specific case of a discrete *time*-varying signal it may be written as $x(nt_{\mathrm{s}})$ and the coefficients of the basis functions may be written $X(mf_{\mathrm{d}})$, where f_{d} is their spacing in Hz. As in Equation (8.29), $f_{\mathrm{d}}=1/Nt_{\mathrm{s}}$. Inserting $1/N=f_{\mathrm{d}}t_{\mathrm{s}}$ into Equation (8.36) we obtain

$$X(mf_{\mathrm{d}}) = \sum_{n=0}^{N-1} x(nt_{\mathrm{s}})\mathrm{e}^{-\mathrm{j}2\pi mf_{\mathrm{d}}nt_{\mathrm{s}}} \qquad 0 \leqslant mf_{\mathrm{d}} \leqslant (N-1)f_{\mathrm{s}}/N \tag{8.37}$$

This has strong similarities in its form with the continuous Fourier transform given by Equation (6.7) which, for a signal $x(t)$, is $X(f)=\int_{-\infty}^{\infty} x(t)\mathrm{e}^{-\mathrm{j}2\pi ft}\,\mathrm{d}t$.

As an example of what Equation (8.37) does, Figure 8.10(a) shows samples of an arbitrary time waveform while Figure 8.10(b) shows the magnitude of the discrete Fourier coefficients and Figure 8.10(c) shows their phases.

An alternative way of conveying the information of these Fourier coefficients would be via their real and imaginary components. Either way, two parameters are needed to characterize each Fourier coefficient, and therefore $2N$ parameters are needed to characterize all N Fourier coefficients. The reason that the number of parameters needed to characterize the Fourier coefficients is twice the number needed to characterize the original time waveform is that the Fourier coefficients are

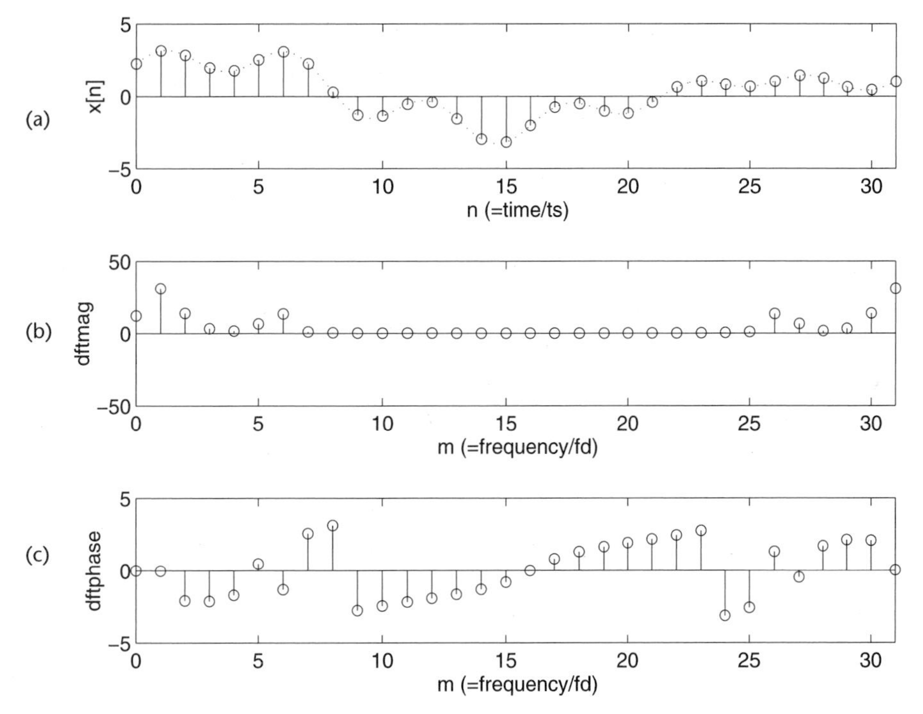

Fig. 8.10 Example of a DFT: (a) samples of an analogue waveform; (b) magnitude of the discrete Fourier coefficients; (c) phase of the discrete Fourier coefficients

not all independent. In accordance with Equation (8.28) there is an even symmetry of the amplitudes of the Fourier coefficients about the one whose index is $N/2$, but an odd symmetry of their phases.

In Chapter 17 it will be found that complex waveforms are more than mathematical abstractions and can have a physical existence. For example, it will be shown that a low-frequency *complex* waveform can be derived from a high frequency narrowband *real* waveform and can contain all of its information, with the advantage that it requires a lower sampling frequency. In the case of a complex waveform the conjugate relationship between the two halves of the DFT is no longer valid. N independent complex signal samples then produce N independent complex frequency coefficients, and vice versa.

8.10 The discrete Fourier transform pair

Repeating the results of Equations (8.36) and (8.27) we have

$$X[m] = \sum_{n=0}^{N-1} x[n]\mathrm{e}^{-\mathrm{j}2\pi mn/N} \qquad 0 \leqslant m \leqslant (N-1) \tag{8.38}$$

$$x[n] = \frac{1}{N}\sum_{m=0}^{N-1} X[m]\mathrm{e}^{\mathrm{j}2\pi mn/N} \qquad 0 \leqslant n \leqslant (N-1) \tag{8.39}$$

These are the DFT and inverse DFT respectively. Together they constitute the discrete Fourier transform pair. It is quite common to present a more compact form of these equations by introducing a parameter W_N defined as

$$W_N = e^{-j2\pi/N} \tag{8.40}$$

Inserting this into Equations (8.38) and (8.39) gives

$$X[m] = \sum_{n=0}^{N-1} x[n]\, W_N{}^{mn} \qquad 0 \leqslant m \leqslant (N-1) \tag{8.41}$$

and

$$x[n] = \frac{1}{N} \sum_{m=0}^{N-1} X[m]\, W_N{}^{-mn} \qquad 0 \leqslant n \leqslant (N-1) \tag{8.42}$$

Sometimes the coefficients of $X[m]$ and $x[n]$ are replaced by subscripts, giving

$$X_m = \sum_{n=0}^{N-1} x_n W_N{}^{mn} \qquad 0 \leqslant m \leqslant (N-1) \tag{8.43}$$

and

$$x_n = \frac{1}{N} \sum_{m=0}^{N-1} X_m W_N{}^{-mn} \qquad 0 \leqslant n \leqslant (N-1) \tag{8.44}$$

Very often, for economy, the subscript N is omitted from the parameter W_N. This will frequently be done in what follows.

The equations above can also be written very compactly using matrix notation. For a 4 point DFT Equation (8.43) can be expressed as

$$\begin{bmatrix} X_0 \\ X_1 \\ X_2 \\ X_3 \end{bmatrix} = \begin{bmatrix} W^0 & W^0 & W^0 & W^0 \\ W^0 & W^1 & W^2 & W^3 \\ W^0 & W^2 & W^4 & W^6 \\ W^0 & W^3 & W^6 & W^9 \end{bmatrix} \begin{bmatrix} x_0 \\ x_1 \\ x_2 \\ x_3 \end{bmatrix} \tag{8.45}$$

Consider $\mathbf{x}_n$ and $\mathbf{X}_m$ to be column vectors and $\mathbf{W}^{mn}$ to be a square matrix, where bold type is used to denote vectors and matrices. Generalizing the notation of Equation (8.45) we obtain

$$\mathbf{X}_m = \mathbf{W}^{mn}\mathbf{x}_n \tag{8.46}$$

Similarly, the inverse DFT of Equation (8.44) may be written compactly as

$$\mathbf{x}_n = \frac{1}{N}\mathbf{W}^{-mn}\mathbf{X}_m \tag{8.47}$$

Some useful insight into DFTs is gained by evaluating one.

Example 8.8

Evaluate the DFT of the 8 point sequence

$$\{\underset{\uparrow}{1}, 1, 1, 0, 0, 0, 0, 0\}$$

shown in Figure 8.11.

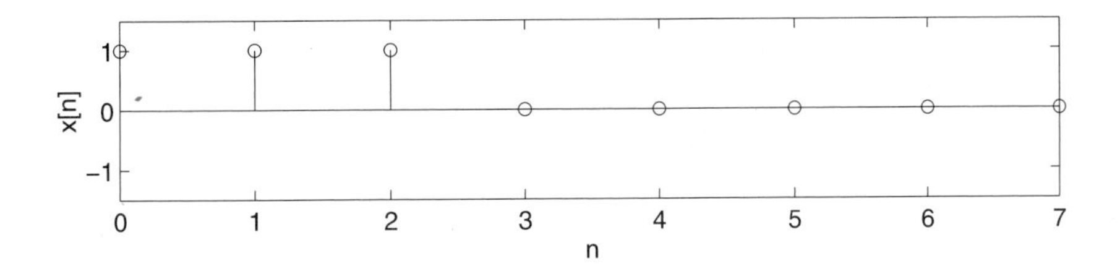

Fig. 8.11 Example of an 8 point sequence

Solution For $N = 8$ Equation (8.43) becomes

$$X_m = \sum_{n=0}^{7} x_n W^{mn} \qquad 0 \leqslant m \leqslant 7$$

where $W = W_8 = e^{-j2\pi/8} = e^{-j\pi/4}$. Inserting the values of x_n

$$\begin{aligned}
X_0 &= 1 \cdot (e^{-j\pi/4})^0 + 1 \cdot (e^{-j\pi/4})^0 + 1 \cdot (e^{-j\pi/4})^0 + 0 + 0 + 0 + 0 + 0 \\
&= 3 \\
X_1 &= 1 \cdot (e^{-j\pi/4})^{1\times 0} + 1 \cdot (e^{-j\pi/4})^{1\times 1} + 1 \cdot (e^{-j\pi/4})^{1\times 2} + 0 + 0 + 0 + 0 + 0 \\
&= 1\angle 0^\circ + 1\angle -45^\circ + 1\angle -90^\circ \\
&= 1 + 0.707 - j0.707 - j \\
&= 1.707 - j1.707 = 2.414\angle -45^\circ \\
X_2 &= 1 \cdot (e^{-j\pi/4})^{2\times 0} + 1 \cdot (e^{-j\pi/4})^{2\times 1} + 1 \cdot (e^{-j\pi/4})^{2\times 2} + 0 + 0 + 0 + 0 + 0 \\
&= 1\angle 0^\circ + 1\angle -90^\circ + 1\angle -180^\circ \\
&= 1 - j - 1 = 1\angle -90^\circ \\
X_3 &= 1 \cdot (e^{-j\pi/4})^0 + 1 \cdot (e^{-j\pi/4})^3 + 1 \cdot (e^{-j\pi/4})^6 + 0 + 0 + 0 + 0 + 0 \\
&= 1\angle 0^\circ + 1\angle -135^\circ + 1\angle 90^\circ \\
&= 1 - 0.707 - j0.707 + j \\
&= 0.293 + j0.293 = 0.414\angle 45^\circ \\
X_4 &= 1 \cdot (e^{-j\pi/4})^0 + 1 \cdot (e^{-j\pi/4})^4 + 1 \cdot (e^{-j\pi/4})^8 \\
&= 1 - 1 + 1 = 1 \\
X_5 &= 1 \cdot (e^{-j\pi/4})^0 + 1 \cdot (e^{-j\pi/4})^5 + 1 \cdot (e^{-j\pi/4})^{10} \\
&= 1\angle 0^\circ + 1\angle 135^\circ + 1\angle -90^\circ \\
&= 0.293 - j0.293 \\
&= 0.414\angle -45^\circ \\
X_6 &= 1 \cdot (e^{-j\pi/4})^0 + 1 \cdot (e^{-j\pi/4})^6 + 1 \cdot (e^{-j\pi/4})^{12} \\
&= 1\angle 0^\circ + 1\angle 90^\circ + 1\angle 180^\circ \\
&= j = 1\angle 90^\circ \\
X_7 &= 1 \cdot (e^{-j\pi/4})^0 + 1 \cdot (e^{-j\pi/4})^7 + 1 \cdot (e^{-j\pi/4})^{14} \\
&= 1\angle 0^\circ + 1\angle 45^\circ + 1\angle 90^\circ \\
&= 1.707 + j1.707 = 2.414\angle 45^\circ
\end{aligned}$$

It should be noted that, in accordance with Equation (8.25), these last three coefficients, X_5, X_6 and X_7, could have been obtained much more easily simply by taking the complex conjugates of X_3, X_2 and X_1 respectively.

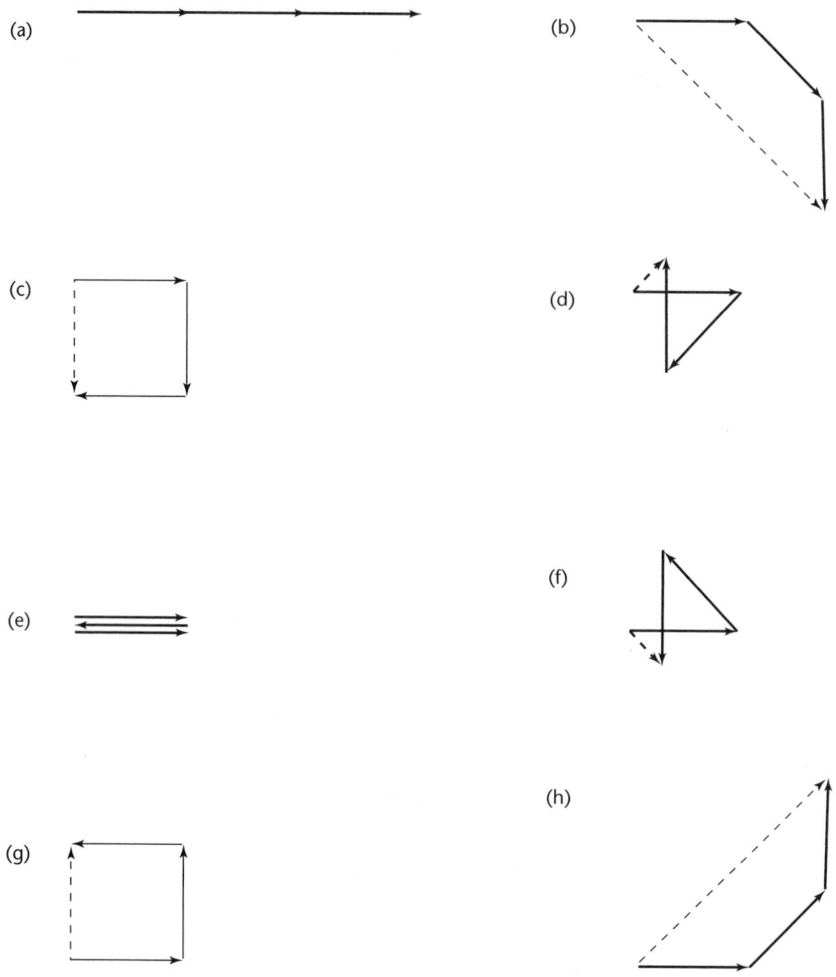

Fig. 8.12 Phasor diagrams for evaluating the DFT of $\{\underset{\uparrow}{1},1,1,0,0,0,0,0\}$ (a) X_0; (b) X_1; (c) X_2; (d) X_3; (e) X_4; (f) X_5; (g) X_6; (h) X_7

Phasor diagrams showing the summation of the three non-zero terms for each of the eight coefficients are given in Figure 8.12. The magnitudes of the eight coefficients are plotted in Figure 8.13.

Example 8.9

Apply the inverse DFT to the coefficients evaluated in Example 8.8 and show that these return the original sequence

$$\{\underset{\uparrow}{1}, 1, 1, 0, 0, 0, 0, 0, \}$$

Solution From Equation (8.37) we have

$$x_n = \frac{1}{8} \sum_{m=0}^{7} X_m W^{-mn} \qquad 0 \leqslant n \leqslant 7$$

Whereas the x_n terms used for evaluating the DFT were real, the X_m terms now needed for evaluating the inverse DFT are complex, and this makes the calculations

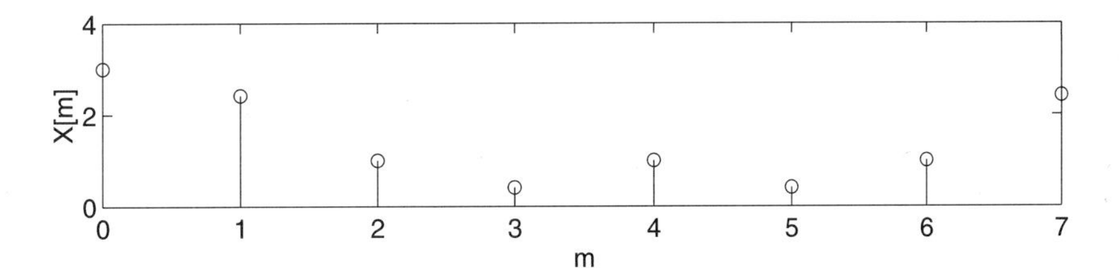

Fig. 8.13 ● The amplitude spectrum of the DFT of $\{\underset{\uparrow}{1},1,1,0,0,0,0,0\}$

more demanding mathematically. For this reason only one x coefficient will be evaluated. This demonstrates the method adequately and confirms that the original value is regained. Choosing x_2 we have

$$\begin{aligned}
x_2 &= \tfrac{1}{8}\{3(\mathrm{e}^{\mathrm{j}\pi/4})^0 + 2.414\angle -45^\circ(\mathrm{e}^{\mathrm{j}\pi/4})^2 + 1\angle 90^\circ(\mathrm{e}^{\mathrm{j}\pi/4})^4 + 0.414\angle 45^\circ(\mathrm{e}^{\mathrm{j}\pi/4})^6 \\
&\quad + 1(\mathrm{e}^{\mathrm{j}\pi/4})^8 + 0.414\angle -45^\circ(\mathrm{e}^{\mathrm{j}\pi/4})^{10} + 1\angle 90^\circ(\mathrm{e}^{\mathrm{j}\pi/4})^{12} + 2.414\angle 45^\circ(\mathrm{e}^{\mathrm{j}\pi/4})^{14}\} \\
&= \tfrac{1}{8}\{3 + 2.414\angle(-45^\circ + 90^\circ) + 1\angle(-90^\circ + 180^\circ) + 0.414\angle(45^\circ + 90^\circ) + 1 \\
&\quad + 0.414\angle(-45^\circ + 90^\circ) + 1\angle(90^\circ - 180^\circ) + 2.414\angle(45^\circ - 90^\circ)\} \\
&= \tfrac{1}{8}\{3 + 1.701 + \mathrm{j}1.707 + \mathrm{j} + 0.293 - \mathrm{j}0.293 + 1 + 0.293 + \mathrm{j}0.293 - \mathrm{j} \\
&\quad + 1.707 - \mathrm{j}1.707\}
\end{aligned}$$

The imaginary terms cancel, leaving $x_2 = 1$ as required.

In Example 8.8 the original signal could be regarded as a sampled version of the rectangular pulse drawn in Figure 8.14(a), for which the continuous Fourier transform produces an amplitude spectrum which is the modulus of a sinc function, as shown in Figure 8.14(b).

It will be noted that this amplitude spectrum looks very different from the amplitude spectrum produced by the DFT that is shown in Figure 8.13. Since one major use of the DFT is as a practical numerical method of obtaining the amplitude spectra of continuous waveforms, this discrepancy clearly needs to be examined further.

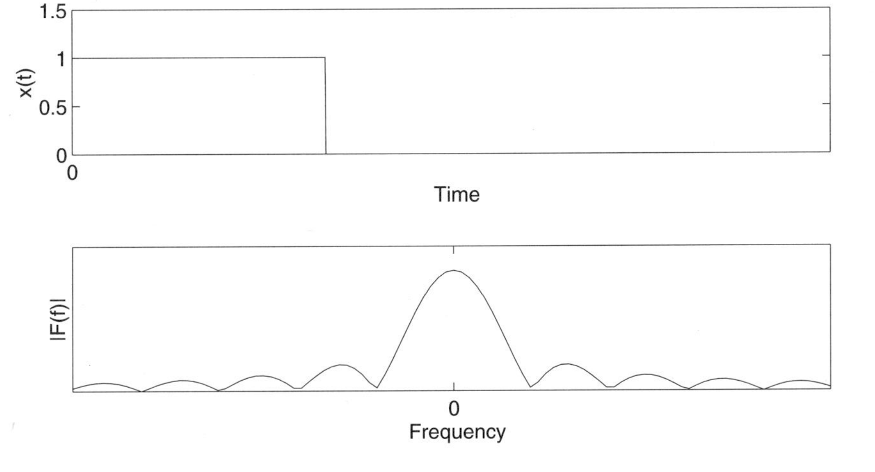

Fig. 8.14 ● Rectangular pulse: (a) waveform; (b) continuous Fourier transform

8.11 Interpretation of the DFT

Most signals that require processing are analogue signals (radar, seismic, cardiac, tomographic and so on). The continuous Fourier transform has many potential applications for such signals, including spectral analysis, convolution and correlation (a discussion of correlation is postponed until Chapter 16). Unfortunately, it is only possible to obtain an analytical solution to the Fourier integral when the signal is a relatively simple deterministic signal, such as one of those listed in Table 6.1. The *practical* way of performing the Fourier transform is to take samples of the signal and to undertake a DFT by means of the fast Fourier transform (FFT) algorithm. It is natural to hope that the DFT will produce a result that is similar to the continuous Fourier transform, and the purpose of this section is to examine whether this is indeed the case. As a cautionary introduction to this, a comparison of Figures 8.13 and 8.14(b) showed that the DFT of a sampled rectangular pulse, as determined by evaluating the DFT of the sequence

$$\{\underset{\uparrow}{1}, 1, 1, 0, 0, 0, 0, 0\}$$

looks substantially different to the continuous Fourier transform of a rectangular pulse.

8.11.1 Differences in the definitions of the continuous and discrete Fourier transforms

The continuous Fourier transform was defined in Equation (6.7) and, applied to a signal $x(t)$, is given by

$$X(f) = \int_{-\infty}^{\infty} x(t)\mathrm{e}^{-\mathrm{j}2\pi ft}\,\mathrm{d}t \tag{8.48}$$

The DFT of a sampled time waveform $x(nt_\mathrm{s})$ was given in Equation (8.37) and is repeated here as

$$X(mf_\mathrm{d}) = \sum_{n=0}^{N-1} x(nt_\mathrm{s})\mathrm{e}^{-\mathrm{j}2\pi mf_\mathrm{d}nt_\mathrm{s}} \qquad 0 \leqslant mf_\mathrm{d} \leqslant (N-1)f_\mathrm{s}/N \tag{8.49}$$

where t_s is the sample spacing and f_d is the separation f_s/N of the frequency coefficients.

There are four major differences between these equations for $X(f)$ and $X(mf_\mathrm{d})$, and these may be summarized as follows.

1. The continuous Fourier transform operates on a continuous waveform $x(t)$, whereas the DFT is applied only to a sampled version $x(nt_\mathrm{s})$ of this waveform.
2. The continuous Fourier transform involves values of $x(t)$ extending over all time, whereas the DFT uses only a finite segment or 'window' of that waveform.
3. The continuous Fourier transform evaluates a continuous frequency function $X(f)$, whereas the DFT evaluates a discrete frequency function $X(mf_\mathrm{d})$ that is constrained to samples f_d apart.
4. The continuous Fourier transform evaluates $X(f)$ over all frequencies, whereas the DFT only evaluates $X(mf_\mathrm{d})$ over a finite frequency window extending from 0 to $(N-1)f_\mathrm{s}/N$.

The next subsection examines the effect of the first two of these differences. Subsection 8.11.3 examines the effect of the third and fourth of these differences.

8.11.2 The effect of sampling and windowing the waveform

Consider the continuous analogue time waveform $x(t)$ shown in Figure 8.15(a) and let us suppose that it has a finite bandwidth such that the magnitude of its Fourier transform (its amplitude spectrum) is as shown in Figure 8.15(b).

The effect of applying a Fourier transform to the windowed $x(nt_s)$ rather than to $x(t)$ is equivalent to applying the Fourier transform to the signal $x(t)p(t)w(t)$, where $p(t)$ is a train of sampling pulses as shown in Figure 8.15(c) and $w(t)$ is a window function as shown in Figure 8.15(f). The product $x(t)p(t)w(t)$ is the waveform shown in Figure 8.15(h). However, with reference to Equation (6.45), the Fourier transform of $x(t)p(t)w(t)$ is the convolution of their individual

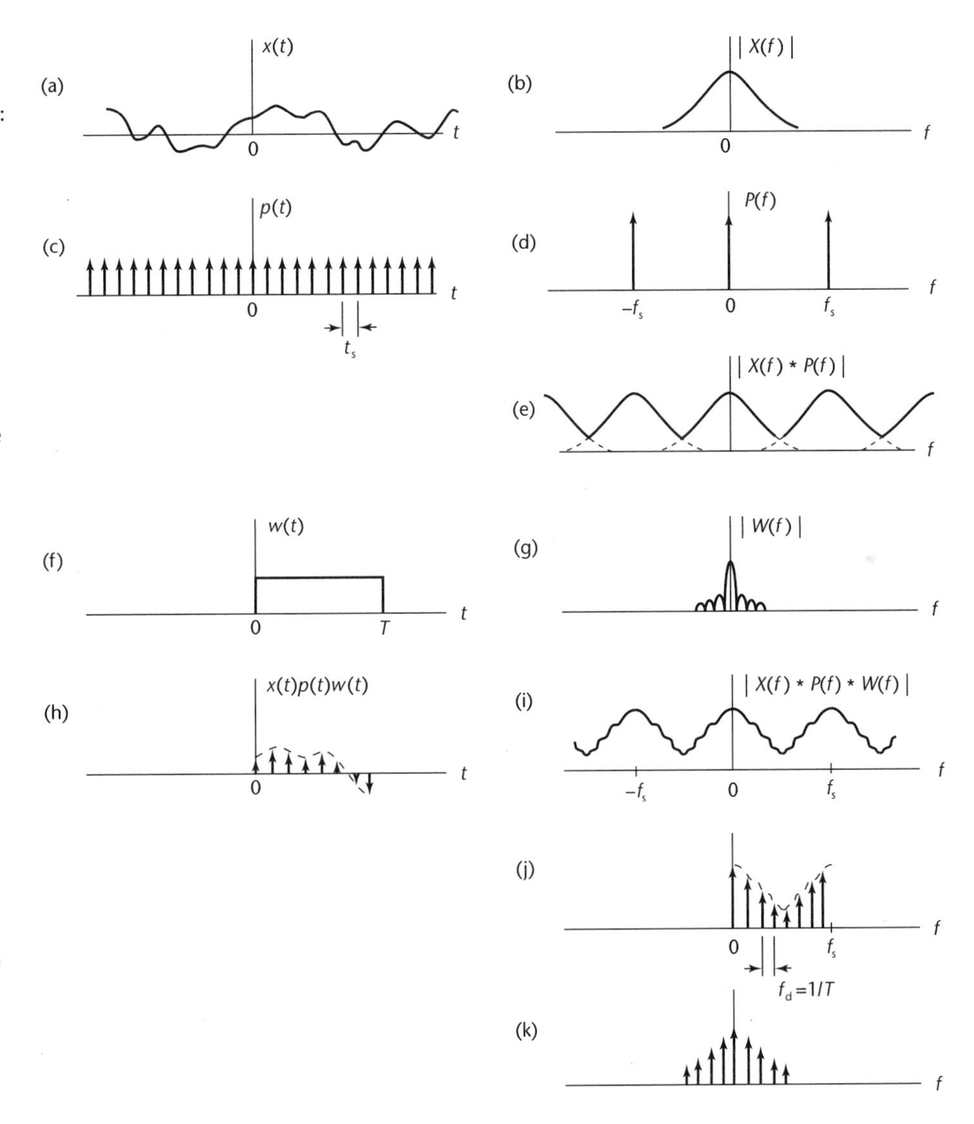

Fig. 8.15 ● Predicting the outcome of a DFT: (a) analogue waveform and (b) its amplitude spectrum; (c) a train of sampling impulses and (d) its amplitude spectrum; (e) amplitude spectrum of sampled analogue waveform; (f) a rectangular window function and (g) its amplitude spectrum; (h) a sampled and windowed analogue waveform and (i) its amplitude spectrum; (j) the same as (i) after sampling and windowing – the amplitude spectrum of the DFT; (k) the same spectrum after reordering of coefficients

Fourier transforms, that is

$$\mathscr{F}\{x(t)p(t)w(t)\} = X(f) * P(f) * W(f) \tag{8.50}$$

A graphical interpretation of this convolution can help reveal the difference between the Fourier transform of $x(t)$ and the Fourier transform of a sampled and windowed version of $x(t)$.

Unfortunately the graphical interpretation of convolution that was applied to time waveforms in Chapter 2 is not rigorously possible because Fourier transforms are complex quantities with real and imaginary components. It is not easy to visualize the effect of multiplication and integration on complex quantities. However, some simplifications and approximations are possible.

The Fourier transform of the train of sampling pulses t_s apart was derived in Example 6.3 and is a set of impulse functions $f_s = 1/t_s$ apart in the frequency domain, as shown by the spectrum $P(f)$ in Figure 8.15(d). This makes the convolution of $X(f)$ with $P(f)$ very simple. We can convolve $X(f)$ with each of these impulse functions separately, and then add the resulting spectra. This is done in Figure 8.15(e). The regions in which there is overlap are regions where there is 'aliasing' and this is where a non-zero phase spectrum in $X(f)$ can complicate the outcome since it means that the summation of aliased components should be done vectorially. Fortunately it is not necessary in this discussion to know the exact shape of the spectrum but only to recognize that aliasing modifies the spectral shapes unless $X(f)$ has decayed to insignificance at half the sampling frequency.

We next require to consider the convolution of $X(f)*P(f)$ with $W(f)$. The Fourier transform of a symmetrical rectangular window of duration τ was shown in Equation (6.9) to be of the form $\sin(\pi f\tau)/\pi f\tau$. Making use of the time shift theorem of Equation (6.51), as done in Example 6.12, the Fourier transform of an offset rectangular window extending from 0 to T (where $T = Nt_s$) is therefore given by

$$W(f) = \frac{\sin(\pi fT)}{\pi fT}\exp(-\mathrm{j}fT) \tag{8.51}$$

The magnitude of this is shown in Figure 8.15(g). Even neglecting the phase spectrum it is difficult to achieve a graphical convolution of $X(f)^*P(f)$ with $W(f)$. However, it is enough in the context of the present discussion to recognize that the effect of the convolution will be to introduce ripple into the resulting spectrum, as shown in Figure 8.15(i). These ripples are commonly referred to as 'leakage' since they arise from the sidelobes of $W(f)$ *leaking out* from its main lobe.

8.11.3 The effect of sampling and windowing the spectrum

The DFT only produces coefficients at the discrete frequencies given by $0 \leqslant mf_d \leqslant (N-1)f_s/N$. Thus the only values of the spectrum in Figure 8.15(i) produced by the DFT are those shown in Figure 8.15(j).

8.11.4 Summary of the relationship between the DFT and the continuous Fourier transform

The following sentence summarizes the similarities and differences between the outcome of a continuous Fourier transform and the outcome of a DFT:

The DFT produces a frequency domain signal that is a sampled and windowed version of what the continuous Fourier transform of a sampled and windowed time domain signal would produce.

In other words the DFT produces Figure 8.15(j) which is a sampled and windowed version of Figure 8.15(i), where Figure 8.15(i) is the continuous Fourier transform of Figure 8.15(h).

8.11.5 Reordering of DFT coefficients

The continuous Fourier transform of Figure 8.15(i) is two-sided and periodic. If we are only able to consider a single spectral period it is more appropriate to consider the region of Figure 8.15(i) lying between $-f_s/2$ and $+f_s/2$ than that lying between 0 and f_s. Therefore, in terms of interpreting the DFT of Figure 8.15(j), it is often useful to imagine the upper half of the frequency coefficients given by $f_s/2 \leqslant mf_d \leqslant (N-1)f_s/N$ to be shifted down in frequency by f_s such that they occupy the frequency range $-f_s/2 \leqslant mf_d \leqslant -f_d$. If the coefficient that was originally at $f_s/2$ is repeated at $f_s/2$, such that it is used at both $-f_s/2$ and $+f_s/2$, there are now $(N+1)$ coefficients occupying the range $-f_s/2 \leqslant mf_d \leqslant +f_s/2$, as shown in Figure 8.15(k). The similarity with the continuous Fourier transform of $x(t)$ shown in Figure 8.15(b) is now much more apparent.

An alternative to this reordering is to consider only the lower half of the DFT, double all the coefficients except X_0, and to interpret it as a *one-sided* spectrum.

8.11.6 An alternative interpretation scheme

The sampled and windowed signal $x(t)p(t)w(t)$ shown in Figure 8.15(h) is repeated in Figure 8.16(a). Let us next imagine that this is made periodic such that it becomes the waveform shown in Figure 8.16(e). Mathematically this can be achieved by

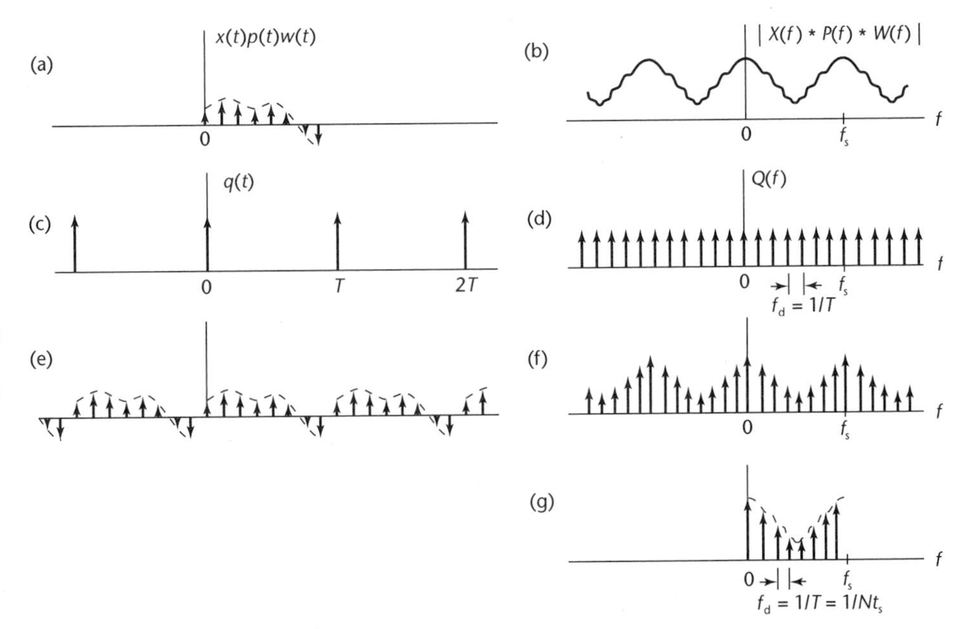

Fig. 8.16 Predicting the outcome of a DFT: (a) a sampled and windowed analogue waveform and (b) its amplitude spectrum; (c) a train of impulses whose period is the duration of the data set and (d) its amplitude spectrum; (e) the data set made periodic and (f) its amplitude spectrum; (g) the outcome of the DFT

convolving $x(t)p(t)w(t)$ with the signal $q(t)$ that is the periodic set of impulse functions T(or Nt_s) apart shown in Figure 8.16(c).

Convolution in the time domain of $x(t)p(t)w(t)$ with $q(t)$ signifies a *multiplication* in the frequency domain of $X(f) * P(f) * W(f)$ by $Q(f)$, where $Q(f)$ is the Fourier transform of $q(t)$. However, $Q(f)$ is a set of impulse functions, $1/T$ or f_d apart in the frequency domain, as shown in Figure 8.16(d). Thus the Fourier transform of the periodic sampled waveform shown in Figure 8.16(e) is the product of Figures 8.16(b) and 8.16(d) and is the sampled spectrum shown in Figure 8.16(f). Compared with Figure 8.15 where we had to sample and window the spectrum of Figure 8.16(i) in order to obtain the DFT, we already have the sampling present in Figure 8.16(f). All that we must do is select the appropriate spectral window. This approach of supposing the finite data set to be made periodic can be very helpful in predicting the outcome of a DFT.

To summarize this:

The DFT produces one period of what is obtained by a continuous Fourier transform acting on the finite data set made periodic.

Example 8.10

Without performing the DFT calculation itself, use the preceding arguments to predict the DFT of 64 samples, 125 μs apart, of a 1000 Hz cosinusoidal signal.

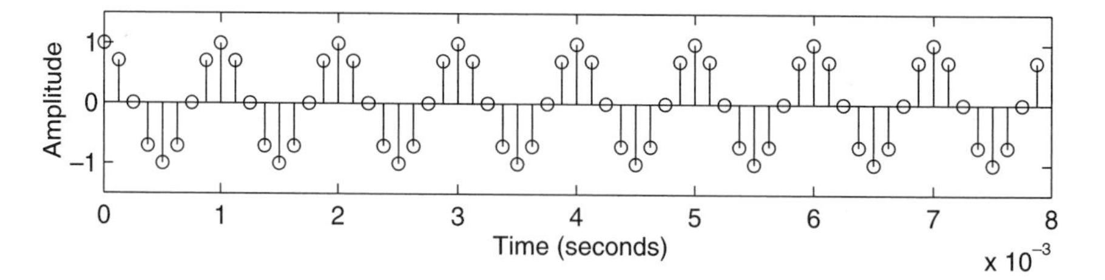

Fig. 8.17 Data set corresponding to eight cycles of a cosinusoid

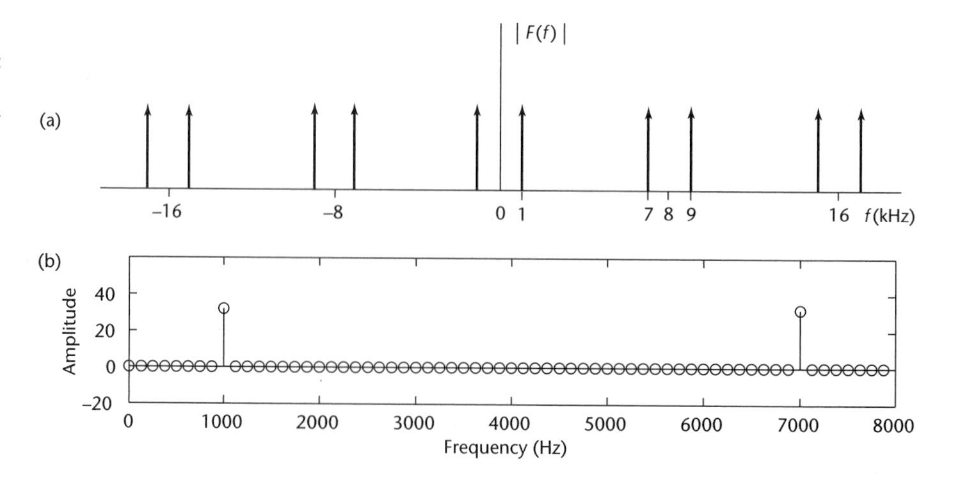

Fig. 8.18 Amplitude spectra: (a) applying the continuous Fourier transform to the data set made periodic; (b) applying the DFT to the finite data set

Solution If the first sample occurs at $t = 0$ and the 64 samples are made to repeat there will be a 65th sample at a time of $(64 \times 0.000125) = 0.008$ s. The period of this periodic data set is thus 0.008 s and is exactly equal to the time of 8 cycles of the 1000 Hz cosinusoid. The data set is thus as shown in Figure 8.17 and means that the effect of making the data set periodic is to produce a sampled 1000 Hz cosinusoid of infinite extent.

However, a sampled 1000 Hz cosinusoid of infinite extent is represented by $p(t)\cos(2000\ \pi t)$, where $p(t)$ is an 8000 Hz sampling signal. Using the frequency translation theorem of Equation (6.52), the continuous Fourier transform of $p(t)\cos(2000\pi t)$ is given by

$$F(f) = P(f) * [\tfrac{1}{2}\delta(f + 1000) + \tfrac{1}{2}\delta(f - 1000)]$$

However, $P(f)$ is a periodic set of impulse functions at the harmonics of the sampling frequency, 8000 Hz apart, and it follows that $F(f)$ is the periodic spectrum shown in Figure 8.18(a). Using the italicized sentence that preceded this example we can deduce that the outcome of a DFT would be the period of Figure 8.18(a) that lies between 0 and 8000 Hz (except that it produces discrete-time impulse functions rather than continuous-time impulse functions). This is shown expanded in Figure 8.18(b) and it is noted that it contains a spectral line without sidelobes (that is without leakage) at 1000 Hz.

An interesting observation from this previous example is that a perfect leakage-free spectrum has been obtained with the DFT. This has been done in spite of having a limited data set. It should be noted, however, that the situation is exceptionally fortuitous in that the waveform resulting from making the finite data set periodic just happens to be identical to the *infinite* data set. This does not normally happen.

Example 8.11

Without performing the DFT calculation itself, use the preceding arguments to predict the DFT of 64 samples, 125 μs apart, of a 1062.5 Hz cosinusoidal signal.

Solution As in Example 8.10 the length of the data set is 0.008 s. For a 1062.5 Hz cosinusoidal signal this signifies that the data set contains exactly 8.5 cycles. If this data set is made periodic we thus obtain Figure 8.19. A close examination shows that this contains discontinuities at 0, 0.008, 0.016 and 0.024 s.

A prediction of the continuous Fourier transform of Figure 8.19 is much less straightforward than that of Figure 8.17 but it can be expected that the resulting spectrum will be much more complicated than that of Figure 8.18. Fortunately there is a simple 'trick' for determining the continuous Fourier transform of Figure 8.19,

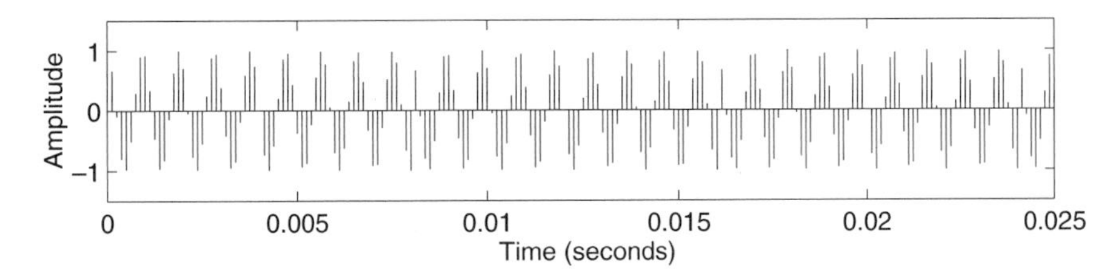

Fig. 8.19 Data set made periodic

and that is to note that Figure 8.19 can be obtained by multiplying the sampled cosine wave of Figure 8.20(a) with the square wave of Figure 8.20(b).

Multiplication in the time domain signifies convolution in the frequency domain and it follows that the continuous Fourier transform of Figure 8.19 is the convolution of $A(f)$ and $B(f)$, where $A(f)$ is the continuous Fourier transform of a sampled 1062.5 Hz cosinusoidal signal as shown in Figure 8.21(a), and $B(f)$ is the continuous Fourier transform of a 62.5 Hz square wave as shown in Figure 8.21(b). Thus the continuous Fourier transform of the data set made periodic is

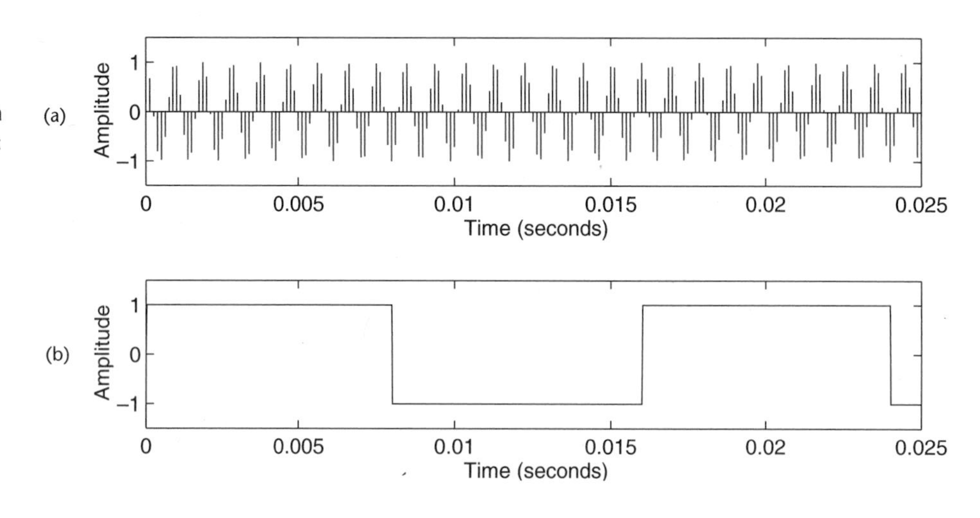

Fig. 8.20 Means of obtaining the data set made periodic: (a) sampled cosinusoid; (b) square wave which multiplied with sampled cosinusoid gives required periodic waveform

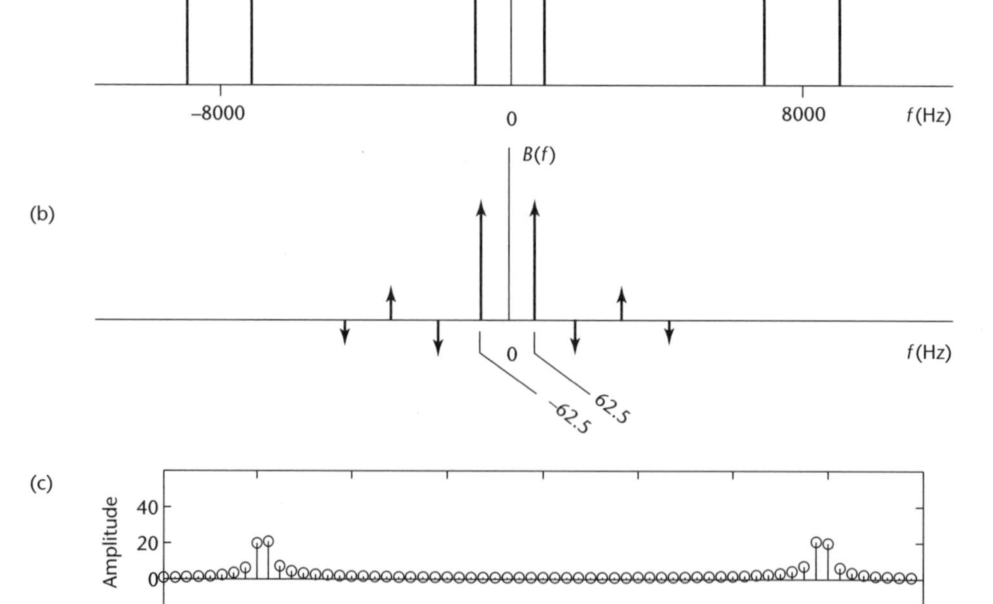

Fig. 8.21 Predicting the outcome of the DFT of 64 samples of a 1062.5 Hz cosinusoid sampled at 8 kHz. (a) The continuous Fourier transform of a sampled 1062.5 Hz cosinusoidal signal; (b) the continuous Fourier transform of a 62.5 Hz square wave on a different frequency scale; (c) the magnitude of the DFT

$A(f) * B(f)$. Using the statement that *the DFT produces one period of what is obtained by a continuous Fourier transform acting on the finite data set made periodic* we thus predict that the DFT is one period of $A(f) * B(f)$. The magnitude of the DFT coefficients is shown in Figure 8.21(c) and confirms the prediction. It is noted that the DFT comes far from producing a clean spectral line at 1062.5 Hz.

A comparison of the previous two examples shows that the outcome of a DFT can be very different in form depending on quite subtle differences in the signal being analysed. These two examples show up very clearly some of the pitfalls when the DFT is used for spectral analysis. The result in Figure 8.18 is perfect, whereas that in Figure 8.21 is quite poor. The subject of spectral analysis will be examined more thoroughly in the next chapter.

8.12 Summary

It has been shown that any discrete-time signal can be considered to be made up of the sum of a set of discrete-time sinusoids and cosinusoids. The set of discrete-time sinusoids and cosinusoids can be considered a set of basis functions. A closely related set of basis functions is a set of discrete-time complex exponentials and the summation expressing the discrete-time signal as the summation of these discrete-time complex exponentials is the inverse discrete Fourier transform. The expression that gives the *coefficients* of these discrete-time complex exponentials in terms of the signal is the discrete Fourier transform itself.

The DFT is frequently used when what is really wanted is the continuous Fourier transform (for example for the spectral analysis of an analogue signal). The DFT is related to the continuous Fourier transform and one of the most useful ways of predicting the outcome of a DFT lies in the statement that 'the DFT produces one period of what is obtained by a continuous Fourier transform acting on the finite data set made periodic'. The outcome of a DFT appears substantially different from that of a continuous Fourier transform. Various methods for obtaining results more similar to true spectra are discussed in the next chapter.

8.13 Problems

1. Evaluate by hand the DFT, $X[m] = \sum_{n=0}^{N-1} x[n]e^{-j2\pi mn/N}$ for the signal $x[n] = \{3, 2, -1, 0, 0, 1, 0, 2\}$. Considering the inverse DFT given by $x[n] = 1/N\sum_{m=0}^{N-1} X[m]e^{j2\pi mn/N}$ combine the terms for which $m = k$ and $m = N - k$ and thus express $x[n]$ as the sum of discrete-time trigonometric basis functions. Plot each of the eight basis functions and show that their sum does indeed equal $x[n]$.

2. A DFT is performed on a signal using 1024 samples 0.1 ms apart. What is the spacing of the frequency coefficients and what frequency range do they cover?

3. A 20 point sequence given by $x[n] = \{1, 1, 1, 0, 0, 0, 0, 0, 1, 1, 1, 1, 1, 0, 0, 0, 0, 0, 1, 1\}$ has its samples 1 ms apart. Based solely on the statement that *the DFT produces one period of what is obtained by a* *continuous Fourier transform acting on the finite data set made periodic*, predict its DFT.

4. It is wished to use the DFT to determine the spectrum of a rectangular pulse 1 ms long. To do this the pulse is sampled at 10 kHz and padded with zeros such that the DFT acts on a sequence of 10 ones followed by 54 zeros. Based solely on the statement that *the DFT produces one period of what is obtained by a continuous Fourier transform acting on the finite data set made periodic*, predict and sketch the approximate amplitude spectrum that would be obtained in the region $0 < f < 2$ kHz.

5. A DFT analysis is made of 512 samples 1 ms apart of the signal $\cos(1890t) + 0.05\cos(1915t)$. Present an argument of whether the DFT coefficients will or will not clearly separate the two cosinusoidal components. Repeat for the signal $\cos(2301t) + 0.05\cos(2325t)$.

6. The signal $\cos(4909t)$ is sampled at 10 kHz and 128 of these samples undergo a DFT in order to produce a display of the estimated amplitude spectrum. State the separation of the frequency coefficients and the range they cover. Sketch the magnitude of the frequency coefficients and show how these should be arranged to obtain the spectrum most similar to the true spectrum of $\cos(4909t)$. Describe how the amplitude spectrum will appear if the frequency of the signal is increased to 5154 rad/s and explain why it is very different in appearance from when the signal frequency is 4909 rad/s.

Chapter nine

The fast Fourier transform and some applications

9.1 ● Preview

The DFT has important applications in signal processing for spectral analysis, convolution and correlation. The fast Fourier transform (FFT) is simply an efficient algorithm for evaluating a DFT. It produces an *identical* result to the DFT but involving fewer calculations. The FFT is used so universally for evaluating the DFT that the term FFT is often used even when the method of calculation is not under consideration.

Many scientific software packages (of which MATLAB is one) include an FFT command that is as easy to apply as the commands for functions such as sines or logs. Thus, just as most users are unconcerned how a software package achieves these latter functions, they are likely to be equally uninterested in the computation of a fast Fourier transform. Hence, although some insight is given to the FFT algorithm itself, the main emphasis of this chapter concerns the use of the FFT and its computational advantages compared with a 'brute force' DFT.

9.2 Computational demands of the discrete Fourier transform

Consider the DFT pair given by Equations (8.41) and (8.42), namely

$$X[m] = \sum_{n=0}^{N-1} x[n]\, W^{mn} \qquad 0 \leqslant m \leqslant (N-1)$$

and

$$x[n] = \frac{1}{N} \sum_{m=0}^{N-1} X[m]\, W^{-mn} \qquad 0 \leqslant n \leqslant (N-1)$$

where W equals $e^{-j2\pi/N}$ and is a complex quantity (the subscript $_N$ is omitted from W_N for simplicity).

It is common for $x[n]$ to be samples of a real waveform and the evaluation of each coefficient $X[m]$ in the DFT therefore concerns the multiplication of N real sample amplitudes by N complex values of W^{mn}, followed by a summation. There are N coefficients to be evaluated. Thus, neglecting the summations because they are computationally less demanding than complex multiplications, the time taken to evaluate a DFT is approximately that needed for N^2 complex multiplications. There is some small scope for computational savings by exploiting the result that $X[N-k] = \overset{*}{X}_k$, and this approximately halves the computational load.

When considering the *inverse* DFT the $X[m]$ values are usually complex and the evaluation of each coefficient $x[n]$ in the inverse DFT therefore requires the multiplication of N complex coefficients by N complex values of W^{mn}, followed by a summation. The multiplication of two complex numbers requires more computation time than the multiplication of a complex number by a real number in the case of the 'forward' DFT. Furthermore, the $x[n]$ values are usually independent and there is no scope for any computational savings based upon conjugate dependencies between them. Thus the inverse DFT tends to be computationally more demanding than the 'forward' DFT. The amount of computation involved in an inverse DFT will therefore be taken as the 'yardstick' against which the computational demands of the FFT will be compared. This yardstick will be taken as being the time taken to compute N^2 complex multiplications.

If N is large the N^2 complex multiplications of the DFT can take a long time on a computer or in a digital processor. The purpose of the FFT is to reduce this number of complex multiplications. (Note: The physical meaning and practical relevance of *complex* waveforms has been introduced at the end of Chapter 8 and will be discussed further in Chapter 17. When they do arise the computational demands of the forward DFT become as great as those of the inverse DFT).

9.3 Computational demands of the fast Fourier transform

The objective of this section is not to explain the thought processes that led Cooley and Tukey to discover the FFT algorithm in 1965, nor to present code for the algorithm. Instead it will confine itself to just one of *many possible* signal flow graphs for an 8 point FFT, confirm that it provides the correct result, show how its concept may be extended to a larger transform, and finally determine the reduced number of complex multiplications now required. Other FFT algorithms have identical or very similar computational demands.

Figure 9.1 is the signal flow graph of an 8 point transform for one of the most common FFT algorithms. On the left-hand side the input coefficients are presented in their natural sequence. On the right-hand side the output coefficients emerge in a scrambled sequence. Each one of the inputs contributes to each one of the outputs and has a path by which this occurs. Wherever the path includes a solid line the input coefficient is multiplied by W^k, where k is the number in the circle at the end of the arrowhead. Wherever the path includes a dotted line no such multiplication is imposed. For example, Figure 9.2 shows the path between the input coefficient $x[5]$ and the output coefficient $X[3]$. It commences with a solid line so that the first circle indicates that a multiplication of W^4 must be applied. The next part of the path

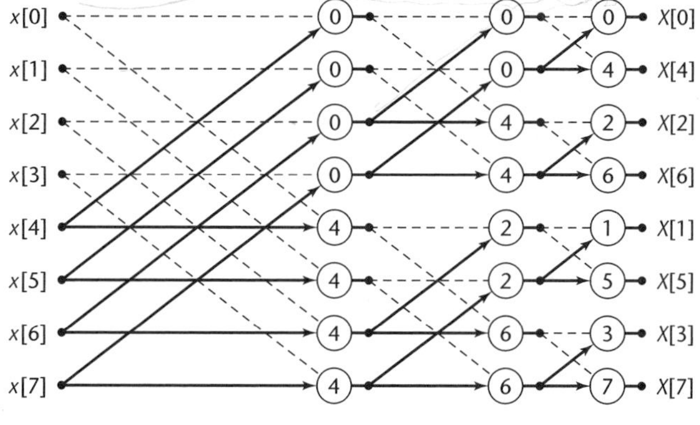

Fig 9.1 Flow graph for FFT

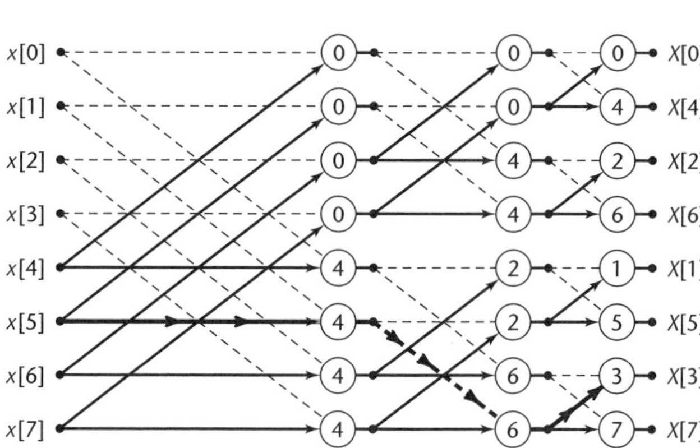

Fig. 9.2 Contribution of input $x[5]$ to output $X[3]$

involves a dotted line so that the multiplication of W^6 within the circle is not applied. The final part is another solid line such that the multiplication of W^3 within the circle is applied. Thus the contribution of $x[5]$ to $X[3]$ is $x[5]W^4W^3$, or simply $x[5]W^7$.

If we consider the contributions of the other seven inputs in a similar way we find that

$$X[3] = x[0] + x[1]W^3 + x[2]W^6 + x[3]W^9 + x[4]W^4 + x[5]W^7 + x[6]W^{10} + x[7]W^{13}$$

To verify that this is the correct answer we can compare it with the result using the definition of the DFT, that $X[m] = \sum_{n=0}^{N-1} x[n]W^{mn}$. This gives

$$X[3] = x[0] + x[1]W^3 + x[2]W^6 + x[3]W^9 + x[4]W^{12} + x[5]W^{15} + x[6]W^{18} + x[7]W^{21}$$

However, $W = e^{-j2\pi/8}$ such that an increase in the exponent of W by an integer multiple of 8 make no difference to its value. For example $W^{15} = W^8W^7 = W^7$. The application of this result shows that the expression for $X[3]$ using the FFT is identical to that using the DFT. The same can be demonstrated for all the other coefficients.

The signal flow graph of Figure 9.1 is for eight coefficients. It will be noted that, working backwards through it, the four sets of neighbouring output coefficients are associated with four FFT 'butterflies', the name arising from the way in which they are connected to the multipliers of the previous column by four lines having a shape akin to a pair of butterfly wings.

Working further backwards, each pair of 'butterflies' may be considered as joined to the previous column of multipliers by a larger 'butterfly'. Finally these two large 'butterflies' interact to form an even larger 'butterfly'.

The concept of butterflies helps reveal the symmetries of the flow graph. It becomes easy to see how the flow graph would be extended to handle $16, 32, \ldots$ or 2^r inputs. Whereas there are three columns of multipliers for 8 inputs, there would be 4 columns for 16 inputs, 5 columns with 32 inputs and r columns for 2^r inputs.

The next aspect to be examined is the number of multiplications involved in the FFT. Returning to Figure 9.2 and considering the passage of $x[5]$ to $X[3]$ it appears that there are *three* multiplications involved: $X[3] = x[5].W^4.1.W^3$. If the contributions of *all eight inputs* to $X[3]$ are included there are a total of 3×8 multiplications that contribute to $X[3]$.

Similarly each of the eight other output coefficients involves 3×8 multiplications, thus apparently making $3 \times 8 \times 8$ multiplications in all. This compares poorly with the 8×8 multiplications required by a conventional DFT. However, what has been overlooked here is that many of the multiplications are common to many of the paths and need not be repeated.

The signal flow graph of Figure 9.1 for an 8 point FFT contains three columns of circles. Let these be applied one column at a time, with each circle producing an output which is the *sum* of the multiplications arising from *both* of its inputs (it is now established that the circles represent multiplier/adders). The results are then *stored* and used for calculating the outputs of the next column.

Let the outputs of the first column of multiplier/adders be denoted by the vector $a[n]$, those from the second column of multiplier/adders be denoted by the vector $b[n]$, and those from the third column of multiplier/adders be denoted by the vector $c[n]$. Thus

$$\begin{aligned}
a[0] &= x[0] + x[4]\,W^0 \\
a[1] &= x[1] + x[5]\,W^0 \\
a[2] &= x[2] + x[6]\,W^0 \\
a[3] &= x[3] + x[7]\,W^0 \\
a[4] &= x[0] + x[4]\,W^4 \\
a[5] &= x[1] + x[5]\,W^4 \\
a[6] &= x[2] + x[6]\,W^4 \\
a[7] &= x[3] + x[7]\,W^4
\end{aligned} \tag{9.1}$$

The stored vector $a[n]$ is now used as the input to the second column. Thus

$$\begin{aligned}
b[0] &= a[0] + a[2]\,W^0 \\
b[1] &= a[1] + a[3]\,W^0 \\
b[2] &= a[0] + a[2]\,W^4 \\
b[3] &= a[1] + a[3]\,W^4 \\
b[4] &= a[4] + a[6]\,W^2 \\
b[5] &= a[5] + a[7]\,W^2 \\
b[6] &= a[4] + a[6]\,W^6 \\
b[7] &= a[5] + a[7]\,W^6
\end{aligned} \tag{9.2}$$

These results are again stored and then used as the inputs to the third column of multiplier/adders. The outputs of the third column are

$$\begin{aligned}
c[0] &= b[0] + b[1]\,W^0 \\
c[1] &= b[0] + b[1]\,W^4 \\
c[2] &= b[2] + b[3]\,W^2 \\
c[3] &= b[2] + b[3]\,W^6 \\
c[4] &= b[4] + b[5]\,W^1 \\
c[5] &= b[4] + b[5]\,W^5 \\
c[6] &= b[6] + b[7]\,W^3 \\
c[7] &= b[6] + b[7]\,W^7
\end{aligned} \tag{9.3}$$

The number of operations needed to derive each of these equations is eight complex multiplications and eight additions. Thus the total number of operations is 24 complex multiplications and 24 additions. An example will now be given to show that $c[n]$ supplies the required transform coefficients, except in a scrambled order as indicated by Figure 9.1.

Supposing that we require $X[3]$, we see from Figure 9.1 that this corresponds to $c[6]$. Working backwards from Equation (9.3), through Equation (9.2), and then Equation (9.1), we find

$$\begin{aligned}
c[6] &= b[6] + b[7]\,W^3 \\
&= (a[4] + a[6]\,W^6) + (a[5] + a[7]\,W^6)\,W^3 \\
&= x[0] + x[4]\,W^4 + (x[2] + x[6]\,W^4)\,W^6 + (x[1] + x[5]\,W^4)\,W^3 \\
&\quad + (x[3] + x[7]\,W^4)\,W^6\,W^3
\end{aligned}$$

This contains one contribution from each of the input coefficients and is readily seen to equal the value of $X[3]$ obtained earlier. The same procedure confirms that the other output coefficients are correct.

The total number of complex multiplications required in this 8 point transform is 3×8, or 24. This is much less than the 8×8 multiplications required by the DFT, or the $3 \times 8 \times 8$ multiplications required if the transform algorithm applies the flow graph of Figure 9.1 *incorrectly* by separately computing the contribution of the path between each input and each output.

If these principles are extended to a signal flow graph with $N = 2^r$ inputs there would now be r columns, each containing N multiplier/adders. The total number of multiplications needed would be $r \cdot N$. Alternatively, writing $r = \log_2 N$, the total number of multiplications is $N \log_2 N$. When N is large this number is considerably less than the N^2 multiplications required by the DFT. Table 9.1 lists N^2 and $N \log_2 N$ for various values of N. For a 2048 point transform the number of multiplications is reduced by a factor of 186. Neglecting the additions, that which might take 3 minutes on a PC using the DFT directly would take only 1 second with the FFT!

The flow graph of Figure 9.1 is for a forward FFT. For an inverse FFT it would be identical except that the multiplying factors would have *negative* exponents and the final sum would need to be divided by N. For example the multiplier corresponding to the circle with a four in it should be changed from W^4 to W^{-4}.

Table 9.1 Multiplications required by DFT and FFT

N	N^2	$N\log_2 N$
8	64	24
32	1 024	160
128	16 384	896
512	262 144	4 608
2048	4 194 304	22 528

This follows directly from the difference in the definition of the DFT with that of the inverse DFT.

Example 9.1

Calculate the 8 point DFT of the sequence

$$\{\underset{\uparrow}{2}, 1, 3, 1, 4, 0, 0, 0\}$$

using the FFT flowchart of Figure 9.1. Do this is in such a way that no more than $N\log_2 N$ complex multiplications are needed (the number is actually less than $N\log_2 N$ because some of the multiplications are by 1 or 0).

Solution For an 8 point sequence we have $W = 1\angle -\pi/4$. Therefore

$$\begin{aligned}
a[0] &= x[0] + x[4]W^0 = 2 + 4 &&= 6 \\
a[1] &= x[1] + x[5]W^0 = 1 + 0 &&= 1 \\
a[2] &= x[2] + x[6]W^0 = 3 + 0 &&= 3 \\
a[3] &= x[3] + x[7]W^0 = 1 + 0 &&= 1 \\
a[4] &= x[0] + x[4]W^4 = 2 + 4\angle -\pi &&= -2 \\
a[5] &= x[1] + x[5]W^4 = 1 + 0 &&= 1 \\
a[6] &= x[2] + x[6]W^4 = 3 + 0 &&= 3 \\
a[7] &= x[3] + x[7]W^4 = 1 + 0 &&= 1
\end{aligned}$$

Including the multiplications by 1s and 0s the operations of this first column require 8 multiplications. Proceeding to the second column

$$\begin{aligned}
b[0] &= a[0] + a[2]W^0 = 6 + 3 &&= 9 \\
b[1] &= a[1] + a[3]W^0 = 1 + 1 &&= 2 \\
b[2] &= a[0] + a[2]W^4 = 6 - 3 &&= 3 \\
b[3] &= a[1] + a[3]W^4 = 1 - 1 &&= 0 \\
b[4] &= a[4] + a[6]W^2 = -2 + 3\angle -90^\circ &&= -2 - 3\mathrm{j} \\
b[5] &= a[5] + a[7]W^2 = 1 + 1\angle 90^\circ &&= 1 - \mathrm{j} &&= 1.414\angle -45^\circ \\
b[6] &= a[4] + a[6]W^6 = -2 + 3\angle 90^\circ &&= -2 + \mathrm{j}3 &&= 3.6\angle 123.7^\circ \\
b[7] &= a[5] + a[7]W^6 = 1 + 1\angle 90^\circ &&= 1 + \mathrm{j} &&= 1.414\angle 45^\circ
\end{aligned}$$

This requires a further 8 multiplications. Proceeding to the third column

$$\begin{aligned}
c[0] &= b[0] + b[1]W^0 = 9 + 2 = 11 = X[0] \\
c[1] &= b[0] + b[1]W^4 = 9 - 2 = 7 = X[4]
\end{aligned}$$

$$c[2] = b[2] + b[3]\,W^2 = 3 + 0 = 3 = X[2]$$
$$c[3] = b[2] + b[3]\,W^6 = 3 + 0 = 3 = X[6]$$
$$c[4] = b[4] + b[5]\,W^1 = -2 - \mathrm{j}3 + 1.414\angle -45°.1\angle -45° = -2 - \mathrm{j}4.414$$
$$= 4.85\angle -114.4° = X[1]$$
$$c[5] = b[4] + b[5]\,W^5 = -2 - \mathrm{j}3 + 1.414\angle -45°.1\angle -135° = -2 - \mathrm{j}1.586$$
$$= 2.55\angle -141.6° = X[5]$$
$$c[6] = b[6] + b[7]\,W^3 = -2 - \mathrm{j}3 + 1.414\angle 45°.1\angle -135° = -2 + \mathrm{j}1.586$$
$$= 2.55\angle 141.6° = X[3]$$
$$c[7] = b[6] + b[7]\,W^7 = -2 - \mathrm{j}3 + 1.414\angle 45°.1\angle 45° = -2 + \mathrm{j}4.414$$
$$= 4.85\angle 114.4° = X[7]$$

This also involves 8 complex multiplications, thus making a total for all three columns of 8×3, or $N \log_2 N$, where $N = 8$. It will be noted that some of these calculations could have been avoided by taking advantage of the result that $X[N-k] = X_k^*$, since this gives us

$$X[7] = X^*[1], \qquad X[6] = X^*[2], \qquad X[5] = X^*[3].$$

As mentioned earlier, there are many variations on the signal flow graph of Figure 9.1, but these will not be studied in this book. There are some differences in their computational efficiency, but only very subtle ones. The important point is that they all avoid the large number of duplicate calculations that occur with a direct DFT. Instead, results of multiplications are stored and then used for the computation of more than one output coefficient.

The purpose of this section has been to provide some insight into the nature of the FFT flowchart and to gain an appreciation of the computational savings that result. It is worth ending with an example of its application in MATLAB.

Example 9.2

Repeat the computation of Example 9.1 using MATLAB.

Solution The following is a suitable program.

```
% fig9_3.m
x=[2 1 3 1 4 0 0 0];
y=fft(x);
% It should be remembered that MATLAB arrays are not numbered from index 0
% but from index 1, and this includes the outcome of the FFT. To overcome this it
% is useful to generate an array of numbers that increase progressively from zero,
% and to plot the FFT against this.
n=[0:7];
subplot(4,1,1)
stem(n,abs(y))
ylabel('fft_mag')
```

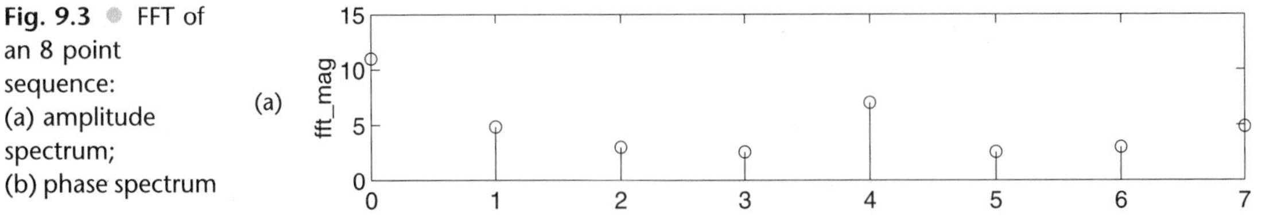

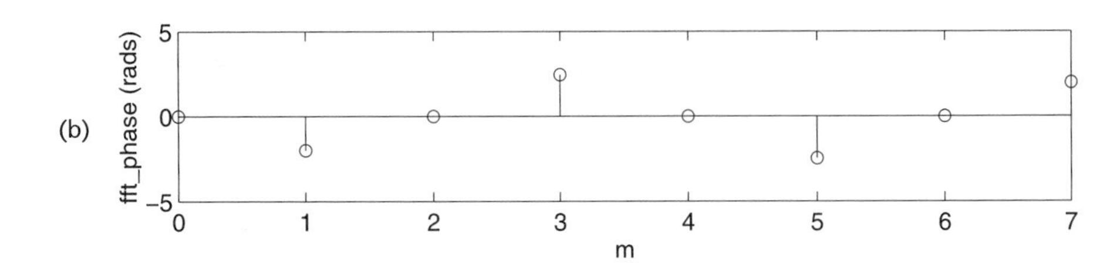

Fig. 9.3 FFT of an 8 point sequence: (a) amplitude spectrum; (b) phase spectrum

```
subplot(2,1,2)
stem(n,angle(y))
xlabel('m'),ylabel('fft_phase(rads)')
```

The outcome is shown in Figure 9.3.

It should be noted that this example represents a total underutilization of the power of MATLAB. A more typical task would be a 1024 point transform and the student edition can even transform 8192 points quite rapidly.

9.4 Spectral analysis

One of the most common applications of the FFT is that of determining the spectral content of power signals. One indication of the spectral content is the amplitude spectrum of a segment of the signal. Much more common, however, is the power spectral density (PSD) defined using Equation (6.38) as

$$S(f) = \lim_{T\to\infty} \frac{1}{T} |X_T(f)|^2$$

where $X_T(f)$ is the Fourier transform of a segment of the signal $x(t)$ of duration T. We face two problems. The first is whether we can get a good estimate of the PSD when T is finite. The second is whether any errors are introduced because of the practical necessity of using the FFT rather than the continuous Fourier transform.

A PSD estimate obtained by means of the DFT/FFT is termed a periodogram, named as such since it is intended to reveal any periodicities in the signal being analysed. Its definition is a discrete-time equivalent of the definition of the PSD and is

$$S[m] = \frac{1}{N} |X_T[m]|^2 \tag{9.4}$$

where the data set of duration T contains N samples and has a DFT $X_T[m]$.

In order to encompass a wide dynamic range of signal amplitudes it is customary to plot PSDs using a decibel scale of power versus a linear scale of frequency, and it should be noted that, apart from an offset due to the division by N in the definition of the PSD estimate, the amplitude spectrum and power spectrum will then be identical. This is because the amplitude spectrum would be plotted as $20 \log_{10} |X_T[m]|$ and the PSD would be plotted as $10 \log_{10} |X_T[m]|^2 - 10 \log_{10} N$, or $20 \log_{10} |X_T[m]| - 10 \log_{10} N$. It is rare to be concerned about the absolute levels of power, in which case the additional $10 \log_{10} N$ term is unimportant. In what follows, there will be an alternation between plots of $|X_T[m]|$, $|X_T[m]|^2$ and $20 \log_{10} |X_T[m]|$. The choice will depend primarily on which has the most suitable dynamic range to show up the features under discussion.

Since the particular choice of signal segment can introduce small artefacts it can be useful to average PSD estimates from several segments, then giving what is termed an averaged periodogram. This can be particularly advantageous if the total number of data points available is excessive for a single FFT. Even if the segment length is amenable to a single FFT it may be beneficial to break it up into shorter segments so that the same strategy can be adopted. This degrades the frequency resolution but reduces the variance of the estimate. If this approach is adopted it is a good idea to have overlapping segments, since, for a given number of segments to be averaged, this extends the duration of each. Segments with windowing and 50% overlaps are particularly effective.

Section 8.11 concerned the interpretation of DFTs and the differences between the DFT and the continuous Fourier transform. The problems concerned the effect of sampling and windowing the time domain signal, and the effect of sampling and windowing the resulting frequency domain signal. To summarize, the effect of sampling in the time domain was that the spectrum could become distorted due to *aliasing*; the effect of windowing in the time domain was that the ideal spectrum became convolved with the Fourier transform of the rectangular window of duration T and this introduced false ripple in the spectrum, an effect known as *leakage*; the effect of sampling in the frequency domain is that the spectrum is only seen at discrete frequencies $f_d = 1/T$ apart and it is rather as if the ideal continuous spectrum is seen through the vertical slits in a picket fence, such that the sampled nature of an FFT spectrum is often referred to as the *picket fence* effect; and the effect of windowing in the frequency domain is that the spectrum is only usefully available between zero frequency and half the sampling frequency.

Concerning the problem of aliasing there are two possible approaches for avoiding spectral distortion. One is that the sampling frequency must be increased sufficiently that the true amplitude spectrum has no significant content above half the sampling frequency, but the problem here is that sampling frequency may be predetermined and, if not, we have the disadvantage that more samples mean a greater computational load. The other approach is that the signal must first be passed through an anti-aliasing lowpass filter to force the tails of $|X_T(f)|$ to be negligible above half the sampling frequency. Clearly this latter approach can truncate the spectrum, but it is generally more acceptable than the interference effects of aliasing.

Before discussing ways of minimizing the leakage and picket fence effects it is worth first considering an example of spectral analysis.

Example 9.3

Sixty-four samples 125 μs apart are available of the supposedly 'unknown' signal $x(t) = \sin(2\pi f_1 t) + 0.05\sin(2\pi f_2 t)$, where $f_1 = 1062.5$ Hz and $f_2 = 1625$ Hz. Use MATLAB to perform an FFT analysis on the data set and plot the amplitude spectrum.

Solution The following is a listing of suitable MATLAB code.

```
ts=0.000125;                              % the sampling interval in seconds
f1=1062.5                                 % the frequency of the first sine wave
f2=1625                                   % the frequency of the second sine wave
t=(0:63)*ts;                              % a vector of 64 sampling instants
x=sin(2*pi*f1*t)+.05*sin(2*pi*f2*t);      % a vector of the digital sequence x(nts)
```

We can now display this waveform in the upper half of the screen using the subplot command. The 64 discrete data values without any interpolation are shown most clearly using the 'stem' command.

```
subplot(3,1,1)
stem(t,x)                                 % plot x(nts) against time.
axis([0 .008 -1.5 1.5])                   % set the scales of the axes.
xlabel('Time (seconds)'),ylabel('x[n]')
```

The outcome of this is shown in Figure 9.4(a).

The FFT generates N frequency coefficients $f_d = 1/Nt_s$ apart. For a 64 point transform the frequencies are derived with

```
f=(0:63)/(64*ts)
```

To plot the magnitude of the FFT against frequency in the second viewing window on the screen we have

```
subplot(3,1,2), stem(f,abs(fft(x)))
axis([0 8000 0 30])                       % this sets the scales of the axes
xlabel('Frequency (Hz)'), ylabel('fftmag')
```

The outcome is shown in Figure 9.4(b) and is the required amplitude spectrum.

Although Figure 9.4(b) is the required FFT, the result is easier to interpret if the upper half of the coefficients are ignored or are translated down in frequency by 8000 Hz to lie in the negative frequency range. This down translation can be done very conveniently in MATLAB using the **'fftshift'** command. Using the instructions:

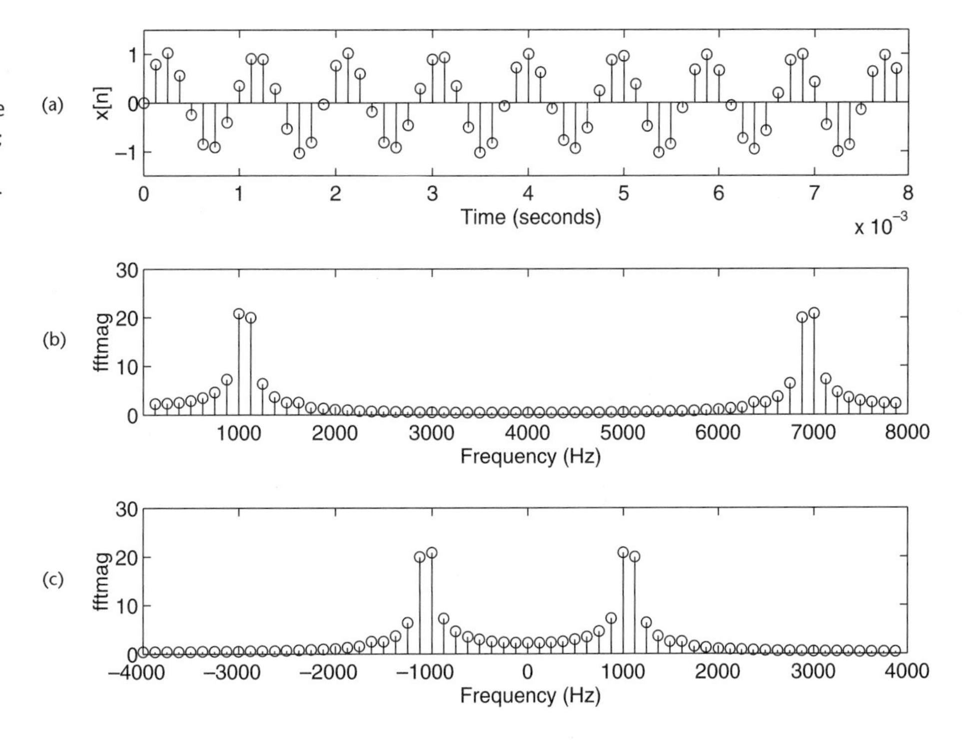

Fig. 9.4 FFT analysis: (a) data set; (b) magnitude of FFT coefficients; (c) coefficients recorded to better show comparison with continuous Fourier transform

```
f1 = f-4000;
subplot(3,1,3), stem(f1,abs(fftshift(fft(x))))
axis([-4000 4000 0 30])
xlabel('Frequency (Hz)'), ylabel('fftmag')
```

the plot of Figure 9.4(c) is obtained.

Several points are worthy of note in Example 9.3.

1. The sampling frequency is 8000 Hz. The final result is easier to interpret as a spectrum if the coefficients above half the sampling frequency are ignored or are translated down in frequency by 8000 Hz to lie in the negative frequency range, as in Figure 9.4(c). This avoids the components at around 7000 Hz that are absent in the analogue signal.
2. The power spectral density can be obtained by squaring the amplitude spectrum but, in this particular example, a plot of amplitude spectrum is better since weaker coefficients are less apparent if squared. (Note that in order to accommodate a wide dynamic range of coefficients it is common in practice to plot spectra on a dB scale: 10 $\log_{10}$(PSD) or equivalently 20 $\log_{10}$(amplitude spectrum).)
3. The frequency coefficients occur at integer multiples of $f_d = f_s/N$ (that is, at multiples of 125 Hz). This means that the frequency of the stronger sinusoid is exactly midway between coefficients at 1000 and 1125 Hz. The picket fence effect

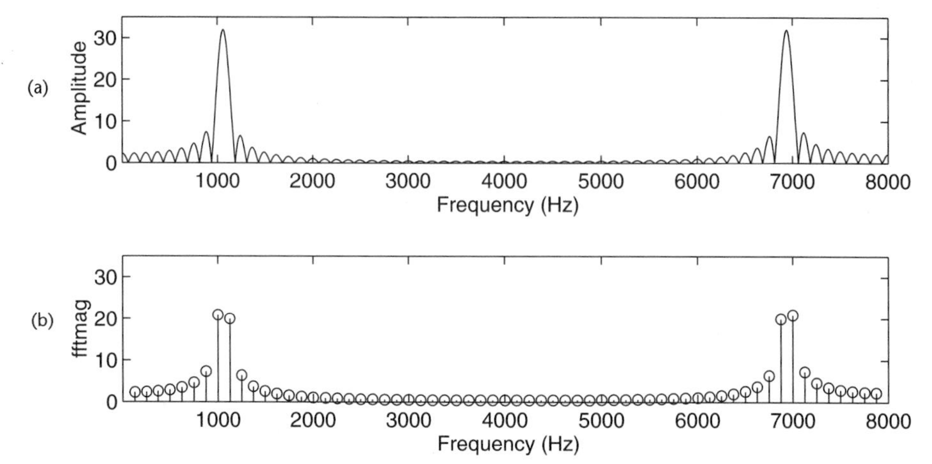

Fig. 9.5 The picket fence effect applied to leakage. (a) Continuous amplitude spectrum of the segment of a sinusoid showing leakage; (b) samples of this amplitude spectrum

thus prevents the true spectral peak from being seen correctly in either amplitude or frequency. However, a visual interpolation of the almost symmetrical frequency coefficients does suggest a peak midway between 1000 and 1125 Hz that is stronger than either individually.

4. Leakage from the main spectral peak is readily apparent, but not in the form of a sinc function shape that might have been anticipated from Figure 8.15(g). The reason for the lack of sinc function sidelobes is that the leakage also encounters the picket fence effect. This is illustrated by Figure 9.5, where Figure 9.5(b) is a sampled version of the continuous spectrum with leakage shown in Figure 9.5(a). Note that Figure 9.5(b) is identical to Figure 9.4(b), except that the weaker component $0.05\sin(2\pi f_2 t)$ has been omitted from the FFT analysis.
5. The leakage causes the weak sinusoid to be masked.

Much can be done to improve spectral estimates by applying special window functions to the data.

9.4.1 Window functions for improving spectral estimates

The unavoidable act of limiting the data used in spectral analysis to a segment of finite duration is equivalent to multiplying the signal by a rectangular window function. As demonstrated in Figure 9.4 a rectangular window function causes false ripple, or leakage, in a spectral estimate. In order to reduce this effect it is beneficial to use window functions whose Fourier transforms have lower sidelobes, and hence less leakage from the main lobe. Two particularly well known and widely used ones are the von Hann window and the Hamming window, titled after men of those names. Probably because of the similarity of the two names 'Hann' and 'Hamming' a von Hann window is also frequently known as a Hanning window (though it is unlikely that its originator would have approved of this).

A von Hann window centred about $t = 0$ is shown in Figure 9.6. Its equation is

$$\begin{aligned} w(t) &= 0.5 + 0.5\cos(2\pi t/T) \qquad && -T/2 < t < T/2 \\ &= 0 && \text{elsewhere} \end{aligned} \tag{9.5}$$

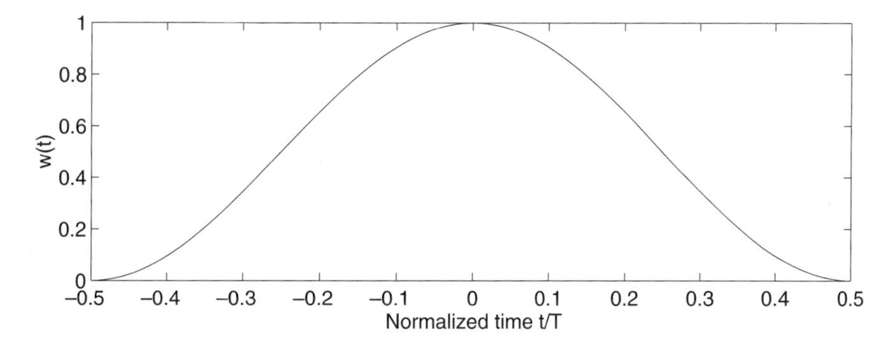

Fig. 9.6 The von Hann window function

It is a segment of a cosine wave superposed on a d.c. level and, in accordance with this, is also known as a raised cosine window. The Fourier transform of the von Hann window produces the amplitude spectrum shown in Figure 9.7. It has a first sidelobe which is −32 dB relative to the main lobe. This compares with a sidelobe level of −13 dB for the rectangular window function. Therefore the von Hann window causes very much less spurious ripple than does the rectangular window function.

The Hamming window is similar to the von Hann window except that it is raised up on a small pedestal such that it is non-zero at its extremities. When centred about the time origin its equation is

$$\begin{aligned} w(t) &= 0.54 + 0.46 \cos(2\pi t/T) \qquad -T/2 < t < T/2 \\ &= 0 \qquad \text{elsewhere} \end{aligned} \tag{9.6}$$

The amplitude spectrum of this has even lower sidelobes than that of the von Hann window. It has a first sidelobe which is −43 dB relative to the main lobe. This compares with a sidelobe level of −32 dB for the von Hann window and −13 dB for the rectangular window. The amplitude spectra of the rectangular, von Hann and Hamming windows are plotted on a dB scale in Figure 9.8.

An important but comparatively minor penalty of using these special window functions is that both their Fourier transforms have a main lobe which is slightly wider than that with the rectangular window, thus increasing the smearing of fine structure and degrading the resolution. Physically this arises because the 'amplitude shading' used in these windows reduces their effective width. The main properties of the amplitude spectra of rectangular, von Hann and Hamming windows are summarized in Table 9.2. This gives the levels of the largest sidelobe relative to the

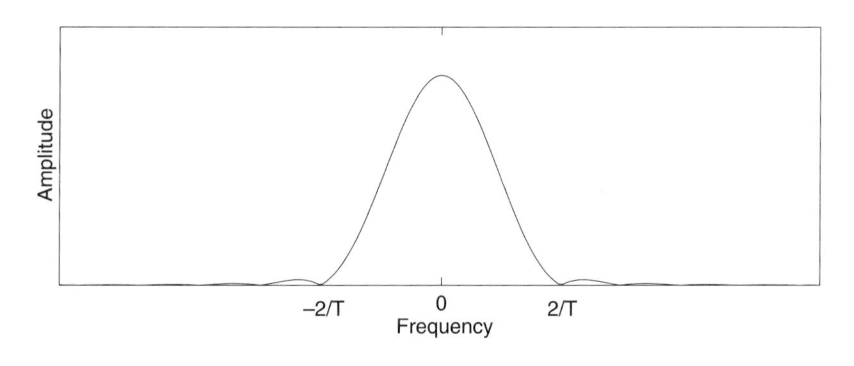

Fig. 9.7 Amplitude spectrum of the von Hann window function

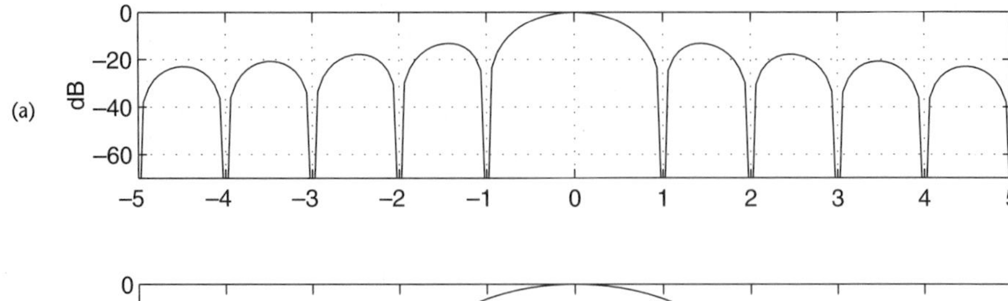

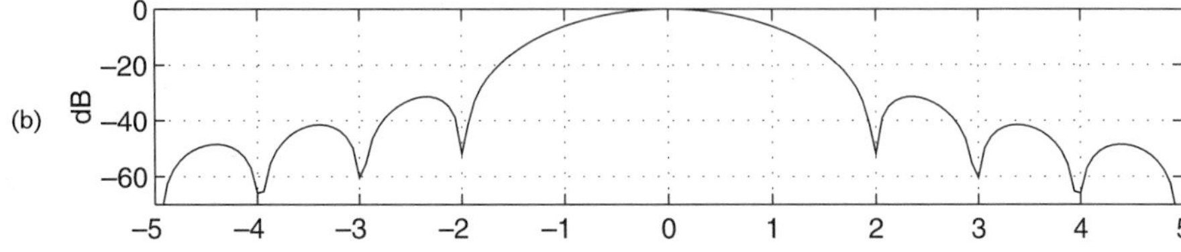

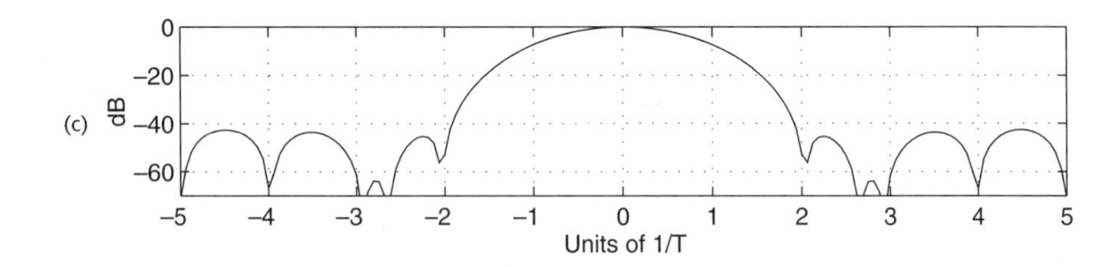

Fig. 9.8 Amplitude spectra of window functions. (a) The rectangular window; (b) the von Hann window; (c) the Hamming window

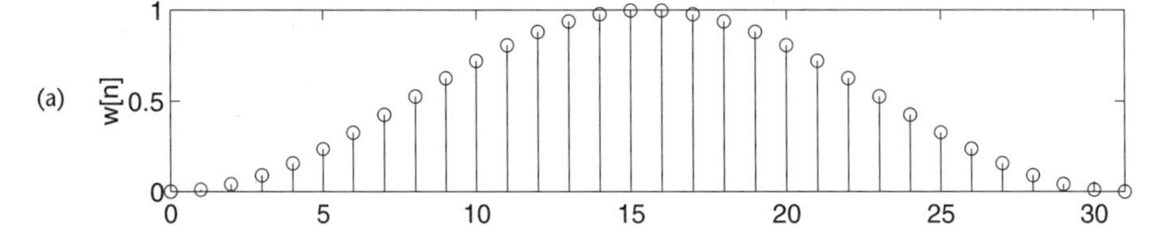

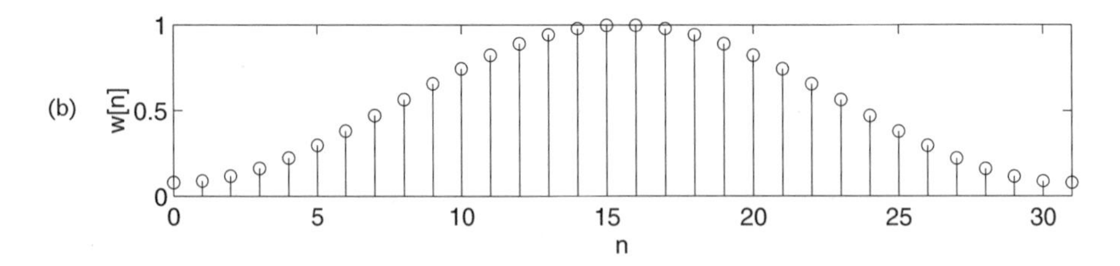

Fig. 9.9 Discrete 32 coefficient window functions: (a) von Hann; (b) Hamming

peak amplitude of the main lobe. It also gives the -3 dB and the null-to-null widths of the main lobe in terms of the window duration T.

The windows discussed are continuous functions but are applied to sequences of discrete samples. Taking this into account, and shifting the windows to commence at the first sample and to end one sample after the last sample, a von Hann window function of duration T with N terms indexed fron 0 to $(N-1)t_s$, where $t_s = T/N$, may be written in discrete form as

$$\begin{aligned} w(nt_s) &= 0.5 - 0.5\cos\left(\frac{2\pi n t_s}{N-1}\right) \qquad 0 < nt_s < (N-1)t_s \\ &= 0 \qquad \text{elsewhere} \end{aligned} \tag{9.7}$$

Table 9.2 Properties of amplitude spectra of selected window functions

Window function	Peak sidelobe level	−3 dB width of mainlobe	Null-to-null width of mainlobe
Rectangular	−13.3 dB	0.88/*T*	2/*T*
von Hann	−31.5 dB	1.44/*T*	4/*T*
Hamming	−42.7 dB	1.31/*T*	4/*T*

Alternatively, in terms solely of the index

$$w[n] = 0.5 - 0.5\cos\left(\frac{2\pi n}{N-1}\right) \qquad 0 < n < (N-1)$$
$$= 0 \qquad \text{elsewhere} \tag{9.8}$$

Similarly the Hamming window function may be written

$$w[n] = 0.54 - 0.46\cos\left(\frac{2\pi n}{N-1}\right) \qquad 0 < n < (N-1)$$
$$= 0 \qquad \text{elsewhere} \tag{9.9}$$

These two functions are illustrated in Figure 9.9 for the case of $N = 32$.

Example 9.4

As in Example 9.3, derive the amplitude spectrum of the signal $x(t) = \sin(2\pi f_1 t) + 0.05\sin(2\pi f_2 t)$ using 64 samples 125 μs apart, but this time apply a Hamming window to the data set.

Solution The following are the necessary MATLAB commands:

```
ts=0.000125;
f1=1062.5;
f2=1625;
t=(0:63)*ts;
x=sin(2*pi*f1*t)+.05*sin(2*pi*f2*t);
w=hamming(64)';                            % a row vector of 64 Hamming weights
xhamm=w.*x;                                % the product of the data and weight
                                           % vectors
subplot(3,1,1)
stem(t,xhamm)                              % plot windowed x(nts) against time
axis([0 .008 -1.5 1.5])                    % set the scales of the axes
xlabel('Time (seconds)'),ylabel('x(n)')
%
f=(0:63)/(64*ts);                          % a vector of the frequencies of the 64
                                           % FFT coefficients.
subplot(3,1,2), stem(f,abs(fft(x1hamm)))   % plot the magnitude of the FFT
axis([0 8000 0 30])                        % set the scales of the axes
xlabel('Frequency (Hz)'), ylabel('fftmag')
```

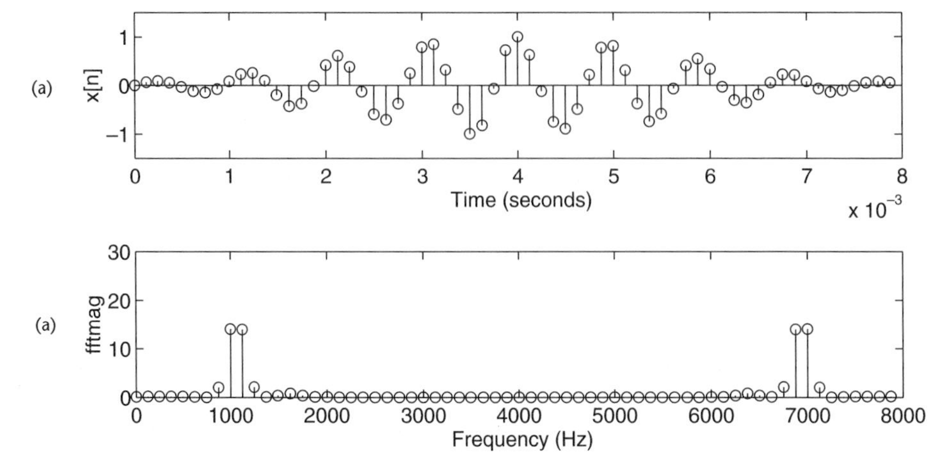

Fig. 9.10 Effect of windowing. (a) A Hamming window applied to the signal segment; (b) the resulting amplitude spectrum

The outcome is shown in Figure 9.10.

When comparing the results of Figure 9.4 and 9.10 it will be noted that the use of a Hamming window has caused a major reduction in leakage from the dominant spectral component at 1062.5 Hz. The level of the spurious ripple is decreased so much that, on close inspection, the weak sinusoid at 1625 Hz is clearly apparent.

There are windows with even lower sidelobes than the Hamming window, and the Kaiser window is one such example. It is appropriate, however, to sound a very strong note of caution about using such windows. The signal itself is likely to contain small measurement errors and these can be compounded by quantization errors after digitization. These errors make the signal equivalent to the true signal plus a noise signal. The noise signal introduces a noise floor into the spectrum and this can easily exceed very low leakage sidelobes arising from spectral components within the signal, thus rendering pointless the attainment of very low sidelobes. For most purposes the Hamming window is totally adequate.

The application of window functions to data sets is very common in spectral analysis and is usually advantageous. There are rare occasions, however, when a rectangular window can give the best results and it is important to have an understanding how this occurs. The key lies in the statement presented in Section 8.11.6 that '*The DFT produces one period of what is obtained by a continuous Fourier transform acting on the finite data set made periodic*'. Example 8.10 considered the DFT of a data set that consisted of samples of exactly eight cycles of a cosinusoid. The result was the ideal spectrum consisting of a finite coefficient at the frequency of the cosinusoid, and zero coefficients elsewhere. The reason for this success was that the data set made periodic became an exact sampled version of the *complete* original signal. However, when considering a totally unknown signal this can be regarded as an exceptionally fortuitous occurrence, and a very different scenario was illustrated by Example 8.11 where a data set containing $8\frac{1}{2}$ cycles of an infinite cosinusoid gave an amplitude spectrum that was somewhat different from that of the continuous Fourier transform of a cosinusoid. The reason is that a data set of $8\frac{1}{2}$ cycles made periodic is very different from the original signal, as shown by Figure 9.11.

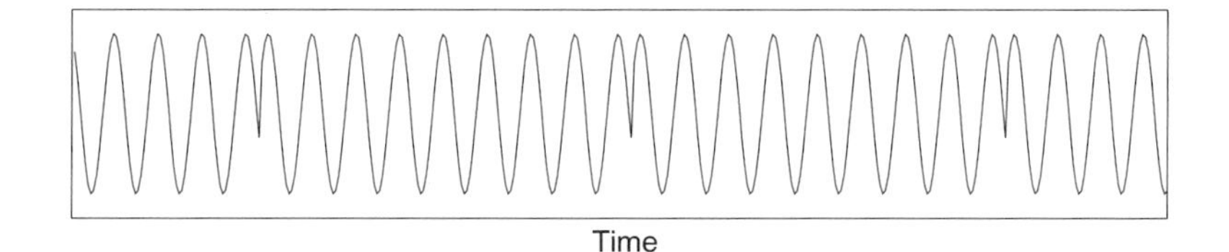

Fig. 9.11 A data set of $8\frac{1}{2}$ cycles of a cosinusoid after being made periodic

In the case of this data set consisting of $8\frac{1}{2}$ cycles of cosinusoid, a Hamming or similar window will eliminate the sharp discontinuities when the data set is made periodic and will therefore improve the spectral estimate. In the case of the data set containing 8 cycles of a cosinusoid the Hamming or similar window can only degrade the spectral estimate.

Consider next a more typical signal for analysis such as the more complex signal shown in Figure 9.12(a) and the data set taken from it shown in Figure 9.12(b). The periodic versions of this data set, with and without a Hamming window, are shown in Figures 9.12(c) and 9.12(d).

Based upon the italicized statement, the FFT of the unmodified data set produces one period of the continuous Fourier transform of the signal of Figure 9.12(c), whereas the FFT of the Hamming-windowed data set produces one period of the Fourier transform of the signal of Figure 9.12(d). The question now reduces to deciding which of these two Fourier transforms is most likely to be similar to the ideal result, which is the continuous Fourier transform of Figure 9.12(a).

It will be noted that *both* of the waveforms of Figures 9.12(c) and 9.12(d) are *totally dissimilar* to the waveform of Figures 9.12(a) *outside* the duration of the finite data set. This situation is the norm and will only be otherwise in very special cases such as when the data set consists of an integral number of periods of a periodic signal.

Throughout the duration of the data set the signal of Figure 9.12(c) is clearly much more like Figure 9.12(a) than is the signal of Figure 9.12(d). Overall however the waveform of Figure 9.12(d) is more likely than the waveform of Figure 9.12(c) to have a PSD similar to that of Figure 9.12(a). The reason is that it does not suffer from the sharp transition at the end of each period that causes ripple in the PSD.

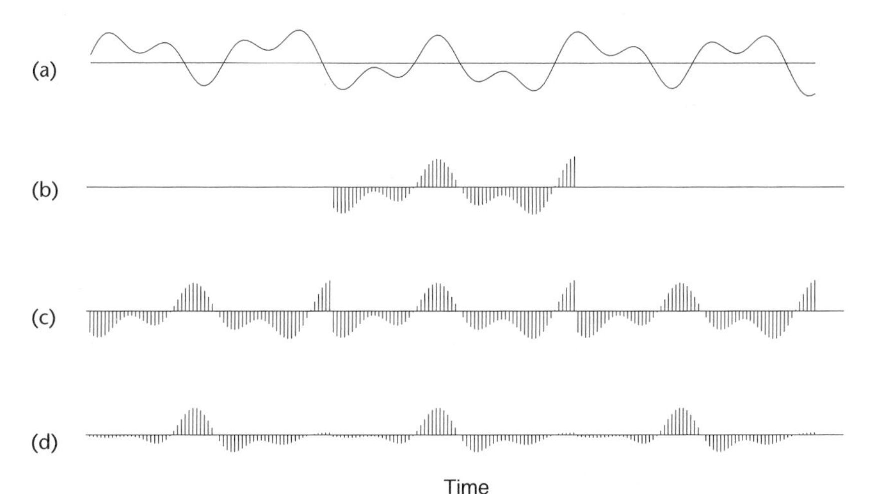

Fig. 9.12 Spectral analysis with the FFT. (a) The waveform to be analysed; (b) the finite data set available for analysis; (c) the data set made periodic; (d) the data set with a Hamming window made periodic

One important observation arises from Figure 9.12. If nothing is known about the signal other than what is available from the finite data set of Figure 9.12(b), there is absolutely no knowledge of the true data outside of this segment. On the basis of our finite data set there is no reason to dismiss the possibility that the signal in Figure 9.12(c) is the true complete signal, *except that this is very, very improbable*. In the unlikely event that Figure 9.12(c) *does* represent the true complete signal, the PSD resulting from the FFT of an unmodified data set would be the *true* PSD in spite of the leakage. (This is what happened in Example 8.11.) If Figure 9.12(c) did represent the true signal, the Hamming window would destroy some of its most important characteristics, namely its sharp transitions, and the PSD would be incorrect. The moral is that power spectral densities obtained from a single segment of a signal that extends outside that segment need to be treated with considerable caution. Since there is no total guarantee that spectral analysis using a smoothing window is better than without, and since the two power spectral densities can be significantly different, it emerges that *any* PSD can be significantly in error. However, it should be recognized that *window functions usually give substantial improvements to the PSD and should in general be used.*

There are many techniques that have been investigated for performing a spectral analysis on a single segment of data other than using an FFT. The maximum entropy method and Prony's method are interesting examples among many, but a treatment of them is outside the scope of this book. However, resulting from the observation that the FFT of an unmodified data set gives a set of complex exponential basis functions that add up to reproduce the data set exactly (which is equivalent to saying that the inverse FFT restores the data set), it can be argued that superior spectral estimates can only be achieved on the basis of the following.

1. Something specific is known about the signal generating the data set, such that there is some *a priori* knowledge of the signal outside of the data set. For example, it might be known that the signal is the sum of a finite number of pure sinusoids, in which case some mathematical method for fitting a small number of sinusoids to the observed data can produce superb results (this is the physical basis behind Prony's method). Alternatively, there may be a model of the process producing the signal which it may be possible to exploit.
2. Enough is known about the signal generating the data set to say that it is very unlikely to contain the sharp transitions of Figure 9.12(c). Effectively this is what is behind the motivation for window functions.
3. If there are several possible spectra, all of which fit the data set, the one with the least detail is the least likely to be misleading and is therefore the most satisfactory. This is the reason for preferring a PSD without sidelobes to one with them, even though there is no certainty that they are not genuine. It is also the physical basis behind maximum entropy methods of spectral analysis.

9.4.2 Zero-padding

As discussed at the beginning of this section the finite spacing of the frequency coefficients obtained by the FFT means that the wanted spectrum is seen as if through a picket fence. A convenient way of interpolating the spectrum between the cofficients is to add zeros to the data set such that its duration is extended from T to aT, where $a > 1$. This causes the spacing of the frequency coefficients to decrease from $1/T$ to $1/aT$ Hz. In general this results in a spectrum that is easier to interpret. In particular it is easier to identify the amplitudes and frequencies of spectral peaks.

It should be noted that, since the duration of the meaningful signal remains at T, the frequency resolution remains unchanged at $1/T$. Thus although the spectrum obtained using zero-padding has a *higher density of spectral lines* to give the *impression* of a better resolution, this better resolution is only an illusion. All that zero-padding does is *interpolate* additional coefficients between those that would be obtained without zero-padding. There are many perfectly satisfactory ways of performing this interpolation other than by using zero-padding. However, zero-padding is often the simplest to implement and is therefore used very frequently.

The merits of some form of interpolation should not be underrated; it can be very helpful for showing up the exact shape of the spectral estimate, thereby revealing more clearly the frequencies of spectral peaks. A useful time domain analogy is that a continuous plot of a time waveform is usually preferable to a plot of a sampled waveform even though, in accordance with the sampling theorem, the samples may contain all the information of that waveform. This is particularly the case when the sampling rate is close to its theoretical minimum for no loss of information. For example, it is not obvious by inspection that a display of a 1 kHz sinewave by means of samples 0.4 ms apart does indeed represent that sinewave. Similar arguments apply to sampled spectra.

Example 9.5

A signal consists of a strong 100 Hz sine wave, a weak 164.25 Hz sine wave and some weak noise. Mathematically it is given by

$$x(t) = 100\sin(638t) + 2\sin(1032t) + n(t)$$

where $n(t)$ is a normally (or Gaussian) distributed random signal with zero mean and unity variance. A data set of this signal is available for analysis and consists of 64 samples 1 ms apart. The noise is such that samples of it 1 ms apart are independent. Derive the PSD of the data set and plot it on a dB scale. Compare the outcomes

(a) using the data set without zero-padding;
(b) using the data set extended to 512 points by zero-padding;
(c) using the extended data set modified by a Hamming window;
(d) using the original data set modified by a Hamming window but then extended to 512 points by zero-padding.

Solution The following MATLAB program is stored as an M-file under the name 'ex9_5.m' and is run in MATLAB command mode by entering the command **'ex9_5'**. The program is divided into four sections, each separated by the **'keyboard'** command which causes the program to stop running until the command **'return'** is typed and entered. Each section plots the waveform used for the FFT, plus the resultant PSD on a dB scale. Each pair of plots can be printed before proceeding to the next section. The data points and frequency coefficients for the first part are plotted as individual values by means of the

'stem' command used previously. The next three parts involve 512 data points and 512 frequency coefficients and this large number makes the individual 'stems' of the stem display too close and the **'stalk'** and/or **'plot'** commands are therefore preferable. The plot command provides a continuous interpolated envelope.

```
% ex9_5.m
ts = .001;
t=(0:63)*ts;                              % the times of 64 samples
n=randn(size(t));                         % this generates the noise
x1=100*sin(638*t)+2*sin(1032*t)+n;        % the waveform equation
subplot(3,1,1)
stem(t,x1),axis([0 .064 -150 150])        % plot the waveform
xlabel('Time(seconds)'),ylabel('Sampled x(t)')
f=(0:63)/(64*ts);                         % frequencies of the 64 coefficients
s1=20*log10(abs(fft(x1)));                % the PSD in dB
subplot(3,1,2)
stem(f,s1),axis([0 1000 0 80])            % plot the PSD
xlabel('Frequency (Hz)'),ylabel('fftmag (dB)')
subplot(3,1,3)
plot(f,s1),axis([0 1000 0 80])
xlabel('Frequency(Hz)'),ylabel('fftmag (dB)')
keyboard                                  % this stops the program at Fig. 9.13
%
clf                                       % clear existing figures
t=(0:511)*ts;                             % the times of 512 samples
x2=1;
x2(65:512)=zeros(size(65:512));           % pad with zeros
subplot(3,1,1)
plot(t,x2), axis([0 .512 -150 150])       % plot the waveform
xlabel('Time(seconds)'),ylabel('x(t)')
f=(0:511)/(512*ts);                       % frequencies of the 512 coefficients
s1=20*log10(abs(fft(x2)));                % the PSD in dB
subplot(3,1,2)
stalk(f,s1),axis([0 1000 0 80])           % plot the PSD
xlabel('Frequency (Hz)'),ylabel('fftmag (dB)')
subplot(3,1,3)
plot(f,s1),axis([0 1000 0 80])            % plot the PSD
xlabel('Frequency (Hz)'),ylabel('fftmag (dB)')
keyboard                                  % this stops the program at Fig. 9.14
%
clf
w=hamming(512)';                          % array of 512 Hamming coefficients
x3=w.*x2;                                 % multiplication of extended data set
                                          % by Hamming window
subplot(3,1,1)
plot(t,x3),axis([0 .512 -150 150])
xlabel('Time(seconds)'),ylabel('x(t)')
```

```
f=(0:511)/(512*ts);
s1=20*log10(abs(fft(x3)));
subplot(3,1,2)
plot(f,s1),axis([0 1000 0 80])
xlabel('Frequency (Hz)'),ylabel('fftmag (dB)')
keyboard                        % this stops the program at Fig. 9.15
%
w=hamming(64)';                 % array of 64 Hamming coefficients
x4=w.*x1;                       % multiplication of original data set
                                % by Hamming window

x4(65:512)=zeros(size(65:512));
subplot(3,1,1)
plot(t,x4),axis([0 .512 -150 150])
xlabel('time(seconds)'),ylabel('x(t)')
f=(0:511)/(512*ts);
s1=20*log10(abs(fft(x4)));
subplot(3,1,2)
plot(f,s1),axis([0 1000 0 80])
xlabel('Frequency (Hz)'),ylabel('fftmag (dB)')
```

Figure 9.13 applies to just the first section of this program, where the data set is used directly. Figure 9.13(a) shows the data set itself and the Figure 9.13(b) shows the resulting PSD. Visual interpolation suggests that the main peak is at around 100 Hz, but the picket fence effect renders the PSD far from ideal. There is some indication of the second sinusoid, but its presence cannot be recognized with any reliability. Certainly it is not possible to assess its magnitude relative to the stronger sinusoid.

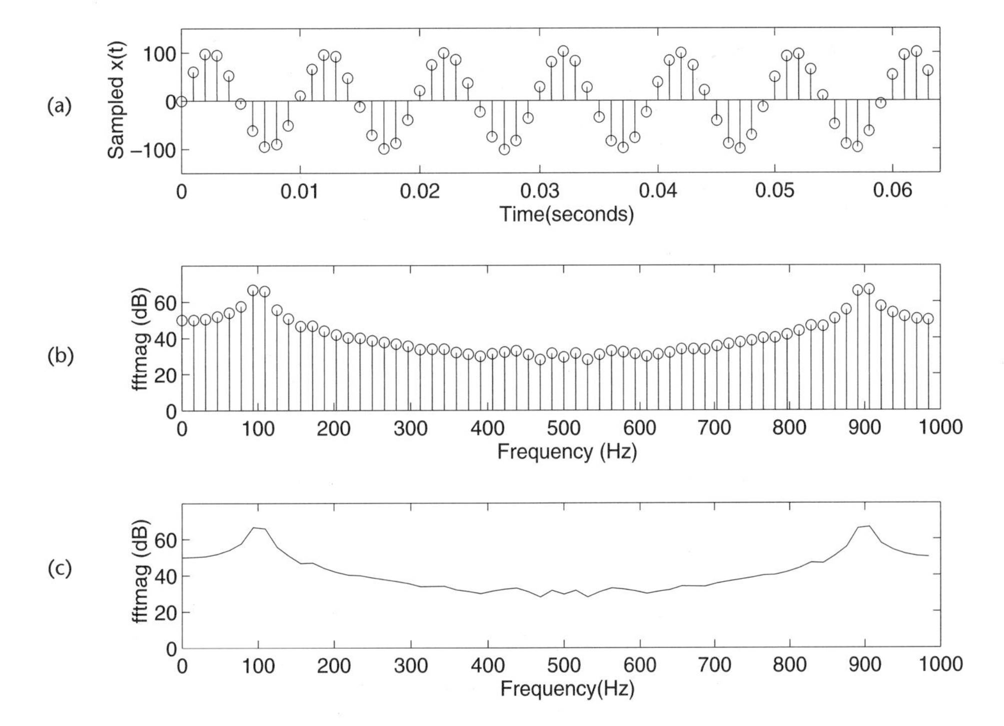

Fig. 9.13 PSD estimation: (a) data set; (b) amplitude spectrum; (c) interpolated amplitude spectrum

It has been stated that the purpose of zero-padding is to provide interpolation between frequency coefficients. *Before* applying zero-padding it is instructive to investigate an alternative form of interpolation. By using the MATLAB **'plot'** command in place of the **'stem'** command an interpolation betweeen data values is made by MATLAB. The result of applying the plot command to the magnitude of the frequency coefficients is shown in Figure 9.13(c).

Consider next the effect of using the data set extended to 512 points by zero-padding. Figure 9.14(a) shows the extended data set, while Figure 9.14(b) shows the resulting PSD estimate. Because the circles on top of the stems using the **'stem'** command are too large for closely spaced data points the modified **'stalk'** function that was created in Section 5.7 for plotting line spectra is used to produce Figure 9.14(b). Because of the high density of frequency coefficients the spectral envelope is easier to interpret here than in Figure 9.13(b), but much of the spectral character is erroneous because leakage from the strong sinusoid has caused misleading sidelobes which now show up. If the **plot** command is used the result of Figure 9.14(c) is obtained and, because of the high density of frequency coefficients, this is effectively a cleaner but otherwise identical result to that of Figure 9.14(b). It is not immediately clear whether Figure 9.14(b) or Figure 9.14(c) are any better than Figure 9.13(c). On the whole Figures 9.14(b) and 9.14(c) are preferable because leakage sidelobes are present in them and at least they are apparent as such. In contrast they are absent from Figures 9.13(b) and 9.13(c) because of the picket fence effect. If an artefact is present it is best to know about it.

In order to recognize the weaker sinusoid with any confidence it is important to eliminate the erroneous sidelobes from Figure 9.14(c), and Figure 9.15(a) shows the zero-padded data set of Figure 9.14(a) after a Hamming window is applied to it. Figure 9.15(b) shows the resulting PSD and it is noted that the sidelobes caused by

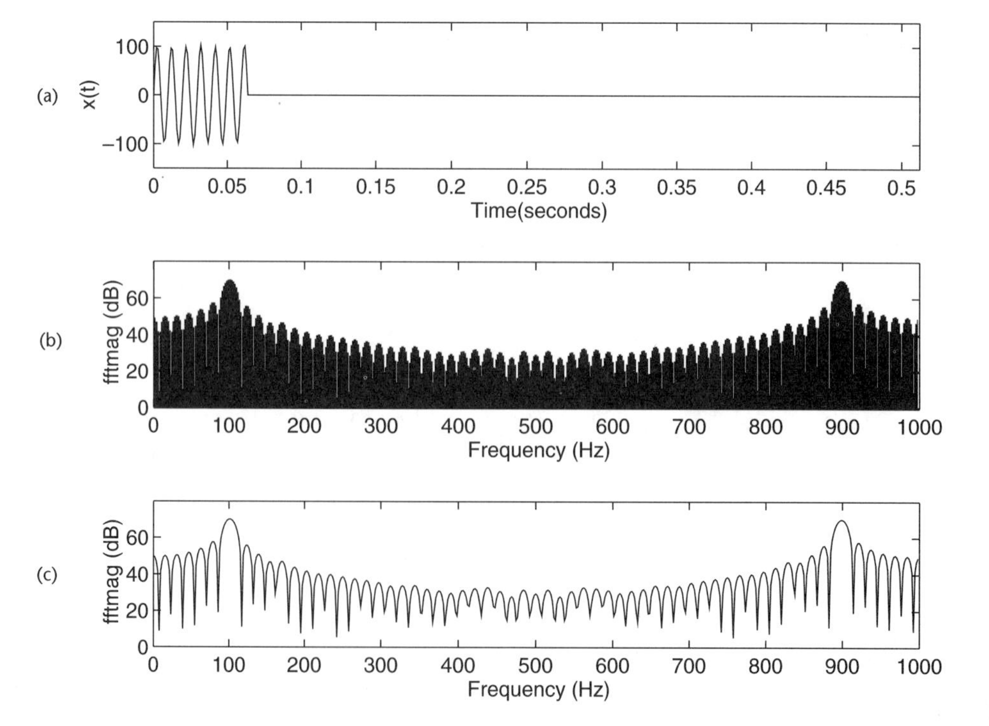

Fig. 9.14 ● PSD estimation: (a) zero-padded data set; (b) amplitude spectrum; (c) interpolated amplitude spectrum

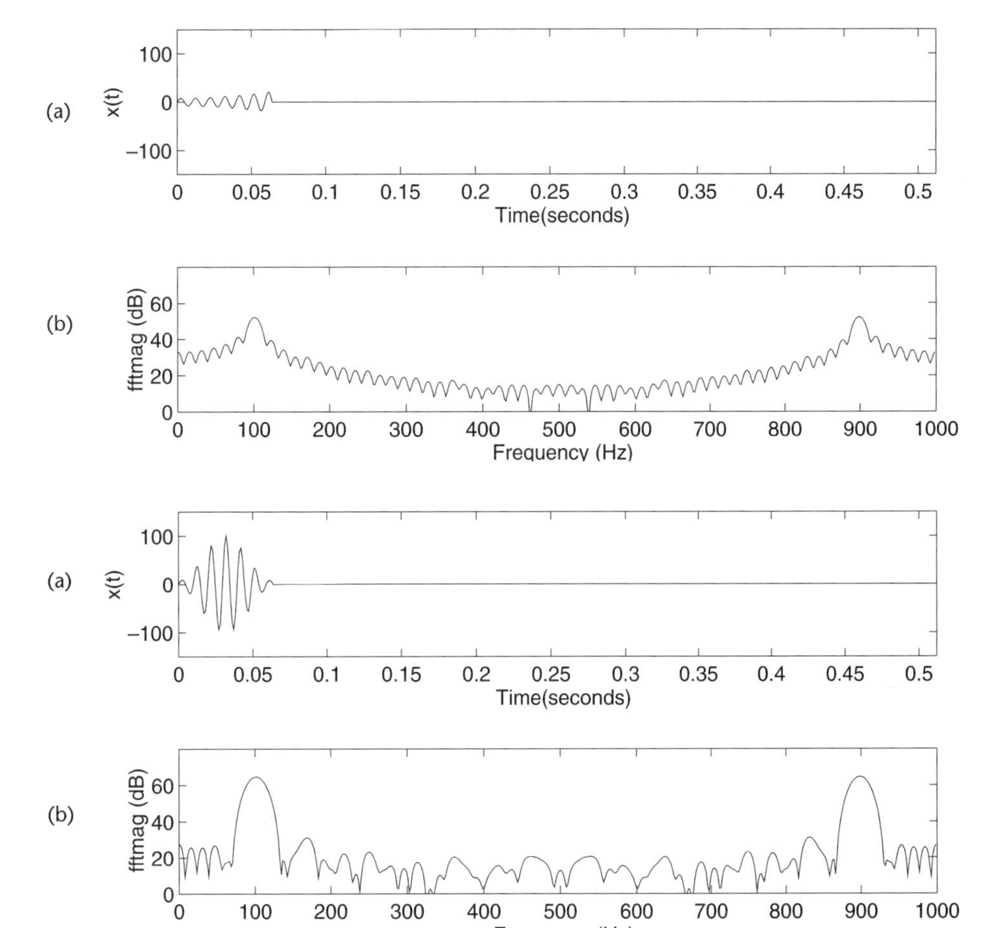

Fig. 9.15 PSD estimation: (a) data set with zero-padding followed by Hamming windowing; (b) interpolated amplitude spectrum

Fig. 9.16 PSD estimation: (a) data set with Hamming windowing followed by zero-padding; (b) interpolated amplitude spectrum

the main spectral component are substantial and that the Hamming window has not proved useful. Physically this failure has arisen because the data set still ends abruptly and has not been rounded by the Hamming window.

A much more satisfactory strategy is adopted in Figure 9.16. Figure 9.16(a) shows the outcome of applying a Hamming window to the original data set of Figure 9.13(a), and then zero-padding it. Figure 9.16(b) shows the resulting PSD and it is noted that the sidelobes caused by the main spectral component are greatly reduced and that the weaker sinusoid can now be identified with reasonable confidence particularly when it is remembered that the sidelobe level for a Hamming window is −43 dB, such that any peaks significantly above a level 43 dB below the main peak can be regarded as probably genuine.

9.5 Spectrograms

One important application of the FFT is for determining and displaying the spectrum of a signal whose frequency content changes with time. To tackle the problem of such a non-stationary signal the waveform is divided up into segments, often overlapping. Each segment is analysed separately by means of the FFT to give a *short-time Fourier transform* (STFT) and then the resulting spectra are displayed side by side to generate a 'spectrogram'.

An example of such a signal with a time-varying spectrum is speech, and an example of a speech spectrogram has already been given in Figure 6.24. When choosing the segment duration for such a signal it should be remembered that the frequency resolution is approximately the reciprocal of the segment duration T, and that T should therefore be chosen long enough to give an adequate frequency resolution. It should, however, be short enough that the spectral features are approximately stationary within that segment. Unfortunately, these requirements may be incompatible, such that some compromise is needed. In the speech example a good choice of segment duration for displaying the strong harmonic content of vowel sounds is around 40 ms, since many of the mechanisms that characterize this type of sound, the shape of the vocal tract for example, change little in that time. A 40 ms segment gives a reasonable frequency resolution of 25 Hz. On the other hand 40 ms is too long to show up the rapid onsets of some other types of speech sound and for these a more suitable choice of segment duration is about 15 ms.

As with any segment the frequency resolution using the short-time Fourier transform (STFT) equals the separation of the frequency coefficients and is given by $f_d = 1/T$, and hence T should be selected to give a suitable resolution. However, some signals change their frequency very rapidly and it may be found that these change their frequency more than $1/T$ Hz within the selected time T. When this happens some technique other than the STFT may prove beneficial. For example the Wigner–Ville distribution produces spectacular spectrograms of isolated fast-changing frequency modulated signals but unfortunately there is a penalty to pay for this. When multiple signal frequencies are present simultaneously the signals interact and often produce a spectrogram that is less satisfactory than that using the STFT. The quest for improved techniques of short-time spectral analysis is a continuing topic of research.

9.6 Linear convolution and periodic (or circular) convolution using the FFT

Whenever a linear system acts on a signal, convolution takes place between the signal and the system's impulse response, and this makes convolution one of the most important operations in signal processing. Often the linear system will be a beneficial one; for example a filter or a control system that translates the movement of a joystick into the signal used to drive the servo motors of an aileron. Sometimes the linear system is detrimental, and an example would be a badly designed transmission line that introduces distortion between the input and output signals.

Some signals and impulse responses are discrete, but a much more common scenario is that the signal at least is analogue. It can be very difficult to evaluate analytically a system output using the convolution integral, $y(\tau) = \int_{-\infty}^{\infty} x(t)h(\tau - t)\mathrm{d}t$, and one justification of discrete convolution is as a practical numerical means of determining the output when a known analogue signal enters a linear time-invariant (LTI) system of known impulse response. The input signal and impulse response would be replaced by sampled waveforms and the discrete convolution would then be evaluated on a computer or digital processor.

A second justification of discrete convolution is that it is a very flexible means of *implementing* an LTI system. For example an analogue signal requiring processing could be digitized, convolved with a suitable impulse response on a computer or digital processor, and then converted back to an analogue signal using a digital to analogue converter.

A particularly important application of convolution is for performing the closely related operation of correlation. This will be discussed in Chapter 16, where it will be shown that the cross-correlation function between $x(t)$ and $y(t)$ can be obtained by convolving $x(t)$ with $y(-t)$.

The convolution of two discrete signals $x[n]$ and $h[n]$ has already been defined in Equation (2.4) as

$$x[n] * h[n] = y[n] = \sum_{k=-\infty}^{\infty} x[k]h[n-k] \tag{9.10}$$

where the asterisk denotes linear convolution and the adjective *linear* has been added to contrast with *periodic* or *circular* convolution to be discussed shortly. By the commutative property of convolution this can also be expressed as

$$x[n] * h[n] = y[n] = \sum_{k=-\infty}^{\infty} h[k]x[n-k] \tag{9.11}$$

Alternatively, expressing the output in terms of a variable k rather than n, as in Equation (2.5), the convolution sum is given by

$$y[k] = \sum_{n=-\infty}^{\infty} x[n]h[k-n] \tag{9.12}$$

For a graphical interpretation of convolution, Equation (9.12) is generally easier to visualize than Equation (9.10) since the reversal of the sequence $h[n]$ to become $h[-n]$ is more obvious when the original variable of n is maintained. Equation (9.12) will generally be preferred in what follows.

A lower-case h is a commonly used symbol for an impulse response, and a common operation is that of convolution between a signal and an impulse response. Note, however, that the definition of Equation (9.12) is just as valid if $h[n]$ is some other form of signal.

It may be useful at this stage to have a reminder of how to evaluate a linear convolution.

Example 9.6

Determine the linear convolution between the sequence

$$x[n] = \{\underset{\uparrow}{3}, 2, 8, 1\}$$

and the sequence

$$h[n] = \{\underset{\uparrow}{2}, 4, 1, 7\}$$

Solution To determine the outcome of $y[k] = \sum_{n=-\infty}^{\infty} x[n]h[k-n]$ let us first consider the result of the summation when $k = 0$. The first term in this summation can be rewritten more fully as

$$x[n] = \{\cdots 0, 0, 0, 0, 0, \underset{\uparrow}{3}, 2, 8, 1, 0, 0, 0, 0, 0, \cdots\}$$

while the second term is now $h[0-n]$ and is simply the original sequence in reverse:

$$h[0-n] = \{\cdots 0, 0, 0, 0, 0, 7, 1, 4, \underset{\uparrow}{2}, 0, 0, 0, 0, 0, \cdots\}$$

Thus the product between the two sequences $x[n]$ and $h[0-n]$ is as shown in Figure 9.17.

The summation of the resulting terms is 6.

Considering next the product $x[n]h[k-n]$ when $k = 1$ the second term becomes $h[1-n]$ and is given by the sequence

$$h[1-n] = \{\cdots 0, 0, 0, 0, 0, 7, 1, \underset{\uparrow}{4}, 2, 0, 0, 0, 0, 0, \cdots\}$$

The sequence $x[n]$ is unchanged and the product between the two sequences $x[n]$ and $h[1-n]$ is as shown in Figure 9.18.

The summation of the resulting terms is 12 + 4 and equals 16. A continuation of this procedure gives the result that $x[n] * h[n] = y[n]$ where, changing the variable

0	0	0	0	0	3	2	8	1	0	0	0	0
					↑							
0	0	7	1	4	2	0	0	0	0	0	0	0
					↑							
0	0	0	0	0	6	0	0	0	0	0	0	0

Fig. 9.17 Product between the two sequences $x[n]$ and $h[0-n]$

0	0	0	0	0	3	2	8	1	0	0	0	0
					↑							
0	0	0	7	1	4	2	0	0	0	0	0	0
					↑							
0	0	0	0	0	12	4	0	0	0	0	0	0

Fig. 9.18 Product between the two sequences $x[n]$ and $h[1-n]$

(a)

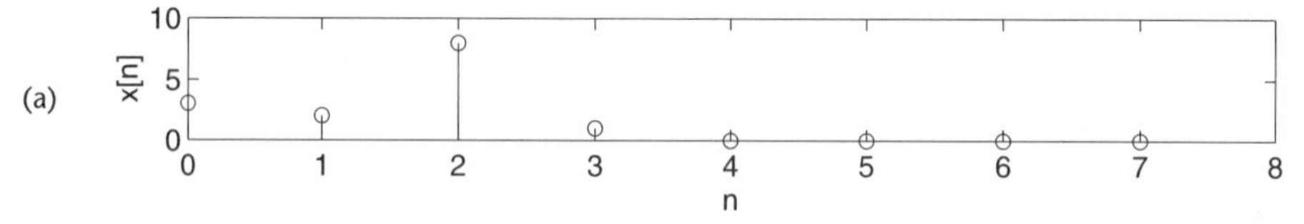

(b)

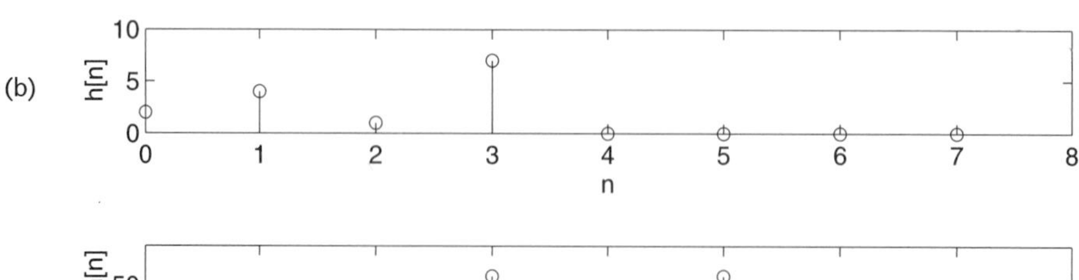

(c)

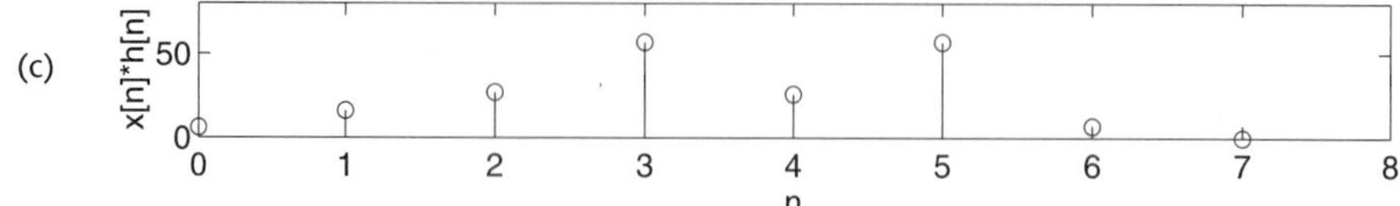

Fig. 9.19 Convolution: (a) first sequence; (b) second sequence; (c) their convolution sum

from n to k,

$$y[k] = \sum_{n=-\infty}^{\infty} x[n]h[k-n] = \{\underset{\uparrow}{6}, 16, 27, 57, 26, 57, 7\}$$

As described in Chapter 2, convolution can be done very simply using MATLAB. The following program performs the same convolution and produces the displays of Figure 9.19.

```
n=(0:7);
x=[3 2 8 1 0 0 0 0];
h=[2 4 1 7 0 0 0 0];
subplot(4,1,1),stem(n,x), axis([0 8 0 10]), xlabel('n'),ylabel('x[n]')
subplot(4,1,2),stem(n,h), axis([0 8 0 10]), xlabel('n'),ylabel('h[n]')
y=conv(x,h);                          % MATLAB convolution command
subplot(4,1,3),stem(n,y(1:8)), axis([0 8 0 80]), xlabel('n'),ylabel('x[n]*h[n]')
```

It is important to establish how many multiplications are required to achieve the convolution of two discrete sequences. Considering the convolution of two 4 point sequences in Example 9.6 and excluding multiplications by zero there was only one multiplication for $k = 0$. There were two multiplications for $k = 1$ and there would have been three for $k = 2$. For $k = 3$ there would have been four non-zero multiplications, but the number declines for $k > 3$ because of the decrease in the overlap of the two sequences. The total requirements are summarized in Table 9.3.

Extending this procedure to the convolution of two 5 point sequences, the number of multiplications would be $(1 + 2 + 3 + 4 + 5 + 4 + 3 + 2 + 1) = 25 = 5^2$. For two six-point sequences the number would be $(1 + 2 + 3 + 4 + 5 + 6 + 5 + 4 + 3 + 2 + 1) = 36 = 6^2$. For two N point sequences the number would be N^2.

If N is very large the computational demands of 'brute force' convolution can be very considerable.

Using the time convolution theorem of Equation (6.41) we have

$$x(t) * h(t) = \mathscr{F}^{-1}[X(f)H(f)] \tag{9.13}$$

For two sampled signals $x(nt_s)$ and $h(nt_s)$ this suggests the procedure of Figure 9.20(a), while the practical implementation of this using the FFT is shown in Figure 9.20(b).

Table 9.3 Multiplications required for convolving two 4 point sequences

k	Number of multiplications
0	1
1	2
2	3
3	4
4	3
5	2
6	1
	Total 16

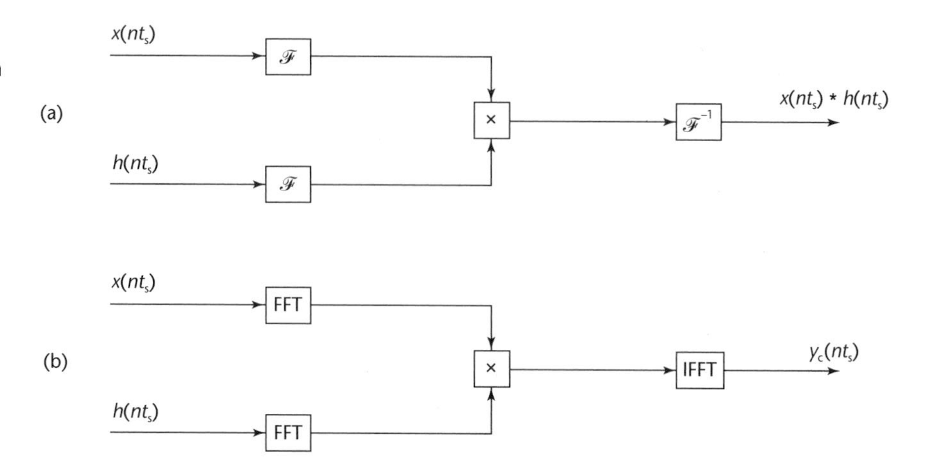

Fig. 9.20 Frequency domain convolution: (a) ideal; (b) practical (but fallible) method using the FFT

Unfortunately, since the discrete or fast Fourier transform of a discrete waveform $x(nt_s)$ does not produce the same result as the *continuous* Fourier transform of $x(nt_s)$, it does not follow that the operation of Figure 9.20(a) can be achieved successfully using Figure 9.20(b). Thus if the output of Figure 9.20(b) is $y_c(nt_s)$ such that

$$y_c(nt_s) = \text{IFFT}[\text{FFT}\{x(nt_s)\}\text{FFT}\{h(nt_s)\}] \tag{9.14}$$

where the abbreviation IFFT is used to denote an inverse FFT, then, in general

$$y_c(nt_s) \neq x(nt_s) * h(nt_s) \tag{9.15}$$

However, the computational advantages of using the FFT are so great that the technique of taking the FFT of each signal, and performing the inverse FFT on their product, is potentially a very attractive method of achieving convolution. What we must do now is to examine how $y_c(nt_s)$ differs from $x(nt_s) * h(nt_s)$ and whether steps can be taken to eliminate any discrepancy.

As a prelude to this we note that if we create periodic versions of $x(nt_s)$ and $h(nt_s)$ that are denoted $x_p(nt_s)$ and $h_p(nt_s)$, and which have *continuous* Fourier transforms $X_p(mf_d)$ and $H_p(mf_p)$, then by applying the time convolution theorem

$$x_p(nt_s) * h_p(nt_s) = \mathscr{F}^{-1}[X_p(mf_d)H_p(mf_d)] \tag{9.16}$$

Thus the convolution of $x_p(nt_s)$ and $h_p(nt_s)$ is achieved by Figure 9.21(a).

However, to repeat the important statement in Section 8.11.6:

The DFT produces one period of what is obtained by a continuous Fourier transform acting on the finite data set made periodic.

Therefore, returning now to the proposed system of Figure 9.20(b) using the FFT and repeated in Figure 9.21(b), it is apparent that the upper FFT produces a spectrum $X(mf_d)$ that is one period of $X_p(mf_d)$, and the lower FFT produces a spectrum $H(mf_d)$ that is one period of $H_p(mf_d)$. It then follows that the multiplier of Figure 9.21(b) produces a spectrum $X(mf_d)X(mf_d)$ that is one period of $X_p(mf_d)H_p(mf_d)$.

The definitions of the DFT and inverse DFT contain a high degree of symmetry and from these we can generate a dual to the previous italicized statement, namely

The inverse DFT produces one period of that predicted by a continuous inverse Fourier transform acting on the finite spectrum made periodic.

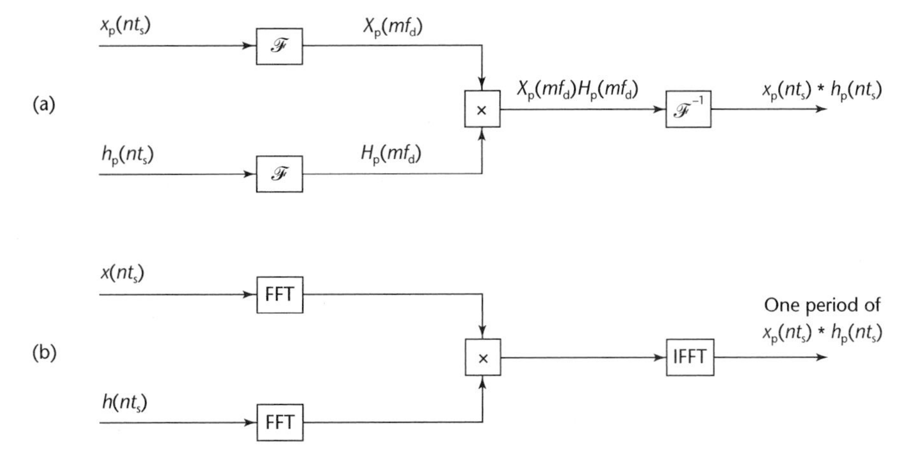

Fig. 9.21 Convolution: (a) of infinite periodic sequences using the continuous Fourier transform; (b) of finite data sets using the FFT

Using this it finally follows that Figure 9.21(b) produces a signal that is one period of $x_p(nt_s) * h_p(nt_s)$.

To summarize, the proposed technique of convolving an N point sequence $x(nt_s)$ with an N point sequence $h(nt_s)$ by means of the FFT actually produces the first N terms of $x_p(nt_s)$ convolved with $h_p(nt_s)$. It now remains to examine the difference between this and the wanted result of $x(nt_s) * h(nt_s)$.

The first N terms of $x_p(nt_s) * h_p(nt_s)$ produce what is termed the *periodic* (or *circular*) convolution between $x(nt_s)$ and $h(nt_s)$. In contrast to the operation of linear convolution between $x(nt_s)$ and $h(nt_s)$ that is denoted by an asterisk, the periodic convolution between two sequences of length N will be given the circular symbol $\bullet$. Thus, referring to Figures 9.20(b) and 9.21(b), $y_c(nt_s) = x(nt_s) \bullet h(nt_s)$.

Some discrete signals are not functions of time, and it is useful to generalize the discussion to data sets that are expressed solely in terms of their index, such as $x[n]$ and $h[n]$ rather than $x(nt_s)$ and $h(nt_s)$. Hence we shall examine the difference between $y_c[n] = x[n] \bullet h[n]$ and $y[n] = x[n] * h[n]$, where

$$y_c[n] = x[n] \bullet h[n] = \sum_{k=0}^{N-1} x_p[k]h_p[n-k] \tag{9.17}$$

and, using Equation (9.10)

$$y[n] = x[n] * h[n] = \sum_{k=-\infty}^{\infty} x[k]h[n-k] \tag{9.18}$$

Alternatively, changing the variable from n to k and considering the permitted possibility of changing the order of convolution, we wish to find the difference between

$$y_c[k] = \sum_{n=0}^{N-1} x_p[n]h_p[k-n] \text{ or } \sum_{n=0}^{N-1} h_p[n]x_p[k-n] \tag{9.19}$$

and

$$y_c[k] = \sum_{n=-\infty}^{\infty} x[n]h[k-n] \text{ or } \sum_{n=-\infty}^{\infty} h[n]x[k-n] \tag{9.20}$$

It is useful to begin by considering an example of periodic convolution and examining how the result differs from that of linear convolution.

Example 9.7

Determine the periodic convolution between the sequence

$$x[n] = \{\underset{\uparrow}{3}, 2, 8, 1\}$$

and the sequence

$$h[n] = \{\underset{\uparrow}{2}, 4, 1, 7\}$$

Compare the result with that obtained by linear convolution.

Solution Making the two sequences periodic we have

$$x_p[n] = \{\cdots 3, 2, 8, 1, \underset{\uparrow}{3}, 2, 8, 1, 3, 2, 8, 1, 3, 2, 8, 1, \cdots\}$$

and

$$h_p[n] = \{\cdots 2, 4, 1, 7, \underset{\uparrow}{2}, 4, 1, 7, 2, 4, 1, 7, 2, 4, 1, 7, \cdots\}$$

To determine the outcome of the operation $y_c[k] = \sum_{n=0}^{N-1} x_p[n]h_p[k-n]$ let us just consider the result of the summation when $k = 0$. The second term in this summation is $h_p[0-n]$ and is given by the sequence

$$h_p[0-n] = \{\cdots 7, 1, 4, 2, 7, 1, 4, 2, 7, 1, 4, \underset{\uparrow}{2}, 7, 1, 4, 2, \cdots\}$$

The multiplication of this by $x_p[n]$ for values of n between 0 and 3 is illustrated in Figure 9.22.

The sum of the four multiplications is $6 + 14 + 8 + 4$ and equals 32. Thus $\sum_{n=0}^{N-1} x_p[n]h_p[0-n] = 32$.

Let us next consider the result of periodic convolution when $k = 1$. We now wish to evaluate $\sum_{n=0}^{3} x_p[n]h_p[1-n]$. The appropriate multiplication is shown in Figure 9.23.

This time the result of the summation is 73.

In a similar manner the periodic convolution sum gives 34 when $k = 2$, and 57 when $k = 3$. Thus

$$x[n] \bullet h[n] = \{\underset{\uparrow}{32}, 73, 34, 57\}$$

2	8	1	3	2	8	1	3	2	8	1	3	2	8	1	3	2	8	1	3
											↑								
7	1	4	2	7	1	4	2	7	1	4	2	7	1	4	2	7	1	4	2
											↑								
											6	14	8	4					

Fig. 9.22 Multiplication of data set made periodic with reversed data set made periodic

2	8	1	3	2	8	1	3	2	8	1	3	2	8	1	3	2	8	1	3
											↑								
2	7	1	4	2	7	1	4	2	7	1	4	2	7	1	4	2	7	1	4
											↑								
											12	4	56	1					

Fig. 9.23 Multiplication of data set made periodic with shifted reversed data set made periodic

It will be noted that this bears little similarity with the result of linear convolution in Example 9.3 which, using the same two sequences, gave

$$x[n] * h[n] = \{\underset{\uparrow}{6}, 16, 27, 57, 26, 57, 7\}$$

There is a more elegant way of obtaining the result of periodic convolution than done in the previous example, and which leads to the alternative terminology for periodic convolution, namely *circular* convolution. In this technique the sequence $x[n]$ is written clockwise around the circumference of a circle, while the sequence $h[n]$ is written anticlockwise around the circumference of a disc that rotates within the first circle. The disc is stepped clockwise by each value of k in turn and, in each position, the N terms opposite one another are multiplied and summed to give $y_p[k]$.

Example 9.8

Determine the periodic convolution between the sequence

$$x[n] = \{\underset{\uparrow}{3}, 2, 8, 1\}$$

and the sequence

$$h[n] = \{\underset{\uparrow}{2}, 4, 1, 7\}$$

by means of circular convolution.

Solution Figure 9.24(a) shows the sequence

$$x[n] = \{\underset{\uparrow}{3}, 2, 8, 1\}$$

written clockwise around the outside of a circle, starting with $x[0]$ located at '12 o'clock'. Inside this circle the sequence

$$h[n] = \{\underset{\uparrow}{2}, 4, 1, 7\}$$

is written anticlockwise, starting with $h[0]$ located at '12 o'clock'. If opposite terms are mutiplied and the resulting four products added the number $[(2 \times 3) + (7 \times 2) + (1 \times 8) \times (4 \times 1)] = 32$ is obtained. This is the same result as obtained in Example 9.6.

Figures 9.24(b), (c) and (d) are similar, except with the inner sequences rotated one, two and three positions clockwise. The results of summing the relevant products

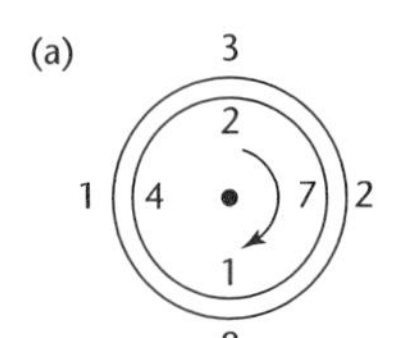

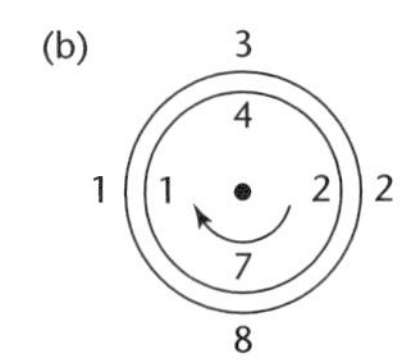

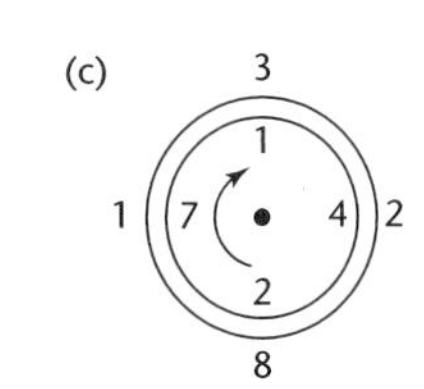

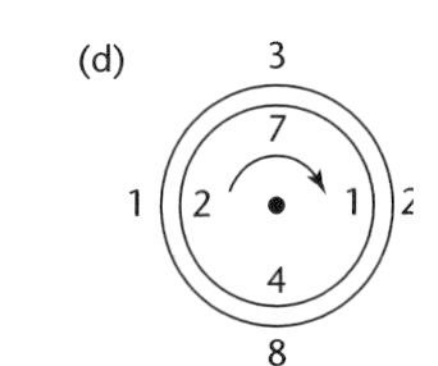

Fig. 9.24 Circular convolution for different shifts between sequences. (a) $k=0$; (b) $k=1$; (c) $k=2$; (d) $k=3$

are $[(4 \times 3) + (2 \times 2) + (7 \times 8) + (1 \times 1)] = 73$, for $k = 1$; $[(1 \times 3) + (4 \times 2) + (2 \times 8) + (7 \times 1)] = 34$, for $k = 2$; $[(7 \times 3) + (1 \times 2) + (4 \times 8) + (2 \times 1)] = 57$, for $k = 3$. This gives the same results as in Example 9.6, that

$$x[n] \bullet h[n] = \{\underset{\uparrow}{32}, 73, 34, 57\}$$

In the examples given the result has been noted, and will now be re-emphasized, that $x[n] \bullet h[n]$ does *not* generally equal $x[n] * h[n]$. The reason for this is clearly apparent from the last example, and is because numbers in the sequence overlap with numbers with which they would not overlap in linear convolution. This is an effect known as *wrap-around* error.

The next task is to examine whether we can somehow avoid this problem, to make periodic (or circular) convolution, realized physically using the FFT, give the same result as linear convolution. The answer is yes. It can be done quite simply by padding the original sequences with extra zeros.

Example 9.9

Let the sequences

$$x[n] = \{\underset{\uparrow}{3}, 2, 8, 1\} \quad \text{and} \quad h[n] = \{\underset{\uparrow}{2}, 4, 1, 7\}$$

be zero-padded to become the sequences

$$x_z[n] = \{\underset{\uparrow}{3}, 2, 8, 1, 0, 0, 0, 0\} \quad \text{and} \quad h_z[n] = \{\underset{\uparrow}{2}, 4, 1, 7, 0, 0, 0, 0\}$$

The z subscripts have been used here to emphasize the zero-padding. Determine the periodic convolution between $x_z[n]$ and $h_z[n]$.

Solution For compactness the graphical construction for circular convolution will be used in preference to that for periodic convolution. Figure 9.25 shows the sequence

$$x_z[n] = \{\underset{\uparrow}{3}, 2, 8, 1, 0, 0, 0, 0\}$$

written clockwise around the outside of a circle, starting with $x_z[0]$ located at '12 o'clock'. Inside this circle the sequence

$$h_z[n] = \{\underset{\uparrow}{2}, 4, 1, 7, 0, 0, 0, 0\}$$

is written anticlockwise, starting with $h_z[0]$ located at '12 o'clock'. Everything is identical to Example 9.8 except that each sequence has undergone zero-padding, such that the number of points is extended from 4 to 8.

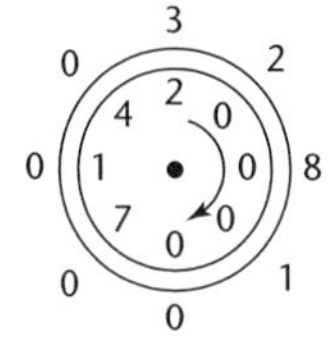

Fig. 9.25 Circular convolution between zero-padded sequences for $k = 0$

If opposite terms are mutiplied and the resulting eight products added the number $(2 \times 3) = 6$ is obtained. If the 'disc' is now rotated clockwise by one position the sum of the products is $[(2 \times 2) + (4 \times 3) = 16$. If it is rotated by two positions the sum of the products is $[(2 \times 8) + (4 \times 2) + (1 \times 3)] = 27$. Continuing this process gives the result that the periodic convolution of $x_z[n]$ and $h_z[n]$ is given by

$$x_z[n] \bullet h_z[n] = \{\underset{\uparrow}{6}, 16, 27, 57, 26, 57, 7\}$$

This is now the same result as obtained in Example 9.6 for the linear convolution between $x[n]$ and $h[n]$, that is:

$$x_z[n] \bullet h_z[n] = x[n] * h[n] = \{\underset{\uparrow}{6}, 16, 27, 57, 26, 57, 7\}$$

To summarize, it has been shown

- that the periodic (or circular) convolution of two sequences can be achieved by taking the FFT of each sequence, multiplying the two FFTs, and taking the inverse FFT of the product;
- that periodic (or circular) convolution produces the same result as linear convolution so long as the two sequences are padded with sufficient zeros (or already have these zeros at the begining and/or end of the sequences).

When two FFTs are multiplied together they must contain the same number of coefficients, and it follows that the two zero-padded sequences must each contain the same number of points. For the sake of the FFT algorithm this should be an integer power of two. In general, the extent of the padding will be different for the two sequences. It is sufficient the the number of *consecutive* zeros at the begining *and* end of the first sequence plus the number of consecutive zeros at the beginning and end of the second sequence is equal to or greater than the total number of all other terms. If, for example, we wish to convolve $\{\underset{\uparrow}{0}, 1, 1, 0, 1, 0, 1\}$ and $\{\underset{\uparrow}{0}, 1, 1\}$ we could zero-pad the two sequences to become $\{\underset{\uparrow}{0}, 1, 1, 0, 1, 0, 1, 0\}$ and $\{\underset{\uparrow}{0}, 1, 1, 0, 0, 0, 0, 0\}$ such that we have $(1 + 1 + 1 + 5)$ relevant zeros and $(6 + 2)$ other terms.

Example 9.10

Show how to convolve the two sequences

$$x[n] = \{\underset{\uparrow}{3}, 6, 2, -2, -4, -1\} \quad \text{and} \quad h[n] = \{\underset{\uparrow}{5}, 1\}$$

by means of the FFT.

Solution Using the FFT it is desirable that the number of terms in each sequence should be equal and should equal 2^r, where r is an integer. Therefore let $x[n]$ be padded by two

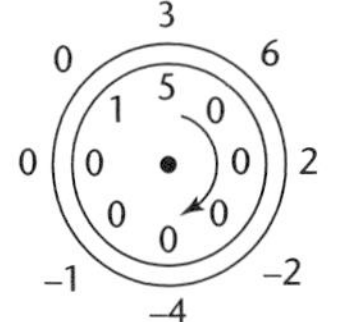

Fig. 9.26 Use of zero-padding to avoid wrap-around error

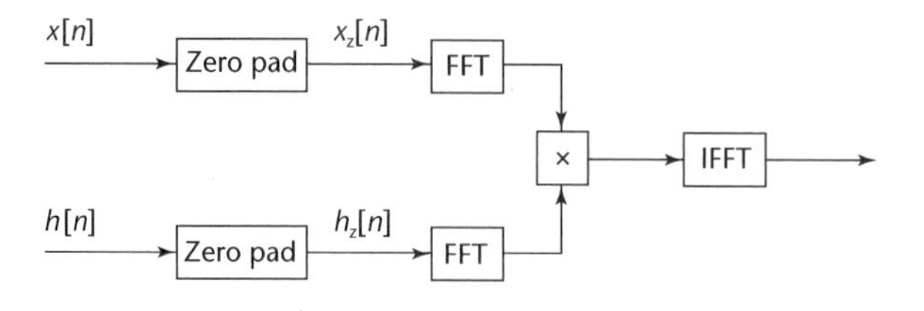

Fig. 9.27 Convolution of data sets using the FFT

zeros to give

$$x_z[n] = \{\underset{\uparrow}{3}, 6, 2, -2, -4, -1, 0, 0\}$$

and $h[n]$ be padded by six zeros to give

$$h_z[n] = \{\underset{\uparrow}{5}, 1, 0, 0, 0, 0, 0, 0\}$$

The total number of zeros is then eight and this equals the the number of non-zeros. There is no problem with wrap-around error as can be easily seen from the graphical construction used for circular convolution and shown in Figure 9.26.

In summary, digital convolution of two sequences $x[n]$ and $h[n]$ is usually achieved using the procedure shown in Figure 9.27, where the sequences are first zero-padded to become sequences $x_z[n]$ and $h_z[n]$.

The computational advantage of this over a brute force convolution is less than the computational advantage of the FFT over the DFT. This is because

- two FFTs and one inverse FFT are required;
- zero-padding increases the number of terms involved in the FFTs.

However, in spite of these two factors the number of computations is very much less than required by a direct application of the convolution sum, particularly when long sequences are involved; hence the FFT is very commonly used in signal processing for performing convolution.

Example 9.11

Sequence $a[n]$ extends from $n = 0$ to $n = 199$ and consists entirely of ones. Sequence $b[n]$ extends from $n = 0$ to $n = 179$ and consists entirely of ones. Using MATLAB convolve the two sequences:

(a) using the **conv(a,b)** command;

(b) extending the each sequence to 256 points and using the FFT technique;

(c) extending the each sequence to 512 points and using the FFT technique.

Solution For the sake of comparison each of the results is best plotted on the same horizontal scale. This is done by extending the outcomes of the three convolutions to 600 points, supplementing with zeros as appropriate. Each result is plotted against n. A suitable program is

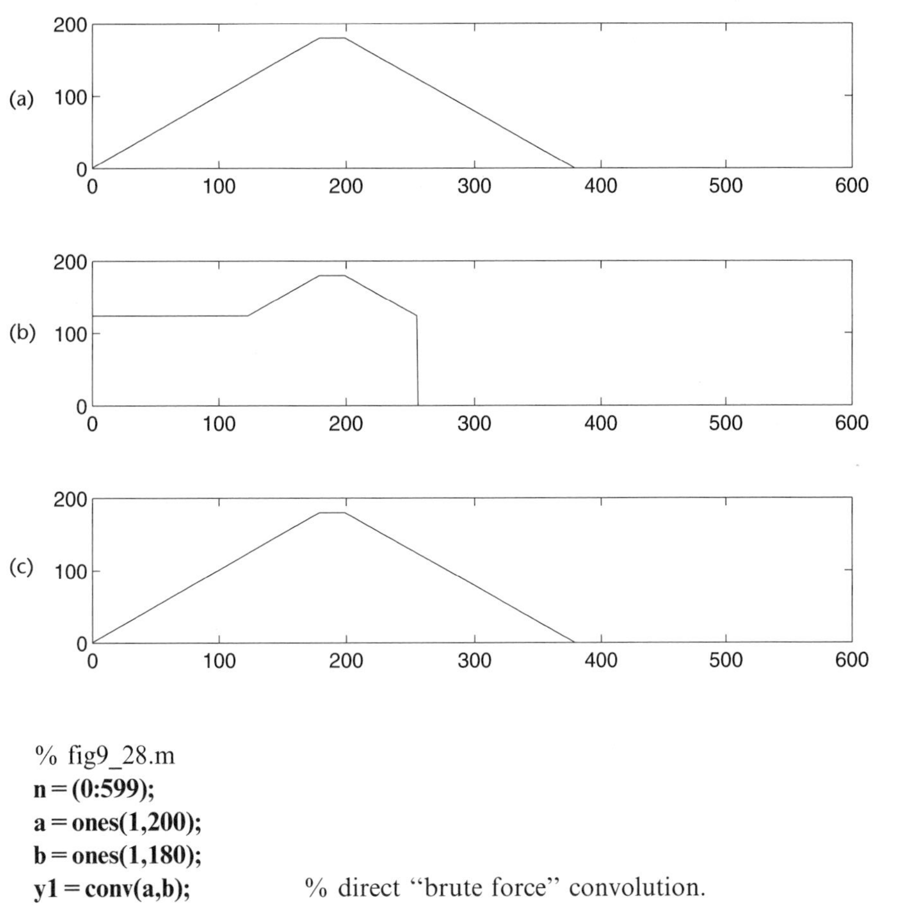

Fig. 9.28 Convolution with MATLAB: (a) time domain; (b) frequency domain, showing wrap-around; (c) frequency domain with zero-padding to avoid wrap-around error

```
% fig9_28.m
n=(0:599);
a=ones(1,200);
b=ones(1,180);
y1=conv(a,b);                 % direct "brute force" convolution.
y1(600)=0;                    % result is extended to 600 points by adding zeros.
subplot(3,1,1)
plot(n,y1)                    % the plot corresponding to part (a).
a(256)=0;                     % extension of vector to become an integer power of 2.
b(256)=0;                     % extension of vector to become an integer power of 2.
y2=real(ifft(fft(a).*fft(b)));  % convolution by frequency domain multiplication.
y2(600)=0;                    % result is extended to 600 points by adding zeros.
subplot(3,1,2)
plot(n,y2)                    % the plot corresponding to part (b).
a(512)=0;                     % zero-padding to avoid wrap-around error.
b(512)=0;                     % zero-padding to avoid wrap-around error.
y3=real(ifft(fft(a).*fft(b)));  % convolution by frequency domain multiplication.
y3(600)=0;                    % result is extended to 600 points by adding zeros.
subplot(3,1,3)
plot(n,y3)                    % the plot corresponding to part (c).
```

The outcome is shown in Figure 9.28. Labelling can be added if required. Note that the FFT technique produces the same result as direct convolution if zero-padding is used to avoid wrap-around error, but not otherwise.

9.7 ● Summary

The example has been given of one of many possible flowgraphs for performing an FFT. By avoiding a large number of duplicate multiplications the FFT is computationally much more efficient than a DFT. Indeed the number of complex multiplications involved in an N point transform is reduced from N^2 to $N \log_2 N$.

The DFT/FFT of a sampled signal does not produce the same result as the continuous Fourier transform of the original signal before sampling. The problems of aliasing, leakage and the picket fence effect have been described, together with the means of minimizing their degradations. A very important application of the FFT is for spectral analysis and application of a suitable window function to the data set is found to be particularly helpful for improving estimates of power spectral density.

The FFT can also reduce the computational time needed for digital convolution. However, the FFT method achieves periodic convolution rather than linear convolution and the two can differ because of wrap-around error. This error can be avoided by applying appropriate zero-padding to the data sets.

A particularly important application of convolution is for performing the closely related operation of correlation. This will be discussed in Chapter 16.

9.8 ● Problems

1. Calculate the 8 point DFT of the sequence $\{3,2,-1,1,-4,-2,0,1\}$ using the FFT flowchart of Figure 9.1 in such a way that $8 \log_2 8$ (or fewer) complex multiplications are required. Confirm the result using MATLAB.

2. A 10.7 MHz sinusoidally modulated FM signal is given by $x(t) = \cos[6.72 \times 10^7 t + 5 \sin(5 \times 10^5 t)]$. Generate in MATLAB a sequence consisting of 8192 samples of this signal each 10 ns apart. Generate the code for plotting the amplitude spectrum of this signal between 10 MHz and 11.4 MHz. Repeat with a Hamming window applied to the sequence. Estimate the bandwidth of the FM signal.

3. It is wished to determine the amplitude spectrum of a pulse that consists of two complete cycles of a 10 kHz square wave. Using MATLAB, determine the outcome if this is done using 64 samples at a spacing of 20 samples/cycle. Repeat for the case where extra zeros are added such that there are 256 samples at a spacing of 20 samples/cycle. Comment on the relative merits of the two results.

4. Repeat Problem 9.3 for the case where the two complete cycles are first modified by a Hamming window.

5. Convolve the two sequences $x[n] = \{3,1,1,0,2,2,0,0\}$ and $y[n] = \{2,4,0,0,3,1,1,0\}$. What would be the result of convolving these two sequences by means of the FFT? What measures can be taken to obtain the correct result?

6. It is wished to convolve 1024 samples 0.1 ms apart of the signal

$$x(t) = \sin(3000t + 4000t^2) \qquad 0 \leqslant t \leqslant 1023s$$

with a signal $y(t)$ that is the same as $x(t)$ except time-reversed, and to then plot the result. Do this in MATLAB: (a) using the **y=fliplr(x)** to produce $y(t)$ and then using the **conv(x,y)** function; (b) using the FFT without zero padding; and (c) using the FFT with zero-padding.

Chapter ten

The steady state response of analogue systems by consideration of the excitation e^{st}

10.1 ● Preview

As an aid for determining the response of a linear system to a cosinusoidal excitation $\cos\omega t$ it was found useful in Chapter 3 to replace the cosinusoid by $e^{j\omega t}$. This was because the system differential equation was much simpler to solve using the complex exponential than using a cosinusoid, and yet the solution could readily be adapted to determine the response to the cosinusoid. The replacement of $\cos\omega t$ by $e^{j\omega t}$ was justified by recognizing that $e^{j\omega t}$ is composed of a *real* (or physical) component $\cos\omega t$ which causes *the real part of the response,* plus an imaginary component that gives rise to an imaginary part of the response, and which may therefore be disregarded.

The complex exponential $e^{j\omega t}$ is a mathematical abstraction that was introduced because of its useful practical applications. Another waveform that is perhaps still more abstract, but which turns out to be even more useful and powerful, is the closely related waveform $e^{\sigma t}e^{j\omega t}$. This may be thought of as a complex exponential time waveform $e^{j\omega t}$ that rises, or decays, exponentially in accordance with the value of σ in $e^{\sigma t}$. This signal $e^{\sigma t}e^{j\omega t}$ reduces to $e^{j\omega t}$ if $\sigma = 0$. Thus, by finding the response of a system to the excitation $e^{\sigma t}e^{j\omega t}$ and then putting $\sigma = 0$ into the result, we can find the response to $e^{j\omega t}$. The real part of this then gives the response to $\cos\omega t$, just as in Chapter 3. In other words an excitation of the form $e^{\sigma t}e^{j\omega t}$ has, *at the very least,* the same potential for producing useful physical results as does the excitation $e^{j\omega t}$. The apparent increased complexity of $e^{\sigma t}e^{j\omega t}$ compared with $e^{j\omega t}$ means, however, that some good reasons are needed before its use can be justified.

The first point to note is that $e^{\sigma t}e^{j\omega t}$ is *not* in fact more complicated to use than $e^{j\omega t}$. The waveform $e^{\sigma t}e^{j\omega t}$ may be written $e^{(\sigma + j\omega)t}$, or simply as e^{st}, where, by definition, $s = \sigma + j\omega$. Because the imaginary component of the exponential is hidden within the complex variable s, this waveform e^{st} is actually *easier to handle* than $e^{j\omega t}$. As is shown in Sections 10.2 and 10.3, impedance functions, admittance functions, transfer functions and so on are of simpler form and are easier to manipulate using e^{st} than when using $e^{j\omega t}$. However, as shown in the subsequent sections, there are even more important reasons for using e^{st}.

10.2 Impedance and admittance when the excitation is e^{st}

If we consider the voltage across an inductance when the current through it is $\tilde{i} = I_0 e^{st}$, we obtain

$$\tilde{v} = L\frac{\mathrm{d}i}{\mathrm{d}t} = sLI_0 e^{st} = sL\tilde{i} \tag{10.1}$$

Hence we can introduce an impedance function $Z(s)$ relating voltage and current, when one of them is of the form e^{st}, that is given by

$$Z(s) = \frac{\tilde{v}}{\tilde{i}} = sL \tag{10.2}$$

This is comparable to the impedance function $\mathbf{Z} = \mathrm{j}\omega L$ of Chapter 3. In a similar manner we find that, in place of the $1/\mathrm{j}\omega C$ obtained for a capacitance when considering the current $I_0 e^{\mathrm{j}\omega t}$, the impedance function for a current $I_0 e^{st}$ is

$$Z(s) = 1/sC \tag{10.3}$$

The impedance function for a resistance is simply

$$Z(s) = R \tag{10.4}$$

An admittance function $Y(s)$ is defined as $\tilde{i}/\tilde{v}$, the reciprocal of $Z(s)$. The values of $Y(s)$ for inductances, capacitances and resistances are $1/sL$, sC and $1/R$.

As in Chapter 3, the impedances of elements add when they are *in series*, that is

$$Z(s) = Z_1(s) + Z_2(s) + Z_3(s) + \cdots \tag{10.5}$$

The admittances add when elements are *in parallel*, that is

$$Y(s) = Y_1(s) + Y_2(s) + Y_3(s) + \cdots \tag{10.6}$$

Example 10.1

Determine the impedance function looking into the left port of the network configuration of Figure 10.1. Hence determine the input current caused by an input voltage e^{st}. Use this result to determine the input current due to a real input voltage $A\cos\omega t$.

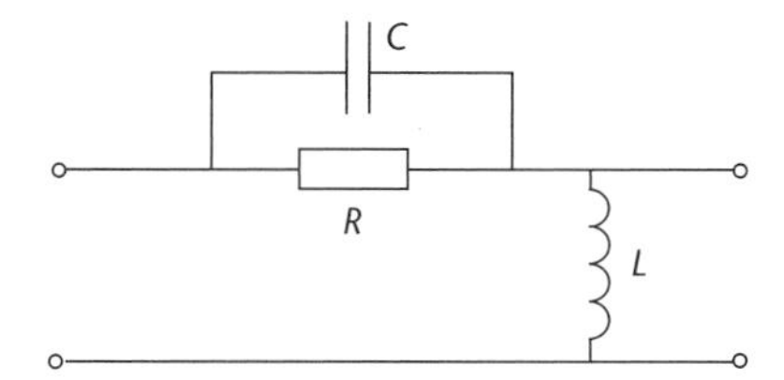

Fig. 10.1 A two-port network

Solution The admittance of the resistance and capacitance in parallel is $(1/R + sC)$. The corresponding impedance is the reciprocal of this, or $R/(1 + sCR)$. The total impedance is therefore

$$Z(s) = sL + \frac{R}{1 + sCR} = \frac{R + sL + s^2LCR}{1 + sCR} \tag{10.7}$$

If we now apply the voltage $\tilde{v}_{\text{in}} = Ae^{st}$ to the network the resulting current is

$$\tilde{i} = \frac{\tilde{v}_{\text{in}}}{Z(s)} = Ae^{st}\frac{1 + sCR}{R + sL + s^2LCR} \tag{10.8}$$

Remembering that $s = \sigma + j\omega$, this result can be used to find the real current resulting from a real applied voltage $A \cos \omega t$ by constraining the solution of Equation (10.8) to the special case of $\sigma = 0$, such that s can be replaced by $j\omega$. This causes the voltage source to be $Ae^{j\omega t}$ and the output current then to be

$$\tilde{i} = Ae^{j\omega t}\frac{1 + j\omega CR}{R + j\omega L - \omega^2 CR} \tag{10.9}$$

The real part of this is then the current that results from the real part of the complex voltage $Ae^{j\omega t}$, namely $A \cos \omega t$, and can be found in the same way as in Chapter 3, which was to express numerator and denominator in polar coordinates and to divide out.

10.3 Transfer functions

A transfer function is the ratio of an output signal to an input signal when the input signal is of the form e^{st}. Admittance functions are particular types of transfer function in which the output refers to the current through a network element and the input refers to the voltage that causes it. Impedance functions are another special type of transfer function in which the output refers to the voltage through a network element and the input refers to the current. With the exception of admittance and impedance functions, the usual symbol for a transfer function is $H(s)$. In electrical networks the most common use of $H(s)$ is for describing the ratio of output *voltage* to input *voltage* when the input voltage is of the form e^{st}.

Example 10.2

Let the voltage across the inductance in Figure 10.1 be the output voltage of the network. Determine the transfer function of the network.

Solution The output voltage equals the input current multiplied by the impedance function of the inductance. Hence the transfer function is given by

$$H(s) = \frac{\tilde{v}_0}{\tilde{v}_{\text{in}}} = \frac{sL\tilde{i}}{\tilde{v}_{\text{in}}} = \frac{sL}{\tilde{Z}(s)} = \frac{sL}{sL + R/(1 + sCR)} = \frac{sL(1 + sCR)}{R + sL + s^2LCR} \tag{10.10}$$

The transfer function $H(s)$ is closely related to the *frequency* transfer function introduced in Chapter 3, which was the ratio of output voltage to input voltage when the input was of the form $Ae^{j\omega t}$. It is easy to go from $H(s)$ to $H(\omega)$ by putting $\sigma = 0$, thus enabling s to be replaced by $j\omega$.

10.4 Further justification for the use of e^{st}

In Section 5.3 it was argued that one of the very special properties of a complex exponential signal $Ae^{j\omega t}$ was that, when applied to any network, the output was of the same complex exponential form but multiplied by a simple modifying factor $H(\omega)$. As can be seen by applying convolution, the same is true for the complex exponential e^{st}. We have the completely general result derived in Section 2.6 that the output of a network whose impulse response is $h(t)$ is

$$y(t) = x(t) * h(t) = h(t) * x(t)$$

where $x(t)$ is the input and $y(t)$ is the output. Expressing this alternatively in terms of τ, using t as a dummy variable, this becomes

$$y(\tau) = \int_{-\infty}^{\infty} h(t)x(\tau - t)\, dt \tag{10.11}$$

If $x(t) = e^{st}$ then

$$y(\tau) = \int_{-\infty}^{\infty} h(t)e^{s(\tau - t)}\, dt$$

This can be rearranged to give

$$y(\tau) = e^{s\tau}\int_{-\infty}^{\infty} h(t)e^{st}\, dt = H(s)e^{s\tau}$$

Alternatively, in terms of the variable t

$$y(t) = H(s)e^{st} \tag{10.12}$$

From this it is noted that a complex exponential input signal e^{st} has the special property that the output is the same complex exponential, except modified by a simple factor $H(s)$. (In mathematical terminology this makes e^{st} an eigenfunction and $H(s)$ an eigenvalue.)

Although e^{st} may be more difficult conceptually than $e^{j\omega t}$, it is seen that the evaluation of impedance functions, transfer functions and so on avoids multiplications involving $j\omega$ and is thus marginally easier to use. However, a much more important reason for using e^{st} extensively in system analysis centres around *pole–zero plots*.

10.5 Pole–zero plots

Consider the results of Example 10.2, in which the transfer function was found to be given by

$$H(s) = \frac{sL(1 + sCR)}{R + sL + s^2LCR} \tag{10.13}$$

This may be manipulated to give

$$H(s) = \frac{LCRs(s + 1/CR)}{LCR(s^2 + s/CR + 1/LC)} = \frac{s(s + 1/CR)}{(s^2 + s/CR + 1/LC)} \tag{10.14}$$

This can be factorized to give

$$H(s) = K\frac{(s - z_1)(s - z_2)}{(s - p_1)(s - p_2)} \tag{10.15}$$

It will be noted from Equation (10.15) that, except for a constant gain factor K which happens to be unity in this particular example, the transfer function of Equation (10.14) is completely defined by the roots of its numerator, z_1 and z_2, and the roots of its denominator, p_1 and p_2.

If s equals z_1 or z_2 the function $H(s)$ equals zero, and hence z_1 and z_2 are known as zeros. If s equals p_1 or p_2 the function $H(s)$ becomes infinite and, to reflect this, p_1 and p_2 are known as poles.

In the example of a transfer function given by Equation (10.14) we have

$$K = 1 \tag{10.16}$$

$$z_1 = 0 \tag{10.17}$$

$$z_2 = -1/CR \tag{10.18}$$

and p_1 and p_2 are the roots of $s^2 + s/CR + 1/LC$, obtained by solving for s in the quadratic equation

$$s^2 + s/CR + 1/LC = 0 \tag{10.19}$$

This gives

$$p_1 = -\frac{1}{2CR} + \sqrt{\left(\frac{1}{2CR}\right)^2 - \frac{1}{LC}} \quad \text{and} \quad p_2 = -\frac{1}{2CR} - \sqrt{\left(\frac{1}{2CR}\right)^2 - \frac{1}{LC}} \tag{10.20}$$

If $1/LC \leqslant (1/2CR)^2$ these roots are real. If however $1/LC > (1/2CR)^2$ we have the square root of a negative number and, by multiplying the inside and outside of the square root by $\sqrt{-1}$, we obtain

$$p_{1,2} = -\frac{1}{2CR} \pm \mathrm{j}\sqrt{\frac{1}{LC} - \left(\frac{1}{2CR}\right)^2} \tag{10.21}$$

In the particular example of a transfer function given by Equation (10.14), z_1 and z_2 are real, whereas p_1 and p_2 can either be real or complex depending on whether $1/LC$ is greater or less than $(1/2CR)^2$.

The values of z_1, z_2, p_1 and p_2 may be plotted in the complex plane. Remembering that poles and zeros are particular values of the variable $s = \sigma + \mathrm{j}\omega$, it will be appreciated that this is a plot in the s-plane of a $\mathrm{j}\omega$ value versus a σ value.

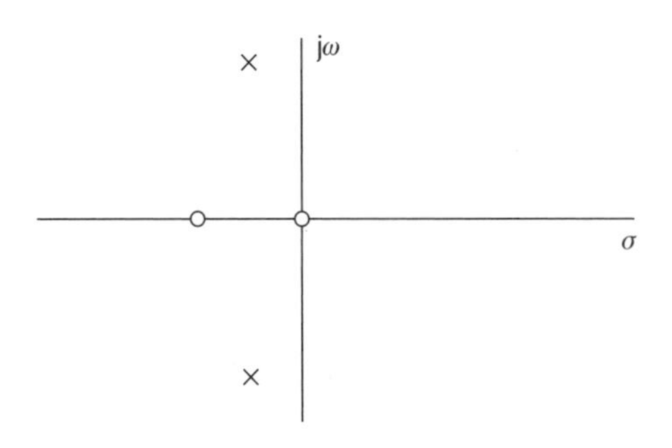

Fig. 10.2 Example of a pole – zero plot

Depending on the values of R, L and C, there are an infinite number of possible pole and zero positions, one of which is shown in Figure (10.2). Here circles are used for zeros and crosses for poles.

All electrical linear systems are composed of impedances that are of the form R, sL or $1/sC$. It follows that all impedance functions, admittance functions, transfer functions and so on are the ratio of two polynomials in s. In other words, a transfer function $H(s)$ is always of the form

$$H(s) = \frac{Y(s)}{X(s)} = \frac{b_m s^m + b_{m-1} s^{m-1} + \cdots + b_0}{a_n s^n + a_{n-1} s^{n-1} + \cdots + a_0} \tag{10.22}$$

Such functions can always be factorized to be expressed as

$$H(s) = K \frac{(s - z_1)(s - z_2) \cdots (s - z_m)}{(s - p_1)(s - p_2) \cdots (s - p_n)} \tag{10.23}$$

This function has a set of poles and zeros which can be plotted on the complex plane.

The most important justifications for the use of e^{st} are centred around this pole–zero plot, and:

- that, with the exception of the constant gain factor K, any impedance function, admittance function or transfer function *can be completely defined by a plot of poles and zeros*;
- that, with practice and experience, *a tremendous amount can be learnt about the response from a direct visual inspection of the pole–zero plot*. The pole–zero plot can provide immediate information about the frequency response. It will be shown in the next chapter that it can also provide immediate information about the response of the system to a sudden change in the excitation (the transient response). In the case of active systems, including control systems, it follows from this that it can give information about the system stability.

The factorization of the numerator and denominator polynomials of the transfer function can be laborious by hand when their order is higher than two. Fortunately, a command in MATLAB is available to perform the necessary computations. Considering as an example a transfer function given by

$$H(s) = \frac{0.0032s^4 + 0.00843s^2 + 0.3031}{s^4 + 0.981s^3 + 1.5204s^2 + 0.8178s + 0.3361}$$

the polynomial coefficients can be entered as row vectors using

```
b=[.0032 0 .0843 0 .3031];
a=[1 .981 1.5204 .8178 .3361];
```

Then, in order to factorize the numerator to determine the zeros, we use the command

z = roots(b)

The result is a column vector of the zeros that is displayed on the screen as

```
z =
      0 + 4.6955i
      0 - 4.6955i
      0 + 2.0727i
      0 - 2.0727i
```

Similarly, to determine the poles we require

p = roots(a)

resulting in the display

```
p =
-0.1244 + 0.9938i
-0.1244 - 0.9938i
-0.3661 + 0.4484i
-0.3661 - 0.4484i
```

These poles and zeros can readily be plotted manually and, apart from a gain factor, this plot provides a complete description of the transfer function. If wished, a function **pzmap** in Version 5 of the student edition of MATLAB can be used for the *automatic* plotting of these s-plane poles and zeros. This function is absent from Version 4, however, and will not be used in this text.

10.6 Frequency response from the pole–zero plot

If a transfer function for an excitation of the form e^{st} is expressed in the general form of Equation (10.23) the *frequency* transfer function for an excitation of the form $e^{j\omega t}$ can be obtained by setting s to $j\omega$. The frequency transfer function becomes

$$H(\omega) = A(\omega)e^{j\theta(\omega)} = K\frac{(j\omega - z_1)(j\omega - z_2)\cdots(j\omega - z_m)}{(j\omega - p_1)(j\omega - p_2)\cdots(j\omega - p_n)} \qquad (10.24)$$

If we now put

$$A_1 = |j\omega - z_1|, \qquad A_2 = |j\omega - z_2| \text{ and so on}$$

and

$$\alpha_1 = \angle(j\omega - z_1), \qquad \alpha_2 = \angle(j\omega - z_2) \text{ and so on}$$

this can be expressed in the form

$$H(\omega) = K\frac{A_1e^{j\alpha_1}A_2e^{j\alpha_2}\cdots A_me^{j\alpha_m}}{B_1e^{j\beta_1}B_2e^{j\beta_2}\cdots B_ne^{j\beta_n}} = K\frac{A_1A_2\cdots A_me^{j(\alpha_1+\alpha_2+\alpha_3\cdots)}}{B_1B_2\cdots B_ne^{j(\beta_1+\beta_2+\beta_3\cdots)}} \qquad (10.25)$$

It follows that the *magnitude* response to the excitation $e^{j\omega t}$ is given by

$$A(\omega) = |H(\omega)| = K\frac{A_1A_2\cdots A_m}{B_1B_2\cdots B_n} = K\frac{|j\omega - z_1||j\omega - z_2|\cdots|j\omega - z_m|}{|j\omega - p_1||j\omega - p_2|\cdots|j\omega - p_n|} \qquad (10.26)$$

and the *phase* response is given by

$$\begin{aligned}\theta(\omega) = \angle H(\omega) &= \alpha_1 + \alpha_2 + \cdots \alpha_m - \beta_1 - \beta_2 \cdots - \beta_n \\ &= \angle(\mathrm{j}\omega - z_1) + \angle(\mathrm{j}\omega - z_2) + \cdots + \angle(\mathrm{j}\omega - z_m) \\ &\quad - \angle(\mathrm{j}\omega - p_1) - \angle(\mathrm{j}\omega - p_2) - \cdots - \angle(\mathrm{j}\omega - p_n)\end{aligned} \tag{10.27}$$

To interpret Equations (10.26) and (10.27), consider the network of Figure 10.1, with its transfer function given by Equation (10.14). This has two zeros and two poles. The two zeros are given by Equations (10.17) and (10.18) and correspond to the points B and C in the complex plane representation of Figure 10.3. If $1/LC > (1/2CR)^2$, Equation (10.21) tells us that there are two complex poles such as those at points D and E in Figure 10.3.

Restricting ourselves to some specific frequency ω_x denoted by point A on the imaginary axis, the magnitude response of Equation (10.26) contains terms such as $|\mathrm{j}\omega_x - z_2|$. If z_2 is point C this can be interpreted as the length AC. Thus, when all the terms in Equation (10.26) are included the magnitude response can be interpreted as the product of the lengths from the point A to the zeros, divided by the product of the lengths from the point A to the poles. In other words the magnitude (or amplitude) response is given by

$$A(\omega_x) = |H(\omega_x)| = K\frac{AB \cdot AC}{AD \cdot AE} \tag{10.28}$$

In general, the amplitude response at any frequency ω_x can be obtained from any pole–zero plot by taking the product of the lengths from point $\mathrm{j}\omega_x$ to the zeros, and dividing it by the product of the lengths from point $\mathrm{j}\omega_x$ to the poles.

In a similar manner, using Equation (10.27), the phase response of a system at any frequency ω_x can be obtained from any pole–zero plot by taking the sum of the

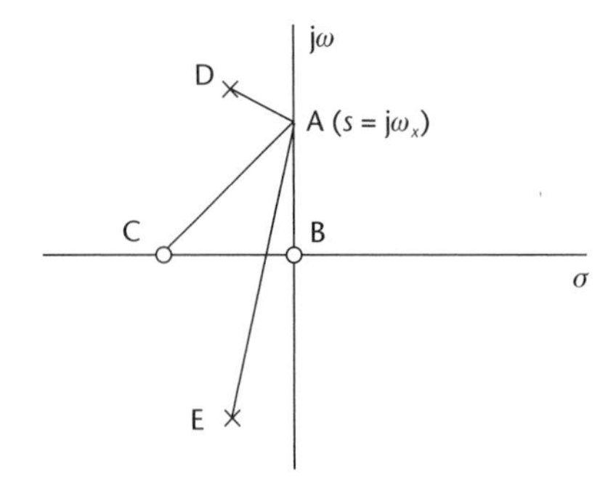

Fig. 10.3 Determination of amplitude response from a pole–zero plot

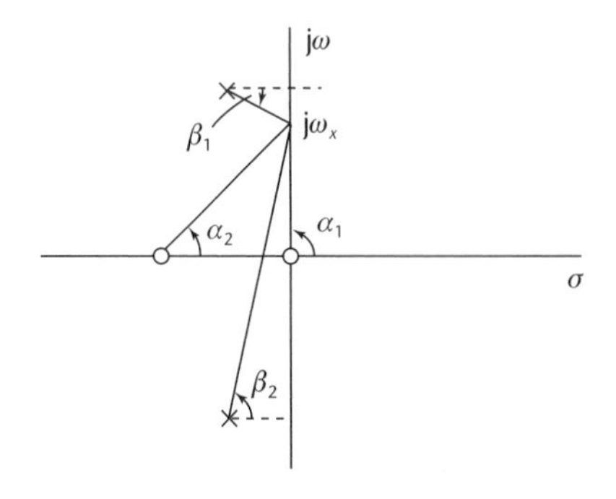

Fig. 10.4 Determination of phase response from a pole–zero plot

angles of the vectors from point $j\omega_x$ to the zeros, and subtracting from it the sum of the angles of the vectors from point $j\omega_x$ to the poles.

In Figure 10.4 for example

$$\theta(\omega_x) = \angle H(\omega_x) = \alpha_1 + \alpha_2 - \beta_1 - \beta_2 \tag{10.29}$$

where α_1, α_1 and β_1 are positive quantities and β_1 is a negative quantity.

Example 10.3

Derive a pole–zero plot for the network of Figure 10.1 for the case where $R = 1\,k\Omega$, $L = 0.2$ mH and $C = 1$ nF. Use it to make an approximate sketch of the amplitude response.

Solution Substituting the values of R, L and C we obtain

$$1/LC = 5 \times 10^{12};\ 1/CR = 10^6;\ 1/(2CR)^2 = 2.5 \times 10^{11}$$

Therefore, from Equations (10.17) and (10.18),

$$z_1 = 0,\ z_2 = -10^6$$

and, from Equation (10.21)

$$p_{1,2} = -0.5 \times 10^6 \pm j\sqrt{5 \times 10^{12} - 2.5 \times 10^{11}}$$
$$= -0.5 \times 10^6 \pm j2.18 \times 10^6$$

Figure (10.5) gives the corresponding pole–zero plot. Proceeding up the imaginary axis of this plot it will be clear that distance AD will reach a minimum value when $\omega = 2.18 \times 10^6$ rads/s and that the product $(AB \cdot AC)/(AD \cdot AE)$ will therefore be a

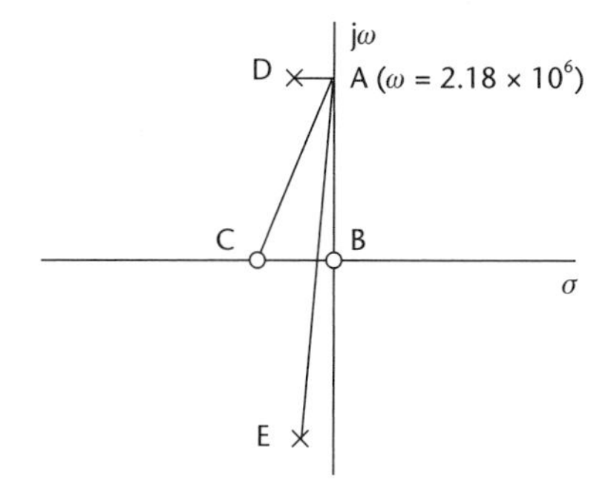

Fig. 10.5 Pole–zero plot

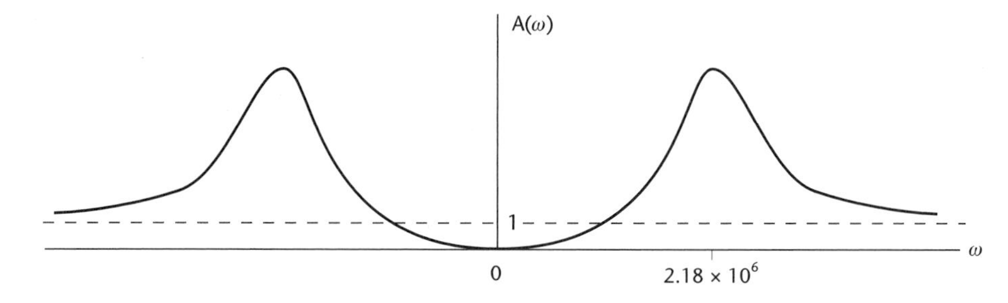

Fig. 10.6 Predicted form of amplitude response

maximum at around this value. It will also be clear that, because of the zero at the origin, the product $(AB \cdot AC)/(AD \cdot AE)$ will be zero when $\omega = 0$. It will also be apparent that as $\omega \rightarrow \infty$ the four lengths AB, AC, AD and AE will all become approximately equal such that, since $K = 1$ in Equation (10.10), the gain will tend to unity. It follows that the amplitude response will be of the broad overall shape shown in Figure 10.6.

Example 10.4

Consider a hypothetical network whose transfer function has complex conjugate poles at $s = 80 \pm j314$ and complex conjugate zeros at $s = \pm j314$ as shown in Figure 10.7(a). Determine its frequency response and suggest an application.

Solution We have

$$A(\omega) = \frac{|j\omega - z_1||j\omega - z_2|}{|j\omega - p_1||j\omega - p_2|} = \frac{|j\omega - z_1|}{|j\omega - p_1|}\frac{|j\omega - z_1^*|}{|j\omega - p_1^*|}$$

With reference to Figure 10.7(b) it is seen that the ratio $|j\omega - z_1^*|/|j\omega - p_1^*|$ is approximately unity for all positive frequencies. At frequencies between 0 and about 290 rad/s, and between 330 rad/s and infinity, the ratio $|j\omega - z_1|/|j\omega - p_1|$ is also very close to unity. In the immediate vicinity of 314 rad/s this latter ratio is significantly less than unity, becoming zero at 314 rad/s itself. The response will be much as shown in Figure 10.8 and it is apparent that we have a notch filter with its notch at 314 rad/s, or 50 Hz. Such a filter would have extensive applications for the rejection of mains interference in the UK. Whether it is physically possible to achieve a filter whose transfer function has such poles and zeros is, however, another matter. Chapter 13 takes a brief look at the design of analogue filters.

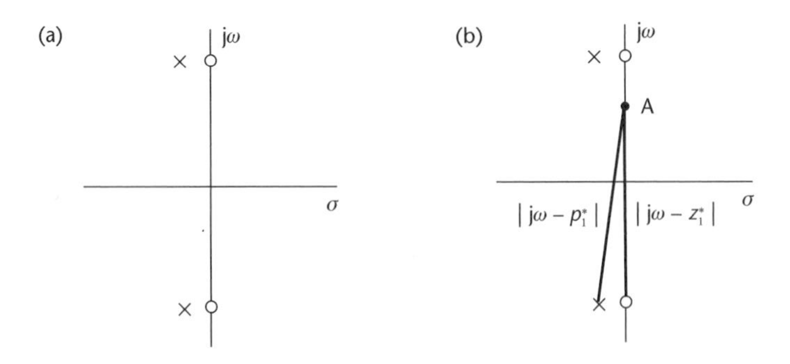

Fig. 10.7 Pole–zero plot

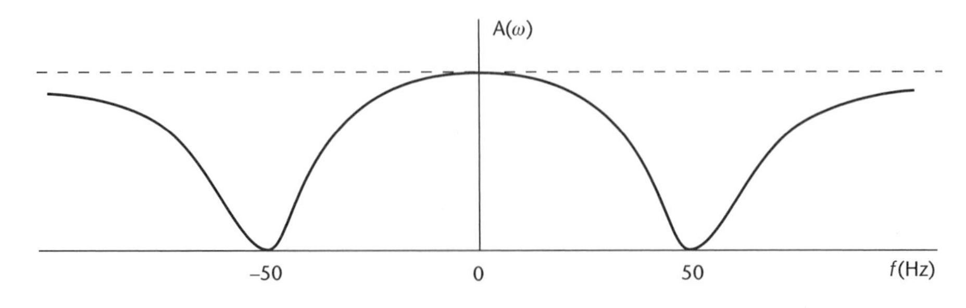

Fig. 10.8 Predicted form of amplitude response

10.7 Frequency response using MATLAB

MATLAB contains two powerful commands for computing and plotting the frequency response corresponding to a known transfer function. If we have a vector b giving the coefficients of the polynomial of the numerator of the transfer function, and a vector a giving the coefficients of the polynomial of the donominator, the command **freqs(b,a)** produces two plots, one of log magnitude versus log frequency and one of phase versus log frequency. A very similar result is obtained using **bode(b,a)** the only significant difference being that the log magnitude is given a dB scale and the phase uses 'phase unwrapping' to avoid some of the 2π discontinuities that occur with **freqs(b,a)** when the phase shift has a range exceeding 2π radians.

Example 10.5

Determine an accurate amplitude response for Example 10.3 using MATLAB.

Solution The transfer function was found to have zeros given by $z_1 = 0$ and $z_2 = -10^6$ and poles given by $p_{1,2} = -0.5 \times 10^6 \pm \text{j}2.18 \times 10^6$. Hence, neglecting any constant gain factor, the transfer function is given by

$$H(s) = \frac{s(s + 10^6)}{(s + 0.5 \times 10^6 - \text{j}2.18 \times 10^6)(s + 0.5 \times 10^6 + \text{j}2.18 \times 10^6)}$$

$$= \frac{s^2 + 10^6 s}{s^2 + 10^6 s + 5.0024 \times 10^{12}}$$

The command **freqs([1 1.e6 0],[1 1.e6 5.0024e12])** produces Figure 10.9.

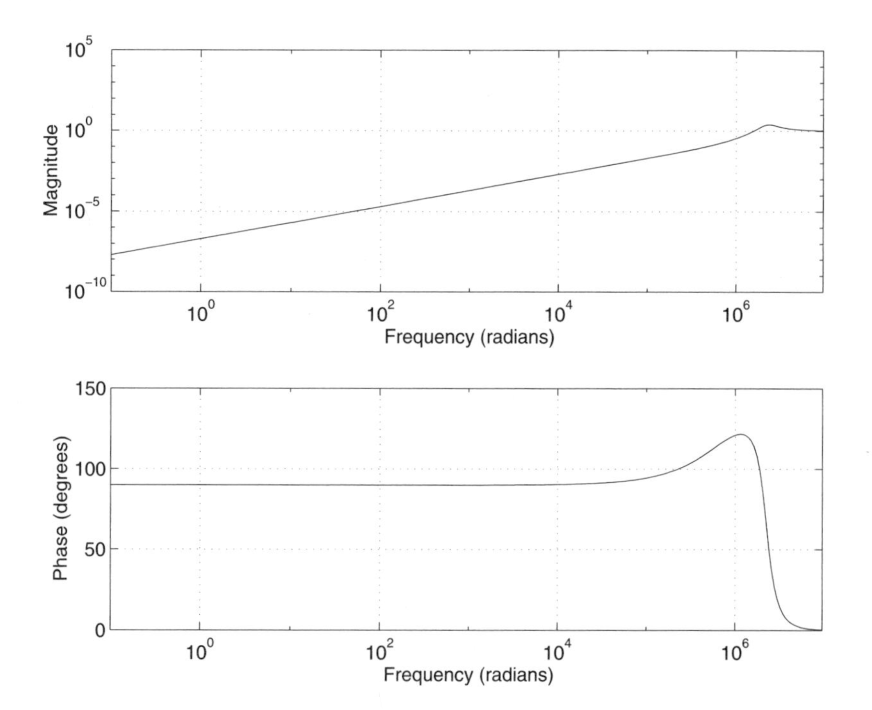

Fig. 10.9 Magnitude and phase response obtained using the **freqs(b,a)** command in MATLAB

The reason why the magnitude plot appears very dissimilar to Figure 10.6 is that the magnitude and frequency are now both on a logarithmic scale. If we want a plot on linear scales we can use the commands **clf; [H,w] = freqs([1 1.e6 0], [1 1.e6 5.0024e12]); subplot(2,1,1); plot(w,abs(H))**. The first of these clears the screen of Figure 10.9. The second calculates the complex frequency response $H(\omega)$ at 200 equispaced frequencies whose range is automatically selected to suit the response and which are given by the vector *w* (denoting ω). The fourth command plots $|H(\omega)|$ versus ω. The outcome is shown in Figure 10.10 which is now similar to Figure 10.6.

If we want to maintain the log frequency axis but to occupy a different range of frequencies we can specify a vector of logarithmically spaced frequencies using the **logspace** command. If, for example, we want to compute the transfer function at 200 logarithmically spaced frequencies between 10^2 and 10^7 rad/s, and then plot the outcome on logarithmic axes we require

```
w=logspace(2,7,200);
H=freqs([1 1.e6 0],[1 1.e6 5.0024e12],w);
loglog(w,abs(H))            % for log-log plots this replaces the plot command.
```

The outcome is shown in Figure 10.11.

If we had preferred a plot of linear magnitude versus log frequency we could have used **semilogx(w,abs(H))**. If we had preferred log magnitude versus linear frequency we could have used **semilogy(w,abs(H))**.

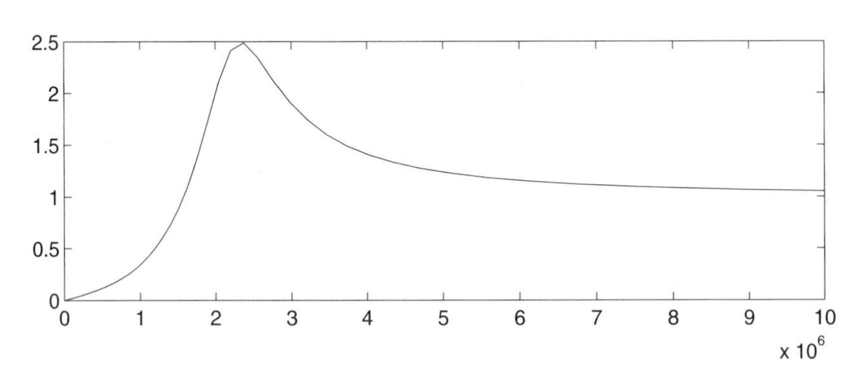

Fig. 10.10 Magnitude plot on linear axes based upon the **freqs(b,a)** command in MATLAB

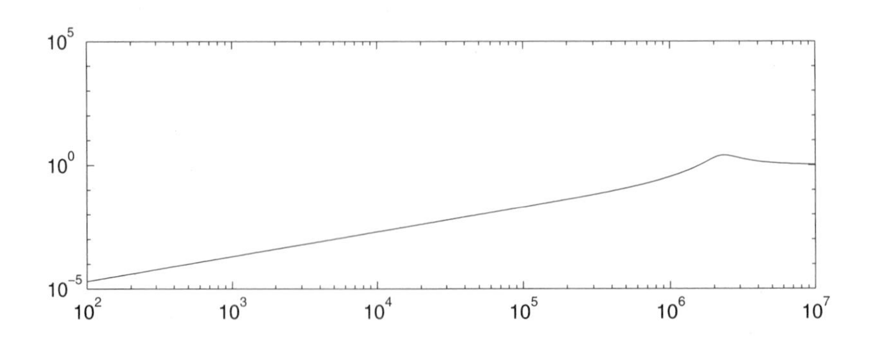

Fig. 10.11 Magnitude plot with selected frequency range

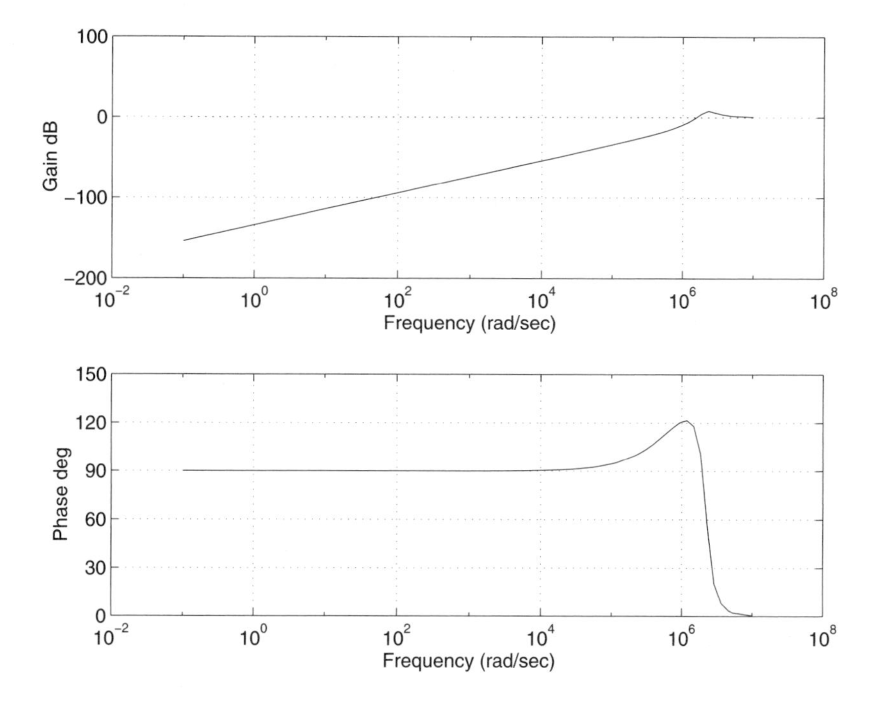

Fig. 10.12 Plots obtained with the **bode(b,a)** command in MATLAB

For comparison with Figure 10.9 the plot obtained with the command **bode([1 1.e6 0],[1 1.e6 5.0024e12])** is shown in Figure 10.12. Apart from the labelling of the axes the two results are very similar. As with **freqs(b,a)** there is considerable scope for flexibility in the range of frequencies that is plotted.

10.8 The order of a transfer function, some examples of second-order systems, and Q

The order of a transfer function is defined as being the highest order of s in the denominator of its transfer function following maximum reduction of terms within the transfer function (this last proviso is added since a transfer function such as $s/(s^2 + as)$ can be simplified to $1/(s + a)$ and would therefore be first-order rather than second-order). It follows that the order also equals the number of poles of the transfer function. If we consider the differential equation relating the output to the input it also follows that the order of a transfer function is the order of the *highest derivative* so long as this cannot be reduced by a change of variable (this last caution is added since, as an example, the differential equation $v = iR + L\,\mathrm{d}i/\mathrm{d}t$ describing the response of a series RL circuit could be rewritten $v = R\,\mathrm{d}q/\mathrm{d}t + L\,\mathrm{d}^2q/\mathrm{d}t^2$, where $\mathrm{d}q/\mathrm{d}t = i$ and yet the order of the system remains as first order.

Often, *but not invariably*, the order of the transfer function equals the number of frequency sensitive (or energy storage) elements in the system that have different currents through them or voltages across them (this condition of *different* currents or

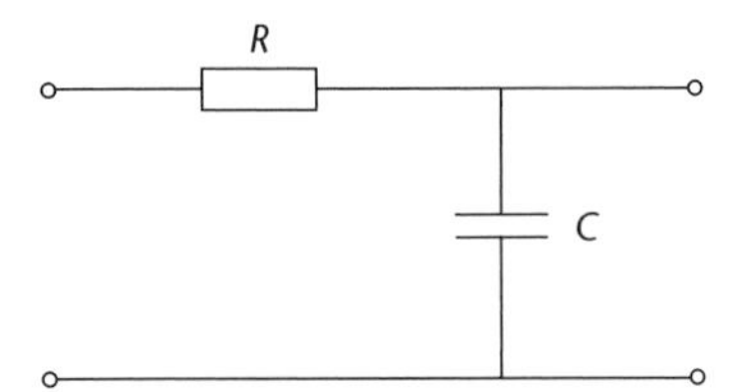

Fig. 10.13 RC network as an example of a first-order (single pole) system

voltages is necessary since otherwise we could, for example, divide an inductance into two in series and change the order of the transfer function).

First-order transfer functions arise from a resistance and a capacitance or from a resistance and an inductance. For example, the transfer function of the *RC* network shown in Figure 10.13 is given by

$$H(s) = \frac{1/sC}{R + 1/sC} = \frac{1}{RC(s + 1/CR)} \tag{10.30}$$

Second-order passive electrical networks involve a capacitance and an inductance. They are frequently used as a very simple means of obtaining bandpass frequency selectivity, and this is best studied by examples.

Example 10.6

It is frequently wished to realize a bandpass filter, and one of the simplest ways to achieve this involves a 'tank' circuit in the collector of a transistor amplifier, as shown in Figure 10.14. Determine the transfer function of the amplifier and plot its poles and zeros. Use this to make an approximate estimate of its bandwidth and fractional bandwidth.

Solution When analysing this circuit there is a practical problem, namely that real inductors are often far from being ideal inductances. They have a significant series resistance due to the finite resistance of the coil winding. This resistance is constant at low frequencies but eventually increases with the square root of frequency due to current being concentrated on the surface of the wire (the 'skin effect'). A second problem is that inductors can have a self-capacitance due to the close proximity of neighbouring turns in the winding. It will be assumed in what follows, however, that this is small compared with the external capacitance. In this example a practical inductor will be considered as comprised of an an ideal inductance L in series with an ideal resistance r.

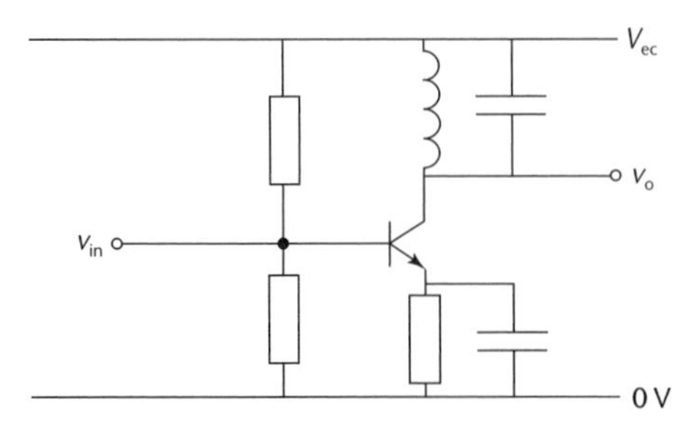

Fig. 10.14 Simple bandpass transistor amplifier

The transistor is close to being an ideal transconductance amplifier, which means that it may be considered to be a current source whose value is proportional to the applied base voltage and is therefore given by $g_m v_{in}$. The equivalent circuit of the transistor amplifier is then as shown in Figure 10.15. To find the response of this circuit to an input sinusoid we consider a complex voltage source, $\tilde{v}_{in} = V_0 e^{st}$ such that the current source becomes $\tilde{i} = g_m \tilde{v}_{in}$. For the tuned circuit the admittance is given by

$$Y(s) = \frac{1}{r + sL} + sC \tag{10.31}$$

such that the impedance $\tilde{v}/\tilde{i}$ is

$$\frac{\tilde{v}_0}{g_m \tilde{v}_{in}} = Z(s) = \frac{r + sL}{1 + sCr + s^2 LC} = \frac{1}{C} \frac{s + r/L}{s^2 + sr/L + 1/LC} \tag{10.32}$$

The transfer function is given by $H(s) = \tilde{v}_0 / \tilde{v}_{in}$. Hence, using Equation (10.32)

$$H(s) = g_m Z(s) = \frac{g_m}{C} \frac{s + r/L}{s^2 + sr/L + 1/LC} \tag{10.33}$$

This may be written as

$$H(s) = K \frac{s - z_1}{(s - p_1)(s - p_2)}$$

where the gain factor K equals g_m/C. The zero z_1 equals $-r/L$ and, since p_1 and p_2 are those values of s which cause the denominator to equal zero, their values correspond to the solution of the quadratic equation

$$s^2 + s\frac{r}{L} + \frac{1}{LC} = 0 \tag{10.34}$$

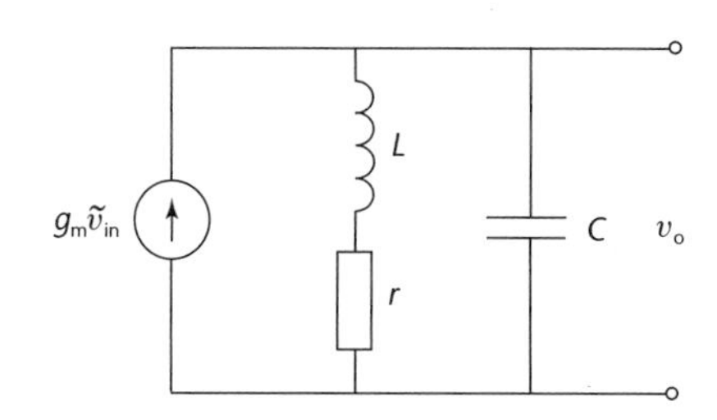

Fig. 10.15 Equivalent circuit of a transistor amplifier

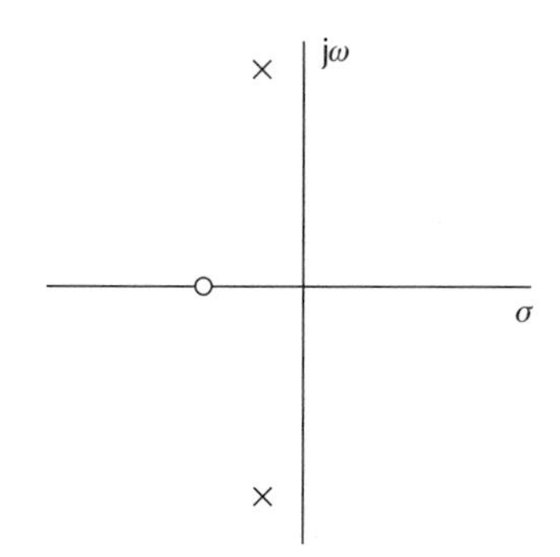

Fig. 10.16 Pole–zero plot for a transistor amplifier

This gives

$$p_{1,2} = -\frac{r}{2L} \pm \sqrt{\left(\frac{r}{2L}\right)^2 - \frac{1}{LC}} \tag{10.35}$$

For a practical inductor the resistance is small and this equation approximates to

$$p_{1,2} = -\frac{r}{2L} \pm \mathrm{j}\sqrt{\frac{1}{LC}} \tag{10.36}$$

The zero and the two conjugate poles are shown in Figure 10.16.

As before, the amplitude response can be obtained by moving up the imaginary axis and, at each frequency, taking the product of the distances to the zeros and dividing this by the product of the distances to the poles. It will be apparent from this procedure that a peak in the amplitude response occurs at, or very close to, the frequency corresponding to the imaginary component of the upper pole. In terms of radian frequency this is given by

$$\omega_0 \equiv \frac{1}{\sqrt{LC}} \tag{10.37}$$

The corresponding cyclic frequency is

$$f_0 \equiv \frac{1}{2\pi\sqrt{LC}} \tag{10.38}$$

This is the frequency at which the inductance and capacitance *resonate* together and is the resonant frequency of the tank circuit.

The real component of the upper pole is $-r/2L$, and a small absolute value of this signifies a narrow absolute bandwidth, since then the length to the upper pole changes substantially with small deviations in frequency away from ω_0. A visual inspection of the pole–zero plot indicates that the bandwidth is approximately r/L rad/s. The fractional bandwidth is given by the ratio of the bandwidth to the resonant frequency. As stated earlier, this 'feel' for the frequency response that is provided by the pole–zero plot is a major justification for using e^{st} as an excitation function.

In order to obtain a more precise and quantitative estimate of bandwidth we can replace s by $\mathrm{j}\omega$ in the expression for $H(s)$ given by Equation (10.33) and thence obtain the amplitude response $|H(\omega)|$. A detailed analysis of this shows that the 3 dB bandwidth is, in fact, exactly r/L rad/s, which is twice the distance of the pole to the imaginary axis. Hence the ratio of bandwidth to centre frequency (the fractional bandwidth) is $r/\omega_0 L$.

In Example 10.5 the fractional bandwidth was $r/\omega_0 L$; the criterion that an inductor is suitable for achieving a very selective filter centred at ω_0 is therefore that the reciprocal of this, $\omega_0 L/r$, should be large. The 'quality factor' of an inductor is defined by

$$Q = \frac{\text{series reactance}}{\text{series resistance}} \tag{10.39}$$

At low frequencies this Q factor increases linearly with frequency due to the increase in reactance with frequency. At higher frequencies, however, the Q flattens out due to an increase in r caused by the skin effect. Eventually it decreases because the self capacitance becomes significant, thus causing the reactance to decline. For any inductor there is a frequency at which its Q is maximum, and this frequency varies with the inductance and construction details of the inductor.

Although the Q factor was originally a term used specifically to specify the suitability of an inductor for achieving a frequency selective circuit with a small fractional bandwidth, it nowadays has an extended definition that is less constrained. It can now refer to any bandpass filter, is termed a selectivity factor rather than a quality factor, and is defined as

$$Q = \frac{f_0}{BW} \equiv \frac{1}{\text{fractional bandwidth}} \tag{10.40}$$

where BW is the filter bandwidth, which usually, but not always, refers to the -3 dB bandwidth. Using the same definition, the use of a selectivity factor Q may also be extended to bandstop filters.

Example 10.7

There are occasions where the combination of a capacitor with a typical inductor in the circuit of Figure 10.14 may result in a fractional bandwidth that is smaller than wanted. In this case the resonance may be damped by placing a resistance in parallel with the two, as shown in Figure 10.17. Determine the new bandwidth, the fractional bandwidth and the Q.

Solution Typical values of the external resistance, R_0, are likely to result in electrical damping that significantly exceeds the damping provided by the self-resistance of the inductor. The transistor circuit and its equivalent circuit become approximately as shown in Figure 10.18.

The admittance of the parallel combination is

$$Y(s) = \frac{1}{R_0} + \frac{1}{sL} + sC = \frac{sL + R_0 + s^2LCR_0}{sLR_0} = C\,\frac{s^2 + s/R_0C + 1/LC}{s} \tag{10.41}$$

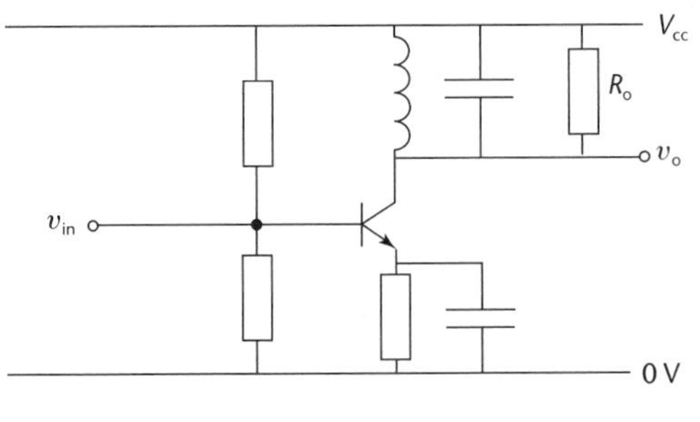

Fig. 10.17 Bandpass transistor amplifier with added damping

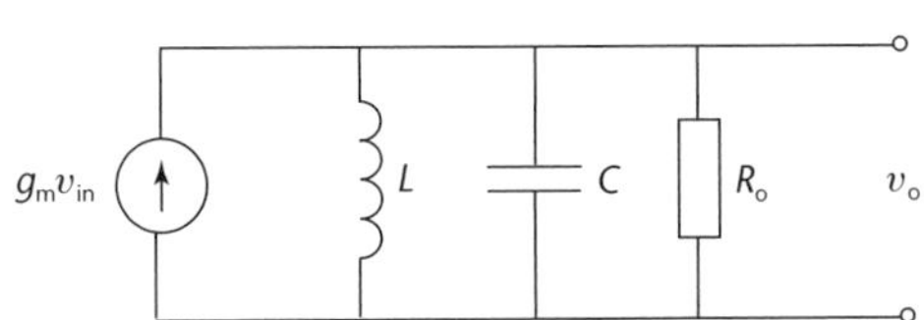

Fig. 10.18 Equivalent circuit of a transistor amplifier with added damping

Hence the transfer function is given by

$$H(s) = \frac{\tilde{v}_0}{\tilde{v}_{in}} = g_m \frac{\tilde{v}_0}{\tilde{i}} = g_m Z(s) = \frac{g_m}{C} \frac{s}{s^2 + s/R_0C + 1/LC} \tag{10.42}$$

This has a zero at the origin and there are two poles with positions given by

$$p_{1,2} = -\frac{1}{2R_0C} \pm \sqrt{\left(\frac{1}{2R_0C}\right)^2 - \frac{1}{LC}} \tag{10.43}$$

As with the tank circuit of Example 10.6 we again use the definition of Equation (10.37) that $\omega_0 \equiv 1/\sqrt{LC}$. Also, a term α_0 is defined as

$$\alpha_0 \equiv 1/2R_0C \tag{10.44}$$

From Equation (10.43) this gives

$$p_{1,2} = -\alpha_0 \pm \sqrt{\alpha_0^2 - \omega_0^2} \tag{10.45}$$

or

$$p_{1,2} = -\alpha_0 \pm \mathrm{j}\sqrt{\omega_0^2 - \alpha_0^2} \tag{10.46}$$

R_0 can have any value, such that ω_0 can be either smaller or greater than α_0. There are three cases of importance:

1. $\omega_0 < \alpha_0$: this gives two poles on the negative real axis, as shown in Figure 10.19(a).
2. $\omega_0 > \alpha_0$: this gives two complex conjugate poles, as shown in Figure 10.19(b).
3. $\omega_0 = \alpha_0$: this gives a double pole on the negative real axis at $-\alpha_0$, as shown in Figure 10.19(c).

From the point of view of a narrowband filter we are only interested in the case of $\omega_0 \gg \alpha_0$. As in Example 10.3 it should be clear that the peak in the amplitude response will be very close to the frequency ω_0, where $\omega_0 = 1/\sqrt{LC}$. By applying the graphical technique of interpreting the pole–zero plot in order to determine the amplitude response it should be apparent that, with reference to Figure 10.20, the bandwidth $(\omega_U - \omega_L)$ will be approximately twice the distance of the pole to the imaginary axis, or $2\alpha_0$ rad/s.

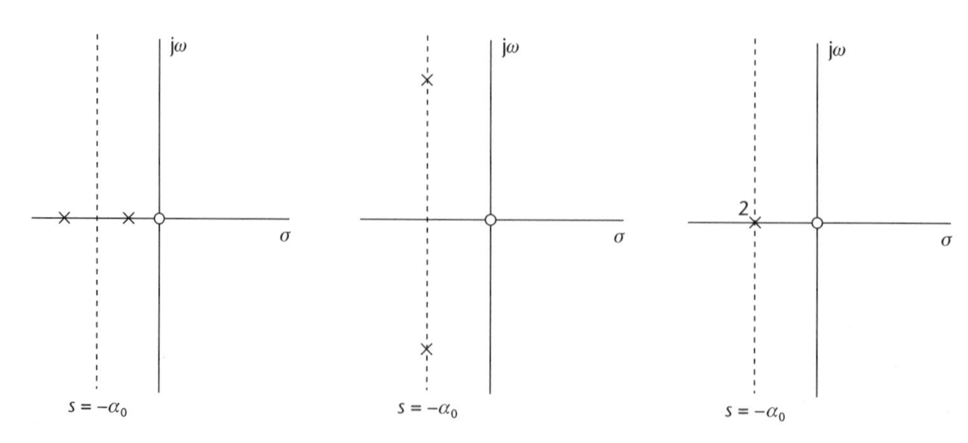

Fig. 10.19 Important cases of pole placement: (a) $\omega_0 < \alpha_0$; (b) $\omega_0 > \alpha_0$; (c) $\omega_0 = \alpha_0$

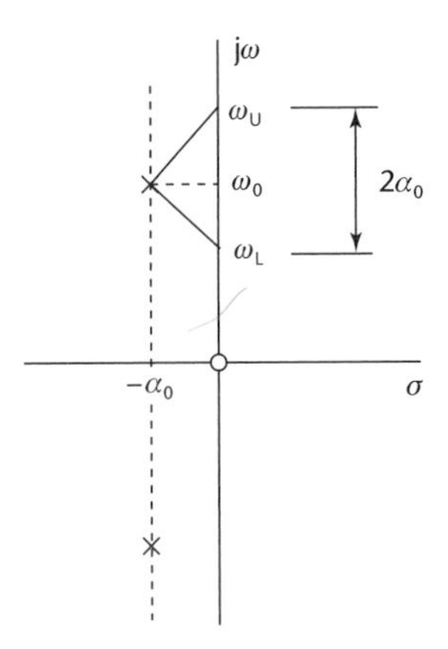

Fig. 10.20 Estimation of bandwidth from a pole–zero plot

Hence

$$\text{fractional bandwidth} = \frac{2\alpha_0}{\omega_0} = \frac{1}{\omega_0 R_0 C} \tag{10.47}$$

Using the definition of Equation (10.40) the selectivity factor is given by

$$Q = \omega_0 R_0 C \tag{10.48}$$

Alternatively, inserting $C = 1/\omega_0^2 L$ from $\omega_0 = \sqrt{1/LC}$

$$Q = R_0/\omega_0 L \tag{10.49}$$

It will be noticed that the form of this Q using a *shunt* resistance is the reciprocal of the form of the inductor quality factor given in Equation (10.39). This difference arises because the resistance there was the equivalent *series* value at the resonant frequency.

Example 10.8

Consider next a *series* resonant circuit as shown in Figure 10.21. Determine the current due to an excitation voltage $\cos \omega t$; also the Q.

Solution Replacing the voltage $\cos \omega t$ by e^{st} we obtain

$$\frac{\tilde{i}}{\tilde{v}} = \frac{1}{Z(s)} = \frac{1}{R + sL + 1/sC} = \frac{1}{L}\frac{s}{s^2 + sR/L + 1/LC} \tag{10.50}$$

It is noted that the current caused by an applied voltage in this series resonant circuit is of the same form as the voltage caused by a current in the parallel resonant circuit of Figure 10.14 (compare Equation (10.50) with the ratio $\tilde{v}/\tilde{i}$ obtained from the reciprocal of $Y(s)$ in Equation (10.41)). Because of this the series and parallel resonant circuits are known as 'duals'. The analysis is essentially identical. As in the previous example the polynomial can be expressed in terms of it poles and zeros, that is

$$\frac{\tilde{i}}{\tilde{v}} = \frac{1}{L}\frac{(s - z_1)}{(s - p_1)(s - p_2)} \tag{10.51}$$

where $z_1 = 0$ and p_1 and p_2 are pole positions corresponding to the solutions of the

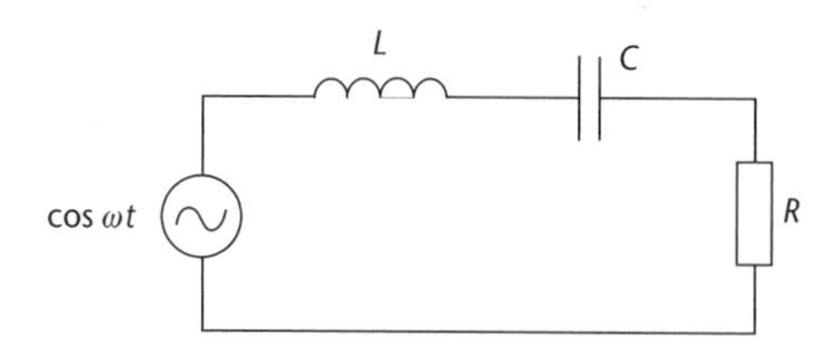

Fig. 10.21 Series resonant circuit

quadratic equation

$$s^2 + s\frac{R}{L} + \frac{1}{LC} = 0 \tag{10.52}$$

These pole positions are thus

$$p_{1,2} = -\frac{R}{2L} \pm \sqrt{\left(\frac{R}{2L}\right)^2 - \frac{1}{LC}} \tag{10.53}$$

These may be written as $p_{1,2} = -\alpha_0 \pm \sqrt{\alpha_0^2 - \omega_0^2}$ which is exactly the same as for the case of R_0, L and C in parallel except that, while $\omega_0 = 1/\sqrt{LC}$ as before, we now have $\alpha_0 \equiv R/2L$. For the case where $\omega_0 \gg \alpha_0$ we again have the fractional bandwidth equal to $2\alpha_0/\omega_0$. This time, however, substitution shows that

$$\text{fractional bandwidth} = \frac{2\alpha_0}{\omega_0} = \frac{R}{\omega_0 L} \tag{10.54}$$

and

$$Q = \frac{\omega_0 L}{R} \tag{10.55}$$

This is of the same form as Equation (10.39) but different from that of Equation (10.49). Care needs to be taken to use the correct formula.

10.9 Conjugate pole and conjugate zero pairs

It will have been noted in all cases so far that, whenever there has been a pole with an imaginary component, this has been accompanied by a conjugate pole; that is, by one having the same real component but with an equal and opposite imaginary component. The same would happen with zeros. The simplest argument that this should always be the case is that the pole-zero plot must be symmetrical about the real axis in order that the amplitude response should be symmetrical about zero frequency, as must always happen.

For a second viewpoint we note that two non-conjugate poles or zeros would signify polynomials within the transfer function of the form $(s - a_1 - jb_1)$ $(s - a_2 - jb_2)$. On expansion these can be shown to contain an imaginary component unless $b_2 = -b_1$ and $a_1 = a_2$. Similarly a *single* pole or zero at $(a_1 + jb_1)$ would signify a polynomial in the transfer function having an imaginary component. However, as stated in Section 10.3, the expressions for transfer functions come about through arithmetic operations on impedances that are of the form R, sL or $1/sC$

and there is no way in which an s transfer function can have numerator or denominator polynomials containing imaginary components.

It follows that *poles and zeros with imaginary components must always exist in conjugate pairs*.

10.10 Impedance functions for non-electrical components and systems

The voltages and currents in the various parts of a linear network are related by differential equations. Because e^{st} remains of the same form after differentiation it follows that a signal e^{st} at one part of the network signifies signals of the form e^{st} at all other parts of the network. For example, if the current through a series RLC circuit is $I_0 e^{st}$ the voltage across the combination is $(R + sL + 1/sC)I_0 e^{st}$.

However, the quantities related by a differential equation need not be constrained to electrical circuit elements for this to apply.

As the first example, the acceleration a of a mass M due to an applied force F is given by $F = Ma$, or by $F = M\,\mathrm{d}u/\mathrm{d}t$ where u is the velocity. If $\tilde{u} = Ae^{st}$ this gives $\tilde{F} = sMAe^{st}$, and hence $\tilde{F}/\tilde{u} = sM$. Thus, just as sL is the impedance to current flow when a voltage is applied to an inductance, so sM may be considered the impedance to velocity when a force is applied to a mass. It follows that the mass may be considered to have a mechanical impedance given by

$$Z_M(s) \equiv \frac{\tilde{F}}{\tilde{u}} = sM \tag{10.56}$$

As a second example, the relationship between the extension x of a spring and the applied force F is $x = KF$, or $F = x/K$, where K is the compliance of the spring (equal to the reciprocal of its stiffness). However, the velocity of the end of the spring is $\mathrm{d}x/\mathrm{d}t$ and, in terms of velocity, this equation for force therefore becomes $F = (1/K)\int_{-\infty}^{t} u\,\mathrm{d}t$. If $\tilde{u} = U_0 e^{st}$ this gives $\tilde{F} = (1/Ks)U_0 e^{st}$ and hence $\tilde{F}/\tilde{u} = 1/sK$. The mechanical impedance of the spring is thus given by

$$Z_M(s) = 1/sK \tag{10.57}$$

This is analogous to the electrical impedance $1/sC$ of a capacitance.

As a third example, the relationship between force and velocity for a viscous damper is $F = f_v u$, where f_v is the coefficient of viscous friction. If $\tilde{u} = U_0 e^{st}$ then $\tilde{F} = f_v U_0 e^{st}$ and $\tilde{F}/\tilde{u} = f_v$. This is another mechanical impedance, analogous this time to an electrical resistance. This time we have

$$Z_M(s) = f_v \tag{10.58}$$

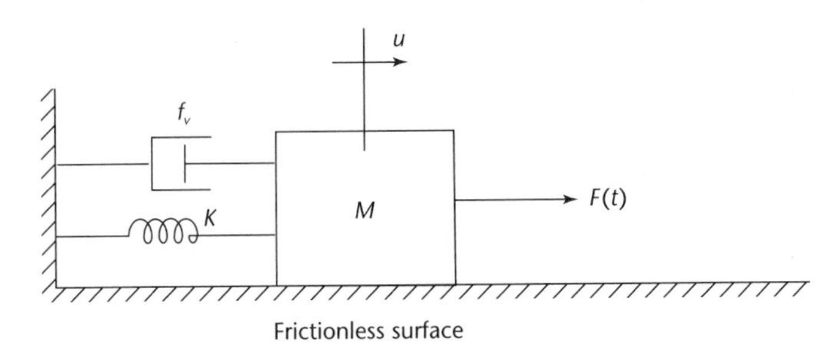

Fig. 10.22 Translational mechanical system

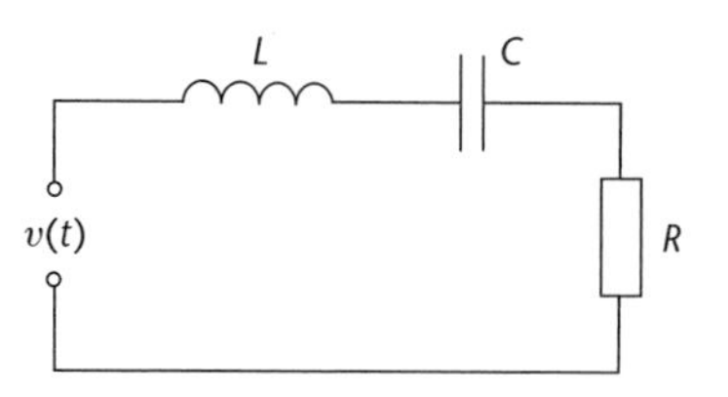

Fig. 10.23 RLC circuit equivalent to a mechanical system

We could combine these results for the system shown in Figure 10.22 to give

$$F(t) = M\frac{\mathrm{d}u}{\mathrm{d}t} + \frac{1}{K}\int u\,\mathrm{d}t + f_{\mathrm{v}}u \qquad (10.59)$$

If $\tilde{u} = U_0 \mathrm{e}^{st}$ this gives

$$\tilde{F} = sM\tilde{u} + \frac{1}{sK}\tilde{u} + f_{\mathrm{v}}\tilde{u} \qquad (10.60)$$

and

$$Z_M(s) = \frac{\tilde{F}}{\tilde{u}} = sM + \frac{1}{sK} + f_{\mathrm{v}} \qquad (10.61)$$

This equation may be considered as giving the impedance function relating force and velocity for this mechanical system. There is an exact analogy between it and the voltage-driven series *RLC* circuit shown in Figure 10.23, for which

$$\frac{\tilde{v}}{\tilde{i}} = Z(s) = sL + \frac{1}{sC} + R \qquad (10.62)$$

Compared with this electrical circuit it is much less likely that the mechanical system would be driven by a sinusoidal forcing function. However, it will be shown in the next chapter that a knowledge of the response of the mechanical system to a driving function of the form e^{st} can be invaluable for predicting the response to other excitations. If, for example, there were a sinusoidal excitation then the resonant frequency would be $1/2\pi\sqrt{MK}$ Hz, just as the resonant frequency of the electrical circuit was shown in Example 10.6 to be given by $1/2\pi\sqrt{LC}$ Hz.

10.11 Transfer functions for electromechanical components and systems

Because of its importance in automatic control systems a particularly good example of the transfer function of an electromechanical system is that of an electric motor. A typical d.c. servo motor uses permanent magnets in its stator to provide a constant magnetic flux density B across the region in which the rotor (or armature) is situated. A d.c. voltage is applied to the armature and causes a current i to circulate around its windings. Each conductor of the armature of length l experiences a force Bli and thus the torque on the armature is proportional to i and is given by

$$T = K_{\mathrm{t}}i \qquad (10.63)$$

where K_{t} is a motor torque constant involving the strength of the magnetic field and the length and number of conductors in the armature.

The motor is applied to a load, and the load and motor armature have a combined moment of inertia J. The rotation of the armature and load result in power being dissipated as mechanical work and wasted heat, and this is associated with a combined coefficient of viscous damping D for rotation defined as the ratio torque/rotational velocity. Thus there is opposition to the torque generated by the armature that is given by

$$T = J\frac{d^2\theta}{dt^2} + D\frac{d\theta}{dt} \tag{10.64}$$

where θ is the angle of rotation.

As the armature rotates, the passing of its windings through the stator field causes a back e.m.f. which opposes the voltage e_a applied to the armature. This back e.m.f. is proportional to the rotational speed. There is also an ohmic and inductive voltage drop in the armature winding. Often the inductive voltage drop is negligible compared with the ohmic loss, in which case we have

$$e_a = K_b\frac{d\theta}{dt} + iR_a \tag{10.65}$$

where K_b is a constant of proportionality called the back e.m.f. constant and R_a is the armature resistance.

Inserting the value of T from Equation (10.64) into Equation (10.63) to obtain an expression for i, and then putting this resulting value of i into Equation (10.65), we obtain

$$\begin{aligned} e_a &= K_b\frac{d\theta}{dt} + \frac{R_a}{K_t}\left(J\frac{d^2\theta}{dt^2} + D\frac{d\theta}{dt}\right) \\ &= \frac{R_a J}{K_t}\frac{d^2\theta}{dt^2} + \left(\frac{R_a D}{K_t} + K_b\right)\frac{d\theta}{dt} \end{aligned} \tag{10.66}$$

If e_a is of the form $\tilde{e}_a = E_a e^{st}$ then θ will be of the form $\tilde{\theta} = \Theta e^{st}$ where Θ may be expected to be a function of s, and substitution of these into Equation 11.66 produces

$$E_a e^{st} = \frac{R_a J}{K_t}s^2\Theta e^{st} + \left(\frac{R_a D}{K_t} + K_b\right)s\Theta e^{st} \tag{10.67}$$

Hence

$$\tilde{e}_a = \frac{R_a J}{K_t}s^2\tilde{\theta} + \left(\frac{R_a D}{K_t} + K_b\right)s\tilde{\theta} \tag{10.68}$$

and

$$\frac{\tilde{\theta}}{\tilde{e}_a} = \frac{1}{(R_a J/K_t)s^2 + (R_a D/K_t + K_b)s} = \frac{K_t/R_a J}{s\{s + (1/J)[D + (K_t K_b/R_a)]\}} \tag{10.69}$$

This transfer function may be written in the simplified form

$$H(s) = \frac{A}{s(s+\alpha)} \tag{10.70}$$

where

$$A = K_t/R_a J \tag{10.71}$$

and

$$\alpha = \frac{1}{J}\left(D + \frac{K_t K_b}{R_a}\right) \tag{10.72}$$

Equation (10.70) could be used to find the response of the motor to a sinusoidal driving voltage. However, this is not likely to be a useful application. The main significance of this transfer function will become evident in Chapter 12 when such a motor is incorporated into an automatic control system and it is wished to determine its step function response.

10.12 Summary

By determining the response of a system to an excitation function of the form e^{st}, where $s = \sigma + j\omega$, the response to a cosinusoidal excitation function $\cos \omega t$ can be deduced by setting σ to zero. Much more importantly though, the complete characteristics of the response can be conveyed by a plot of the s values for which the response is infinite or zero (a pole–zero plot). A great deal can be learnt about the frequency response by a simple visual inspection of this pole–zero plot. Accurate responses can be readily obtained using MATLAB. The techniques can be applied to electromechanical or mechanical systems as well as to electrical systems. The next chapter will show that the pole–zero plot can further be interpreted to give information on the response of the system to abrupt changes in excitation, and consequently on the stability of the system.

Because inductors are widely used in frequency-selective analogue systems, a sideline to the main thrust of the chapter has been a discussion of how inductors deviate in practice from ideal inductances and how their performance is characterized by their Q factor.

10.13 Problems

1. A system has a pair of conjugate poles at $(-412 \pm j8020)$ and no zeros. Sketch its approximate amplitude response. What would be the effect of adding a zero at the origin?
2. A network has a transfer function given by

$$H(s) = \frac{0.0032s^6 + 0.0591s^4 + 0.2000s^2 + 0.1818}{s^6 + 1.1454s^5 + 2.3499s^4 + 1.7969s^3 + 1.5044s^2 + 0.6395s + 0.1926}$$

Determine the poles and zeros using MATLAB. Show the results on a pole–zero plot and use this to sketch an approximate amplitude response. Check the result by using MATLAB to produce a Bode plot. Produce an amplitude plot using MATLAB in which the axes are linear to make the result more easily comparable with the sketch.

3. The tank circuit of a bandpass transistor amplifier uses an inductor in parallel with a 1.5 nF capacitance. If the equivalent circuit of the inductor is a 3 mH inductance in series with a resistance, determine the centre frequency of the amplifier. If the inductor has Q of 180 at the resonant frequency, estimate the amplifier bandwidth and the equivalent series resistance of the inductor at resonance. Estimate the centre frequency and bandwidth of the amplifier. Determine what extra resistance should be added in parallel with the inductor and capacitance in order to give a fractional bandwidth of 0.05. Without undertaking any network analysis, deduce the pole positions for the two cases.
4. The equivalent circuit of a piezoelectric hydrophone is as shown in Figure 10.24, where $L = 13$ mH, $C = 52$ pF, $R = 1.6$ kΩ and $C_0 = 250$ pF. Plot its frequency response.

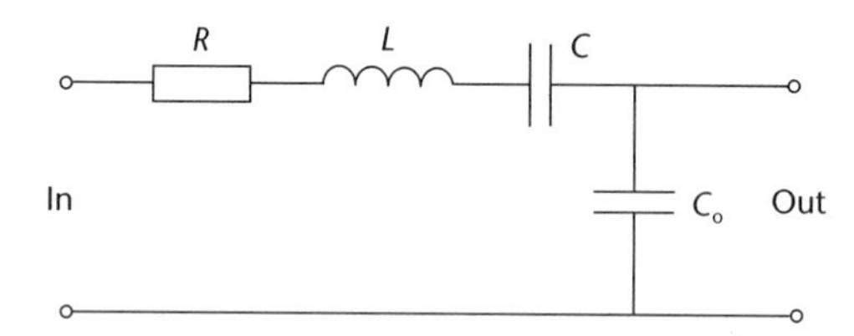

Fig. 10.24

5. The output displacement $x(t)$ of an electromechanical system due to a voltage input $v(t)$ is given by the system differential equation

$$6\frac{d^3x(t)}{dt^3} + 54\frac{d^2x(t)}{dt^2} + 17\frac{dx(t)}{dt} + 3x(t) = 2.2\frac{dv(t)}{dt} + 0.7v(t)$$

Using the methods described in this chapter, find the transfer function; hence plot the frequency response using MATLAB.

Chapter eleven

Natural responses, transients and stability

11.1 ● Preview

Chapter 10 was concerned with the steady state response of analogue systems to cosinusoidal excitations. The next two chapters are concerned mainly with the responses of analogue systems to other types of excitation.

The analytical techniques used in this chapter for determining system responses are somewhat laborious and the methods of the next chapter involving the Laplace transform are usually preferred. Alternatively, it is extremely easy to determine a system response using MATLAB – often with just a single command. In spite of these reservations the approach given in this present chapter does have value. The system response is separated into a natural response and a forced response, and this provides valuable *physical insight* into the behaviour of the system under a wide range of excitation conditions. It was shown in Section 10.5 that the system is characterized by the poles and zeros of the system transfer function, and it follows that the poles and zeros are directly related to the natural response. A very important consequence is that certain pole positions are an indication of instability.

11.2 ● Natural response

The *natural* response of a system concerns the response that is *a property of the system itself*, rather than what is caused by any input to the system. It is most easily visualized as the response of that system due to the release of internal energy in the absence of any external excitation. For example, a guitar string gains energy when pulled but, on being released, undergoes its natural response – it dissipates its internal energy by means of decaying vibrations whose characteristics are a property of the guitar. A comparable scenario in an electrical system would be the response of a network as a charged capacitor within it discharges through the other components of the network.

Internal energy can be acquired in a number of ways. The main one considered in this chapter is where a continuous source has been applied to the system for a very long time such that the system has reached its steady state condition. The natural response then concerns the release of energy acquired by the system when this source is suddenly disconnected. However, another way of causing a system to acquire energy is to apply an impulsive excitation to it. This causes energy to be assimilated over the infinitesimal duration of the impulse and to be dissipated thereafter in accordance with the natural response. It follows that, except at the instant of the impulse itself, the impulse response is identical to the natural response. (The proviso arises because there are some systems for which the impulse response contains an impulse at the instant of the input impulse function, and this is not part of the natural response.)

Somewhat less directly related and less easily visualized than the response due to a *release* of energy is the fact that the natural response is also relevant to the way in which a system *acquires* energy. If, for example, air is suddenly forced through an organ pipe, the way in which the sound builds up is determined by the natural response of the pipe. Similarly, the buildup of current when a voltage is first applied to an electrical circuit is affected by the natural response of the circuit.

If we consider a system such as a mass suspended by a spring it is much easier to visualize what happens if the mass is pushed down and then released than to visualize what happens if an impulsive force is applied to the mass. Because impulses are infinitesimal in duration we are very unfamiliar with them in everyday life. We find it intuitively difficult to predict an impulse response. For this reason the impulse response approach will be avoided in this chapter but will be delayed to the next chapter, where further mathematical tools become available to help in the analysis.

11.3 The natural response of first-order systems

Two cases will be considered; a series RC network and a series RL network.

11.3.1 The natural current response of a series *RC* network

Consider the electrical network shown in Figure 11.1. Prior to $t = 0$ it is assumed that the network is in its settled or steady state condition such that $i = 0$ and the capacitance is charged up to the voltage E_0. At $t = 0$ the external forcing function is disconnected; thus the voltage applied to the resistance and capacitance is as shown in Figure 11.2.

Following this disconnection of the voltage source, the energy stored in the capacitance is discharged through the resistance and the resulting behaviour therefore reflects the *natural*, or free, response of the network. We could consider the natural response of the voltage across the capacitance, but what follows concerns the natural response of the *current*. The two are related but different in form.

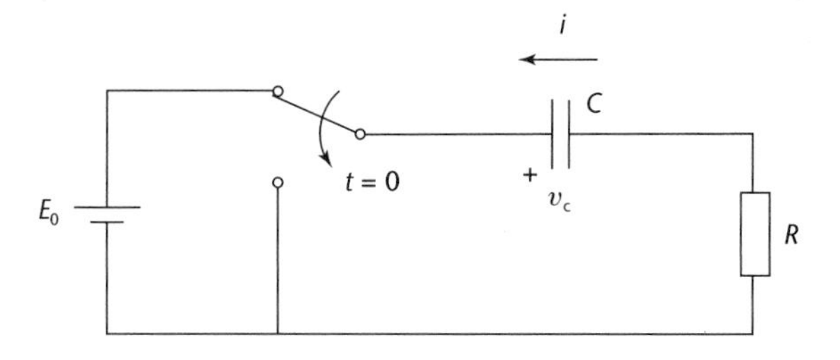

Fig. 11.1 A series RC network

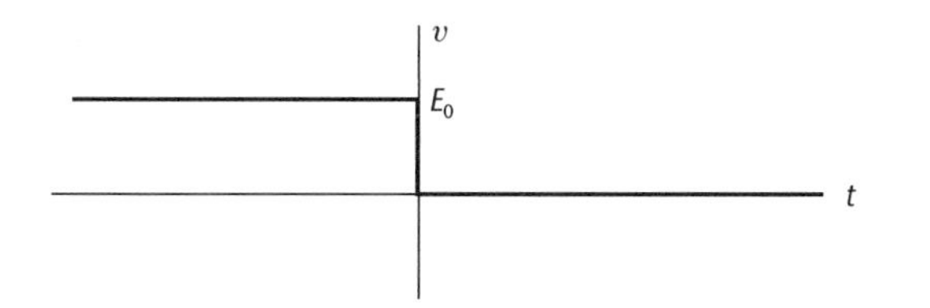

Fig. 11.2 Voltage applied to the RC network

In order to determine the natural current response it is necessary to formulate and then solve the differential equation describing the network behaviour for $t \geqslant 0$. Adopting the sign convention of Figure 11.1 for the directions of positive i and positive v_C we can apply Kirchhoff's voltage law after the switch is thrown to obtain

$$0 = v_C - iR \tag{11.1}$$

The capacitance discharges such that the voltage across it decreases from its initial value of E_0 according to the relationship

$$v_C = E_0 - \frac{1}{C}\int_0^t i \, dt \tag{11.2}$$

Using this to eliminate v_C from Equation (11.1) we obtain

$$0 = E_0 - \frac{1}{C}\int_0^t i \, dt - iR \tag{11.3}$$

In order to solve this it is helpful to differentiate each term so that the integro-differential equation becomes a straightforward differential equation. Doing this eliminates the E_0 term and produces

$$\frac{1}{C} i + R\frac{di}{dt} = 0 \tag{11.4}$$

The *highest derivative* in Equation (11.4) is the first, making this is a *first-order* differential equation and meaning that we are dealing with a *first-order system.*

Because there is no forcing function within the differential equation of Equation (11.4), this differential equation is known as a *homogeneous* equation. Because e^{st} remains of the same form when differentiated we begin by assuming a solution of the form

$$i = Ae^{st} \tag{11.5}$$

Putting this into Equation (11.4) we obtain

$$\frac{A}{C} e^{st} + RAse^{st} = 0 \tag{11.6}$$

such that, dividing through by $(A/C)e^{st}$,

$$1 + sCR = 0 \tag{11.7}$$

The differential equation of Equation (11.4) has been reduced to a very much simpler equation, one that is an *algebraic* equation in terms of s. Such an equation is known as the *characteristic equation*.

In our example the solution of the characteristic equation is $s = -1/RC$ and substitution of this into Equation (11.5) gives

$$i = Ae^{-t/CR} \tag{11.8}$$

To evaluate A we consider the *initial* conditions of the network. Prior to the switch being thrown, the voltage on the capacitance is E_0 and, at the instant when the switch is thrown at $t = 0$, this appears across the resistance to produce an initial current of $i = E_0/R$. Inserting this initial current into Equation (11.8) for $t = 0$ we obtain

$$A = E_0/R \tag{11.9}$$

and hence Equation (11.8) becomes

$$i = \frac{E_0}{R} e^{-t/CR} \qquad t \geqslant 0 \tag{11.10}$$

This is the natural response and is shown in Figure 11.3. The hatching for $t < 0$ is included to emphasize that the result of Equation (11.10) is only valid for $t \geqslant 0$.

The current can also be expressed in the form

$$i = \frac{E_0}{R} e^{-t/\tau} \qquad t \geqslant 0 \tag{11.11}$$

where τ is the time taken for the current to decrease to e^{-1} (36.8%) of its original value and is known as the time constant. In this example $\tau = CR$. Differentiation of Equation (11.11) shows that the initial gradient is $-E_0/R\tau$, thus making the initial gradient intersect the time axis at time $\tau = CR$, as shown in Figure 11.3.

It should be noted that the initial current depends on what happened in the circuit prior to $t = 0$, but that *the form of the response subsequent to* $t = 0$ (a decaying exponential in this example) *is solely a function of the network itself*, thus confirming the appropriateness of the term *natural response*.

11.3.2 The natural current response of a series *RL* network

Consider the electrical network shown in Fig. 11.4. The differential equation describing the current for $t \geqslant 0$ is

$$L\frac{di}{dt} + Ri = 0 \tag{11.12}$$

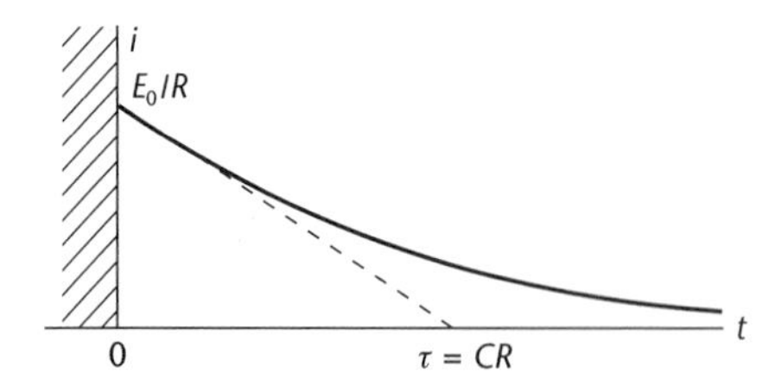

Fig. 11.3 Natural response of current in series *RC* network

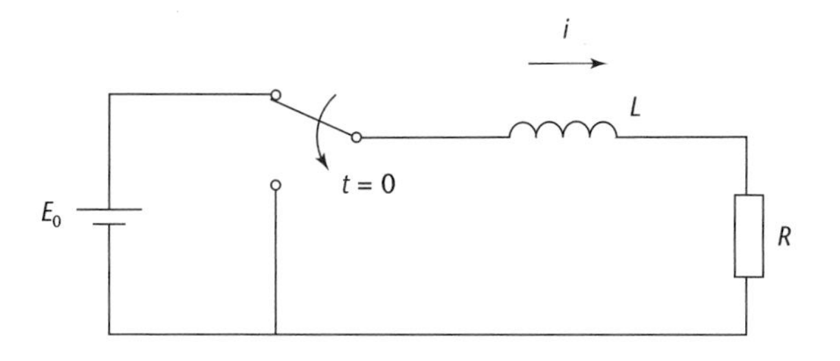

Fig. 11.4 A series *RL* network

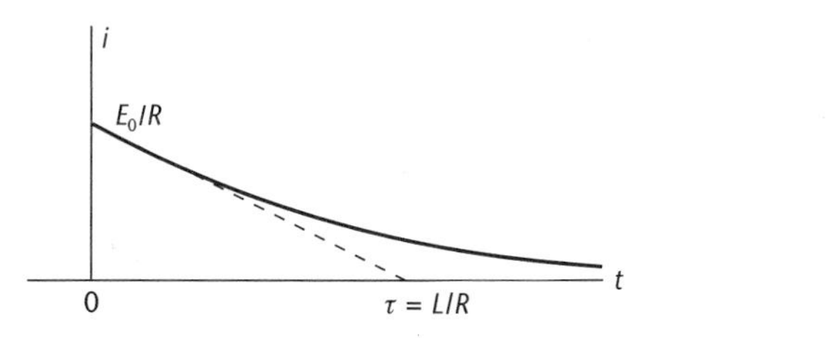

Fig. 11.5 Natural response of current in series *RL* network

Assuming the solution $i = Ae^{st}$ we obtain

$$LAse^{st} + RAe^{st} = 0 \tag{11.13}$$

Therefore

$$s = -R/L \tag{11.14}$$

such that

$$i = Ae^{-(R/L)t} \tag{11.15}$$

For $t < 0$ the current is E_0/R . This current cannot change instantaneously at $t = 0$ because of the inductance. Therefore, putting this current into Equation (11.15) for $t = 0$ we find $A = E_0/R$ and hence

$$i = \frac{E_0}{R} e^{-(R/L)t} \qquad t \geqslant 0 \tag{11.16}$$

This is the natural response of the circuit and is shown in Figure 11.5. The time constant is now L/R.

11.4 The complete response

The natural responses of the systems considered have concerned the release of internal energy following the *disconnection* of an excitation by means of a switch. By the *complete* response we mean the response of the system following some change to the system input, but such that there is an excitation source present after the change.

Consider for example the same network as Figure 11.1 but consider this time that the voltage source is *connected* at $t = 0$, rather than being disconnected. A suitable switching arrangement is shown in Figure 11.6. For $t \geqslant 0$ there is now a steady state component of current due to the excitation and there is another component of current which is due to the *sudden change* in the circuit and which is therefore of the same form as the natural response. In other words, the complete response for $t \geqslant 0$ is $i_F + i_N$, where i_F is the forced current and i_N is the current associated with the

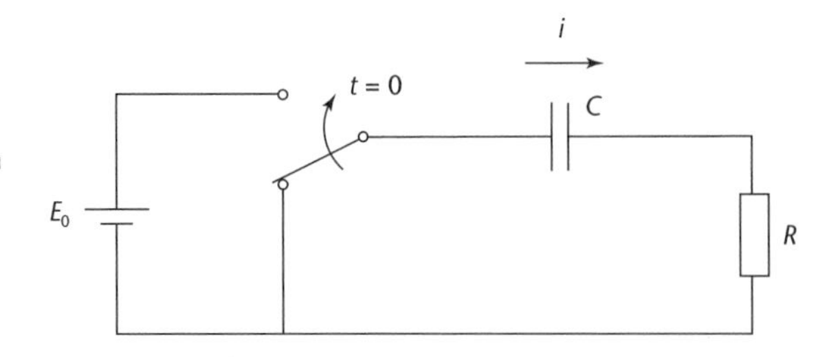

Fig. 11.6 Sudden application of a voltage source to a series *RC* network

natural response. The current associated with the natural response eventually dies away to leave just the forced current, and this is readily obtained by network analysis. The form of the natural response is known from Equation (11.8) and all that remains therefore is to determine its *magnitude*. This is done by imposing the initial conditions on the response.

Example 11.1

Determine the complete current response for the series *RC* network of Figure 11.6.

Solution Because of the capacitance we have $i_F = 0$ and, from Equation (11.8), $i_N = Ae^{-t/CR}$. Hence

$$i = i_F + i_N = Ae^{-t/CR} \tag{11.17}$$

Immediately after the switch is thrown (at a time denoted by $t = 0^+$) the voltage across the capacitance is zero and the current is therefore E_0/R. Inserting this initial condition into Equation (11.17) we find that the complete response is given by

$$i = \frac{E_0}{R}\,e^{-t/CR} \qquad t \geqslant 0 \tag{11.18}$$

This then is the current in a series *RC* network in response to a step function of voltage. It is the same equation as when a voltage E_0 is suddenly removed (Figure 11.1 and Equation (11.10)) except that the currents in the two cases are in opposite directions.

It is interesting to note that, if a mathematician were asked to solve Example 11.1 he or she would start by writing down the differential equation for the current in the network and would then find a specific or *particular solution* to this differential equation which, in this example, would be the final or steady state current. The mathematician would then set the forcing function to zero to produce a homogeneous equation and determine what is termed the *complementary solution*. This gives the *form*

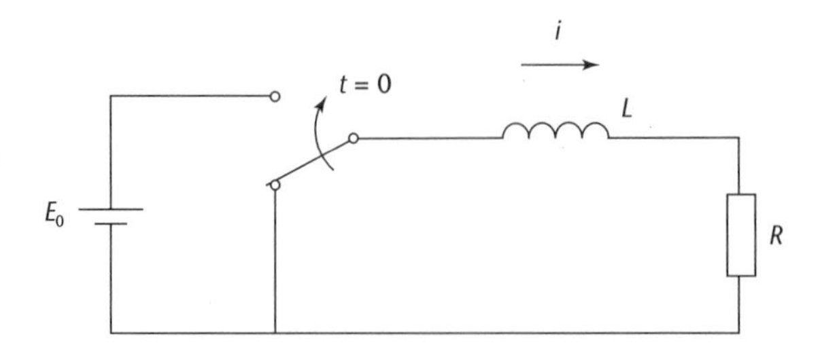

Fig. 11.7 Sudden application of a voltage source to a series *RL* network

of a current response but not its magnitude. This solution corresponds to the *natural response*. The sum of the particular solution and the complementary solution is the *complete solution* and it is finally necessary to determine the magnitude of the complementary solution such that the boundary conditions are satisfied.

In effect this procedure is identical to what has been described earlier; it is just the terminology that is different.

Example 11.2

Determine the complete current response for the series RL network shown in Figure 11.7.

Solution The differential equation describing the current for $t \geqslant 0$ is given by

$$E_0 = L\frac{\mathrm{d}i}{\mathrm{d}t} + Ri \tag{11.19}$$

By inspection of the network the final or steady state current is given by

$$i = E_0/R \tag{11.20}$$

This is the *particular* solution.

The homogeneous equation is

$$L\frac{\mathrm{d}i}{\mathrm{d}t} + Ri = 0 \tag{11.21}$$

Putting $i = A\mathrm{e}^{st}$ into this we obtain

$$sLA\mathrm{e}^{st} + RA\mathrm{e}^{st} = 0 \tag{11.22}$$

Hence $s = -R/L$ and

$$i = A\mathrm{e}^{-(R/L)t} \tag{11.23}$$

This is the *complementary* solution.

The complete solution is the sum of the particular solution and the complementary solution and, using Equations (11.20) and (11.23) is therefore

$$i = \frac{E_0}{R} + A\mathrm{e}^{-(R/L)t} \tag{11.24}$$

Prior to the switch being thrown the current is zero. The inductance does not allow the current to be changed instantly and hence the initial condition is that $i = 0$ when $t = 0^+$. Substituting this into Equation (11.24) gives

$$A = -E_0/R \tag{11.25}$$

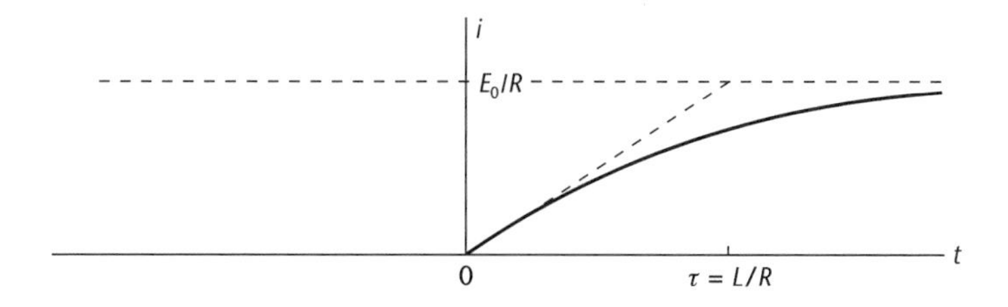

Fig. 11.8 Step function response of a series *RL* network

and hence Equation (11.24) becomes

$$i = \frac{E_0}{R}(1 - e^{-(R/L)t}) \tag{11.26}$$

This is the current response of the RL network to a step voltage function and is plotted in Figure 11.8.

The last two examples have concerned the sudden *introduction* of an excitation. *An interesting modification* to the approaches used is to consider the excitation to be the superposition of a continuous excitation *and one which ceases at* $t = 0$. Because this approach involves the *removal* of an excitation rather than the introduction of one it can be somewhat easier to visualize the role and magnitude of the natural response. This is best understood by examples.

Example 11.3

Solve the problem of Example 11.2 by considering the applied voltage as the sum of a continuous voltage and a voltage that cancels it for $t < 0$.

Solution The voltage applied to the series RL circuit of Figure 11.7 is the step function shown in Figure 11.9(a). This may be thought of as the continuous d.c. voltage of Figure 11.9(b) superposed on the excitation function of Figure 11.9(c), and it is noted that these two waveforms cancel for $t < 0$. Because there is no net excitation for $t < 0$, the two waveforms of Figure 11.9(b) and 11.9(c) could be regarded as fictitious. However, it is very helpful to consider them as both existing. This is because the total response is the sum of the two and yet the response to each excitation waveform separately is easy to predict and visualize.

The response to the d.c. voltage of Figure 11.9(b) is clearly given by

$$i_F = E_0/R \tag{11.27}$$

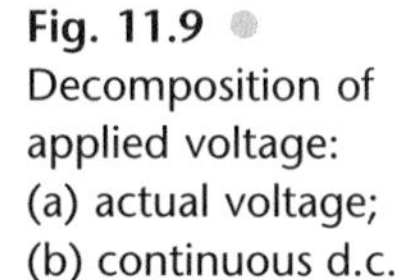

(a)

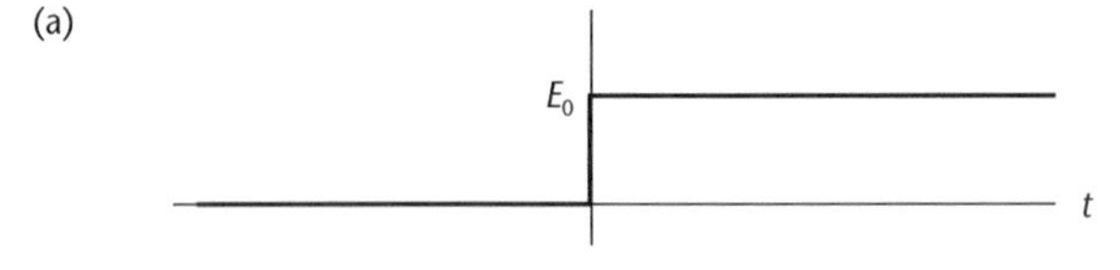

(b)

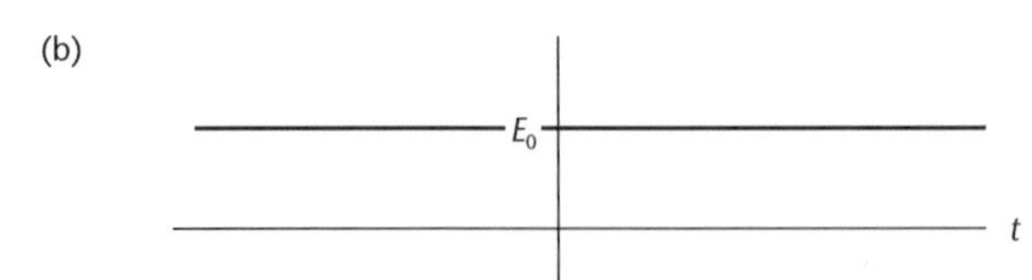

(c)

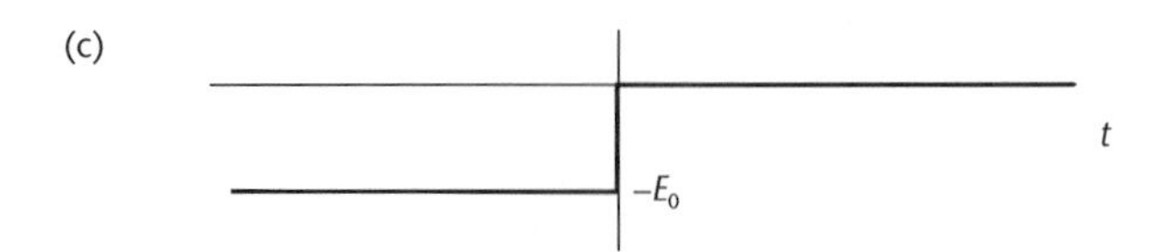

Fig. 11.9 Decomposition of applied voltage: (a) actual voltage; (b) continuous d.c. voltage; (c) fictitious opposing voltage prior to $t = 0$

The response to the voltage of Figure 11.9(c) is obtained by recognizing that this voltage is the exact negative of that in Figure 11.4, thus causing the current to be the negative of that given by Equation (11.16). Thus this component of current is given by

$$i_{\mathrm{N}} = -\frac{E_0}{R}\,\mathrm{e}^{-(R/L)t} \qquad t \geqslant 0 \tag{11.28}$$

Thus the step function current response is given by

$$i = i_{\mathrm{F}} + i_{\mathrm{N}} = \frac{E_0}{R}\,(1 - \mathrm{e}^{-(R/L)t}) \qquad t \geqslant 0 \tag{11.29}$$

This is the same result as obtained in Equation (11.26) and illustrated in Figure 11.8. The approach has the advantage over the approach of Example 11.2 that it is easier to visualize and evaluate the natural component of current since it arises from the *cessation* of a voltage rather than from the *introduction* of one.

Example 11.4

In the circuit of Figure 11.10 the 230 V 50 Hz alternating source $325.3\cos(100\pi t + 40°)$ is connected into the circuit at $t = 0$. Determine the response of the voltage across the capacitance.

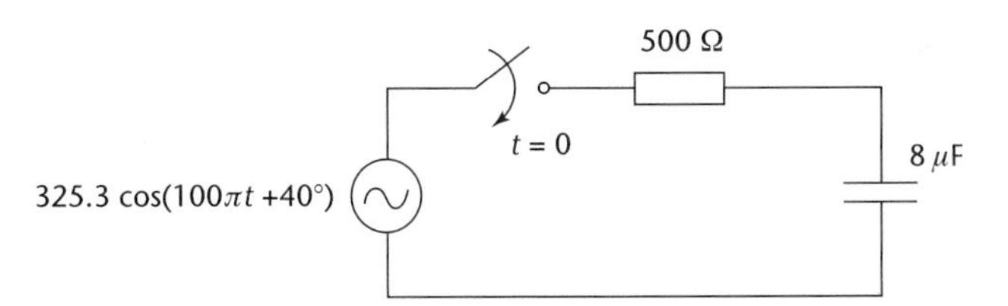

Fig. 11.10 Sudden application of an a.c. voltage source to a series *RC* network

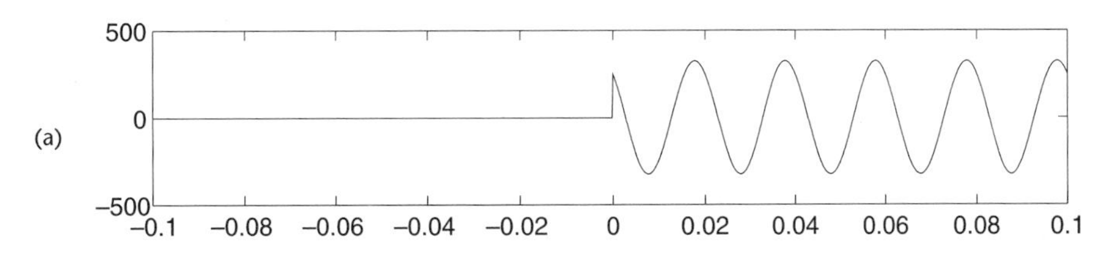

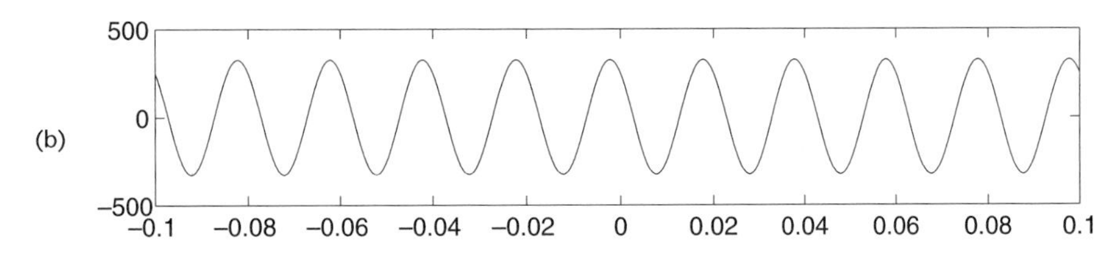

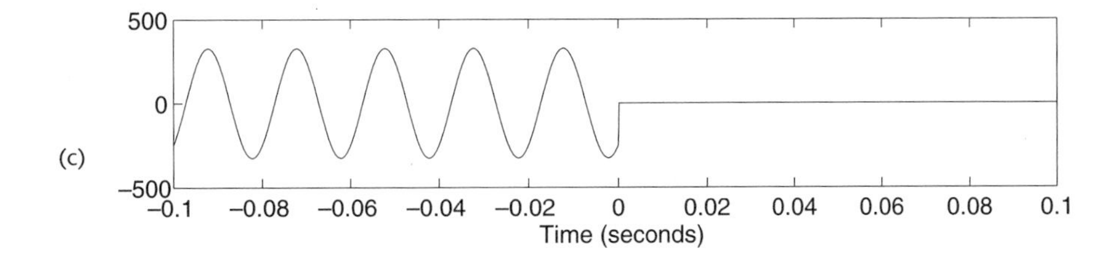

Fig. 11.11 Decomposition of applied voltage: (a) actual voltage; (b) continuous a.c. voltage; (c) fictitious opposing voltage prior to $t = 0$

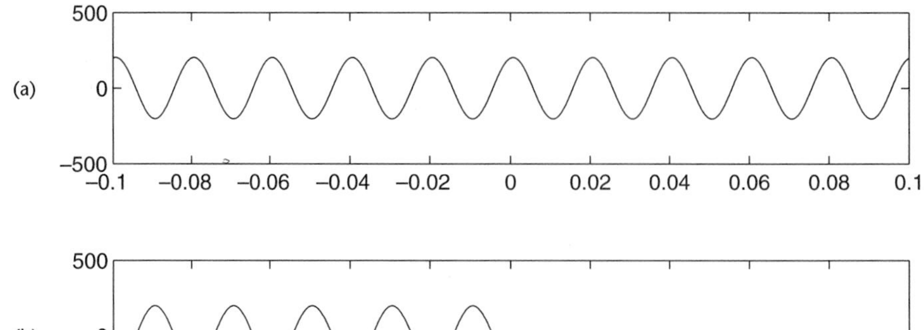

Fig. 11.12 Components of voltage across capacitance: (a) due to continuous a.c. voltage; (b) due to fictitious opposing voltage prior to $t = 0$

Solution The applied voltage is shown in Figure 11.11(a). This may be considered to be composed of the continuous a.c. voltage shown in Figure 11.11(b) and the voltage shown in Figure 11.11(c) that cancels it for $t < 0$.

The continuous a.c. voltage of Figure 11.11(b) can be written in phasor form as

$$\mathbf{V} = 230\angle 40^\circ \tag{11.30}$$

Hence

$$\mathbf{V}_{\mathrm{CF}} = \mathbf{V}\frac{1/\mathrm{j}\omega C}{R + 1/\mathrm{j}\omega C} = \frac{\mathbf{V}}{1 + \mathrm{j}\omega CR} = \frac{230\angle(40^\circ - \tan^{-1}(\omega CR))}{\sqrt{1 + (\omega CR)^2}} \tag{11.31}$$

For $\omega = 100\pi$, $R = 500\ \Omega$ and $C = 8\ \mu\mathrm{F}$ this becomes $\mathbf{V}_{\mathrm{CF}} = 143.2\angle{-11.5^\circ}$ and hence

$$v_{\mathrm{CF}} = 202.5\cos(100\pi t - 11.5^\circ) \tag{11.32}$$

This is shown in Figure 11.12(a)

Prior to $t = 0$ the fictitious opposing voltage of Figure 11.11(c) must cause a voltage across the capacitance that is equal and opposite to that of Figure 11.12(a), as shown in Figure 11.12(b). At $t = 0$ the value of this voltage is $-202.5\cos(-11.5^\circ)$ or -198.5 V. For $t \geqslant 0$ the charge associated with this voltage is discharged with a time constant of RC and the natural response is therefore

$$v_{\mathrm{CN}} = -198.5\mathrm{e}^{-t/RC} = -198.5\mathrm{e}^{-250t} \tag{11.33}$$

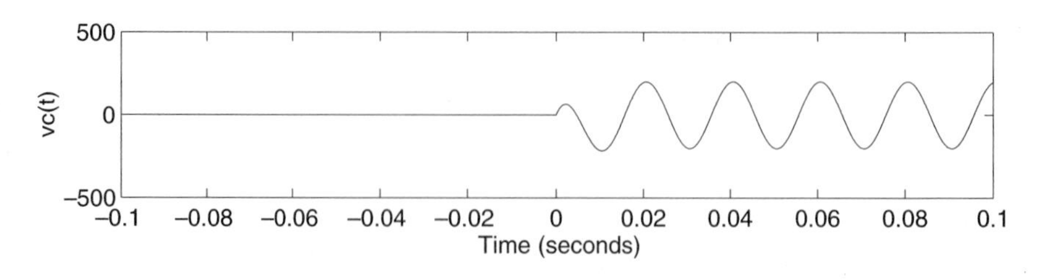

Fig. 11.13 Complete response of voltage across capacitance

This is also shown in Figure 11.12(b). The complete response is the sum of the voltages in Figures 11.12(a) and 11.12(b) and is therefore given by

$$v_C = 202.5\cos(100\pi t - 11.5°) - 198.5e^{-250t} \tag{11.34}$$

This is shown in Figure 11.13.

11.5 Transients

When the excitation to a system is changed, transients occur. Section 11.3 considered cases where a forcing function was *disconnected* from a first-order linear system and Section 11.4 considered cases where a forcing function was *connected* into a first-order linear system. In all cases the period during which the response goes from one steady state condition to another steady state condition is known as the *transient period.*

Consider a system in which a forcing function is changed at $t = 0$. The response during a transient period is the sum of a natural response and a steady state response due to the new forcing function (zero in the case where the source is disconnected). The transient period lasts as long as the natural response has a significant effect on the total response.

What is termed *the transient response* is the *complete response within the transient period.*

The term *transient* on its own is often used synononomously with transient response. However, the term tends to be used somewhat loosely and is sometimes used to refer solely to the component of the response which is transient in character (that is, *the natural response*). Since the complete response is the sum of the forced response and the natural response this leads to two different possible interpretations of *transient*, either the complete response *or* the natural response within the transient period. Care needs to be taken.

So far *electrical* transients have been considered. A *mechanical* example of a transient would be the oscillation of a car after it passes over a pothole. An electromechanical example would be the deflection of the pointer of an analogue voltmeter when it is first switched into a circuit. An understanding of the transient response can be useful in the design of a system. When designing a voltmeter, for example, it may well be wished to know the relationship between the overshoot and the mechanical parameters of inertia, spring constant and damping.

11.6 Relationship between the natural response and the system poles and zeros

Consider the series RC network of Figure 11.14 to be driven by a continuous forcing function that is the previously encountered mathematical abstraction of the form

$$\tilde{v} = Be^{st} \tag{11.35}$$

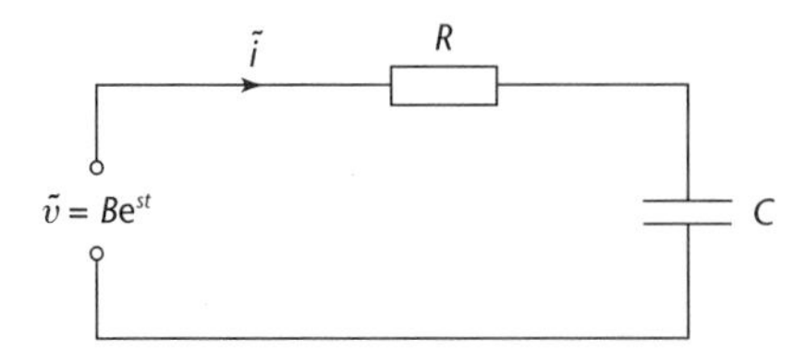

Fig. 11.14 Series RC network

The current is given by

$$\tilde{i} = \frac{\tilde{v}}{Z(s)} = Be^{st}Y(s) \tag{11.36}$$

where

$$Y(s) = \frac{1}{Z(s)} = \frac{1}{R + 1/sC} = \frac{1}{R}\frac{s}{s + 1/RC} = \frac{1}{R}\frac{(s - z_1)}{(s - p_1)} \tag{11.37}$$

where $z_1 = 0$ and the pole p_1 equals $-1/RC$.

If $s = -1/RC$, $Y(s)$ is infinite and the current given by Equation (11.36) can be non-zero even if the magnitude B of the complex exponential voltage is zero. It follows that a complex exponential current of the form

$$i = Ae^{-t/RC} \tag{11.38}$$

can exist, *even with a zero applied voltage.*

In physical terms, however, the only way the current can be non-zero in the absence of an applied voltage is because of energy already stored in the circuit. This corresponds with the mechanism of the natural response as described in Section 11.2.

It follows that *the natural current response of the circuit is of exactly the same form as the current for which the admittance is infinite.* This conclusion is confirmed by noting that the current of Equation (11.38) is identical to the natural response of the series RC circuit, as given by Equation (11.8).

If we have an admittance function containing multiple poles, each pole signifies that a current of that complex exponential form can exist in the absence of an applied voltage. In other words, if

$$Y(s) = \frac{(s - z_1)(s - z_2)(s - z_3)\cdots}{(s - p_1)(s - p_2)(s - p_3)\cdots} \tag{11.39}$$

there can be natural current components $A_1e^{p_1t}$, $A_2e^{p_2t}$, $A_3e^{p_3t}$ and so on. It also follows that a natural current can exist which is the sum of these terms, that is

$$i_n = A_1e^{p_1t} + A_2e^{p_2t} + A_3e^{p_3t} + \cdots \tag{11.40}$$

If one of the poles has an imaginary component then there must be another pole which is its conjugate in order that their separate natural responses can combine to give a natural response that is real. (This provides a sound *physical* argument to confirm the conclusion of Section 10.8 that poles with imaginary components must exist in conjugate pairs.) An admittance function with poles at $-\alpha_0 + j\omega_1$ and $-\alpha_0 - j\omega_1$ signifies a natural current response of the form $A_1e^{(-\alpha_0 + j\omega_1)t}$ $+A_2e^{(-\alpha_0 - j\omega_1)t}$. We must have $A_1 = A_2$ and the response reduces to the form $(A_1/2)e^{-\alpha_0 t}\cos(\omega_1 t + \theta)$, which is an exponentially decaying sinusoid.

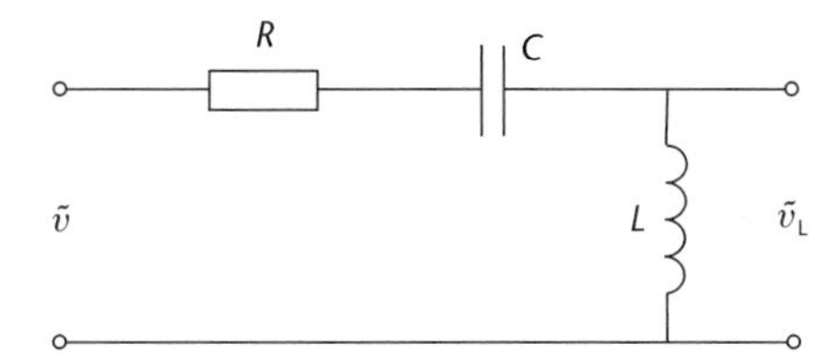

Fig. 11.15 Two-port *RLC* network

A pole without an imaginary component, such as $p_1 = -\alpha_0$, signifies a natural response of the form $e^{-\alpha_0 t}$ and is a non-oscillatory decaying exponential.

If currents can exist in a circuit without any applied voltage it follows that voltages across components can also exist in the circuit in the absence of any applied voltage. Therefore, just as poles in an admittance function indicate the nature of the natural response, so do the poles in a transfer function. As an example, consider the transfer function in the *RLC* circuit of Figure 11.15.

We have

$$\frac{\tilde{v}_L}{\tilde{v}} = \frac{sL}{R + sL + 1/sC} = \frac{s^2}{s^2 + (R/L)s + 1/LC} = \frac{(s-0)^2}{(s-p_1)(s-p_2)} \tag{11.41}$$

The same argument can be applied as before that, if we suppose $\tilde{v} = Be^{st}$ and s has the value of one of the poles, the voltage across the inductance can be non-zero even if $B = 0$. In other words there are natural response components of the voltage across the inductance given by $A_1 e^{p_1 t}$ and $A_2 e^{p_2 t}$, and the total natural response can have contributions from both, namely

$$v_L = A_1 e^{p_1 t} + A_2 e^{p_2 t} \tag{11.42}$$

Methods exist for calculating the constants A_1 and A_2 directly from the s plane plot of poles and zeros but an examination of these will be delayed until the next chapter.

11.7 The natural response of second-order systems

Second-order passive electrical networks involve an inductance and a capacitance. There are many possible configurations but the case of the series *RLC* network of Figure 11.16 will be the one considered. The switch takes the d.c. source out of the circuit at time $t = 0$ and subsequently the energies stored in the capacitance and in the inductance move around the series *RLC* circuit to be dissipated ultimately in the resistance. There is no forcing function for $t \geqslant 0$ and therefore the response for $t \geqslant 0$ is the natural response. Based on the previous section its natural response will be deduced from the poles of its admittance function.

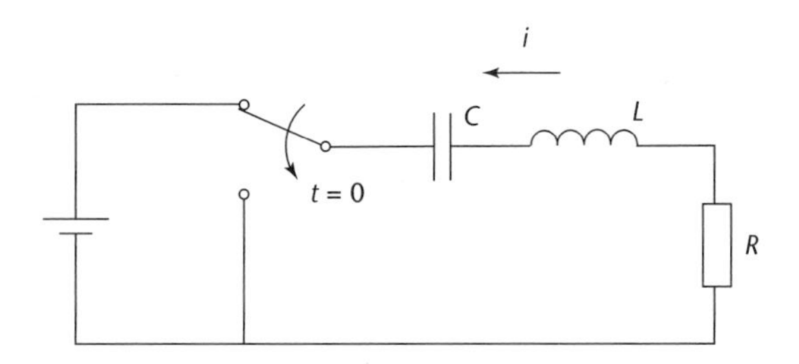

Fig. 11.16 Series *RLC* network as a second-order system

We have

$$Y(s) = \frac{1}{Z(s)} = \frac{1}{R + sL + 1/sC} = \frac{1}{L}\frac{s}{s^2 + sR/L + 1/LC}$$
$$= \frac{1}{L}\frac{(s - z_1)}{(s - p_1)(s - p_2)} \tag{11.43}$$

where the two poles reaffirm that this is a second-order system. The roots of the denominator give

$$p_{1,2} = -\frac{R}{2L} \pm \sqrt{\left(\frac{R}{2L}\right)^2 - \frac{1}{LC}} \tag{11.44}$$

or

$$p_{1,2} = -\alpha_0 \pm \sqrt{\alpha_0^2 - \omega_0^2} \tag{11.45}$$

where

$$\alpha_0 = R/2L \tag{11.46}$$

and

$$\omega_0 = 1/\sqrt{LC} \tag{11.47}$$

It follows from Equation (11.40) that the natural response is of the form

$$i = A_1 e^{p_1 t} + A_2 e^{p_2 t} \tag{11.48}$$

There are three important cases of the current given by Equation (11.48), corresponding to $\omega_0 > \alpha_0$, $\omega_0 < \alpha_0$ and $\omega_0 = \alpha_0$. The three cases are considered in turn.

Case 1: $\omega_0 < \alpha_0$, or $\dfrac{1}{LC} < \left(\dfrac{R}{2L}\right)^2$

This time the poles are real and negative and are given by

$$p_1 = -\alpha_0 + \sqrt{\alpha_0^2 - \omega_0^2} = -\alpha_1 \tag{11.49}$$

and

$$p_2 = -\alpha_0 - \sqrt{\alpha_0^2 - \omega_0^2} = -\alpha_2 \tag{11.50}$$

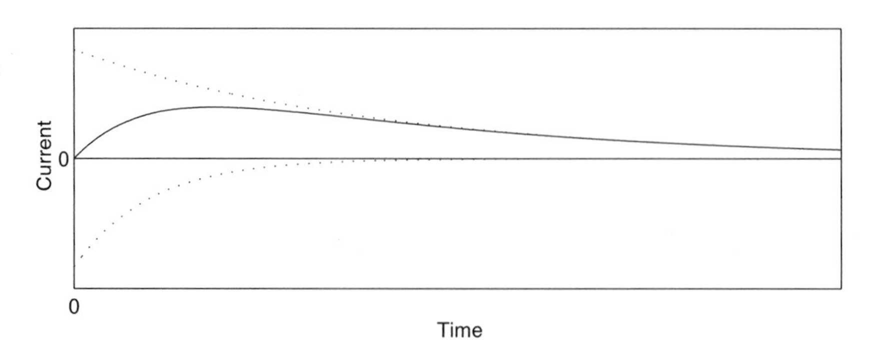

Fig. 11.17 The natural response of an over-damped second-order system

Therefore, from Equation (11.48)

$$i = A_1 e^{-\alpha_1 t} + A_2 e^{-\alpha_2 t} \tag{11.51}$$

We can find numerical values for α_1 and α_2 using their definitions and knowing R, L and C. The values of A_1 and A_2 are determined by imposing the initial conditions of the circuit. Based upon particular values of A_1 and A_2, the two components of current in Equation (11.51) might be as shown by the dotted lines in Figure 11.17. The combined current is then as shown by the solid line. There is no oscillation in this current subsequent to $t = 0$ and the condition is known as the over-damped case.

Case 2: $\omega_0 > \alpha_0$, or $\dfrac{1}{LC} > \left(\dfrac{R}{2L}\right)^2$

Here Equation (11.45) gives us

$$p_{1,2} = -\alpha_0 \pm \mathrm{j}\sqrt{\omega_0^2 - \alpha_0^2} = -\alpha_0 \pm \mathrm{j}\omega_1 \tag{11.52}$$

where

$$\omega_1 \equiv \sqrt{\frac{1}{LC} - \left(\frac{R}{2L}\right)^2} \tag{11.53}$$

Therefore, from Equation (11.48)

$$i = A_1 e^{-\alpha_0 t} e^{\mathrm{j}\omega_1 t} + A_2 e^{-\alpha_0 t} e^{-\mathrm{j}\omega_1 t} \tag{11.54}$$

It is permissible for A_1 and A_2 to be complex but the current i must be real. This occurs if $A_1 = (A/2)e^{\mathrm{j}\beta}$ and $A_2 = (A/2)e^{-\mathrm{j}\beta}$, where A is a real constant, since then Equation (11.54) gives

$$i = (A/2)e^{-\alpha_0 t} e^{\mathrm{j}(\omega_1 t + \beta)} + (A/2)e^{-\alpha_0 t} e^{-\mathrm{j}(\omega_1 t + \beta)} = Ae^{-\alpha_0 t}\cos(\omega_1 t + \beta)$$

or, defining a phase θ as $(\beta + 90°)$,

$$i = Ae^{-\alpha_0 t}\sin(\omega_1 t + \theta) \tag{11.55}$$

As in Case 1 the values of A and θ are determined by the initial conditions. For the configuration of Figure 11.16 the current is initially zero, such that $\theta = 0°$. A plot of Equation (11.55) for particular values of A, α_0 and ω_1 is given in Figure 11.18. There is a damped oscillation, or ringing, and this condition of $\omega_0 > \alpha_0$ is known as the under-damped case. The frequency ω_1 is the natural, or characteristic, frequency of the system.

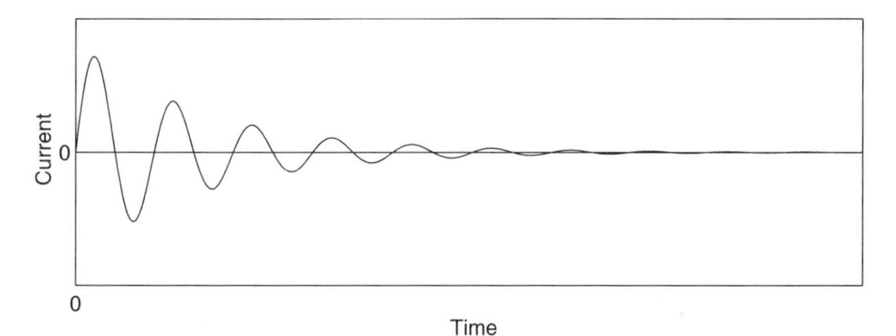

Fig. 11.18 The natural response of an under-damped second-order system

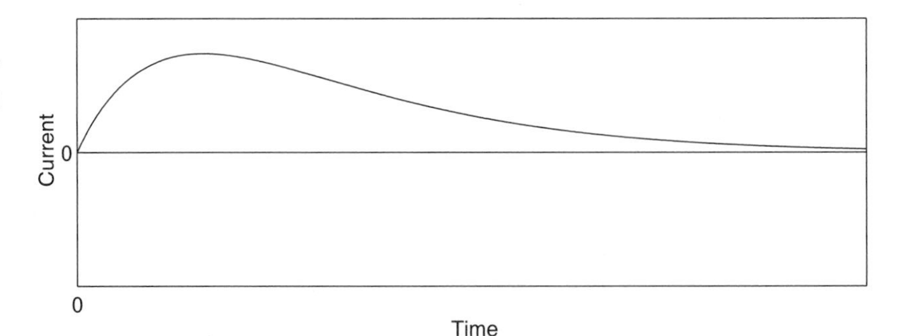

Fig. 11.19 The natural response of a critically damped second-order system

Case 3: $\omega_0 = \alpha_0$, that is $\dfrac{1}{LC} = \left(\dfrac{R}{2L}\right)^2$

This case represents the borderline between the other two. This time $p_1 = p_2$, and Equation (11.48) implies incorrectly

$$i = (A_1 + A_2)\mathrm{e}^{p_1 t} = A_0 \mathrm{e}^{p_1 t} \tag{11.56}$$

The single arbitrary constant in Equation (11.56) is insufficient to satisfy two independent initial conditions determined by the amplitude of the alternating source and the instant in its cycle that the switch is thrown. For this special case it can be shown that the solution takes the form

$$i = K_1 t\mathrm{e}^{-\alpha_0 t} + K_2 \mathrm{e}^{-\alpha_0 t} \tag{11.57}$$

This is known as the critically damped condition. For the configuration of Figure 11.16, $K_2 = 0$ and the current is as shown in Figure 11.19.

11.8 The natural response of hybrid electrical and mechanical systems

As an example of a hybrid system that is partly electrical and partly mechanical consider the d'Arsonval movement of an ordinary d.c. ammeter for converting current into an angular deflection. As shown in Figure 11.20, it consists of a coil supported on bearings, located between the pole pieces of a permanent magnet. The current in the coil causes a torque that is proportional to the current: $T = k_M i$. The angular deflection causes an opposing torque in the spring: $T = k\theta$. As the coil turns there are additional opposing torques due to inertia and damping that affect the dynamics of the rotation. If J is the inertia of the rotating components and d is the rotational damping coefficient these torques are $J\,\mathrm{d}^2\theta/\mathrm{d}t^2$ and $D\,\mathrm{d}\theta/\mathrm{d}t$. Thus the equation of motion is

$$J\frac{\mathrm{d}^2\theta}{\mathrm{d}t^2} + D\frac{\mathrm{d}\theta}{\mathrm{d}t} + k\theta = k_M i \tag{11.58}$$

To find the natural response we can remove the forcing function $k_M i$, consider the deflection $\tilde{\theta} = \Theta\mathrm{e}^{st}$, where Θ can be expected to be complex, and then solve for Θ and s. Alternatively, we can apply a forcing current $\tilde{i} = I_0\mathrm{e}^{st}$, suppose a resulting deflection, $\tilde{\theta} = \Theta\mathrm{e}^{st}$, thence find the system transfer function, and finally deduce the natural response from the poles, as in Section 11.6. The two methods are equivalent but, adopting the latter, we obtain

$$Js^2\Theta\mathrm{e}^{st} + Ds\Theta\mathrm{e}^{st} + k\Theta\mathrm{e}^{st} = k_M I_0 \mathrm{e}^{st} \tag{11.59}$$

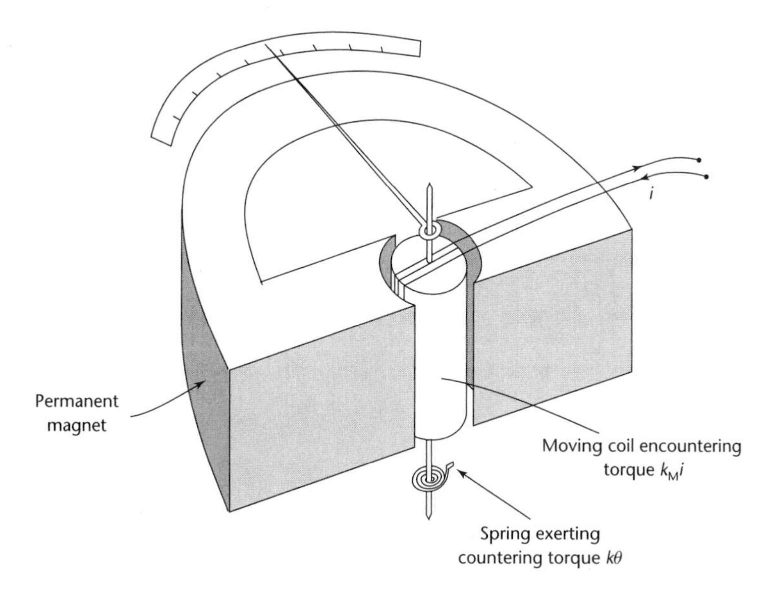

Fig. 11.20 The d'Arsonval movement of a d.c. ammeter

The term 'transfer function' is defined as the ratio of output signal to input signal. For the d'Arsonval movement, where current is the cause and angular deflection is the effect, the transfer function of the system is therefore given by the expression for $\tilde{\theta}/\tilde{i}$. We have

$$\frac{\tilde{\theta}}{\tilde{i}} = \frac{\Theta e^{st}}{I_0 e^{st}} = \frac{\Theta}{I_0} = \frac{k_M/J}{s^2 + sD/J + k/J} \tag{11.60}$$

It will be noted that if I_0 is real, complex values of s will result in complex values of Θ; hence the absence of a zero subscript from Θ.

The roots of the denominator of Equation (11.60) correspond to the poles of the system transfer function and are given by

$$p_{1,2} = -\alpha_0 \pm \sqrt{\alpha_0^2 - \omega_0^2} \tag{11.61}$$

where

$$\alpha_0 = D/2J \tag{11.62}$$

and

$$\omega_0 = \sqrt{\frac{k}{J}} \tag{11.63}$$

The natural response of this system is given by

$$\theta = A_1 e^{p_1 t} + A_2 e^{p_2 t} \tag{11.64}$$

where, depending on the values of J, D and k, this will correspond to a second-order overdamped or underdamped response.

11.9 Stability

The treatment so far has been restricted to first- and second-order systems (ones leading to first- and second-order differential equations and therefore yielding one or two poles respectively). Higher order systems merely lead to a greater number of poles, the number equalling the order of the system. Each pair of conjugate poles occurring at $-\alpha_0 \pm j\omega_1$ signifies a natural response of the form $e^{-\alpha_0 t}\sin(\omega_1 t + \theta)$, which, for $\alpha_0 > 0$, is an exponentially decaying sinusoid. If, however, we somehow had poles with positive real parts, that is, lying in the right-hand half of the s-plane, say at $+\alpha_R \pm j\omega_1$, this would imply a natural response that is an exponentially *increasing* sinusoid of the form $e^{\alpha_R t}\sin(\omega_1 t + \theta)$. Such a system must be unstable. *Therefore poles in the right-hand half-plane indicate instability.*

For a series *RLC* circuit it will be noted from Equation (11.45) that $p_{1,2} = -\alpha_0 \pm \sqrt{\alpha_0^2 - \omega_0^2}$ where $\theta_0 = R/2L$, and hence the poles always have negative real parts. It follows that the series *RLC* circuit is inherently stable.

Example 11.5

The circuit of Figure 11.21 has some interesting properties. Determine the ratio between the output voltage $\tilde{v}_0$ and an input current $\tilde{i} = I_0 e^{st}$. Find the poles and zeros and make conclusions about the properties of the circuit.

Solution It is useful to recognize that part of this circuit is the standard non-inverting operational amplifier shown in Figure 11.22. It will be noted that the inverting input of the operational amplifier is at a voltage $\tilde{v}_0 R_4/(R_3 + R_4)$.

Because of the near infinite gain of the operational amplifier it follows that the non-inverting input $\tilde{v}_{in}$ is at a very similar voltage. Hence, equating these two voltages, the gain of the amplifier, $\tilde{v}_0/\tilde{v}_{in}$, is given by K, where

$$K \approx \frac{R_3 + R_4}{R_4} \tag{11.65}$$

Returning to Figure 11.21, we again have the non-inverting input at a voltage $\tilde{v}_0/K$ and, recognizing that the sum of the three components of current into this node must equal zero, we obtain

$$\tilde{i} + \frac{\tilde{v}_0 - \tilde{v}_0/K}{R + 1/sC} - \frac{\tilde{v}_0/K}{R/(1 + sCR)} = 0 \tag{11.66}$$

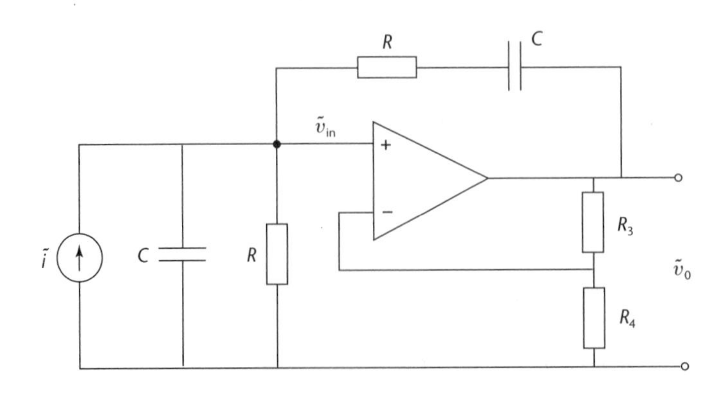

Fig. 11.21 Circuit for stability analysis

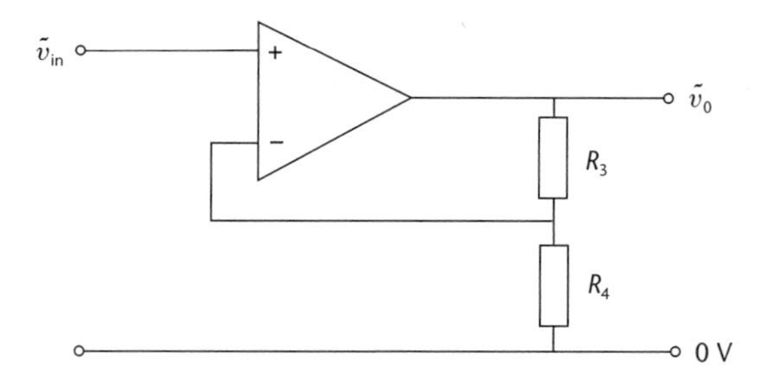

Fig. 11.22 Non-inverting operational amplifier with an input $\tilde{v}_{in}$ in the form e^{st}

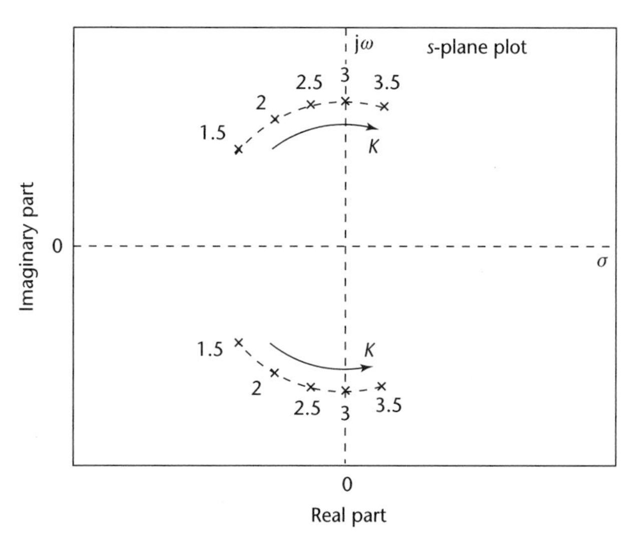

Fig. 11.23 Pole–zero plot for the circuit

This simplifies to give

$$\frac{\tilde{v}_0}{\tilde{i}} = \frac{(1 + sCR)KR}{s^2C^2R^2 + sCR(3 - K) + 1} \tag{11.67}$$

The poles are found by equating the denominator to zero, that is:

$$s^2 + \frac{3 - K}{CR}s + \frac{1}{C^2R^2} = 0 \tag{11.68}$$

This gives

$$p_{1,2} = -\frac{3 - K}{2CR} \pm \sqrt{\left(\frac{3 - K}{2CR}\right)^2 - \frac{1}{C^2R^2}} \tag{11.69}$$

These poles are plotted in Figure 11.23 for different values of K. It is noted that the poles cross the imaginary axis from left to right as K rises past the value 3. If K is slightly less than 3 the circuit will act as a bandpass filter centred on $1/CR$. If $K > 3$, however, the circuit is unstable. This implies that, even with the current source disconnected, there will be oscillation. If K is exactly 3, or minutely greater, we can expect a sinusoidal oscillation at radian frequency $1/CR$. In fact, the circuit of Figure 11.21 without the current source is the well-known Wien Bridge oscillator.

If one operates at the frequency corresponding to a pole, where the gain is infinite by definition, one may ask why the output is not infinite. The answer of course is that an operational amplifier can never support an infinite output signal, if only because of its finite d.c. supply. If the output rises too high the output will begin to clip. At large but finite output levels there will, in practice, be some non-linearity in the amplifier such that its gain K is somewhat less than the $(R_3 + R_4)/R_4$ given by Equation (11.65). The gain will decrease with increasing output levels. If, for example, the values of R_3 and R_4 are such that $(R_3 + R_4)/R_4$ equals 3.01 there are output amplitudes above which the amplifier gain is less than three and, since oscillation cannot be maintained when $K < 3$, the output cannot build up to these levels. If $(R_3 + R_4)/R_4$ is made significantly larger than 3, the output amplitude will be larger, because K can now be maintained at 3 or greater at these larger levels of oscillation. The consequence is that the amplifier is operating in a region of greater non-linearity. This means that raising $(R_3 + R_4)R_4$ much above 3 may result in an output that deviates significantly from a perfect sine wave. Depending on the application there may be an unacceptable level of distortion. For low distortion we make $(R_3 + R_4)/R_4$ only slightly greater than 3 and accept a small output amplitude.

11.10 ● Summary

Various methods have been described for determining the response of a system to a sudden change in excitation. There is one component whose form depends on the nature of the excitation, the forced response, and one component which is solely dependent on the system itself, the natural response. One way of determining the natural response is to solve the system differential equation, but another is from the poles of the transfer function. Poles in the right-hand half-plane signify system instability.

The networks and systems examined in this chapter have been very simple ones. As they become more complex the techniques presented for determining their response become difficult to apply. The next chapter introduces a much more sophisticated and elegant method that is based on the Laplace transform.

11.11 ● Problems

1. For the network of Figure 11.24 determine the response after the switch is thrown to position 2.

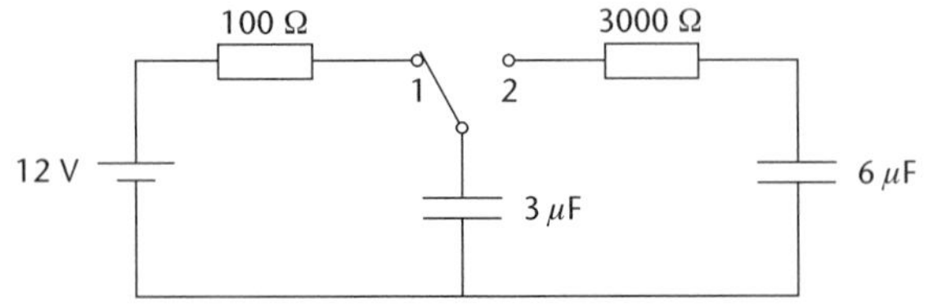

Fig. 11.24

2. In the series *RLC* circuit of Figure 11.25 the 110 V 60 Hz alternating source $155.6\cos(377t + 25°)$ is

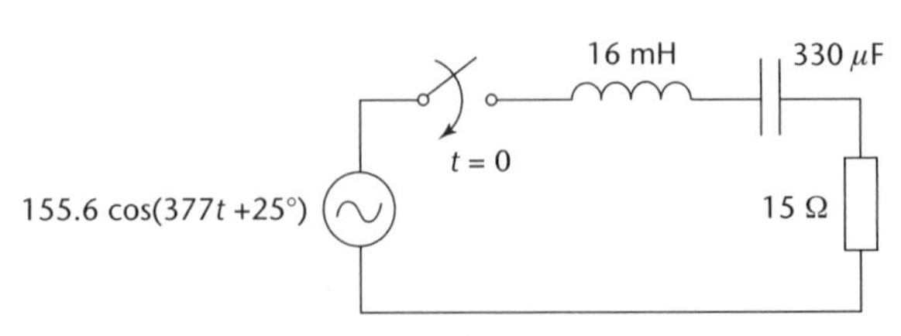

Fig. 11.25

connected at $t = 0$. Using the methods of this chapter, determine the current in the circuit following the closure of the switch. Plot the result using MATLAB. (Note: A very much easier way of obtaining the solution is possible using the Laplace transform techniques of the next chapter. Also, a special MATLAB command for determining system responses is available and, as demonstrated in Problem 10 of Chapter 12, this whole problem can be solved with just three lines of code.)

3. A boat has a transfer function relating its angle of roll to a torque exerted by the sea that is given by

$$H(s) = \frac{8.3}{s^2 + 2.3s + 7.1}$$

Determine the form of the natural response of the boat. Draw a pole–zero plot.

4. Due to an excitation force $f(t)$ a mass, spring and damper system undergoes a displacement $x(t)$ that is given by the differential equation

$$M\frac{d^2x(t)}{dt^2} + f_v\frac{dx(t)}{dt} + Kx(t) = f(t)$$

where $M = 200$ kg, f_v is in Ns/m, and $K = 9000$ N/m. Determine the transfer function and determine the resonant frequency in the absence of damping. Determine the pole positions in terms of M, f_v and K.

5. An automatic control system used for adjusting the angle of a telescope in response to an applied voltage has a transfer function given by

$$H(s) = \frac{0.06K(s + 0.2)}{s^3 + 0.6s^2 + 0.08s + 0.06K}$$

where K is a gain parameter which the system designer can adjust. Use MATLAB to find the poles for values of K between 0.2 and 1. Deduce the value of K at which the system becomes unstable.

Chapter twelve

The Laplace transform

12.1 ● Preview

The Laplace transform has less physical significance than the Fourier transform in terms of indicating spectral content and it is less suited to evaluation by numerical techniques, as done using the FFT. In return, it provides a very powerful and elegant method for solving differential equations and thereby determining system responses. As an added bonus the Laplace transforms of signals can be represented pictorially by means of poles and zeros, and this can be integrated with the pole–zero plots already introduced for portraying system transfer functions.

12.2 The bilateral Laplace transform

Consider multiplying a waveform $f(t)$ by $e^{-\sigma t}$ before taking its Fourier transform. The result is then a function of σ as well as of frequency. Nothing has been lost by this multiplication since the multiplying factor of $e^{-\sigma t}$ is unity if $\sigma = 0$, and hence the Fourier transform $f(t)$ can be obtained simply by setting $\sigma = 0$ in the result. There is, however, some gain because the Fourier transform of $f(t)e^{-\sigma t}$ covers a larger class of waveforms than is covered by just considering the Fourier transform of $f(t)$. The Fourier transform of $f(t)e^{-\sigma t}$ is known as the bilateral Laplace transform of $f(t)$, that is

$$\mathscr{L}\,[f(t)] = \mathscr{F}\,[f(t)e^{-\sigma t}] = \int_{-\infty}^{\infty} f(t)e^{-\sigma t}e^{-j\omega t}\,dt \tag{12.1}$$

where the symbol $\mathscr{L}$ is one way of denoting the Laplace transform operation.

The two exponential terms $e^{-\sigma t}$ and $e^{-j\omega t}$ can be merged to become $e^{-(\sigma + j\omega)t}$, and it is seen that the Laplace transform of $f(t)$ is thus a function of s, where $s = \sigma + j\omega$. Denoting the bilateral Laplace transform $f(t)$ as $F_b(s)$ we thus have the most usual way of expressing the definition of the bilateral Laplace transform, which is

$$\mathscr{L}\,[f(t)] = F_b(s) = \int_{-\infty}^{\infty} f(t)e^{-st}\,dt \tag{12.2}$$

Example 12.1

Determine the Fourier transform of $u(t)e^{-\sigma t}$, where $u(t)$ is the unit step function.

Solution Using the definition of the Fourier transform given by Equation (6.14) we obtain

$$F(\omega) = \int_{-\infty}^{\infty} u(t)e^{-\sigma t}e^{-j\omega t}\,dt = \int_{0}^{\infty} e^{-\sigma t}e^{-j\omega t}\,dt = \int_{0}^{\infty} e^{-(\sigma + j\omega)t}\,dt$$

$$= \frac{-1}{\sigma + j\omega}\,e^{-(\sigma + j\omega)t}\Big|_0^{\infty} \tag{12.3}$$

So long as $\sigma > 0$, the value of $e^{-(\sigma + j\omega)t}$ is zero at the upper limit of $t = \infty$ and the definite integral then simplifies to

$$F(\omega) = \frac{1}{\sigma + j\omega} \tag{12.4}$$

Example 12.2

Determine the bilateral Laplace transform of $u(t)$.

Solution Using Equation (12.2)

$$F_b(s) = \int_{-\infty}^{\infty} u(t)e^{-st}\,dt = \int_{0}^{\infty} e^{-st}\,dt = -\frac{1}{s}\,e^{-st}\Big|_0^{\infty}$$

When considering the value of the integral at the upper limit it is tempting to forget that s is a complex quantity and simply to assume that $e^{-st} = 0$ when $t = \infty$. In fact this is only the case when the same constraint as in Example 11.1 is applied, namely that σ must be positive. With this proviso

$$F_b(s) = \frac{1}{s} \tag{12.5}$$

As expected, the results of Examples 12.1 and 12.2 are the same, although the necessary constraint that σ must be greater than zero is less obvious in Example 12.2. This constraint is discussed further in Section 12.3.

Consider next what happens if $f(t)$ is a signal such as a continuous cosine wave $A\cos(2\pi f_0 t)$. The Fourier transform of this was shown in Section 6.7 to be infinite at $f = \pm f_0$, but the problem was overcome by the use of impulse functions, that is

$$\mathscr{F}[A\cos(2\pi f_0 t)] = \frac{A}{2}\,\delta(f - f_0) + \frac{A}{2}\,\delta(f + f_0)$$

If, however, we consider the Fourier transform of $Ae^{-\sigma t}\cos(2\pi f_0 t)$ we are now dealing with a signal which, depending on whether σ is positive or negative, has an infinite amplitude at either $t = -\infty$ or $t = +\infty$. Thus the integral of Equation (12.1) does not converge. This signifies that the bilateral Laplace transform is unsuitable for continuous signals such as $A\cos(2\pi f_0 t)$.

12.3 The unilateral Laplace transform

To overcome the problem that the Fourier transform of $f(t)e^{-\sigma t}$ cannot be found for a signal such as $f(t) = A\cos(2\pi f_0 t)$ unless $\sigma = 0$, it is common to restrict use of the Laplace transform to signals which are zero for $t < 0$. For such signals Equation (12.1) becomes

$$\mathscr{L}[f(t)] = \mathscr{F}[f(t)e^{-\sigma t}] = \int_{-\infty}^{\infty} f(t)e^{-\sigma t}e^{-j\omega t}\,dt = \int_{0}^{\infty} f(t)e^{-\sigma t}e^{-j\omega t}\,dt \tag{12.6}$$

The *unilateral* Laplace transform applies *only* to signals which are zero for $t < 0$ and is defined as

$$\mathscr{L}[f(t)] = F(s) = \int_{0-}^{\infty} f(t)e^{-st}\,dt \tag{12.7}$$

where the meaning and significance of 0– in the lower limit of the integral becomes apparent in Example 12.3. Very often the *unilateral* Laplace transform is written more simply without this as

$$\mathscr{L}[f(t)] = F(s) = \int_{0}^{\infty} f(t)e^{-st}\,dt \tag{12.8}$$

Henceforth this book deals exclusively with the unilateral Laplace transform; indeed the term 'Laplace transform' on its own will be taken as signifying the *unilateral* Laplace transform. For this reason the subscript b used in Equation (12.2) for the bilateral Laplace transform has been omitted from the notation $F(s)$.

For signals which are zero for $t < 0$ the unilateral Laplace transform gives the same result as the bilateral Laplace transform. Hence, from Equation (12.5), the

Laplace transform of the unit step function is

$$\mathscr{L}\,[u(t)] = 1/s \tag{12.9}$$

Example 12.3

Determine the Laplace transform of a unit impulse at $t = 0$, as shown in Figure 12.1.

Solution By its definition the impulse function is a function having zero duration but finite area. So long as this constraint is satisfied its exact shape is unimportant. As stated in Section 6.5 an assumption which is as good as any is to consider it the limiting case, as $\tau \to 0$, of a rectangular pulse of duration τ and amplitude $1/\tau$. This is shown in Figure 12.2.

It is required that, after multiplying this rectangular pulse by e^{-st}, the whole of the pulse should be involved in the integration: hence the need to extend the lower limit of the integration to just before $t = 0$ as indicated by the symbol $0-$. Therefore, using Equation (12.9)

$$\mathscr{L}\,[\delta(t)] = \lim_{\tau \to 0} \int_{0-}^{\infty} \mathrm{rect}(t/\tau)\mathrm{e}^{-st}\,\mathrm{d}t \tag{12.10}$$

At $t = 0$, $\mathrm{e}^{-st} = \mathrm{e}^{0} = 1$. Hence

$$\mathscr{L}\,[\delta(t)] = \lim_{\tau \to 0} \int_{-\tau/2}^{\tau/2} \mathrm{rect}(t/\tau)\,\mathrm{d}t \tag{12.11}$$

This then gives

$$\mathscr{L}\,[\delta(t)] = 1 \tag{12.12}$$

It has been shown in this last example that, strictly, the lower limit of integration in the definition of the Laplace transform should be written as $0-$ in order that the full area of any impulse function centred at the origin should be included. For simplicity of writing, however, this rigour will be dispensed with in this text and the notation of Equation (12.8) will be used. It must be remembered though that, if an impulse is present at $t = 0$, it should be included within the integral.

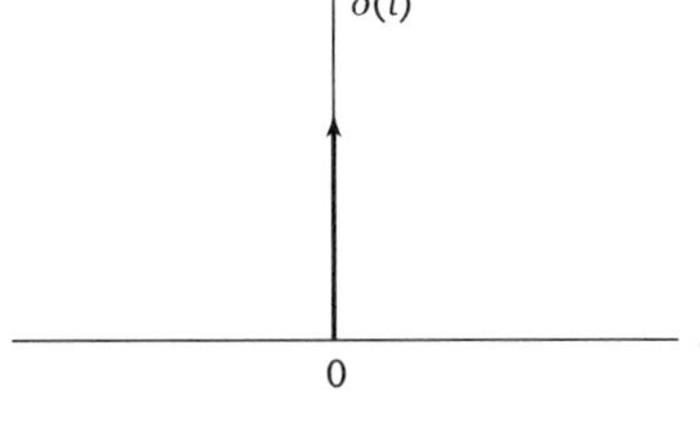

Fig. 12.1 The unit impulse function

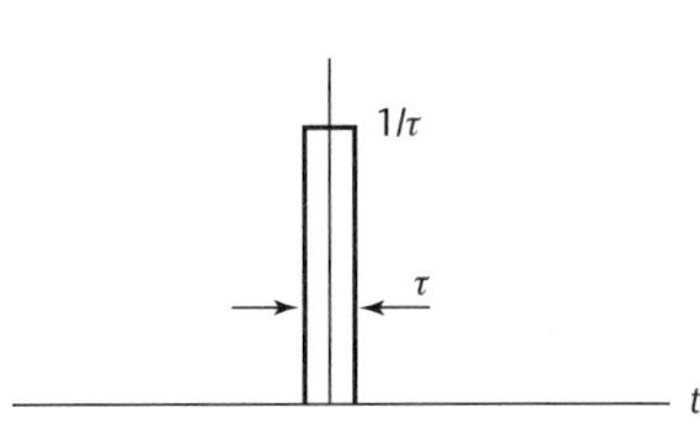

Fig. 12.2 Rectangular pulse of unit area

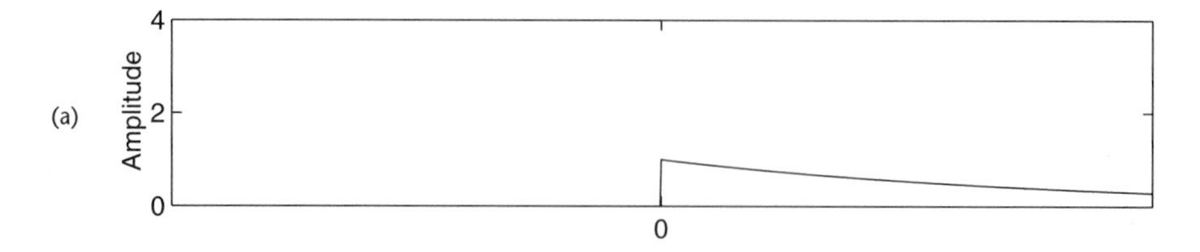

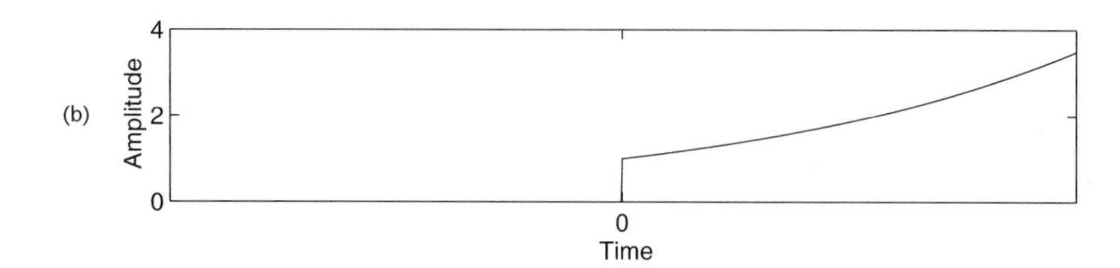

Fig. 12.3 $u(t)e^{-st}$: (a) for $\sigma > 0$; (b) for $\sigma < 0$

12.4 Convergence

Even with the constraint that the Laplace transform is restricted to those signals which are zero for $t < 0$ there are values of s for which the result has little meaning. Returning to the interpretation that the Laplace transform of $f(t)$ is the Fourier transform of $f(t)e^{-\sigma t}$, this difficulty is because the Fourier integral does not converge for certain values of σ. For example Figure 12.3(a) shows the unit step function $u(t)$ multiplied by $e^{-\sigma t}$ for the case where σ is positive, while Figure 12.3(b) shows the unit step function multiplied by $e^{-\sigma t}$ for the case where σ is negative. It will be apparent that the Fourier transform of $u(t)e^{-\sigma t}$, and hence the Laplace transform of $u(t)$, does not converge if $\sigma < 0$.

Example 12.4

Evaluate the Laplace transform of the exponentially decaying step function $u(t)e^{-at}$ and determine the values of σ for which the result is meaningful.

Solution Using Equation (12.7) we have

$$\mathscr{L}\left[u(t)e^{-at}\right] = \int_0^\infty u(t)e^{-at}e^{-st}\,dt = \int_0^\infty e^{-(s+a)t}\,dt = -\frac{1}{s+a}\,e^{-(s+a)t}\Big|_0^\infty$$
$$= -\frac{1}{s+a}\,e^{-(\sigma+j\omega+a)t}\Big|_0^\infty = -\frac{1}{s+a}\,e^{-(\sigma+a)t}e^{-j\omega t}\Big|_0^\infty$$

So long as $a + \sigma > 0$ the value of $e^{-(\sigma+a)t}$ is zero at the upper limit of $t = \infty$ and the result simplifies to become

$$\mathscr{L}\left[u(t)e^{-at}\right] = \frac{1}{s+a} \tag{12.13}$$

Thus the convergence condition is that $\sigma > -a$.

A more visual approach to convergence arises by considering the Laplace transform of $u(t)e^{at}$ to be the same as the Fourier transform of $u(t)e^{-at}e^{-\sigma t}$, where $u(t)e^{-at}$ is shown in Figure 12.4, together with the function $e^{-\sigma t}$ for the case of

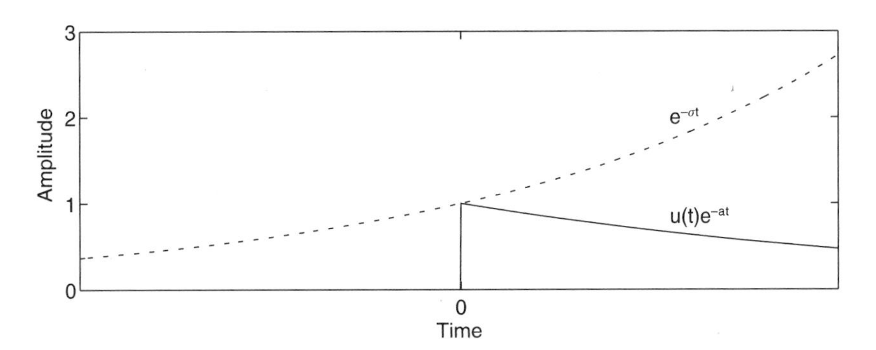

Fig. 12.4 Example of a σ value for which the Laplace transform does not converge

$\sigma < -a$. It is apparent in Figure 12.4 that the value of σ is such that $u(t)\mathrm{e}^{-at}\mathrm{e}^{-\sigma t}$ increases exponentially with time, such that its Fourier transform will not converge. Hence, as already stated the condition for the convergence of the Laplace transform of $u(t)\mathrm{e}^{-at}$ is that $a + \sigma > 0$ or $\sigma > -a$.

For most *practical* applications of the Laplace transform the problem of convergence is of no consequence and henceforth the conditions of convergence will rarely be mentioned. It is possible to find extensive and important uses of the Laplace transform without ever being aware of the values of σ for which the tabulated transform, such as $1/(s + \alpha)$ above, is valid.

12.5 A physical interpretation of the Laplace transform

It is useful to begin with a brief reminder of Fourier analysis. The concept of the Fourier series is that a periodic waveform $f_{\mathrm{p}}(t)$ of period T can be expressed as the sum of a d.c. term, a fundamental cosinusoid, and an infinite number of harmonics. As in Equation (5.18), we have

$$f_{\mathrm{p}}(t) = c_0 + \sum_{n=1}^{\infty} c_n \cos(2\pi n f_0 t + \theta_n), \tag{12.14}$$

where $f_0 = 1/T$. By using the relationship $\cos x = \frac{1}{2}(\mathrm{e}^{\mathrm{j}x} + \mathrm{e}^{-\mathrm{j}x})$, Equation (12.14) can be written in terms of complex exponential terms having positive and negative frequencies as

$$f_{\mathrm{p}}(t) = \sum_{n=-\infty}^{\infty} F_n \mathrm{e}^{\mathrm{j}2\pi n f_0 t} \tag{12.15}$$

This is the exponential Fourier series of Equation (5.22). A non-periodic signal $f(t)$ may be thought of as a periodic signal whose period T tends to infinity. This causes the complex exponential terms to become infinitely close together such that their individual frequencies become meaningless and it is necessary to consider a *continuous* spectrum, $F(f)$. The discrete coefficient F_n applies to a frequency nf_0 and it is necessary that its strength should equal the strength of the continuous spectrum within a frequency band f_0 wide centred at nf_0. This led to the result of

Equation (6.12) that $F_n = \lim_{f_0 \to 0} f_0 F(nf_0)$. Inserting this into Equation (12.15) we obtain the result of Equation (6.13) that

$$f(t) = \lim_{f_0 \to 0} \sum_{n=-\infty}^{\infty} f_0 F(nf_0) \mathrm{e}^{\mathrm{j}2\pi n f_0 t} \tag{12.16}$$

As argued in Chapter 6 this leads on to the definition of the inverse Fourier transform namely

$$f(t) = \int_{-\infty}^{\infty} F(f) \mathrm{e}^{\mathrm{j}2\pi f t} \, \mathrm{d}f$$

Working in terms of radian frequency rather than cyclic frequency the equivalent to Equation (12.16) is

$$f(t) = \lim_{\omega_0 \to 0} \frac{1}{2\pi} \sum_{n=-\infty}^{\infty} \omega_0 F(n\omega_0) \mathrm{e}^{\mathrm{j}n\omega_0 t} \tag{12.17}$$

To summarize so far then, the basis of the Fourier transform is to suppose that a signal $f(t)$ can be regarded as composed of an infinite number of cosinusoidal waveforms, each of which can be split into a pair of complex exponential waveforms. The amplitude of the complex exponential waveform at frequency $n\omega_0$ is $(\omega_0/2\pi)F(n\omega_0)$, where $F(\omega)$ is the Fourier transform of $f(t)$.

A physical interpretation of the Laplace transform is based on the supposition that a signal can be regarded as composed of an infinite number of *exponentially rising or decaying* cosinusoidal waveforms, each of which can be split up into a pair of exponentially rising or decaying complex exponential waveforms. There are an infinite number of possible decay constants for this supposition to be valid. If we choose some specific value σ_0 the amplitude of the complex exponential waveform of frequency $n\omega_0$ at time $t = 0$ is $(\omega_0/2\pi)F(\sigma_0 + \mathrm{j}n\omega_0)$, where $F(\sigma_0 + \mathrm{j}n\omega_0)$ is available from the Laplace transform $F(s)$. In this way the signal $f(t)$ can be written, as an alternative to Equation (12.17), as

$$f(t) = \lim_{\omega_0 \to 0} \frac{1}{2\pi} \sum_{n=-\infty}^{\infty} \omega_0 F(\sigma_0 + \mathrm{j}n\omega_0) \mathrm{e}^{(\sigma_0 + \mathrm{j}n\omega_0)t} \tag{12.18}$$

In Section 8.2 the terms within the summation of Equation (12.16) were described as basis functions. The terms within the summation of Equation (12.18) can similarly be described as basis functions.

This interpretation of Equation (12.18) is best clarified with an example.

Example 12.5

Equation 12.9 gave the Laplace transform of a unit step function as $1/s$. Confirm that a unit step function can be interpreted as an infinite sum of exponentially rising or decaying complex exponential time waveforms of the form e^{st}, whose complex amplitudes are given by $1/s$.

Solution The variable s is defined as $\sigma + \mathrm{j}\omega$ and hence we wish to interpret the Laplace transform of $1/(\sigma + \mathrm{j}\omega)$. Hence, on the basis of Equation 12.18, we wish to show that

$$u(t) = \lim_{\omega_0 \to 0} \frac{1}{2\pi} \sum_{n=-\infty}^{\infty} \omega_0 \frac{1}{\sigma_0 + \mathrm{j}n\omega_0} \mathrm{e}^{(\sigma_0 + \mathrm{j}n\omega_0)t} \tag{12.19}$$

In order to demonstrate this interpretation of the Laplace transform we

1. choose some specific, though arbitrary, value of σ_0;
2. constrain the summation to a finite (though large) number of complex exponential components whose frequencies are equally spaced by some specific, though arbitrary, small value ω_0.

Let us try $\sigma_0 = 40$ and $\omega_0 = 5$, and let the number of terms in the summation be limited to 401. Hence we wish to show that

$$\frac{1}{2\pi} \sum_{n=-200}^{200} \frac{5}{40 + \mathrm{j}5n} \mathrm{e}^{(40 + \mathrm{j}5n)t} \approx u(t) \tag{12.20}$$

Taking $5\mathrm{e}^{40t}$ outside the summation and expressing $40 + \mathrm{j}5n$ by its amplitude and phase terms, the left-hand side of this equation can be written as

$$\begin{aligned} f(t) &= \frac{5}{2\pi} \mathrm{e}^{40t} \sum_{n=-200}^{200} \frac{\mathrm{e}^{\mathrm{j}5nt}}{\sqrt{40^2 + (5n)^2}\ \mathrm{e}^{\mathrm{j}\tan^{-1}(5n/40)}} \\ &= 0.796\mathrm{e}^{40t} \sum_{n=-200}^{200} \frac{1}{\sqrt{1600 + 25n^2}} \mathrm{e}^{\mathrm{j}[5nt - \tan^{-1}(5n/40)]} \end{aligned} \tag{12.21}$$

Separating the $n = 0$ term from the rest, to be placed outside the summation, and combining each positive frequency term with the corresponding negative frequency term using the relationship $\mathrm{e}^{\mathrm{j}x} + \mathrm{e}^{-\mathrm{j}x} = 2\cos x$ we obtain

$$f(t) = 0.796 \frac{\mathrm{e}^{40t}}{40} + 0.796\mathrm{e}^{40t} \sum_{n=1}^{200} \frac{2}{\sqrt{1600 + 25n^2}} \cos(5nt - \tan^{-1}(0.125n)) \tag{12.22}$$

Because the number of terms is finite it can only be expected that this would be a reasonable approximation to $u(t)$ over a limited timespan. For the values allocated to σ_0 and ω_0 a suitable timespan is $-0.12 < t < 0.12$. The following program in MATLAB calculates and displays the result of Equation (12.22).

```
% fig12_5.m
t=(-0.12:0.001:0.12)                    % 241 time points 0.001s apart
f0=0.796*exp(40*t)/40;                  % the n=0 term outside the summation
f=zeros(size(t));                       % initialize the function to zero
for n=1:200                             % the number of terms in the summation
a=0.796*exp(40*t)*2/sqrt(1600+25*n^2).*cos(5*n*t-atan(0.125*n));
f=f+a;                                  % summation of terms
end
f=f+f0;                                 % add the n=0 term
plot(t,f)
axis([-.12 .12 -1.5 1.5])
xlabel('Time (seconds)'),ylabel('Amplitude')
```

The result of this program is shown in Figure 12.5. It is seen to give a reasonable approximation to a unit step function. The ripple and its rise with time is reduced if ω_0 is decreased and if the number of terms in the summation is increased. In the

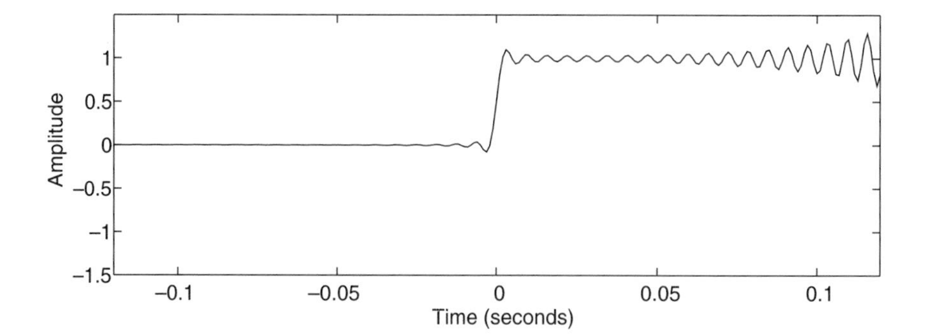

Fig. 12.5 Realization of a step function as the sum of exponentially increasing cosinusoids

limiting case of an infinite number of terms that are infinitely closely spaced in frequency, Equation (12.19) can produce a perfect step function whatever the value of σ_0, so long as it satisfies the covergence condition (see Section 12.4) that $\sigma > 0$.

It is instructive to examine some of the individual terms in Equation (12.22). The following MATLAB program generates and plots the term outside the summation (corresponding to $n = 0$), and the terms inside the summation corresponding to $n = 40$, 80, 120 and 160.

```
% fig12_6.m
t=(-.12: .001: .12);
f0=.796*exp(40*t)/40;
subplot(5,1,1)
plot(t,f0), axis([-.12 .12 -.2 .2])
f40=.796*exp(40*t)*2/sqrt(1600+25*40^2).*cos(5*40*t-atan(.125*40));
subplot(5,1,2)
plot(t,f40), axis([-.12 .12 -.2 .2])
```

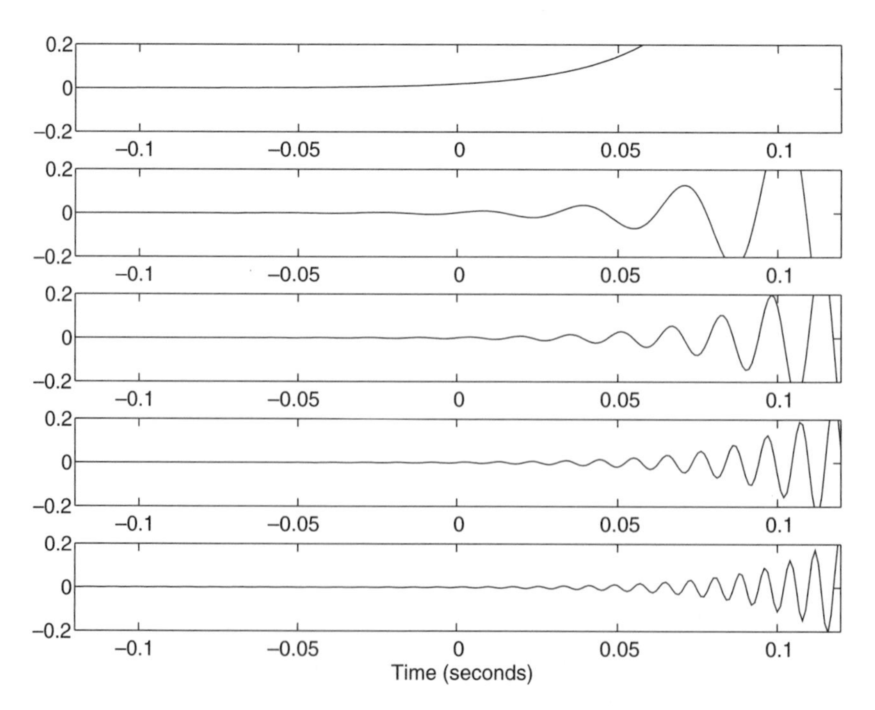

Fig. 12.6 Some of the exponentially increasing cosinusoids that sum to approximate a step function, thus interpreting its Laplace transform of $1/s$

```
f80=.796*exp(40*t)*2/sqrt(1600+25*80^2).*cos(5*80*t-atan(.125*80));
subplot(5,1,3)
plot(t,f80), axis([-.12 .12 -.2 .2])
f120=.796*exp(40*t)*2/sqrt(1600+25*120^2).*cos(5*120*t-atan(.125*120));
subplot(5,1,4)
plot(t,f120), axis([-.12 .12 -.2 .2])
f160=.796*exp(40*t)*2/sqrt(1600+25*160^2).*cos(5*160*t-atan(.125*160));
subplot(515)
plot(t,f160), axis([-.12 .12 -.2 .2])
xlabel('Time (seconds)')
```

The outcome is shown in Figure 12.6. It will be noted that all of the terms increase exponentially and yet, because of their phase relationships, they reinforce each other at certain times and cancel each other at other times; this occurs in such a way as to produce the required step function.

The example above demonstrates how a step function can be realized by a set of exponentially increasing cosinusoids (or complex exponentials) of appropriate amplitudes and phases. Specifically, its Laplace transform is $1/s$ and it can therefore be thought of as an infinite sum of basis functions of the form $(1/s_1)e^{s_1 t} + (1/s_2)e^{s_2 t} + (1/s_3)e^{s_3 t} + \cdots$. Extending this, *most signals* can be broken down into an infinite set of basis functions of the form e^{st}, all of whose strengths are given by the Laplace transform of the signal. *It is this concept that forms the basis of a physical interpretation of the Laplace transform.*

The greatest difficulty with a physical understanding of a Laplace transform is that, in contrast with the Fourier transform, *two* variables are involved. Besides frequency we also have σ. However, this conceptual difficulty can be overcome by simply regarding σ as adding flexibility to the nature of the complex exponentials. Whereas the complex exponential functions do not change their amplitude with time in the case of a Fourier transforms, they can have an exponential rise (and sometimes decay) in the case of a Laplace transform.

12.6 Applications of the Laplace transform to system analysis

If we accept that a signal is the sum of an infinite number of terms of the form e^{st} and that we know the response of a system to each of these terms, then we can sum these responses to determine the output. For example, if $F(s)$ is considered as representing the strengths of the e^{st} basis functions that sum together to constitute the input signal and $H(s)$ is the response of the system to an input e^{st}, as a function of s, then $F(s)H(s)$ must represent the strengths of the e^{st} basis functions that sum together to constitute the output signal. Hence, by finding the signal that has $F(s)H(s)$ as its transform, we can determine the time waveform that is the output of the system. This latter process is that of performing an inverse Laplace transform. A block diagram of this procedure is given in Figure 12.7.

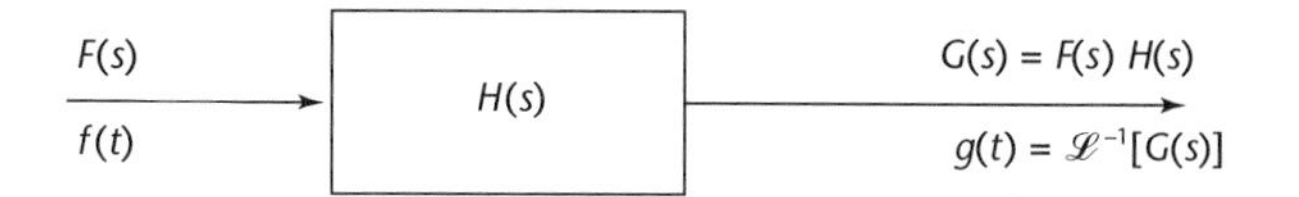

Fig. 12.7 Determining a system output by means of the Laplace transform

In practice, the inverse Laplace transform of $F(s)H(s)$ is usually achieved by breaking it up into the sum of a simpler set of terms, each of which is a standard and well-known Laplace transform available from a table of Laplace transforms. The associated time waveforms can then be summed. If, for example, $F(s)H(s)$ contains high-order polynomials, the procedure would be to break it up into partial fractions, each of which has a simple inverse Laplace transform that is available from the table.

Example 12.6

Use Laplace transforms to determine the response of the simple *RC* network of Figure 12.8 to a unit step function.

Solution We have

$$H(s) = \frac{1/sC}{R + 1/sC} = \frac{1}{1 + sCR} \tag{12.23}$$

In network analysis it is more usual to denote a voltage waveform by $v(t)$ than by $f(t)$. Hence, using the symbol $V_1(s)$ to denote the Laplace transform of the input, we have $V_1(s) = 1/s$ for a unit step.

Letting $V_2(s)$ denote the Laplace transform of the output, we therefore have

$$V_2(s) = V_1(s)H(s) = \frac{1}{s(1 + sCR)} \tag{12.24}$$

This can be separated into partial fractions as

$$V_2(s) = \frac{A}{s} + \frac{B}{1 + sCR}$$

where $A(1 + sCR) + Bs = 1$, thus giving $A = 1$, $B = -CR$. Hence

$$V_2(s) = \frac{1}{s} - \frac{CR}{1 + sCR} = \frac{1}{s} - \frac{1}{s + 1/CR} \tag{12.25}$$

We know from Equation (12.9) that $\mathscr{L}\,[u(t)] = 1/s$ and from Equation (12.13) that $\mathscr{L}\,[u(t)\mathrm{e}^{-at}] = 1/(s + a)$. Using these results, it follows that the inverse Laplace transform of $V_2(s)$ is

$$v_2(t) = u(t) - u(t)\mathrm{e}^{-t/CR} \tag{12.26}$$

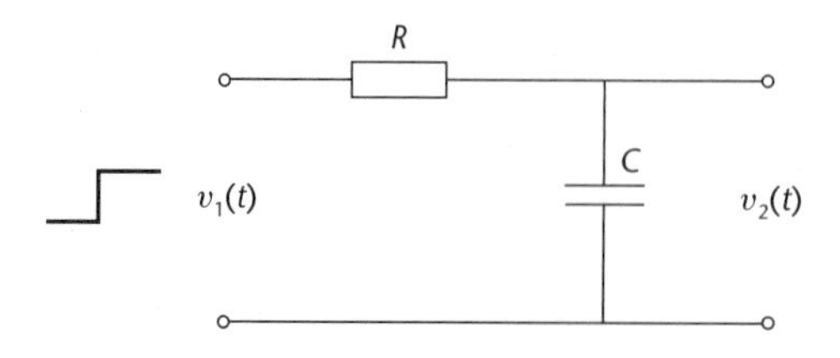

Fig. 12.8 Application of unit step function to an *RC* network

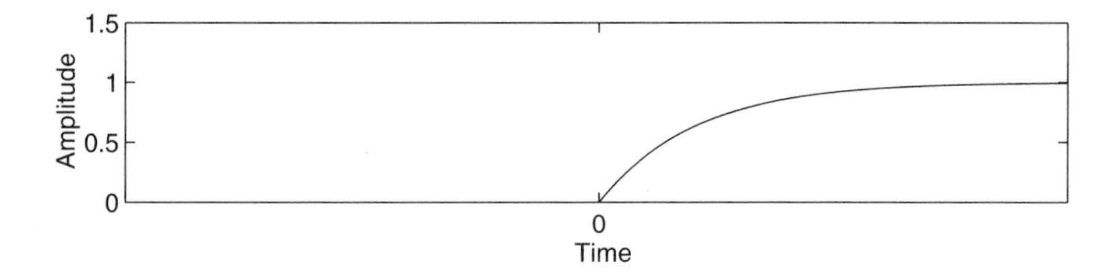

Fig. 12.9 Step function response of an *RC* network

or

$$v_2(t) = 1 - \mathrm{e}^{-t/CR} \tag{12.27}$$

with it being implicitly understood that this applies only for $t \geqslant 0$, and that $v_2(t) = 0$ for $t < 0$. This step function response is shown in Figure 12.9.

12.7 Justification of the Laplace transform and a comparison with the Fourier transform

It has been shown in the previous section that the Laplace transform can be used for determining the response of a system to an input signal. The following are some of its advantages compared to the Fourier transform.

1. As argued in Chapter 10, it is generally easier to determine the transfer function of a system to an input e^{st} than to an input $\mathrm{e}^{\mathrm{j}\omega t}$. In other words, by avoiding the j operator, it is easier to determine $H(s)$ than $H(\omega)$.
2. Tables are available giving the Fourier transforms and Laplace transforms of many important signals. These are also used for the *inverse* operation of determining what signal corresponds to a given transform. When the transform is too complex to be listed in the tables it can be expanded into partial fractions, each of which has a simple inverse transform. In this way the signal can be determined. *By avoiding the* j *operator it is very much easier to expand a Laplace transform into partial fractions than it is to expand a Fourier transform into partial fractions*. This method of performing inverse transforms is examined more fully in Section 12.9.
3. It will be shown in Section 12.15 that it is possible to use Laplace transforms as a direct method of solving differential equations without resource to transfer functions.
4. It will also be shown in Section 12.15 that Laplace transforms can readily cover the situation where the system has energy stored in it before the signal is applied, caused perhaps by a capacitor that is already charged, or by an inductor which already carries current. These non-zero initial conditions, caused by events in the system prior to $t = 0$, present little difficulty to the Laplace transform method.
5. The Laplace transform can be described concisely and effectively in pictorial form by a plot of poles and zeros in the s-plane. For example, the Laplace transform of the step function in Example 12.6 was $1/s$ and this can be represented by a pole in the s-plane at the origin, as shown in Figure 12.10(a). The transfer function of the

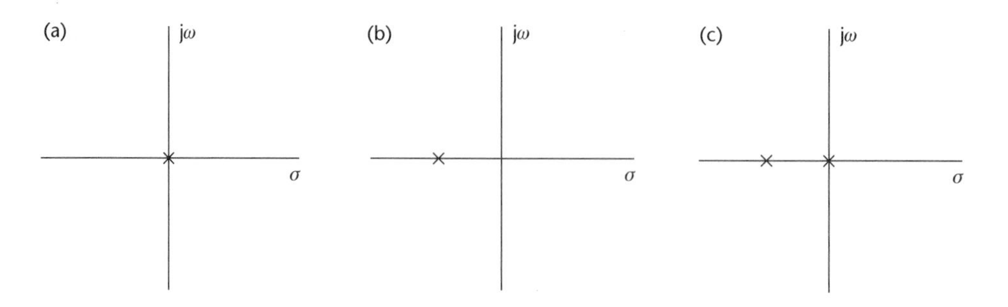

Fig. 12.10 Use of pole-zero plots: (a) for an input signal; (b) for a transfer function; (c) for the output signal

RC network in Example 12.6 was $(1/CR)[1/(s + 1/CR)]$ and, apart from a gain factor, can be represented by a pole in the *s*-plane at $s = -1/CR$, as shown in Figure 12.10(b). The transform of the output signal is given by the product of the input transform with the transfer function and, apart from a gain factor, can be represented by two poles in the *s*-plane, one at the origin and one at $s = -1/CR$, as shown in Figure 12.10(c).

The converse of this operation is possible in that it is possible to recognize what signal corresponds to an *s*-plane plot. For example, it is possible with experience to recognize that Figures 12.10(a), (b) and (c) represent, respectively, a step function, a network whose transfer function is of the form $1/(s + 1/CR)$, and an output of the form $u(t) - u(t)\mathrm{e}^{-t/CR}$. The interpretation of these *s*-plane plots in this way will be discussed further in Chapter 13. Such an interpretation is *particularly important in control systems analysis* where the variation of a parameter such as gain can vary pole positions along a 'root locus'. *By being able to recognize what are desirable pole positions it is possible to optimize the control system parameters.*

The many advantages of the Laplace transform over the Fourier transform should not be taken to suggest that it supersedes the Fourier transform. The following are some of the advantages of the Fourier transform.

1. The Fourier transform of a signal has the advantage that it has a very important physical significance, namely that it conveys the spectral content of the signal. Similarly, a frequency transfer function $H(\omega)$ conveys the frequency response directly. (Note, however, that *s* domain signals and transfer functions can be converted to frequency domain signals and transfer functions by simply replacing *s* with $\mathrm{j}\omega$.)
2. For complicated signals whose Fourier transform cannot be evaluated analytically (such as speech or engine noise) computational techniques can be used to derive the closely related discrete Fourier transform by means of the FFT. This makes spectral analysis possible. The FFT also enables a signal to be convolved with a system impulse response, thus enabling a system response to be computed. Because of the complication introduced by the extra parameter σ, the Laplace transform has no computational counterpart to the FFT.

12.8 *s*-plane plots and some more Laplace transforms

12.8.1 The step function

The Laplace transform of the unit step function has already been shown in Equation (12.9) to be given by $\mathscr{L}\,[u(t)] = 1/s$. Because the unilateral Laplace transform is only applied to signals which are zero for $t < 0$ the symbol $u(t)$ for the unit step may be omitted and we can write

$$\mathscr{L}\,[1] = 1/s \tag{12.28}$$

This can be represented in the s-plane by a single pole at the origin, as shown in Figure 12.11.

As discussed in Section 12.4, the Laplace transform of the unit step does not converge if $\sigma < 0$, and this leads to the right-hand half-plane being known as the 'region of convergence'. However, an awareness of this fact and what it means is rarely necessary.

12.8.2 The exponentially decaying step function

Equation (12.13) showed that $\mathscr{L}\, u(t)\mathrm{e}^{-at} = 1/(s+a)$. Because the step function symbol is not necessary we can write

$$\mathscr{L}\,[\mathrm{e}^{-at}] = 1/(s+a) \tag{12.29}$$

This can be represented in the s-plane by a single pole at $s = -a$, as shown in Figure 12.12. Here the region of convergence is the area to the right of the pole but, once again, this is rarely mentioned.

12.8.3 The ramp function *At*

This is a common test signal for automatic control systems. Using the definition of the Laplace transform gives

$$\mathscr{L}\,[At] = \int_0^\infty At\mathrm{e}^{-st}\,\mathrm{d}t$$

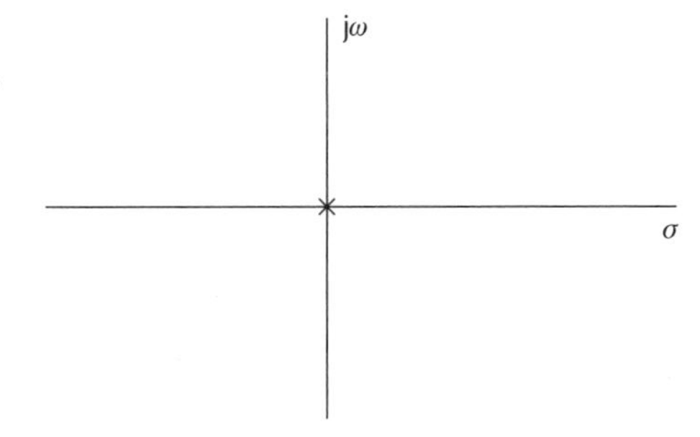

Fig. 12.11 *s*-plane plot of step function

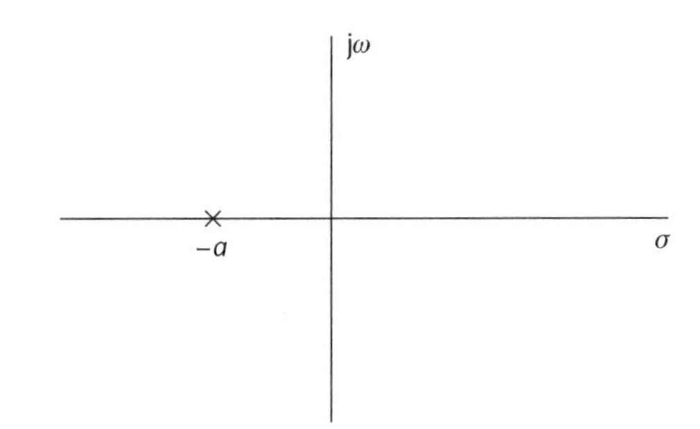

Fig. 12.12 *s*-plane plot of exponentially decaying step function

Integration by parts yields

$$\mathcal{L}\,[At] = A/s^2 \tag{12.30}$$

This has a double pole at the origin, as shown in Figure 12.13. Note that this is shown as a single cross with the multiplier of 2 beside it.

12.8.4 Sine and cosine functions

The Laplace transform of the sine and cosine functions can be derived by extending the result for the Laplace transform of e^{-at}. We have $\sin \omega_0 t = (1/2j)(e^{j\omega_0 t} - e^{-j\omega_0 t})$. By using the result of Equation (12.29) we obtain

$$\mathcal{L}\,[\sin(\omega_0 t)] = \frac{1}{2j}\left\{\frac{1}{s - j\omega_0} - \frac{1}{s + j\omega_0}\right\} = \frac{\omega_0}{(s - j\omega_0)(s + j\omega_0)} = \frac{\omega_0}{s^2 + \omega_0^2} \tag{12.31}$$

This has conjugate poles at $s = \pm j\omega_0$ as shown in Figure 12.14.

Similarly $\cos(\omega_0 t) = \frac{1}{2}(e^{j\omega_0 t} + e^{-j\omega_0 t})$, giving

$$\mathcal{L}\,[\cos(\omega_0 t)] = \frac{1}{2}\left\{\frac{1}{s - j\omega_0} + \frac{1}{s + j\omega_0}\right\} = \frac{s}{(s - j\omega_0)(s + j\omega_0)} = \frac{s}{s^2 + \omega_0^2} \tag{12.32}$$

This has conjugate poles at $s = \pm j\omega_0$ plus a zero at the origin, as shown in Figure 12.15.

12.8.5 Functions modified by an exponential decay

Consider the Laplace transform of $e^{-at}f(t)$ when the Laplace transform of $f(t)$ is $F(s)$. Using the definition of the Laplace transform we have

$$\mathcal{L}\,[e^{-at}f(t)] = \int_0^\infty e^{-at}f(t)e^{-st}\,dt = \int_0^\infty f(t)e^{-(s+a)t}\,dt$$

$$\therefore \mathcal{L}\,[e^{-at}f(t)] = F(s + a) \tag{12.33}$$

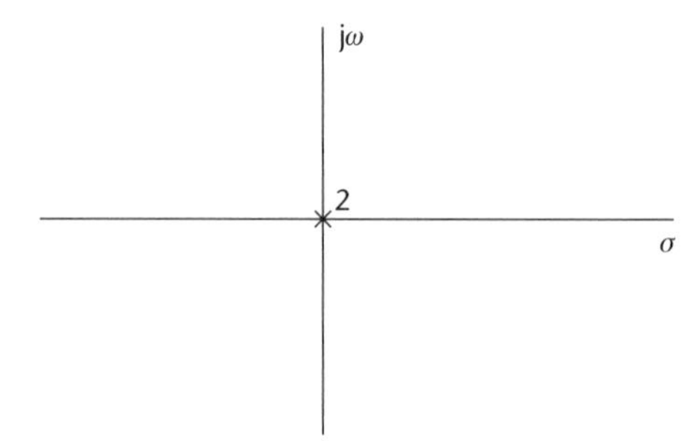

Fig. 12.13 *s*-plane plot of a ramp function

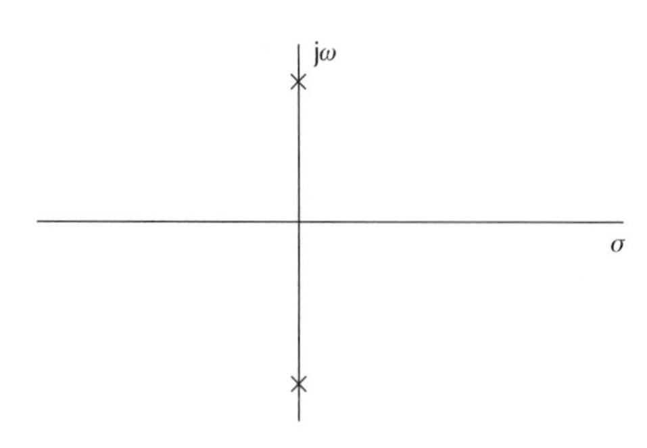

Fig. 12.14 *s*-plane plot of a sinusoidal function

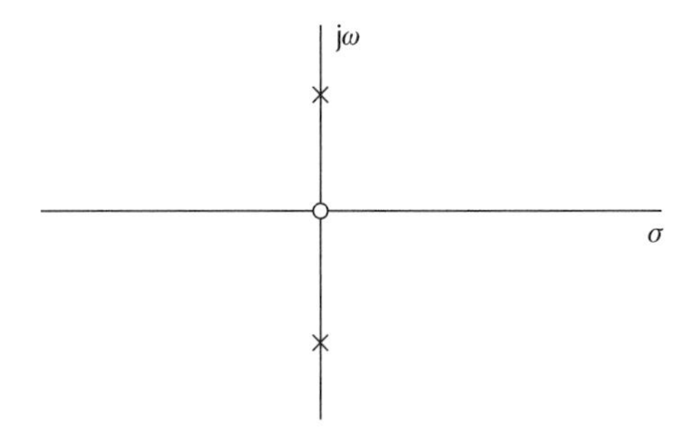

Fig. 12.15 s-plane plot of a cosinusoidal function

This signifies that the poles and zeros are all moved to the left by a. The change of Figure 12.11 to Figure 12.12 has already provided one example of this.

12.8.6 Exponentially decaying sines and cosines

Exponentially decaying sines and cosines occur commonly, and the result of Equation (12.33) can be applied directly to the transforms of Equations (12.31) and (12.32) to derive their Laplace transform. We obtain

$$\mathscr{L}\,[\mathrm{e}^{-at}\sin\omega_0 t] = \frac{\omega_0}{(s+a)^2+\omega_0^2} \tag{12.34}$$

An alternative way of expressing this is as

$$\mathscr{L}\,[\mathrm{e}^{-at}\sin\omega_0 t] = \frac{\omega_0}{(s+a+\mathrm{j}\omega_0)(s+a-\mathrm{j}\omega_0)} \tag{12.35}$$

and is seen to have a pair of conjugate poles at $s = -a \pm \mathrm{j}\omega_0$. These are shown in Figure 12.16. Similarly

$$\mathscr{L}\,[\mathrm{e}^{-at}\cos\omega_0 t] = \frac{s+a}{(s+a)^2+\omega_0^2} \tag{12.36}$$

An alternative way of expressing this is as

$$\mathscr{L}\,[\mathrm{e}^{-at}\cos\omega_0 t] = \frac{s+a}{(s+a+\mathrm{j}\omega_0)(s+a-\mathrm{j}\omega_0)} \tag{12.37}$$

This has the same two poles as for $\mathrm{e}^{-at}\sin\omega_0 t$, but also a zero at $-a$ as shown in Figure 12.17.

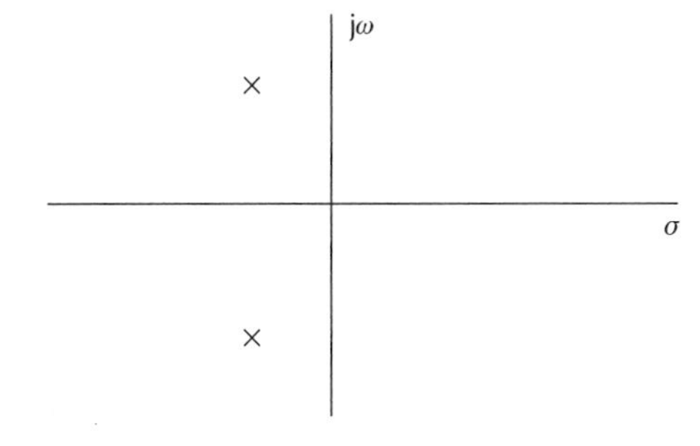

Fig. 12.16 s-plane plot of an exponentially decaying sinusoidal function

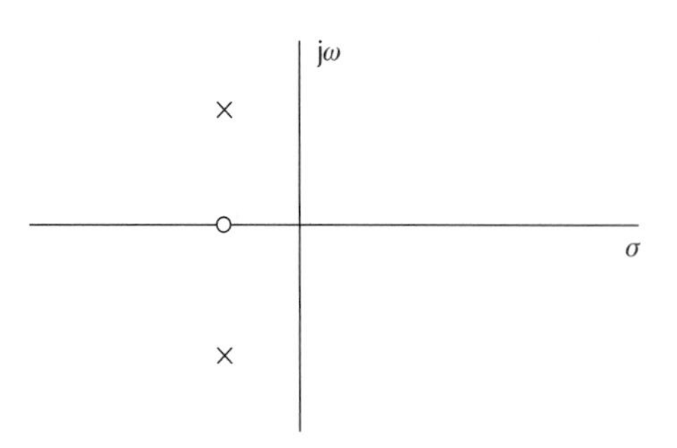

Fig. 12.17 s-plane plot of an exponentially decaying cosinusoidal function

12.8.7 Delayed functions

Using the definition of the Laplace transform

$$\mathscr{L}\,[f(t-t')] = \int_0^\infty f(t-t')\mathrm{e}^{-st}\,\mathrm{d}t$$

Substituting $x = t - t'$, this becomes $\int_{-t'}^{\infty} f(x)\mathrm{e}^{-s(x+t')}\,\mathrm{d}x$ and thence, changing the variable from x back to t, we obtain $\int_{-t'}^{\infty} f(t)\mathrm{e}^{-s(t+t')}\,\mathrm{d}t$. But, since $f(t)$ is zero for $t < 0$, the lower limit of integration may be changed to 0.

$$\therefore \mathscr{L}\,[f(t-t')] = \int_0^\infty f(t)\mathrm{e}^{-s(t+t')}\,\mathrm{d}t = \mathrm{e}^{-st'}\int_0^\infty f(t)\mathrm{e}^{-st}\,\mathrm{d}t$$

$$\therefore \mathscr{L}\,[f(t-t')] = \mathrm{e}^{-st'}F(s) \tag{12.38}$$

A particularly important application of this result is for determining the Laplace transform of a delayed impulse function, $\delta(t-t')$. Equation (12.12) gave the result that $\mathscr{L}\,[\delta(t)] = 1$. It follows from Equation (12.38) that

$$\mathscr{L}\,[\delta(t-t')] = \mathrm{e}^{-st'} \tag{12.39}$$

This result will prove particularly important in discussions on discrete-time systems, as there the sampled waveform is closely akin to a sequence of delayed impulse functions, each of appropriate weight (strictly this concept applies to the continuous-time signal arising from ideal sampling, whereas a discrete-time signal is a sequence of *discrete-time* impulse functions, each of appropriate amplitude).

12.9 Inverse Laplace transformation

If we have an s domain description of a signal, $F(s)$, the inverse Laplace transform is generally performed by expanding $F(s)$ into partial fractions. The inverse Laplace transform of $F(s)$ is then the sum of these constituent inverse Laplace transforms.

Consider $F(s)$ to be the ratio of two polynomials such that the multiplying factor of the highest order term in the denominator is unity, that is

$$F(s) = \frac{b(s)}{a(s)} = \frac{b(s)}{s^n + a_{n-1}s^{n-1} + a_{n-2}s^{n-2} + \cdots + a_0} \tag{12.40}$$

The first step of a partial fraction expansion is to factor the denominator of $F(s)$ to give

$$F(s) = \frac{b(s)}{(s-p_1)(s-p_2)(s-p_3)\cdots(s-p_n)} \tag{12.41}$$

Assuming that the order of the numerator is equal to or less than that of the denominator and that all the poles are distinct (that is, different from one another) this can be expressed as

$$F(s) = \frac{r_1}{s - p_1} + \frac{r_2}{s - p_2} + \frac{r_3}{s - p_3} + \cdots + \frac{r_n}{s - p_n} + k \tag{12.42}$$

Here k is the direct term and, if present, has an inverse Laplace transform $k\delta(t)$. The coefficients r_1, r_2, r_3 and so on are known as *residues* and it is necessary next to find their values.

To determine r_1, both sides of the equation should be multiplied by $(s - p_1)$ to give

$$\frac{b(s)}{(s - p_2)(s - p_3)\cdots} = r_1 + \frac{r_2(s - p_1)}{s - p_2} + \frac{r_3(s - p_1)}{s - p_3} + \cdots + k(s - p_1)$$

Now, by putting $s = p_1$

$$r_1 = \left.\frac{b(s)}{(s - p_2)(s - p_3)\cdots}\right|_{s = p_1} \tag{12.43}$$

Having evaluated all the residues in this way, the inverse Laplace transform of each partial fraction can be determined.

If a partial fraction contains a negative real pole its inverse Laplace transform, from Equation (12.29), is a decaying exponential, for example

$$\mathscr{L}^{-1}\left[\frac{r}{s + a}\right] = re^{-at} \tag{12.44}$$

If a partial fraction contains a complex pole it will be accompanied by a second partial fraction containing a conjugate pole and having a residue which is the conjugate of the first. Thus a very important inverse Laplace transform is that of

$$\frac{r}{s - p} + \frac{r^*}{s - p^*}$$

Let us write this as

$$\frac{c + jd}{s + a - jb} + \frac{c - jd}{s + a + jb} = \frac{2cs + 2ca - 2bd}{(s + a)^2 + b^2} = 2c\frac{s + a}{(s + a)^2 + b^2} - 2d\frac{b}{(s + a)^2 + b^2}$$

However, these last terms can be inverse transformed using Equations (12.34) and (12.36). Hence

$$\mathscr{L}^{-1}\left[\frac{r}{s - p} + \frac{r^*}{s - p^*}\right] = 2ce^{-at}\cos bt - 2de^{-at}\sin bt \tag{12.45}$$

where a, b, c and d are obtained from $p = -a + jb$ and $r = c + jd$

Equation (12.45) can be expressed in the form $Ae^{-at}\cos(bt + \gamma)$ but this is best avoided since γ involves an arctangent and careless calculations can easily make it end up in the wrong quadrant.

In the event of repeated pole the partial fraction expansion must be modified. Considering a second-order pole for example, we have

$$F(s) = \frac{b(s)}{(s - p_1)^2(s - p_3)\cdots(s - p_n)} = \frac{r_1}{s - p_1} + \frac{r_2}{(s - p_1)^2} + \frac{r_3}{s - p_3} + \cdots + \frac{r_n}{s - p_n}$$

Using the results of Equations (12.30) and (12.33) the inverse Laplace transform of the second term gives rise to a signal of the form

$$\mathscr{L}^{-1}\left[\frac{r_2}{(s-p_1)^2}\right] = r_2 t e^{p_1 t} \tag{12.46}$$

In practice the greatest problem with the method of partial fractions is factorizing the denominator when its order is greater than two. This can be quite laborious using manual techniques. Fortunately MATLAB comes to the rescue and its 'residue' command determines the residues, the poles, and the direct term if present, in the partial fraction expansion

$$\frac{b(s)}{a(s)} = \frac{r_1}{s-p_1} + \frac{r_2}{s-p_2} + \cdots + \frac{r_n}{s-p_n} + k$$

Example 12.7

Network analysis shows that a third-order Butterworth lowpass filter has a transfer function given by

$$H(s) = \frac{1}{s^3 + 2s^2 + 2s + 1}$$

Determine its step function response by means of the Laplace transform.

Solution The Laplace transform of a unit step function input is $F(s) = 1/s$. Therefore the Laplace transform of the output is given by

$$G(s) = F(s)H(s) = \frac{1}{s(s^3 + 2s^2 + 2s + 1)} = \frac{1}{s^4 + 2s^3 + 2s^2 + s + 0}$$

In order to find the partial fraction expansion of this the MATLAB commands needed are:

```
b=1                    % the coefficients of the numerator
a=[1 2 2 1 0]          % the coefficients of the denominator
[r p k]=residue(b,a)   % the residues, the poles, any direct term
```

The resulting printout is

```
r =
   -1.0000
    0.0000 + 0.5774i
    0.0000 - 0.5774i
    1.0000 - 0.0000i

p=
   -1.0000
   -0.5000 + 0.8660i
   -0.5000 - 0.8660i
        0

k=
    []
```

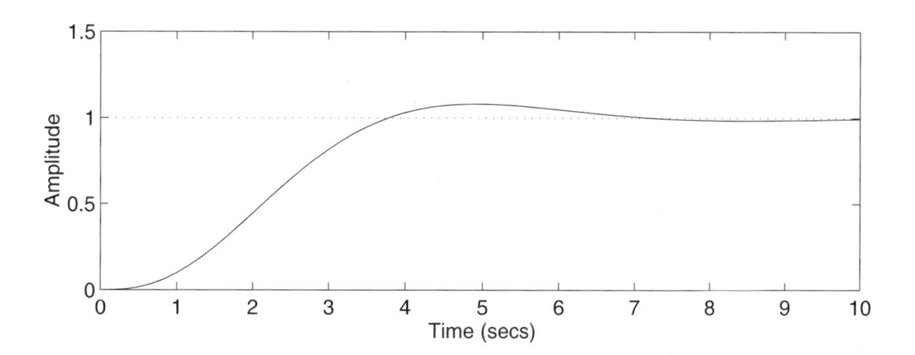

Fig. 12.18 Step function response of a third-order Butterworth filter

In other words the residues are $r_1 = -1$, $r_2 = \text{j}0.5774$, $r_3 = -\text{j}0.5774$, $r_4 = 1$; the poles are $p_1 = -1$, $p_2 = -0.5 + \text{j}0.866$, $p_3 = -0.5 - \text{j}0.866$, $p_4 = 0$; there is no direct term. That is,

$$G(s) = \frac{1}{s(s^3 + 2s^2 + 2s + 1)} = \frac{-1}{s+1} + \frac{\text{j}0.5774}{s + 0.5 - \text{j}0.866} - \frac{\text{j}0.5774}{s + 0.5 + \text{j}0.866} + \frac{1}{s}$$

The inverse Laplace transform of the two terms with conjugate poles can be obtained by inserting $a = 0.5$, $b = 0.866$, $c = 0$ and $d = 0.5774$ into Equation (12.45). Making use of Equation (12.44) for the other two terms we find that the inverse Laplace transform, and hence the required step function response, is given by

$$\begin{aligned} g(t) &= -\text{e}^{-t} + 0 - 1.1548\text{e}^{-0.5t}\sin(0.866t) + 1 \\ &= 1 - \text{e}^{-t} - 1.1548\text{e}^{-0.5t}\sin(0.866t) \end{aligned}$$

Since Laplace transforms are used it is assumed, without the necessity for writing it, that this is only valid for $t \geqslant 0$, and that $g(t) = 0$ for $t < 0$. The result is plotted in Figure 12.18. (In fact, this plot was obtained using a specific MATLAB command for obtaining step responses that will be described in Section 12.13.)

12.10 Interpreting signal waveforms from pole – zero plots

As shown in Example 12.7 the Laplace transform of a system response can be determined by multiplying the Laplace transform of the input signal by the system transfer function. A very compact way of representing the resulting Laplace transform is by an s-plane plot of its poles and zeros, and it can be very useful to be able to interpret the significance of the pole and zero positions. This can be particularly beneficial when there are opportunities for manipulating the poles and zeros of the transfer function.

It is readily apparent from Equation (12.44) that a pole at $-a$ signifies an inverse Laplace transform of the form e^{-at}, and from Equation (12.45) that a pair of conjugate poles at $(-a \pm \text{j}b)$ will give rise to an inverse Laplace transform of the form $\text{e}^{-at}\cos(bt + \gamma)$. Much less obvious are the magnitudes of these terms. They can be found from a full partial fraction expansion of the Laplace transform, but the objective of this section is to achieve an approximate estimate of their values from a visual inspection of the s-plane plot.

When zeros exist we can expand the numerator of Equation (12.41) to obtain

$$F(s) = \frac{b(s)}{a(s)} = K\frac{(s-z_1)(s-z_2)(s-z_3)\cdots(s-z_m)}{(s-p_1)(s-p_2)(s-p_3)\cdots(s-p_n)} \tag{12.47}$$

Equation (12.43) then gives the residue r_1 as

$$\begin{aligned} r_1 &= K\frac{(s-z_1)(s-z_2)(s-z_3)\cdots}{(s-p_2)(s-p_3)\cdots}\bigg|_{s=p_1} \\ &= K\frac{(p_1-z_1)(p_1-z_2)(p_1-z_3)\cdots}{(p_1-p_2)(p_1-p_3)\cdots} \end{aligned} \tag{12.48}$$

Expressed in words

$$\text{residue at } p_1 = K\frac{\text{product of vectors from each zero to } p_1}{\text{product of vectors from each other pole to } p_1} \tag{12.49}$$

The same procedure applies to all the other poles.

In the absence of zeros the numerator of Equation (12.47) is unity and we obtain

$$\text{residue at } p_1 = K\frac{1}{\text{product of vectors from each other pole to } p_1} \tag{12.50}$$

Example 12.8

The Laplace transform of a signal is given by

$$F(s) = \frac{2(s+3)}{s(s+1)(s+6)}$$

Determine the signal waveform from an inspection of the s-plane plot.

Solution The poles and zeros are shown in Figure 12.19.

Applying Equation (12.49) to the pole at the origin

$$\text{residue} = 2\frac{3}{(1)(6)} = 1$$

Applying Equation (12.49) to the pole at -1

$$\text{residue} = 2\frac{2}{(-1)(5)} = -0.8$$

Applying Equation (12.49) to the pole at -6

$$\text{residue} = 2\frac{-3}{(-5)(-6)} = -0.2$$

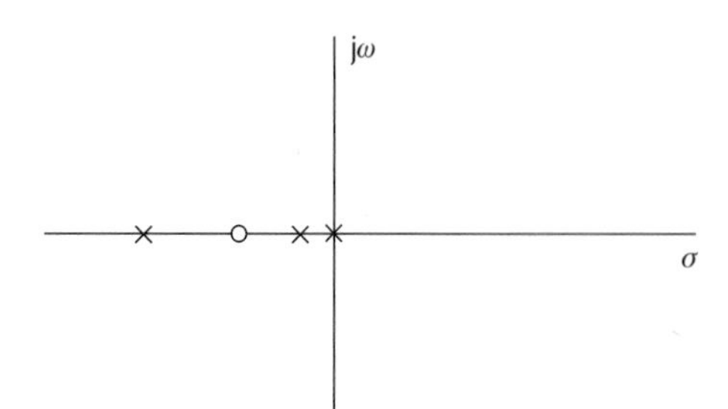

Fig. 12.19 Poles and zeros of the Laplace transform of a signal

Hence

$$F(s) = \frac{1}{s} - \frac{0.8}{s+1} - \frac{0.2}{s+6}$$

Therefore

$$f(t) = 1 - 0.8e^{-t} - 0.2e^{-6t}$$

Noting that the Laplace transform of a unit step function is $1/s$ it is apparent that $F(s)$ may have arisen as the result of applying a step function to a system whose transfer function is

$$H(s) = \frac{2(s+3)}{(s+1)(s+6)}$$

If we wish we could determine the impulse response of this system.

Example 12.9

Determine the impulse response corresponding to the transfer function

$$H(s) = \frac{2(s+3)}{(s+1)(s+6)}$$

Solution The Laplace transform of a unit impulse is unity and the impulse response is thus the inverse Laplace transform of $H(s)$. The poles and zeros of this are shown in Figure 12.20.

Applying Equation 12.49 to the pole at -1

$$\text{residue} = 2\,\frac{2}{5} = 0.8$$

Applying Equation (12.49) to the pole at -6

$$\text{residue} = 2\,\frac{-3}{(-5)} = 1.2$$

Hence

$$H(s) = \frac{0.8}{s+1} + \frac{1.2}{s+6}$$

Therefore

$$h(t) = 0.8e^{-t} + 1.2e^{-6t}$$

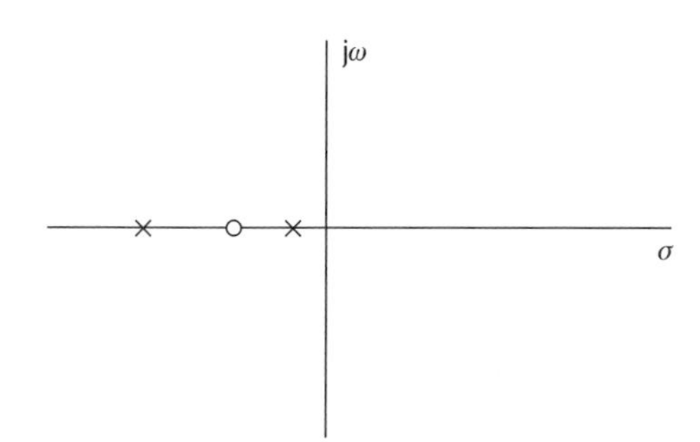

Fig. 12.20 Poles and zeros of a transfer function

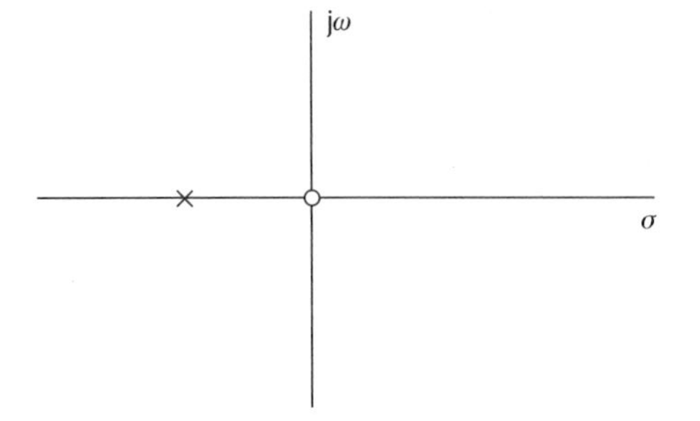

Fig. 12.21 Pole–zero plot of signal

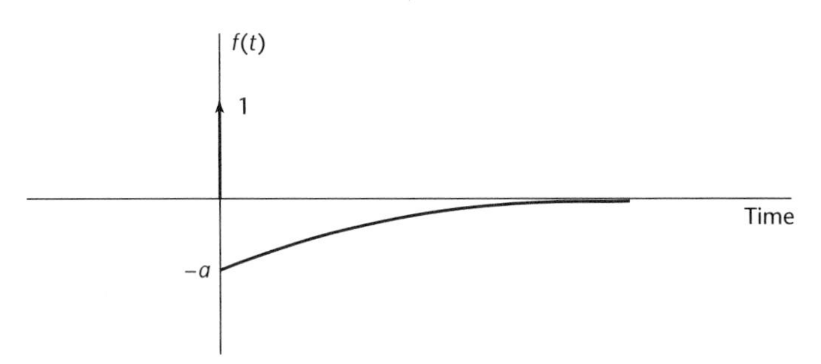

Fig. 12.22 Signal corresponding to the pole–zero plot

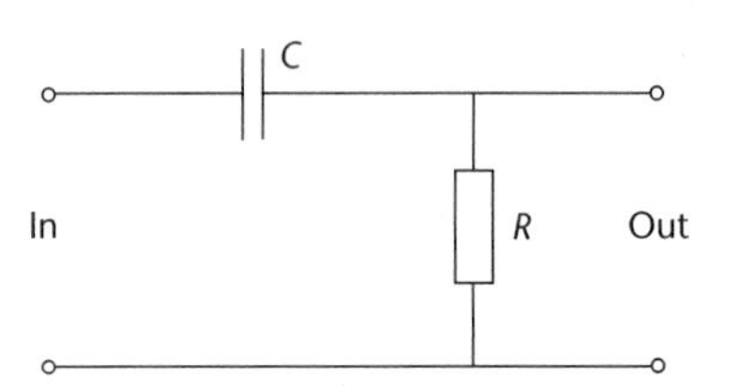

Fig. 12.23 Network whose impulse response is of the form shown in Figure 12.22

Example 12.10

Determine the form of the signal $f(t)$ whose pole–zero plot is that shown in Figure 12.21.

Solution Neglecting any constant gain term, the Laplace transform is $F(s) = s/(s + a)$. Since this has as many zeros as poles, the expression can be expanded to be of the form $k + r/(s + a)$, where $k = 1$ and $r = -a$. Hence $f(t) = \delta(t) - a\mathrm{e}^{-at}$. This signal is shown in Figure 12.22.

One way in which such a signal can arise is as the impulse response of a network whose transfer function has the pole–zero plot of Figure 12.20. Indeed, the network shown in Figure 12.23 has the transfer function $H(s) = s/(s + 1/CR)$ and its impulse response is therefore $h(t) = \delta(t) - (1/CR)\mathrm{e}^{-t/CR}$.

It will be noted from the examples that, the further a pole is from the imaginary axis, the greater is the rate of exponential decay of the component that it represents. Poles close to the imaginary axis have longer lasting effects than poles far from the imaginary axis. *For this reason poles close to the imaginary axis are generally the most important in their effect on a system response.* If, for example, we have a third-order system whose

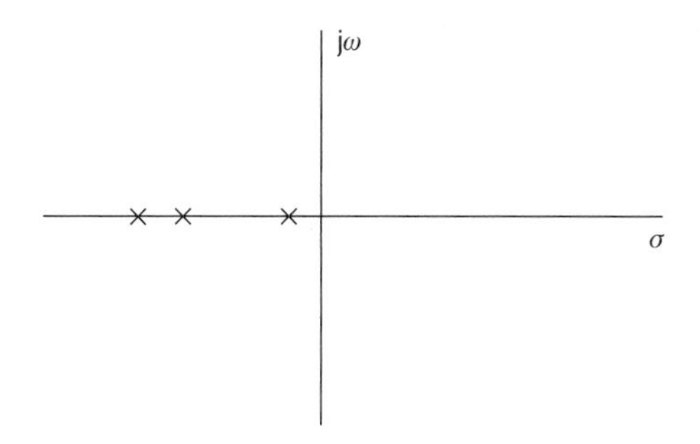

Fig. 12.24 Third-order system with negative real poles

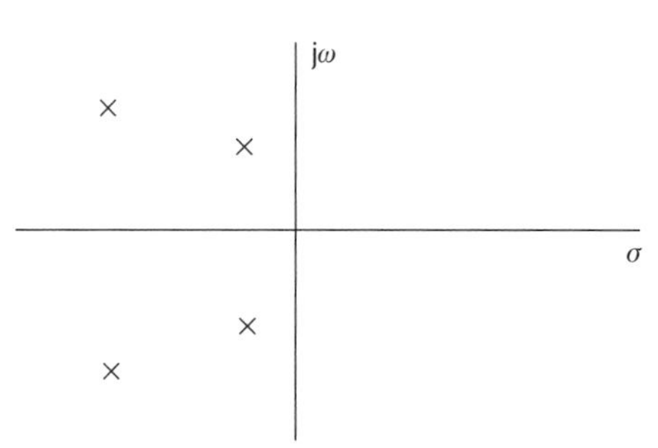

Fig. 12.25 Fourth-order system with two pairs of conjugate poles

transfer function contains the three negative real poles shown in Figure 12.24, the most important part of its response will be that caused by the pole nearest the origin. The most significant part of the response will therefore be that of a first-order system.

Similarly, if we have a fourth-order system whose transfer function contains the four poles shown in Figure 12.25, the most important part of the response will be that due to the two poles nearest the origin. The most significant part of the response will therefore be that of a second-order system.

The s-plane plot of the response of a system to a step function differs from that of the system transfer function by the addition of a pole at the origin. Step function responses are frequently relevant and, because the poles closest to the origin have the most long-lasting effect, it is very important to have a good understanding of the step function responses of first- and second-order systems.

12.11 The step function response of first-order systems

This case has already been examined in Example 12.6, but is repeated to link it up with an s-plane plot and its interpretation. The RC network in Figure 12.26 is again taken as a simple example of a first-order electrical system (*one* energy storage element).

Its transfer function is given by

$$H(s) = \frac{1/sC}{R + 1/sC} = \frac{1/CR}{s + 1/CR}$$

This is an example of a first-order system without zeros. If the resistance and capacitance of Figure 12.26 are interchanged the transfer function becomes

$$H(s) = \frac{s}{s + 1/CR}$$

and we now have the first-order system with a zero that was considered in Example 12.10.

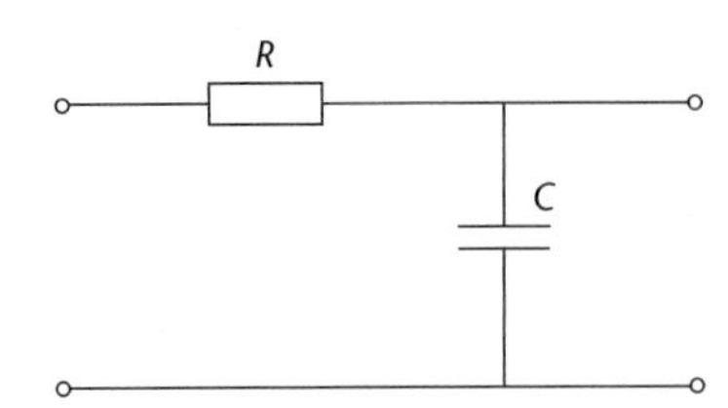

Fig. 12.26 Example of first-order system

The Laplace transform of a unit step function is $F(s) = 1/s$, such that the step function response for the system of Figure 12.26 is described in the s domain by

$$C(s) = F(s)H(s) = \frac{1/CR}{s(s + 1/CR)}$$

Generalizing this to any first-order system without zeros we have

$$H(s) = \frac{a}{(s + a)} \tag{12.51}$$

$$C(s) = \frac{a}{s(s + a)} \tag{12.52}$$

$C(s)$ has two poles, as shown in Figure 12.27.

From Equation (12.50) the residue for the pole at the origin is given by

$$r_1 = \frac{a}{a} = 1$$

and the residue for the pole at $-a$ is given by

$$r_2 = \frac{a}{-a} = -1$$

Hence

$$C(s) = \frac{1}{s} - \frac{1}{s + a} \tag{12.53}$$

and

$$c(t) = 1 - \mathrm{e}^{-at} \tag{12.54}$$

This response is shown by the solid line in Figure 12.28. In contrast to Example 12.6 the residues have been obtained by a graphical interpretation of the pole plot.

In the case of a first-order system with zeros (as for example if R and C are interchanged) we find

$$H(s) = \frac{s}{(s + a)} \tag{12.55}$$

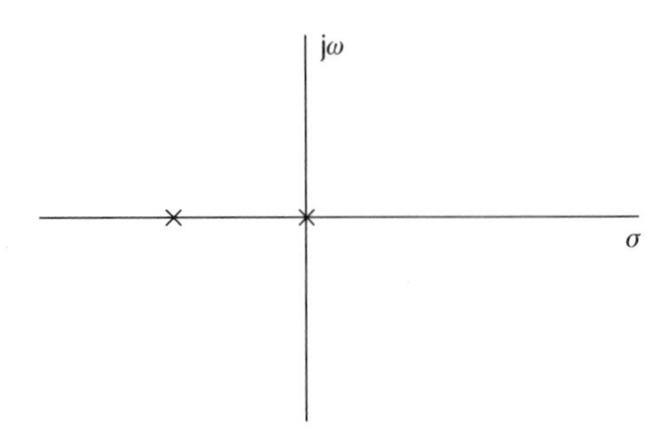

Fig. 12.27 Pole plot of step function response of first-order system without zeros

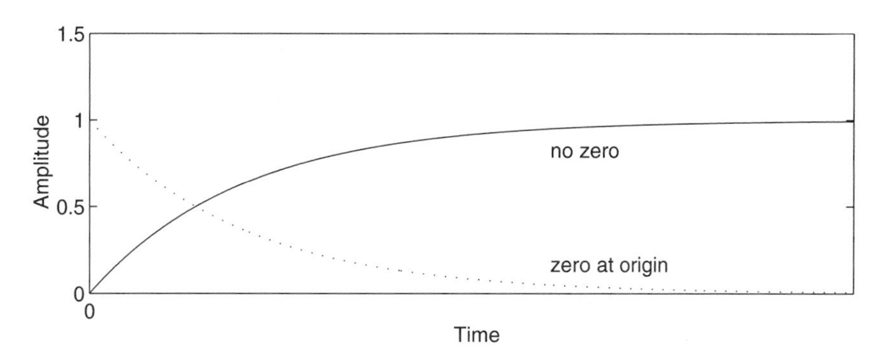

Fig. 12.28 Step function response of first-order system with and without a zero

such that $C(s) = 1/(s+a)$ and hence

$$c(t) = \mathrm{e}^{-at} \tag{12.56}$$

This is shown by the dotted line in Figure 12.28, and is a decaying exponential rather than a rising one.

12.12 The step function response of second-order systems

The series *RLC* network in Figure 12.29 is taken as a simple example of a second-order electrical system (*two* totally separate energy storage elements). Its transfer function is given by

$$H(s) = \frac{1/sC}{R + sL + 1/sC} = \frac{1/LC}{s^2 + s(R/L) + (1/LC)}$$

The Laplace transform of a unit step function is $F(s) = 1/s$, such that the step function response is described in the s domain by

$$C(s) = F(s)H(s) = \frac{1/LC}{s[s^2 + s(R/L) + (1/LC)]}$$

It is useful to generalize this to any second-order system and, in doing this, it is common to introduce two parameters ζ and ω_n defined by

$$\omega_n = \frac{1}{\sqrt{LC}} \tag{12.57}$$

and

$$\zeta = \frac{R}{2\sqrt{LC}} \tag{12.58}$$

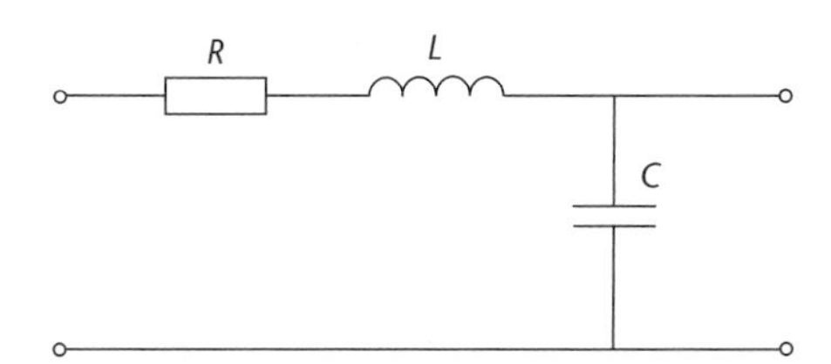

Fig. 12.29 Example of a second-order system

Using these parameters the transfer function of a general second-order system is

$$H(s) = \frac{\omega_n^2}{s^2 + 2\zeta\omega_n s + \omega_n^2} \tag{12.59}$$

and the step function response is

$$C(s) = \frac{\omega_n^2}{s(s^2 + 2\zeta\omega_n s + \omega_n^2)} \tag{12.60}$$

The physical significance of the parameters ω_n and ζ for a general second-order system is that ω_n represents the natural frequency in rad/s in the absence of damping, while ζ is a useful normalized measure of the damping and is known as the damping ratio.

Factorization of the denominator of Equation (12.59) shows the transfer function to have the real poles

$$p_{1,2} = -\zeta\omega_n \pm \omega_n\sqrt{\zeta^2 - 1} \tag{12.61}$$

if $\zeta \geqslant 1$, or complex poles given by

$$p_{1,2} = -\zeta\omega_n \pm j\omega_n\sqrt{1 - \zeta^2} \tag{12.62}$$

if $0 \leqslant \zeta < 1$. The poles of the step function response are thus as shown in Figure 12.30 for three different conditions of the damping ratio. Repeating the terms introduced in Section 11.9, these correspond to the under-damped, critically damped and over-damped conditions.

For the under-damped case the radial distances from the origin to the system poles are given by ω_n, and the separation of the two conjugate poles is given by $2\omega_n\sqrt{1-\zeta^2}$. Also, with reference to Figure 12.30(a), it is useful to introduce two angles, the angle θ of the poles relative to the negative real axis, and the angle ϕ of the uppermost pole to the imaginary axis. We have

$$\cos\theta = \frac{\omega_n\zeta}{\omega_n} = \zeta \tag{12.63}$$

and

$$\phi = \tan^{-1}\frac{\zeta}{\sqrt{1-\zeta^2}} \tag{12.64}$$

θ is useful as a direct indication of the damping ratio, but ϕ is more relevant to what follows. Applying Equation (12.50) to the pole–zero plot, and noting from Equation (12.60) that $K = \omega_n^2$, we see that the residue of the pole at the origin is

$$r_1 = \frac{\omega_n^2}{\omega_n^2} = 1 \tag{12.65}$$

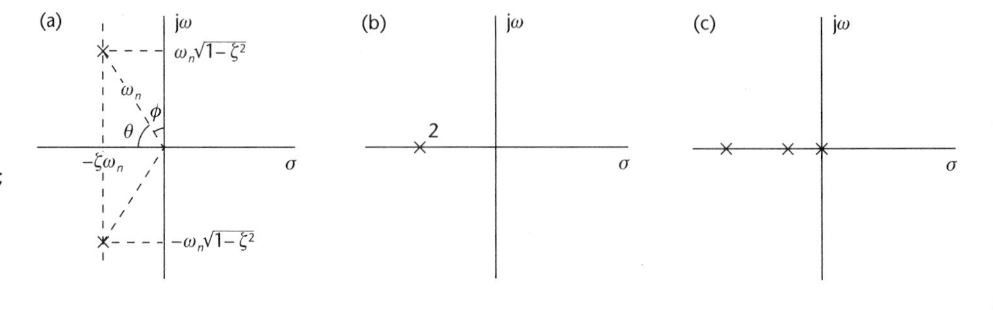

Fig. 12.30 Poles of the step function response of a second-order system: (a) under-damped; (b) critically damped; (c) over-damped

and the residue at the upper pole is

$$r_2 = \frac{\omega_n^2}{\omega_n e^{j(\pi/2+\phi)} 2\omega_n \sqrt{1-\zeta^2} e^{j\pi/2}} = \frac{-e^{-j\phi}}{2\sqrt{1-\zeta^2}}$$

$$= -\frac{\cos\phi}{2\sqrt{1-\zeta^2}} + j\frac{\sin\phi}{2\sqrt{1-\zeta^2}} \tag{12.66}$$

Thus we can write $C(s)$ in the form

$$C(s) = \frac{1}{s} + \frac{r_2}{s-p_2} + \frac{r_2^*}{s-p_2^*} = \frac{1}{s} + \frac{c+jd}{s+a-jb} + \frac{c-jd}{s+a+jb} \tag{12.67}$$

where

$$a = \zeta\omega_n, \quad b = \omega_n\sqrt{1-\zeta^2}, \quad c = -\frac{\cos\phi}{2\sqrt{1-\zeta^2}} \quad \text{and} \quad d = \frac{\sin\phi}{2\sqrt{1-\zeta^2}}$$

Hence, using Equations (12.44) and (12.45)

$$c(t) = \mathscr{L}^{-1}C(s) = 1 - \frac{\cos\phi}{\sqrt{1-\zeta^2}} e^{-\zeta\omega_n t}\cos\left(\omega_n\sqrt{1-\zeta^2}t\right)$$
$$- \frac{\sin\phi}{\sqrt{1-\zeta^2}} e^{-\zeta\omega_n t}\sin\left(\omega_n\sqrt{1-\zeta^2}t\right)$$

Manipulation simplifies this to

$$c(t) = 1 - \frac{1}{\sqrt{1-\zeta^2}} e^{-\zeta\omega_n t}\cos\left(\omega_n\sqrt{1-\zeta^2}t - \phi\right) \tag{12.68}$$

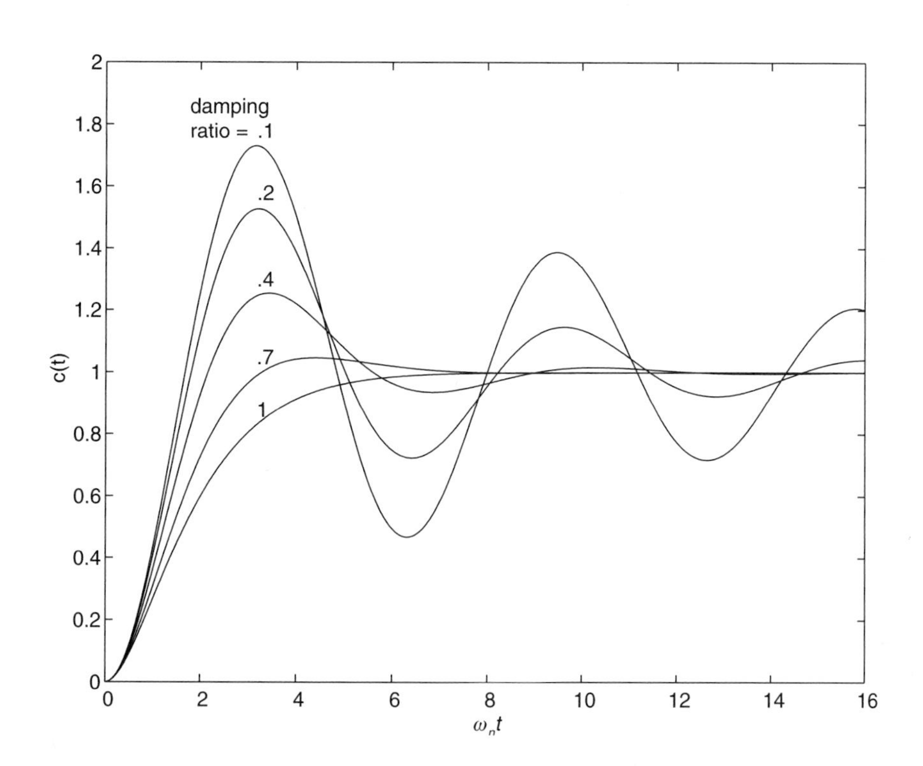

Fig. 12.31* Step function response of under-damped second-order system as a function of damping ratio

where

$$\phi = \tan^{-1} \frac{\zeta}{\sqrt{1 - \zeta^2}}$$

Figure 12.31 shows this step function response for various values of the damping ratio.

One can use Equation (12.68) to draw several broad conclusions about the effect of pole positions on the step function response.

1. The frequency of the under-damped oscillations depends solely on the imaginary components of the system poles, and not on their real components. In other words, conjugate poles lying anywhere on the dotted line of Figure 12.32 signify the same oscillatory frequency. The ones nearest the imaginary axis correspond to the step function response that is the least quickly damped out.
2. The time constant of the exponential damping depends solely on the real components of the system poles, and not on their imaginary components. In other words, conjugate poles lying anywhere on the dotted line of Figure 12.33 have the same rate of decay. The ones furthest from the real axis have the highest oscillatory frequency.

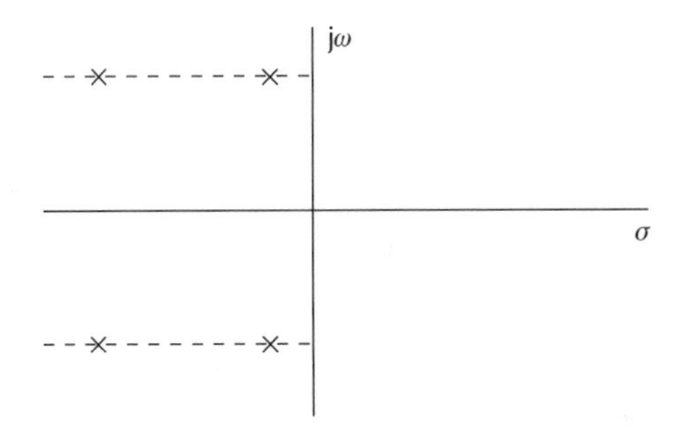

Fig. 12.32 Pole positions giving the same oscillatory frequency

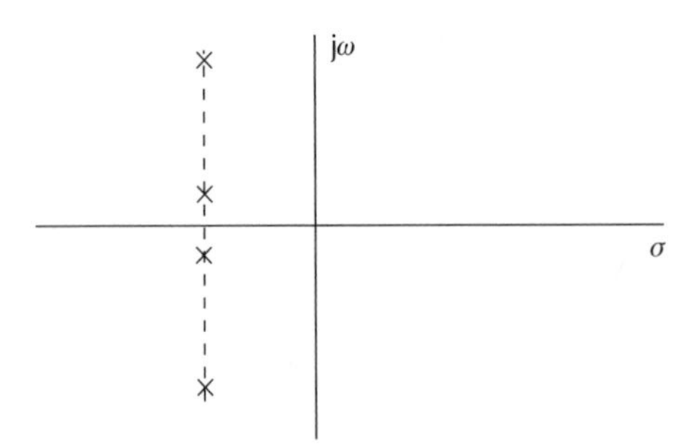

Fig. 12.33 Pole positions giving the same rate of decay

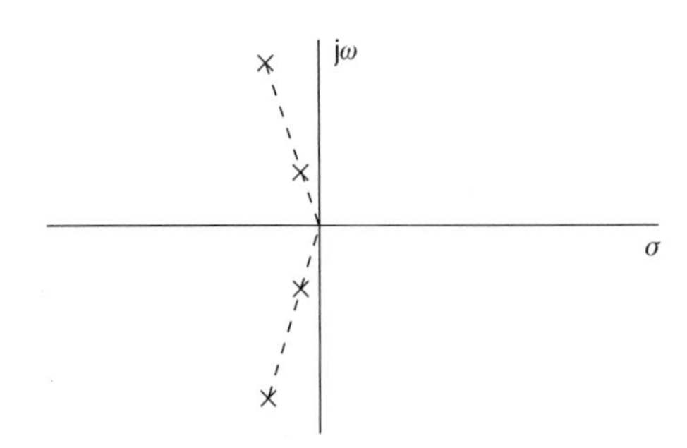

Fig. 12.34 Pole positions giving the same shape of decaying oscillations

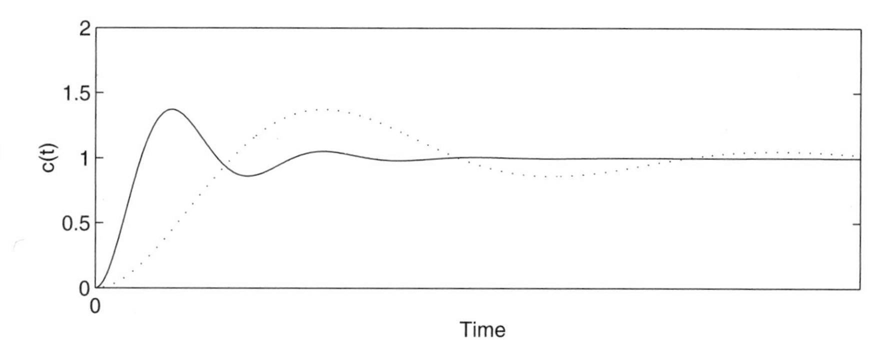

Fig. 12.35 Second-order under-damped responses when the radial distances from the origin of the poles are changed

3. The effect of changing the radial distances from the origin of the system poles while keeping the ϕ value constant is purely to expand or contract the step function response in time. If, for example, the outermost set of system poles in Figure 12.34 gives the step function response using the solid line in Figure 12.35, the innermost set of system poles in Figure 12.34 gives the step function response shown dotted in Figure 12.35.
4. Using Equation (12.64) a ϕ value of 45° gives $\phi = 0.707$. From Figure 12.31 this is ideal for a fast rise time without significant overshoot.

12.13 System responses using MATLAB

Powerful commands exist in MATLAB for determining the responses of systems with a known transfer function. The command **step(b,a)** produces the step function response of a system whose transfer function is defined by a vector b giving the polynomial coefficients of its numerator, and a vector a giving the polynomial coefficients of its denominator. The command **impulse(b,a)** produces the corresponding impulse response.

Example 12.11

Use MATLAB to plot the step function response and impulse response of the Butterworth filter described in the next chapter for which

$$H(s) = \frac{1}{s^4 + 2.6131s^3 + 3.4142s^2 + 2.613s + 1}$$

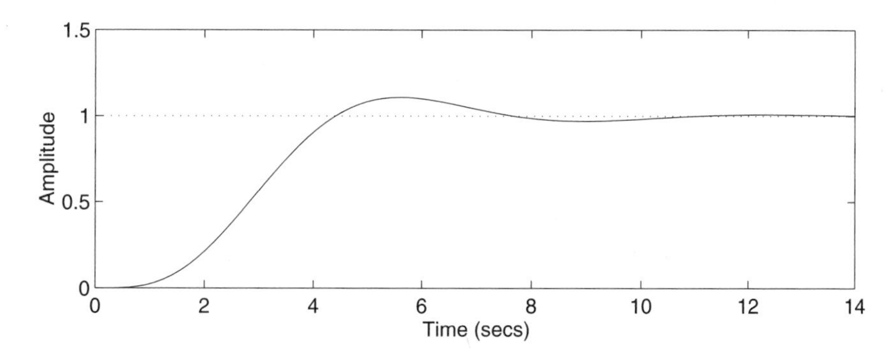

Fig. 12.36 Step function response obtained using MATLAB

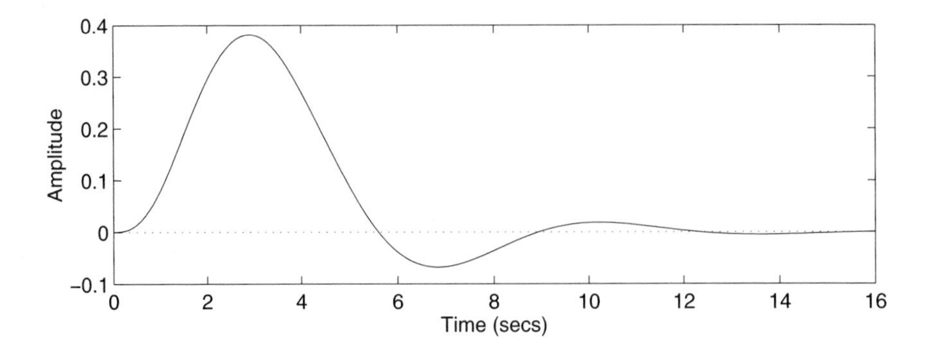

Fig. 12.37 Impulse response obtained using MATLAB

Solution The command **step(1,[1 2.6131 3.4142 2.6131 1])** produces Figure 12.36. The command **impulse(1,[1 2.6131 3.4142 2.6131 1])** produces Figure 12.37.

An even more powerful MATLAB command computes and plots the response of a system to an arbitrary input. If we have a time vector t and an input signal $x(t)$ the command **lsim(b,a,x,t)** plots the output response.

Example 12.12

A sonar echo consisting of 6 cycles of a 50 kHz sine wave enters a receiver consisting of the frequency-selective bandpass transistor amplifier shown in Figure 10.17. The

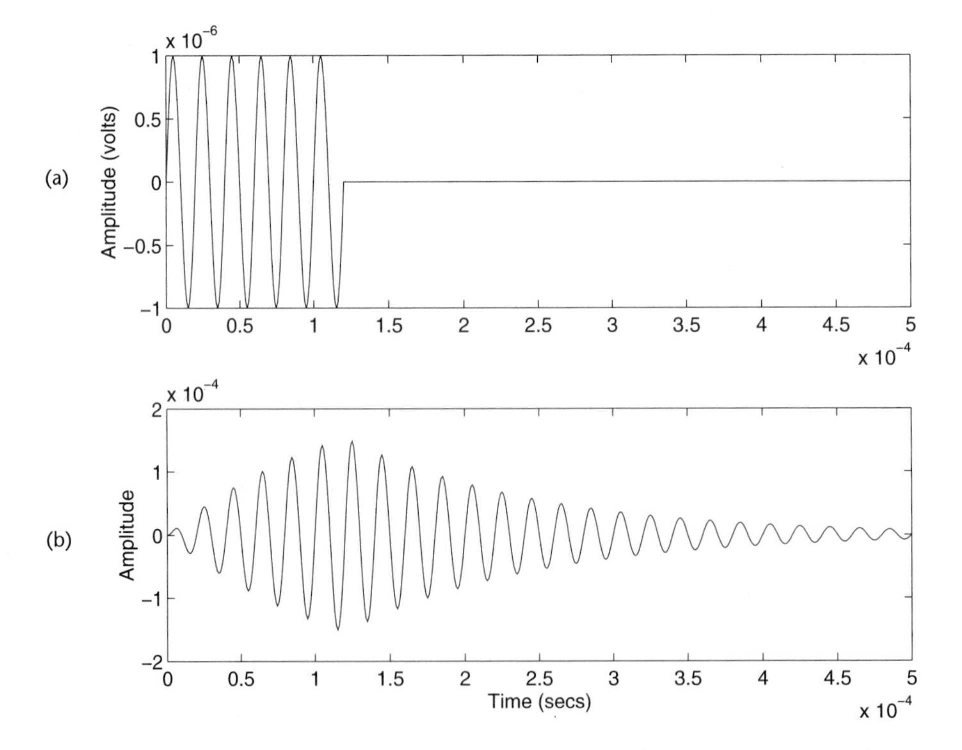

Fig. 12.38 System response using MATLAB: (a) input signal; (b) output signal

tank circuit is resonant at 50 kHz and has a Q of 20 such that, making use of Equations (10.42) and (10.48), the transfer function can be shown to be given by

$$H(s) = 4 \times 10^6 \frac{s}{s^2 + 15708s + 9.87 \times 10^{10}}$$

If the peak amplitude of the echo is 1 μV, plot the input and output signals.

Solution The duration of the pulse is $6/50000 = 120\,\mu s$. The pulse and a zero amplitude period following will be approximated using 500 samples 1 μs apart. The following commands are appropriate.

```
t=(0:500)*1.e-6;
x=l.e-6*sin(2*pi*50000*t(1:120));      % this generates the pulse.
x(501)=0;                    % this extends the signal with zero amplitude samples.
subplot(2,1,1),plot(t,x),ylabel('Amplitude')
subplot(2,1,2), lsim([4e6 0],[1 15708 9.87e10],x,t)
```

The outcome is shown in Figure 12.38.

12.14 An introduction to automatic control

Consider the situation where the armature-controlled servo motor described in Section 10.11 is used to pan a closed-circuit television camera used for surveillance. If an operator wishes to rotate the camera to some specific new position, one possibility is to apply a voltage to the armature of the motor and to switch it off just before the camera arrives at the new required angle, hoping that the camera will stop at the required position following a small amount of further rotation due to inertia. Such a technique involves human *feedback* based on a monitoring of the camera angle and an awareness of the inertia of the camera.

An alternative arrangement with many advantages is to use the typical automatic control system shown in Figure 12.39. Here the operator changes the camera direction to a desired angle θ_R by adjusting a knob on a calibrated potentiometer to a position that is known to correspond to the required angle θ_R. This potentiometer generates a voltage $v_R = k\theta_R$. The actual angle of the camera, θ_C, is measured by a second potentiometer and converted to a voltage $v_C = k\theta_C$ that is fed back to a summing point. Here it is subtracted from the input to generate a difference or 'error' voltage given by $v_E = k(\theta_R - \theta_C)$. This error voltage is amplified by an amplifier of

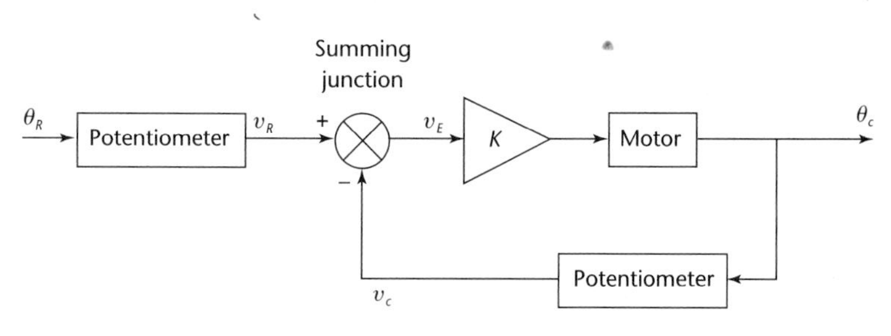

Fig. 12.39 Very simple automatic control system for adjusting camera angle

gain K to become $Kk(\theta_R - \theta_C)$ and used to drive the motor. When the camera attains its correct position, $\theta_C = \theta_R$ such that the error voltage is zero and there is no further drive on the motor.

Simple intuition might suggest that the amplifier acting on the error signal should have a very high gain in order that the signal $Kk(\theta_R - \theta_C)$ should be large, thus driving the camera quickly to its correct equilibrium position. Such intuition is very dangerous. Too much gain can cause a major overshoot of the camera. A clarification of the phenomenon involves a knowledge of the camera dynamics and some analysis.

An improved response can often be obtained by introducing a suitably designed frequency-dependent network into the feedback loop, as shown in Figure 12.40. It will be assumed that its transfer function $H(s)$ incorporates the factor k that converts output angle to voltage by means of a potentiometer. Besides this introduction of $H(s)$ there are, however, several other important differences in the ways in which Figures 12.39 and 12.40 are portrayed. Firstly, Figure 12.40 begins with the signal into the summing junction, namely the output of the input potientiometer. Secondly, because we wish to work with the s domain descriptions of the motor dynamics and the feedback network (that is, their responses to signals of the form e^{st}) we label system components by their transfer functions, and we prefer to describe the signals by their Laplace transforms rather than by their time waveforms. Thus the input in Figure 12.40 is labelled $V_R(s)$ and the output $\theta_C(s)$.

In Figure 12.40, $G(s)$ is the 'forward' or 'direct' transfer function relating the angle of the camera to the voltage from the amplifier when the voltage is of the form e^{st}. It involves the dynamics of the camera and of any additional networks inserted between the amplifier and the servo motor (none in Figure 12.39). In other words

$$G(s) = \tilde{\theta}_C / \tilde{v}_M \tag{12.69}$$

where $\tilde{v}_M$ is a voltage of the the form e^{st} applied to the motor and $\tilde{\theta}_C$ is the resulting angular response.

It would be very common to determine $G(s)$ by a *theoretical* study of electrical and mechanical properties of the motor. An alternative is to deduce $G(s)$ from *experimental* measurements of its response to appropriate excitation signals, perhaps based on the more general equation

$$G(s) = \theta_C(s)/V_M(s) \tag{12.70}$$

Much can be learnt if sinusoidal excitations with a wide range of frequencies are used, but other possibilities are to use step, impulse or pseudo-noise inputs. Unfortunately, there can be difficulties in the procedure and the problem of *system identification* is outside the scope of this text.

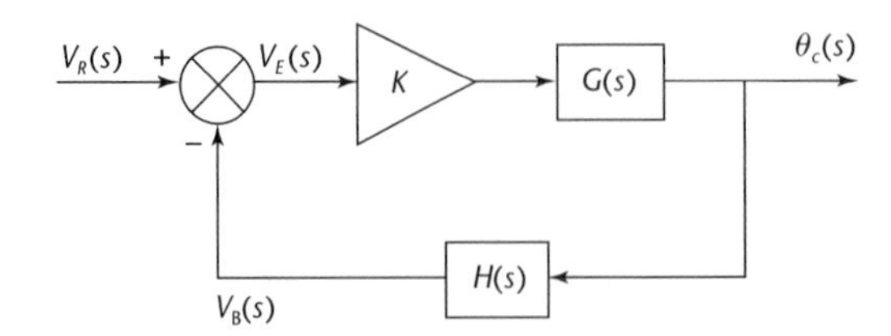

Fig. 12.40 Block diagram of more flexible control system for adjusting camera angle

Following the summing unit in Figure 12.40 the error signal is given by

$$V_{\mathrm{E}}(s) = V_{\mathrm{R}}(s) - V_{\mathrm{B}}(s) \tag{12.71}$$

Hence

$$\theta_{\mathrm{C}}(s) = (V_{\mathrm{R}}(s) - V_{\mathrm{B}}(s))KG(s) \tag{12.72}$$

We have $H(s) = V_{\mathrm{B}}(s)/\theta_{\mathrm{C}}(s)$ and using this to eliminate $V_{\mathrm{B}}(s)$ from Equation (12.72) we obtain

$$\theta_{\mathrm{C}}(s) = KV_{\mathrm{R}}(s)G(s) - K\theta_{\mathrm{C}}(s)G(s)H(s) \tag{12.73}$$

The ratio of $\theta_{\mathrm{C}}(s)$ to $V_{\mathrm{R}}(s)$ is known as the *closed loop* transfer function and will be given the symbol $T(s)$. Thus

$$\frac{\theta_{\mathrm{C}}(s)}{V_{\mathrm{R}}(s)} = T(s) = \frac{KG(s)}{1 + KG(s)H(s)} \tag{12.74}$$

The term $KG(s)H(s)$ in the denominator is known as the *open loop* transfer function since it gives the transfer function between the signal fed back to the summer and the input if the feedback loop is broken at the summer.

Without any feedback at all the transfer function would be $KG(s)$ and this will have a set of poles that depends on the dynamics of the camera. With feedback new poles are introduced into the system which, from Equation (12.74), are given by

$$KG(s)H(s) = -1 \tag{12.75}$$

These *closed loop* poles control the system response and it is important to determine their values. Depending on $G(s)$ and $H(s)$ there may be many poles. Considering the example of the panned closed-circuit TV camera a good result would be that the poles nearest the imaginary axis are a conjugate pair giving a second-order step function response corresponding to a ζ value of about 0.6 (see Figure 12.31). Most important of all, it is essential to ensure that the closed loop poles are not in the right-hand half of the s-plane, as this indicates instability. It can be very useful to evaluate how the pole positions vary as a function of the variable K, as then this can be adjusted to an optimal value. This variation of pole positions is known as the *root locus*.

Example 12.13

The armature-controlled d.c. motor discussed in Section 10.11 is used to pan a closed-circuit TV camera and the electromechanical and mechanical constants are such that the transfer function of the motor/camera combination is given by

$$G(s) = \frac{0.06}{s(s + 0.4)}$$

Determine the closed loop poles if $H(s) = 1$ and $K = 2$ and deduce whether the system is satisfactory.

Solution Using Equation (12.75), the closed loop poles are given by

$$KG(s)H(s) = \frac{0.12}{s(s + 0.4)} = -1$$

Hence

$$s^2 + 0.4s + 0.12 = 0$$

This yields

$$p_{1,2} = -0.2 \pm j0.283$$

It may be noted that these pole positions correspond to $\phi = \tan^{-1}(0.2/0.283)$. Hence, using Equation (12.64),

$$\frac{\zeta}{\sqrt{1-\zeta^2}} = \frac{0.2}{0.283}$$

to give $\zeta = 0.58$. From Figure 12.31 this is seen to give a fast response without excessive overshoot. It would probably be advantageous to reduce K very slightly.

Because of MATLAB the obvious alternative to the procedure given is to find the closed loop transfer function and to plot the step function response using the **step(b,a)** command.

In the example above $KG(s)H(s) = -1$ produced a second-order equation. For higher order equations it can be difficult to find the roots. Various graphical methods exist, but these are laborious, and a much simpler way using the **roots(a)** command in MATLAB was first described in Section 10.5. For example, the roots to the equation $s^3 + 12s^2 + 121s + 1560 = 0$ can be determined by entering the coefficients of the polynomial by the command, **a=[1 12 121 1560]**, and following this with the command, **r=roots(a)**.

12.15 Laplace transforms used directly for the solution of differential equations

Let us consider again the response of the simple RC network to a unit step. The approach adopted so far has been to find the transfer function of the network to an input of the form e^{st}, namely $H(s) = 1/(1 + sCR)$, to multiply this by the Laplace transform of the input, and then to evaluate the inverse Laplace transform.

An alternative which is closely related, but even more powerful *because it can cope with non-zero initial conditions*, is to apply the Laplace transform technique directly to differential equation that relates output to input. However, before demonstrating this, it is necessary first to determine the relationship between the Laplace transform of a signal and the Laplace transform of the derivatives of that signal.

If the Laplace transform of $f(t)$ is $F(s)$ we wish to determine the Laplace transforms of $df(t)/dt$, $d^2f(t)/dt^2$ and so on. Using the rigorous definition of the Laplace transform given in Equation (12.7), where the lower limit is 0– we have

$$\mathscr{L}\left[\frac{df(t)}{dt}\right] = \int_{0-}^{\infty} \frac{df(t)}{dt} e^{-st}\, dt \qquad (12.76)$$

Integrating by parts gives

$$\mathscr{L}\left[\frac{df(t)}{dt}\right] = e^{-st} f(t)\Big]_{0-}^{\infty} - \int_{0-}^{\infty} - se^{-st}f(t)\, dt$$

Taking the s outside the integral of the second term and changing the order of the terms gives

$$\mathscr{L}\,\frac{\mathrm{d}f(t)}{\mathrm{d}t} = sF(s) - f(0-) \tag{12.77}$$

where $f(0-)$ is the value of $f(t)$ immediately before $t = 0$. Because it is quite common for $f(t)$ to change its value abruptly at $t = 0$ such that $f(0+)$ is very different from $f(0-)$ the minus sign has been included in $f(0-)$ even though the convention of this text has been usually to omit it from the lower limit of integration in the definition of the Laplace transform (see discussion of the Laplace transform of an impulse function in Section 12.3).

The significance of $f(0-)$ is that it represents a possible initial condition. Neglecting this possibility just for the moment, let us repeat Example 12.6 for finding the response of an RC network to a unit step, but now using the result of the Laplace transform of a derivative.

Example 12.14

For the RC network of Figure 12.41, derive the differential equation that relates the output voltage, $v_2(t)$, to the input voltage, $v_1(t)$. Apply the Laplace transform to this differential equation to determine the output caused by a unit step function input.

Solution The charge across the capacitance equals $Cv_2(t)$. The current through the series combination of resistance and capacitance is therefore $C\,\mathrm{d}v_2(t)/\mathrm{d}t$, and the voltage drop across the resistance is $CR\,\mathrm{d}v_2(t)/\mathrm{d}t$. But this voltage drop is also given by $v_1(t) - v_2(t)$.

$$\therefore \quad v_1(t) - v_2(t) = CR\,\frac{\mathrm{d}v_2(t)}{\mathrm{d}t}$$

or

$$v_2(t) + CR\,\frac{\mathrm{d}v_2(t)}{\mathrm{d}t} = v_1(t) \tag{12.78}$$

Taking the Laplace transform of both sides, and assuming $v_C(0-) = 0$, gives

$$V_2(s) + sCRV_2(s) = V_1(s) \tag{12.79}$$

$$\therefore \quad V_2(s) = \frac{V_1(s)}{1 + sCR}$$

If the input is a unit step, $V_1(s) = 1/s$, and we have

$$V_2(s) = \frac{1}{s(1 + sCR)} \tag{12.81}$$

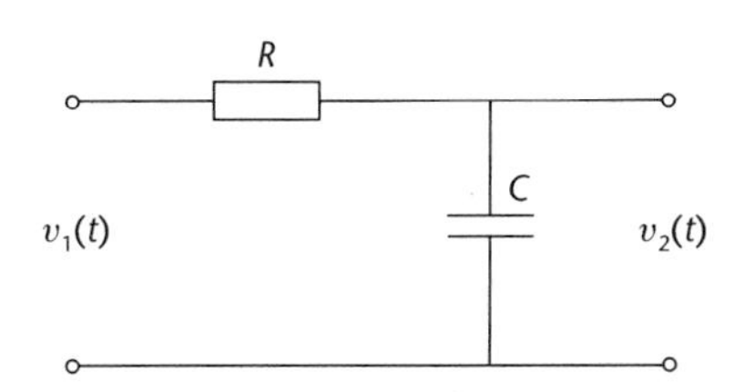

Fig. 12.41 RC network

But, as in Example 12.6, this can be expanded as

$$V_2(s) = \frac{1}{s} - \frac{1}{s + 1/CR}$$

Therefore

$$v_2(t) = 1 - e^{-t/RC} \tag{12.82}$$

As expected, this is the same result as obtained in Example 12.6.

It will be noted that the solution in Example 12.14 has been obtained without using the impedance functions of circuit elements (R, sL and $1/sC$). Instead it derives the differential equation relating output to input and solves this directly using Laplace transforms. It will be noted that the differential equation of Equation (12.78) has been replaced by the much simpler *algebraic* equation of Equation (12.79).

In this example there was no energy stored in the system prior to $t = 0$. Let us next consider an example in which there is.

Example 12.15

Repeat Example 12.14 for the case where the capacitance is initially charged to 0.6 V.

Solution As in Example 12.14,

$$v_2(t) + CR\frac{dv_2(t)}{dt} = v_1(t) \tag{12.83}$$

Taking the Laplace transform of both sides but recognizing that $v_2(t)$ has an initial condition

$$V_2(s) + CR - [sV_2(s) - v_2(0-)] = V_1(s) \tag{12.84}$$

where $v_2(0-) = 0.6$. Hence

$$V_2(s) = \frac{V_1(s) + 0.6CR}{1 + sCR} \tag{12.85}$$

Putting $V_1(s) = 1/s$ gives

$$V_2(s) = \frac{1}{s(1 + sCR)} + \frac{0.6CR}{1 + sCR} \tag{12.86}$$

The first of these terms is the same as in the previous example and may again be expanded into partial fractions. Thus

$$V_2(s) = \frac{1}{s} - \frac{CR}{1 + sCR} + \frac{0.6CR}{1 + sCR} = \frac{1}{s} - \frac{0.4CR}{1 + sCR} \tag{12.87}$$

The inverse Laplace transform gives

$$v_2(t) = 1 - 0.4e^{-t/CR} \qquad t \geqslant 0 \tag{12.88}$$

It is interesting to note that, in this particular example, there is a simpler and more physical way of obtaining the same result. One could recognize that the initial

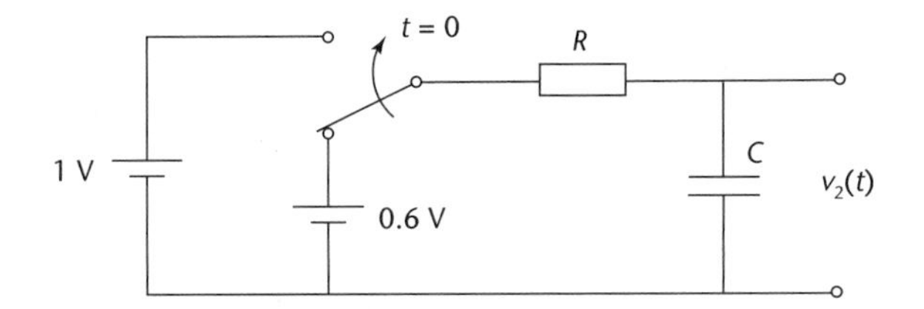

Fig. 12.42 Circuit causing initial condition

condition could have arisen because of a situation such as shown in Figure 12.42, where the input to the network is changed from 0.6 V for $t < 0$ to 1 V for $t \geqslant 0$. The input voltage may be considered to be a 0.4 V step superposed on a d.c. level of 0.6 V. By superposition the output is 0.6 V due to the d.c. input component plus the response of Equation (12.82) multiplied by 0.4, that is

$$v_2(t) = 0.6 + 0.4(1 - \mathrm{e}^{-t/RC}) = 1 - 0.4\mathrm{e}^{-t/RC}$$

as before.

The procedure given at the beginning of this section for calculating the Laplace transform of the first derivative of $f(t)$ can be repeated for the second derivative of $f(t)$ to give

$$\mathscr{L}\left[\frac{\mathrm{d}^2 f(t)}{\mathrm{d}t^2}\right] = s^2 F(s) - sf(0-) - \frac{\mathrm{d}f(0-)}{\mathrm{d}t} \tag{12.89}$$

Extending this to the nth derivative, but restricting it to the case where the initial conditions of $f(t)$ and its derivatives are zero, yields

$$\mathscr{L}\left[\frac{\mathrm{d}^n f(t)}{\mathrm{d}t^n}\right] = s^n F(s) \tag{12.90}$$

In practice it is very common to have zero initial conditions.

12.16 Laplace transforms used directly for the solution of integro-differential equations

When solving integro-differential equations we often encounter a term of the form $\int_{-\infty}^{t} f(\tau)\,\mathrm{d}\tau$. Before demonstrating the application of Laplace transforms to solving integro-differential equations it is necessary first to determine the Laplace transform of such an integral.

The first point is that the unilateral Laplace transform is confined to signals which are zero for $t < 0$. Hence

$$\int_{-\infty}^{t} f(\tau)\,\mathrm{d}\tau = \int_{0}^{t} f(\tau)\,\mathrm{d}\tau \tag{12.91}$$

Using the definition of the Laplace transform we have

$$\mathscr{L}\left[\int_0^t f(\tau)\,\mathrm{d}\tau\right] = \int_0^{\infty}\left[\int_0^t f(\tau)\,\mathrm{d}\tau\right]\mathrm{e}^{-st}\,\mathrm{d}t \tag{12.92}$$

Integrating this by parts gives

$$\mathscr{L}\left[\int_0^t f(\tau)\,\mathrm{d}\tau\right] = -\frac{1}{s}\,\mathrm{e}^{-st}\left[\int_0^t f(\tau)\,\mathrm{d}\tau\right]_0^\infty - \int_0^\infty -\frac{1}{s}\,\mathrm{e}^{-st} f(\tau)\,\mathrm{d}\tau \tag{12.93}$$

Considering the first term, we have $\mathrm{e}^{-st} = 0$ when $t = \infty$ at the upper limit of integration, and $\int_0^t f(\tau)\,\mathrm{d}\tau = 0$ when $t = 0$ at the lower limit of integration (an exception arises if $f(\tau)$ is an impulse function, but this is a case that will be ignored, since it is not usual to want to know the Laplace transform of its integral). It follows that the first term of Equation (12.93) is zero.

The $1/s$ term can be taken outside the integral of the second term to give $F(s)/s$. Hence, combining Equations (12.91) and (12.93), we obtain

$$\mathscr{L}\left[\int_{-\infty}^t f(\tau)\,\mathrm{d}\tau\right] = \frac{1}{s}\,F(s) \tag{12.94}$$

Example 12.16

Considering once again the simple *RC* network of Figure 12.41, determine now the *current* when the input voltage excitation function, $v(t)$, is a unit step function.

Solution We have

$$v(t) = Ri(t) + 1/C\int_{-\infty}^t i(\tau)\,\mathrm{d}\tau \tag{12.95}$$

Taking the Laplace transforms of each term gives

$$V(s) = RI(s) + I(s)/Cs \tag{12.96}$$

$$\therefore\ I(s) = \frac{V(s)}{R + 1/Cs} \tag{12.97}$$

If $V(s) = 1/s$ for a unit step function this becomes

$$I(s) = \frac{1}{s(R + 1/Cs)} = \frac{1}{R}\,\frac{1}{s + 1/RC} \tag{12.98}$$

$$\therefore\ i(t) = \frac{1}{R}\,\mathrm{e}^{-t/RC} \tag{12.99}$$

12.17 A comparison of Laplace transforms and phasors for system analysis

Consider the example of needing to determine the current in the inductance L_2 of the network shown in Figure 12.43. If the input signal is v we can introduce the loop currents i_1 and i_2 and write down the differential equations for the network, expressing Kirchhoff's voltage law that the sum of all voltages around a closed circuit is zero. The equations for the two loops are

$$v = L_1\frac{\mathrm{d}i_1}{\mathrm{d}t} + \frac{1}{C}\int_{-\infty}^t i_1\,\mathrm{d}t - \frac{1}{C}\int_{-\infty}^t i_2\,\mathrm{d}t \tag{12.100}$$

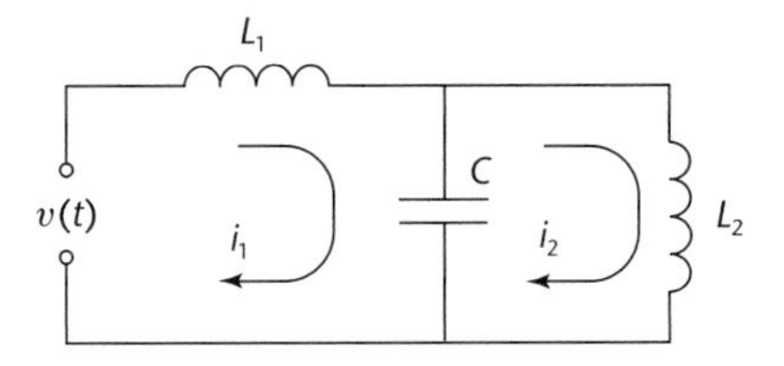

Fig. 12.43 Network requiring determination of current

and

$$0 = -\frac{1}{C}\int_{-\infty}^{t} i_1 \, dt + L_2 \frac{di_2}{dt} + \frac{1}{C}\int_{-\infty}^{t} i_2 \, dt \tag{12.101}$$

The phasor approach to network analysis is to assume an excitation $v = Ae^{j(\omega t + \alpha)}$, or $\mathbf{V}e^{j\omega t}$ where $\mathbf{V}$ is a phasor given by $\mathbf{V} = Ae^{j\alpha}$. For this excitation it can be assumed that i_1 is of a similar complex exponential form, namely $Be^{j(\omega t + \beta)}$, or $\mathbf{I}_1 e^{j\omega t}$ where $\mathbf{I}_1$ is a phasor given by $\mathbf{I}_1 = Be^{j\beta}$. Similarly, i_2 is of the form $Ce^{j(\omega t + \gamma}$, or $\mathbf{I}_2 e^{j\omega t}$ where $\mathbf{I}_2$ is a phasor given by $\mathbf{I}_2 = Ce^{j\gamma}$. Substituting these values into Equations (12.100) and (12.101) and dividing through by $e^{j\omega t}$ produces the standard form of phasor network equations, namely

$$\mathbf{V} = \mathbf{I}_1 \left(j\omega L_1 + \frac{1}{j\omega C} \right) - \mathbf{I}_2 \frac{1}{j\omega C} \tag{12.102}$$

$$0 = -\mathbf{I}_1 \frac{1}{j\omega C} + \mathbf{I}_2 \left(j\omega L_2 + \frac{1}{j\omega C} \right) \tag{12.103}$$

These two simultaneous equations can be solved in the usual way to eliminate $\mathbf{I}_1$ and thus find the relationship between $\mathbf{I}_2$ and $\mathbf{V}$.

An alternative procedure is to take the Laplace transforms of Equations (12.100) and (12.101) to produce

$$V(s) = I_1(s)\left(sL_1 + \frac{1}{sC} \right) - I_2(s) \frac{1}{sC} \tag{12.104}$$

$$0 = -I_1(s) \frac{1}{sC} + I_2(s)\left(sL_2 + \frac{1}{sC} \right) \tag{12.105}$$

Again these are simultaneous equations which can be solved to eliminate $I_1(s)$ and thus find the relationship between $I_2(s)$ and $V(s)$. However, whereas the phasor equations are only meaningful when dealing with excitations of the form $e^{j\omega t}$ or $\cos(\omega t)$, this time we are not constrained in the form of the excitation function. We can determine the Laplace transform $V(s)$ of *any input signal*, solve the two equations to determine $I_2(s)$, and finally perform an inverse Laplace transform on $I_2(s)$ to determine the current.

For the sake of brevity it is very common practice when using Laplace transforms to take it *as understood* that each quantity is a function of s and to write Equations (12.104) and (12.105) simply

$$V = I_1\left(sL_1 + \frac{1}{sC} \right) - I_2 \frac{1}{sC} \tag{12.106}$$

$$0 = -I_1 \frac{1}{sC} + I_2 \left(sL_2 + \frac{1}{sC} \right) \tag{12.107}$$

Note here that the bold notation often used for phasors is not adopted.

Network equations such as those of Equations (12.106) and (12.107) can be obtained from the phasor equations by simply replacing $j\omega$ by s. However, they have very much more potential in that they are not constrained to sinusoidal excitations. The Laplace transform technique is very much more general and powerful than that of phasors.

12.18 Summary

It was shown in Chapter 10 that the transfer function of a linear system to an excitation of the form e^{st} was a very compact way of conveying all the information of that system, with the added advantage that it could be portrayed pictorially with a plot of its poles and zeros. It was easy to use this transfer function to determine the steady state response to cosinusoids. The Laplace transform is applicable to a much wider class of signals than cosinusoids. Effectively, it breaks down signals into an infinite set of basis functions, each of the form e^{st}, such that each one separately can be modified by the transfer function. An inverse Laplace transform effectively recombines the modified basis functions and gives the system response. Although this sounds complicated, the use of known transforms and the existence of techniques for simplifying transforms into these known transforms makes the procedure fairly simple in practice. Laplace transforms are very powerful and are used very extensively in system analysis, and particularly for the design and understanding of control systems. It has also been shown that MATLAB is very easy to use and effective for computing and plotting system responses.

12.19 Problems

1. Determine the Laplace transform of the following signals by considering each one to be the superposition of simpler signals whose Laplace transforms are known:

(a) a rectangular pulse of unit amplitude commencing at $t = 0$ and ending at $t = 2\,s$;

(b) a tone burst of the sinusoid $\sin(10\pi t)$ commencing at $t = 0$ and ending at $t = 2\,s$;

(c) a triangular pulse of unit peak amplitude commencing at $t = 0$ and ending at $t = 2\,s$.

2. Find the inverse Laplace transforms of:

(a) $X(s) = \dfrac{2s+1}{(s+2)(s+3)}$

(b) $Y(s) = \dfrac{5s^2 - 3s + 2}{(s+1)(s^2+2s+7)}$

(c) $Z(s) = \dfrac{1 + e^{-3s}}{s(2s+5)}$

3. A system has a transfer function given by

$$H(s) = \frac{(s+2)}{(s+1)(s+5)}$$

What is the Laplace transform $C(s)$ of its step function response? Based upon an inspection of the s-plane plot of $C(s)$ make an approximate sketch of the step function response. What happens to the response if the zero at -2 is moved to -1.2? Confirm the accuracy of the sketches by applying the **step(b,a)** command of MATLAB.

4. A system has a transfer function given by

$$H(s) = \frac{(s+3)}{(s+0.5+j2.3)(s+0.5-j2.3) \times (s+3.1+j1.4)(s+3.1-j1.4)}$$

Which pair of conjugate poles is likely to have the greatest effect on the system response? What is the natural frequency and damping ratio corresponding to this pair of poles? Based upon Figure 12.31, sketch an approximate step function response. Compare with the result obtained using the **step(b,a)** command in MATLAB.

5. When a voltage $v(t)$ is applied to an electromechanical system the resulting displacement $x(t)$ is governed by the differential equation

$$2\frac{d^3x(t)}{dt^3}+5\frac{d^2x(t)}{dt^2}+2.9\frac{dx(t)}{dt}+1.7x(t) = v(t)+0.4\frac{dv(t)}{dt}$$

Use MATLAB to determine the step function response.

6. A rectangular pulse 15 s long is applied to a filter whose transfer function is given by

$$H(s)=\frac{0.01s^6+0.1217s^4+0.2982s^2+0.2026}{s^6+0.9877s^5+2.2944s^4+1.6068s^3+1.499s^2+0.6169s+0.2222}$$

Use the **lsim(b,a,x,t)** command in MATLAB to plot the output.

7. A signal $x(t)$ has a Laplace transform given by

$$X(s)=\frac{2.2s^4+0.6s^3+1.4s^2+2.4s+0.74}{s^5+2.7s^4+4.2s^3+4.32s^2+1.83s+0.57}$$

Use MATLAB to plot $x(t)$. (Hint: think of $X(s)$ as a transfer function and determine the corresponding impulse response.)

8. A signal $x(t)$ has a Laplace transform given by

$$X(s)=\frac{1-e^{-3s}}{s^5+2.7s^4+4.2s^3+4.32s^2+1.83s+0.57}$$

Use MATLAB to plot $x(t)$. Suggest a likely way in which such a signal would have arisen.

9. As in Problem 5 of Chapter 11, an automatic control system used for adjusting the angle of a telescope in response to an applied voltage has a transfer function given by

$$H(s)=\frac{0.06K(s+0.2)}{s^3+0.6s^2+0.08s+0.06K}$$

where K is a gain parameter which the system designer can adjust. Use MATLAB to plot the step function response for $K=0.02$, $K=0.04$, $K=0.08$ and $K=0.16$. Suggest which of these values is likely to be the most satisfactory.

10. Determine and plot the current in Problem 2 of Chapter 11 using the **lsim(b,a,x,t)** command of MATLAB.

Chapter thirteen

Synthesis of analogue filters

13.1 ● Preview

The purpose of a filter is to enhance a wanted signal relative to unwanted signals, interference or noise. In some radar and sonar applications this is done on the basis of searching the received signal for echoes whose *time waveform* matches that of the transmit pulse. Such a system uses a *matched filter* for which the output may have a very different shape from that of the input signal, but with a greatly enhanced ratio of peak signal to noise. This topic will be covered in Chapter 16. More usually, however, filtering is done on the basis of emphasizing the *frequencies* of the wanted signal relative to those of unwanted signals, while avoiding any distortion in the wanted signal. This chapter will confine itself to this latter situation – namely the enhancement or suppression of selected frequencies. It also confines itself to the case where the filters are analogue.

13.2 Basic approach to filter design

There are several ways of categorizing analogue filters:

- By whether they are low-pass, high-pass, band-pass, or band-stop. (Note that these and many similar terms have become words in their own right so that, in accordance with common practice, they will henceforth be written without the hyphens.)
- By their cutoff frequencies. (A cutoff frequency is defined as that where the gain has changed by some specified amount relative to the mean mid-band gain.)
- By the shape of their amplitude response.
- By the shape of their phase response.
- By whether or not the filter contains active devices with gain, such as operational amplifiers – in other words by whether the filter is active or passive.

Apart from resistances in the terminations, passive filters are usually composed of inductors and capacitors. In contrast, active filters use resistors and capacitors. Many factors must be considered when making the decision whether to design a passive *LC* filter or an active *RC* filter. The main argument against an *LC* filter centres around the size and cost of its inductors, particularly when the filter is to be used at sub-audio frequencies where the inductance values are likely to be large. Another major problem with inductors is that practical *inductors* do not have the properties of ideal *inductances*. They suffer from losses (as described by their finite Q) and can also be degraded sometimes by significant stray capacitance arising from the close proximity of adjacent turns in their winding. In their favour they do not require power supplies, rarely encounter saturation problems when large signals are involved, and are easy to adjust due to the tuning slugs of most inductors. For certain specialist applications very high performance passive filters can be realized by converting the electrical input into a mechanical displacement, and then exploiting *mechanical* resonances; for example piezoelectric crystal, piezoelectric ceramic, surface acoustic wave and electromechanical filters. The resulting mechanical output is then converted back into an electrical signal. As an example, ceramic filters the size of a 16 pin integrated circuit are commercially available for use in the intermediate frequency stages of communications receivers and have extremely impressive frequency-selective properties.

The main argument against active filters centres around the requirement that the operational amplifiers used within them should have a high open-loop gain over the frequency range of the filter. This tends to limit their use to a few MHz, whereas *LC* filters can operate at hundreds of megahertz.

The usual way of designing a filter is to:

1. Select an amplitude response that is known to have desirable properties.
2. Use available sets of tables to determine the network configuration and element values for a lowpass filter of the selected amplitude response, but having a cutoff frequency of 1 rad/s. The resulting filter is termed a *lowpass prototype*.
3. Consider adding a delay equalizer in cascade if the phase characteristic of the frequency selective filter is unsatisfactory for the application (though this is rarely

the case). A delay equalizer is an allpass network having a frequency dependent delay but a constant amplitude response.

4. Transform the lowpass prototype filter (or filter and equalizer if applicable) into a lowpass filter *of the required bandwidth.*
5. If a lowpass filter is not what is wanted, transform this lowpass filter into the required highpass, bandpass or bandstop filter.

In most cases there is also some impedance scaling to be done, since the filter from the design tables, the so-called 'normalized' lowpass filter, typically assumes that the filter is driven from a source with a source impedance of 1 Ω, and that the filter is terminated by 1 Ω. These impedance values are usually inappropriate for a real design.

13.3 Amplitude and delay distortion

If the waveform leaving a network is of identical *shape* to the wanted component of the input signal the network is considered to be distortionless.

An output waveform can be amplitude-scaled and delayed relative to the input waveform and yet have the same shape. Thus, with reference to Figure 13.1, the input signal $f(t)$ is considered to pass undistorted through a network if the output $g(t)$ is related by

$$g(t) = Kf(t - t_0) \tag{13.1}$$

Performing the Fourier transform on Equation (13.1) using the time shift theorem of Equation (6.51) gives

$$G(\omega) = KF(\omega)\exp(-j\omega t_0) \tag{13.2}$$

where, for compactness of notation, it is now preferred to work in terms of radian frequency rather than cyclic frequency.

Since $G(\omega) = F(\omega)H(\omega)$, where $H(\omega)$ is the frequency transfer function of the network, this signifies that

$$H(\omega) = K\exp(-j\omega t_0) \tag{13.3}$$

However, since $\exp(\pm j2n\pi) = 1$, Equation (13.3) should be changed to encompass this possibility, giving

$$H(\omega) = K\exp[j(-\omega t_0 \pm 2n\pi)] \tag{13.4}$$

Clearly this condition is necessary only over the bandwidth of the signal. Hence the required condition for no distortion is that, over the bandwidth of the signal,

$$A(\omega) = K \tag{13.5}$$

$$\theta(\omega) = -\omega t_0 \pm 2n\pi \tag{13.6}$$

The first of these conditions, that the gain should be constant over the signal bandwidth, is obvious. The second is much less so. Equation (13.6) states that there is

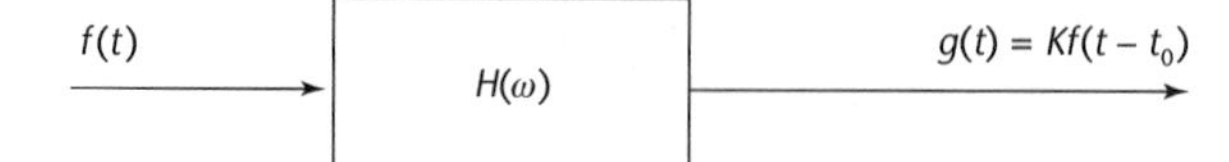

Fig. 13.1 Distortionless transmission

a requirement for a linear relationship between phase and frequency, but also that this straight line law should pass through $\pm 2n\pi$. For example, considering a signal passing through a bandpass filter whose amplitude characteristics are as shown in Figure 13.2(a), a suitable phase response for distortionless transmission would be that shown in Figure 13.2(b).

To test this, consider the signal $f(t) = \cos(50t) + \cos(43t)$ and the case of $A(\omega) = 1$ and $\theta(\omega) = -0.03\omega + 6\pi$. The output is then given by

$$g(t) = \cos(50t - 50 \times 0.03 + 6\pi) + \cos(43t - 43 \times 0.03 + 6\pi)$$

In Figure 13.3 the solid line shows $f(t)$ while the dotted line shows $g(t)$. The two waveforms are seen to be identical except for a delay between them. The physical explanation for this similarity is that the phase component of -0.03ω in $\theta(\omega)$ signifies an identical delay of 0.03 s for the two components and the phase component of term 6π in $\theta(\omega)$ signifies that each sinusoidal component is shifted by an integral number of cycles, thus leaving it unchanged.

If the phase response of the network is given by

$$\theta(\omega) = -\omega t_0 + \alpha, \qquad \text{where} \quad \alpha \neq \pm 2n\pi \tag{13.7}$$

we encounter what is known as *intercept* distortion. The phenomenon of intercept distortion will be clarified by two examples.

Consider again the signal $f(t) = \cos(50t) + \cos(43t)$, but this time the case of $A(\omega) = 1$ and $\theta(\omega) = -0.03\omega + 6.8\pi$. The output is then given by

$$g(t) = \cos(50t - 50 \times 0.03 + 6.8\pi) + \cos(43t - 43 \times 0.03 + 6.8\pi)$$

The solid line in Figure 13.4 shows $f(t)$ while the dotted line shows the new waveform $g(t)$. Now the two waveforms are dissimilar, demonstrating that intercept distortion is present.

The phase shift of lowpass filters is always zero at zero frequency and hence lowpass filters never encounter any problem with intercept distortion. The only requirement is a linear phase shift.

For bandpass filters intercept distortion is often present *but can usually be tolerated.* This is because the most common application of a bandpass filter is that of extracting a *modulated* signal from a background of unwanted signals, and it turns

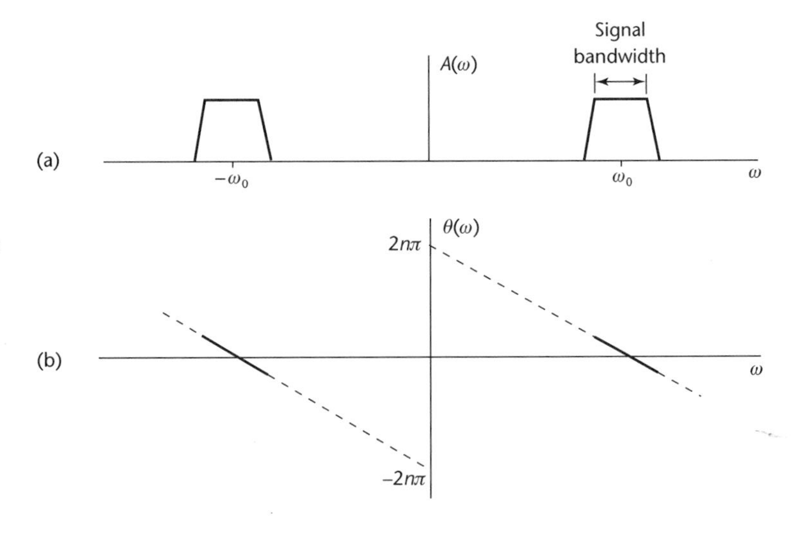

Fig. 13.2 Conditions for distortionless transmission of a bandpass signal: (a) amplitude response; (b) phase response

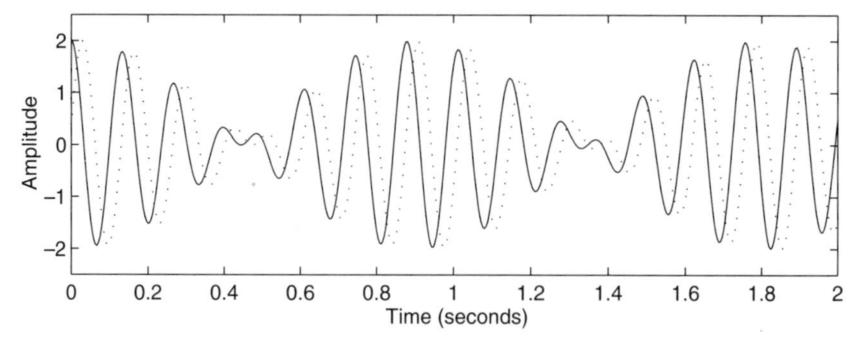

Fig. 13.3 Input and output signals with distortionless phase response

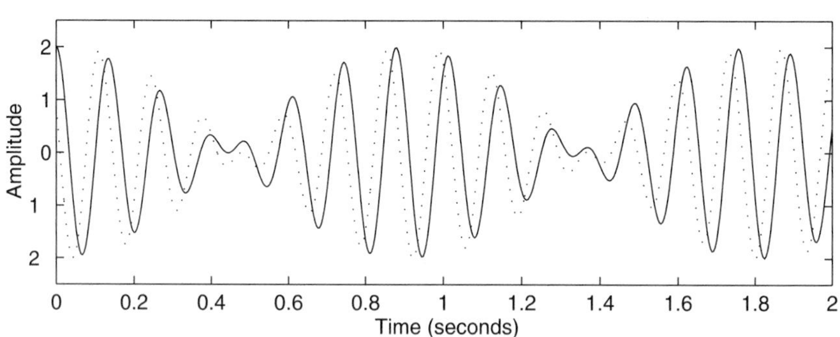

Fig. 13.4 Input and output signals with linear phase response but showing intercept distortion

out that the modulation itself is unaffected by intercept distortion. As an example, consider the situation in which a carrier of frequency ω_0 is amplitude modulated by a cosinusoid $\cos pt$. We have

$$f(t) = (1 + m \cos pt) \cos \omega_0 t \tag{13.8}$$

where m is the modulation index. An example of such a waveform is shown in Figure 13.5, where $(1 + m \cos pt)$ is the envelope of the carrier $\cos \omega_0 t$.

Using the trigonometric identity $\cos A \cos B = \frac{1}{2} \cos(A - B) + \frac{1}{2} \cos(A + B)$, we can express $f(t)$ as a carrier and two sidebands, that is

$$f(t) = (1 + m \cos pt) \cos \omega_0 t = \cos \omega_0 t + \frac{m}{2} \cos(\omega_0 + p)t + \frac{m}{2} \cos(\omega_0 - p)t \tag{13.9}$$

Passing this through a network for which $\theta(\omega) = -\omega t_0 + \alpha$ the output is given by

$$g(t) = \cos(\omega_0 t - \omega_0 t_0 + \alpha) + \frac{m}{2} \cos[(\omega_0 + p)t - (\omega_0 + p)t_0 + \alpha] + \frac{m}{2} \cos[(\omega_0 - p)t - (\omega_0 - p)t_0 + \alpha]$$

However, a reverse use of the same trigonometric identity reduces this to

$$g(t) = [1 + m \cos(pt - pt_0)] \cos[\omega_0 t - \omega_0 t_0 + \alpha] \tag{13.10}$$

A comparison of Equations (13.8) and (13.10) shows that the envelope is shifted relative to the carrier and that $g(t)$ is therefore different from $f(t)$. *Usually, however, this shift between envelope and carrier does not matter and, so long as this is the case, the intercept distortion is unimportant.*

Extending this analysis to amplitude modulation by the sum of several cosinusoids, the same procedure shows that a signal

$$f(t) = (1 + m_1 \cos p_1 t + m_2 \cos p_2 t + \cdots) \cos \omega_0 t \tag{13.11}$$

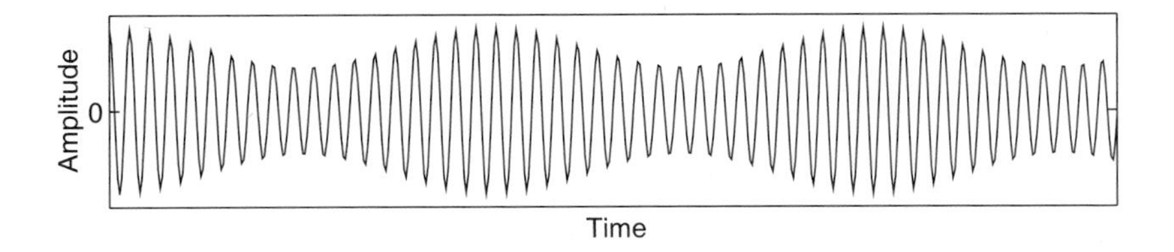

Fig. 13.5 Example of amplitude-modulated signal

will be changed by the network to

$$g(t) = [1 + m_1 \cos(p_1 t - p_1 t_0) + m_2 \cos(p_2 t - p_2 t_0) + \cdots] \cos[\omega_0 t - \omega_0 t_0 + \alpha]$$

or

$$g(t) = \{1 + m_1 \cos[p_1(t - t_0)] + m_2 \cos[p_2(t - t_0)] + \cdots\} \cos[\omega_0 t - \omega_0 t_0 + \alpha]$$

Relative to the input the envelope is delayed by t_0. This delay is known as the *envelope* or *group* delay. Relating this to the network phase response, $\theta(\omega) = -\omega t_0 + \alpha$, it is seen that this group delay is given by

$$t_g = -\frac{d\theta}{d\omega} \tag{13.12}$$

The example has been given of a simple amplitude-modulated signal but the result can be extended to any type of modulated signal. *The modulation of a signal is undistorted by the phase response if the group delay is constant across the signal bandwidth.* For modulated signals intercept distortion is usually acceptable and a sufficient condition for distortionless transmission is usually taken to be that the group delay $-d\theta/d\omega$ is constant.

Although introduced in the context of bandpass networks, constant group delay is commonly applied to lowpass networks as the criterion for no distortion. This is perfectly acceptable since the phase shift is always zero at zero frequency, and the condition of constant group delay therefore signifies a truly distortionless system having a linear phase shift passing through the origin.

When the group delay is not constant across the signal bandwidth, different frequency components will undergo different delays and distortion will result.

13.4 Selecting a frequency response

The amplitude response of a general lowpass filter is shown in Figure 13.6. Here the passband is the region of almost constant gain extending up to the cutoff frequency, where the cutoff frequency is the frequency at which the gain has dropped from the

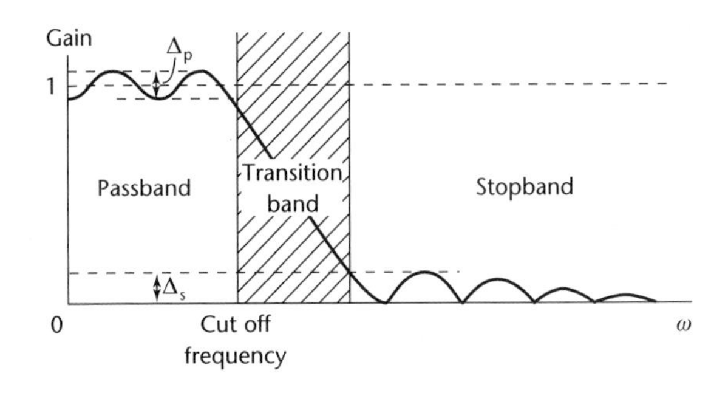

Fig. 13.6 Ideal brickwall amplitude response

mean passband gain by some specified amount. Above the cutoff frequency is the transition band, where we want a rapid decrease in gain. This 'roll-off rate' is normally expressed in dB/decade. Finally, we reach the stopband, which is the region of very low gain. As shown in Figure 13.6, there may be ripple in the pass band and stopband. Assuming that the mean passband gain is unity this is defined as $-20\log_{10}(1-\Delta)$ dB, where Δ is the deviation of the gain. For example, the passband (or stopband) ripple might be $-20\log_{10}(1-0.05)=0.446$ dB.

The ideal lowpass filter is a 'brickwall' filter but, as shown in Example 6.3, this has an impulse response with infinite tails. Even if a linear phase response causes this response to be delayed relative to the input impulse function, as shown in Figure 13.7, these tails extend into the time region *before* the impulse (making it *non-causal*), thus demonstrating that this ideal filter cannot be physically realized. It is not possible to obtain an ideal brickwall filter response and it is necessary to accept some degree of approximation to it.

There are many frequency responses that have been studied, but the best known of these are the Butterworth, Chebyshev and elliptic (or Cauer) function responses. In their lowpass form these are intended to give a near constant attenuation in most of the band up to the specified cutoff frequency, followed by a rapid increase in attenuation above this frequency.

Another well-known response is that of the Bessel–Thomson filter, since this gives a much more constant delay throughout the passband. Unfortunately, the slope of its attenuation above cutoff is significantly less than that of a Butterworth, Chebyshev or elliptic filter of comparable complexity. For this reason a more usual alternative to the Bessel–Thomson filter, on the somewhat rare occasions when delay distortion is important, is to cascade one of these other filters with a delay equalizer.

The Butterworth magnitude response for a normalized lowpass filter having a 1 rad/s cutoff frequency is given by

$$|H(\omega)|^2=\frac{1}{1+\omega^{2n}} \tag{13.13}$$

where n is the *order* of the filter. In the case of a passive Butterworth filter n also corresponds to the number of reactive elements.

The amplitude response of Equation (13.13) is reduced to half its zero-frequency value when $\omega^{2n}=1$ and hence when $\omega=1$. Thus, whatever the value of n, the amplitude response is down by 3 dB at a frequency of 1 rad/s. For $\omega\gg1$ the response approximates to $1/\omega^{2n}$ such that, taking the logarithm of both sides

$$10\log_{10}|H(\omega)|^2=-20n\log_{10}\omega \qquad \text{for} \quad \omega\gg1 \tag{13.14}$$

This signifies that the slope of the high-frequency asymptote (also known as the roll-off or fall-off) is $20n$ dB/decade, or $6n$ dB/octave, where a decade represents a 10 : 1 range of frequencies, and an octave a 2 : 1 range of frequencies.

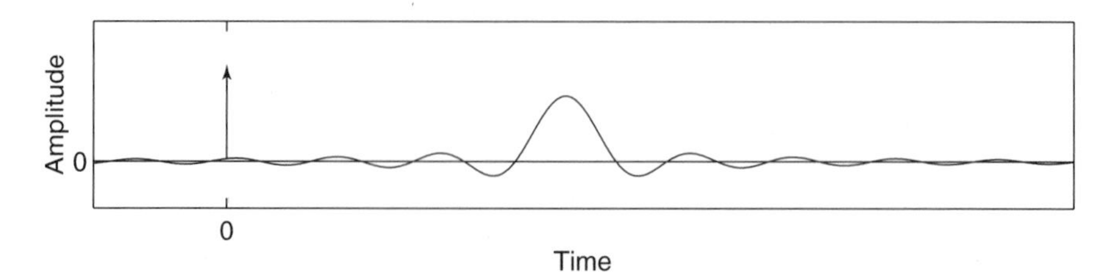

Fig. 13.7 Input impulse and the impulse response of a brickwall filter with linear phase

The main merits of these Butterworth responses are their flatness in the passband. It can be shown by successive differentiation that all derivatives of $|H(\omega)|^2$ with respect to ω are zero at zero frequency and, because of this, the response is known as *maximally flat*.

If the order of a Butterworth filter is made large the filter takes on a sharp rate of cutoff, thus becoming a closer approximation to the brickwall response. Unfortunately the phase response becomes less satisfactory as the order of the filter is increased. The Butterworth filter does, however, have a more satisfactory phase response than many of the alternatives.

The Chebyshev response permits a small amount of ripple in the passband and, in return, gives a sharper cutoff than the Butterworth response for the same number of filter elements. The Chebyshev response is

$$|H(\omega)|^2 = \frac{1}{1+\epsilon^2 C_n^2(\omega)} \tag{13.15}$$

where ϵ is a constant and

$$C_n(\omega) = \cos(n \cos^{-1} \omega) \qquad \text{for} \quad |\omega| \leqslant 1 \tag{13.16}$$

This function $C_n(\omega)$ can be expressed as a polynomial of degree n such that, as with the Butterworth filter, n gives the order of the filter. $C_n(\omega)$ is known as the Chebyshev polynomial.

The Chebyshev polynomial has special properties such that $|H(\omega)|^2$ oscillates between 1 and $1/(1+\epsilon^2)$ within the passband, equalling $1/(1+\epsilon^2)$ at a frequency of 1 rad/s. The designer specifies what amount of ripple is acceptable in the passband, bearing in mind that the greater the allowed ripple the sharper will be the cutoff for a given order of filter. The attenuation in the stopband of the Chebyshev response increases monotonically with increasing frequency.

In the elliptic function (or Cauer) filter, ripple is allowed in the stopband as well as in the passband. The consequence is that, for a given order of filter, the roll-off rate of the filter becomes even greater than that of the Chebyshev filter. The designer specifies the permissible ripple within the passband (that is, for the band of ω values between 0 and 1 rad/s) and also the minimum attenuation in the stopband.

13.5 Transfer functions and frequency responses of analogue filters using MATLAB

MATLAB can be used to determine the transfer function of a Butterworth, Chebyshev or elliptic filter of specified order, and then to plot the resulting frequency response. The frequency response can then be examined to see if it meets the requirement. For some applications it can also be useful to plot the impulse response or step response. Examples of the appropriate commands are presented in this section for these three types of filter. By maintaining the same arbitrary order of filter (order 4 in all three cases) the differences between these filters become apparent. The section constrains itself to lowpass filters with a cutoff frequency of 1 rad/s but the commands can be adapted to other cutoff frequencies and to highpass, bandpass or bandstop filters. Beginning with a Butterworth filter the first command is

[b,a] = butter(4,1,'s')

where **4** specifies the order of the filter, **1** specifies the cutoff frequency of the filter in rad/s, and **s** specifies that we are concerned with an *analogue* filter and its s domain transfer function. The command returns a row vector **b** giving the coefficients in the numerator of the transfer function, and a row vector **a** giving the coefficients in the denominator of the transfer function. The printout resulting from the command is

```
b =
     0     0.0000     0.0000     0     1.0000

a =
     1.0000   2.6131   3.4142   2.6131   1.0000
```

This tells us that the necessary transfer function is given by

$$H(s) = \frac{1}{s^4 + 2.6131s^3 + 3.4142s^2 + 2.6131s + 1}$$

In order to plot the frequency response of this transfer function we can either use the command **freqs(b,a)** or **bode(b,a)**, where **b** and **a** are row vectors giving the coefficients of the numerator and denominator polynomials, but do not need to be entered since we have already derived them.

The outcome of **bode(b,a)** is shown in Figure 13.8.

For the Chebyshev filter the necessary commands are

[b,a] = cheby1(4,.9,1,'s')
bode(b,a)

where **4**, **1** and **s** are as before but **.9** is added to specify the allowable passband ripple in dB. Note also that the cutoff frequency of 1 rad/s now refers to the frequency where the gain is reduced from the zero frequency gain by the specified ripple of 0.9 dB; it does not refer to the −3 dB frequency. **cheby1** signifies that we are

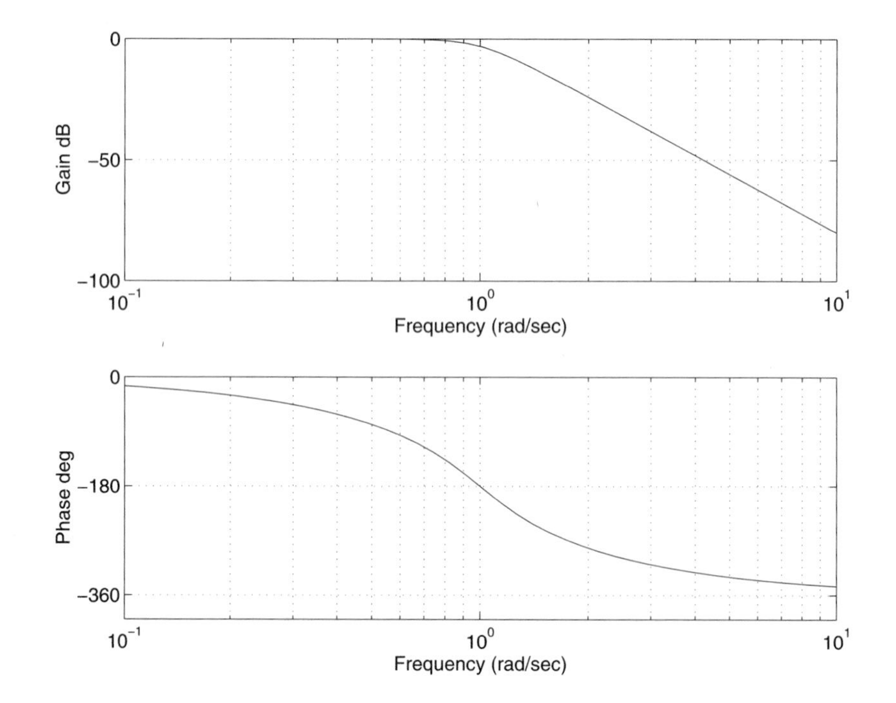

Fig. 13.8 Frequency response of fourth-order Butterworth filter

undertaking a Chebyshev type I filter design in which the *passband* ripple is controlled (and contrasts with **cheby2** which is for a type II design where the *stopband* ripple is controlled). The outcome is shown in Figure 13.9.

For the elliptic function filter the necessary commands are

```
[b,a]=ellip(4,.9,50,1,'s')
bode(b,a)
```

where **4**, **.9**, **1** and **s** are as for the Chebyshev filter but **50** is added to specify that, after the initial monotonic increase of attenuation in the stopband to 50 dB, stopband ripple is then acceptable so long as the attenuation never falls below 50 dB. The outcome is shown in Figure 13.10.

If the amplitude responses of Figures 13.8, 13.9 and 13.10 are compared it is clear that the rate of descent into the stopband increases progressively. It should also be noted that the passband phase responses are remarkably similar. In view of these observations it is not surprising that elliptic filters are probably the most widely used.

If we require the step or impulse responses of any of these filters we can obtain these using the **step(b,a)** or **impulse(b,a)** commands described in Chapter 12.

With a small modification of the commands so far presented MATLAB can be used very effectively for deriving the transfer functions and plotting the frequency responses of highpass, bandpass and bandstop filters of specified cutoff frequencies. As an example, consider the need for a bandpass filter with a passband from 1 kHz to 1.2 kHz. Our investigation might begin with one based on a third-order Butterworth lowpass filter. To determine the resulting frequency response the following commands are needed

```
[b,a]=butter(3,[1000*2*pi 1200*2*pi],'s')
bode(b,a)
```

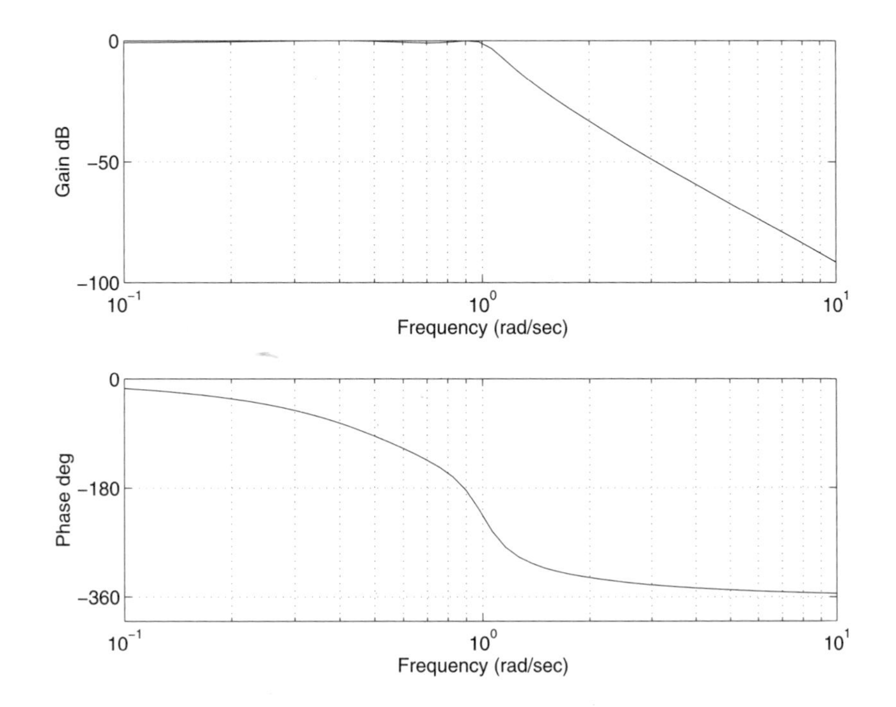

Fig. 13.9 Frequency response of fourth-order Chebyshev filter with 0.9 dB ripple

where the only difference from the commands leading to Figure 3.8 is that the order of the prototype filter is reduced from 4 to 3 and the cutoff frequency of 1 rad/s is now specified by a vector of the lower and upper cutoff frequencies in rad/s. The outcome of these commands is shown in Figure 13.11 and the designer can decide whether the filter is suitable for the task in hand. If not the order or type of response can be

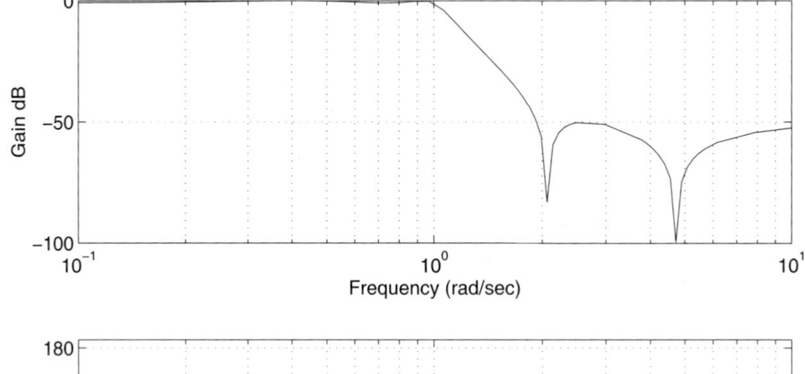

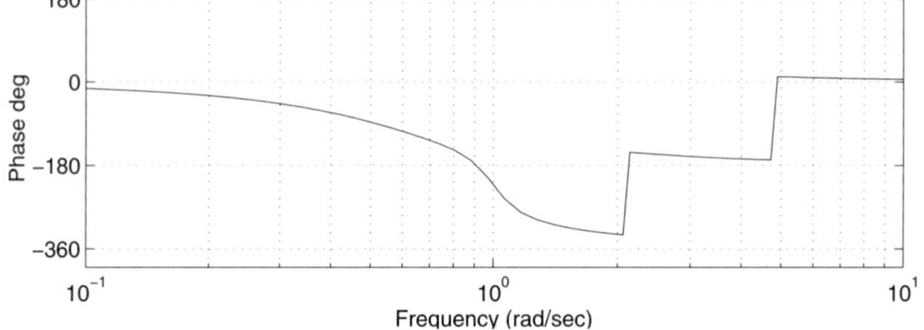

Fig. 13.10 Frequency response of fourth-order elliptic filter with 0.9 dB ripple and 50 dB stopband attenuation

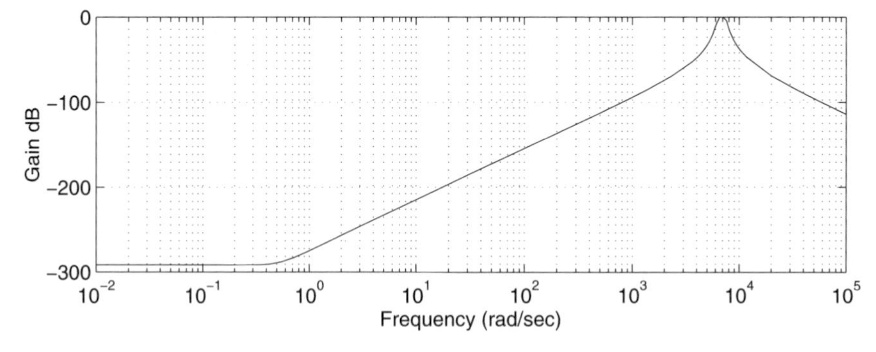

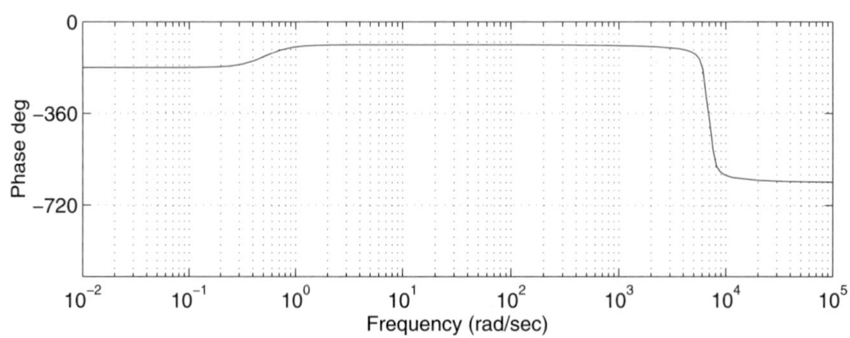

Fig. 13.11 Frequency response of bandpass filter based on third-order Butterworth lowpass response

changed. Similar commands are also available for highpass and bandstop filters. Note in Figure 13.11 that the phase shift and the gain at the low frequencies where the gain is exceedingly small should not be trusted; they are likely to have suffered from computational errors.

13.6 Obtaining the transfer function corresponding to a desired frequency response

It has been shown in the previous section how MATLAB can produce transfer functions corresponding to well-known types of frequency response. This section outlines the principles which make this possible. These are not essential to filter design and the reader may well decide it appropriate to omit this section and to proceed to the more practical sections that follow.

The technique is centred around the result (to be proved) that, if we replace ω in the required expression for $|H(\omega)|^2$ by s/j, we obtain $H(s)H(-s)$, that is

$$|H(\omega)|^2_{\omega=s/\mathrm{j}} = H(s)H(-s) \tag{13.17}$$

Then, since the poles of a stable filter with transfer function $H(s)$ lie entirely in the left-hand half of the s-plane, and those of $H(-s)$ therefore lie entirely in the right-hand half of the s-plane, it becomes obvious which poles of $H(s)H(-s)$ belong to $H(s)$. From this it is possible to derive an expression for $H(s)$.

The key to this procedure is the validity of Equation (13.17). It is useful to prove this for a simple second-order network. The arguments can readily be extended to all networks.

Example 13.1

An all-pole network (one with no zeros) has one set of conjugate poles such that

$$H(s) = \frac{1}{(s-p_1)(s-p_1{}^*)}$$

Demonstrate that $|H(\omega)|^2$ is obtained by replacing s by $\mathrm{j}\omega$ in $H(s)H(-s)$. Hence deduce that $H(s)H(-s)$ is obtained by replacing ω by s/j in $|H(\omega)|^2$.

Solution If

$$H(s) = \frac{1}{(s-p_1)(s-p_1{}^*)}$$

then $H(-s)$ is given by

$$H(-s) = \frac{1}{(-s-p_1)(-s-p_1{}^*)} = \frac{1}{(s+p_1)(s+p_1{}^*)}$$

It will be noted that, since the two conjugate poles of $H(s)$ must be in the left-hand half of the s-plane for $H(s)$ to be physically realizable, the two poles of $H(-s)$ are in the right-hand half of the s-plane and $H(-s)$ does not therefore represent a physically realizable transfer function. However, this does not affect the derivation. We have

$$H(s)H(-s) = \frac{1}{(s-p_1)(s-p_1{}^*)(s+p_1)(s+p_1{}^*)}$$

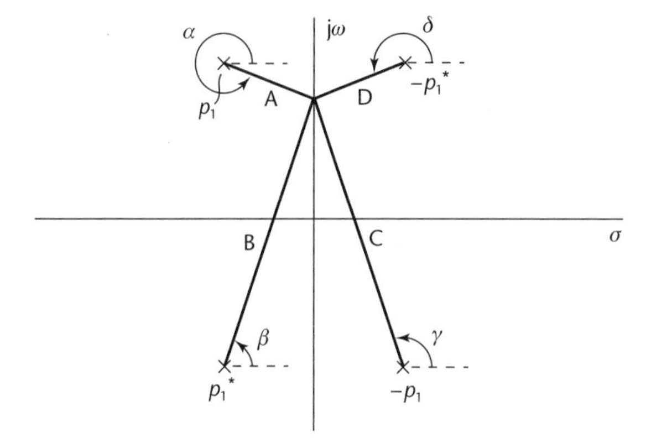

Fig. 13.12 Pole plot of $H(s)H(-s)$

Substituting $s = j\omega$ we obtain

$$\{H(s)H(-s)\}_{s=j\omega} = \frac{1}{(j\omega - p_1)(j\omega - p_1^*)(j\omega + p_1)(j\omega + p_1^*)}$$

However, with reference to Figure 13.12, this may be expressed as

$$\{H(s)H(-s)\}_{s=j\omega} = \frac{1}{Ae^{j\alpha}Be^{j\beta}Ce^{j\gamma}De^{j\delta}}$$

But we see from Figure 13.12 that

$$D = A,\ C = B,\ \alpha + \delta = 3\pi,\ \beta + \gamma = \pi$$

Hence

$$\{H(s)H(-s)\}_{s=j\omega} = \frac{1}{A^2B^2}$$

However, using the geometrical method of interpreting pole–zero plots for obtaining the frequency response that was presented in Section 10.6 we see that this last ratio equals $|H(\omega)|^2$. Hence we have shown that the replacement of s by $j\omega$ in $H(s)H(-s)$ gives $|H(\omega)|^2$. Conversely therefore the replacement of ω by s/j in $|H(\omega)|^2$ gives $H(s)H(-s)$, that is

$$|H(\omega)|^2_{\omega = s/j} = H(s)H(-s)$$

The above example confirms Equation (13.17) for the specific case of an all-pole transfer function. The key point is that $H(s)$ and $H(-s)$ have poles which are diametrically opposite but that, because poles are always real or occur in conjugate pairs, the poles of $H(-s)$ are the mirror image about the imaginary axis of the poles of $H(s)$. Based upon this the reasoning may be extended to any network to confirm the general applicability of Equation (13.17).

The next task is to show how Equation (13.17) can be applied to determine the transfer function corresponding to a required frequency response.

Example 13.2

From Equation (13.13) the amplitude response of a third-order Butterworth filter is

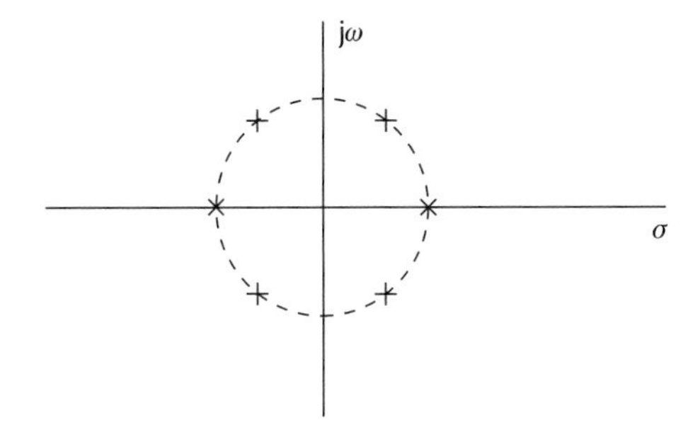

Fig. 13.13 Poles of $H(s)H(-s)$ for a third-order Butterworth filter

given by

$$|H(\omega)|^2 = \frac{1}{1+\omega^6}$$

Determine its transfer function.

Solution In accordance with Equation (13.17) ω is replaced by s/j in $|H(\omega)|^2$ to give

$$H(s)H(-s) = \frac{1}{1+(s/\mathrm{j})^6}$$

To find the poles we equate the denominator to zero, that is $1+(s/\mathrm{j})^6 = 0$, or $s^6 = -\mathrm{j}^6 = 1$. However, unity is also represented by $\mathrm{e}^{\mathrm{j}2\pi k}$, where k is an integer. Hence

$$s = (\mathrm{e}^{\mathrm{j}2\pi k})^{1/6}$$

From this we see that s has unity magnitude and an angle θ_k given by

$$\theta_k = \frac{2\pi k}{6} = 0,\ \pi/3,\ 2\pi/3,\ \pi,\ 4\pi/3,\ 5\pi/3$$

as is shown in Figure 13.13. The poles lie equispaced around a unit circle, and the half that lie in the left half-plane belong to $H(s)$; that is

$$H(s) = \frac{1}{(s-1\angle 180°)(s-1\angle 120°)(s-1\angle 240°)}$$

Multiplying this out gives

$$H(s) = \frac{1}{s^3+2s^2+2s+1} \qquad (13.18)$$

This then is the transfer function required for a third-order Butterworth filter with a cutoff frequency of 1 rad/s.

The principles of Example 13.2 can be applied to find the transfer functions corresponding to other filter responses, say Chebyshev, elliptic or Bessel. For these the poles do not simply lie on such a simple contour as the unit circle in the s-plane.

13.7 Translating a transfer function into a filter design

The usual configuration for a passive LC filter is the ladder network. The number of frequency-dependent elements required equals the order of the filter and, continuing

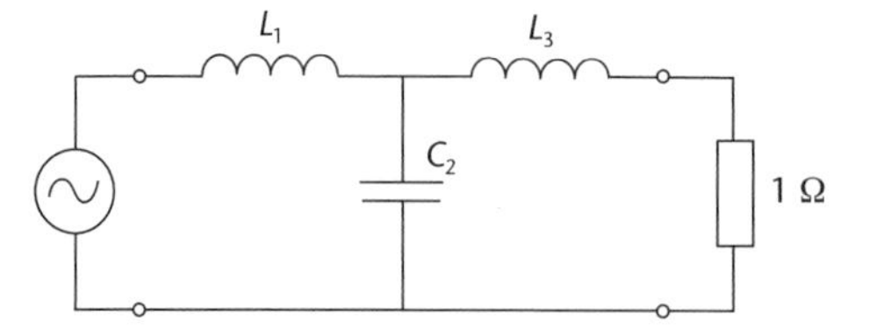

Fig. 13.14 Possible configuration of third-order Butterworth filter

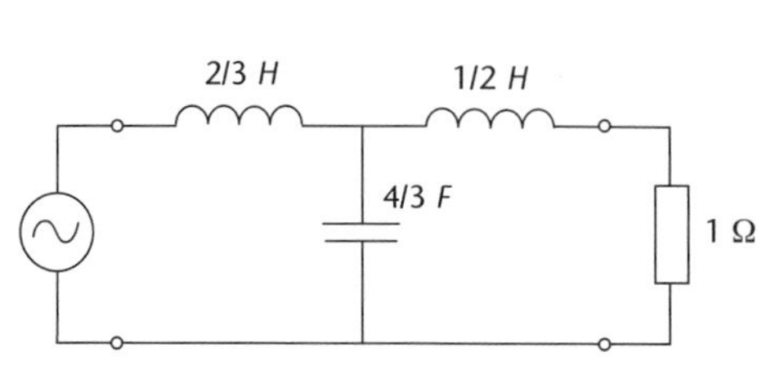

Fig. 13.15 Third-order Butterworth filter

with the example of a third-order Butterworth filter, the network of Figure 13.14 is appropriate. The filter is driven by a voltage source and terminated by a 1 Ω resistive load.

Network analysis shows that

$$H(s) = \frac{1}{s^3 L_1 L_3 C_2 + s^2 L_1 C_2 + s(L_1 + L_3) + 1} \tag{13.19}$$

By comparing this with Equation (13.18) we see that $L_1 L_3 C_2 = 1$, $L_1 C_2 = 2$, $L_1 + L_3 = 2$. This gives $L_1 = \frac{3}{2} H$, $C_2 = \frac{4}{3} F$, $L_3 = \frac{1}{2} H$. Thus the network of Figure 13.15 is a Butterworth normalized lowpass filter.

Once the network configuration is decided, component values may be determined in the same way for other orders of filter and for other types of response.

13.8 Design of passive lowpass filters using tables

Tables are available giving the network configuration and component values for the normalized lowpass filter of each response type (see, for example Geffe (1963); Williams and Taylor (1988)). For this reason it is not necessary or advisable to synthesize filters from first principles using the techniques of the previous two sections.

An integral part of any filter is its terminations and the *LC* ladder network may be used in any of the following ways:

- between a source with an output impedance of 1 Ω and a load with an impedance of 1 Ω;
- between a source with an output impedance of 1 Ω and a load with an infinite impedance;

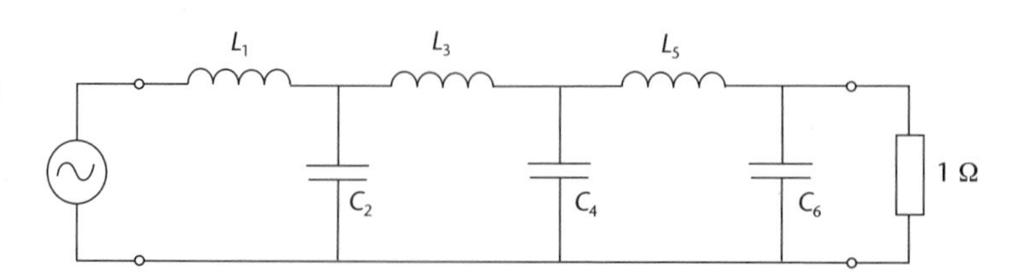

Fig. 13.16 Filter configuration

Table 13.1 Component values for Butterworth filter driven by a voltage source and terminated in 1 Ω

n	L_1	C_2	L_3	C_4	L_5	C_6
3	1.500	1.333	0.500			
4	1.531	1.577	1.802	0.383		
5	1.545	1.694	1.382	0.894	0.309	
6	1.533	1.759	1.553	1.202	0.758	0.259

Table 13.2 Component values for Chebyshev filter with 0.9 dB ripple driven by a voltage source and terminated in 1 Ω

n	L_1	C_2	L_3	C_4	L_5	C_6
3	1.477	1.332	0.977			
4	1.288	1.879	1.420	1.011		
5	1.640	1.602	1.960	1.445	1.029	
6	1.357	2.023	1.665	1.992	1.464	1.039

- between a voltage source and a load with an impedance of 1 Ω.

What follows applies only to the last of these cases and for ideal lossless reactive components in the configuration shown in Figure 13.16, where the voltage source might well be provided by the output of an operational amplifier circuit.

Tables 13.1 and 13.2 give the component values for Butterworth filters and for Chebyshev filters with 0.9 dB of ripple in the passband. More extensive tables, such as those by Geffe, include alternative terminations, higher order filters and other response types. Geffe also presents tables of component values for situations where the components are lossy. The values given in Tables 13.1 and 13.2 should give satisfactory frequency responses so long as the inductors have Qs exceeding about 150. Such Qs are perfectly realistic, but it should be noted that the conditions of component dissipation become more stringent as the order of the filter rises.

13.9 Impedance scaling

The normalized lowpass filter is based upon 1 Ω terminations. If it is wished to change these terminations to R Ω, it is necessary only to change all other impedances in the same proportion; that is, to multiply them by R. Since the impedance of an inductance is proportional to the inductance, this means multiplying the normalized inductance values by R. Since the impedance of a capacitance is *inversely* proportional to the capacitance this requires the normalized capacitance values to be *divided* by R.

13.10 Frequency scaling of lowpass filters

The impedance-scaled lowpass filter has a cutoff frequency of 1 rad/s. If we want a lowpass filter with a cutoff frequency of B Hz we need the reactances in the new filter at this new frequency to be the same as the reactances in the lowpass filter at 1 rad/s. This arises because, if the impedances are the same, the gain must be the same. Considering an inductance of value L_k in the normalized (or prototype) filter, and one of value L'_k in the new filter, we require the reactance ωL to be the same in both cases; that is, $2\pi BL'_k = L_k$, or

$$L'_k = \frac{L_k}{2\pi B} \tag{13.20}$$

By a similar argument a new capacitance C'_m is related to the capacitance C_m in the prototype by $1/(2\pi BC'_m) = 1/C_m$, or

$$C'_m = \frac{C_m}{2\pi B} \tag{13.21}$$

13.11 Lowpass to bandpass transformation

Suppose we require a *bandpass* filter that has a bandwidth of B Hz centred about a frequency f_0. The first step is to evaluate the component values that are needed for a *lowpass* filter of bandwidth B Hz.

A capacitance is then placed *in series* with each resulting inductance, of a value that gives a series resonance at f_0, and an inductance is placed *in parallel* with each capacitance of the lowpass filter, of a value that gives a parallel resonance at f_0.

To see the effect of this let us compare the reactance of the original inductance L'_k with the reactance of the series resonant circuit that replaces it. The capacitance C'_k that is put in series with L'_k is given by the relationship $2\pi f_0 L'_k = 1/2\pi f_0 C'_k$. Hence

$$C'_k = \frac{1}{(2\pi f_0)^2 L'_k} \tag{13.22}$$

The reactance of the series combination of L'_k and C'_k at some frequency y Hz is $2\pi y L'_k - (1/2\pi y C'_k)$. However, this is the same as the reactance of the inductance L'_k of the scaled lowpass filter at some frequency x Hz so long as we satisfy

$$2\pi y L'_k - \frac{1}{2\pi y C'_k} = 2\pi x L'_k$$

Eliminating C'_k by use of Equation (13.22) gives

$$2\pi y L'_k - \frac{1}{2\pi y}(2\pi f_0)^2 L'_k = 2\pi x L'_k$$

or, dividing through by $2\pi L'_k$

$$y - \frac{f_0^2}{y} = x \tag{13.23}$$

We find an identical relationship when we replace the original capacitance C'_m with a parallel resonant circuit. Here the inductance L'_m that is put in parallel with C'_m is given by $2\pi f_0 L'_m = (1/2\pi f_0)^2 C'_m$. Hence

$$L'_m = \frac{1}{(2\pi f_0)^2 C'_m} \tag{13.24}$$

This time it is easiest to prove Equation (13.23) by working with the reciprocal of reactance, the susceptance. The susceptance of the parallel resonant circuit at y Hz is

$2\pi y C'_{\mathrm{m}} - (1/2\pi y L'_{\mathrm{m}})$. The susceptance of C'_{m} in the lowpass filter at a frequency of x Hz is $2\pi x C'_{\mathrm{m}}$, and is the same as the susceptance of the parallel combination at y Hz if

$$2\pi y C'_{\mathrm{m}} - \frac{1}{2\pi y L'_{\mathrm{m}}} = 2\pi x C'_{\mathrm{m}}$$

Eliminating L'_{m} by use of Equation (13.24) gives

$$2\pi y C'_{\mathrm{m}} - \frac{1}{2\pi y}(2\pi f_0)^2 C'_{\mathrm{m}} = 2\pi x C'_{\mathrm{m}}$$

and, dividing through by $2\pi C'_{\mathrm{m}}$, we again obtain Equation (13.23).

The significance of this is that the impedances in the lowpass filter components at frequency x Hz are the same as those of their bandpass filter replacement at y Hz if x and y are related by Equation (13.23). It follows that the *gain* (or *attenuation*) of the lowpass filter at x Hz is the same as that of their bandpass filter replacement at y Hz if x and y are related by Equation (13.23). We can therefore perform the frequency mapping illustrated by Figure 13.17. Here the left-hand section shows the gain of the lowpass filter in a form where the plot is rotated 90° anticlockwise from what is normally encountered, such that its frequency axis is parallel with the x-axis. The central section displays the relationship between x and y given by Equation (13.23). Each point on the lowpass response is projected across to this $x = f(y)$ curve and down to the new frequency where a point is plotted of the same gain as that of the original lowpass filter. In other words, the mapping is such that the new filter has the same gain at y Hz as the lowpass filter at x Hz. Figure 13.17 confirms that the new filter is indeed a bandpass filter.

It is relevant to compare the bandwidth of the bandpass filter with the bandwidth of the lowpass filter from which it is derived. Let the lowpass filter have an attenuation of 3 dB at a frequency x_1. Then, using Equation (13.23), the bandpass filter has an attenuation of 3 dB at a frequency y_1, where

$$y_1 - \frac{f_0^2}{y_1} = x_1 \tag{13.25}$$

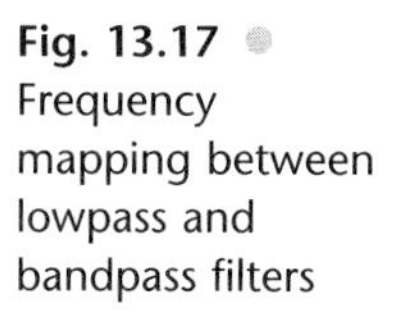

Fig. 13.17 Frequency mapping between lowpass and bandpass filters

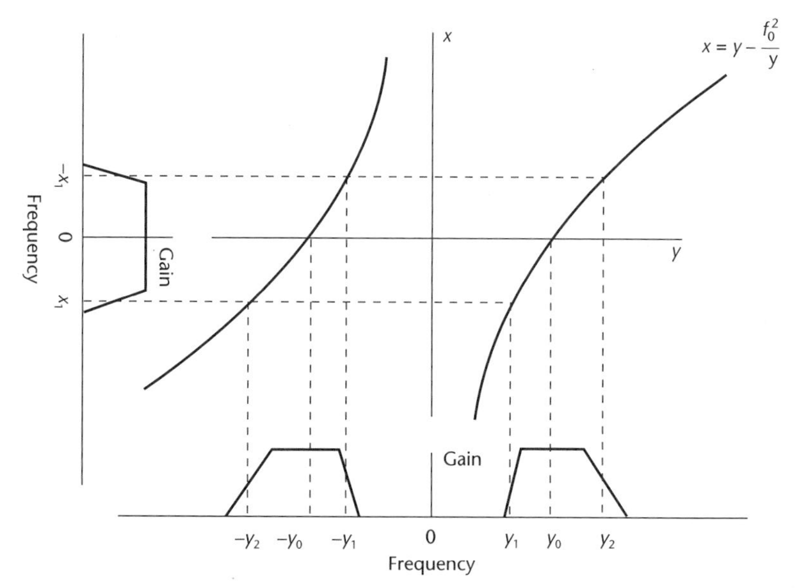

However, the lowpass filter also has the attenuation of 3 dB at the negative frequency $-x_1$ and it is this which gives the lower cutoff frequency of the bandpass filter. Therefore the bandpass filter also has an attenuation of 3 dB at the frequency y_2, where

$$y_2 - \frac{f_0^2}{y_2} = -x_1 \tag{13.26}$$

To find the relationship between the cutoff frequencies of the bandpass filter and its centre frequency we add Equations (13.25) and (13.26) to eliminate x_1. This gives

$$y_1 - \frac{f_0^2}{y_1} = -y_2 + \frac{f_0^2}{y_2}$$

which reduces to

$$f_0 = \sqrt{y_1 y_2} \tag{13.27}$$

It has emerged that f_0 is the *geometric* mean of the the two cutoff frequencies and not their arithmetic mean.

To find the 3 dB bandwidth $(y_1 - y_2)$ of the bandpass filter we eliminate f_0 from Equation (13.25) using Equation (13.27). This gives

$$y_1 - y_2 = x_1 \tag{13.28}$$

Thus we find that the bandwidth of the bandpass filter is the same as that of the lowpass filter.

It is relevant to examine the relationship between the roll-off of a bandpass filter and the roll-off of the lowpass filter from which it is derived. Differentiating the frequency mapping law of Equation (13.23) with respect to y we obtain

$$\frac{\mathrm{d}x}{\mathrm{d}y} = 1 + \frac{f_0^2}{y^2} \tag{13.29}$$

At the centre frequency f_0 where $y = f_0$ this slope equals 2.

Considering the quite common situation where the fractional bandwidth of the bandpass filter is small, the skirts of this filter will be close to this region where the slope of the transformational relationship is 2. It follows from Figure 13.17 that the bandpass filter has the same gain change over c Hz as the lowpass filter does over $2c$ Hz. This means that, in units of dB/Hz, the roll-off of the bandpass filter is twice that of the lowpass filter. However, because of the way in which the lowpass response is translated up in frequency to produce the bandpass response, the *fractional* change in frequency for a given change in gain is usually very much greater for the bandpass filter than for the lowpass filter.

In practice, roll-off is usually expressed in units of dB/decade rather than dB/Hz and, since a frequency change of x Hz about some high frequency represents less in decades than the difference in decades between zero and x Hz, the roll-off of a bandpass filter in dB/decade *depends on its centre frequency*. If the centre frequency of a bandpass filter is much greater than the bandwidth, this roll-off in dB/decade *can be exceedingly high*, very much higher for example than $20n$ dB/decade of a Butterworth filter. It is largely for this reason that bandpass filters are rarely realized by cascading lowpass and highpass filters.

It should be emphasized that, as done in Figure 13.10, MATLAB provides a simple and accurate way of portraying the frequency response of bandpass filters (or indeed of most other filters).

Note: a lowpass to bandpass transformation causes the number of energy storage elements to be doubled and an nth-order lowpass filter produces a bandpass transfer function of order $2n$. However, it is common practice to describe the filter as being of the same order as the filter from which it is derived. Because this leaves scope for confusion it is probably best to describe a bandpass filter by its number of poles. Thus a bandpass filter derived from a second-order lowpass filter is less ambiguously described as a 4 pole bandpass filter than as a second-order bandpass filter.

Example 13.3

Design a bandpass filter with a bandwidth of 7 kHz centred at 80 kHz. Assume lossless elements and that the filter is driven from a voltage source and terminated by 4.7 kΩ. Let the filter be a 6 pole filter based on a third-order Butterworth lowpass prototype filter. Plot the frequency response between 10^5 and 10^6 rad/s.

Solution From Table 13.1 the component values for the normalized lowpass prototype are $L_1 = \frac{3}{2}$ H, $C_2 = \frac{4}{3}$ F and $L_3 = \frac{1}{2}$ H. After impedance scaling from 1 Ω to 4.7 kΩ these become $L_1 = 1.5 \times 4700 = 7050$ H, $C_2 = 1.333/4700$ F $= 283.68$ μF and $L_3 = 0.5 \times 4700 = 2350$ H. The next step is to apply the frequency scaling of Equations (13.20) and (13.21) to obtain the component values for a lowpass filter of 7 kHz bandwidth. This gives $L_1' = 7050/(2\pi.7000) = 0.160$ H, $C_2' = 283.68/(2\pi.7000)$ μF $= 6.45$ nF and $L_3' = 2350/(2\pi.7000) = 0.053$ H.

We now place a capacitance in series with each inductance of a value that gives a series resonance at 80 kHz, and an inductance in parallel with the capacitance of a value that gives a parallel resonance at 80 kHz. Using Equations (13.22) and (13.24) the added component values are

$$C_1' = \frac{1}{0.160(2\pi.80000)^2} \text{ F}$$

$$L_2' = \frac{1}{(2\pi.80000)^2(6.45 \times 10^{-9})} \text{ H}$$

$$C_3' = \frac{1}{0.053(2\pi.80000)^2} \text{ F}$$

The design values are therefore $L_1' = 0.160$ H, $C_1' = 24.7$ pF, $L_2' = 0.61$ mH, $C_2' = 6.45$ nF, $L_3' = 0.053$ H and $C_3' = 74.6$ pF. This gives us the configuration of Figure 13.18(a). In this example the 0.160 H inductance is a rather higher value than is easily obtained from a winding on a small ferrite core. More importantly though, a practical inductor of this inductance value is likely to have a self-capacitance that is comparable with or greater than the 24.7 pF that is in parallel with it. When this happens it is advisable to try changing the specification in some acceptable way, in particular by changing the terminating impedance. If the terminating impedance is reduced to 470 Ω in this example the much more satisfactory design of Figure 13.18(b) arises.

In order to plot the theoretical response we must first determine the lower and upper −3 dB frequencies, y_1 and y_2. Converting to radian frequency, 80 kHz corresponds to a centre frequency of 502 655 rad/s and 7 kHz corresponds to a bandwidth of 43 982 rad/s. Remembering that the centre frequency is the geometric

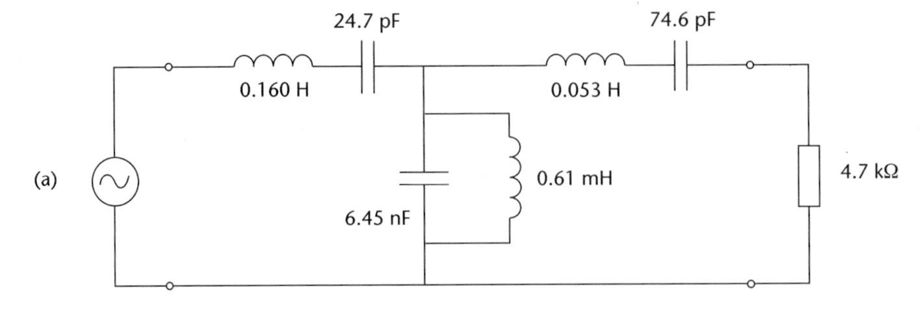

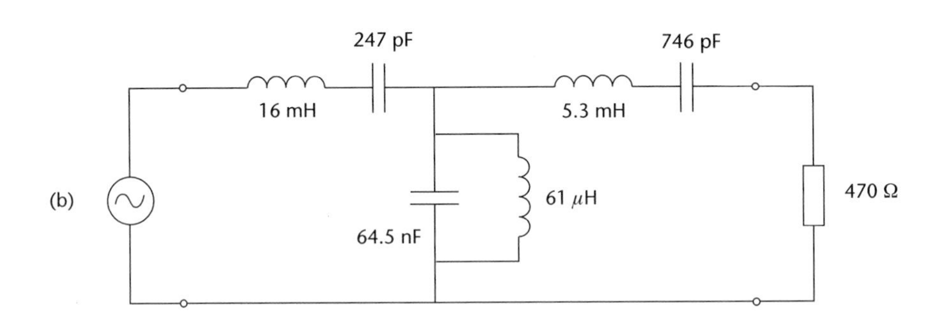

Fig. 13.18 Bandpass filter: (a) original design; (b) modified design

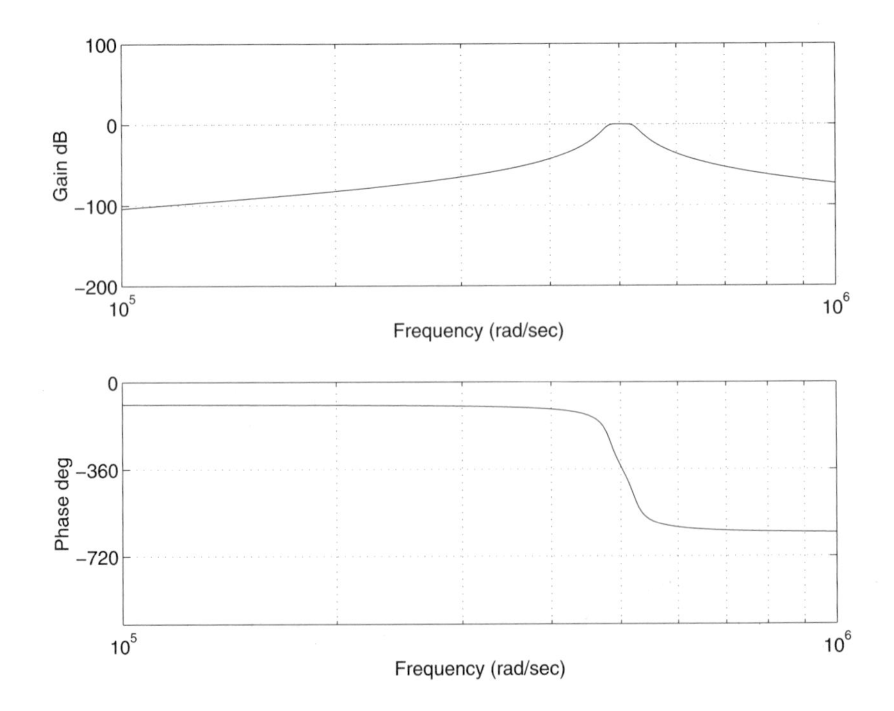

Fig. 13.19 Bode plot of bandpass filter response

mean we must therefore solve the equations

$$y_2 - y_1 = 43\,982$$
$$\sqrt{y_1 y_2} = 502\,655$$

This leads to $y_1 = 480\,145$ rad/s and $y_2 = 525\,127$ rad/s. Hence to determine the coefficients of the polynomials in the transfer function we require

```
[b,a]=butter(3,[480145 525127],'s');
```

In order to achieve a Bode plot with the frequencies confined to the specified range we must define a frequency vector. The two commands

```
w=(100:1000)*1.e3;
bode(b,a,w)
```

produce Figure 13.19.

It was noted in Section 13.2 that a major problem with passive filters concerned the losses within practical inductors. This applies to all forms of passive filter but tends to be a particular problem in bandpass filters with very sharp rates of roll-off. Because of the difficulty in obtaining high Q inductors, mechanical resonances are exploited in many commercially available high-performance bandpass filters.

13.12 Lowpass to highpass transformation

Consider a lowpass prototype filter with a cutoff frequency of 1 rad/s and let us replace any inductance L by a capacitance $1/L$. The reactance of the original inductance in the lowpass filter at a frequency x Hz is then the same as that of the capacitance in the highpass filter at a frequency of y Hz if we satisfy the relationship

$$2\pi x L = -\frac{1}{2\pi y(1/L)} \tag{13.30}$$

Similarly, let us replace any capacitance C in the lowpass filter by an inductance $1/C$. The reactance of the original capacitance in the lowpass filter at a frequency x Hz is then the same as that of the inductance in the highpass filter at a frequency of y Hz if we satisfy the relationship

$$-\frac{1}{2\pi x C} = 2\pi y(1/C) \tag{13.31}$$

Both of these equations give the same result, that the reactance of an element in the lowpass filter at x Hz is the same as that of the replacement element in the highpass filter at y Hz, if

$$x = -\frac{1}{4\pi^2 y} \tag{13.32}$$

It follows that the gains of the two filters are the same when this relationship is satisfied.

As with the lowpass to bandpass case of Figure 13.17 we again have a frequency transformation, but this time as shown in Figure 13.20. It is seen that the

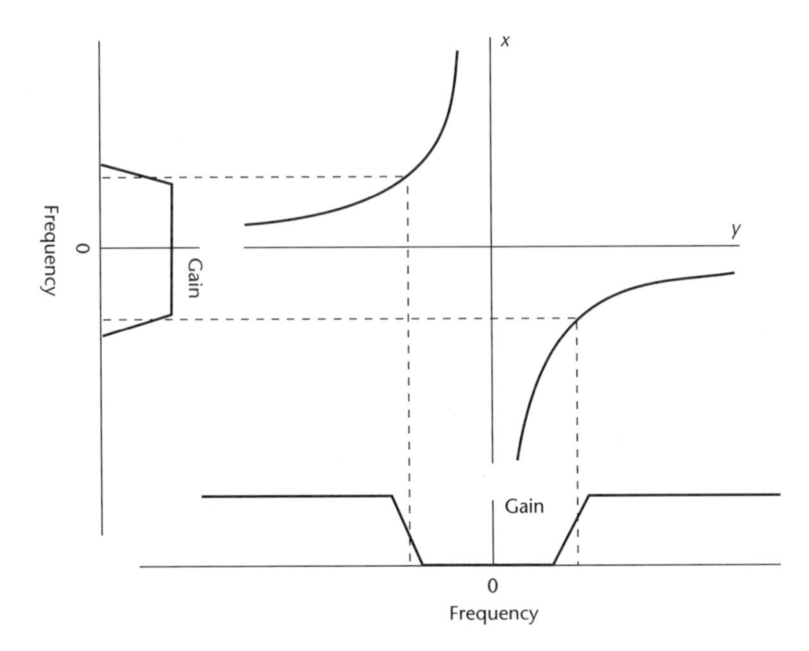

Fig. 13.20 Frequency mapping between lowpass and highpass filters

transformation causes the lowpass filter to become a highpass filter. In terms of radian frequency they both have the same cutoff of 1 rad/s. For terminating impedances other than 1 Ω and for cutoff frequencies other than 1 rad/s, impedance and frequency scaling can be applied just as with the lowpass filter.

13.13 Passive bandstop filters

Bandstop filters are obtained from highpass filters in the same way that bandpass filters are obtained from lowpass filters. Specifically, a capacitance is placed in series with each resulting inductance of a value that gives a series resonance at f_0; also an inductance is placed in parallel with each capacitance of the lowpass filter of a value that gives a parallel resonance at f_0. The frequency mapping is identical, and a construction comparable to Figure 13.17 shows that a highpass filter maps to a bandstop one.

13.14 Active lowpass filters

Whereas a high-order passive lowpass filter is a composite network in which all parts interact, it is normal in an active filter to cascade buffered low-order sections. Two configurations which are very useful are shown in Figures 13.21 and 13.22.

Network analysis shows that the circuit of Figure 13.21 has a transfer function given by

$$H(s) = \frac{1}{C_1C_2s^2 + 2C_2s + 1} \tag{13.33}$$

such that it has two poles.

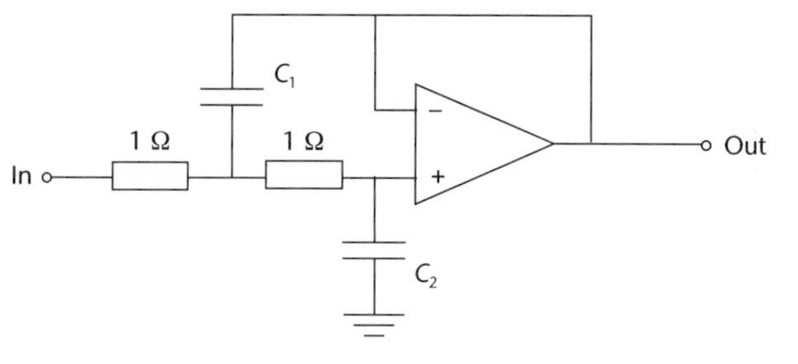

Fig. 13.21 Unity-gain two-pole active lowpass filter

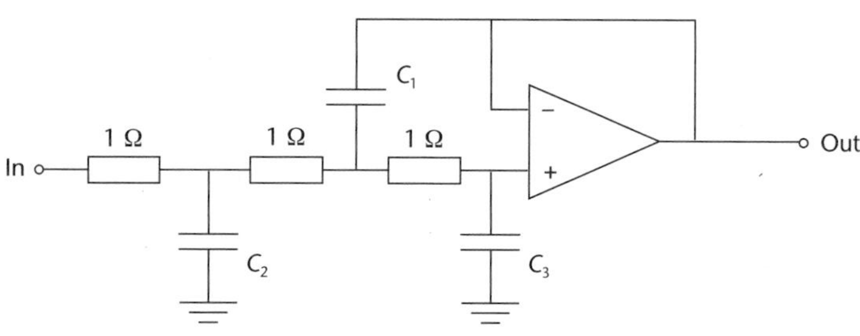

Fig. 13.22 Unity-gain three-pole active lowpass filter

Similarly, network analysis shows that the circuit of Figure 13.22 has a transfer function given by

$$H(s) = \frac{1}{As^3 + Bs^2 + Cs + 1} \tag{13.34}$$

where

$$A = C_1C_2C_3, \; B = 2C_3(C_1 + C_2), \; C = C_2 + 3C_3 \tag{13.35}$$

This transfer function has three poles and, because poles that do not lie on the real axis must exist in conjugate pairs, the presence of three poles signifies that one or all three must be real.

Let us suppose, as an example, that we wish to design an active lowpass filter with a fifth-order Butterworth or Chebyshev response. We can convert the required amplitude response into a transfer function using the techniques of Section 13.5. Consider the case of a transfer function given by

$$H(s) = \frac{1}{(s - p_1)(s - p_2)(s - p_2{}^*)(s - p_3)(s - p_3{}^*)} \tag{13.36}$$

Our technique would be to cascade the network of Figure 13.21 with that of Figure 13.22 and to select values of C_1, C_2 and C_3 in the network of Figure 13.22 such that

$$\frac{1}{As^3 + Bs^2 + Cs + 1} = \frac{1}{(s - p_1)(s - p_2)(s - p_2{}^*)} \tag{13.37}$$

and to select values of C_1 and C_2 of the network of Figure 13.21 such that, if Equation (13.37) is applied,

$$\frac{1}{C_1C_2s^2 + 2C_2s + 1} = \frac{1}{(s - p_3)(s - p_3{}^*)} \tag{13.38}$$

An alternative would be to make

$$\frac{1}{As^3 + Bs^2 + Cs + 1} = \frac{1}{(s - p_1)(s - p_3)(s - p_3{}^*)} \tag{13.39}$$

and

$$\frac{1}{C_1 C_2 s^2 + 2C_2 s + 1} = \frac{1}{(s - p_2)(s - p_2{}^*)} \tag{13.40}$$

The above procedure is straightforward but laborious. Fortunately, as for passive lowpass filters, the necessary calculations have been done by others and presented in tables. Tables are available (for example in Williams and Taylor (1988)) which assume that a standard filter response of a required order is achieved by cascading sections of the form shown in Figures 13.21 and 13.22, and give the necessary capacitor values assuming a cutoff of 1 rad/s. For example, the tables in Williams and Taylor state that a seventh-order Butterworth normalized lowpass filter is realized by a third-order section having $C_1 = 1.531$ F, $C_2 = 1.336$ F and $C_3 = 0.4885$ F, followed by two second-order sections, the first having $C_1 = 1.604$ F and $C_2 = 0.6235$ F and the second having $C_1 = 4.493$ F and $C_2 = 0.225$ F. There is therefore no need to derive values from first principles using the procedures given above. All that does need to be done is to apply impedance and frequency scaling. If it is deemed appropriate to change the resistance values in Figure 13.22 from 1 Ω to R Ω the capacitances must be *decreased* by a factor R in order that the relative impedances of resistances and capacitances should remain unchanged. If the cutoff frequency is *increased* by a factor of x from 1 rad/s the capacitances must be *decreased* by a factor x in order to keep their impedances the same at the cutoff frequency.

13.15 Active highpass filters

It was originally shown by Mitra that a lowpass active filter could be transformed into a highpass filter by changing each resistance R into a capacitance of value $1/R$ farads, and changing each capacitance C into a resistance of value $1/C$ ohms. This is often known as the RC–CR transformation. Following this, impedance and frequency scaling can be imposed as before.

13.16 Active bandpass filters

The procedure for designing an active bandpass filter begins by determining the pole positions of the corresponding lowpass filter. For example, Figure13.13 showed that the poles of a third-order Butterworth filter with a cutoff frequency of 1 rad/s are at $1\angle 120°$, $1\angle 180°$ and $1\angle 240°$. A lowpass filter with a bandwidth of x rad/s would have poles at $x\angle 120°$, $x\angle 180°$ and $x\angle 240°$, as shown in Figure 13.23.

The geometric technique of evaluating a frequency response from an s-plane plot of poles and zeros suggests that the corresponding bandpass filter centred at ω_0 should have the pole–zero plot of Figure 13.24, where the three poles have been displaced upwards and downwards by ω_0 and zeros have been introduced at the origin, one for each conjugate pole pair. We can see that this produces the wanted amplitude response by selecting a point $j\omega'$ on the imaginary axis close to $j\omega_0$ and drawing lines to all poles and zeros as shown in Figure 13.25. It is observed that if ω' changes slightly, the product of the distances to the zeros divided by the product of

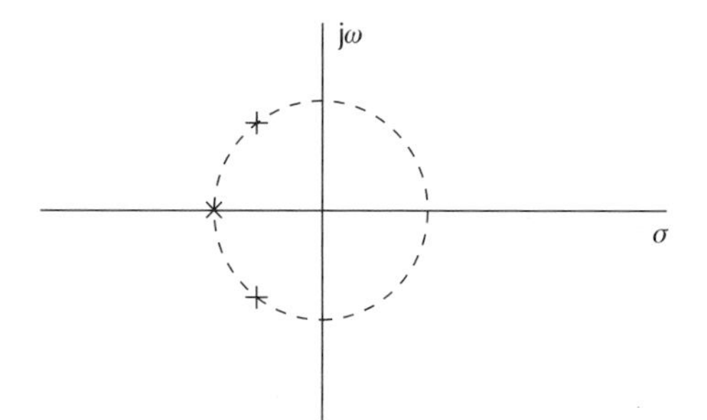

Fig. 13.23 Pole plot for Butterworth lowpass filter

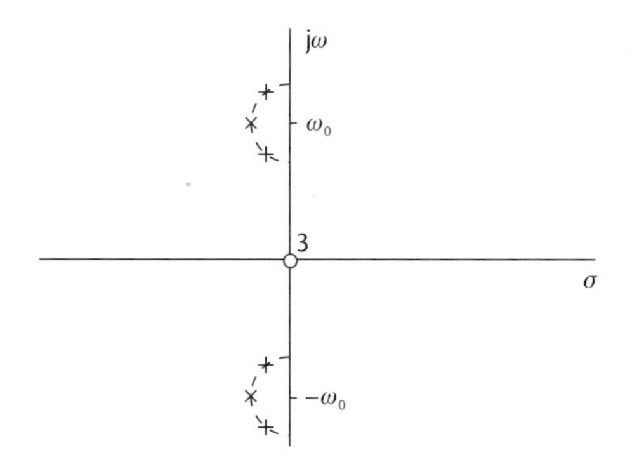

Fig. 13.24 Pole-zero plot for Butterworth bandpass filter

the distances to the lower poles is little altered. This means that the amplitude response is determined primarily by the relation of the point $j\omega'$ to the upper three poles. So long as the centre frequency of the bandpass filter is much larger than the cutoff frequency of the lowpass filter, its amplitude response will have a similar shape to that of the lowpass filter except for being shifted up in frequency by ω_0.

There are three conjugate pole pairs and one possibility is to cascade three circuits, each one providing one of these plus a zero at the origin. One of several possible configurations is the multiple feedback bandpass circuit (also known as the Delyiannis-Friend circuit) shown in Figure 13.26. This has the transfer function

$$H(s) = \frac{sC/R_1}{s^2C^2 + 2sC/R_2 + 1/R_1R_2} \tag{13.41}$$

The component values can be chosen to give one of the pole pairs. Similar sections can give the other pole pairs. In effect, the overall response is obtained by 'stagger tuning' sections.

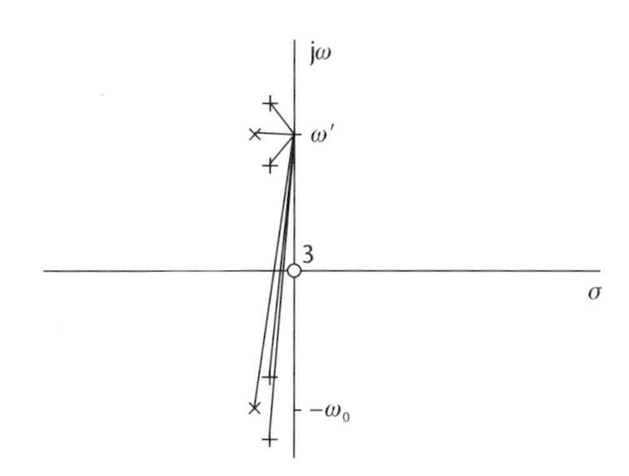

Fig. 13.25 Geometrical construction for determining amplitude response of bandpass filter

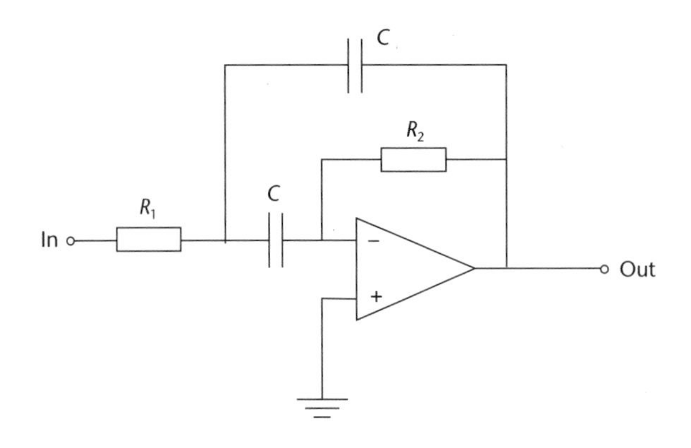

Fig. 13.26 Multiple feedback bandpass circuit

This treatment of active bandpass filters is very superficial and is only intended to give an introduction to the subject. For a more comprehensive coverage of these and other active filters including bandstop filters, the reader is referred to more specialized texts (such as Williams and Taylor (1988)).

13.17 Delay equalizers

Active and passive circuits are available for achieving group delay equalization. However, this chapter is deliberately constrained to a very brief overview of filter design methods and only the active network shown in Figure 13.27 will be considered. The main objective is to demonstrate that such a network can have a constant gain while modifying the phase.

Equating the voltage on the non-inverting input of the op-amp to that on the inverting input we have

$$V_1(s)\,\frac{R}{R+1/sC} = V_1(s) + \frac{V_2(s)-V_1(s)}{2} = \frac{V_1(s)}{2} + \frac{V_2(s)}{2} \tag{13.42}$$

Rearranging and simplifying the notation by taking it as implicitly understood that V_1 and V_2 are functions of s, this becomes

$$V_2 = V_1\left(\frac{2}{1+1/sRC} - 1\right) = V_1\,\frac{1-1/sRC}{1+1/sRC}$$

$$\therefore \frac{V_2}{V_1} = \frac{s-1/RC}{s+1/RC} \tag{13.43}$$

The transfer function has a pole at $-1/RC$ and a zero at $+1/RC$, as shown in Figure 13.28. The gain is given by

$$\left|\frac{V_2}{V_1}\right| = \left|\frac{\mathrm{j}\omega - 1/RC}{\mathrm{j}\omega + 1/RC}\right| \tag{13.44}$$

This does not vary with frequency, demonstrating that Figure 13.27 is an allpass network. The phase shift is given by

$$\begin{aligned}\theta(\omega) &= \angle(\mathrm{j}\omega - 1/RC) - \angle(\mathrm{j}\omega + 1/RC)\\ &= -2\,\tan^{-1}(\omega RC)\end{aligned} \tag{13.45}$$

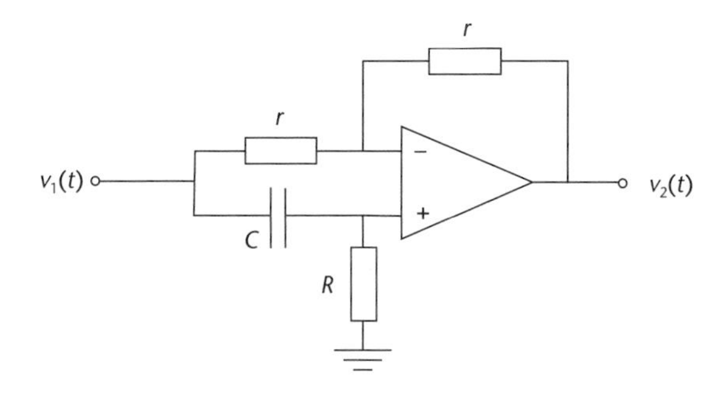

Fig. 13.27 Delay equalizer circuit

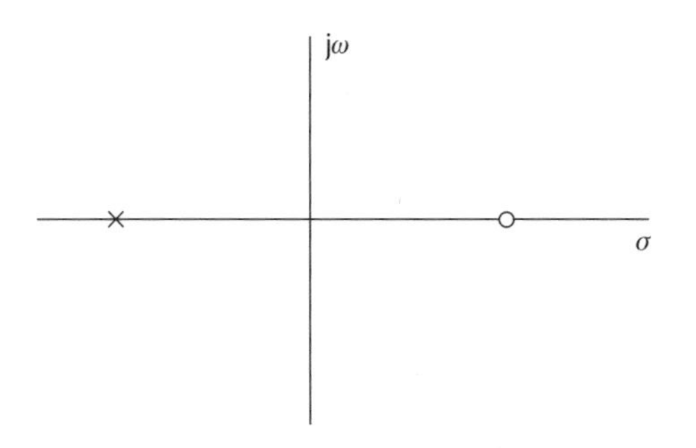

Fig. 13.28 Pole–zero plot for delay equalizer

Using Equation (13.11) the group delay is thus

$$\tau(\omega) = -\frac{\mathrm{d}}{\mathrm{d}\omega}\,\theta(\omega) = \frac{2RC}{(1+\omega^2R^2C^2)} \tag{13.46}$$

Table 13.3 applies this equation to frequencies between 0 and 1 rad/s for a small range of RC values. These group delays are in seconds and the table shows that a suitable choice of RC can provide a substantial amount of compensation for the group delay characteristics of a lowpass prototype filter. The compensation can be shown to be particularly suitable for a filter with a Butterworth response.

For compensating filters with cutoff frequencies other than 1 rad/s the capacitance value is frequency-scaled in accordance with Equation (13.21). If RC is decreased to compensate a filter with a higher cutoff frequency the group delay decreases in accordance with Equation (13.46).

For the case of providing group delay equalization for a bandpass filter an inductance can be placed in parallel with the capacitance, of a value that resonates with the capacitance at the centre frequency of the filter.

The circuit of Figure 13.27 has a first-order transfer function given by Equation (13.43). Such circuits may be cascaded to add further flexibility in obtaining good

Table 13.3 Group delay (seconds) of group delay equalizer

	ω										
RC	**0**	**0.1**	**0.2**	**0.3**	**0.4**	**0.5**	**0.6**	**0.7**	**0.8**	**0.9**	**1.0**
0.5	1.0	0.997	0.990	0.978	0.962	0.941	0.917	0.891	0.862	0.832	0.800
1.0	2.0	1.98	1.92	1.83	1.72	1.60	1.47	1.34	1.22	1.10	1.00
2.0	4.0	3.85	3.45	2.94	2.44	2.00	1.64	1.35	1.12	0.943	0.800
4.0	8.0	7.90	4.88	3.28	2.25	1.60	1.18	0.905	0.712	0.573	0.471

delay compensation. Circuits with higher order allpass transfer functions are also feasible and add still further flexibility.

13.18 Summary

The design of active and passive filters usually begins with a prototype lowpass filter that is normalized to 1 Ω terminations and to a cutoff frequency of 1 rad/s. Frequency responses for different types of filter can be readily plotted using MATLAB. Network configurations and the component values of prototype filters are best determined with the aid of tables (found in books such as those by Geffe (1963) or Williams and Taylor (1988)) or specialized software. The main purpose of this chapter has been to show how the component values for lowpass prototype filters may be converted into lowpass, highpass, bandpass or bandstop filters with the required cutoff frequencies and having the specified terminating impedances. There has also been a discussion on phase distortion where it was shown that a constant group delay can cause intercept distortion but that this is usually acceptable. If constant group delay is an adequate but necessary criterion a group delay equalizer can be incorporated.

13.19 Problems

1. A lowpass filter is required to have a 3 dB cutoff frequency of 12 kHz and an attenuation of 30 dB at 16 kHz. Determine analytically what order of Butterworth filter is needed. Check the result by plotting the frequency response using MATLAB.
2. A speech signal filtered to occupy the frequency range 300 Hz to 3400 Hz is denoted by $f(t)$. This is multiplied by a 455 kHz cosine wave to produce $f(t)\cos\omega_s t$. Considering only a 300 Hz component of the speech signal we thus have $\cos(2\pi\times 300t)\cos(2\pi\times 455\,000t)$. Using the trigonometric identity $\cos A\cos B=\frac{1}{2}\cos(A-B)+\frac{1}{2}\cos(A+B)$ this can be expanded as $\cos A\cos B=\frac{1}{2}\cos(2\pi\times 454\,700t)+\frac{1}{2}\cos(2\pi\times 455\,300t)$. In order to produce a single sideband signal for communication purposes, $f(t)\cos\omega_c t$ is passed through a bandpass filter whose 3 dB bandwidth is 3100 Hz centred at 456.85 kHz such that the upper 'sideband' is attenuated by 3 dB (note that, $456\,850-3100/2=455\,300$). Determine analytically what order of Butterworth filter is needed to attenuate the lower of the two sidebands by 40 dB.
3. Use MATLAB to plot the frequency response of a fifth-order Chebyshev lowpass filter having 0.9 dB ripple in the passband and a cutoff frequency of 3.4 kHz. Sketch a passive filter configuration and determine component values assuming it is driven from a voltage source and terminated by 200 Ω.
4. Use MATLAB to plot the frequency response of a fourth-order Butterworth bandpass filter (8 pole) having a lower cutoff frequency of 12 kHz and an upper cutoff frequency of 15 kHz. Sketch a passive filter configuration and determine component values assuming it is driven from a voltage source and terminated by 1 kΩ.
5. Determine the transfer function of a fifth-order Butterworth lowpass filter with a cutoff frequency of 1 rad/s. Use the **[mag,phase,w]=bode(b,a)** command in MATLAB as a convenient method of determining the filter's phase response in degrees, as a function of a specified frequency vector. Derive and plot the group delay as a function of frequency. By referring to Table 13.3, show that the variation of group delay over most of the passband can be considerably reduced by the addition of two first-order delay equalizers in cascade, each having $RC=1$.

Chapter fourteen

An introduction to digital networks and the z transform

14.1 ● Preview

Although most signals that require processing are analogue it is becoming more and more common to digitize such signals and then process them digitally on a computer, microprocessor or special DSP chip. This allows enormous flexibility with the type of processing that is possible. Earlier chapters have described the use of the FFT, particularly for spectral analysis and convolution, but, as with analogue processing, the enhancement of certain frequencies relative to others by filtering is another very important signal processing procedure. This chapter introduces digital filters and shows how their frequency responses may be determined.

As with analogue networks it is possible to describe digital networks by their s transfer functions, but it emerges that these have an infinite number of poles and zeros and this makes them rather unsatisfactory. Similarly, the Laplace transforms of important discrete-time signals such as step functions or cosinusoids have an infinite number of poles and zeros. Everything becomes very much simpler if the s variable is replaced by a different but related variable z. This leads to the z transform. As one example of its application, digital control systems are briefly examined.

14.2 Structure of digital networks and the difference equation

Digital networks can contain both feedforward and feedback paths. Those containing only feedforward paths are termed non-recursive networks. Those containing only feedback paths, or containing feedforward *and* feedback paths, are termed recursive networks.

Figure 14.1 is an example of a very simple recursive digital filter. As drawn, it might be thought of as representing an analogue network consisting of one analogue delay and an adder. In that event the relationship between input and output would be given by

$$y(t) = x(t) + 0.9y(t - t_s) \tag{14.1}$$

However, this is not the way it which it is implemented in practice. Figure 14.1 is the symbolic representation of an algorithm that is applied exclusively to *sampled* data, and is implemented on a computer, microprocessor or special DSP chip. The delay represents the delay t_s between data samples. With this understood, the relationship between input and output would be written as

$$y[n] = x[n] + 0.9y[n - 1] \tag{14.2}$$

As an example of appying this relationship, consider two MATLAB files, the first containing some data, the second one acting on it. Let us imagine the data to be 200 measurements of a voltage that is sinusoidal but with random noise superposed on it. An artificial data set can be generated and loaded into a data file with the name C14data.mat using the commands

```
for n=1:200
x(n)=sin(n/12)+.6*randn;  % this generates the artificial data set of a random
                          % Gaussian distributed random variable added
                          % to a sinusoid.
end
save C14data.mat x
```

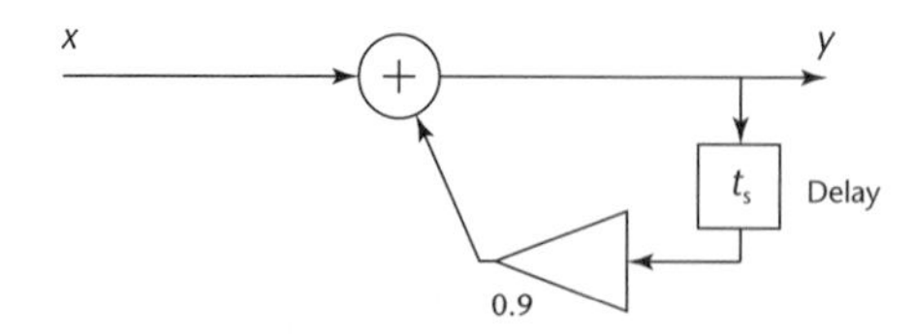

Fig. 14.1 Example of a simple recursive digital filter

In order to process this data file using the relationship of Equation (14.2) a suitable program stored in an M-file named sec14_2 would be

```
% sec14_2.m
load C14data.mat          % this calls up the data file of this name and hence
                          % places the variable x in the MATLAB workspace
y=zeros(size(x));         % this initializes the output values to be zero.
for n=2:200               % we start at n=2 so that the index of the first
                          element
                          % of the output array to be addressed is one.
y(n)=x(n)+.9*y(n-1);
end
subplot(2,1,1)
plot(x)
ylabel('x[n]')
subplot(2,1,2)
plot(y)
ylabel('y[n]'),xlabel('n')
```

The outcome is shown in Figure 14.2, where Figure 14.2(a) plots the input data and Figure 14.2(b) plots the processed output data. The results have been labelled as discrete signals using square brackets for the *y*-axis labels but, because of the large number of points, are presented as continuous signals using the **plot** command.

A comparison of Figures 14.2(a) and 14.2(b) shows that the noise is reduced and the sinusoid is enhanced. What follows in later sections will clarify these two changes.

A generalized configuration that can give all possible linear relationships between input and output is shown in Figure 14.3, where the input $x[n]$ is assumed to be a sampled data signal and again the rectangular boxes represent delays of one sample period.

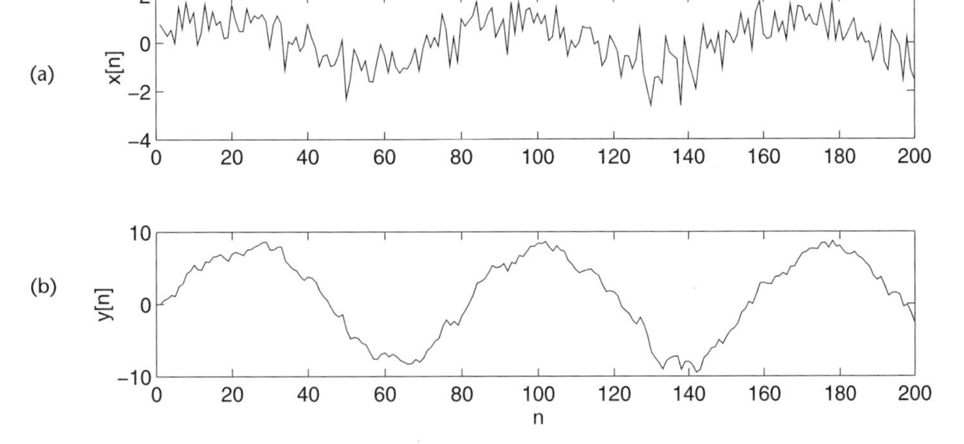

Fig. 14.2 Effect of simple recursive filter: (a) input signal; (b) output signal

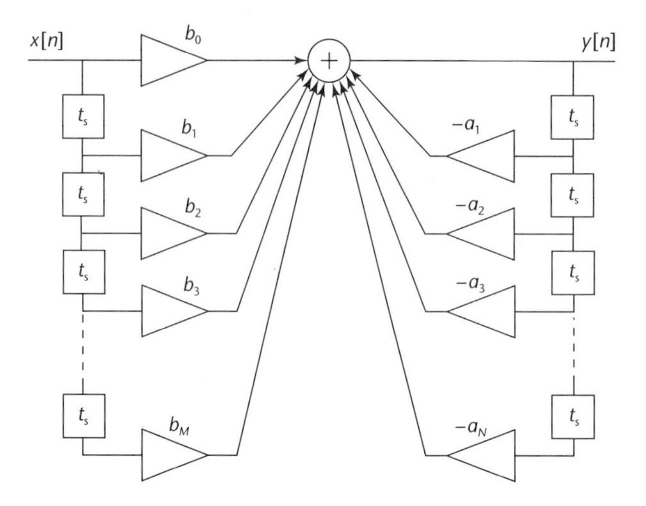

Fig. 14.3 Generalized configuration of a digital network

The output signal $y[n]$ may be written down by inspection and is

$$\begin{aligned} y[n] &= b_0x[n] + b_1x[n-1] + b_2x[n-2] + \cdots + b_Mx(n-M) \\ &\quad -a_1y[n-1] - a_2y[n-2] - \cdots - a_Ny(n-N) \end{aligned} \tag{14.3}$$

This may be simplified as

$$y[n] = \sum_{i=0}^{M} b_ix[n-i] - \sum_{i=1}^{N} a_iy[n-i] \tag{14.4}$$

or as

$$\sum_{i=0}^{N} a_iy[n-i] = \sum_{i=0}^{M} b_ix[n-i] \tag{14.5}$$

where $a_0 = 1$. This is the difference equation (or recurrence formula) of the digital network. The integer N, *referring to the recursive part of this equation*, is known as the *order* of the difference equation. It should be noted, however, that some authors take the order of the equation to be the higher of the two values M and N. This convention will not be adopted here.

The symmetry of Equation (14.5) justifies the convention of negative coefficients for the feedback terms. It should be noted, however, that great care is needed when referring to the literature as there are considerable variations in the symbol and sign conventions used. It is not unusual to find texts:

- that use the *a*s and *b*s the other way around (the *a*s for the non-recursive coefficients, and the *b*s for the recursive coefficients). The argument of this is that it is logical to have the *a*s occurring first when reading the structure of Figure 14.3 from left to right.
- that use positive coefficients for the feedback (or recursive) terms, the argument being that a sign convention that is consistently positive throughout Figure 14.3 is more logical.
- that use different symbols altogether, typically *c*s and *d*s or *l*s and *k*s.

However, the convention adopted here is the most widely accepted.

The configuration of Figure 14.3 is a *causal* structure, which means that, as with analogue networks, the output depends entirely on the present and past values of the input sequence. However, it is quite feasible to have *non-causal* digital systems in which the output is also affected by future values of the input sequence. For example, stored data can be smoothed off-line by an averaging process that includes past, present and future data values. The difference equation for a simple non-causal five point averager could be

$$y[n] = \tfrac{1}{5}(x[n-2] + x[n-1] + x[n] + x[n+1] + x[n+2]) \tag{14.6}$$

The key requirement for a non-causal system is the storage of input samples (requiring memory), such that when a particular sample is presented to the network, subsequent samples are also available. Since much data processing is done off-line, with the entire data sequence available for processing, non-causal systems are perfectly feasible. This book, however, will put the emphasis on causal systems.

14.3 The impulse responses of digital networks – FIR and IIR systems

The simple five point, causal, non-recursive averager of Figure 1.11 is repeated in Figure 14.4. As then, the difference equation is

$$y[n] = \tfrac{1}{5}(x[n] + x[n-1] + x[n-2] + x[n-3] + x[n-4]) \tag{14.7}$$

By definition the impulse response $h[n]$ is the output when the input is a unit impulse function $\delta[n]$. Therefore

$$h(n) = \tfrac{1}{5}(\delta[n] + \delta[n-1] + \delta[n-2] + \delta[n-3] + \delta[n-4]) \tag{14.8}$$

It contains a *finite* number of terms, equal to the number of non-recursive coefficients, and therefore represents a finite impulse response, or FIR, filter. Because it takes the average of successive sets of five values it may also be termed a moving average (MA) filter. Other terms are feedforward filter, tapped delay line filter and transversal filter.

Figure 14.5 is a repeat of Figure 14.1 except that the one recursive coefficient now has the generalized value of $-a_1$. The initial response when a unit impulse is applied

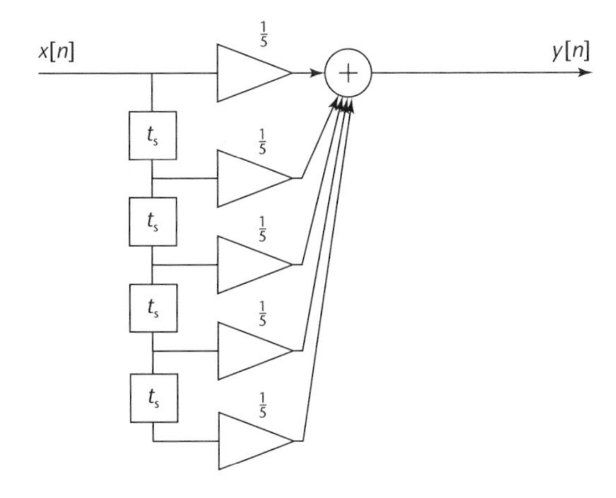

Fig. 14.4 Five point non-recursive digital filter

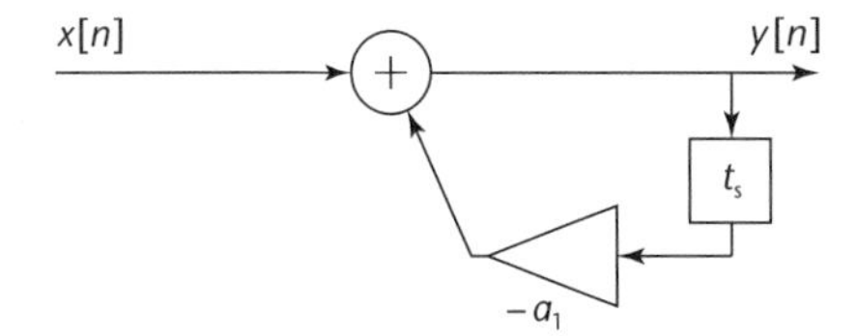

Fig. 14.5 Simple IIR filter

to the input is a unit impulse at the output. However, this output impulse then circulates around the feedback loop producing an infinite number of impulses at the output. The impulse response is thus

$$h[n] = \delta[n] - a_1\delta[n-1] + (-a_1)^2\delta[n-2] + (-a_1)^3\delta[n-3] + \cdots \tag{14.9}$$

There are an infinite number of terms and this is thus an *infinite* impulse response, or IIR, filter. Another way of describing the filter of Figure 14.5 would be as a feedback or autoregressive (AR) filter.

Non-recursive filters are always FIR. This is because their finite number of non-recursive coefficients cannot allow them an infinite impulse response. An infinite impulse response can only arise when there is feedback, and it follows that IIR filters must be recursive.

Although non-recursive filters are always FIR the converse does not necessarily apply. An FIR filter can be recursive. For example, it was shown in Chapter 1 that the five point averager of Figure 14.4 could be realized more economically by the recursive configuration of Figure 14.6.

The relationships between non-recursive, recursive, FIR and IIR are summarized by Figure 14.7. Because it is unusual for a recursive filter to be FIR it is common (but somewhat dangerous) practice to use the terms 'recursive filter' and 'IIR filter' synonymously, and to use the terms 'non-recursive filter' and 'FIR filter' synonymously.

Filters with both feedforward and feedback paths are also referred to as autoregressive moving average (ARMA) filters.

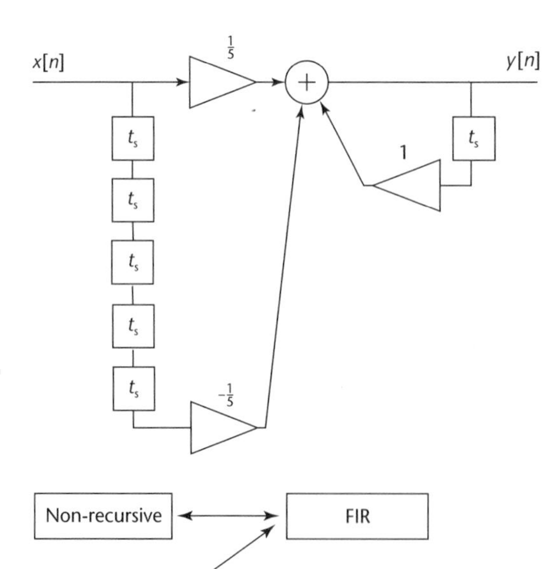

Fig. 14.6 Recursive realization of an FIR filter

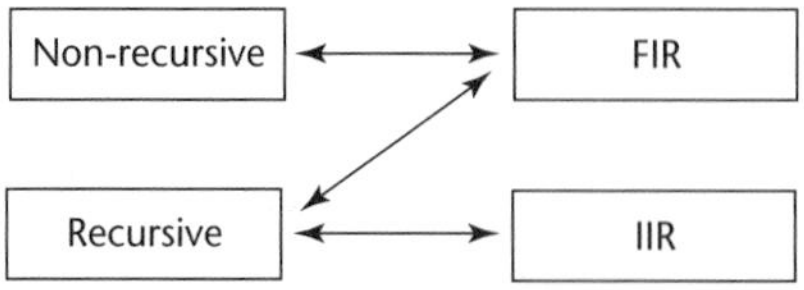

Non-recursive
FIR
Recursive
IIR

Fig. 14.7 Interdependence of filter terminology

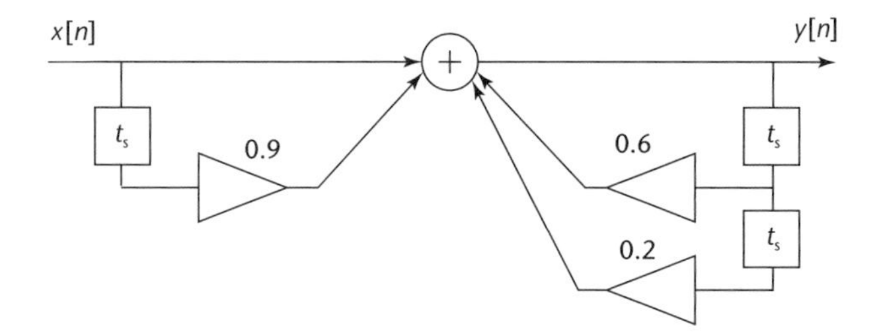

Fig. 14.8 Example of a digital network

If it is wished to find the output $y[n]$ of a digital filter due to an input $x[n]$ it is generally easiest to do so directly from the difference equation.

Example 14.1

Derive the difference equation of the digital network in Figure 14.8 and determine its output when the input is given by

$$x[n] = \{\underset{\uparrow}{3}, 1, 2\}$$

Solution By inspection the difference equation is given by

$$y[n] = x[n] + 0.9x[n-1] + 0.6y[n-1] + 0.2y[n-2]$$

The easiest manual way of determining numerical values is to draw up a table in which terms are evaluated successively for each value of n.

n	$x[n]$	$x[n-1]$	$y[n-1]$	$y[n-2]$	$y[n]$
0	3	0	0	0	3
1	1	3	3	0	1 + 0.9(3) + 0.6(3) + 0.2(0) = 5.5
2	2	1	5.5	3	2 + 0.9(1) + 0.6(5.5) + 0.2(3) = 6.8
3	0	2	6.8	5.5	0 + 0.9(2) + 0.6(6.8) + 0.2(5.5) = 6.98
4	0	0	6.98	6.8	0 + 0 + 0.6(6.98) + 0.2(6.8) = 5.548
5	0	0	5.568	6.98	0 + 0 + 0.6(5.548) + 0.2(6.98) = 4.725
etc.					

An alternative procedure for finding the output, $y[n]$, due to an input $x[n]$ is to determine the impulse response of the system and to apply the convolution sum introduced in Equation (2.4), that is

$$y[n] = x[n] * h[n] = \sum_{k=-\infty}^{\infty} x[k]h[n-k] \tag{14.10}$$

In general this is less easy.

14.4 Other configurations of digital networks

The configuration of Figure 14.3 is the most straightforward and obvious way of realizing the difference equation of Equation (14.3). It is known as the Direct Form I

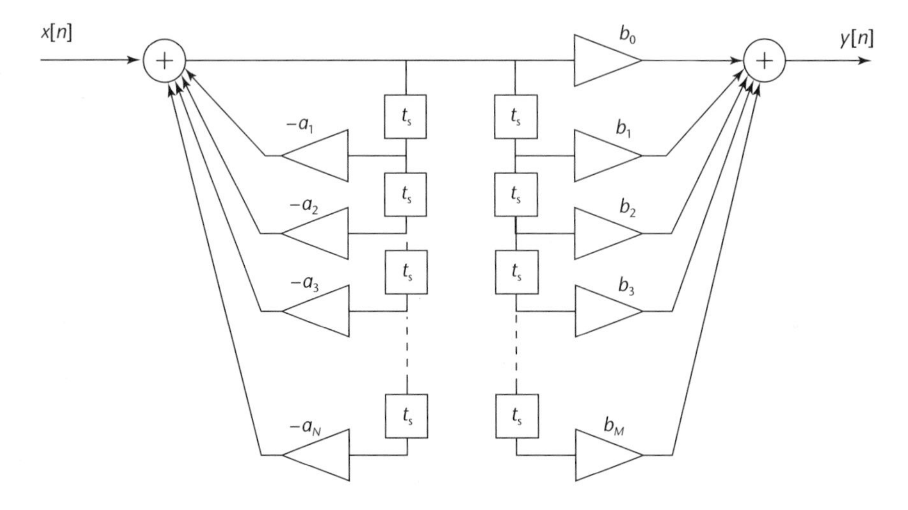

Fig. 14.9 Interchange of non-recursive and recursive sections

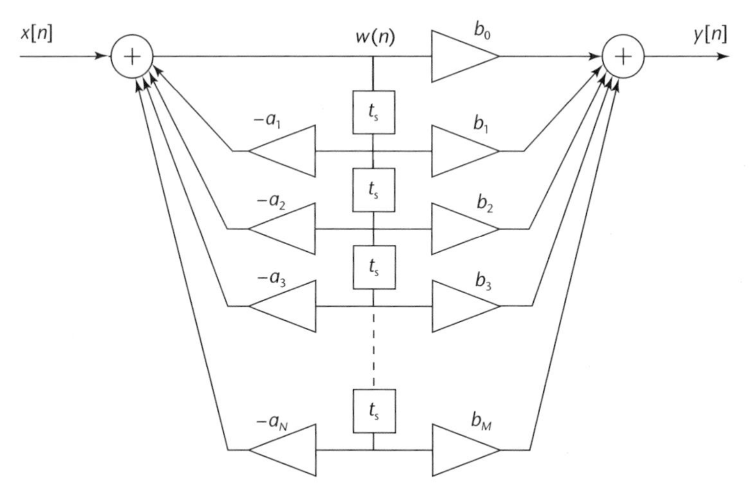

Fig. 14.10 Direct Form II realization of a digital filter

realization. Other configurations can achieve the same difference equation, often with fewer delays. For example, the non-recursive and recursive sections can be interchanged, as shown in Figure 14.9. The delays can then be shared between the two sections to give a configuration with a minimum number of delays, thus making it the 'canonical' realization of the network of Figure 14.3. This is shown in Figure 14.10. This is also sometimes known as the Direct Form II realization of the network. Its characteristics are clearly the same as those of Figure 14.3, but are most easily described now by a *pair* of difference equations, namely

$$w[n] = x[n] - a_1 w[n-1] - a_2 w[n-2] - \cdots$$
$$y[n] = b_0 w[n] + b_1 w[n-1] + b_2 w[n-2] - \cdots \tag{14.11}$$

It is also possible to break down a high-order network into a set of lower order networks *in parallel*, or into a set of lower order networks *in cascade*. As will be shown in Chapter 15, such measures can be very beneficial because they make the performance of the network much less sensitive to inaccuracies in the coefficients.

14.5 The hardware realization of digital networks

It has already been shown how a difference equation can be implemented by a computer algorithm. The objective now is to give an introduction to how fast on-line processing can be achieved using special-purpose single-chip devices.

As an example, consider the need to implement an IIR filter whose difference equation is given by

$$y[n] = b_0x[n] + b_1x[n-1] + b_2x[n-2] - a_1y[n-1] - a_2y[n-2] - a_3y[n-3]$$

Figure 14.11 shows the general principle of a typical processor. The filter coefficients are stored in a coefficient memory with storage cells CM1, CM2, CM3, CM4, CM5 and CM6. The required input and output data are stored in a data memory with storage cells DM1, DM2, DM3, DM4, DM5 and DM6.

Let us assume that some of the input data have already been filtered and that we are about to evaluate $y[7]$. Let us also assume that the digital word corresponding to $x[7]$ has just gone from the ADC to storage cell DM1. The subsequent sequence of events might be as follows:

1. contents of accumulator set to zero
2. DM1 and CM1 sent to multiplier/accumulator to give b_0x_7
3. DM2 and CM2 sent to multiplier/accumulator to give $b_0x_7 + b_1x_6$
4. DM3 and CM3 sent to multiplier/accumulator to give $b_0x_7 + b_1x_6 + b_2x_5$
5. DM4 and CM4 sent to multiplier/accumulator to give $b_0x_7 + b_1x_6 + b_2x_5 - a_1y_6$
6. DM5 and CM5 sent to multiplier/accumulator to give $b_0x_7 + b_1x_6 + b_2x_5 - a_1y_6 - a_2y_5$
7. DM6 and CM6 sent to multiplier/accumulator to give $y_7 = b_0x_7 + b_1x_6 + b_2x_5 - a_1y_6 - a_2y_5 - a_3y_4$
8. contents of accumulator, y_7, sent to DAC and output
9. contents of DM5 written to DM6 to put y_5 in DM6
10. contents of DM4 written to DM5 to put y_6 in DM5
11. contents of accumulator written to DM4 to put y_7 in DM4
12. contents of DM2 written to DM3 to put x_6 in DM3
13. contents of DM1 written to DM2 to put x_7 in DM2
14. output of ADC written to DM1 to put x_8 in DM1

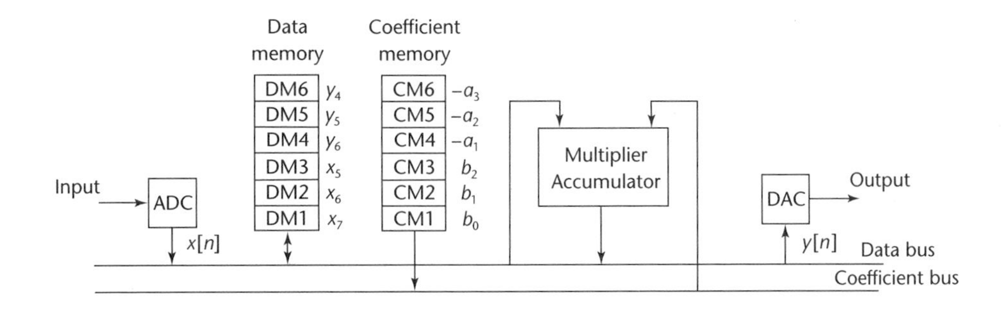

Fig. 14.11 Architecture of hardware digital filter

Now the sequence of events repeats itself, that is

1. contents of accumulator set to zero
2. DM1 and CM1 sent to multiplier/accumulator to give b_0x_8
3. DM2 and CM2 sent to multiplier/accumulator to give $b_0x_8 + b_1x_7$
4. DM3 and CM3 sent to multiplier/accumulator to give $b_0x_8 + b_1x_7 + b_2x_6$
5. DM4 and CM4 sent to multiplier/accumulator to give $b_0x_8 + b_1x_7 + b_2x_6 - a_1y_7$
6. DM5 and CM5 sent to multiplier/accumulator to give $b_0x_8 + b_1x_7 + b_2x_6 - a_1y_7 - a_2y_6$
7. DM6 and CM6 sent to multiplier/accumulator to give $y_8 = b_0x_8 + b_1x_7 + b_2x_6 - a_1y_7 - a_2y_6 - a_3y_5$
8. contents of accumulator, y_8, sent to DAC and output
9. contents of DM5 written to DM6 to put y_6 in DM6
10. contents of DM4 written to DM5 to put y_7 in DM5
11. contents of accumulator written to DM4 to put y_8 in DM4
12. contents of DM2 written to DM3 to put x_7 in DM3
13. contents of DM1 written to DM2 to put x_8 in DM2
14. output of ADC written to DM1 to put x_9 in DM1

The description above is a simplification, but it does establish the main principles of the hardware realization of a digital filter. Of particular note are the following observations:

- The coefficients are applied by means of a digital multiplier, and not by an amplifier or attenuator acting on an analogue signal as might be supposed from the symbols used in schematics such as Figure 14.3.
- The necessary multiplications and additions are done serially.
- The memories can be very large. This is particularly relevant to FIR filters, where very many coefficients are sometimes required.
- Various organizations of memory space are possible, as demonstrated by the difference between Direct Form I and Direct Form II realizations. These can vary in their memory requirements and in their consequent processing speeds.
- The serial processing described may be too slow for processing signals requiring very high sampling rates. However, there are special-purpose chips now available that do the necessary multiplications in parallel.

14.6 An *s*-plane treatment of digital networks

If the first-order recursive network of Figure 14.1 is treated as an analogue network and presented with an impulse function $\delta(t)$, the output is the impulse response $h(t)$. From Equation (14.1), this is given by

$$h(t) = \delta(t) + 0.9h(t - t_s) \tag{14.12}$$

The transfer function may be determined by taking the Laplace transform, to obtain

$$H(s) = 1 + 0.9H(s)e^{-st_s} \tag{14.13}$$

This simplifies to give

$$H(s) = \frac{1}{1 - 0.9e^{-st_s}} \tag{14.14}$$

An alternative way of arriving at the same result is to consider an input signal $x(t)$ equal to e^{st} and to substitute this and an output signal of the form $H(s)e^{st}$ into Equation (14.1) to obtain

$$H(s)e^{st} = e^{st} + 0.9H(s)e^{s(t-t_s)} \tag{14.15}$$

The e^{st} term divides out, again producing

$$H(s) = \frac{1}{1 - 0.9e^{-st_s}}$$

It is instructive to examine the s-plane poles and zeros of this digital network.

Example 14.2

Plot the s-plane poles and zeros of the transfer function of the network of Figure 14.1 and deduce the nature of its frequency response.

Solution Equation (14.14) gives the transfer function as $H(s) = 1/(1 - 0.9e^{-st_s})$. We have $s = \sigma + j\omega$ and there is a zero when $\sigma = -\infty$, since then the denominator equals $(1 - 0.9e^{\infty}e^{-j\omega t_s})$ and this is minus infinity. There is a pole when the denominator equals zero, that is when $(1 - 0.9e^{-st_s}) = 0$ or when $0.9e^{-st_s} = 1$. Taking the natural logarithm of both sides gives

$$\ln(0.9) - st_s = \ln 1$$

or

$$s = [(\ln(0.9) - \ln(1))]/t_s = -0.105/t_s$$

However, this is not the only pole. If we consider a much more general possibility, that $s = -0.150/t_s \pm j2n\pi/t_s$, we find now that $e^{-st_s} = e^{(0.105 \pm j2n\pi)} = e^{0.105 \pm} e^{j2n\pi}$. This simplifies to $e^{0.105}$, whatever the value of n. Thus all the values of s given by $s = -0.105/t_s \pm j2n\pi/t_s$ are poles, and it is seen that there are an infinite number of them. They all have the same real component, namely $-0.105/t_s$, but have different

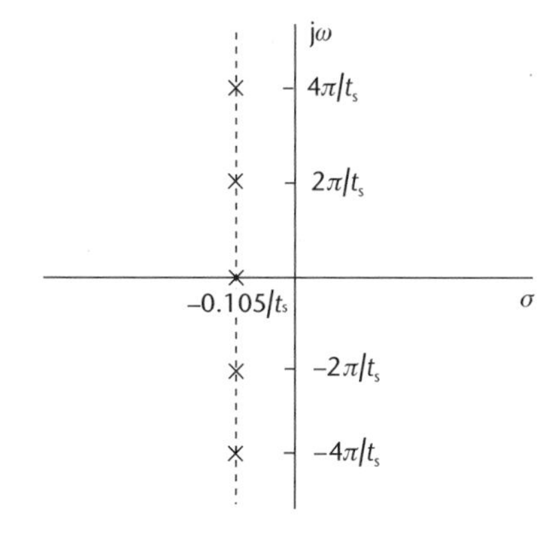

Fig. 14.12 s-plane plot of poles and zeros for a digital network

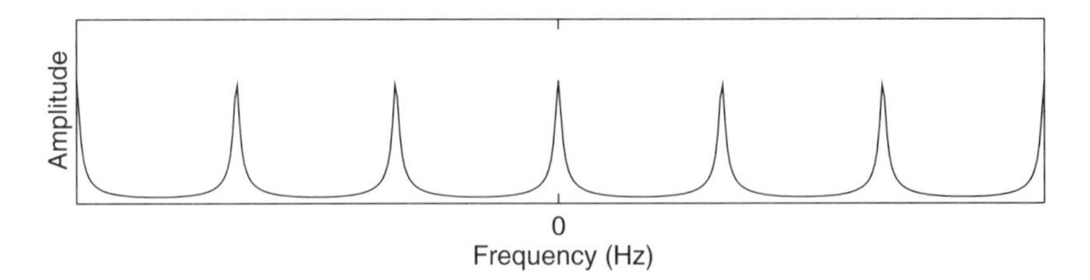

Fig. 14.13 Amplitude response corresponding to pole plot

imaginary components, namely $2n\pi/t_s$, where n is any positive or negative integer. A few of these poles are plotted in Figure 14.12. Being at $-\infty$ it is not possible to plot the zero.

Based upon the graphical technique introduced in Section 10.6 for deducing the frequency response from an s-plane plot of poles and zeros, it will be apparent that the periodic nature of the poles causes the amplitude response to be periodic in frequency, of the form shown in Figure 14.13.

A plot of poles and zeros provides invaluable insight into the frequency response, transient response and stability of a digital network in just the same way as for an analogue network. However, it is far from satisfactory to have s-plane plots such as that of Figure 14.12 in which we have a zero at $s = -\infty$ and an infinite number of poles; some alternative is desirable. First, though, it is important to confirm that the problem is a general one and not just a peculiarity of the network of Figure 14.1.

Considering the generalized digital signal processor of Figure 14.3 and treating it as an analogue delay network with an input $x(t)$, the input/output equation equivalent to the difference equation of Equation (14.4) is

$$y(t) = \sum_{i=0}^{M} b_i x(t - it_s) - \sum_{i=1}^{N} a_i y(t - it_s) \tag{14.16}$$

Putting $x(t) = \mathrm{e}^{st}$ and assuming $y(t) = H(s)\mathrm{e}^{st}$ gives

$$H(s)\mathrm{e}^{st} = \sum_{i=0}^{M} b_i \mathrm{e}^{s(t - it_s)} - \sum_{i=1}^{N} a_i H(s)\mathrm{e}^{s(t - it_s)}$$

The e^{st} terms divide out, leading to

$$H(s) = \frac{\sum_{i=0}^{M} b_i \mathrm{e}^{-sit_s}}{1 + \sum_{i=1}^{N} a_i \mathrm{e}^{-sit_s}} \tag{14.17}$$

(An alternative way of obtaining this same result would be to perform a Laplace transform on Equation (14.16) and to take the ratio $Y(s)/X(s)$).

It can now be shown that this generalized network has an *infinite number* of poles and zeros which are *periodic in frequency*.

Suppose that there is a pole at $s = -a + \mathrm{j}b$ such that the denominator of Equation (14.17) equals zero at this value of s, that is

$$1 + \sum_{i=1}^{N} a_i \mathrm{e}^{ait_s} \mathrm{e}^{-\mathrm{j}bit_s} = 0$$

Considering other values of s given by $s = -a + j(b \pm 2n\pi/t_s)$, the denominator becomes $1 + \sum_{i=1}^{N} a_i \mathrm{e}^{ait_s} \mathrm{e}^{-\mathrm{j}bit_s} \mathrm{e}^{\mp \mathrm{j}2n\pi i}$ and, because $\mathrm{e}^{\mp \mathrm{j}2n\pi i} = 1$, this again equals zero. This shows that, besides the pole at $s = -a + jb$, there are also poles at $s = -a + j(b \pm 2n\pi/t_s)$. It follows that there are an infinite number of periodic poles. By a similar argument any zeros arising from the numerator of Equation (14.17) will be periodic in frequency and infinite in number.

The consequence of the analysis above is that it is not possible to include all the poles and zeros of a digital network in an s-plane plot. This situation is far from satisfactory and it is worth looking for some better alternative.

14.7 A z-plane treatment of digital networks

Suppose we introduce a new variable z, defined by

$$z \equiv e^{st_s} \tag{14.18}$$

Considering the periodic poles just discussed in the previous section and given by $s = -a + j(b \pm 2n\pi/t_s)$, application of Equation (14.18) shows that the corresponding z values are $e^{-at_s}e^{j(bt_s \pm 2n\pi)}$. But, since $e^{\pm j2n\pi} = 1$, these reduce to $e^{-at_s}e^{jbt_s}$ and it can be seen the the periodic s-plane poles correspond to a *single* value of z having a magnitude e^{-at_s} and an angle e^{jbt_s}. The same simplification applies to zeros.

It follows that poles expressed in terms of z are not periodic in the way that they are when expressed in terms of s. If we plot the real and imaginary coordinates of the z values of the poles and zeros of a digital network we shall end up with a much simpler graphical representation of the network than if we plot the s-plane poles and zeros.

Example 14.3

Plot the z-plane poles and zeros of the transfer function of the network of Figure 14.1.

Solution Equation (14.14) gave the transfer function as $H(s) = 1/(1 - 0.9e^{-st_s})$. If we change the variable from s to z, where $z \equiv e^{st_s}$, we obtain a transfer function in terms of z, namely

$$H(z) = \frac{1}{1 - 0.9z^{-1}} = \frac{z}{z - 0.9}$$

$H(z)$ has a zero at $z = 0$ and a single pole at $z = 0.9$. These are plotted in Figure

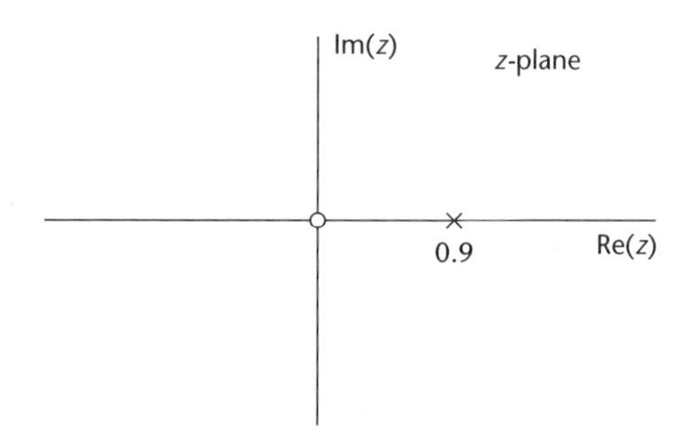

Fig. 14.14 z-plane plot of poles and zeros

14.14 and it is seen how much simpler and more satisfactory this is than the s-plane plot of Figure 14.12.

The use of the variable z may be usefully extended to all digital networks. If we substitute $z = e^{st_s}$ into the transfer function $H(s)$ given by Equation (14.17), we obtain a function of z, namely

$$H(z) = \frac{\sum_{i=0}^{M} b_i z^{-i}}{1 + \sum_{i=1}^{N} a_i z^{-i}} \tag{14.19}$$

This can be factorized. Multiplication of numerator and denominator by z^M if $M \geqslant N$, or by z^N if $N \geqslant M$, causes both to become polynomials in positive powers of z. Factorization then allows $H(z)$ to be expressed in the form

$$H(z) = \frac{K(z - z_1)(z - z_2)(z - z_3)\cdots}{(z - p_1)(z - p_2)(z - p_3)\cdots} \tag{14.20}$$

The terms z_1, z_2 and so on are the values of z which cause this expression to be zero, and p_1, p_2 and so on are the values of z which cause it to be infinite. Thus $z_1, z_2, \ldots$ are the zeros of the transfer function, while $p_1, p_2, \ldots$ are the poles. In general they have complex values, and a plot of their real and imaginary values in the complex plane produces a z-plane plot of the poles and zeros. It will be noted that the same symbols $z_1, z_2, \ldots$ and $p_1, p_2, \ldots$ have been used for the z-plane zeros and poles as for the s-plane zeros and poles. This is only because z and p are very appropriate symbols for zeros and poles and not because they have the same values in the two planes. They do not.

14.8 Mapping between the s-plane and the z-plane

A mapping exists between the s-plane and the z-plane. Points on the s-plane may be transformed into points on the z-plane using the relationship $z = e^{st_s}$. Points on the z-plane may be transformed to points on the s-plane using the inverse of this, namely $s = (\ln z)/t_s$. However, this latter relationship is incomplete. Because $e^{\pm j2n\pi} = 1$, we must permit a more flexible relationship to give all possible s values, which is

$$s = (\ln z)/t_s \pm j2n\pi \tag{14.21}$$

In the s-plane the imaginary axis has been shown to be of particular significance, since moving along it corresponds to changing the frequency. Substituting $j\omega$ for s in the expression $z = e^{st_s}$ gives $z = e^{j\omega t_s}$. In Steinmetz notation this is $z = 1\angle\omega t_s$ and a change of frequency signifies a movement around a circle of unity radius in the z-plane. Each increase of $2\pi/t_s$ rad/s, or $1/t_s$ Hz, corresponds to one revolution in the z-plane. It follows that the imaginary axis in the s-plane transforms to the 'unit circle' in the z-plane. These concepts are illustrated in Figures 14.15(a) and 14.15(b).

The point in the z-plane corresponding to zero frequency is where z equals $1\angle 0°$; that is, where the unit circle intercepts the real axis at $z = 1$. The same point can be described by $1\angle 2n\pi$ and therefore also corresponds to the radian frequency values $2\pi/t_s$, $4\pi/t_s$, $6\pi/t_s$, ... (or cyclic frequency values $1/t_s$, $2/t_s$, $3/t_s$, ...).

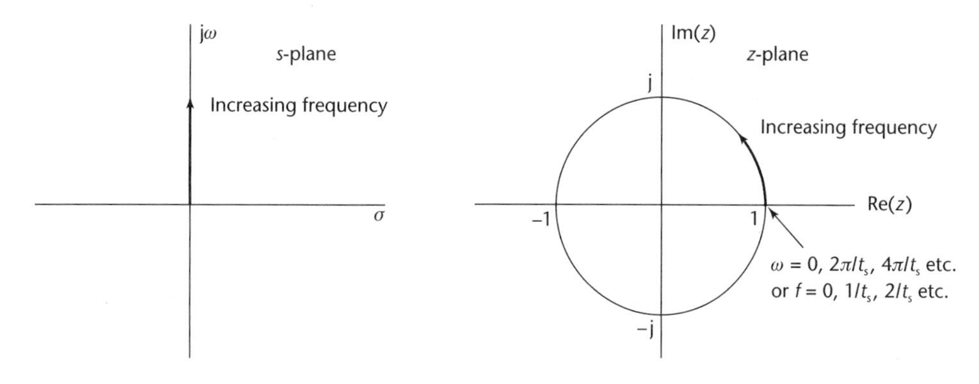

Fig. 14.15 Mapping of the imaginary axis in s-plane to the unit circle in z-plane

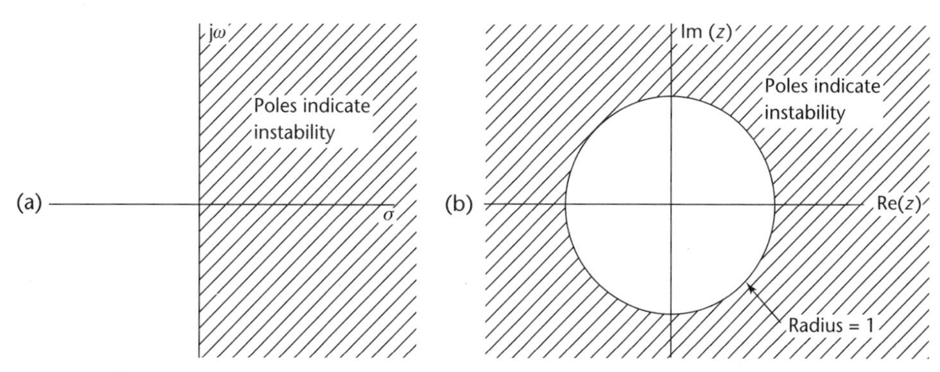

Fig. 14.16 Pole positions indicating instability: (a) s-plane; (b) z-plane

Inserting $s = \sigma + \mathrm{j}\omega$ into the definition $z = \mathrm{e}^{st_s}$ gives $z = \mathrm{e}^{\sigma t_s}\mathrm{e}^{\mathrm{j}\omega t_s} = \mathrm{e}^{\sigma t_s}\angle\omega t_s$. A negative value of σ gives a radius that is less than unity and the left-hand half of the s-plane is thus seen to map to the inside of the unit circle in the z-plane. The right-hand half of the s-plane, for which $\sigma > 0$, maps to the outside of the unit circle. Therefore, just as poles in the right-hand half of the s-plane indicate instability, so do poles outside the unit circle in the z-plane. These concepts are illustrated by Figures 14.16(a) and 14.16(b).

14.9 The frequency response of a digital network from a geometrical interpretation of z-plane poles and zeros

We can find the value of the transfer function at a frequency ω by replacing s with $\mathrm{j}\omega$ in Equation (14.17), or by replacing z by $\mathrm{e}^{j\omega t_s}$ in Equations (14.19) or (14.20). Doing the latter to Equation (14.20) we obtain

$$H(\omega) = \frac{K(\mathrm{e}^{\mathrm{j}\omega t_s} - z_1)(\mathrm{e}^{\mathrm{j}\omega t_s} - z_2)(\mathrm{e}^{\mathrm{j}\omega t_s} - z_3)\cdots}{(\mathrm{e}^{\mathrm{j}\omega t_s} - p_1)(\mathrm{e}^{\mathrm{j}\omega t_s} - p_2)(\mathrm{e}^{\mathrm{j}\omega t_s} - p_3)\cdots} \tag{14.22}$$

The amplitude response at frequency ω is obtained by taking the magnitude of this transfer function. Hence

$$|H(\omega)| = \frac{K|\mathrm{e}^{\mathrm{j}\omega t_s} - z_1||\mathrm{e}^{\mathrm{j}\omega t_s} - z_2||\mathrm{e}^{\mathrm{j}\omega t_s} - z_3|\cdots}{|\mathrm{e}^{\mathrm{j}\omega t_s} - p_1||\mathrm{e}^{\mathrm{j}\omega t_s} - p_2||\mathrm{e}^{\mathrm{j}\omega t_s} - p_3|\cdots} \tag{14.23}$$

This may be evaluated numerically. However, a very useful application of a z-plane plot of poles and zeros arises from the way in which it can be used to provide a graphical means of obtaining the same result.

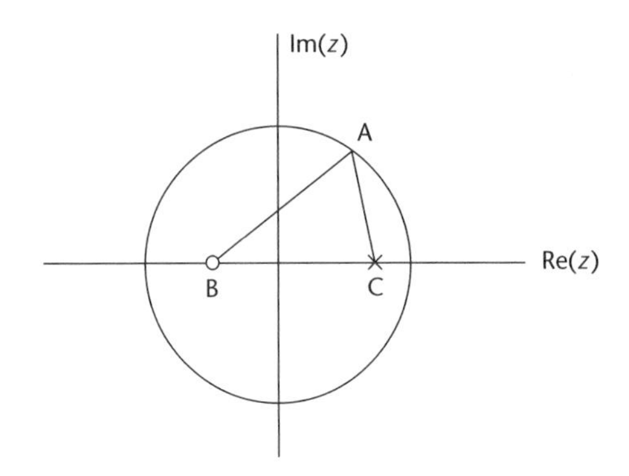

Fig. 14.17 Determination of amplitude response from a pole–zero plot

Considering a specific frequency point on the unit circle, Equation (14.23) can be interpreted as the product of the distances of the zeros to this point, divided by the product of the distances of the poles to this point and multiplied by the constant K. This ratio changes as one moves around the unit circle, thus enabling the amplitude response to be determined. For example, in the z-plane plot of $H(z) = (z + 0.5)/(z - 0.7)$ shown in Figure 14.17, point A corresponds to some specific frequency ω_1, and the amplitude response at this frequency is given by the ratio AB/AC.

Example 14.4

Determine the approximate amplitude response of a network which has z-plane poles at $0.95\angle \pm 45°$ and a zero at the origin, as shown in Figure 14.18. Assume that the samples are 1 ms apart, such that the sampling frequency is 1 kHz.

Solution One revolution around the unit circle corresponds to changing the frequency by 1 kHz. By considering the distance to the poles and zeros as one progresses around the unit circle it is apparent that the denominator of Equation (14.23) will become very small whenever the frequency becomes close to that of any of the poles. However, since one revolution around the unit circle is 1 kHz and the angle of the poles is one eighth of a revolution, this happens when $f = \pm 125$ Hz. It also happens if f is increased or decreased by any multiple of the sampling frequency; that is, when $f = (\pm 125 + 1000n)$ Hz. It follows that the amplitude response will correspond to that of a periodic bandpass filter and be of the form shown in Figure 14.19.

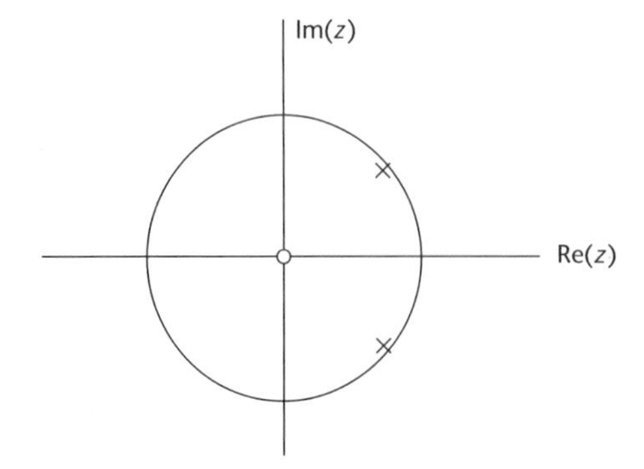

Fig. 14.18 Pole–zero plot

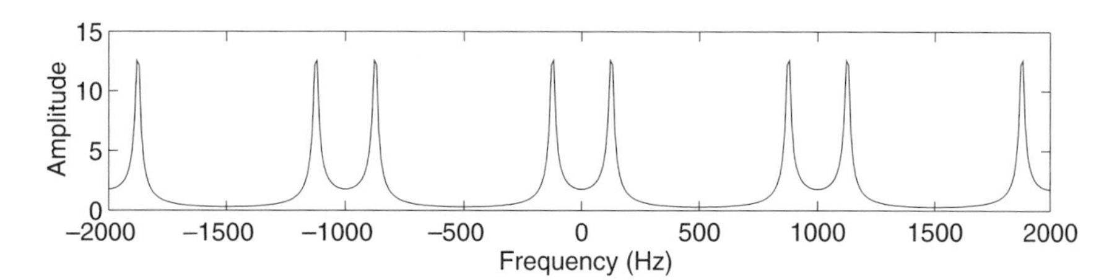

Fig. 14.19 Predicted amplitude response

Example 14.5

Derive the z-plane plot that represents the following DSP algorithm in MATLAB.

```
% ex14_5.m
load C14data.mat          % The data file of x that was created in Section 14.2
y=zeros(1,200);           % As in previous examples we initialize y to zero
for n=3:200
y(n)=x(n)+1.45*y(n-1)-1.04*y(n-2);
end
```

If the input data is obtained by sampling at 8 kHz, make any obvious deductions regarding frequency response, stability and so on. Show how minor changes to the algorithm can change its usefulness.

Solution The algorithm is readily seen to accord with the difference equation $y[n] = x[n] + 1.45y[n-1] - 1.04y[n-2]$, and this corresponds with the configuration of Figure 14.20.

The coefficients are given by $-a_1 = 1.45$, $-a_2 = -1.04$. Therefore, using Equation (14.19),

$$H(z) = \frac{\sum_{i=0}^{M} b_i z^{-i}}{1 + \sum_{i=1}^{N} a_i z^{-i}} = \frac{1}{1 - 1.45z^{-1} + 1.04z^{-2}} = \frac{z^2}{z^2 - 1.45z + 1.04} \tag{14.24}$$

(An alternative and probably easier way of obtaining this arises later when the z transform is introduced in Section 14.12. It will be shown that the z transform of $y[n-k]$ is $z^{-k}Y(z)$. Thus one simply performs a z transform on the difference equation, to obtain $Y(z) = X(z) + 1.45z^{-1}Y(z) - 1.04z^{-2}Y(z)$, and then takes the ratio $Y(z)/X(z)$).

The roots of the denominator may be found by equating it to zero and solving the resultant quadratic equation.

$$p_{1,2} = \frac{1.45 \pm \sqrt{1.45^2 - 4.16}}{2} = 0.725 \pm \mathrm{j}0.717 \tag{14.25}$$

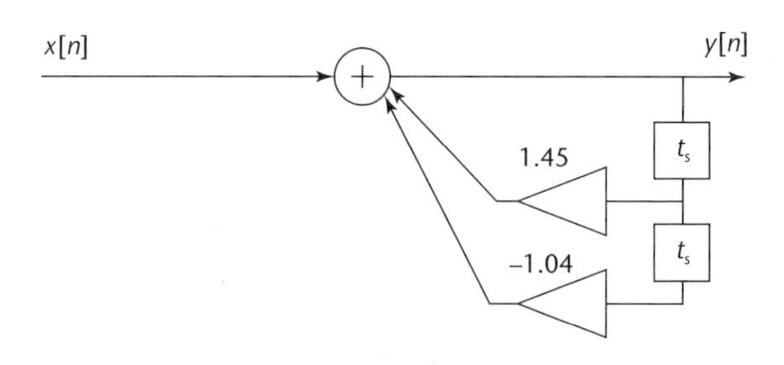

Fig. 14.20 Realization structure of digital filter

Hence

$$H(z) = \frac{(z - z_1)(z - z_2)}{(z - p_1)(z - p_2)} \tag{14.26}$$

where

$$z_1 = z_2 = 0; \quad p_1 = 0.725 + \text{j}0.717; \quad p_2 = 0.725 - \text{j}0.717 \tag{14.27}$$

These poles and zeros are plotted in Figure 14.21, together with the unit circle. Because the poles lie outside the unit circle it is immediately apparent that the system is unstable. This is confirmed by running the program. Figure 14.22 shows the variable **y** plotted by following the program with the instructions

plot(y), ylabel('y[n]'), xlabel('n')

If the poles were, however, to fall just inside the unit circle, the system would behave like a very sharp bandpass filter, just as in Example 14.4. Since the frequency change for one complete revolution of the unit circle is $1/t_s$, or 8 kHz in this example, and since the upper pole is an eighth of the way around, it is seen that the centre of the bandpass filter would have been close to 1 kHz.

It does not require a major change to make the system act as a stable bandpass filter. Let the algorithm be changed to

```
% ex14_5.m
load C14data.mat          % The data file of x that was created in Section 14.2
y=zeros(1,200);           % As in previous examples we initialize y to zero
for n=3:200
y(n)=x(n)+1.358*y(n-1)-0.922*y(n-2);
end
```

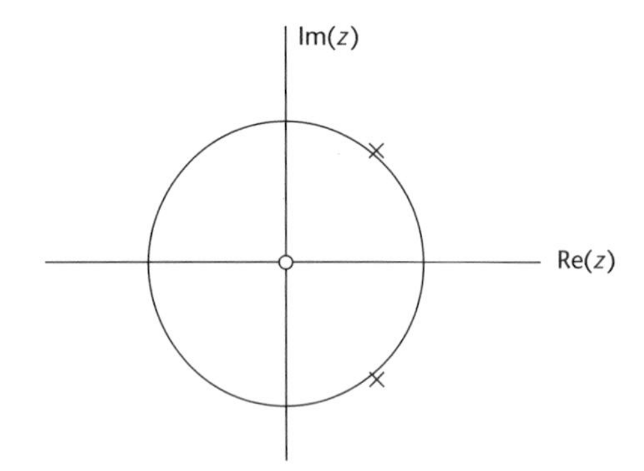

Fig. 14.21 Pole–zero plot

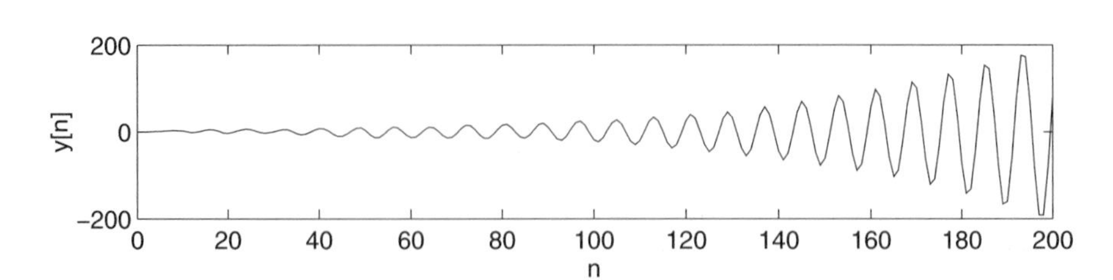

Fig. 14.22 Outcome of unstable filter

A repeat of the above calculation shows that the poles are now at $0.96\angle \pm 45°$ and this constitutes a sharp but stable bandpass filter with a centre frequency of 1 kHz. The output is not presented since 1 kHz bandpass filtering is inappropriate to separating the sinusoid from noise in this particular data set.

A procedure similar to that given for predicting the amplitude response can be applied for determining the phase response. This is done by taking the phase of Equation (14.22). This gives

$$\begin{aligned}\angle H(\omega) &= \angle(\mathrm{e}^{\mathrm{j}\omega t_s} - z_1) + \angle(\mathrm{e}^{\mathrm{j}\omega t_s} - z_2) + \angle(\mathrm{e}^{\mathrm{j}\omega t_s} - z_3) + \cdots \\ &\quad - \angle(\mathrm{e}^{\mathrm{j}\omega t_s} - p_1) - \angle(\mathrm{e}^{\mathrm{j}\omega t_s} - p_2) - \angle(\mathrm{e}^{\mathrm{j}\omega t_s} - p_3) - \cdots\end{aligned} \tag{14.28}$$

This can be expressed alternatively as

$$\theta(\omega) = \alpha_1 + \alpha_2 + \alpha_3 + \cdots - \beta_1 - \beta_2 - \beta_3 - \cdots \tag{14.29}$$

where Figure 14.23 shows the angles appropriate to a specific pole–zero plot and a frequency ω_1. It should be noted that the angle to the origin in this example should be used twice since there is a double zero at the origin.

14.10 An analytical approach to the frequency response of a digital network

Equation (14.22) derived the frequency transfer function from the factorized z transfer function of Equation (14.20). Alternatively, we can replace z by $\mathrm{e}^{\mathrm{j}\omega t_s}$ in Equation (14.19) to obtain

$$H(\omega) = \frac{\sum_{i=0}^{M} b_i \mathrm{e}^{-\mathrm{j}\omega t_s i}}{1 + \sum_{i=1}^{N} a_i \mathrm{e}^{-\mathrm{j}\omega t_s i}} \tag{14.30}$$

In terms of the digital radian frequency defined in Section 8.5 by $\Omega = \omega / f_s = \omega t_s$ radians/sample this gives

$$H(\Omega) = \frac{\sum_{i=0}^{M} b_i \mathrm{e}^{-\mathrm{j}\Omega i}}{1 + \sum_{i=1}^{N} a_i \mathrm{e}^{-\mathrm{j}\Omega i}} \tag{14.31}$$

We can determine the amplitude and phase responses from these equations.

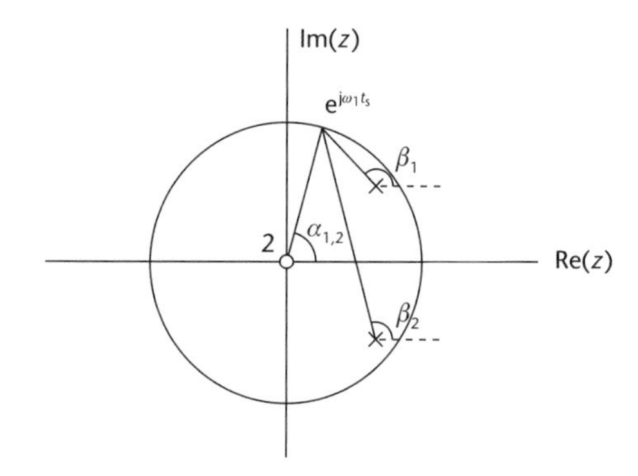

Fig. 14.23 Geometrical construction for determining phase response

Alternatively, we can use the digital cyclic frequency defined in Section 8.5 by $F = f/f_s = ft_s$ cycles/sample, and this gives

$$H(F) = \frac{\sum_{i=0}^{M} b_i e^{-j2\pi Fi}}{1 + \sum_{i=1}^{N} a_i e^{-j2\pi Fi}} \tag{14.32}$$

Example 14.6

Determine the amplitude response of the five point averager of Figure 14.4.

Solution The non-recursive coefficients are given by $b_0 = b_1 = b_2 = b_3 = b_4 = 0.2$. The recursive cofficients are all zero. Hence Equation (14.31) gives

$$H(\Omega) = \sum_{i=0}^{4} 0.2e^{-j\Omega i} = 0.2(e^{-j0} + e^{-j\Omega} + e^{-j2\Omega} + e^{-j3\Omega} + e^{-j4\Omega})$$

This expression acquires a symmetry if $e^{-j2\Omega}$ is taken outside the bracket, to give

$$\begin{aligned} H(\Omega) &= 0.2e^{+j2\Omega}(e^{+j2\Omega} + e^{+j\Omega} + e^{j0} + e^{-j\Omega} + e^{-j2\Omega}) \\ &= 0.2e^{-j2\Omega}(2 \cos 2\Omega + 2 \cos \Omega + 1) \end{aligned}$$

The magnitude of $e^{-j2\Omega}$ is unity and hence the amplitude response is given by

$$|H(\Omega)| = 0.4(\cos 2\Omega + \cos \Omega + 0.5)$$

or by

$$|H(F)| = 0.4(\cos 4\pi F + \cos 2\pi F + 0.5)$$

This is plotted in Figure 14.24 and the periodicity is noted. The 3 dB bandwidth (corresponding to $|H(F)| = 1/\sqrt{2}$) occurs when $F = 0.0907$. Using the relationship that $f = F/t_s = Ff_s$ this signifies a 3 dB bandwidth of 9.07 Hz if the sampling frequency is 100 Hz, 90.7 Hz if the sampling frequency is 1000 Hz, and so on.

Example 14.7

In Example 14.3 it was shown that the z transfer function of Figure 14.1 is given by $H(z) = z/(z - 0.9)$. Assuming a sampling frequency of 10 kHz, demonstrate how the frequency response may be obtained (a) analytically and (b) graphically.

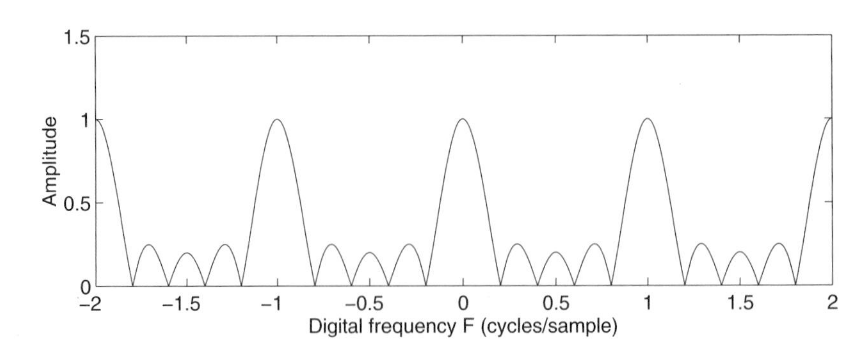

Fig. 14.24 Amplitude response of five point averager

Table 14.1 Sample values of amplitude response

f(Hz)	$\vert H(f)\vert$
0	10
100	8.59
200	6.43
300	4.89
600	2.71
1000	1.68
2000	0.89
3000	0.65
4000	0.55
5000	0.53

Solution Replacing z in $H(z)$ by $e^{j\omega t_s}$ and taking the magnitude we obtain

$$|H(\omega)| = \frac{|e^{j\omega t_s}|}{|e^{j\omega t_s} - 0.9|} = \frac{1}{|\cos(\omega t_s) + j\sin(\omega t_s) - 0.9|}$$

$$= \frac{1}{\sqrt{[\cos(\omega t_s) - 0.9]^2 + \sin^2(\omega t_s)}}$$

Using the result that $\sin^2 A + \cos^2 A = 1$ this simplifies to

$$|H(f)| = \frac{1}{\sqrt{1.81 - 1.8\cos(\omega t_s)}} = \frac{1}{\sqrt{1.81 - 1.8\cos(2\pi f/10000)}}$$

This is evaluated for different values of frequency in Table 14.1.

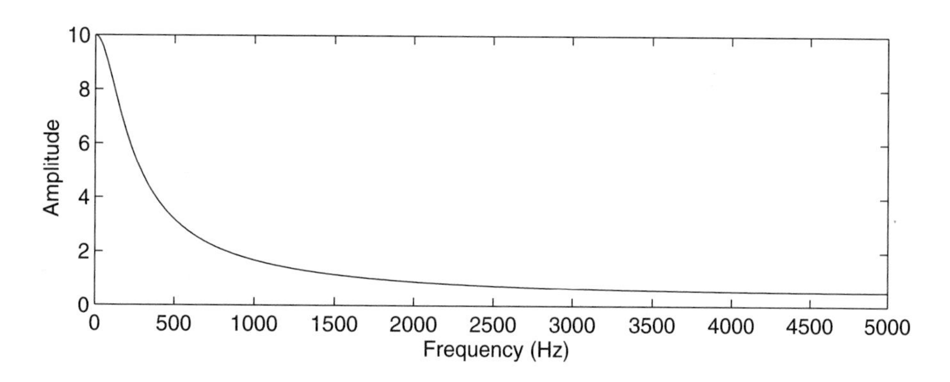

Fig. 14.25 Amplitude response of filter

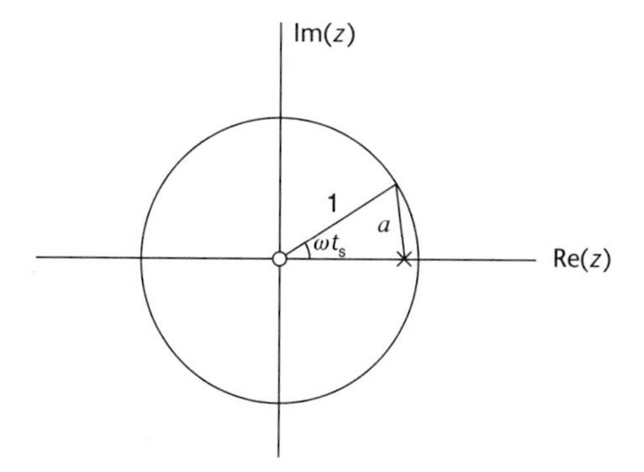

Fig. 14.26 Geometric construction for evaluating amplitude response

The results for these and other frequencies are plotted in Figure 14.25.

The alternative is to obtain an approximate answer graphically using the construction of Figure 14.26.

In this example the distance to the pole changes with frequency, but the distance to the zero is constant. The amplitude response is thus $|H(\omega)| = 1/a$ and Figure 14.25 is clearly appropriate.

14.11 Frequency responses and pole–zero plots using MATLAB

If we recognize that $a_0 = 1$ the transfer function of Equation (14.19) can be rewritten as

$$H(z) = \frac{\sum_{i=0}^{M} b_i z^{-i}}{1 + \sum_{i=1}^{N} a_i z^{-i}} = \frac{\sum_{i=0}^{M} b_i z^{-i}}{\sum_{i=0}^{N} a_i z^{-i}} \tag{14.33}$$

If we express the coefficients in the numerator by a vector **b** and those in the denominator by a vector **a** we can use the MATLAB command **freqz(b,a)** to plot the frequency response and the MATLAB command **zplane(b,a)** to produce a pole–zero plot. It should be noted that both vectors should include coefficients down to the most negative power in z in either one, and that any coefficients that are missing based on this criterion should be entered as zeros. (A good alternative is to express $H(z)$ as the ratio of two polynomials in *positive* powers in z and then to specify the coefficients of these polynomials.)

Example 14.8

Use MATLAB to produce a pole–zero plot and to display the frequency response of the filter examined in Example 14.7 for which the transfer function was given by $H(z) = z/(z - 0.9)$.

Solution We can enter the two vectors directly from the coefficients in the numerator and denominator of the given expression for $H(z)$. Alternatively, dividing through by z to express the transfer function in the form of Equation (14.33) we obtain $H(z) = 1/(1 - 0.9z^{-1})$ with the most negative power in z being z^{-1}. Whichever approach is adopted, a suitable set of MATLAB commands is thus

```
b=[1 0]; a=[1 -0.9];   % Note that is incorrect to enter b=[1];
subplot(2,2,1)         % This is an easy way of reducing the size of the plot to a
                       % quarter of the screen while maintaining the full size of
                       % the labelling
zplane(b,a)
```

The outcome is shown in Figure 14.27. Clearly this is a trivial example, since the poles and zeros are apparent from the original expression for $H(z)$. In more difficult cases the z-plane command performs the necessary factorization of numerator and denominator before producing the plot.

Continuing to now produce frequency response curves we enter

```
freqz(b,a)
```

This produces Figure 14.28.

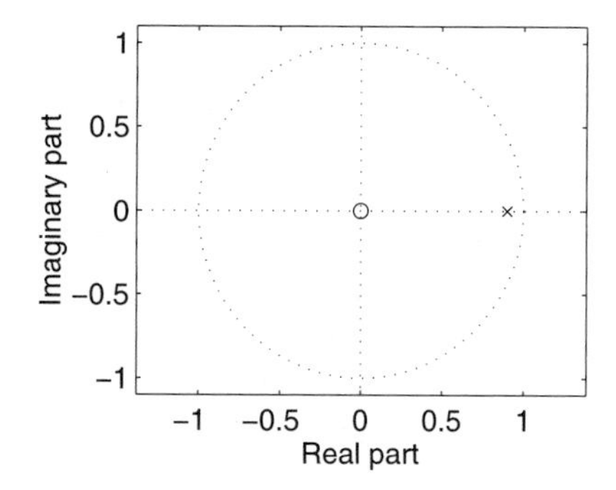

Fig. 14.27 Use of MATLAB to produce pole–zero plots

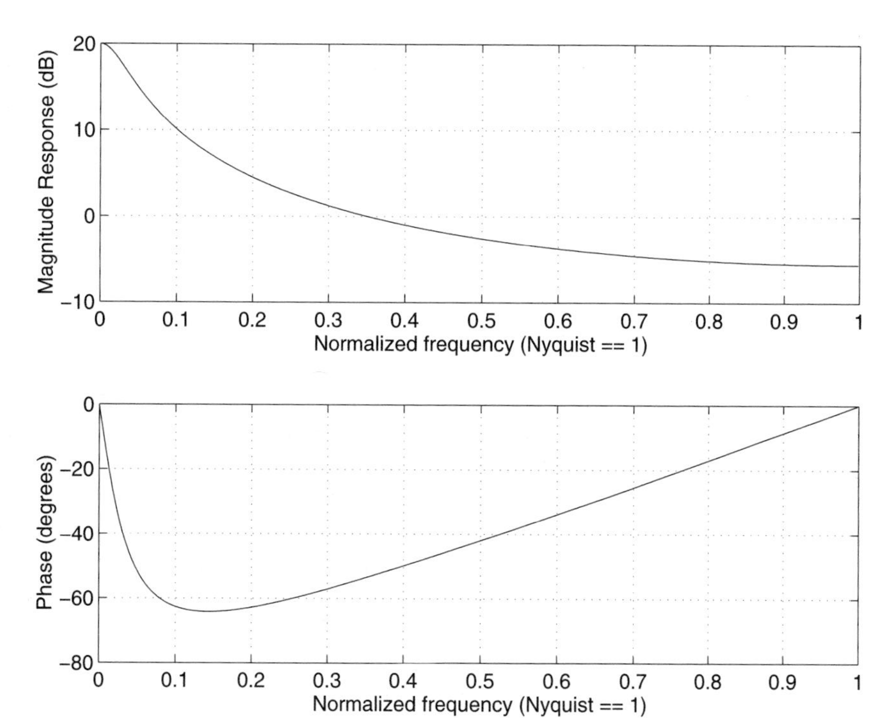

Fig. 14.28 Frequency response plot using MATLAB

Clearly the power of the software is not fully exploited in this very simple example. It should be noted that, in accordance with the definitions given in Section 8.5, a normalized or Nyquist frequency value of 1 in these plots corresponds to a digital frequency of 0.5 cycles/sample (or π radians/sample).

14.12 The *z* transform

So far this chapter has introduced z transfer functions. For digital networks it has been shown that these are simpler in form than s domain transfer functions; also that their resulting z-plane plots are much more compact and easier to interpret than the corresponding s-plane plots.

A *transform* is an operation that is normally applied to *signals* rather than to networks. We could take the Laplace transform of the impulse response of a network (in which case we would obtain the transfer function $H(s)$ of the network), but the Laplace transform is more general than this and can be applied to a much broader class of signals than just impulse responses. Similarly, the z transform of the impulse response of a digital network will now be shown to equal the z transfer function of the network, but can just as well be applied to other signals.

Let us consider the example of the non-recursive network in Figure 14.29. By inspection the response to a discrete-time impulse $\delta[n]$ can be expressed as

$$h[n] = \{\underset{\uparrow}{3}, 5, 7, 9, 2, 4\} \tag{14.34}$$

and the response to a continuous-time impulse $\delta(t)$ can be expressed as

$$h(t) = 3\delta(t) + 5\delta(t - t_s) + 7\delta(t - 2t_s) + 9\delta(t - 3t_s) + 2\delta(t - 4t_s) + 4\delta(t - 5t_s) \tag{14.35}$$

In the latter case the Laplace transform of $h(t)$ is given by

$$H(s) = 3 + 5e^{-st_s} + 7e^{-2st_s} + 9e^{-3st_s} + 2e^{-4st_s} + 4e^{-5st_s} \tag{14.36}$$

Making the substitution $z = e^{st_s}$, this gives

$$H(z) = 3 + 5z^{-1} + 7z^{-2} + 9z^{-3} + 2z^{-4} + 4z^{-5} \tag{14.37}$$

This can be interpreted either as the z transform of the continuous-time impulse response given in Equation (14.35) or as the z transform of the discrete-time impulse response given in Equation (14.34). The latter is the more usual.

The z transform of an impulse response can be regarded as a power series in z^{-1} whose coefficients correspond to the successive values of the impulse response. On this basis Equation (14.37) could have been obtained *directly* from Equation (14.34) and, except for demonstrating the relationship between z transforms and Laplace transforms, it is not necessary or helpful to involve the intermediate stage of the Laplace transform.

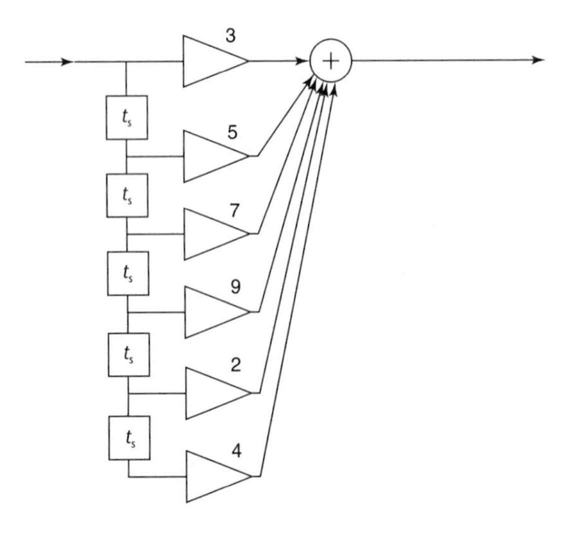

Fig. 14.29 Example of non-recursive network

It is useful to note that the same result as Equation (14.37) could have been obtained from Equation (14.19), which was that

$$H(z) = \frac{\sum_{i=0}^{M} b_i z^{-i}}{1 + \sum_{i=1}^{N} a_i z^{-i}}$$

thus demonstrating that the z transform of the impulse response of the network does indeed equal the z transfer function of the network.

The application of the z transform is not constrained to impulse responses but can apply to any signal. As with an impulse response, the z transform of any sequence $x[n]$ commencing at $n \geqslant 0$ can be regarded as a power series in z^{-1} whose coefficients correspond to the successive values of the signal. Thus

$$X(z) = x[0] + x[1]z^{-1} + x[2]z^{-2} + \cdots = \sum_{n=0}^{\infty} x[n]z^{-n} \tag{14.38}$$

where z is a complex variable.

Although the approach in this chapter has been to define z as equalling e^{st_s} and then to show that the Laplace transform leads to Equation (14.38) when this change of variable is made, a common alternative is to regard Equation (14.38) as the definition of the z transform.

Just as the Laplace transform was constrained in Equation (12.7) to signals which were zero for $t < 0$, so this z transform is constrained to discrete signals which are zero for $n < 0$. It is for this reason that, strictly, the definition of Equation (14.38) should be termed the *unilateral* z transform. Just as there is a *bilateral* Laplace transform, there is also a *bilateral* z transform which can cover the case of signals existing for $n < 0$. The *bilateral* z transform is defined as

$$X(z) = \sum_{n=-\infty}^{\infty} x[n]z^{-n} \tag{14.39}$$

However, the bilateral z transform is rarely necessary, can give problems, and will not be covered in this book except briefly in Section 14.16 in the context of convolution.

It will be noted that z can be thought of simply as a time shift operator. Multiplication by z corresponds to advancing a sample by one sampling interval. Multiplication by z^{-1} corresponds to delaying a sample by one sampling interval. Hence, if the z transform of $\delta[n]$ is 1, the z transform of $\delta[n-1]$ is z^{-1}, the z transform of $\delta[n-2]$ is z^{-2} and so on. If the z transform of a sequence $x[n]$ is $X(z)$ this concept can be extended further to give the z transform of $x[n-1]$ as $z^{-1}X(z)$, and the z transform of $x[n-2]$ as $z^{-2}X(z)$ and so on. Generalizing this we have

$$x[n-k] \leftrightarrow z^{-k}X(z) \tag{14.40}$$

Example 14.9

Determine the z transform of the sequence

$$y[n] = \{\underset{\uparrow}{0}, 3, 5, 7, 9, 2, 4\}.$$

Solution Using the definition of Equation (14.38) we can write down directly that

$$Y(z) = 3z^{-1} + 5z^{-2} + 7z^{-3} + 9z^{-4} + 2z^{-5} + 4z^{-6}$$

Alternatively, we can recognize that $y[n]$ is the same as the sequence $h[n]$ in Equation (14.34) except that all the samples are delayed by one sampling interval. Therefore, applying Equation (14.40) to the result of Equation (14.37) that $H(z) = 3 + 5z^{-1} + 7z^{-2} + 9z^{-3} + 2z^{-4} + 4z^{-5}$ we have

$$Y(z) = z^{-1}H(z) = 3z^{-1} + 5z^{-2} + 7z^{-3} + 9z^{-4} + 2z^{-5} + 4z^{-6}$$

Elsewhere in this book the attempt has been made to give the Fourier and Laplace transforms a *physical* interpretation as representing the coefficients of basis functions. In the case of the Fourier transform these basis functions are sinusoids and cosinusoids, which are combined in pairs as complex exponential waveforms. There is an infinite set of these basis functions, each of a different frequency. In the case of the Laplace transform these basis functions have an extra degree of freedom in that the complex exponential waveforms can increase or decay exponentially.

It is not easy to give the z transform the same physical interpretation. The approach adopted in this book has been to regard the z transform as being the same as the Laplace transform, except that it entails a change of variable from s to z based on the relationship $z = e^{st_s}$. The purpose of this change of variable is that it produces transforms which are simpler in form for discrete signals.

14.13 Some important z transforms

For some important signals the z transform is a geometric series that can be replaced by a compact closed form representation. Some examples follow.

14.13.1 The unit step function

The discrete unit step function is given by

$$\begin{aligned} u[n] &= 1 \qquad 0 \leqslant n \leqslant \infty \\ &= 0 \qquad n < 0 \end{aligned} \tag{14.41}$$

Using Equation (14.38) its z transform is given by

$$U(z) = 1 + z^{-1} + z^{-2} + z^{-3} + z^{-4} + \cdots \tag{14.42}$$

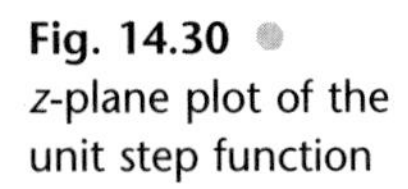

Fig. 14.30 z-plane plot of the unit step function

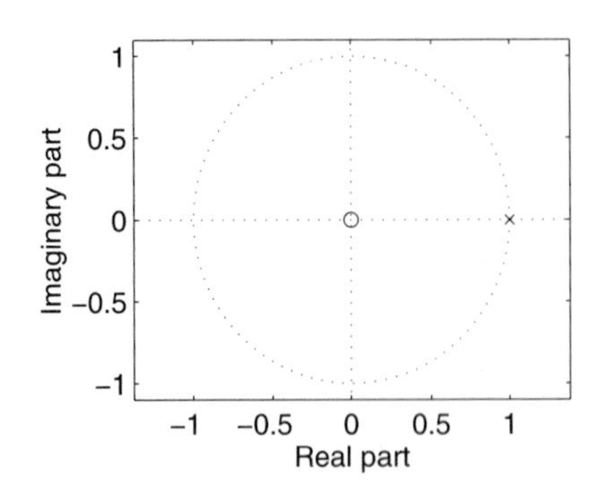

However, this is a well-known geometric series and may be rewritten in closed form as

$$U(z) = \frac{1}{1 - z^{-1}} = \frac{z}{z - 1} \tag{14.43}$$

Just as it has been shown earlier that a z transfer function of a network can be portrayed by a plot of z-plane poles and zeros, so also can the z transform of a sequence. In this example, where $U(z) = z/(z - 1)$ for the sequence $u[n]$, there is a zero at the origin and a pole at $z = 1$. These are shown in Figure 14.30.

14.13.2 The unit ramp

The z transform of a unit ramp $r[n]$ is given by

$$Z(r[n]) = R(z) = \frac{z}{(z - 1)^2} \tag{14.44}$$

where the symbol Z has been used to denote a z transform. This result is readily confirmed by long division, which gives $z^{-1} + 2z^{-2} + 3z^{-3} + 4z^{-4} + \cdots$ such that this clearly corresponds to a sequence

$$\{\underset{\uparrow}{0}, 1, 2, 3, 4, \ \ldots\}$$

The z-plane plot is shown in Figure 14.31, where, as before and in accordance with the MATLAB convention, the double pole is indicated by the '2'.

14.13.3 The decaying exponential

The z transform of a decaying exponential given by $a^n u[n]$ is given by

$$Z(a^n u[n]) = \frac{z}{(z - a)} \tag{14.45}$$

Again this result is confirmed by long division, which gives $1 + az^{-1} + a^2 z^{-2} + a^3 z^{-3} + \cdots$. The z-plane plot is shown in Figure 14.32. The unit step function is a special case of this in which $a = 1$.

14.13.4 The damped cosinusoid

The z transform of a damped cosinusoid $u[n]a^n \cos(2\pi Fn)$ is given by

$$Z\{u[n]a^n \cos(2\pi Fn)\} = \frac{z(z - a\cos(2\pi F))}{(z^2 - 2az\cos(2\pi F) + a^2)} \tag{14.46}$$

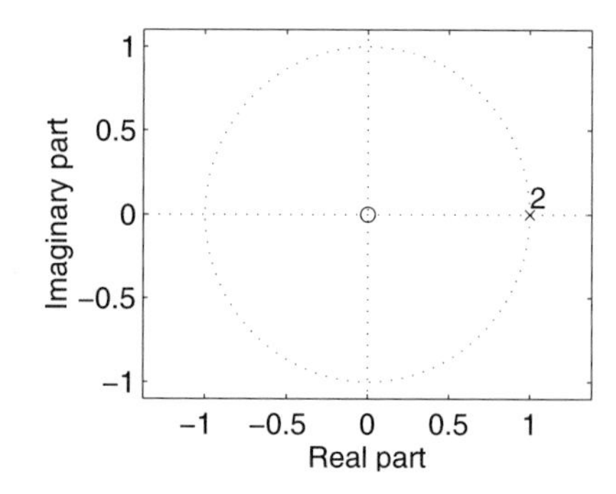

Fig. 14.31 z-plane plot of the unit ramp

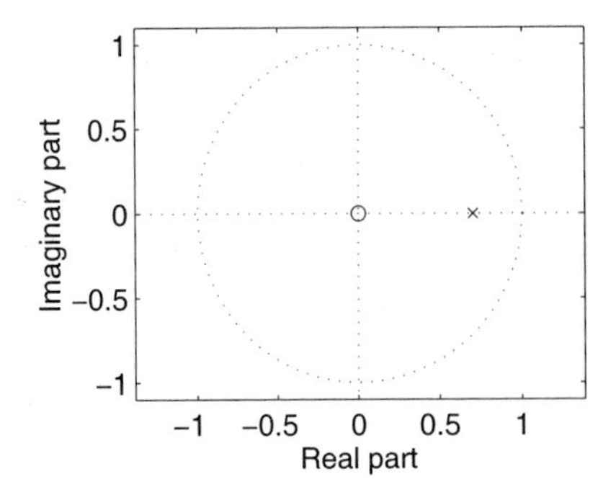

Fig. 14.32 z-plane plot of an exponentially decaying step function

Figure 14.33 shows the corresponding z-plane plots for the undamped case where $a = 1$ and for the damped case where $a < 1$.

14.13.5 The damped sinusoid

The z transform of a damped sinusoid $u[n]a^n \sin(2\pi Fn)$ is given by

$$Z\{u[n]a^n \sin(2\pi Fn)\} = \frac{az \sin(2\pi F)}{(z^2 - 2az \cos(2\pi F) + a^2} \tag{14.47}$$

Figure 14.34 shows the corresponding z-plane plots for the undamped case where $a = 1$ and for the damped case where $a < 1$.

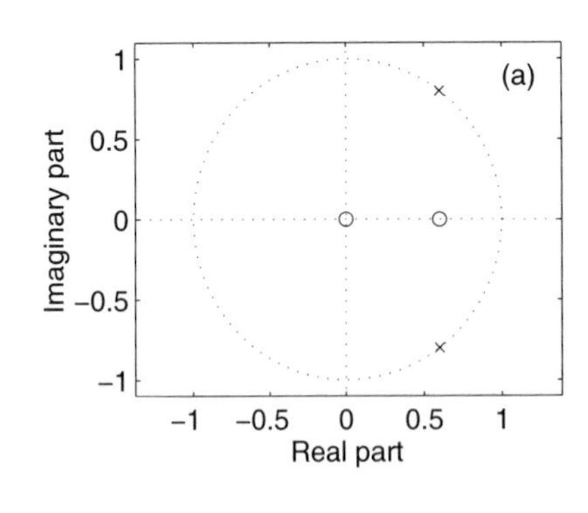

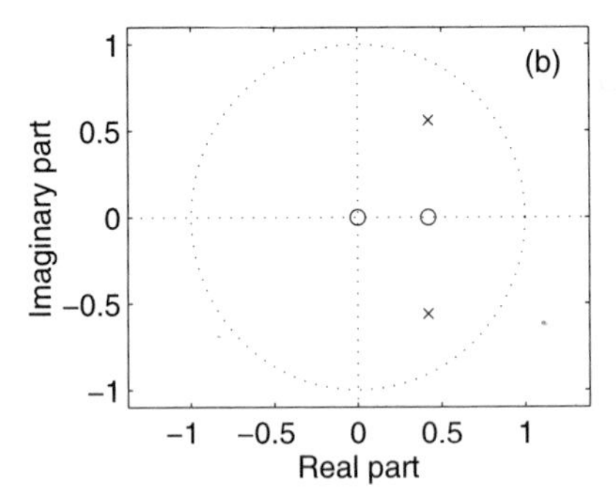

Fig. 14.33 z-plane plot of a cosinusoid: (a) undamped; (b) damped

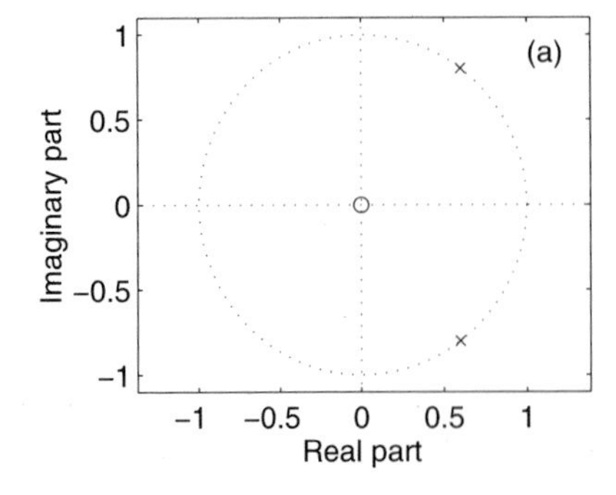

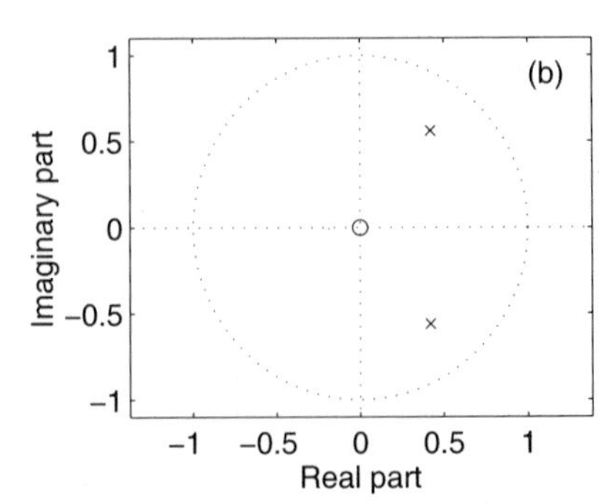

Fig. 14.34 z-plane plot of a sinusoid: (a) undamped; (b) damped

14.14 A direct derivation of the z transfer function

Section 14.6 derived the s transfer function of the generalized digital network of Figure 14.3 and then Section 14.7 obtained the z transfer function by replacing e^{st_s} by z. A better method now is to take the z transform of the difference equation directly.

Equation (14.4) gave us the difference equation

$$y[n] = \sum_{i=0}^{M} b_i x[n-i] - \sum_{i=1}^{N} a_i y[n-i]$$

If the z transform of $x[n]$ is $X(z)$ and we consider a single term in the first summation, such as $b_k x[n-k]$, it will be clear from Equation (14.40) that its z transform is $b_k z^{-k} X(z)$. Extending this treatment to taking the z transform of all terms in the difference equation, we obtain

$$Y(z) = \sum_{i=0}^{M} b_i z^{-i} X(z) - \sum_{i=1}^{N} a_i z^{-i} Y(z)$$

$$= X(z) \sum_{i=0}^{M} b_i z^{-i} - Y(z) \sum_{i=1}^{N} a_i z^{-i}$$

Hence

$$\frac{Y(z)}{X(z)} = H(z) = \frac{\sum_{i=0}^{M} b_i z^{-i}}{1 + \sum_{i=1}^{N} a_i z^{-i}} \tag{14.48}$$

This is the same result as Equation (14.19). Because of the connection established between the z transform and the Laplace transform it follows automatically that this z transfer function is also the z transform of the impulse response of the network.

14.15 The inverse z transform

If a signal is described by its z transform, and if this transform is in the form of a power series, the inverse transform may be performed directly from the definition of the z transform.

Example 14.10

Determine the inverse z transform of $X(z) = 3z^{-1} + 5z^{-2} + 2z^{-5} + 8z^{-7}$.

Solution Using the definition of the z transform given in Equation (14.38), but applying the inverse process, we obtain

$$x[n] = \{\underset{\uparrow}{0}, 3, 5, 0, 0, 2, 0, 8\}$$

Very often a z transform is in the form of the ratio between two polynomials. In the simplest case the z transform may be a well-known one (such as in Section 14.13), and then the inverse transform may be read off directly. For example, because

Section 14.13.1 gives the unilateral z transform of the step function $u[n]$ as $U(z) = z/(z-1)$, the inverse transform of $X(z) = kz/(z-1)$ is given by $x[n] = k$.

In more complicated cases one possible technique is to expand the z transform into a sum of simple partial fractions each of whose inverse transforms are known ones. This was the procedure adopted in Section 12.9 for performing inverse Laplace transforms. However, the method is less useful with inverse z transforms than with inverse Laplace transforms for several reasons:

- We can describe *any* sampled signal by its z transform, whereas we can only describe an analogue signal of *simple analytical form* by its Laplace transform. This means that, having applied a discrete signal to a digital network, we are interested in an infinite number of possible inverse z transforms. With analogue systems our analysis is constrained to much simpler input signals (such as a step function input to an automatic control system) and hence the inverse Laplace transforms needed are more limited.
- High-order digital networks are more common than high-order analogue networks because of the inherent simplicity of high-order digital networks such as the generalized configuration of Figure 14.3. (Note, however, that even low-order digital networks can have involved impulse responses.)
- Alternative methods are available that are usually simpler. They are based on the fact that the inverse z transform of a power series in z is straightforward, whereas the inverse Laplace transform of a power series in s is not.

In real life applications it is rare that an inverse z transform has a simple analytical form, and numerical methods are needed. The first one examined reduces the ratio of the two polynomials into a power series by long division.

Example 14.11

Determine the first four terms of the inverse transform of

$$F(z) = \frac{3z+2}{5z^2+4z+1}$$

Solution Using long division we have

$$\begin{array}{r|lllll}
 & 0.6z^{-1} & -\ 0.08z^{-2} & -\ 0.056z^{-3} & +\ 0.0608z^{-4} & \\
\hline
5z^2+4z+1\ ) & 3z & +2 & & & \\
 & 3z & +2.4 & +0.6z^{-1} & & \\
\cline{2-4}
 & & -0.4 & -0.6z^{-1} & & \\
 & & -0.4 & -0.32z^{-1} & -0.08z^{-2} & \\
\cline{3-5}
 & & & -0.28z^{-1} & +0.08z^{-2} & \\
 & & & -0.28z^{-1} & -0.224z^{-2} & -0.056z^{-3} \\
\cline{4-6}
 & & & & 0.304z^{-2} & +0.056z^{-3}
\end{array}$$

Hence $F(z) = 0.6z^{-1} - 0.08z^{-2} - 0.056z^{-3} + 0.0608z^{-4}$ ... and therefore

$$f[n] = \{\underset{\uparrow}{0}, +0.6, -0.08, -0.056, +0.0608, \cdots\}$$

It will be noted that there is no indication that subsequent terms are insignificant. Indeed there is some hint from the last term that they may have begun to diverge. In any practical application of this problem it would be advisable to evaluate a few more terms.

Example 14.12

Repeat Example 14.11 by means of partial fraction expansion.

Solution As in Example 12.7 we can use MATLAB to determine the partial fraction expansion. The program needed is

```
b=[3 2];
a=[5 4 1];
[r p k]=residue(b,a)
```

The outcome of this is that the residues are $r_1 = 0.3 - \text{j}0.4$, $r_2 = 0.3 + \text{j}0.4$; the poles are $p_1 = -0.4 + \text{j}0.2$, $p_2 = -0.4 - \text{j}0.2$; that is

$$F(z) = \frac{0.3 - \text{j}0.4}{z + 0.4 - \text{j}0.2} + \frac{0.3 + \text{j}0.4}{z + 0.4 + \text{j}0.2}$$

Unfortunately, this has no easy inverse transform. There is no simple analytical expression for the impulse response, thus confirming the advantages of the power series approach. It is interesting to note, however, that the poles lie within the unit circle and that the result of the inverse transform is therefore stable and non-divergent.

A method which is closely related mathematically to the power series long division approach, but which tends to be much simpler, is most easily understood by regarding the transform as a transfer function (even though it may not be) and determining the corresponding impulse response. This impulse response is then the required inverse z transform. The technique for doing this is to find the difference equation associated with the transfer function and then to determine the ouput resulting from an input impulse function.

Example 14.13

Repeat Example 14.11 by considering $F(z)$ as a transfer function and thence determining the impulse response from the associated difference equation.

Solution In keeping with the conventional nomenclature for a transfer function the transform will be relabelled $H(z)$; that is, we require to find the inverse z transform of

$$H(z) = \frac{3z + 2}{5z^2 + 4z + 1}$$

Using the symbols $x[n]$ and $y[n]$ for the input and output of a digital network, with transforms $X(z)$ and $Y(z)$, we thus have

$$H(z) = \frac{Y(z)}{X(z)} = \frac{3z + 2}{5z^2 + 4z + 1} = \frac{3z^{-1} + 2z^{-2}}{5 + 4z^{-1} + z^{-2}}$$

Hence, cross-multiplying

$$Y(z)(5+4z^{-1}+z^{-2}) = X(z)(3z^{-1}+2z^{-2})$$

and, following an inverse z transform,

$$y[n] = \tfrac{3}{5}x[n-1] + \tfrac{2}{5}x[n-2] - \tfrac{4}{5}y[n-1] - \tfrac{1}{5}y[n-2]$$

To find the impulse response corresponding to this difference equation we must put $x[n] = \delta[n]$. We can then determine the output by drawing up a table of input and output samples for different values of n.

n	$x[n]$	$y[n] = 0.6x[n-1] + 0.4x[n-2] - 0.8y[n-1] - 0.2y[n-2]$	
0	1		0
1	0	$0.6(1)+0.4(0)-0.8(0)-0.2(0)=$	0.6
2	0	$0.6(0)+0.4(1)-0.8(0.6)-0.2(0)=$	−0.08
3	0	$0.6(0)+0.4(0)-0.8(-0.08)-0.2(0.6)=$	−0.056
4	0	$0.6(0)+0.4(0)-0.8(-0.056)-0.2(-0.08)=$	0.0608
⋮	⋮	⋮	

From the right-hand column the impulse response associated with $H(z)$ is

$$h[n] = \{\underset{\uparrow}{0}, +0.6, -0.08, -0.056, +0.0608, \ldots\}$$

Hence the inverse z transform of $F(z)$ is

$$f[n] = \{\underset{\uparrow}{0}, +0.6, -0.08, -0.056, +0.0608, \ldots\}$$

as in Example 14.11.

A simple computer program is an ideal alternative to the table for determining the output predicted by the difference equation when the input is a unit impulse. In order to ensure that the two input and output samples prior to the unit impulse are zero, it is easiest to delay the unit impulse. For the case of a unit impulse at $n = 3$ a program in MATLAB for determining the impulse response for $1 \leqslant n \leqslant 20$ is given by

```
% ex14_13.m
x=zeros(1,20);
x(3)=1;
y=zeros(1,20);
for n=3:20
y(n)=.6*x(n-1)+.4*x(n-2)-.8*y(n-1)-.2*y(n-2);
end
y
```

The resulting printout is

```
y=
Columns 1 through 7
     0     0     0    0.6000    -0.0800    -0.0560    0.0608
Columns 8 through 14
    -0.0374    0.0178    -0.0067    0.0018    -0.0001    -0.0003    0.0002
```

Columns 15 through 20

-0.0001	0.0001	0.0000	0.0000	0.0000	0.0000

If this response is advanced by two samples the same result is obtained as before.

14.16 Discrete-time-convolution using the z transform

Equation (2.3) showed that, if a signal $x[n]$ is applied to a system whose impulse response is $h[n]$, the output $y[n]$ is given by the convolution sum, repeated here as

$$y[n] = \sum_{k=-\infty}^{\infty} x[k]h[n-k] \tag{14.49}$$

However, the concept of convolution need not be restricted to system responses. The convolution of two signals $x_1[n]$ and $x_2[n]$ is given by

$$y[n] = \sum_{k=-\infty}^{\infty} x_1[k]x_2[n-k] \tag{14.50}$$

For a signal $x_1[n]$ that has zero values for $n < 0$ this may be expanded as

$$y[n] = x_1[0]x_2[n-0] + x_1[1]x_2[n-1] + x_1[2]x_2[n-2] + \cdots \tag{14.51}$$

Taking the unilateral z transform of this, using the definition of Equation (14.35) and assuming $x_2[n]$ has zero values for $n < 0$, we obtain

$$\begin{aligned} Y(z) &= x_1[0]X_2(z) + x_1[1]z^{-1}X_2(z) + x_1[2]z^{-2}X_2(z) + \cdots \\ &= X_2(z)(x_1[0] + x_1[1]z^{-1} + x_1[2]z^{-2} + \cdots) \end{aligned}$$

Hence

$$Y(z) = X_1(z)X_2(z) \tag{14.52}$$

Equation (14.52) demonstrates that the convolution of two discrete-time sequences becomes a multiplicative process in the z domain. It says that we can convolve two sequences by multiplying their z transforms and then performing an inverse z transform. It adds further justification to the technique of discrete convolution demonstrated in Example 2.3, where the process of convolving two sequences was shown to be achievable by replacing the sequences with appropriate polynomials in z, multiplying them, and then transforming the resulting polynomial back into a sequence.

Example 14.14

Convolve the two sequences

$$x_1[n] = \{\underset{\uparrow}{4}, 2, 5\} \qquad \text{and} \qquad x_2[n] = \{\underset{\uparrow}{2}, 1, 3, 6\}$$

by means of the z transform.

Solution We have $X_1(z) = 4 + 2z^{-1} + 5z^{-2}$ and $X_2(z) = 2 + z^{-1} + 3z^{-2} + 6z^{-3}$. The product of these two polynomials can be achieved by long multiplication as follows:

$$
\begin{array}{cccccccccccc}
4 & + & 2z^{-1} & + & 5z^{-2} & & & & & & \\
2 & + & z^{-1} & + & 3z^{-2} & + & 6z^{-3} & & & & \\
\hline
8 & + & 4z^{-1} & + & 10z^{-2} & & & & & & \\
 & & 4z^{-1} & + & 2z^{-2} & + & 5z^{-3} & & & & \\
 & & & & 12z^{-2} & + & 6z^{-3} & + & 15z^{-4} & & \\
 & & & & & & 24z^{-3} & + & 12z^{-4} & + & 30z^{-5} \\
\hline
8 & + & 8z^{-1} & + & 24z^{-2} & + & 35z^{-3} & + & 27z^{-4} & + & 30z^{-5} \\
\hline
\end{array}
$$

By performing the inverse z transform it follows that the convolution sum gives

$$y[n] = \{\underset{\uparrow}{8}, 8, 24, 35, 27, 30\}$$

The technique of convolving two sequences by multiplying their z transforms and inverse transforming the product can be extended to sequences having non-zero values for $n < 0$. The bilateral z transform defined by Equation (14.39) is needed for this, but this merely means that positive powers in z are needed. The method was demonstrated in Example 2.3.

14.17 The responses of digital systems using MATLAB

In general it is much easier to determine the response of a digital system than that of an analogue system. This is because the digital system can be described by its difference equation and this is directly amenable to a solution on a computer using a 'for' loop, as was done in Section 14.2. There is nothing comparable for solving the differential equations of analogue systems.

In spite of the comparative simplicity of determining digital system responses, there are special functions in MATLAB for doing this and it is useful to be familiar with them. They are aimed at giving the system response when the system is described by its z transfer function rather than when it is described by its difference equation. For graphical plots the relevant functions are **dimpulse(b,a)** for an impulse response, **dstep(b,a)** for a step function response, and **dlsim(b,a,x)** for the response to an arbitrary but defined sequence $x[n]$. As was the case when MATLAB was used in Section 14.11 for plotting frequency responses and producing pole–zero plots, vectors **b** and **a** represent the numerator and denominator coefficients respectively of the z transfer function as given in

$$H(z) = \frac{\sum_{i=0}^{M} b_i z^{-i}}{1 + \sum_{i=1}^{N} a_i z^{-i}} = \frac{\sum_{i=0}^{M} b_i z^{-i}}{\sum_{i=0}^{N} a_i z^{-i}}$$

As there, it should be noted that both vectors should include coefficients down to the most negative power of z present in $H(z)$ and that any coefficients missing based on this criterion should be entered as zeros. Again, an alternative is to express $H(z)$ as the ratio of two polynomials in positive powers of z and then to specify the coefficients of these polynomials. If, for example, we have $H(z) = (2z^{-1} + z^{-2})/(1 + 3z^{-1} + 4z^{-2} + 5z^{-3})$ we could alternatively write this as $H(z) = (2z^2 + z)/(z^3 + 3z^2 + 4z + 5)$. Either way we would define the vectors as **b=[2 1 0]; a=[1 3 4 5];**.

Example 14.15

Determine the impulse response, the step function response and the response to an input sequence $x[n] = \{3, 1, -2, 4, 2, 0, -3, -5, -1, 2, 1, 5, \}$, of a digital network whose transfer function is given by

$$H(z) = \frac{3 + z^{-1} + 4z^{-2}}{1 - 2z^{-1} + 1.75z^{-2} - 0.75z^{-3} + 0.125z^{-4}}$$

Solution For a plot of the impulse response we enter

```
b=[3 1 4 0 0]; a=[1 -2 1.75 -0.75 0.125];
dimpulse(b,a)
```

This produces Figure 14.35. It will be noted that it is in the form of a 'staircase' type waveform rather than a true impulse response.

For a plot of the step function response we follow this with

```
dstep(b,a)
```

This produces Figure 14.36. Again this is in the form of a 'staircase' type waveform rather than a set of impulses.

For a plot of the response of the system to the sequence $x[n] = \{3, 1, -2, 4, 2, 0, -3, -5, -1, 2, 1, 5, \}$ we require

```
x=[3 1 -2 4 2 0 -3 -5 -1 2 1 5];
dlsim(b,a,x)
```

This produces Figure 14.37, where once again it is in the form of a 'staircase' type waveform. It should be noted that this display only covers the timespan of the input signal.

In order to obtain more conventional plots of *impulses* from these three commands we can compute the appropriate outputs $y[n]$ for each case, generate an appropriate

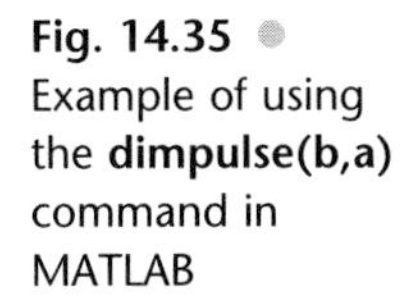

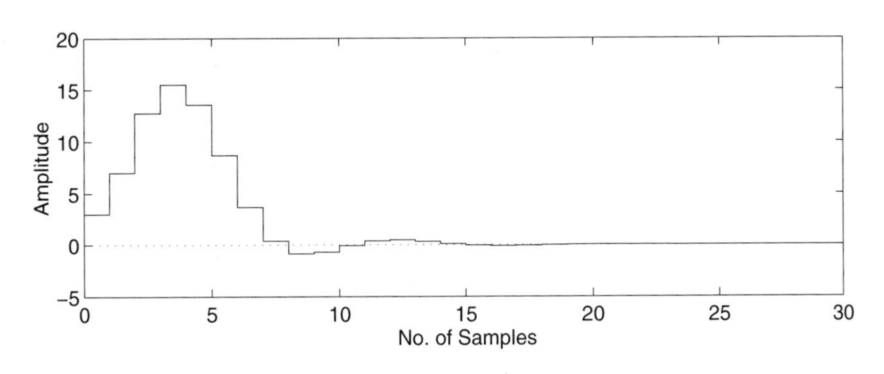

Fig. 14.35 Example of using the **dimpulse(b,a)** command in MATLAB

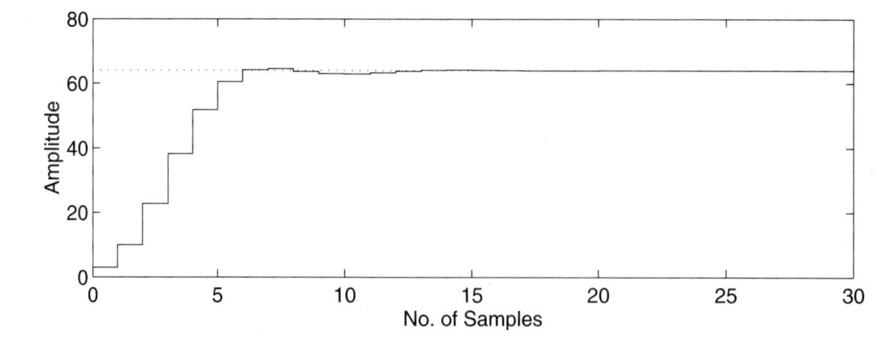

Fig. 14.36 Example of using the **dstep(b,a)** command in MATLAB

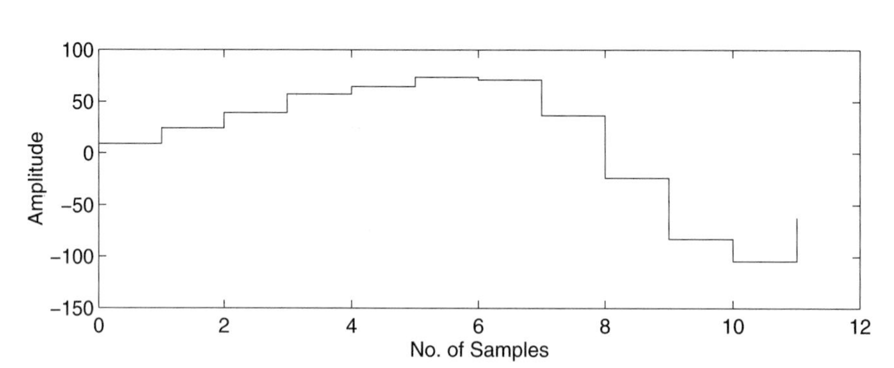

Fig. 14.37 Example of using the **dlsim(b,a,x)** command in MATLAB

set of timing indices n and then plot $y[n]$ using the **stem(n,y)** command. A suitable set of commands for the impulse response is

```
[y,z] = dimpulse(b,a);          % note that the second term evaluated by this
                                % command, z, is not of concern in this book.
% We now determine the number k of elements in y[n] and generate a timing
% index vector in which the first element is zero and the last is (k-1).
n = (0:length(y)-1);
stem(n,y)
set(gca,'XLim',[0 max(n)])  % this causes the computed y[n] values to fill the
                                % display
```

This produces Figure 14.38.

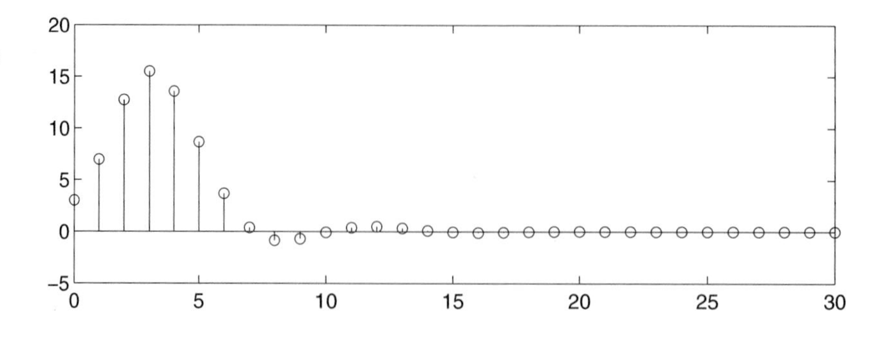

Fig. 14.38 Modified impulse response plot using MATLAB

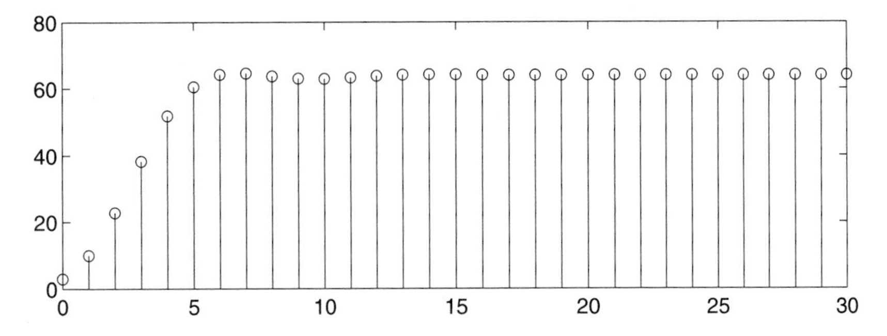

Fig. 14.39 Modified step function response plot using MATLAB

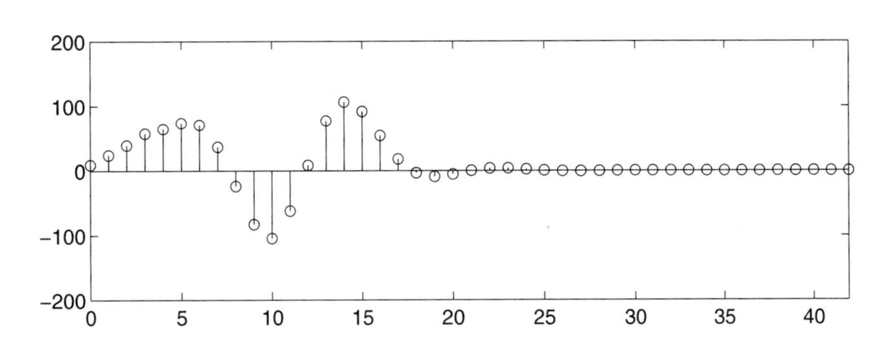

Fig. 14.40 Modified system response plot using MATLAB

Similarly, for a step function response we can have

```
[y,z]=dstep(b,a);
n=(0:length(y)-1);
stem(n,y)
set(gca,'XLim',[0 max(n)])
```

This produces Figure 14.39.

Finally, for the system response we would ideally like to extend the plot beyond the range of the input signal so that the complete response is included. To do this we must extend beyond the duration of the input sequence by the duration of the impulse response. Thus we can have

```
x=[3 1 -2 4 2 0 -3 -5 -1 2 1 5];
[u,v]=dimpulse(b,a);
k=length(x)+length(u);
x(k)=0;                    % this extends x[n] with an appropriate number of zeros
[y,z]=dlsim(b,a,x);
n=(0:length(y)-1);         % this generates a vector of timing indices
stem(n,y)
set(gca,'XLim',[0 max(n)])
```

This produces Figure 14.40.

14.18 Transfer functions of sampled analogue systems

As shown in Chapters 10–12, one very important way of describing an analogue system is by its s transfer function. One of the very simplest examples is an RC network, and for this

$$H(s) = \frac{1/sC}{R + 1/sC} = \frac{1}{CR}\frac{1}{s + 1/CR} \tag{14.53}$$

The corresponding impulse response is $h(t) = (1/CR)e^{-t/CR}$.

If, however, we have an analogue system whose output is *sampled* it is easier to describe the overall transfer function of the analogue system plus sampler by a z transfer function. Taking the case where a continuous-time impulse $\delta(t)$ is applied to the RC network and the output is sampled with a sampling period t_s, the *discrete-time* sequence arising is given by $1/CR\{1, e^{-t_s/CR}, e^{-2t_s/CR}, e^{-3t_s/CR}, \dots\}$. It is readily seen that this is identical to the response of the digital network in Figure 14.41 to a discrete-time impulse $\delta[n]$. The network of Figure 14.41 has been met several times before and, by taking the z transform of its difference equation, its transfer function is readily shown to be given by

$$H(z) = \frac{1}{CR}\frac{1}{1 - e^{-t_s/CR}z^{-1}} \tag{14.54}$$

It has thus been shown that the s transfer function of Equation (14.53) and the z transfer function of Equation (14.54) are strongly related. One relates to an analogue system and the other relates to the analogue system followed by a sampler. Based upon the same criterion of relatedness and using the Laplace transforms given in Section 12.8 and the z transforms given in Section 14.13 we can draw up a table of some other related transforms. This is done in Table 14.2.

What follows next is a prelude to considering in the next section how a particular device, perhaps a motor, can be controlled by digital feedback. In order to avoid confusion of what is meant by the term 'system', whether it be the motor in this case, or the complete system with feedback, the device being controlled will henceforth be termed the 'plant'. This is common control system terminology and arises because the device being controlled is very often more complex than a motor, and may indeed be an industrial plant.

The plant itself will be assumed to be analogue. It is now supposed that this plant (whose output is to be sampled) is itself controlled by an ideally sampled signal consisting of continuous-time impulses but that, in order to have a suitable analogue signal for actuating the analogue plant, this train of impulses is first passed through a zero order hold (ZOH). This means such that the waveform presented to the plant is in fact a staircase waveform. A block diagram containing some hypothetical waveforms

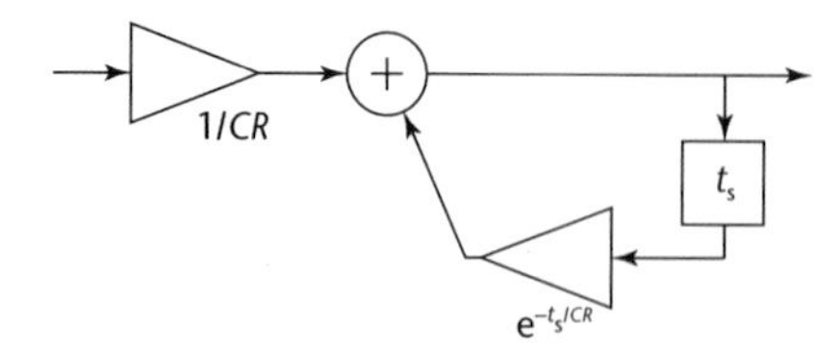

Fig. 14.41 Network for which $h[n] = 1/CR\{1, e^{-t_s/CR}, e^{-2t_s/CR}, e^{-3t_s/CR}, \dots\}$

Table 14.2 Some of the most important related Laplace and z transforms

$h(t)$	$H(s)$	$h[n]$	$H(z)$
$u(t)$	$\frac{1}{s}$	$u[n]$	$\frac{z}{z-1}$
$tu(t)$	$\frac{1}{s^2}$	$nu[n]$	$\frac{t_s z}{(z-1)^2}$
$e^{-at}u(t)$	$\frac{1}{s+a}$	$e^{-ant_s}u[n]$	$\frac{z}{z-e^{-at_s}}$
$e^{-at}\sin(\omega_0 t)u(t)$	$\frac{\omega_0}{(s+a)^2+\omega_0^2}$	$e^{-ant_s}\sin(\omega_0 nt_s)u[n]$	$\frac{ze^{-at_s}\sin(\omega_0 t_s)}{z^2-2ze^{-at_s}\cos(\omega_0 t_s)+e^{-2at_s}}$
$e^{-at}\cos(\omega_0 t)u(t)$	$\frac{s+a}{(s+a)^2+\omega_0^2}$	$e^{-ant_s}\cos(\omega_0 nt_s)u[n]$	$\frac{z^2-ze^{-at_s}\cos(\omega_0 t_s)}{z^2-2ze^{-at_s}\cos(\omega_0 t_s)+e^{-2at_s}}$

is shown in Figure 14.42. The system shown in Figure 14.42 has a discrete-time output such that, even though the plant is analogue, a z domain description of the overall system is the most appropriate. The next task is to find out what this is.

Beginning first with the ZOH and plant without the output sampler, an s domain description is easiest. The impulse response of the ZOH is a rectangular pulse of duration t_s, where t_s is the sampling period that is used throughout the system. Recognizing that this rectangular pulse can also be thought of as a unit step function delayed by t_s subtracted from an undelayed unit step function we can write the impulse response of the ZOH as

$$g_0(t) = u(t) - u(t - t_s)$$

Taking the Laplace transform of this we thus find that the s transfer function of the ZOH is given by

$$G_0(s) = \frac{1}{s} - \frac{e^{-st_s}}{s} = \frac{1 - e^{-st_s}}{s} \tag{14.55}$$

It follows that if the transfer function of the plant is given by $G(s)$, the combined transfer function of this and the ZOH is given by $G'(s) = G_0(s)G(s) = [(1 - e^{-st_s})/s]G(s)$.

When determining the equivalent z transfer function when the output sampler *is included* it is easiest to regroup this as $(1 - e^{-st_s})[G(s)/s]$. Using Table 14.2 we then seek the z transform that is equivalent to $G(s)/s$ and multiply it by the z transform that is equivalent to $(1 - e^{-st_s})$ which, using the definition of the z transform, is simply $(1 - z^{-1})$.

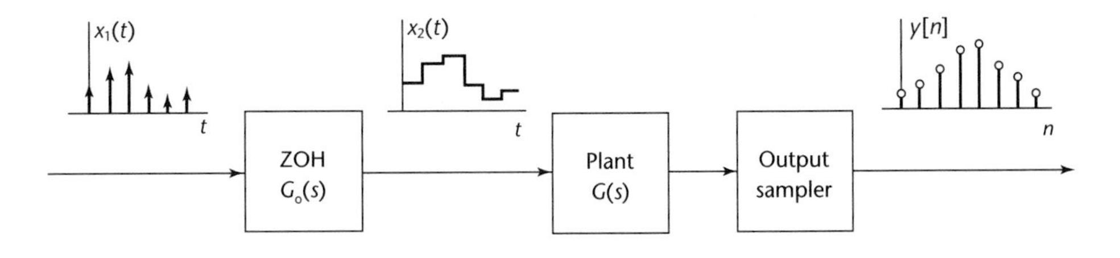

Fig. 14.42 Block diagram showing a plant being driven by a discrete-time signal and having its output monitored by a sampler

Example 14.16

A plant has the s transfer function $G(s) = 6/(s+a)$. It is preceded by a ZOH and is followed by a sampler. Derive the overall sampled data transfer function $G(z)$ if the sampling period is t_s. Determine the step function response if $a = 0.25$ and $t_s = 0.5$ s. Repeat for the case of $a = 0.25$ and $t_s = 4$ s.

Solution Without the final sampler the s transfer function is given by

$$G'(s) = \frac{(1 - e^{-st_s})}{s}\frac{6}{s+a} = (1 - e^{-st_s})\frac{6}{s(s+a)}$$

The next step is to find the z transfer function of a hypothetical system of transfer function $6/[s(s+a)]$ followed by a sampler. Expanding by partial fractions we obtain

$$\frac{6}{s(s+a)} = \frac{6}{a}\left[\frac{1}{s} - \frac{1}{s+a}\right]$$

Using Table 14.2 the corresponding z transfer function is

$$\frac{6}{a}\left[\frac{z}{z-1} - \frac{z}{z - e^{-at_s}}\right]$$

If we now include the $(1 - e^{-st_s})$ term we thus find that the required z transfer function is given by

$$G(z) = \frac{6}{a}(1 - z^{-1})\left[\frac{z}{z-1} - \frac{z}{z - e^{-at_s}}\right]$$

$$= \frac{6}{a}\left[1 - \frac{z-1}{z - e^{-at_s}}\right] = \frac{6(1 - e^{-at_s})}{a(z - e^{-at_s})}$$

(a)

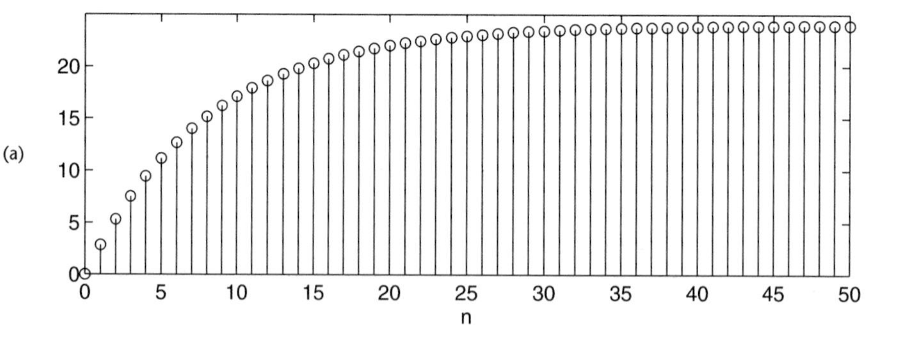

(b)

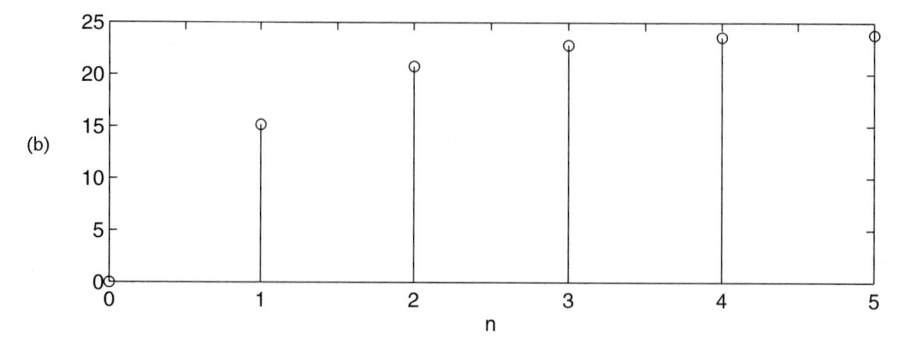

Fig. 14.43 Step function response of plant with transfer function $G(s) = 6/(s+0.25)$ when preceded with a ZOH and followed by a sampler. (a) 0.5 s sampling interval; (b) 4 s sampling interval

If $a = 0.25$ and $t_s = 0.5$ s this becomes

$$G(z) = \frac{24(1 - 0.8825)}{z - 0.8825} = \frac{2.82}{z - 0.8825}$$

To plot the corresponding step function response in MATLAB we can have

```
[y,z]=dstep(2.82,[1 -0.8825]);
n=(0:length(y)-1);
stem(n,y)
set(gca,'XLim',[0 max(n)])
xlabel('n')
```

This produces Figure 14.43(a). Similarly, if $a = 0.25$ and $t_s = 4$ s we obtain $G(z) = 15.17/(z - 0.368)$ and the corresponding step function response is shown in Figure 14.43(b). It is noted that this is sampled less frequently than that of Figure 14.43(a), but both have the same sample amplitudes at the times when both produce an output. This is because the output of a ZOH to a sampled step function is the same, irrespective of the sampling rate.

14.19 An introduction to digital control systems

The block diagram of a digital control system is shown in Figure 14.44. It is essentially the same as the analogue control system discussed in Section 12.14, the main difference being

- The feedback signal is a *digitized* version of the plant output.
- This feedback signal is *processed digitally* before being subtracted from a digitized input signal (though in other systems the processing might occur after the subtraction).
- Before being presented to the plant the digitized error signal undergoes a digital to analogue conversion in which the output analogue values are held constant between conversions.

Assuming that quantization errors are sufficiently small the coding aspects of digitization are not relevant and Figure 14.45 shows a block diagram that is equivalent to Figure 14.44 except that it operates with sampled signals (also some of the blocks are slightly rearranged).

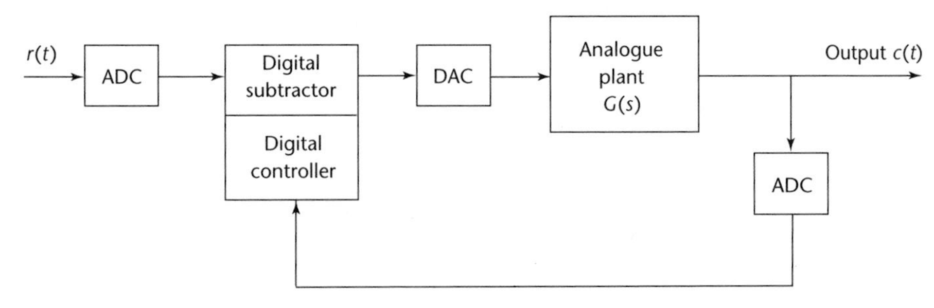

Fig. 14.44 Block diagram of a digital control system

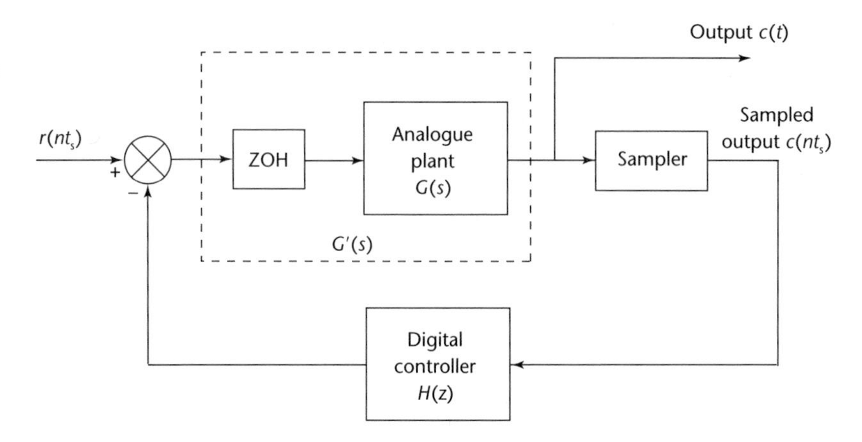

Fig. 14.45 Block diagram of a control system using sampled signals

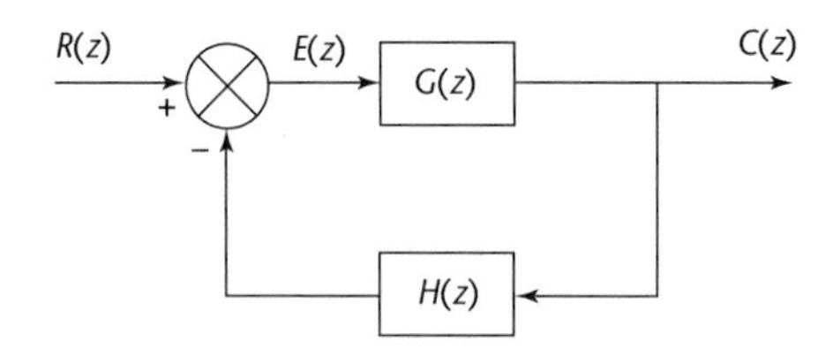

Fig. 14.46 *z* domain block diagram of a digital control system

Finally, working in the z domain rather than in the time domain, we obtain the greatly simplified block diagram of Figure 14.46. Here $G(z)$ replaces the ZOH, plant and sampler of Figure 14.45 in the way that was discussed in Section 14.18.

We can now write down the z domain equations describing the control system. We have

$$E(z) = R(z) - C(z)H(z) \tag{14.56}$$

and

$$C(z) = E(z)G(z) \tag{14.57}$$

Substituting Equation (14.56) into Equation (14.57) we obtain

$$C(z) = [R(z) - C(z)H(z)]G(z)$$

Hence the overall transfer function $T(z)$ is given by

$$T(z) = \frac{C(z)}{R(z)} = \frac{G(z)}{1 + (G(z)H(z)} \tag{14.58}$$

We can examine this to determine the performance of a digital control system.

Example 14.17

Example 12.13 used an armature-controlled d.c. motor in an analogue feedback system for controlling the angle of a camera. Its transfer function relating output angle to input voltage was $G(s) = 0.06/[s(s+0.4)]$. If the analogue feedback is replaced by digital feedback, using the configuration of Figure 14.44, determine the step function response if the sampling interval is 0.2 s and $H(z) = 10$.

Solution The s transfer function of the ZOH and motor is given by

$$G'(s) = \frac{(1-e^{-st_s})}{s}\frac{0.06}{s(s+0.4)} = (1-e^{-st_s})\frac{0.06}{s^2(s+0.4)}$$

Expanding the last part of this into partial fractions we obtain

$$\frac{0.06}{s^2(s+0.4)} = \frac{-0.375s+0.15}{s^2} + \frac{0.375}{(s+0.4)} = \frac{-0.375}{s} + \frac{0.15}{s^2} + \frac{0.375}{s+0.4}$$

Hence, using Table 14.2, the sampled data transfer function that replaces $G'(s)$ is given by

$$G(z) = (1-z^{-1})\left(\frac{-0.375z}{z-1} + \frac{0.15t_s z}{(z-1)^2} + \frac{0.375z}{z-e^{-0.4t_s}}\right)$$
$$= -0.375 + \frac{0.15t_s}{z-1} + \frac{0.375(z-1)}{z-e^{-0.4t_s}}$$

Putting this over a common denominator this reduces, after some manipulation, to

$$G(z) = \frac{0.15t_s(z-e^{-0.4t_s}) - 0.375(z-1)(1-e^{-0.4t_s})}{(z-1)(z-e^{-0.4t_s})}$$

If $t_s = 0.2$ this gives

$$G(z) = \frac{0.0011686z + 0.0011379}{(z-1)(z-0.92312)}$$

If $H(z) = K$ the overall transfer function becomes

$$T(z) = \frac{G(z)}{1+KG(z)} = \frac{0.001168\,6z + 0.0011379}{z^2 + (0.0011686K - 1.92312)z + (0.0011379K + 0.92312)}$$

If $H(z) = 10$ this gives

$$T(z) = \frac{0.001686z + 0.0011379}{z^2 - 1.91143z + 0.93450}$$

To plot the step function response in MATLAB we use

```
b=[0.0011686 0.0011379];a=[1 -1.91143 0.93450];
dstep(b,a)
```

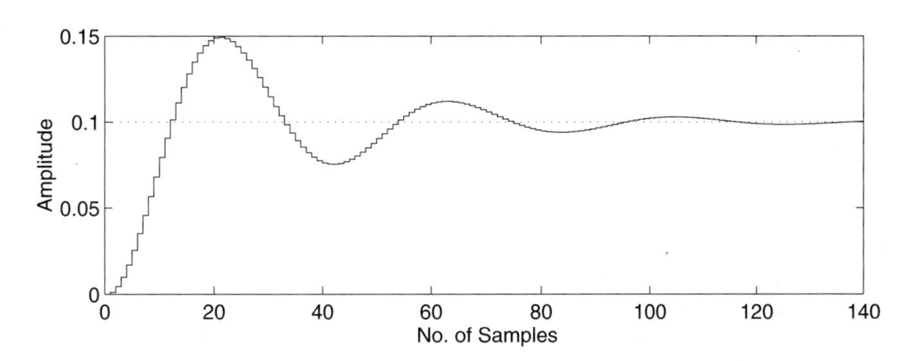

Fig. 14.47 Step function response of a digital control system

This produces Figure 14.47. It is noted that the system is substantially underdamped and is likely to be unsatisfactory. A reduction in K increases the damping and a K value of one is readily shown to be about optimum.

14.20 The relationship between the z transform, the Fourier transform and the DFT

Equation (14.20) put the z transfer function of a digital network in the form of the ratio between two polynomials in z and factorized them, to give

$$H(z) = \frac{K(z - z_1)(z - z_2)(z - z_3)\cdots}{(z - p_1)(z - p_2)(z - p_3)\cdots}$$

Section 14.9 determined the value of this transfer function at a frequency ω by constraining the value of z, defined as $z = \mathrm{e}^{st_\mathrm{s}}$, to the value $\mathrm{e}^{\mathrm{j}\omega t_\mathrm{s}}$. This led to the frequency transfer function being given in Equation (14.22) as

$$H(\omega) = \frac{K(\mathrm{e}^{\mathrm{j}\omega t_\mathrm{s}} - z_1)(\mathrm{e}^{\mathrm{j}\omega t_\mathrm{s}} - z_2)(\mathrm{e}^{\mathrm{j}\omega t_\mathrm{s}} - z_3)\cdots}{(\mathrm{e}^{\mathrm{j}\omega t_\mathrm{s}} - p_1)(\mathrm{e}^{\mathrm{j}\omega t_\mathrm{s}} - p_2)(\mathrm{e}^{\mathrm{j}\omega t_\mathrm{s}} - p_3)\cdots}$$

There is no reason why this procedure for determining a frequency response should be limited to transfer functions. It can apply also to signals.

If, for example, a discrete signal $x[n]$ has the z transform $X(z)$, its frequency content can be determined by constraining z, defined as $z = \mathrm{e}^{st_\mathrm{s}}$, to the value $\mathrm{e}^{\mathrm{j}\omega t_\mathrm{s}}$. Thus if a signal $x[n]$ has the z transform

$$X(z) = \frac{K(z - z_1)(z - z_2)(z - z_3)\cdots}{(z - p_1)(z - p_2)(z - p_3)\cdots} \tag{14.59}$$

then its Fourier transform is given by

$$X(\omega) = \frac{K(\mathrm{e}^{\mathrm{j}\omega t_\mathrm{s}} - z_1)(\mathrm{e}^{\mathrm{j}\omega t_\mathrm{s}} - z_2)(\mathrm{e}^{\mathrm{j}\omega t_\mathrm{s}} - z_3)\cdots}{(\mathrm{e}^{\mathrm{j}\omega t_\mathrm{s}} - p_1)(\mathrm{e}^{\mathrm{j}\omega t_\mathrm{s}} - p_2)(\mathrm{e}^{\mathrm{j}\omega t_\mathrm{s}} - p_3)\cdots} \tag{14.60}$$

If $X(z)$ is described by its poles and zeros in the z-plane this means determining the values of $X(z)$ on the unit circle. More specifically, if the signal samples are t_s apart, the Fourier transform of the signal $x[n]$ at a radian frequency ω is the value of the z transform of $x[n]$ at position $1\angle\omega t_\mathrm{s}$. It will be noted that this Fourier transform is periodic, each period corresponding to one complete revolution around the unit circle.

If only the magnitude of the Fourier transform at frequency ω is required then, as dealt with in Section 14.9, this is given by

$$|X(\omega)| = K\frac{\text{product of the distances from } 1\angle\omega t_\mathrm{s} \text{ to the zeros}}{\text{product of the distances from } 1\angle\omega t_\mathrm{s} \text{ to the poles}} \tag{14.61}$$

Example 14.18

If a signal sampled at 10 kHz has the z transform $z/(z - 0.9)$, determine its amplitude spectrum.

Solution This problem is identical to Example 14.7 except that $z/(z-0.9)$ applies here to a signal rather than to a transfer function. Therefore the amplitude spectrum of this signal is the same as the amplitude response of the network in Example 14.7 and is that plotted in Figure 14.25. Looked at differently, the amplitude response of the network in Example 14.7 is the amplitude spectrum of its impulse response, and this impulse response is identical to the signal of this present example.

Closed forms of the z transform, such as $z/(z-0.9)$, arise with infinite sequences. However, the z transform is also very often applied to finite length sequences such as

$$x[n] = \{\underset{\uparrow}{1}.7, 2.3, 1.5, 5.6, 2.1, 3.9\}$$

The z transform of this is $X(z) = 1.7 + 2.3z^{-1} + 1.5z^{-2} + 5.6z^{-3} + 2.1z^{-4} + 3.9z^{-5}$. Using MATLAB we can find the pole–zero plot using

```
b=[1.7 2.3 1.5 5.6 2.1 3.9];
a=[ 1 0 0 0 0 0 ];
subplot(2,2,1)                % used for some reason as on p. 372
zplane(b,a)
```

The result is shown in Figure 14.48. By moving around the unit circle and applying Equation (14.61) we can use a pole–zero plot such as Figure 14.48 to make a graphical estimate of the spectrum of the signal. The periodicity caused by the signal being sampled is readily apparent, since everything repeats itself after one revolution around the unit circle.

Equations (14.59) and (14.60) demonstrate that, even though their definitions are very different, the z transform and the continuous Fourier transform of a discrete signal are strongly related. If, however, we compare the z transform with the *discrete* Fourier transform we now find that the *definitions* themselves are similar in form. For example, the definition of the unilateral z transform was given by Equation (14.38) as

$$X(z) = \sum_{n=0}^{\infty} x[n]z^{-n} \tag{14.62}$$

while that of the DFT, for an N point sequence, was given in Equation (8.36) by

$$X[m] = \sum_{n=0}^{N-1} x[n]e^{-j2\pi mn/N} \qquad 0 \leqslant m \leqslant (N-1) \tag{14.63}$$

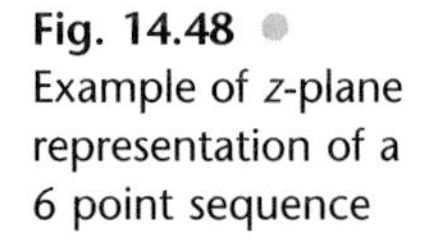

Fig. 14.48 Example of z-plane representation of a 6 point sequence

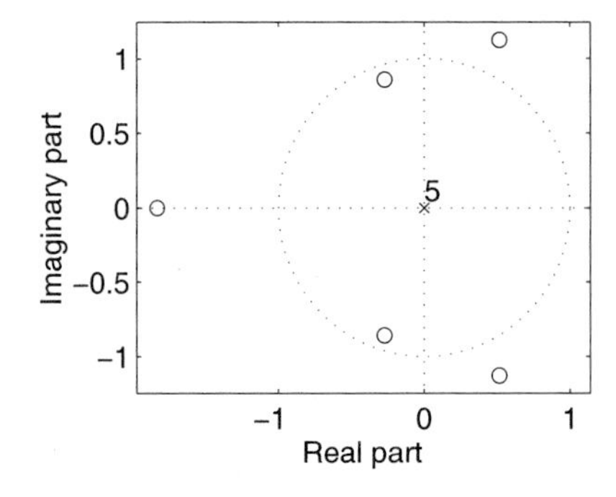

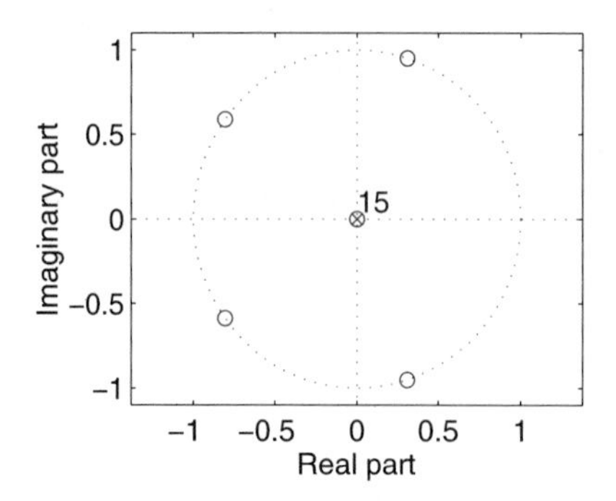

Fig. 14.49 z-plane plot of a discrete approximation to a rectangular pulse

The important differences between these two equations are:

- The value of z is constrained in the DFT equation to the value $e^{j2\pi m/N}$; that is, to points on the unit circle.
- Only N terms are evaluated by the DFT equation.

In other words, whereas the continuous spectrum of a signal can be evaluated by applying the graphical method continuously around the unit circle, the spectrum produced by an N point DFT can be evaluated by applying the graphical method at N equispaced discrete points around the unit circle. Consider for example a rectangular pulse to be approximated by the 16 point sequence $x[n] = \{1, 1, 1, 1, 1, 0, 0, 0, 0, 0, 0, 0, 0, 0, 0, 0, \}$. The z-plane plot of poles and zeros can be obtained with the commands

```
b=[1 1 1 1 1 0 0 0 0 0 0 0 0 0 0 0];
a=[1 0 0 0 0 0 0 0 0 0 0 0 0 0 0 0];
subplot(2,2,1)
zplane(b,a)
```

The result is shown in Figure 14.49. The spectrum produced by the DFT is then obtained by applying the graphical method to 16 points on the unit circle $2\pi/16$ radians, or $22\frac{1}{2}^{\circ}$, apart. The outcome may be compared with the computational result obtained using the MATLAB commands

```
n=(0:15);
stem(n,abs(fft(b)))
ylabel('Amplitude'),xlabel('n')
```

This produces Figure 14.50.

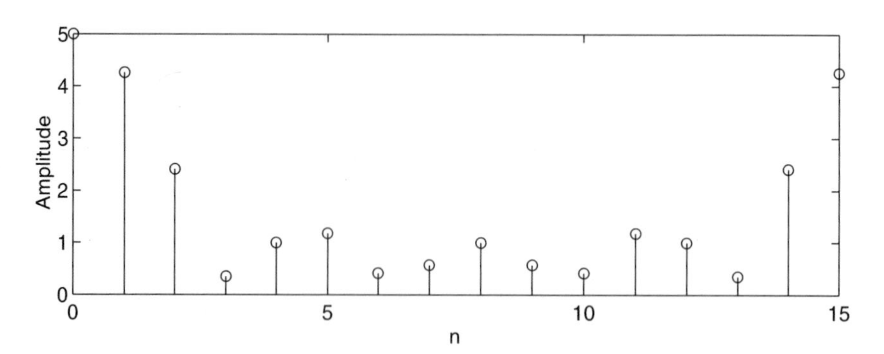

Fig. 14.50 Amplitude spectrum of a discrete approximation to a rectangular pulse

14.21 ● Summary

Whereas Laplace transforms are ideal for analysing analogue signals and systems they produce polynomials with an infinite number of periodic poles and zeros when applied to digital signals and systems. By introducing a new variable, given by $z = e^{st_s}$, much simpler signal transforms and system transfer functions are produced. The z transform is the cornerstone of digital systems analysis and its application to digital control systems is one important example that has been investigated. The next chapter studies the application of z transforms to digital filters.

14.22 ● Problems

1. Derive the difference equation of the digital network shown in Figure 14.51 and determine its transfer function. Find the first six terms of the output due to an input sequence $x[n] = \{4, 2, 3\}$. (a) manually and (b) using the **dlsim(b,a,x)** command in MATLAB.

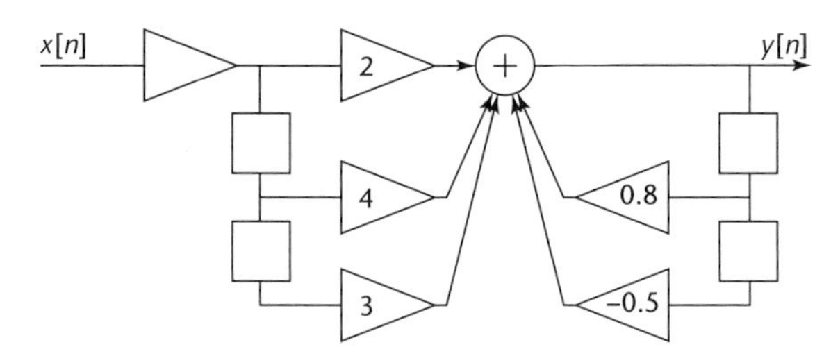

Fig. 14.51

2. A digital filter is based on a sampling frequency of 100 Hz and has z-plane poles at $0.9\angle 40°$, $0.9\angle -40°$, $0.9\angle 50°$ and $0.9\angle -50°$. It has four zeros at the origin.

   Sketch the approximate amplitude response between 0 and 200 Hz. Derive the difference equation for the filter and hence draw the filter configuration, labelling all coefficients. How would the configuration be changed if the poles were unchanged but the zeros were removed? Discuss the repercussions of leaving the poles unchanged but having a fifth zero at the origin.

3. A digital filter is based on a sampling frequency of 1 kHz and has z-plane zeros at $1\angle 21.6°$ and $1\angle -21.6°$. It has z-plane poles at $0.9\angle 21.6°$, and $0.9\angle -21.6°$.

   Describe and justify a graphical means of deriving the amplitude response. Using this method, sketch the appropriate amplitude response, labelling any frequencies of particular significance. Show how the exact amplitude response may be obtained, and calculate the ratio of the gain at 55 Hz to that at 0 Hz. Derive the difference equation for the filter and hence draw the filter configuration, labelling all coefficients.

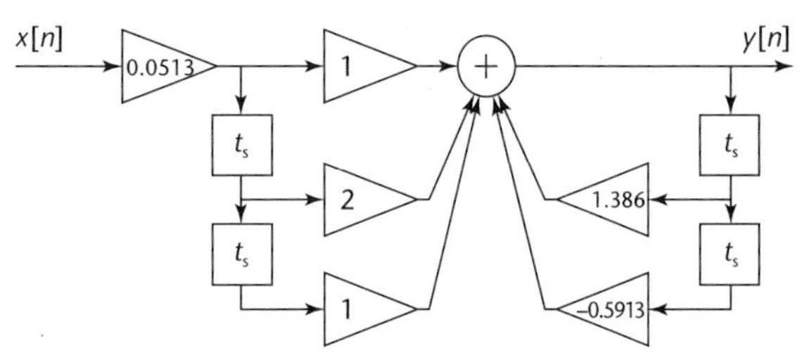

Fig. 14.52

4. Figure 14.52 shows a digital filter in which the delays are 0.5 ms. Write down the difference equation and from this derive the z transfer function. From the z transfer function derive a general expression for the amplitude response and evaluate it for frequencies of 0, 200 and 400 Hz. Derive and plot the poles and zeros of the z transfer function in the z-plane. Describe how such plots can be usefully used to provide approximate estimates of the amplitude and phase response by graphical means. Based upon this method, apply trigonometry to determine the precise gain at frequencies of 0 and 200 Hz, and confirm that the results agree with those obtained previously.

5. A digital filter has four poles, at $z = 0.95\angle \pm 34°$ and $0.95\angle \pm 38°$, and four zeros, all at $z = 0$. The

sampling frequency is 10 kHz. Make an approximate sketch of the amplitude response of the filter between 0 and 10 kHz, labelling any frequencies of special significance. Derive the z transfer function and the difference equation of the digital filter. Draw a block diagram representation of the filter labelling all filter coefficients. By graphical or geometric means estimate the ratio of the maximum gain of the filter to the gain at 0 Hz.

6. It is wished to determine the unit step function response of a system whose z transfer function is

$$H(z) = \frac{1 - 0.4z^{-1}}{1 - z^{-1} + 0.6z^{-2}}$$

(a) Find the first five terms by performing an inverse z transform following a long division;

(b) Find the first five terms by determining the difference equation and tabulating the outputs resulting from successive samples of the unit step function;

(c) Find the result by using MATLAB.

Using MATLAB find the first 17 terms of the response of the system to the input signal $x[n] = \{2, -1, 3, 1, -4, 5, 4, -2, -3\}$.

7. Convolve the two sequences $x[n] = \{2, -1, 3, 1, -4, 5, 4, -2, -3\}$ and $y[n] = \{3, -2, -3, 6, 1, 2, 4\}$ by multiplication of their z transforms. Confirm the result using MATLAB.

8. Determine the unit step function response of the digital control system of Figure 14.53 assuming that the sampling interval is 0.4 s.

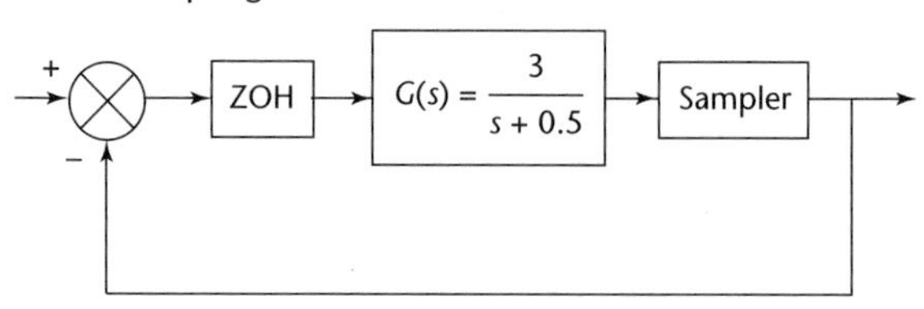

Fig. 14.53

Chapter fifteen

Synthesis of digital filters

15.1 ● Preview

The purpose of filtering a signal is to enhance its wanted features. Digital filtering is exceptionally flexible and can easily incorporate non-linear operations such as the clipping or removal of data samples that appear to be incompatible with neighbouring samples, and therefore erroneous. One of the most important operations, however, is that of linear filtering, in which a chosen frequency band is enhanced (or attenuated) relative to others. As with analogue filters it is common to want low ripple in the passband, high attenuation in the stopband and a high roll-off rate. It is the achievement of these objectives that forms the main subject of this chapter. In contrast with analogue filters it is very easy to translate a chosen transfer function into a practical design.

15.2 Filter design by pole–zero placement

There are some situations where the response characteristics of a filter are not critical. Under these circumstances a simple low-order filter can be satisfactory, and a very speedy method of design is to place poles and zeros at sensible positions that can be seen to give a reasonable frequency response; from these the difference equation that implements this filter can be derived. The technique is based on the result of Section 14.9 that, apart from a constant factor K, the gain at a particular frequency is the product of the distances from the corresponding point on the unit circle to the zeros, divided by the product of the distances from that point to the poles.

Example 15.1

A sequence has samples 1 ms apart. Design a bandpass filter that has a bandwidth of 40 Hz centred at 200 Hz.

Solution The frequency response of a digital filter is given by the values of its z transfer function when z lies on the unit circle. One revolution around the unit circle in the z-plane plot of the filter transfer function signifies changing the frequency from zero frequency up to the sampling frequency of 1 kHz. It follows that the centre frequency of 200 Hz corresponds to a position on the unit circle one fifth of a revolution anticlockwise from the horizontal axis, at $1\angle 72°$. Similarly, the frequency of -200 Hz corresponds to $1\angle -72°$. For many applications a single pair of conjugate poles placed close to these z-plane positions, accompanied possibly by some zeros at the origin, will produce an adequate frequency response. All that remains is to select a suitable radius for these conjugate poles.

The -3 dB points of the filter correspond to the frequencies of 180 Hz and 220 Hz, and hence to the z-plane positions of $1\angle 64.8°$ and $1\angle 79.2°$. The gain at these positions, B and D in Figure 15.1, should be $1/\sqrt{2}$ of what it is at the centre frequency point C. If we consider the approximate right angle triangle ACD it follows that $AC = CD$. But $CD = r\alpha$ where $r = 1$ and $\alpha = C\hat{O}D = 2\pi(72 - 64.8)/360 = 0.1257$ radians. Hence $AC = 0.1257$ and the radius of the pole is $(1 - 0.1257) = 0.874$.

Therefore

$$
\begin{aligned}
H(z) &= \frac{1}{(z - 0.874\angle 72°)(z - 0.874\angle -72°)} \\
&= \frac{1}{z^2 - 1.748z\cos(72°) + 0.764}
\end{aligned}
\tag{15.1}
$$

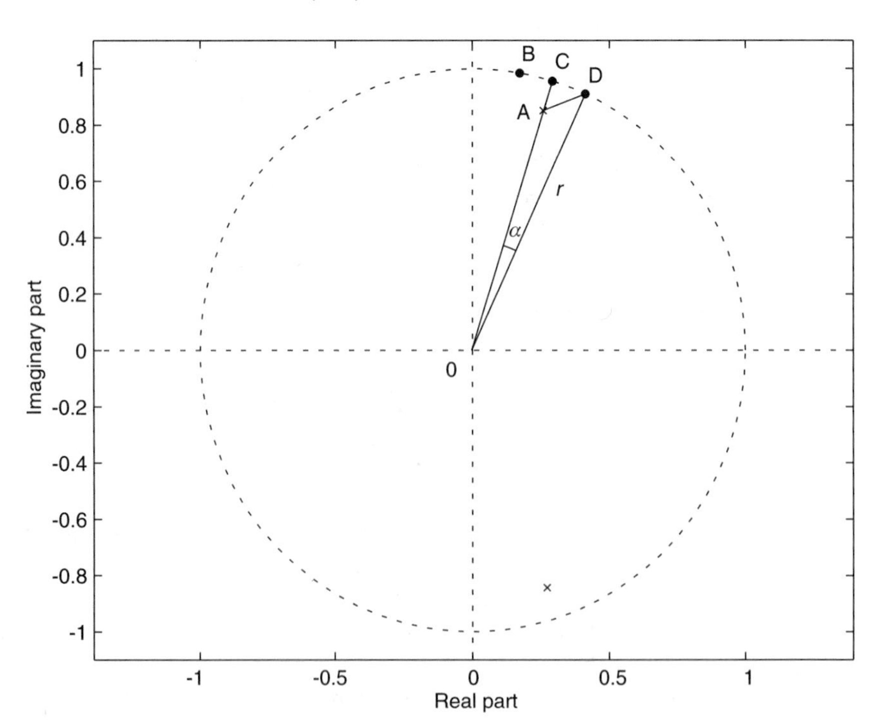

Fig. 15.1 *z*-plane plot for simple bandpass filter design

or

$$\frac{Y(z)}{X(z)} = \frac{z^{-2}}{1 - 0.540z^{-1} + 0.764z^{-2}} \tag{15.2}$$

Cross-multiplying and performing the inverse z transform gives the difference equation. The numerator causes a delay of two samples and is best avoided. We can do this without affecting the frequency response by adding a double zero at the origin, thereby realizing the transfer function

$$H(z) = \frac{Y(z)}{X(z)} = \frac{z^2}{z^2 - 0.540z + 0.764} = \frac{1}{1 - 0.540z^{-1} + 0.764z^{-2}} \tag{15.3}$$

Cross-multiplying now and performing the inverse z transform gives the difference equation of the required bandpass filter.

$$y[n] = x[n] + 0.540y[n-1] - 0.764y[n-2] \tag{15.4}$$

The realization is thus the recursive filter shown in Figure 15.2(a). A common alternative convention is to label delays of one sample as z^{-1}, and this is shown in Figure 15.2(b). This has the advantage that it does not imply a restriction to temporal signals (for example, we might wish to filter data that is amplitude as a function of distance). Both are acceptable, though this book adopts the t_s labelling since it has an easier physical interpretation.

It is always a good idea to check the frequency response. We can do this by substituting $e^{j\omega t_s}$ for z in Equation (15.3) to obtain

$$|H(\omega)| = \frac{1}{|e^{j2\omega t_s} - 0.54e^{j\omega t_s} + 0.764|}$$

$$= \frac{1}{\sqrt{[\cos(2\omega/1000) - 0.54\cos(\omega/1000) + 0.764]^2 + [\sin(2\omega/1000) - 0.54\sin(\omega/1000)]^2}}$$

This can be plotted.

Alternatively, and much easier, we can use the MATLAB commands

```
b=[1  0  0];              % the non-recursive coefficients
a=[1 -.54 .764];          % the recursive coefficients
freqz(b,a)
```

The outcome is shown in Figure 15.3, and the required bandpass response is seen to be achieved.

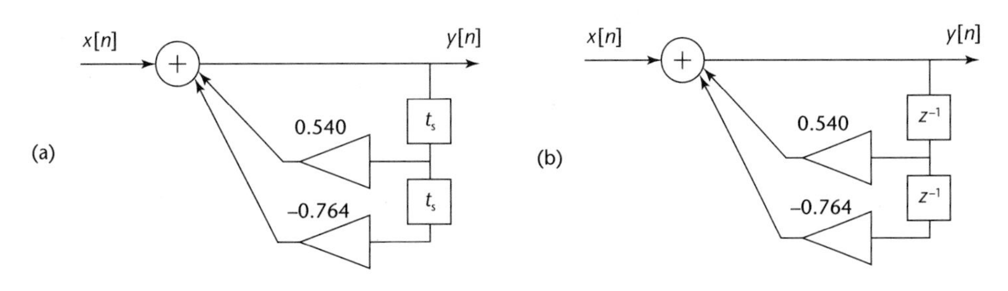

Fig. 15.2 Realization of a bandpass filter: (a) labelling delays as t_s; (b) labelling delays as z^{-1}

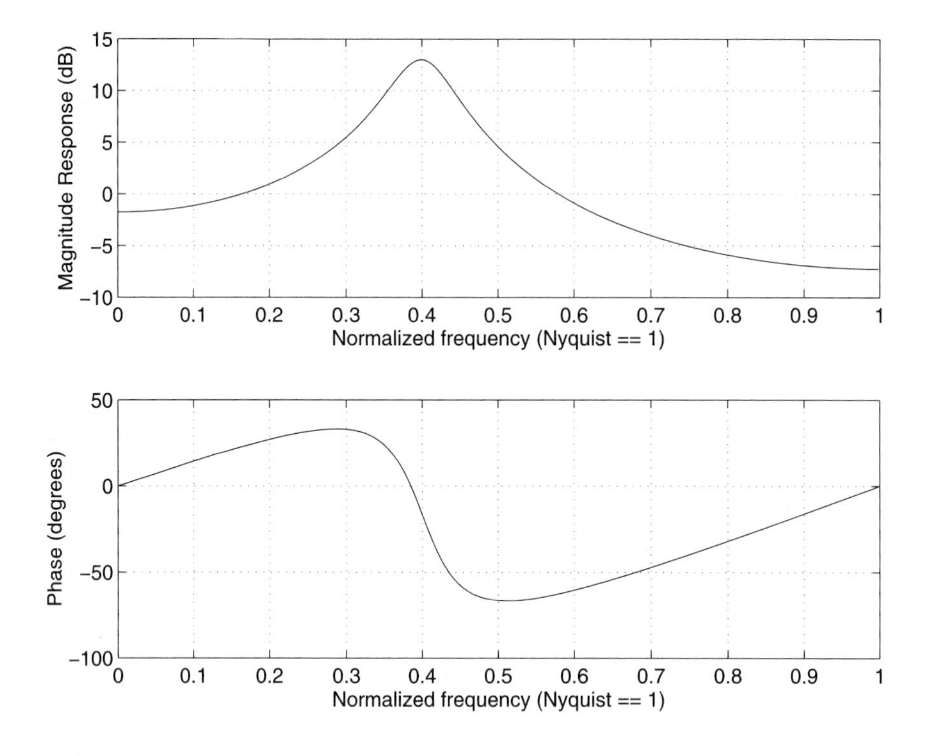

Fig. 15.3 Frequency response of a simple bandpass filter

15.3 Design of recursive digital filters based on the frequency response characteristics of analogue filters

As discussed in Chapter 13 on analogue filters, certain frequency responses have been found to have very desirable properties. The Butterworth response is maximally flat in the passband. The Chebyshev response permits a small amount of ripple in the passband and, in return, gives a sharper cutoff than the Butterworth response for a given order of the filter. The elliptic (or Cauer) filter permits ripple in both pass- and stopbands and gives an even sharper cutoff. Much past effort has gone into determining the s domain transfer functions that achieve these desirable amplitude characteristics, and it is sensible to see whether z transfer functions giving similar amplitude characteristics can be obtained from them. The resulting z transfer function, $H(z)$, can then be used to derive a difference equation which achieves the required filter.

The most obvious way of converting a transfer function $H(s)$ into a transfer function $H(z)$ might appear to be to replace s, using the relationship $z = e^{st_s}$. Unfortunately, this results in a z transfer function which cannot be converted into a difference equation. For example, the very simple first-order filter $H(s) = 1/(s - p_1)$ would become $H(z) = 1/[(1/t_s)\ln(z) - p_1]$, but this is not in the form of the ratio of two polynomials in z, and cannot therefore be transformed into a difference equation. Some other approach is needed.

Two techniques of going from a transfer function $H(s)$ to a transfer function $H(z)$ will be considered. These are the impulse invariant technique and the bilinear transformation technique.

15.3.1 The impulse invariant technique

The impulse invariant technique begins with a partial fraction expansion of $H(s)$. For example the transfer function

$$H(s) = K\frac{(s-z_1)(s-z_2)(s-z_3)\cdots}{(s-p_1)(s-p_2)(s-p_3)\cdots(s-p_n)}$$

is replaced by

$$H(s) = \frac{r_1}{(s-p_1)} + \frac{r_2}{(s-p_2)} + \frac{r_3}{(s-p_3)} + \cdots + \frac{r_n}{(s-p_n)} \tag{15.5}$$

This suggests that the transfer function could be realized by a bank of n single-pole filters in parallel whose outputs are added. The principle of the impulse invariant technique is to replace each of these terms with the transfer function of a digital filter having the same but sampled impulse response and then to recombine them in a form which is more easily implemented.

By taking the inverse Laplace transform of Equation (15.5) it is seen that each of these single-pole filters has an impulse response which is of a simple exponential form. For example, the impulse response of the ith filter is given by

$$\begin{aligned} h_i(t) &= r_i e^{p_i t} \qquad && t \geqslant 0 \\ &= 0 && t < 0 \end{aligned} \tag{15.6}$$

A practical problem arises with this in that some or all of the residues may be complex, as may be some or all of the poles. When this happens the single-pole filter is not physically realizable. However, ignoring this difficulty for the moment and considering the case where r_i is real and where p_i is real and negative, the impulse response of Equation (15.6) is as shown in Figure 15.4(a). Figure 15.4(b) shows the impulse response of Figure 15.4(a) approximated by segments t_s apart, and Figure 15.4(c) shows a sampled version of the impulse response in which each sample has an amplitude equal to the area of the segment that it replaces.

If we now consider the response of the simple digital filter in Figure 15.5 to a discrete-time impulse at the input it can be seen that an impulse function of magnitude $r_i t_s$ appears instantly at the output and then circulates around the feedback loop being successively modified by the recursive coefficient, $\exp(p_i t_s)$, such that

$$h[n] = r_i t_s\{\underset{\uparrow}{1}, e^{p_i t_s}, e^{2p_i t_s}, e^{3p_i t_s}, \cdots\} \tag{15.7}$$

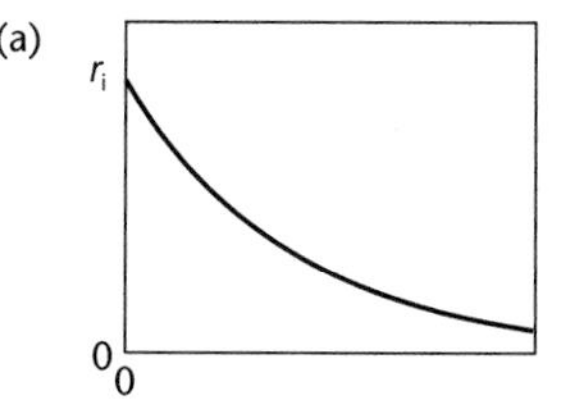

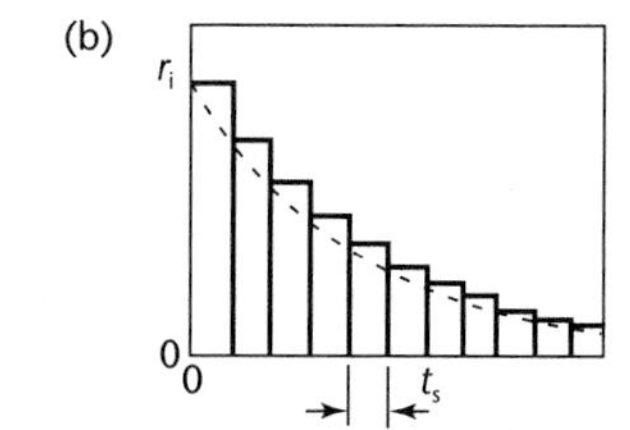

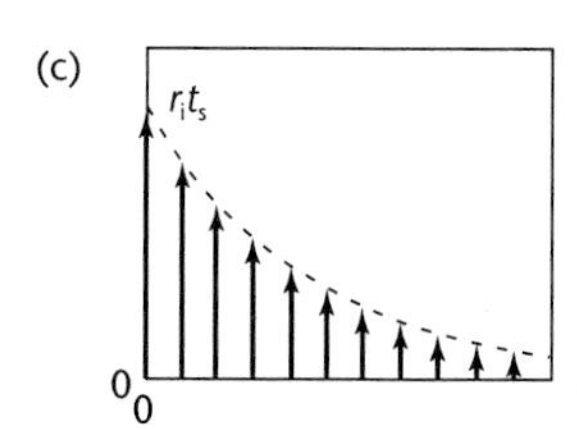

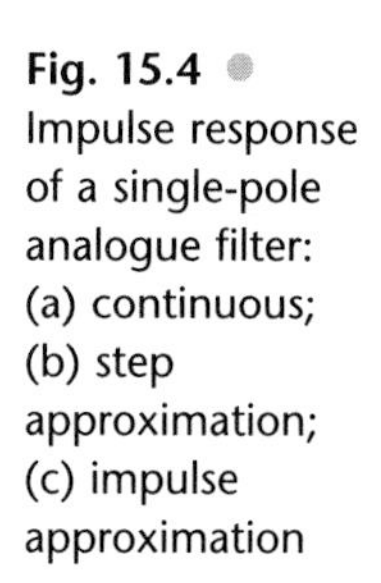

Fig. 15.4 Impulse response of a single-pole analogue filter: (a) continuous; (b) step approximation; (c) impulse approximation

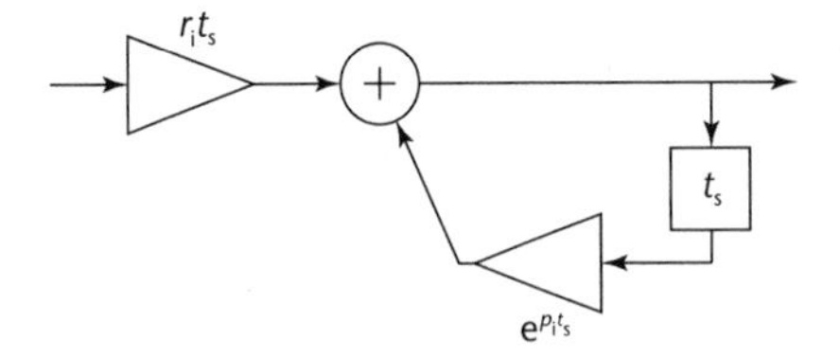

Fig. 15.5 Conceptual building block for the synthesis of a filter by the impulse invariant method

This impulse response is the same as the required sampled impulse response of Figure 15.4(c). The difference equation of the digital filter in Figure 15.5 is given by

$$y[n] = r_i t_s x[n] + \exp(p_i t_s) y[n-1] \tag{15.8}$$

such that, taking the z transform of both sides, and rearranging

$$\frac{Y(z)}{X(z)} = H(z) = \frac{r_i t_s}{1 - \exp(p_i t_s) z^{-1}} \tag{15.9}$$

The concept of the impulse invariant technique is then based upon replacing each term of the form $r/(s-p)$ in Equation (15.5) by one of the form $rt_s/[1 - \exp(pt_s)z^{-1}]$ and then recombining the terms to be of the form

$$H(z) = \frac{\sum_{i=0}^{M} b_i z^{-i}}{1 + \sum_{i=1}^{N} a_i z^{-i}} \tag{15.10}$$

This is the same as obtained in Equation (14.19) for the generalized digital network configuration of Figure 14.3, and from it the filter coefficients may be read off directly.

Just as Equation (15.5) suggests that the s transfer function might be realized by a bank of n single-pole analogue filters in parallel whose outputs are added, so one might suppose that the z transfer function might be realized by a bank of n single-pole digital filters in parallel whose outputs are added. In general, however, this is not feasible since some of the n single-pole filters would contain complex coefficients. When, however, the transfer functions are combined to give Equation (15.10) the coefficients are always real.

Example 15.2

A third-order Butterworth filter with a cutoff frequency of 1 rad/s has the transfer function

$$H(s) = \frac{1}{(s-p_1)(s-p_2)(s-p_3)}$$

where $p_1 = -0.5 + \mathrm{j}0.866$, $p_2 = -1$, and $p_2 = -0.5 - \mathrm{j}0.866$. Design the corresponding digital filter using the impulse invariant technique. Assume a sampling frequency of 2 Hz.

Solution If we undertake a partial fraction expansion of the s domain transfer function we obtain

$$H(s) = \frac{r_1}{s-p_1} + \frac{r_2}{s-p_2} + \frac{r_3}{s-p_3} \tag{15.11}$$

where $r_1 = -0.5 - j0.2287$, $r_2 = 1$ and $r_3 = -0.5 + j0.2287$. From Equation (15.9) the z domain replacement requires terms of the form e^{pt_s}. We have $t_s = 0.5$ s and hence

$$\exp(p_1 t_s) = \exp(-0.25 + j0.433) = 0.7788 \exp(j0.433)$$
$$\exp(p_2 t_s) = \exp(-0.5) = 0.6065$$
$$\exp(p_3 t_s) = \exp(-0.25 - j0.433) = 0.7788 \exp(-j0.433)$$

Hence, replacing the three terms in Equation (15.11) by the corresponding z domain terms required by Equation (15.10), we obtain

$$H(z) = 0.5\left[\frac{-0.5 - j0.2287}{1 - 0.7788\exp(j0.433)z^{-1}} + \frac{1}{1 - 0.6065z^{-1}} + \frac{-0.5 + j0.2287}{1 - 0.7788\exp(-j0.433)z^{-1}}\right]$$

We now wish to combine these terms by putting them over a common denominator. It is easiest to combine the first and third terms first, and then to include the centre term. After some arithmetic we obtain

$$H(z) = 0.5\,\frac{0.08701z^{-1} + 0.0636z^{-2}}{1 - 2.0203z^{-1} + 1.464z^{-2} - 0.3678z^{-3}}$$

Remembering that z^{-1} indicates a delay of one sample and that an overall delay has no effect on the amphitude response, the common factor of z^{-1} in the numerator may be ignored. Thus the objective is to determine a filter configuration that produces the transfer function

$$H(z) = \frac{Y(z)}{X(z)} = \frac{0.0435 + 0.0318z^{-1}}{1 - 2.0203z^{-1} + 1.464z^{-2} - 0.3678z^{-3}}$$

We can now cross-multiply the equation to give

$$Y(z)(1 - 2.0203z^{-1} + 1.464z^{-2} - 0.3678z^{-3}) = X(z)(0.0435 + 0.0318z^{-1})$$

An inverse z transform gives the difference equation and hence the filter coefficients (or alternatively we can read off the coefficients directly using the relationship of Equation (15.10)). Either way we achieve the filter configuration shown in Figure 15.6.

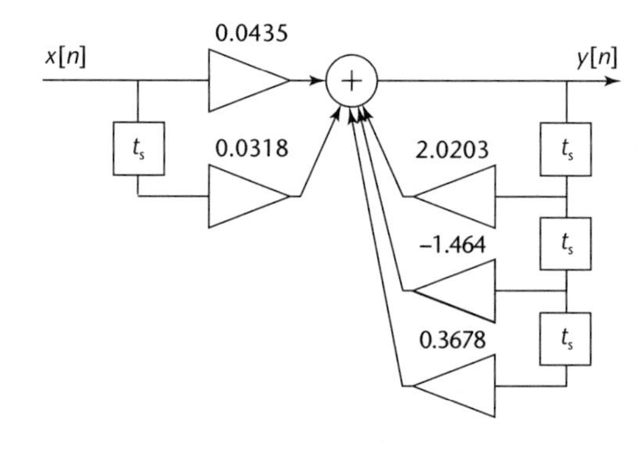

Fig. 15.6 Realization of a third-order Butterworth lowpass filter by the impulse invariant method

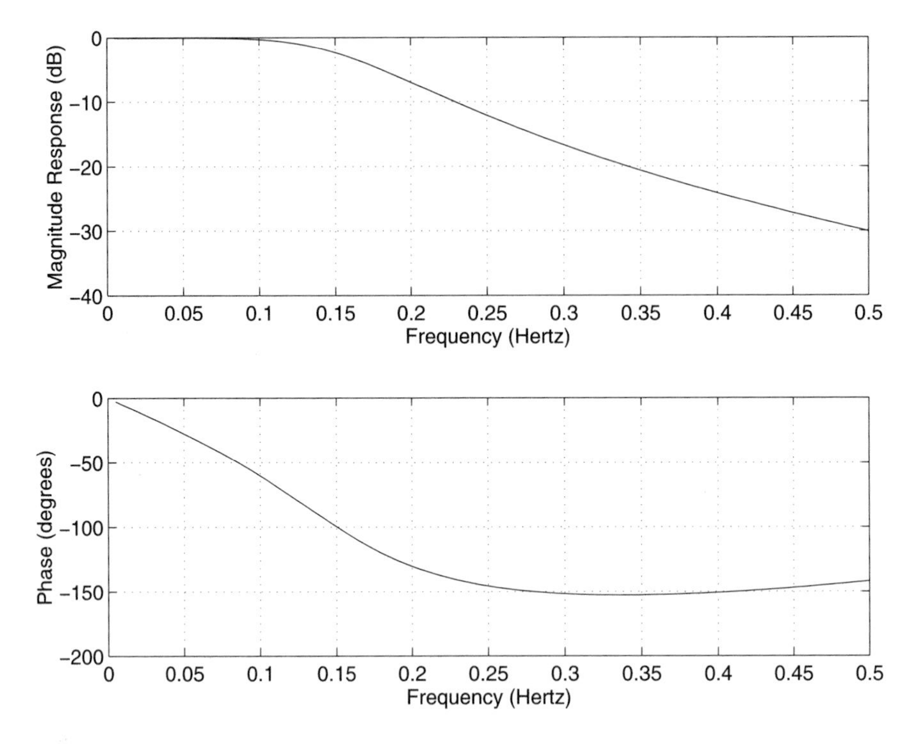

Fig. 15.7 Frequency response of a third-order Butterworth digital filter

As in Example 15.1 it is possible to check the frequency response by substituting $e^{j\omega t_s}$ for z in the expression for the z transfer function. Using MATLAB, however, we simply need the commands

```
f=(0:200)/200;                 % this produces a frequency scale in Hz up to
                               % the folding frequency

b=[.0435 .0318 0 0];
a = [1 -2.0203 1.464 -.3678];
freqz(b,a,f,2)                 % this produces a plot that is not normalized, in
                               % which the horizontal axis is f. The '2' is the
                               % sampling frequency.
```

The outcome is shown in Figure 15.7.

The impulse response of a digital filter designed by the impulse invariant technique is the same as that of the original analogue filter, except that it is sampled. It follows that the frequency response of the digital filter is the same as that of the analogue filter, except that it becomes periodic. This periodicity introduces aliasing, and the consequence is that the attenuation of the digital filter in the region around half the sampling frequency (often known as the folding frequency) is less than for the analogue filter. This problem is illustrated by Figure 15.8. Since the filter of Figure 15.6 is a lowpass filter this reduction in attenuation due to aliasing is clearly undesirable. The next technique does not suffer from this problem because it introduces a *null* at the folding frequency.

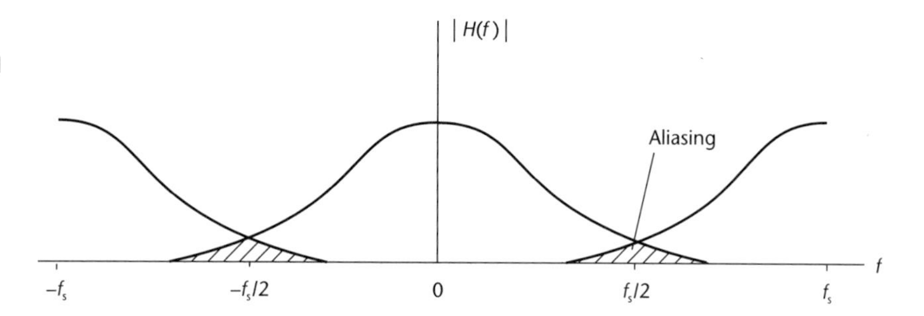

Fig. 15.8 Aliasing in a digital filter designed by the impulse invariant technique

15.3.2 The bilinear transformation technique

It was suggested earlier in this section that the most obvious way of converting a transfer function $H(s)$ into a transfer function $H(z)$ would appear to be to replace each s in $H(s)$ with a function of z based on the relationship $z = e^{st_s}$, but that this unfortunately results in a z transfer function which cannot be converted into a difference equation. The concept behind the bilinear transformation technique is to modify $H(s)$ into a new function $H(s')$ which has a different (but very similar) frequency response, and which is such that the replacement of s' using the relationship $z = e^{s't_s}$ now gives a z transfer function $H(s)$ that *can* be readily converted into a difference equation.

In practice the technique is implemented without involving the intermediate stage of $H(s')$. An awareness of $H(s')$ is, however, very useful for predicting any difference between the frequency response of the original analogue filter and that of the resulting digital filter.

The technique is best explained by an example. Consider the very simple first-order analogue filter shown in Figure 15.9 having the transfer function

$$H(s) = \frac{\alpha}{s + \alpha} \tag{15.12}$$

The objective is to achieve a physically realizable digital filter that has a similar frequency response. The transfer function that we choose to replace this is given by

$$H(s') = \frac{\alpha}{2f_s \tanh(s'/2f_s) + \alpha} \tag{15.13}$$

and we now face two tasks:

1. We must confirm that the frequency response associated with $H(s')$ is indeed a satisfactory alternative to the frequency response associated with $H(s)$.
2. We must obtain the z transfer function corresponding to $H(s')$ by replacing s' with the function of z required by the relationship $z = e^{s't_s}$ and then demonstrate that the resulting $H(z)$ corresponds to a difference equation that can be easily implemented.

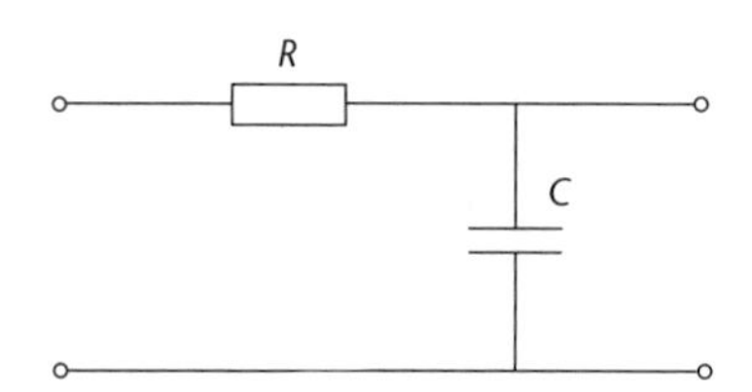

Fig. 15.9 First-order analogue filter

Beginning with the first of these tasks, the frequency response associated with $H(s')$ at a frequency ω' is obtained by substituting $j\omega'$ for s'. From Equation (15.13) and making use of the relationship $\tanh(jx) = j\tan(x)$ we obtain

$$H(\omega') = \frac{\alpha}{2f_s \tanh(j\omega'/2f_s) + \alpha} = \frac{\alpha}{2jf_s \tan(\omega'/2f_s) + \alpha}$$

$$\therefore \quad |H(\omega')| = \frac{\alpha}{\sqrt{(2f_s \tan(\omega'/2f_s))^2 + \alpha^2}} \tag{15.14}$$

This contrasts with the original analogue filter which, from Equation (15.12), had the amplitude response

$$|H(\omega)| = \frac{\alpha}{\sqrt{\omega^2 + \alpha^2}} \tag{15.15}$$

A comparison of Equations (15.14) and (15.15) shows that the new filter has the same gain at a frequency ω' as the original filter has at a frequency ω if ω is related to ω' by the equation

$$\omega = 2f_s \tan \frac{\omega'}{2f_s} \tag{15.16}$$

This inequality of ω and ω' means that the difference in frequency response between $H(s)$ and $H(s')$ can be accounted for by a frequency *warping* given by Equation (15.16). The effect of this frequency warping is very clearly illustrated by Figure 15.10, where $|H(\omega')|$ is obtained by mapping $|H(\omega)|$ onto the curves $\omega = 2f_s \tan(\omega'/2f_s)$.

One observation from this construction is that the value of $|H(\omega')|$ when $\omega' = \pm\omega_s/2, \pm 3\omega_s/2, \pm 5\omega_s/2, \ldots$ is the same as the value of $|H(\omega)|$ when $\omega = \infty$. Hence $|H(\omega')|$ is periodic with a null at the folding frequency.

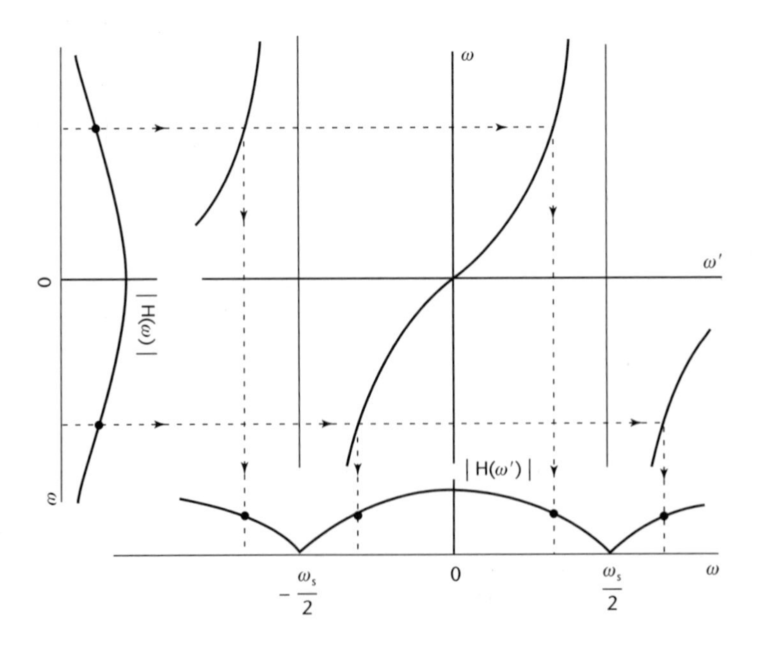

Fig. 15.10 Frequency warping

For *small* values of ω', however, Equation 15.16 reduces to

$$\omega \approx \omega' \tag{15.17}$$

such that $|H(\omega')|$ is then of the same shape as $|H(\omega)|$.

What has emerged is exactly the type of response we want from a digital filter. It is periodic; it is similar to that of the original analogue filter for much of the region up to the folding frequency; and it has zero gain at the folding frequency.

Now that we have found that $H(s')$ is a satisfactory alternative to $H(s)$ we must undertake the second task, which is to convert it into a digital filter. Using the relationship

$$\tanh(x) = \frac{(e^x - e^{-x})}{(e^x + e^{-x})} = \frac{e^x(1 - e^{-2x})}{e^x(1 + e^{-2x})} = \frac{(1 - e^{-2x})}{(1 + e^{-2x})}, \tag{15.18}$$

and using $t_s = 1/f_s$, Equation (15.13) becomes

$$H(s') = \frac{\alpha}{2f_s\,(1 - e^{-s't_s})/(1 + e^{-s't_s}) + \alpha} \tag{15.19}$$

We can obtain the digital filter corresponding to this by substituting z for $e^{s't_s}$. Because $H(s')$ contains terms of the form $e^{s't_s}$ this gives the relatively simple expression

$$H(z) = \frac{\alpha}{2f_s\,(1 - z^{-1})/(1 + z^{-1}) + \alpha} \tag{15.20}$$

This can easily be converted into the form of one polynomial in z divided by another polynomial in z. This signifies a simple difference equation, and hence a simple physical realization of the required digital filter.

A comparison of the original transfer function of Equation (15.13) with the z transfer function of Equation (15.20) shows that all that has been done is to replace s with $2f_s(1 - z^{-1})/(1 + z^{-1})$. It is not necessary to pass through the intermediate stage involving $H(s')$, although an awarenes of $H(s')$ and the mapping procedure of Figure 15.10 is very useful in visualizing the frequency response of the resulting digital filter.

Although demonstrated for one specific filter, it is possible to extend the derivation to show that this substitution is all that is needed for converting *any* analogue filter into a digital filter. Therefore the bilinear transformation is simply to replace s in the s transfer function using the substitution

$$s \rightarrow 2f_s \frac{1 - z^{-1}}{1 + z^{-1}} \tag{15.21}$$

Multiplying numerator and denominator by z, an alternative that may be regarded as simpler arithmetically is to make the substitution

$$s \rightarrow 2f_s \frac{z - 1}{z + 1} \tag{15.22}$$

The subsequent examples prefer the application of Equation (15.21).

Note: if $s = 2f_s(z - 1)/(z + 1)$, then $(sz/2f_s) + (s/2f_s) - z + 1 = 0$ and we have an equation which is linear in s and linear in z. Hence the equation is *bilinear* in s and z, which is what generates the term *bilinear transformation*).

Example 15.3

A third-order Butterworth filter with a cutoff frequency of 1 rad/s has the transfer function

$$H(s) = \frac{1}{(s+1)(s^2+s+1)} \tag{15.23}$$

Assuming a sampling frequency of 2 Hz, design the corresponding digital filter using the bilinear transformation technique.

Solution The transfer function is the same as that of Example 15.2, except expressed in a less factorized form. This time, using Equation (15.21) with $f_s = 2$, we simply replace each value of s by $4(1-z^{-1})/(1+z^{-1})$.

It is best to first multiply out the expression for $H(s)$, producing

$$H(s) = \frac{1}{(s^3+2s^2+2s-1)}$$

The substitution of s by $4(1-z^{-1})/(1+z^{-1})$ gives

$$H(z) = \frac{1}{64[(1-z^{-1})/(1+z^{-1})]^3 + 32[(1-z^{-1})/(1+z^{-1})]^2 + 8[(1-z^{-1})/(1+z^{-1})] + 1}$$

$$= \frac{(1+z^{-1})^3}{64(1-z^{-1})^3 + 32(1-z^{-1})^2(1+z^{-1}) + 8(1-z^{-1})(1+z^{-1})^2 + (1+z^{-1})^3}$$

Hence

$$H(z) = \frac{1+3z^{-1}+3z^{-2}+z^{-3}}{64-192z^{-1}+192z^{-2}-64z^{-3}+32-32z^{-1}-32z^{-2}+32z^{-3}+8+8z^{-1}-8z^{-2}-8z^{-3}+1+3z^{-1}+3z^{-2}+z^{-3}}$$

$$= \frac{1+3z^{-1}+3z^{-2}+z^{-3}}{105-213z^{-1}+155z^{-2}-39z^{-3}}$$

$$= \frac{0.00952+0.02857z^{-1}+0.02857z^{-2}+0.00952z^{-3}}{1-2.0286z^{-1}+1.4762z^{-2}-0.3714z^{-3}}$$

Using Equation (15.10) the coefficients of the filter may be read off directly to give the configuration of Figure 15.11. It will be noted that the recursive part of this filter

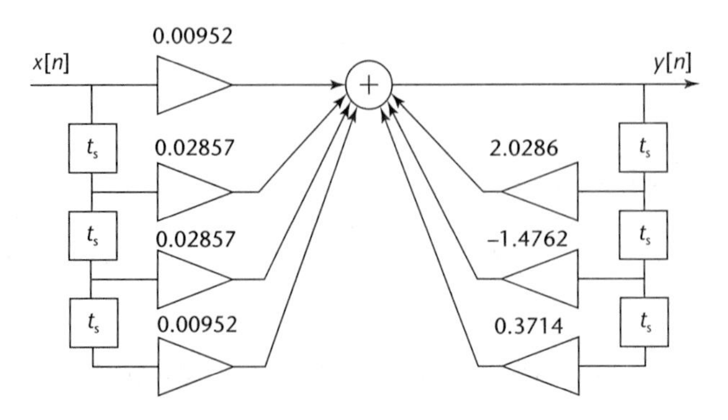

Fig. 15.11 Realization of a third-order Butterworth lowpass filter by the bilinear transformation method

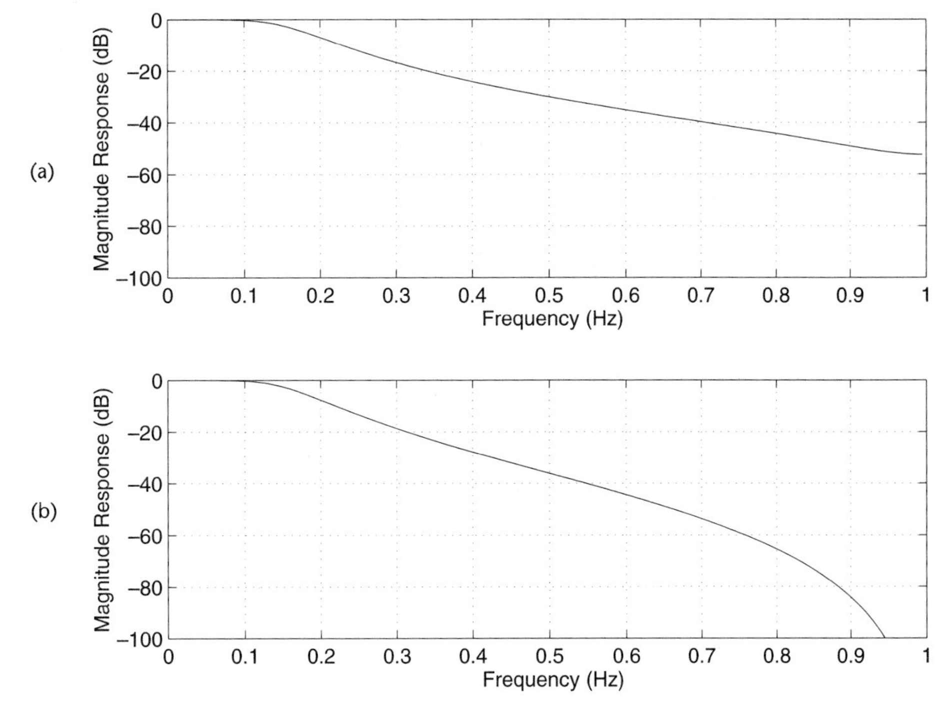

Fig. 15.12* Amplitude response of filters up to the folding frequency; (a) designed by the impulse invariant technique; (b) designed by the bilinear transformation technique

is almost identical to that of the filter in Figure 15.6 designed by the impulse invariant method. Because that designed by the bilinear transformation method has a null at the folding frequency the magnitude response of the two filters is very different as this frequency is approached. At much lower frequencies, however, the two frequency responses are very similar, and Figure 15.12 shows the results for the two filters up to the folding frequency of 1 Hz.

15.3.3 Highpass, bandpass and bandstop filters

A lowpass filter having a cutoff frequency other than that of the lowpass analogue prototype is achieved by transforming the analogue prototype filter into an analogue filter of the correct cutoff frequency, and then applying the impulse invariant or bilinear transformation technique. In a similar manner, one way of designing a highpass, bandpass or bandstop digital filter is to use the frequency transformation techniques described in Chapter 13 for obtaining the appropriate highpass, bandpass or bandstop analogue filter, and then to apply the impulse invariant or bilinear transformation technique to convert it into a digital filter.

As an alternative, there are extensions of the substitution used in the bilinear transformation method whereby s in the lowpass analogue prototype transfer function may be replaced by an expression in z that converts it *directly* to the transfer function of the required digital filter. The transformation formula can include the wanted cutoff frequencies as well as the type of filter wanted.

With both these methods the arithmetic becomes quite complicated and, in view of the powerful software approach for filter design discussed in the next sub-section, these methods will not be elaborated further.

15.4 Designing recursive digital filters with MATLAB

It is useful to have an understanding of the techniques described in Section 15.3, but the design procedures are arithmetically laborious and prone to error, particularly with high-order filters. Simple instructions in MATLAB lead to instant designs for Butterworth, Chebyshev and elliptic filters based upon the bilinear transformation method. This is best appreciated with examples.

Example 15.4

Use MATLAB to design a third-order Butterworth filter with a cutoff frequency of 1 rad/s, assuming a sampling frequency of 2 Hz. Give a z-plane plot of the poles and zeros and plot the frequency response up to 0.5 Hz.

Solution This is the same problem as in Example 15.3, but now the design is done using MATLAB. The following code is entered:

```
[b,a] = butter(3, 0.15915/1)
```

where **b** is a vector of the wanted non-recursive coefficients, **a** is a vector of the wanted recursive coefficients, **butter** indicates that a Butterworth filter is wanted, **3** specifies that the filter should be third-order, and **0.15915/1** is the required cutoff frequency in Hz divided by the folding frequency in Hz (that is, divided by half the sampling frequency). The instruction is almost the same as the instruction used in Section 13.5 for determining the s transfer function of an analogue Butterworth filter, the main difference now being the absence of **'s'** within the brackets of the command. The resulting printout is

```
b =
      0.0100   0.0301   0.0301   0.0100
a =
      1.0000   -2.0090   1.4531   -0.3638
```

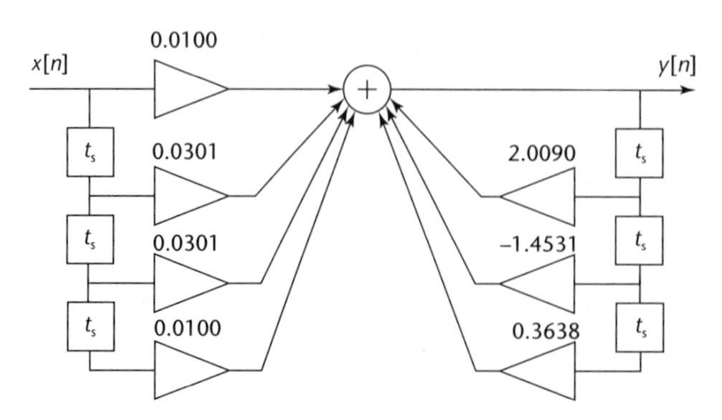

Fig. 15.13 Realization of a third-order Butterworth lowpass filter using MATLAB

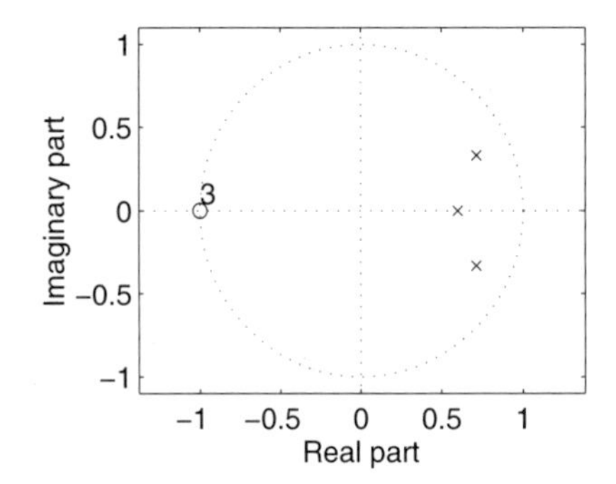

Fig. 15.14 Pole–zero plot of a Butterworth filter

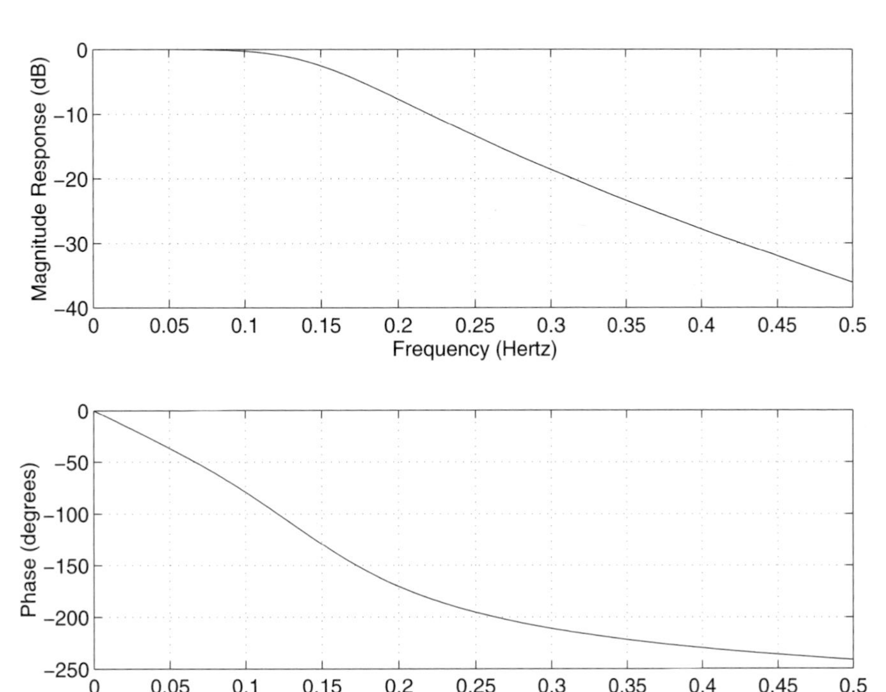

Fig. 15.15 Frequency response of a third-order Butterworth filter designed using MATLAB

In keeping with the sign conventions of Section 14.2 the recursive coefficients are the negatives of these **a** values. The resulting design is that shown in Figure 15.13.

In order to produce a z-plane plot of poles and zeros the following line of code is next entered:

```
subplot(2,2,1)
zplane(b,a)
```

This produces Figure 15.14.

Finally, to determine the frequency response we simply enter

```
f=(0:200)/400;          % a vector of 200 frequencies from 0 to 0.5Hz
freqz(b,a,f,2)          % 2 is the sampling frequency
```

This produces Figure 15.15.

It will be noted that the design effort in Example 15.4 using MATLAB is trivial compared with the manual calculations of Example 15.3. The two sets of coefficients are very slightly different but both are satisfactory. The small discrepancy arises because the **butter** command in MATLAB involves a replacement of s by $2kf_s(1-z^{-1})/(1+z^{-1})$ rather than by $2f_s(1-z^{-1})/(1+z^{-1})$. This changes the frequency warping from $\omega = 2f_s \tan(\omega'/2f_s)$ to $\omega = 2kf_s \tan(\omega'/2f_s)$ such that, with k slightly different from unity and suitably chosen, we can obtain an exact match in the frequency responses of the analogue and digital filters at a selected frequency. In the case of the Butterworth filter this exact match is made to occur at the frequency where the amplitude response is -3 dB.

Single commands in MATLAB can produce designs for lowpass, highpass, bandpass or bandstop filters giving Butterworth, Chebyshev or elliptic function responses. The advantages of using MATLAB over manual methods increase with the increased complexity of the task, as seen in the next example.

Example 15.5

A sequence has 1 kHz sampling. Design an 8 pole Chebyshev bandpass filter that has a bandwidth of 40 Hz centred at 200 Hz, with 0.2 dB of ripple in the passband. Plot the poles and zeros; plot the frequency response between 100 Hz and 300 Hz.

Solution The task is the same as that of Example 15.1 except that now a very much higher order filter is required. Using MATLAB the following single line of code is required

[b,a] = cheby1(4, .2, [180/500 220/500])

where **b** and **a** are vectors giving us the non-recursive and recursive coefficients, as before; **cheby1** denotes that a Chebyshev filter with ripple in the passband but not on the stopband is required; **4** is the number of pole *pairs* and is the order of the lowpass filter from which the bandpass filter is derived; **.2** is the passband ripple in dB; and **[180/500 220/500]** is a two-component vector containing the lower cutoff frequency of 180 Hz divided by half the sampling frequency and the higher cutoff frequency of 220 Hz divided by half the sampling frequency.

The outcome is

```
b =
     1.0e–003*

Columns 1 through 7

 0.1190  0  0.4762  0.0000  0.7143  0.0000  0.4762

Columns 8 through 9

 0.0000  0.1190
 a =

Columns 1 through 7

 1.0000  –2.3536  5.6406  –7.1455  8.6952  –6.4903  4.6544
```

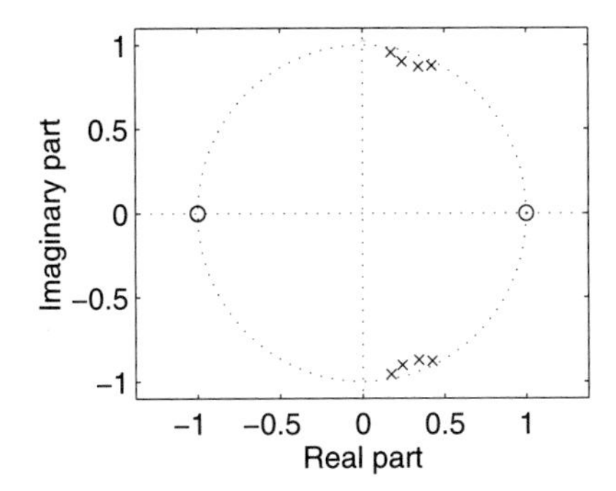

Fig. 15.16 z-plane plot of an 8 pole Chebyshev bandpass filter

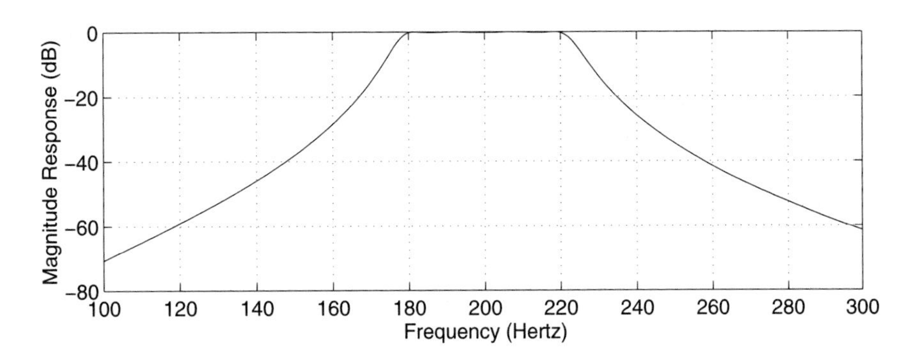

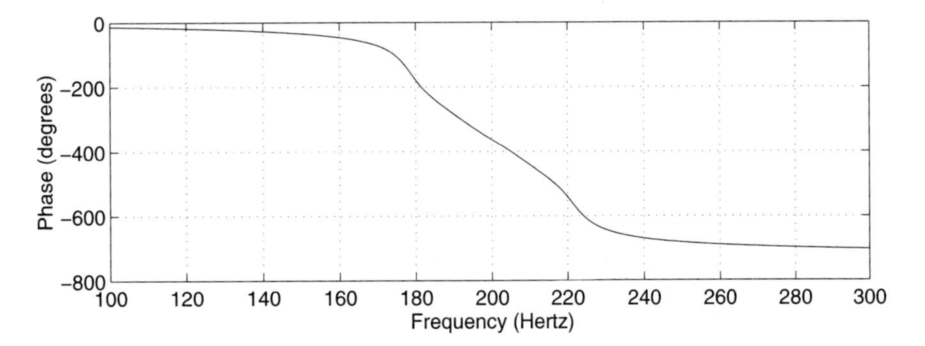

Fig. 15.17 Frequency response of an 8 pole Chebyshev bandpass filter

Columns 8 through 9

−1.7621 0.6802

This signifies the difference equation given by

$$\begin{aligned} y[n] =& 0.001\{0.1190x[n] + 0.4762x[n-2] + 0.7143x[n-4] + 0.4762x[n-6] \\ &+ 0.1190x[n-8]\} + 2.3536y[n-1] - 5.6406y[n-2] + 7.1455y[n-3] \\ &- 8.6952y[n-4] + 6.4903y[n-5] - 4.6544y[n-6] + 1.7621y[n-7] \\ &- 0.6802y[n-8] \end{aligned}$$

The pole–zero plot is obtained by the code

```
subplot(2,2,1)
zplane(b,a)
```

This produces Figure 15.16.

The code

```
f = (100:300);            % the frequency range of the plot
freqz(b,a,f,1000)         % 1000 is the sampling frequency
```

produces the frequency response shown in Figure 15.17.

15.5 Effect of coefficient quantization in IIR filters

When designing a digital filter by hand using the impulse invariant or bilinear transformation technique, it is natural to impose some degree of round-off to the numbers in the calculations. This can cause the filter coefficients to be slightly inaccurate. Even if computer techniques are used to obtain very accurate coefficients, it is tempting to impose some degree of approximation on these coefficients when realizing the filter in software. If the digital filter is realized in hardware the filter coefficients are represented by an n-bit binary word and there is some approximation involved that depends on the value of n.

Unfortunately, very small inaccuracies in the filter coefficients sometimes cause major deviations from the wanted frequency response, and even instability. It is always wise to confirm the properties of a filter using the rounded coefficients actually involved. The problem is best illustrated by an example.

Example 15.6

Use MATLAB to design an eighth-order elliptic lowpass filter with a cutoff frequency of 300 Hz, 0.5 dB of ripple in the passband and a minimum stopband attenuation of 50 dB for use with a signal that is sampled at 4 kHz. Plot the amplitude response

(a) using 16 digit accuracy for the filter coefficients;
(b) rounding off the filter coefficients to six decimal places;
(c) rounding off the filter coefficients to five decimal places.

Show that the filter has become unstable in the last case, such that its frequency response is invalid!

Solution The design of the filter is obtained using the command

```
[b,a]=ellip(8, .5, 50, 300/2000)   % 8th order, .5dB ripple, 50dB attenuation,
                                   % 300Hz cutoff, 2000Hz folding frequency.
```

The outcome is

```
b =

Columns 1 through 7

    0.0046  -0.0249  0.0655  -0.1096  0.1289  -0.1096  0.0655
```

```
Columns 8 through 9

     -0.0249   0.0046

 a =

Columns 1 through 7

     1.0000  -6.9350  21.5565  -39.1515  45.3884  -34.3665  16.5896

Columns 8 through 9

     -4.6673   0.5860
```

It should be noted that these coefficients are calculated and known with an accuracy of 16 digits but that, using the default number display format of MATLAB, they are only displayed with an accuracy of four decimal places.

The following program repeats this design command but then changes the accuracy of the coefficients between three sections of the program and plots the different amplitude responses for each case.

```
% fig15_18.m - Section 1
f=(0:500)*4;                         % frequencies from 0 to 2000Hz
[b,a]=ellip(8, .5, 50, 300/2000);    % 8th order, .5 dB ripple, 50 dB attenuation,
                                     % 300 Hz cutoff, 2000 Hz folding frequency
[h,f]=freqz(b, a, 512, 4000);        % h is the frequency transfer function
m=20*log10(abs(h));                  % magnitude of h in dB
subplot(3,1,1)
plot(f,m)
axis([0 2000 -80 20])
ylabel('Ampl. (dB)')
% Section 2
b2=round(b*1.e6)/1.e6;               % decimal place is moved 6 positions to the
                                     % right; 'round' discards decimals; decimal
                                     % place is then moved back 6 positions to the
                                     % left to give accuracy of 6 decimal places for
                                     % b coefficients.
a2=round(a*1.e6)/1.e6;               % the same treatment is given to the a
                                     % coefficients
[h,f]=freqz(b2, a2, 512, 4000);
m=20*log10(abs(h));
subplot(3,1,2)
plot(f,m)
axis([0 000 -80 20])
ylabel('Ampl. (dB)')
% Section 3
b3=round(b*1.e5)/1.e5;               % b coefficients given accuracy of 5 decimal
                                     % places
a3=round(a*1.e5)/1.e5;               % a coefficients given accuracy of 5 decimal
                                     % places
[h,f]=freqz(b3, a3, 512, 4000);
```

```
m=20*log10(abs(h));
subplot(3,1,3)
plot(f,m)
axis([0 2000 -80 20])
ylabel('Ampl. (dB)'),xlabel('Frequency (Hz)')
```

The outcome is shown in Figure 15.18. However, some caution is needed since frequency responses obtained by replacing z by $e^{j\omega t_s}$ in the transfer function equation

$$H(z) = \frac{\sum_{i=0}^{M} b_i z^{-i}}{1 + \sum_{i=1}^{N} a_i z^{-i}}$$

are invalid if this equation gives poles outside the unit circle, as then the system is unstable. We can check the radii of the poles for each of the three designs by determining the roots of the denominator polynomial in each case. The necessary command for the first design is

```
r = abs(roots(a))'          % this determines the roots of the denominator
                            % of H(z), and hence the poles, when the
                            % coefficients have 16 digit accuracy
```

This produces the printout

```
r =

Columns 1 through 7

  0.9941  0.9941  0.9741  0.9741  0.9259  0.9259  0.8538
```

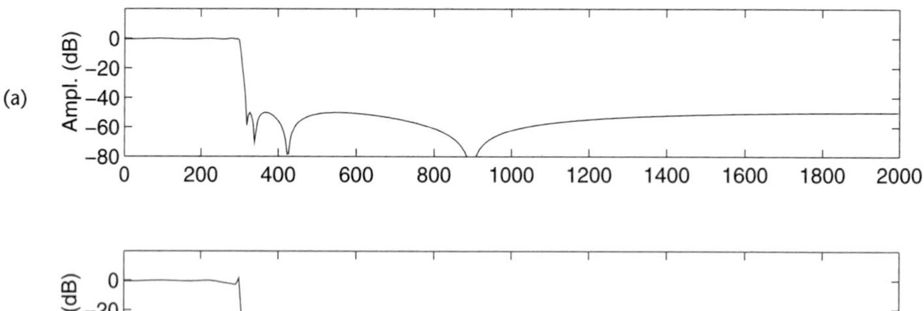

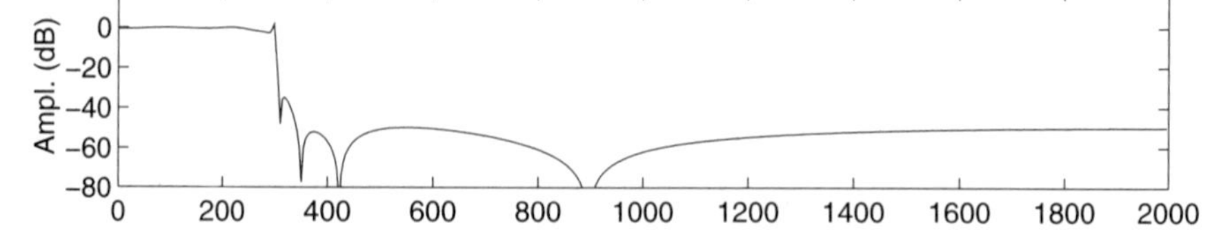

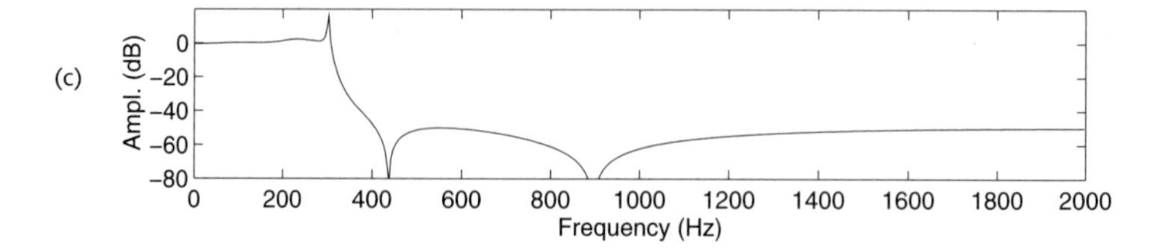

Fig. 15.18 Amplitude responses of an elliptic filter. (a) 16 digit accuracy for coefficients; (b) 6 decimal place accuracy for coefficients; (c) 5 decimal place accuracy for coefficients

Column 8

0.8538

Next, for the second design, we use the command

```
r = abs(roots(a2))'          % this gives the poles when the coefficients are
                             % accurate to 6 decimal places
```

This produces the printout

```
r =

Columns 1 through 7

   0.9974   0.9974   0.9688   0.9688   0.9284   0.9284   0.8533

Column 8

   0.8533
```

Finally, for the third design we use the command

```
r = abs(roots(a3))'          % this gives the poles when the coefficients are
                             % accurate to 5 decimal places
```

This produces

```
r =

Columns 1 through 7

   1.0013   1.0013   0.9587   0.9587   0.9362   0.9362   0.8518

Column 8

   0.8518
```

It is seen that some of the poles in this last case have a radius exceeding unity, thus indicating instability. Even though the filter coefficients are accurate to five decimal places, the design is worthless!

When the filter coefficients are accurate to six decimal places the radii of the poles are all less than one, such that the filter is stable. Even here, however, there is a significant degradation of frequency response compared with the theoretical response, as seen by a comparison of Figure 15.18(b) with Figure 15.18(a).

Although the amplitude response of Figure 15.18(c) is sufficiently degraded relative to the theoretical response to suggest there might be a stability problem, it is not definitive. An informative alternative to examining the poles is to examine the impulse response.

The following program continues from the previous program to plot the impulse responses for the three filters.

```
% R fig15_19.m
subplot(3,1,1)
dimpulse(b,a,1000)
subplot(3,1,2)
dimpulse(b2,a2,1000)
```

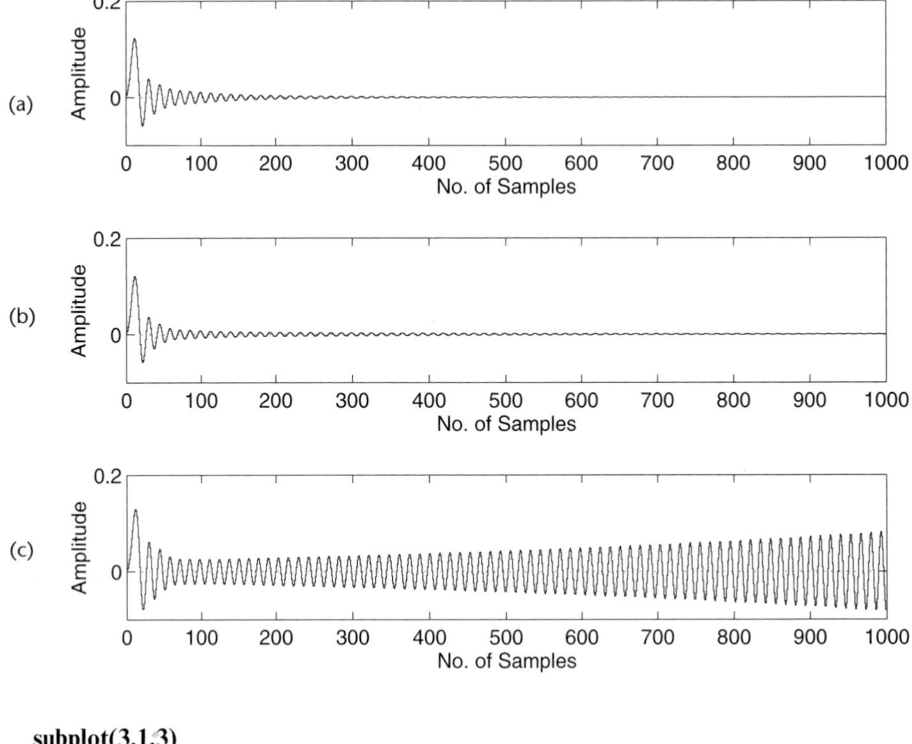

Fig. 15.19 Impulse response of an elliptic filter. (a) 16 digit accuracy for coefficients; (b) 6 decimal place accuracy for coefficients; (c) 5 decimal place accuracy for coefficients

```
subplot(3,1,3)
dimpulse(b3,a3,1000)
```

The result is shown in Figure 15.19.

The last example has shown that an exceedingly small percentage inaccuracy in filter coefficients can cause instability in an IIR filter. Such inaccuracies can easily arise if the filter is designed manually, or if accurate computer-generated coefficients are quantized to enable them to be stored in the memory of a hardware filter. The obvious solution for the quantization problem is to have large word sizes, perhaps 16 or even 32 bits, but the penalty compared with smaller words of 8 or 12 bits is the requirement for a greater memory size and more complicated multiplications. It is better to find ways of making the filter less sensitive to small errors in the coefficients.

15.6 Cascading low-order IIR filters for improved stability

The higher the order of the filter the greater is the problem with instability due to inaccuracies in the coefficients. The most commonly adopted solution is to realize the high-order filter by a cascade of second-order filters. For example, the transfer function given by

$$H(z) = \frac{\sum_{i=0}^{M} b_i z^{-i}}{1 + \sum_{i=1}^{N} a_i z^{-i}} = \frac{K(z-z_1)(z-z_2)(z-z_3)(z-z_4)(z-z_5)(z-z_6)}{(z-p_1)(z-p_2)(z-p_3)(z-p_4)(z-p_5)(z-p_6)} \tag{15.24}$$

can be rewritten as

$$H(z) = KH_1(z)H_2(z)H_3(z) \tag{15.25}$$

where one possible grouping of poles and zeros requires

$$H_1(z) = \frac{(z - z_1)(z - z_2)}{(z - p_1)(z - p_2)}$$

$$H_2(z) = \frac{(z - z_3)(z - z_4)}{(z - p_3)(z - p_4)}$$

$$H_3(z) = \frac{(z - z_5)(z - z_6)}{(z - p_5)(z - p_6)}$$

There are, however, many other groupings, the only constraint being that conjugate poles and conjugate zeros must be assigned to the same second-order filter, since otherwise the filter coefficients would be complex. When making the choice we face two decisions.

1. We must decide which pair of poles should be paired with which pair of zeros (bearing in mind that conjugate poles go together and conjugate zeros go together).
2. We must decide on the sequence of the cascaded sections. If the cascade sections are implemented with high precision arithmetic (as effectively happens on a computer), each one can be regarded as a linear system and their order does not matter. However, with finite precision arithmetic (as happens in the hardware realization of a filter), the round-off errors do affect the optimal sequence. This last point is clarified by an example that is unrelated to filters.

Example 15.7

As a simple example of how sequence can matter, show that the outcome of multiplying the decimal numbers 3.2, 2.1 and 9.8 is affected by the order of multiplication if the result of each multiplication is rounded off to one decimal place.

Solution Let us compare the multiplications $(3.2 \times 2.1) \times 9.8$ and $(3.2 \times 9.8) \times 2.1$. In the first case we have

$$3.2 \times 2.1 = 6.72 \rightarrow 6.7$$

$$6.7 \times 9.8 = 65.66 \rightarrow 65.7$$

In the second case we have

$$3.2 \times 9.8 = 31.36 \rightarrow 31.4$$

$$31.4 \times 2.1 = 65.94 \rightarrow 65.9$$

The two answers are seen to be different.

In order to obtain the best results from cascaded sections the solution to the two problems that have been identified is that the output of each multiplier section should be well matched to the dynamic range of the digital words at all parts of the cascaded filter. To achieve this, two useful rules that are not entirely obvious are:

1. Among all the pairs of conjugate poles, the pair closest to the unit circle is that which has the greatest effect on the amplitude response. This effect should be minimized by associating the nearest pair of zeros with this pole pair. This procedure should continue with remaining pairings.
2. The sub-filters should be situated in the cascade such that the least influential ones occur first. Thus the last sub-filter is the one with poles closest to the unit circle, since this has the greatest influence on the amplitude response.

Example 15.8

Replace the filter of Example 15.6 by four two-pole filters in cascade. Demonstrate that the two-pole filters are stable even when the coefficients are rounded off to two decimal places.

Solution It is instructive to begin by examining the pole–zero plot. Using MATLAB we require

```
[b,a] = ellip(8, .5, 50, 300/2000);
zplane(b,a)
```

or, since we shall require numerical values of the poles and zeros, an alternative program whose first command generates the zeros, poles and gain of the z transfer function is

```
[z,p,k] = ellip(8, .5, 50, 300/2000);
zplane(z,p)
```

Either way the outcome is as shown in Figure 15.20. Using the proposed rules, the optimal pairings for the first and last two-pole filters in the cascade are also indicated.

We can print out numerical values for the zeros and poles by the commands **z** and **p** and then group them as suggested. This leads to

$$H_4(z) = \frac{(z - 1\angle 28.90°)(z - 1\angle -28.90°)}{(z - 0.994\angle 27.06°)(z - 0.994\angle -27.06°)} = \frac{1 - 1.75z^{-1} + z^{-2}}{1 - 1.77z^{-1} + 0.99z^{-2}}$$

$$H_3(z) = \frac{(z - 1\angle 30.69°)(z - 1\angle -30.69°)}{(z - 0.974\angle 25.78°)(z - 0.974\angle -25.78°)} = \frac{1 - 1.72z^{-1} + z^{-2}}{1 - 1.75z^{-1} + 0.95z^{-2}}$$

$$H_2(z) = \frac{(z - 1\angle 38.14°)(z - 1\angle -38.14°)}{(z - 0.926\angle 21.33°)(z - 0.926\angle -21.33°)} = \frac{1 - 1.57z^{-1} + z^{-2}}{1 - 1.73z^{-1} + 0.86z^{-2}}$$

$$H_1(z) = \frac{(z - 1\angle 80.64°)(z - 1\angle -80.64°)}{(z - 0.854\angle 9.28°)(z - 0.854\angle -9.28°)} = \frac{1 - 0.33z^{-1} + z^{-2}}{1 - 1.69z^{-1} + 0.73z^{-2}}$$

Using the command **k** the gain factor k can be read off from MATLAB as equalling 0.0046. This should be incorporated into the cascade.

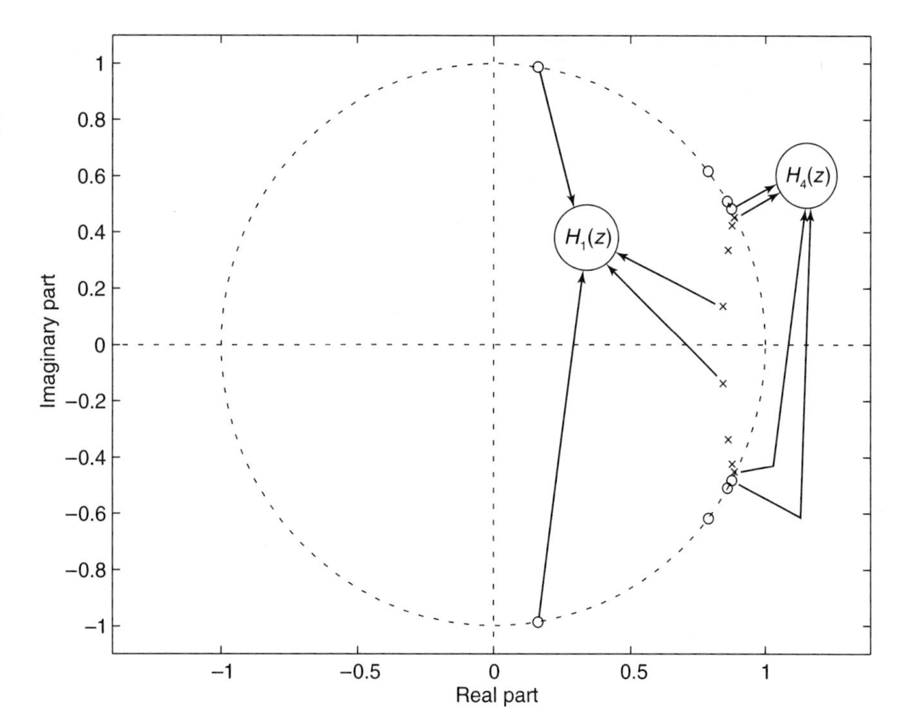

Fig. 15.20 Pole–zero plot for an elliptic function filter, with optimal pairings

It will be noted that all the coefficients have been rounded off to a modest accuracy of two decimal places. We can check the stability of the most critical one of these, namely $H_4(z)$, by determining its poles. To do this we enter the coefficients of the denominator polynomial and find its roots, using

```
x=[1 -1.77 .99];
r=abs(roots(x))'
```

The outcome is

```
r =

0.9950    0.9950
```

These are close to the original values of 0.9941 and are inside the unit circle in spite of the considerable round-off error in the coefficients. In contrast to the six decimal place accuracy required by the filter in the previous example, a two decimal place accuracy is sufficient for this second-order section.

In fact, MATLAB offers an even greater short-cut for producing the design. There is a special command **zp2sos** for breaking down a transfer function defined by its poles and zeros into a cascade of second-order sections and providing the coefficients of these sections. The output is in matrix form, where each row lists the b coefficients and then the a coefficients of one of the second-order sections. In its default mode the sections should be cascaded in the same sequence as the rows. Thus for a complete design of the elliptic filter we need only two commands:

```
[z,p,k] = ellip(8, .5, 50, 300/2000);
m=zp2sos(z,p,k)
```

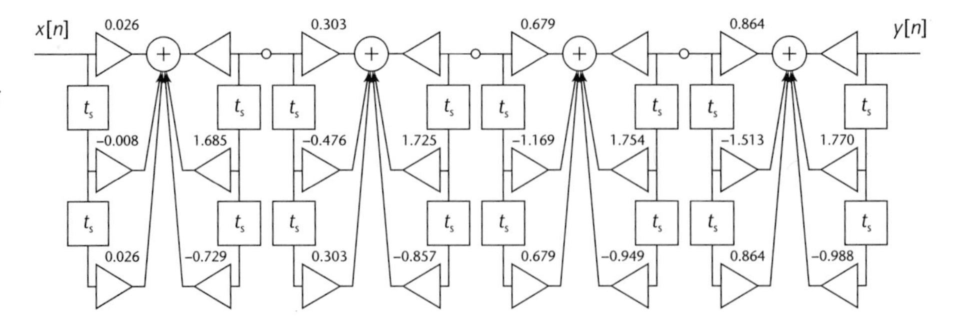

Fig. 15.21 Realization of an elliptic filter by four two-pole sections in cascade

The outcome is

```
m =
  0.0261  -0.0085  0.0261  1.0000  -1.6853  0.7290
  0.3027  -0.4761  0.3027  1.0000  -1.7250  0.8573
  0.6795  -1.1687  0.6795  1.0000  -1.7542  0.9488
  0.8641  -1.5131  0.8641  1.0000  -1.7705  0.9882
```

Apart from automatically including an optimal distribution of the gain factor through the cascade the design is identical to the one suggested by the previously derived functions $H_1(z)$, $H_2(z)$, $H_3(z)$ and $H_4(z)$. The complete filter is shown in Figure 15.21 and, although an accuracy of two decimal places has been shown to be sufficient for the last and most critical section, an accuracy of three or four decimal places is a better choice since it provides a greater safety margin and yet remains economical. In general it is always wise to check the response with the rounded coefficients actually used, whatever their accuracy.

15.7 Design of non-recursive digital filters

The recursive filters described in the preceding sections require only a few coefficients to be stored in memory; also, by making use of previous output samples, they require few multiplications for producing new output samples. This makes them computationally efficient and fast in operation. One problem is their propensity to instability, but, as was shown, this difficulty can be minimized by breaking down the filter into a cascade of second-order filters. There is, however, a problem that cannot be so easily overcome, and that is the one of their phase response. Although Butterworth, Chebyshev and elliptic filters have very desirable amplitude responses, their phase characteristics are non-linear, and this signifies waveform distortion. Sometimes this does not matter. For example, the ear is very insensitive to phase distortion, and the two-tone signal $\cos(1500t) + 0.6\cos(1800t)$ would be indistinguishable from the signal $\cos(1500t) + 0.6\sin(1800t)$, even though the two waveforms are different. In some other situations, however, phase distortion is not acceptable.

In the case of analogue filters it was suggested in Chapter 13 that excessive phase distortion can be reduced by adding a delay equalizer in cascade. In the case of digital filters a much simpler solution is to use a non-recursive filter. This can give a perfect linear phase response, as will now be shown.

Consider the non-recursive configuration shown in Figure 15.22. By inspection, the impulse response is given by

$$h[n] = \{b_0,\ b_1,\ b_2,\ b_3,\ b_4, \cdots\} \tag{15.26}$$

It is seen that there is total flexibility regarding the impulse response except that it must be causal (that is, zero for $t < 0$). The impulse response is the same as the distribution of the non-recursive filter coefficients and, so long as enough coefficients are used, *any causal impulse response can be achieved.*

It is useful to begin by forgetting causality and to show that the transfer function of a system having an impulse response with even symmetry about the origin introduces no phase distortion. An example of an impulse response with even symmetry is shown in Figure 15.23 and the corresponding transfer function is given by its Fourier transform; that is

$$H(\omega) = \int_{-\infty}^{\infty} h(t)\mathrm{e}^{-\mathrm{j}\omega t}\,\mathrm{d}t$$

This can be written as

$$H(\omega) = \int_{-\infty}^{\infty} h(t)(\cos\,\omega t - \mathrm{j}\,\sin\,\omega t)\,\mathrm{d}t$$

$$= \int_{-\infty}^{\infty} h(t)\cos\,\omega t\,\mathrm{d}t - \mathrm{j}\int_{-\infty}^{\infty} h(t)\sin\,\omega t\,\mathrm{d}t \tag{15.27}$$

The last term is zero when $h(t)$ is an even function, signifying that $H(\omega)$ is entirely real with a phase response of either 0° or 180°. It follows that any system having an impulse response which is symmetric about the origin can only introduce two possible phase shifts, either 0° or 180°. It can be shown that the transition from 0° to 180°, or vice versa, can only occur when there is a discontinuity in the amplitude

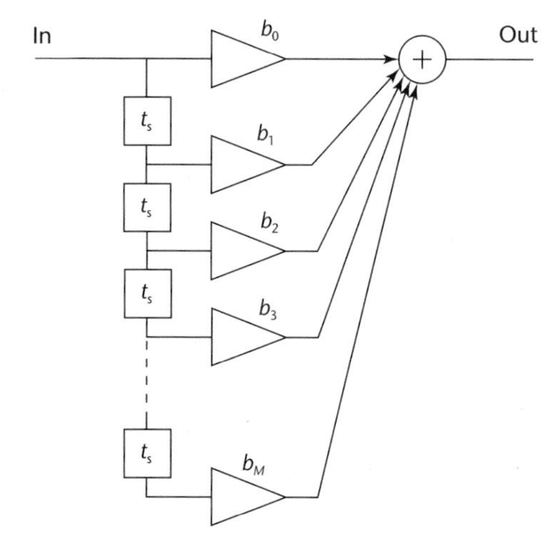

Fig. 15.22 Non-recursive filter configuration

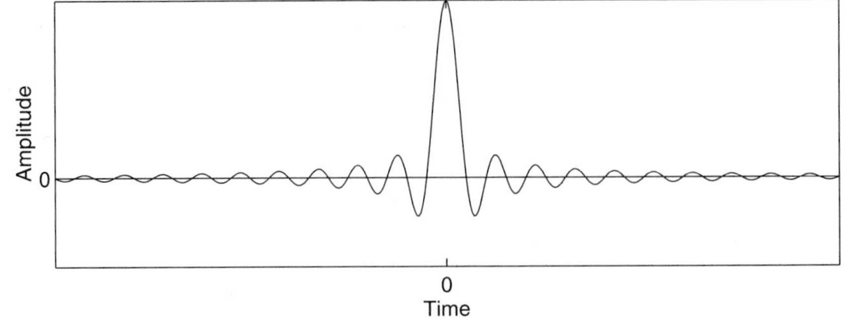

Fig. 15.23 An impulse response with even symmetry

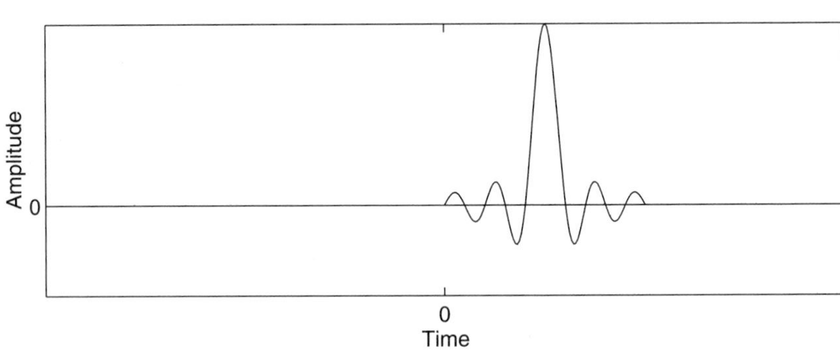

Fig. 15.24 Truncated, offset, symmetrical and causal impulse response that signifies a linear phase response

response, such as at a null. It follows that any system having an impulse response which is symmetric about the origin introduces zero phase shift in the passband.

The problem that Figure 15.23 represents a non-causal impulse response can be removed if the response can be truncated symmetrically and then delayed such that there is no response prior to $t = 0$. From Section 6.18.4 the effect of a constant delay is equivalent to introducing a phase shift proportional to frequency. It is thus seen that a non-recursive filter with a symmetrical but offset impulse response, such as in Figure 15.24, will have a perfect linear phase response within the passband. Since the designer has control over the impulse response this absence of phase distortion is a very appealing feature of non-recursive filters.

The two main methods of non-recursive filter design are the Fourier method and the optimal method. The *Fourier* method is the easiest to understand and can give excellent results. The *optimal* method is more complicated, but can achieve a closer approximation of the desired frequency response using the same number of coefficients. It is the ideal choice when a software package for its implementation is available to the designer.

15.7.1 The Fourier (or window) method

The Fourier design technique begins by selecting a brickwall amplitude response (lowpass, highpass, bandpass or bandstop) and a zero phase response. It then accepts the inevitable fact that all digital filters have *periodic* transfer functions (with a period equal to the sampling frequency) and applies the inverse Fourier transform to a periodic version of the brickwall response in order to determine the associated impulse response. The filter is implemented by selecting coefficients that give a close approximation to a delayed (and hence causal) version of this impulse response.

The main problem with this method of filter design is that the impulse responses corresponding to typical wanted frequency responses are *infinite in extent*. This means that a very large number of coefficients are needed to produce a good approximation of the frequency response and to make the filter causal. In practice, the number of coefficients is constrained for reasons of economy, and the associated truncation of the impulse response signifies a deviation from the intended frequency response.

Let us first consider the effect of making the wanted transfer function periodic such that it repeats itself every sampling frequency. This means that a digital filter intended to realize the amplitude response $A(f)$ shown in Figure 15.25(a) will actually achieve the amplitude response shown in Figure 15.25(b).

We require the inverse Fourier transform of the transfer function corresponding to Figure 15.25(b) in order to determine what impulse response must be realized by the digital filter.

Previously we have dealt with the Fourier transforms of periodic functions of *time* but not with the *inverse* Fourier transforms of periodic functions of *frequency*. To recap on periodic functions of time we found in Equation (5.23) that the coefficients of the exponential Fourier *series* of a time waveform $f_{\mathrm{p}}(t)$ of period T, and hence fundamental frequency $f_0 = 1/T$, were given by

$$F_n = \frac{1}{T}\int_{-T/2}^{T/2} f_{\mathrm{p}}(t)\mathrm{e}^{-\mathrm{j}2\pi n f_0 t}\,\mathrm{d}t \tag{15.28}$$

and Section 6.11 told us that the Fourier *transform* of $f_{\mathrm{p}}(t)$ is the same as its Fourier series, except that the lines of the Fourier series, are replaced by impulse functions whose weights are the same as the amplitudes of the lines.

Proceeding next to a periodic frequency function of period $f_{\mathrm{s}} = 1/t_{\mathrm{s}}$ Hz (rather than a time function of period T seconds), the symmetry of the Fourier and inverse Fourier transforms makes us deduce that the inverse Fourier transform of the periodic amplitude response of Figure 15.25(b) is a set of impulse functions t_{s} apart and whose weights are given by

$$b_i = t_{\mathrm{s}}\int_{-f_{\mathrm{s}}/2}^{f_{\mathrm{s}}/2} A(f)\mathrm{e}^{\mathrm{j}2\pi i t_{\mathrm{s}} f}\,\mathrm{d}f \tag{15.29}$$

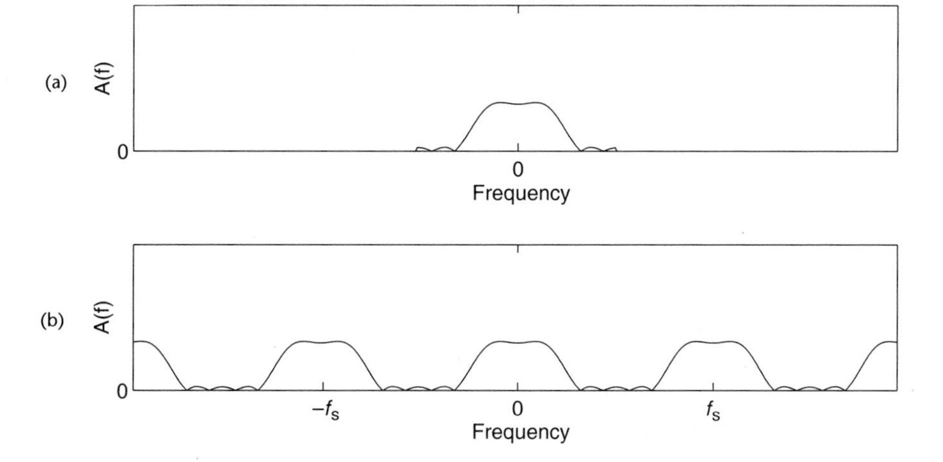

Fig. 15.25 Amplitude response of a digital filter: (a) wanted; (b) actual

It follows that this is the fundamental design equation for determining the ith coefficient of a non-recursive filter using the Fourier method.

Example 15.9

Use Equation (15.29) to design a 15-coefficient causal lowpass filter with a bandwidth of B Hz, assuming a sampling rate of $4B$ Hz. Calculate the actual amplitude response.

Solution For a zero phase filter we have

$$\begin{aligned} A(f) = H(f) &= 1 \qquad -B < f < B \\ &= 0 \qquad \text{elsewhere} \end{aligned} \tag{15.30}$$

Since $B < f_s/2$, we can constrain the limits of integration in Equation (15.29) to $\pm B$ Hz. Hence

$$b_i = t_s \int_{-B}^{B} e^{j2\pi i t_s f}\, df \tag{15.31}$$

Putting $t_s = 1/(4B)$, this becomes

$$b_i = \frac{1}{4B}\int_{-B}^{B} e^{j\pi i f/2B}\, df = \frac{1}{4B}\frac{2B}{j\pi i}\, e^{j\pi i f/2B}\Big|_{-B}^{B} = \frac{\sin(\pi i/2)}{\pi i} \tag{15.32}$$

This equation gives the coefficients of a non-causal filter whose impulse response is that shown in Figure 15.26(a).

In order to make the design into a causal filter with 15 coefficients we must shift the indices of the coefficients by 7 and truncate the coefficients to obtain

$$b_i = \frac{\sin[\pi(i-7)/2]}{\pi(i-7)} \qquad 0 \leqslant i \leqslant 14 \tag{15.33}$$

that is

$$\begin{aligned} &b_0 = -\frac{1}{7\pi},\ b_1 = 0,\ b_2 = \frac{1}{5\pi},\ b_3 = 0,\ b_4 = -\frac{1}{3\pi}, \\ &b_5 = 0,\ b_6 = \frac{1}{\pi},\ b_7 = 0.5,\ b_8 = \frac{1}{\pi},\ b_9 = 0,\ b_{10} = -\frac{1}{3\pi}, \\ &b_{11} = 0,\ b_{12} = \frac{1}{5\pi},\ b_{13} = 0,\ b_{14} = -\frac{1}{7\pi} \end{aligned} \tag{15.34}$$

The corresponding impulse response is shown in Figure 15.26(b). The filter realization is shown in Figure 15.27, where, because some of the coefficients are zero, only nine of the fifteen are actually required.

We can now determine the transfer function using Equation (14.31), which, in terms of digital radian frequency for a non-recursive filter, is

$$H(\Omega) = \sum_{i=0}^{14} b_i e^{-j\Omega i} \tag{15.35}$$

Substituting the filter coefficients and taking $e^{-j7\Omega}$ outside of the summation to

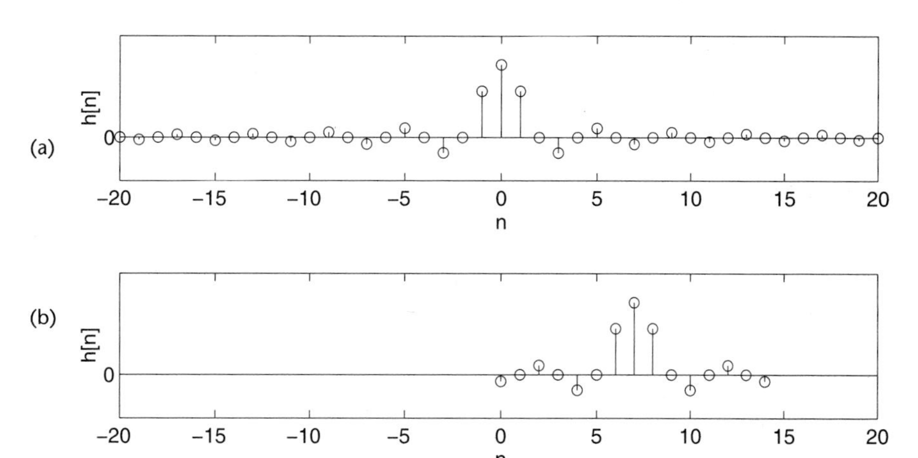

Fig. 15.26 Impulse response of a non-recursive filter: (a) non-causal with an infinite number of coefficients; (b) causal with 15 coefficients

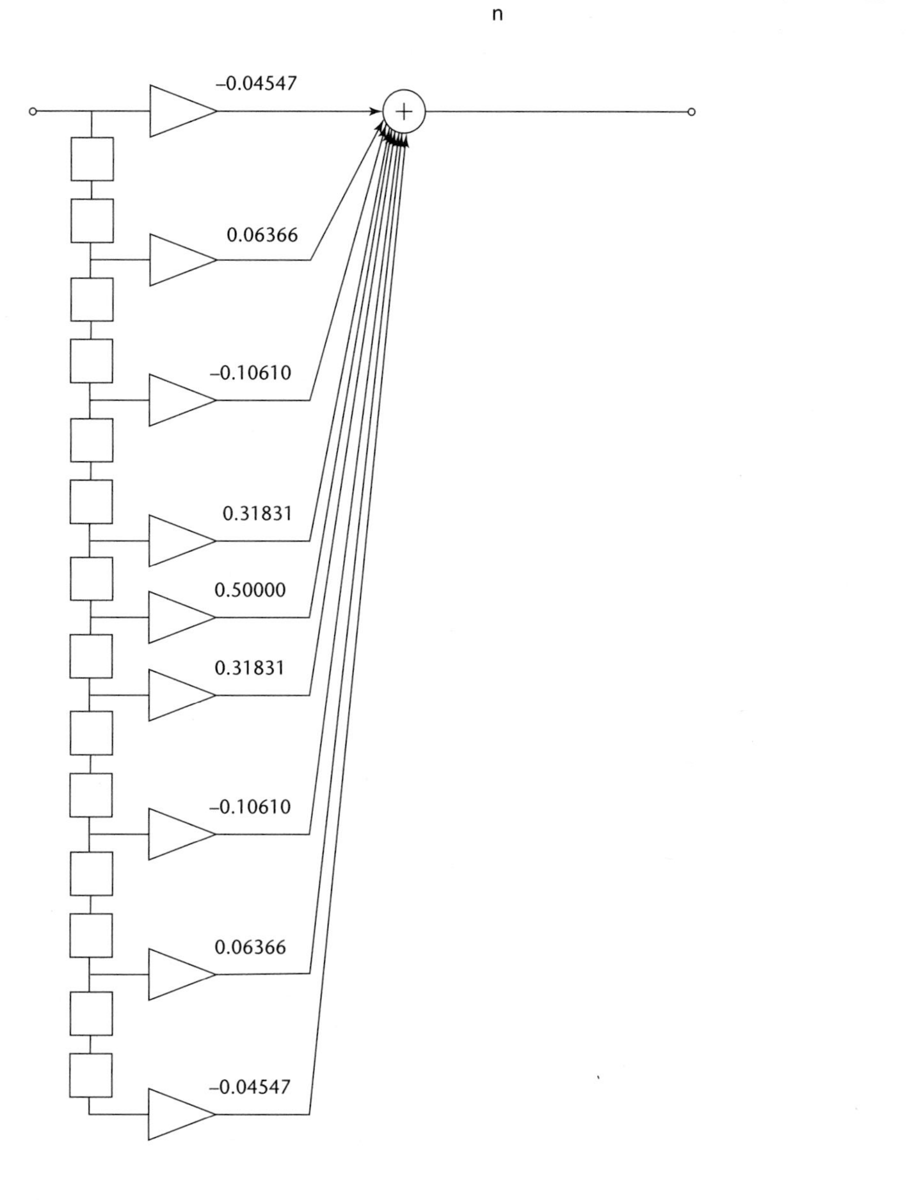

Fig. 15.27 Realization of a non-recursive filter

better reveal symmetry in the expression we obtain

$$H(\Omega) = e^{-j7\Omega}(-0.0455e^{j7\Omega} + 0.0637e^{j5\Omega} - 0.1061e^{j3\Omega} + 0.3183e^{j\Omega} + 0.5 + 0.3183e^{-j\Omega} - 0.1061e^{-j3\Omega} + 0.0637e^{-j5\Omega} - 0.0455e^{-j7\Omega}) \tag{15.36}$$

This simplifies to

$$|H(\Omega)| = 0.5 + 0.6366\cos(\Omega) - 0.2122\cos(3\Omega) + 0.1273\cos(5\Omega) - 0.091\cos(7\Omega) \tag{15.37}$$

This can be evaluated as a function of frequency and plotted. Alternatively, using MATLAB, we can enter the coefficients and use the command for obtaining a frequency response, that is

```
b=[-.04547 0 .06366 0 -.10610 0 .31831 .5 .31831 0 -.10610 0 .06366 0 -.04547];
a=[1 0 0 0 0 0 0 0 0 0 0 0 0 0 0];
freqz(b,a)
```

This produces Figure 15.28. It will be noted that within the passband there is a perfectly linear phase shift and hence no phase distortion.

It should be noted that it would not have been necessary to take $e^{-j7\Omega}$ outside the summation of Equation (15.35) to give Equation (15.36) if we had considered a non-causal filter with coefficients given by Equation (15.32) and then used the expression

$$H(\Omega) = \sum_{i=-7}^{7} b_i e^{-j\Omega i} \tag{15.38}$$

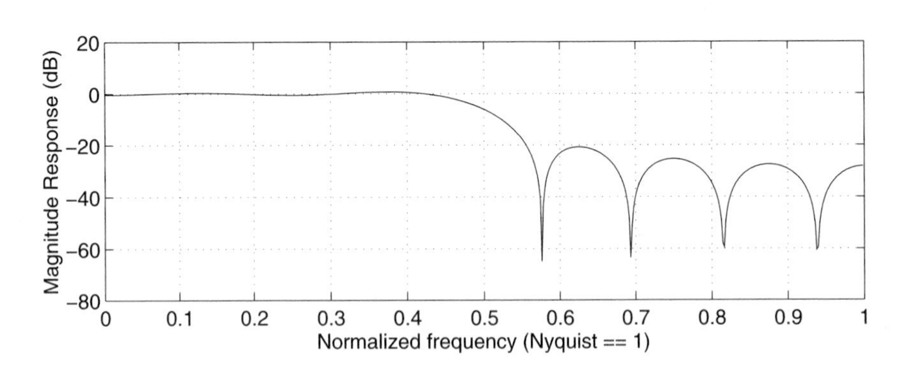

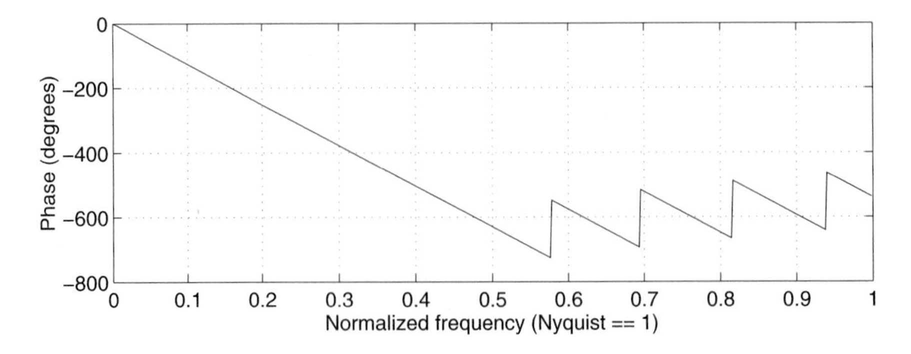

Fig. 15.28 Frequency response of a 15-coefficient lowpass non-recursive (FIR) filter

The amplitude responses of the causal and non-causal filters are identical; it is just their phase responses that are different.

Let us next consider the effect of multiplying the filter coefficients by a weighting function w_i, such that they become $b_i w_i$ where the b_i coefficients are those obtained from Equation (15.29). The effect is to change the discrete-time impulse response of the filter from $h[n]$ to $h[n]w[n]$, where $w[n]$ is a sequence of the weighting function coefficients.

It is instructive to use continuous Fourier transform theory and therefore to consider the response of the filter to a continuous-time impulse function rather than a discrete-time one. The effect of the weighting function is to change the continuous-time impulse response of the filter from $h(t)$ to $h(t)w(t)$, where $w(t)$ is a continuous version of $w[n]$. Using the frequency convolution theorem of Equation (6.45), the effect of changing the impulse response of the filter from $h(t)$ to $h(t)w(t)$ is to change the transfer function from $H(f)$ to $H(f) * W(f)$, where $W(f)$ is the Fourier transform of the weighting function. As will be shown, this modification to $H(f)$ can have a very beneficial effect on the frequency response.

It is useful to begin by applying this theory to one weighting function that has already been considered, that where the filter coefficients are truncated for reasons of economy. This is equivalent to $w(t)$ being a rectangular window function, and for this $W(f)$ is a sinc function as shown in Figure 15.29(a). For the case where a brickwall lowpass filter is wanted, such that the periodic transfer function of the intended filter $H(f)$ is that of Figure 15.29(b), the result of the convolution with $W(f)$ is shown in Figure 15.29(c). The degradation of the filter response due to the windowing of the impulse response is clearly apparent.

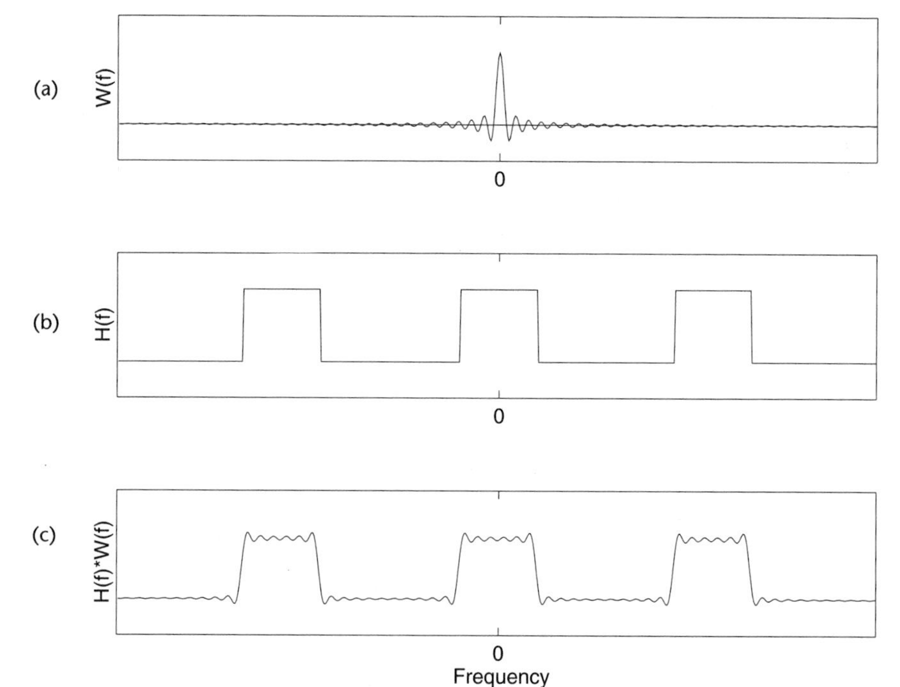

Fig. 15.29 Effect of windowing the coefficients of an FIR filter: (a) Fourier transform of a rectangular window function; (b) transfer function with no windowing; (c) transfer function with windowing

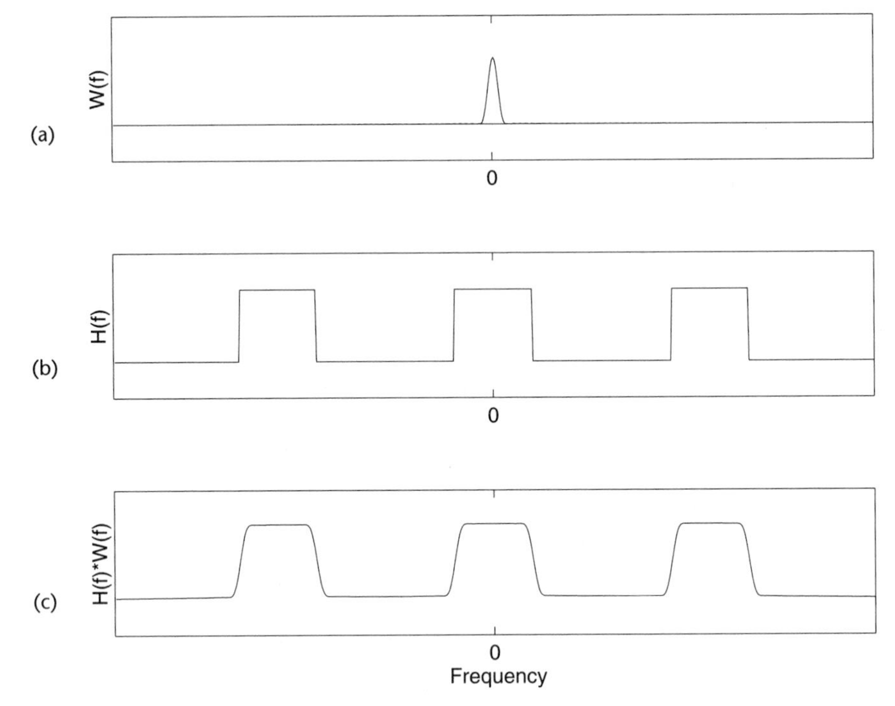

Fig. 15.30 Effect of windowing the coefficients of an FIR filter: (a) Fourier transform of a Hamming window function; (b) transfer function with no windowing; (c) transfer function with windowing

The arguments for imposing a *non*-rectangular window function on the impulse response of a filter is the same as that for imposing a non-rectangular window function on a waveform in spectral analysis, as discussed in Section 9.4. In both cases $W(f)$ can have greatly reduced sidelobes and hence the convolution of $H(f)$ with $W(f)$ causes much less ripple than when a rectangular window is used. As in Section 9.4, the use of von Hann, Hamming and similar windows can be very beneficial. Figure 15.30(a) shows the function $W(f)$ resulting from a Hamming window and Figure 15.30(c) shows the effect of convolving this with the response of the ideal digital lowpass filter response given in Figure 15.30(b). This final response is seen to be much more satisfactory than the comparable response using a rectangular window that was presented in Figure 15.29(c).

Example 15.10

Apply a Hamming window to the filter design of Example 15.9 and determine the resulting frequency response.

Solution Equation (9.9) gives us that the Hamming window is

$$w[n] = 0.54 - 0.46 \cos\left(\frac{2\pi n}{N-1}\right) \qquad 0 < n < N-1$$

$$= 0 \qquad \text{elsewhere}$$

For $N = 15$ this gives

$$w_0 = 0.08,\ w_1 = 0.1255,\ w_2 = 0.2532,\ w_3 = 0.4376,\ w_4 = 0.6424,$$
$$w_5 = 0.8268,\ w_6 = 0.9544,\ w_7 = 1.0000,\ w_8 = 0.9544,\ w_9 = 0.8268,$$
$$w_{10} = 0.6424,\ w_{11} = 0.4376,\ w_{12} = 0.2532,\ w_{13} = 0.1255,\ w_{14} = 0.08$$

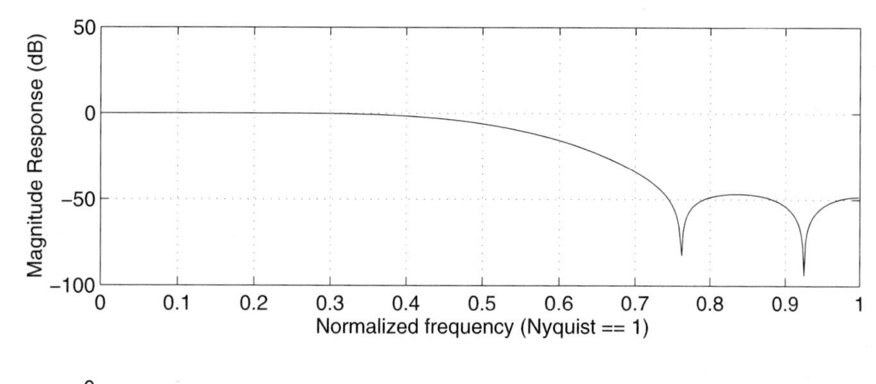

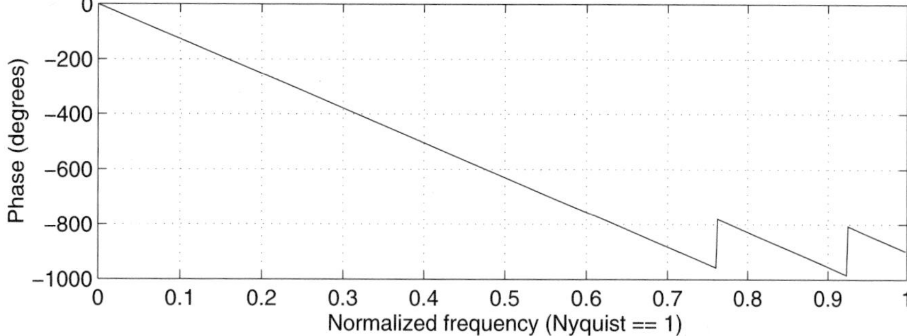

Fig. 15.31 Frequency response of a 15-coefficient lowpass FIR filter with a Hamming window

Applying these weights to the filter coefficients of Example 15.10 we obtain

$$b_0 = -0.00364,\ b_1 = 0,\ b_2 = 0.01612,\ b_3 = 0,\ b_4 = -0.06816,\ b_5 = 0,$$
$$b_6 = 0.30379,\ b_7 = 0.5,\ b_8 = 0.30379,\ b_9 = 0,\ b_{10} = -0.06816,\ b_{11} = 0,$$
$$b_{12} = 0.01612,\ b_{13} = 0,\ b_{14} = -0.00364$$

Proceeding in the same way as in the previous example we obtain

$$|H(\Omega)| = 0.5 + 0.6076\cos(\Omega) - 0.1363\cos(3\Omega) + 0.0322\cos(5\Omega) - 0.0073\cos(7\Omega)$$

However, as in Example 15.19, the easiest way of plotting the frequency response is to apply MATLAB using the commands

```
b=[-.00364 0 .01612 0 -.06816 0 .30379 .5 .30379 0 -.06816 0 .01612 0 -.00364];
a=[1 0 0 0 0 0 0 0 0 0 0 0 0 0 0];
freqz(b,a)
```

This produces Figure 15.31, and it is noted how the amplitudes of the unwanted sidelobes in the amplitude response are decreased compared with those of Figure 15.28.

The design of a linear phase FIR filter by the Fourier method can be done quite quickly in the way presented, but Version 5 of MATLAB (not the student edition of Version 4) contains a function **fir1** that streamlines the procedure even more. In its default mode it imparts a Hamming window (other windows can be specified) and the coefficients can be obtained with a single command. To design the filter of Example 15.10 all that is required is

```
b = fir1(14, 0.5)
```

where **14** is the index of the highest coefficient and is the number of coefficients less one since the lowest coefficient is b_0. The **0.5** is the 6 dB cutoff frequency normalized to half the sampling frequency. The resulting coefficients are almost identical to those obtained in Example 15.10, with any differences arising because Example 15.10 specified the bandwidth of the ideal brickwall filter that formed the starting point of the design, whereas the MATLAB command specifies the 6 dB bandwidth that actually emerges. Probably the reason that the **fir1** function was not included in the student edition of Version 4 of MATLAB is that another method of synthesizing FIR filters exists that produces an 'optimal' design, and this is generally preferable.

15.7.2 The design of optimal linear phase FIR filters

When frequency selectivity is required the main justification for FIR filters is that they give a linear phase characteristic. The most widely accepted method for designing FIR filters is based on an algorithm devised by Parks and McClellan based on the Remez exchange algorithm for which the explanation is complicated and outside the scope of this book. Fortunately, it can be achieved using the **remez** command in MATLAB without requiring much understanding of the working of the algorithm. The method is considered optimal in that a desired frequency response is specified, and the method produces a design that gives the smallest possible error between the actual and desired responses for the number of coefficients specified. Its implementation in MATLAB is best illustrated by an example.

Example 15.11

Assuming a 5000 Hz sampling frequency, use MATLAB to design an optimal 32-coefficient FIR bandpass filter that aims to achieve unity gain between 600 and 900 Hz, and zero gain below 400 Hz and above 1100 Hz.

Solution From the specification, our ideal objective is to realize the amplitude response shown in Figure 15.32. We first enter a vector of the six frequencies shown in Figure 15.32, each normalized to half the sampling, or folding, or Nyquist frequency.

```
f=[0 400/2500 600/2500 900/2500 1100/2500 1];
```

We next enter a vector of the desired amplitude response at the six specified frequencies:

```
m=[0 0 1 1 0 0];
```

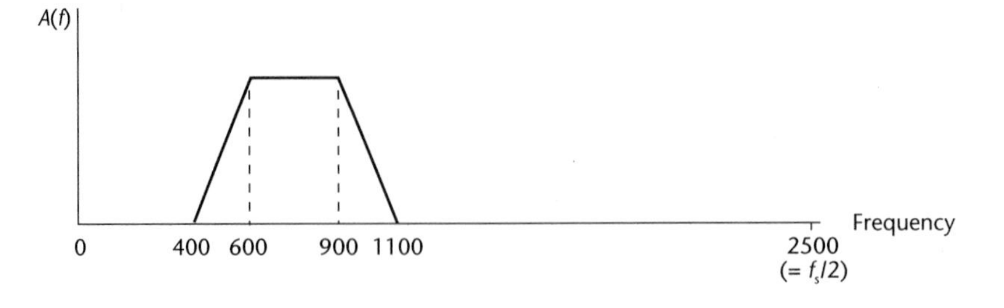

Fig. 15.32 Desired amplitude response

Now we apply the **remez** command to derive the non-recursive coefficients. In this we specify the number of coefficients less one (this is considered in MATLAB to be the *order* of the filter – although in conflict with the definition used in this book (see Section 14.2), where the order is taken to be determined solely by the number of *recursive* coefficients in the filter, and hence by the denominator of the z transfer function). We also include the frequency and amplitude vectors such that the command is

b = remez(31,f,m);

A printout of the coefficients has been suppressed here with the usual semicolon because it is useful to first confirm that the frequency response is satisfactory since, if not, we can change the number of coefficients. We can do this with

freqz(b,1,400,5000)

where **b** is the vector of the coefficients, **1** represents the non-existent recursive coefficients, **400** is the number of frequencies at which the response is calculated, and **5000** is the sampling frequency. The outcome is shown in Figure 15.33. Assuming this is satisfactory we can now printout the coefficients by entering the command **b**. The outcome is

b =

Columns 1 through 7

0.0186 −0.0185 −0.0265 -0.0179 0.0001 0.0036 −0.0091

Columns 8 through 14

−0.0049 0.0373 0.0791 0.0497 −0.0606 −0.1581 −0.1283

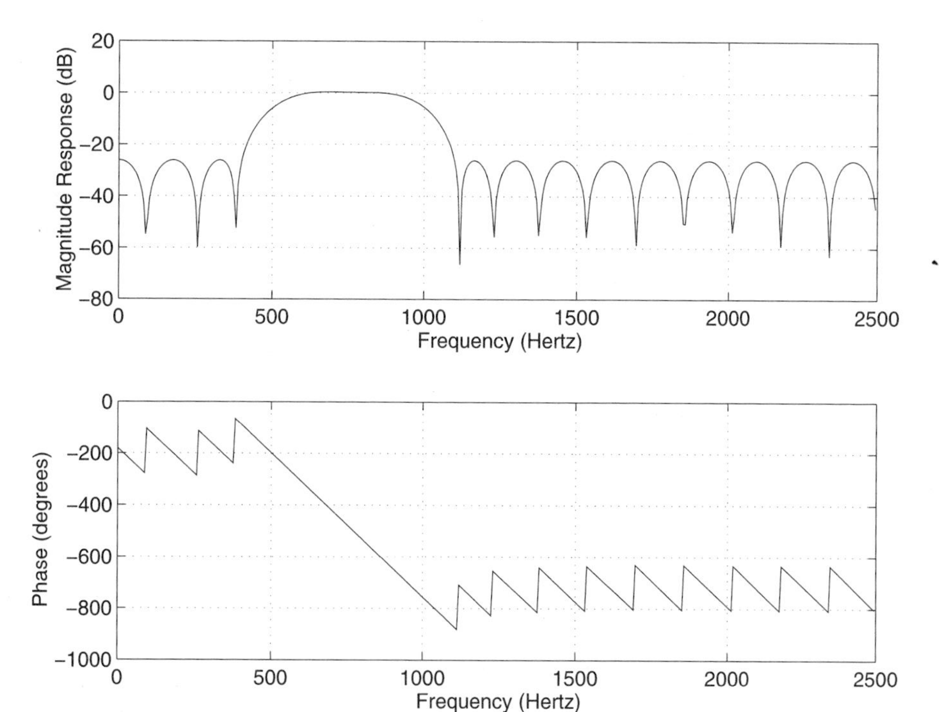

Fig. 15.33 Frequency response of an optimal 32-coefficient bandpass FIR filter

Columns 15 through 21

0.0301 0.1809 0.1809 0.0301 −0.1283 −0.1581 −0.0606

Columns 22 through 28

0.0497 0.0791 0.0373 −0.0049 −0.0091 0.0036 0.0001

Columns 29 through 32

−0.0179 −0.0265 −0.0185 0.0186

Two points are of particular note in the last example:

- Compared with the Fourier method there is a greater ability to specify and thereby attain the frequency response that we really want. The result is that a much sharper transition region is obtained than using the Fourier method, although at the penalty of more ripple. (If we were prepared to relax the sharpness of the transition region to that achieved using the Fourier method we would actually achieve less ripple).
- Because the design is optimal, the criterion that the maximum error between the desired and actual amplitude response outside of the transition region is minimized allows this maximum error to occur more than once. The procedure therefore leads to an equiripple filter.

As stated earlier, this method of designing an FIR filter is widely used and is of considerable importance.

15.8 Summary

A variety of techniques for synthesizing digital filters have been described. IIR filters require fewer coefficients than FIR filters for a specified amplitude response but have a non-linear phase response. Small errors in the values of the coefficients in high-order IIR filters can easily cause a major degradation to the frequency response, or even instability. The best way of overcoming this is to cascade low-order sections. FIR filters require more coefficients but can give a perfect linear phase response. The most important method of designing FIR filters is based on the Parks–McClellan algorithm. This produces a filter giving the optimal fit to a specified amplitude response. The design of both IIR and FIR filters requires only a few lines of code in MATLAB.

15.9 Problems

1. A cardiac signal is sampled at 500 Hz but is found to be badly contaminated by 60 Hz interference. Design a digital notch filter for removing the interference that has one pair of conjugate poles with a radius of 0.95 and one pair of conjugate zeros.

2. A second-order analogue Butterworth lowpass filter has the transfer function

$$H(s) = \frac{400}{s^2 + 20\sqrt{2}s + 400}$$

What is its 3 dB cutoff frequency? Choosing a sampling frequency of 30 Hz use the bilinear transformation technique to determine (by manual calculation) the coefficients of the corresponding digital filter. Sketch the filter, labelling the coefficients. What is the ratio of its gain at 10 Hz to that at 0 Hz? What is this ratio for the analogue filter? Explain why a discrepancy between these two ratios is to be expected.

3. An analogue Chebyshev lowpass filter with a cutoff frequency of 1 rad/s and 0.9 dB passband ripple has a transfer function given by

$$H(s) = \frac{0.521}{s^3 + 1.0276s^2 + 1.278s + 0.521}$$

Assuming a sampling rate of 2 samples/s use the bilinear transformation technique to determine (manually) the z transfer function of the corresponding digital filter. Draw a block diagram representation of the filter, labelling all filter coefficients. Calculate the gain of the digital filter at frequencies of 0 and 1 rad/s. Note that the latter is slightly different to 0.9 dB.

4. Use MATLAB to determine the difference equation of a 6 pole Butterworth bandpass filter having a passband between 800 Hz and 900 Hz that is used for processing a signal sampled at 5 kHz. Plot the frequency response between 500 Hz and 1200 Hz.

5. Redesign the filter of Problem 4 as three two-pole filters in cascade. Derive a set of three difference equations corresponding to the design.

6. Assuming a 10 kHz sampling frequency use MATLAB to design an optimal 40-coefficient FIR lowpass filter that aims to achieve unity gain up to 600 Hz and zero gain above 1300 Hz. Plot the frequency response.

If an acceptable objective is to achieve a minimum attenuation of 40 dB above 1300 Hz, rather than seeking infinite attenuation, determine the minimum number of coefficients that are needed. Plot the frequency response.

Chapter sixteen

Correlation

16.1 ● Preview

The correlation coefficient is used in statistics as a measure of the dependence of two variables. In signal processing, however, the cross-correlation function is more important and reveals the degree of similarity between two signals as a function of the time shift between them. A distinct peak in the cross-correlation function indicates that the two signals are matched for that particular time shift. This has important applications in signal detection.

16.2 Covariance and the correlation coefficient

In statistics correlation is used to determine the extent to which different sets of things depend on one another. For example, we would intuitively expect the weight of humans to be related in some way to their height, and thus to be correlated. The extent to which this premise is true is revealed by the correlation coefficient.

An intermediate step towards the correlation coefficient is the covariance. If we have two sets of measurements

$$x_1, x_2, x_3, \cdots x_N$$

and

$$y_1, y_2, y_3, \cdots y_N$$

the *observed covariance* is defined by the equation

$$\text{observed covariance} = \frac{\sum_{i=1}^{N}(x_i - \bar{x})(y_i - \bar{y})}{N} \tag{16.1}$$

where $\bar{y}$ and $\bar{x}$ are the observed mean values of x and y. Note that, in contrast to what has been done with time sequences, the usual convention in stastics is to commence the indexing at number 1.

If the number of observations used to evaluate the covariance is small the observed covariance will have little statistical significance. The *theoretical covariance* is the limiting value of the observed covariance as the number of observations becomes large.

Example 16.1

Ten adults have the heights and weights given by the following table. Determine the observed covariance.

	1	2	3	4	5	6	7	8	9	10
Height (m)	1.73	1.59	1.31	1.53	1.38	1.90	1.87	1.84	1.26	1.71
Weight (kg)	94.8	70.1	50.2	70.7	63.0	118.1	85.3	70.1	60.8	76.3

Solution The mean height is calculated as

$$0.1(1.73 + 1.59 + 1.31 + 1.53 + 1.38 + 1.90 + 1.87 + 1.84 + 1.26 + 1.71) = 1.612 \text{ m}$$

The mean weight is calculated as

$$0.1(94.8 + 70.1 + 50.2 + 70.7 + 63.0 + 118.1 + 85.3 + 70.1 + 60.8 + 76.3) = 75.94 \text{ kg}$$

$$\therefore \text{covariance} = \frac{(1.73 - 1.612)(94.8 - 75.94) + (1.59 - 1.612)(70.1 - 75.94) + \cdots}{10}$$

$$= 3.124$$

It will be noted in this last example that a change in the units of height from metres to cm would increase the covariance by a factor of 100, thus indicating that the covariance *fails* to reveal the dependence between weight and height. To achieve a satisfactory indicator the covariance must be normalized. This is done by dividing the observed covariance by the population standard deviations of weight and height. This leads to the observed correlation coefficient ρ_{xy}, defined as

$$\rho_{xy} = \frac{\sum_{i=1}^{N}(x_i - \bar{x})(y_i - \bar{y})}{N\sigma_x\sigma_y} \tag{16.2}$$

where σ_x is the population standard deviation of x, defined as

$$\sigma_x = \left(\frac{1}{N} \sum_{n=1}^{\infty} (x_i - \bar{x})^2 \right)^{1/2} \tag{16.3}$$

and similarly σ_y is the population standard deviation of y, defined as

$$\sigma_y = \left(\frac{1}{N} \sum_{n=1}^{\infty} (y_i - \bar{y})^2 \right)^{1/2} \tag{16.4}$$

The squares of the standard deviations, namely σ_x^2 and σ_y^2 are known as the variances of x and y.

Note: there is also a statistical measure known as the sample standard deviation. In contrast to the divisor of N for the *population* standard deviation, the divisor for the *sample* standard deviation is $N - 1$.

If we have the special situation where y is proportional to x such that $y_i = kx_i + c$, substitution into Equation (16.2) shows that the correlation coefficient is then equal to $+1$ if $k > 1$ and equal to -1 if $k < 1$. A correlation coefficient of plus or minus one between two quantities indicates an exact linear relationship between them, while a correlation coefficient of zero indicates no linear dependence.

Example 16.2

Determine the standard deviations of height and weight for the adults tabulated in Example 16.1 and apply them to determine the correlation coefficient.

Solution

$$\sigma_{\mathrm{h}}^2 = 0.1[(1.73 - 1.612)^2 + (1.59 - 1.612)^2 + \cdots] = 0.05018$$

$$\therefore \sigma_{\mathrm{h}} = 0.224 \text{ m}$$

$$\sigma_{\mathrm{w}} = 0.1[(94.8 - 75.94)^2 + (70.1 - 75.94)^2 + \cdots] = 337.57$$

$$\therefore \sigma_w = 18.373 \text{ kg}$$

$$\therefore \rho_{xy} = \frac{3.214}{(0.224)(18.373)} = 0.781$$

It will be noted that a change in the units of height would change the standard deviation of height by the same factor as it would change the covariance between weight and height, thus causing the correlation coefficient to be unchanged. The correlation coefficient does therefore give a valid indication of the dependence between weight and height. Its significance in this example is limited only by the rather small number of observations involved (the sample size is small). The correlation coefficient quantifies what is apparent from the graph of weight versus height shown in Figure 16.1: that weight and height do have some genuine interdependence, even though it is not particularly strong.

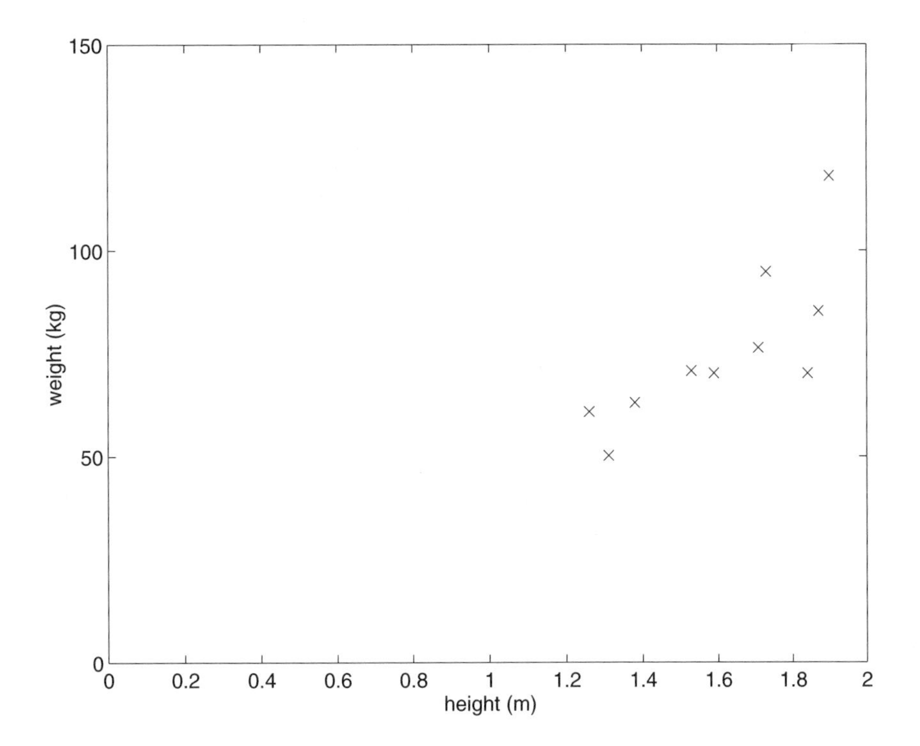

Fig. 16.1 Graph of weight versus height

16.3 The cross-correlation function for finite-duration signals

Consider two microphones mounted in the open air such that their outputs $x(t)$ and $y(t)$ due to some sound source are as shown in Figure 16.2. A set of samples of the two signals might be as shown in Figure 16.3.

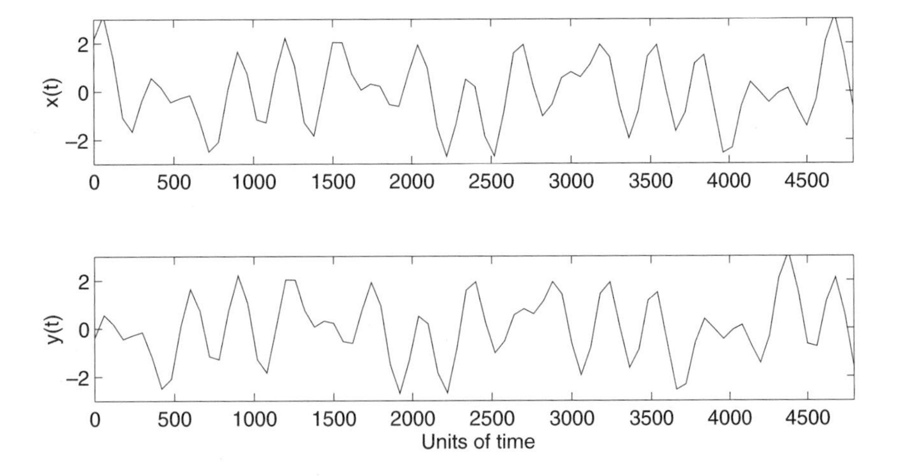

Fig. 16.2 Signals on two microphones

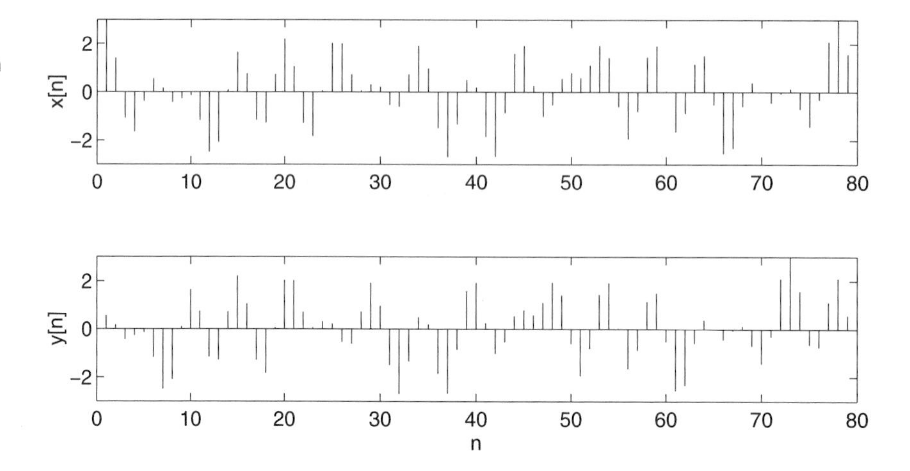

Fig. 16.3 Sampled signals on two microphones

Inspection shows that the two sets of samples are identical except for being out of step in time. This displacement would cause the correlation coefficient to be very small. To indicate the similarity of the signals we require some parameter that indicates the similarity between two signals as a function of the delay between them. Such a function is the cross-correlation function (CCF). The cross-correlation of two sequences $x[n]$ and $y[n]$ is a third sequence $r_{xy}[k]$ defined as

$$r_{xy}[k] = \sum_{n=-\infty}^{\infty} x[n]]y[n-k] \tag{16.5}$$

where it is seen that the second sequence $y[n]$ is delayed by k units relative to $x[n]$, and the sum of the product terms is then evaluated. This is done for all values of k. In the case of Figure 16.3 it is apparent by inspection that the cross-correlation is very small if $k = 0$, but is substantial if $k = 5$. This is because $x[n]$ *lags* behind $y[n]$ by five sample intervals in Figure 16.3 and a match therefore occurs between $x[n]$ and $y[n]$ if $y[n]$ is delayed by five units. The values of k are often referred to as lags.

If *both* the functions in the summation of Equation (16.5) are advanced by k the result of the summation will not change. This leads to an alternative equation for the cross-correlation function, namely

$$r_{xy}[k] = \sum_{n=-\infty}^{\infty} x[n+k]y[n] \tag{16.6}$$

This same equation can also be obtained, but in a less intuitive way, by making a change of variable in Equation (16.5).

Example 16.3

Determine the cross-correlation between the two sequences

$$x[n] = \{\underset{\uparrow}{2}, 1, 4, 3, -2, -3, 2, 4, -1, -3\} \text{ and } y[n] = \{\underset{\uparrow}{3}, -1, 2, 1, 4\}$$

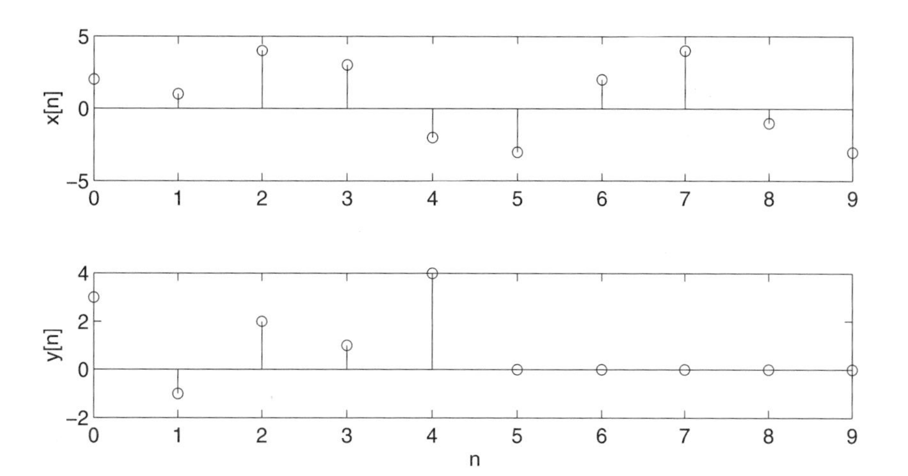

Fig. 16.4 Two sequences to be cross-correlated

Solution The two sequences are illustrated in Figure 16.4. If $k = 0$, the product $x[n]y[n-k]$ within the summation is given by

$$x[n]y[n] = \{\underset{\uparrow}{6}, -1, 8, 3, -8\}$$

and the summation of the terms gives us that $r_{xy}[0] = 8$. If $k = 1$, such that $y[n]$ is delayed by 1 unit, we have

$$y[n-1] = \{\underset{\uparrow}{0}, 3, -1, 1, 2, 1, 4\}$$

and hence

$$x[n]y[n-1] = \{\underset{\uparrow}{0}, 3, -4, 6, -2, -12\}$$

Summation of the terms gives us that $r_{xy}[1] = -9$. If $k = -1$ such that $y[n]$ is advanced by 1 unit we have

$$y[n+1] = \{3, \underset{\uparrow}{-1}, 2, 1, 4\}$$

and hence

$$x[n]y[n+1] = \{\underset{\uparrow}{-2}, 2, 4, 12\}$$

Summation of the terms gives us that $r_{xy}[-1] = 16$. A continuation of this process gives the result that

$$r_{xy}[k] = \{8, 6, 21, 16, \underset{\uparrow}{8}, -9, 10, 23, 1, -16, -3, 7, 0, -9\}$$

It is relevant to make the following observations about the cross-correlation function.

- If one of the sequences is doubled in magnitude the values of the cross-correlation sequence will be doubled. Thus the normalization that was imposed in going from the covariance of Equation (16.1) to the correlation coefficient of Equation (16.2) is missing from the cross-correlation function as defined by Equations (16.5) and (16.6). However, this does not necessarily matter. The usual purpose of the cross-correlation function is to examine how the degree of match varies with the lag, rather than to generate a measure of the exact extent of the match. For cases where the exact extent of the match does matter a normalizing factor can be imposed and this is discussed in Section 16.6.

- In general, the cross-correlation function is affected by the order in which two sequences are correlated. Following the definition of the cross-correlation function between $x[n]$ and $y[n]$ given by Equation (16.5), the cross-correlation function between $y[n]$ and $x[n]$ is given by

$$r_{yx}[k] = \sum_{n=-\infty}^{\infty} y[n]x[n-k] \tag{16.7}$$

A comparison with Equation (16.6) shows that

$$r_{yx}[k] = r_{xy}[-k] \tag{16.8}$$

- There is a connection between correlation and convolution. In order to *convolve* two sequences $x[n]$ and $y[n]$ we found in Section 2.4 that we time reverse $y[n]$ to become $y[-n]$, shift it to the right (or delay it) by k to become $y[-n+k]$, multiply by $x[n]$, and sum the resulting terms. Thus

$$x[n] * y[n] = \sum_{n=-\infty}^{\infty} x[n]y[k-n] \tag{16.9}$$

Similarly, to *convolve* two sequences $x[n]$ and $y[-n]$ we time reverse $y[-n]$ to become $y[n]$, delay it by k to become $y[n-k]$, multiply by $x[n]$, and sum the resulting terms. Thus the convolution of two sequences $x[n]$ and $y[-n]$ is given by

$$x[n] * y[-n] = \sum_{n=-\infty}^{\infty} x[n]y[n-k] \tag{16.10}$$

However, the right-hand side of this equation is identical to the cross-correlation between $x[n]$ and $y[n]$ given in Equation (16.5). It follows that

$$r_{xy}[n] = x[n] * y[-n] \tag{16.11}$$

Thus the cross-correlation function between $x[n]$ and $y[n]$ may be obtained by convolving $x[n]$ with $y[-n]$.

The cross-correlation function can be applied to continuous functions as well as to discrete functions. The relationship for two finite-energy functions, $x(t)$ and $y(t)$, that is equivalent to Equation (16.5) is

$$r_{xy}(\tau) = \int_{-\infty}^{\infty} x(t)y(t-\tau)\mathrm{d}t \tag{16.12}$$

Alternatively, by advancing both $x(t)$ and $y(t-\tau)$ by τ, we can have

$$r_{xy}(\tau) = \int_{-\infty}^{\infty} x(t+\tau)y(t)\mathrm{d}t \tag{16.13}$$

As with discrete signals, the cross-correlation and convolution of continuous signals are related. The comparable equation to Equation (16.11) is

$$r_{xy}(t) = x(t) * y(-t) \tag{16.14}$$

Thus the cross-correlation function between $x(t)$ and $y(t)$ may be obtained by convolving $x(t)$ and $y(-t)$.

Note: Some textbooks give expressions for $r_{xy}[k]$ and $r_{xy}(\tau)$ which would apply to $r_{yx}[k]$ and $r_{yx}(\tau)$ using the conventions adopted in this book. They are thus time reversed compared with what are obtained here; that is they produce $r_{xy}[-k]$ and $r_{xy}(-\tau)$. Fortunately the time reversal is often of little significance and is likely to be corrected unconsciously from a physical understanding of the problem under consideration. Indeed, it is common to omit the subscripts and to write the cross-correlation function simply as

$$r(\tau) = \int_{-\infty}^{\infty} x(t)y(t+\tau)\mathrm{d}t \tag{16.15}$$

The comparable equation for discrete signals is

$$r[k] = \sum_{n=-\infty}^{\infty} x[n]y[n+k] \tag{16.16}$$

In accordance with the conventions of this book these equations refer to $r_{yx}(\tau)$ and $r_{yx}[k]$ but, as discussed, it is common to avoid specifying the order of the correlation.

16.4 Correlation using MATLAB

In the student edition of MATLAB Version 5 there is a specific function **xcorr(a,b)** for cross-correlation. This, however, does not exist in the student edition Version 4, and it is necessary to time reverse the second function and then apply MATLAB's convolution function. Thus in order to cross-correlate

$$x[n] = \{\underset{\uparrow}{2}, 1, 4, 3, -2, -3, 2, 4, -1, -3\}$$

with

$$y[n] = \{\underset{\uparrow}{3}, -1, 2, 1, 4\}$$

we can generate a new variable

$$z[n] = y[-n] = \{4, 1, 2, -1, \underset{\uparrow}{3}\}$$

and then evaluate $x[n] * z[n]$. Thus a suitable set of instructions is

```
x=[2 1 4 3 -2 -3 2 4 -1 -3];
z=[4 1 2 -1 3];
conv(x,z)
```

This produces the output

```
ans=
```

```
Columns 1 through 12

    8   6   21   16   8   -9   10   23   1   -16   -3   7

Columns 13 through 14

    0   -9
```

This is the same result as obtained in Example 16.3. The only problem is that it does not contain timing information. To do this and to produce displays of the original signals and of the cross-correlation function it is useful to create our own special function comparable to the **convstem(x1,n1,x2,n2)** function that was devised in Section 2.5. This new function will be designated **corrstem(x1,n1,x2,n2)**. The following is a suitable M-file.

```
% corrstem.m  This is for plotting two discrete sequences and their cross-
% correlation function such that all three plots have the same scale of indices and
% the index scale extends from the lowest index amongst the three to the highest
% index amongst the three. What follows is very similar to convstem.m except that
% the second function is time reversed. This is done by reversing the order of the
% elements in the second sequence using the fliplr(x) command, and by reversing
% the order and changing the signs of its indices.
function corrstem(x1, n1, x2, n2)
x3=fliplr(x2);
n3=-fliplr(n2);
y=conv(x1,x3);
% We need to find the lowest index kmin of the CCF.
kmin=n1(1)+n3(1);
% We next want to determine the lowest index amongst x1, x2 and the CCF.
kmin_plot = min([n1(1) n2(1) kmin]);
% We need to find the the highest index kmax of the CCF.
kmax= max(n1)+max(n3);
% We now want the highest index amongst x1, x2 and the CCF.
kmax_plot = max([max(n1) max(n2) kmax]);
% We next generate a vector of indices that is appropriate to plots of x1, x2 and
% the convolution sum. We extend beyond kmax_plot by one because we will
% need to assign a zero value to the last element of these variables and we do not
% wish to destroy any data.
k_plot=(kmin_plot : kmax_plot+1);
% Next come the plots themselves. Each one must be padded with zeros outside
% the range of the original data. Following the procedure in Section 1.2.8 this can
% be done with
xx1=[zeros(1,n1(1)-kmin_plot) x1];
xx1(length(k_plot))=0;      % This assignment is why we extended k_plot by one.
xx2=[zeros(1,n2(1)-kmin_plot) x2];
xx2(length(k_plot))=0;
yy=[zeros(1,kmin-kmin_plot) y];
yy(length(k_plot))=0;
% Now plot xx1
```

```
subplot(3, 1, 1)
stem(k_plot, xx1)
% We now set the limits of the horixontal axis of the plots avoiding stems on the
% left or right boundaries. We also reduce the font size and specify a tick mark for
% every lag defined in the k_plot vector.
a=length(k_plot)/50;          % The horizontal axis will be extended by this much.
set(gca,'XLim',[kmin_plot - a  kmax_plot+1+a],'fontsize',10,'XTick',[k_plot])
xlabel('n'),ylabel('x1[n]')
% Now plot xx2
subplot(3 1, 2)
stem(k_plot,xx2)
set(gca,'XLim',[kmin_plot - a  kmax_plot+1+a],'fontsize',10,'XTick',[k_plot])
xlabel('n'),ylabel('x2[n]')
% Now plot the CCF yy
subplot(3, 1, 3)
stem(k_plot,yy)
set(gca,'XLim',[kmin_plot - a  kmax_plot+1+a],'fontsize',10,'XTick',[k_plot])
xlabel('Lag k of x2[n] relative to x1[n]'),ylabel('CCF')
```

Example 16.4

Use the specially created MATLAB function **corrstem(x1,n1,x2,n2)** for displaying the two sequences

$$x[n] = \{2, 1, 4, \underset{\uparrow}{3}, -2, -3, 2, 4, -1, -3\}$$

and

$$y[n] = \{3, -1, 2, \underset{\uparrow}{1}, 4\}$$

and their cross-correlation function. Note that $y[n]$ has a different origin to $y[n]$ in Example 16.3.

Solution Because of the axis labelling incorporated into the **corrstem(x1,n1,x2,n2)** function we need to rename the two sequences $x_1[n]$ and $x_2[n]$. The necessary MATLAB instructions are simply

```
x1=[2 1 4 3 -2 -3 2 4 -1 -3];n1=(-3:6);
x2=[3 -1 2 1 4];n2=(-3:1);
corrstem(x1,n1,x2,n2)
```

The outcome is shown in Figure 16.5.

16.5 The cross-correlation function for power signals

Section 6.2 discussed the difference between energy signals and power signals. Examples of power signals are noise and periodic signals such as sine waves and

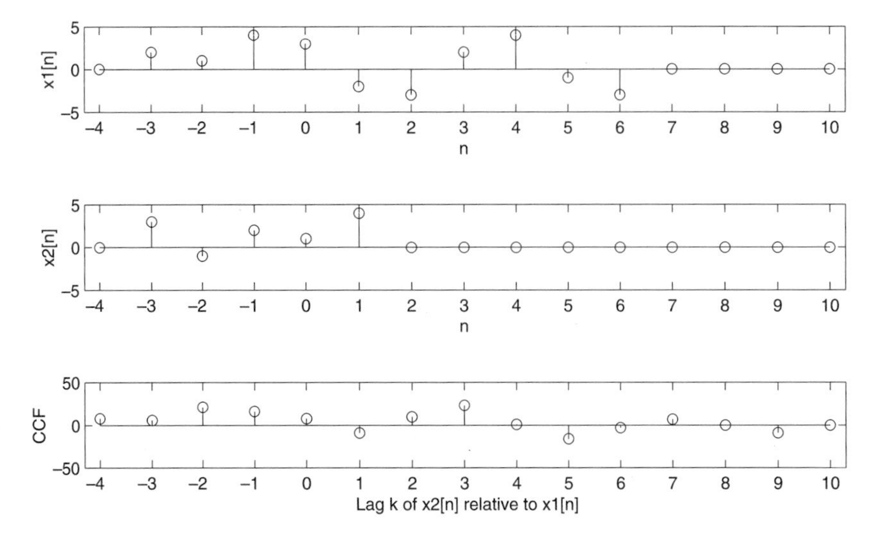

Fig. 16.5 Cross-correlation plot using MATLAB

square waves. For such signals the summations and integrations involved in the definitions of cross-correlation used in the last section would usually produce infinite results. To avoid this, the cross-correlation functions of power signals are redefined such that the summations and integrations are replaced by averages. For two discrete power signals $x[n]$ and $y[n]$ the cross-correlation function is defined as

$$r_{xy}[k] = \lim_{M\to\infty} \frac{1}{2M+1} \sum_{n=-M}^{M} x[n]y[n-k] \tag{16.17}$$

If the signals are periodic, each with period N, the average over an infinite interval is the same as the average over a single period. A simpler alternative is thus

$$r_{xy}[k] = \frac{1}{N} \sum_{n=0}^{N} x[n]y[n-k] \tag{16.18}$$

An alternative way of denoting the averaging operation, applicable to both periodic and non-periodic power signals, is

$$r_{xy}[k] = \overline{x[n]y[n-k]} \tag{16.19}$$

where the bar denotes an average. Alternatively, using the less rigorous nomenclature of Equation (16.16) that does not distinguish the order of the cross-correlation but actually applies to $r_{yx}[k]$, we could simply write

$$r[k] = \overline{x[n]y[n+k]} \tag{16.20}$$

For two *continuous* power signals $x(t)$ and $y(t)$ the cross-correlation function is defined as

$$r_{xy}(\tau) = \lim_{T\to\infty} \frac{1}{T} \int_{-T/2}^{T/2} x(t)y(t-\tau)\mathrm{d}t \tag{16.21}$$

which for a periodic signal of period T, simplifies to

$$r_{xy}(\tau) = \frac{1}{T} \int_{-T/2}^{T/2} x(t)y(t-\tau)\mathrm{d}t \tag{16.22}$$

Again we can use the bar notation for denoting the averaging operation to give

$$r_{xy}(\tau) = \overline{x(t)y(t-\tau)} \tag{16.23}$$

As before, we can omit the subscripts from r if we do not require the cross-correlation function to indicate the order of the cross-correlation. Although strictly applicable to $r_{yx}(\tau)$ rather than $r_{xy}(\tau)$ we can thus write a commonly encountered formula for the cross-correlation function as

$$r(\tau) = \overline{x(t)y(t+\tau)} \tag{16.24}$$

16.6 Autocorrelation functions

The autocorrelation function arises when a signal is correlated with itself. Clearly the order does not matter. If the signal is a finite energy sequence $x[n]$, Equation (16.6) therefore gives the autocorrelation function (or ACF) as

$$r_{xx}[k] = \sum_{n=-\infty}^{\infty} x[n]x[n+k] \tag{16.25}$$

If the discrete time signal is a power signal, Equation (16.20) gives the ACF as

$$r_{xx}[k] = \overline{x[n]x[n+k]} \tag{16.26}$$

From Equation (16.13) the ACF for a finite energy *continuous* signal is

$$r_{xx}(\tau) = \int_{-\infty}^{\infty} x(t)x(t+\tau)\mathrm{d}t \tag{16.27}$$

and, for a continuous *power* signal, Equation (16.24) gives it as

$$r_{xx}(\tau) = \overline{x(t)x(t+\tau)} \tag{16.28}$$

Example 16.5

Determine the autocorrelation function of a sine wave $\sin(\omega_0 t)$.

Solution The period of the sine wave is given by $T = 1/f_0 = 2\pi/\omega_0$. Using Equation (16.28) we obtain

$$\begin{aligned} r_{xx}(\tau) &= \frac{1}{T}\int_0^T \sin(\omega_0 t)\sin[\omega_0(t+\tau)]\mathrm{d}t \\ &= \frac{1}{T}\int_0^T \sin(\omega_0 t)\sin(\omega_0 t + \omega_0\tau)\mathrm{d}t \end{aligned}$$

Using the trigonometric identity that $\sin A \sin B = \frac{1}{2}\cos(A-B) - \frac{1}{2}\cos(A+B)$ we obtain

$$\begin{aligned} r_{xx}(\tau) &= \frac{1}{2T}\int_0^T [\cos\omega_0\tau - \cos(2\omega_0 t + \omega_0\tau)]\mathrm{d}t \\ &= \tfrac{1}{2}\cos\omega_0\tau \end{aligned}$$

The sine wave and its autocorrelation function are shown in Figure 16.6.

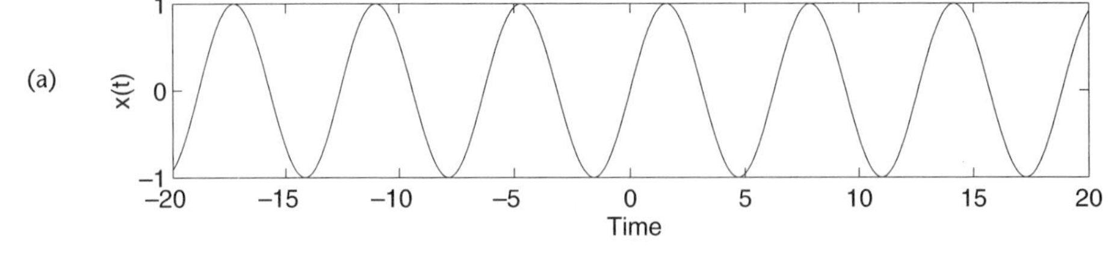

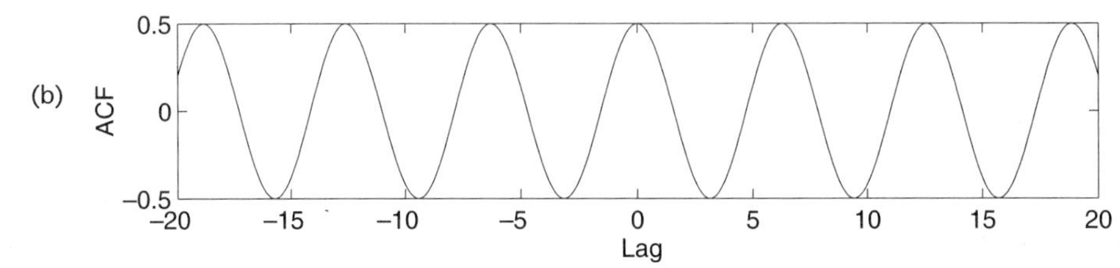

Fig. 16.6 Example of autocorrelation: (a) a sine wave; (b) its ACF

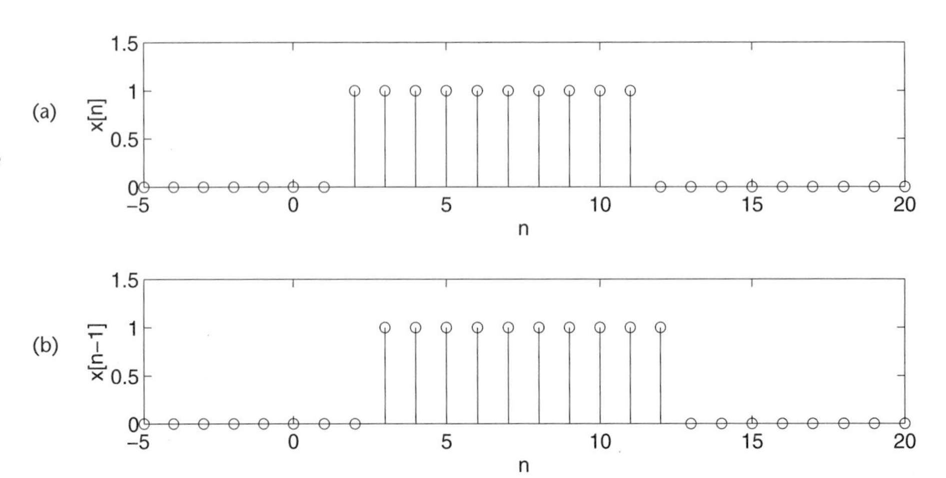

Fig. 16.7 Samples of a rectangular pulse: (a) original; (b) delayed by one sample

Example 16.6

Determine the autocorrelation function of a set of ten unit samples that represent a rectangular pulse.

Solution Figure 16.7(a) shows the sequence of ten unit samples and Figure 16.7(b) shows the same sequence displaced by one sample interval. The product of the two sequences is clearly a sequence of nine unit samples. The sum of these samples is nine.

In a similar way the result for a lag of two sample intervals is eight, that for three sample intervals is seven, and so on. Thus the autocorrelation function is a triangular sequence, as shown in Figure 16.8. It should be noted that, as always, the ACF has its peak at zero lag in spite of the offset of the signal.

A signal without lag is perfectly matched to itself. It can be very useful to have an autocorrelation function that is normalized to give a unity value when the lag is zero.

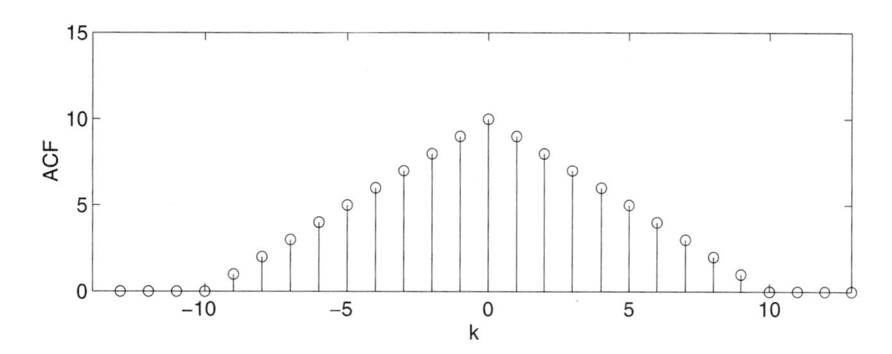

Fig. 16.8 Autocorrelation function of sampled rectangular pulse

The normalized autocorrelation function for a discrete-time signal is defined as

$$\rho_{xx}[k] = \frac{r_{xx}[k]}{r_{xx}[0]} \tag{16.29}$$

and for a continuous-time signal as

$$\rho_{xx}(\tau) = \frac{r_{xx}(\tau)}{r_{xx}(0)} \tag{16.30}$$

Normalizing factors can also be applied to cross-correlation functions. The normalized cross-correlation functions for discrete and continuous signals are defined as

$$\rho_{xy}[k] = \frac{r_{xy}[k]}{\sqrt{r_{xx}[0]r_{yy}[0]}} \tag{16.31}$$

and

$$\rho_{xy}(\tau) = \frac{r_{xy}(\tau)}{\sqrt{r_{xx}(0)r_{yy}(0)}} \tag{16.32}$$

Example 16.7

Determine the autocorrelation function and normalized autocorrelation function of a sequence of 60 random numbers having a normal distribution with a standard deviation of one.

Solution The result is shown in Figure 16.9 and is obtained with the following MATLAB program.

```
% fig16_9.m
N=60;                   % the number of random numbers
x=randn(1,N);           % Generate a vector of n random numbers
acf=conv(x,fliplr(x));  % Convolve x with a time reversed x. The indices of
                        % the acf vector extend from 1 to (2N-1) but are
                        % different from the lags which extend between
                        % -(N-1) and (N-1).
lag=(-(N-1):(N-1));     % the lags resulting from correlation
```

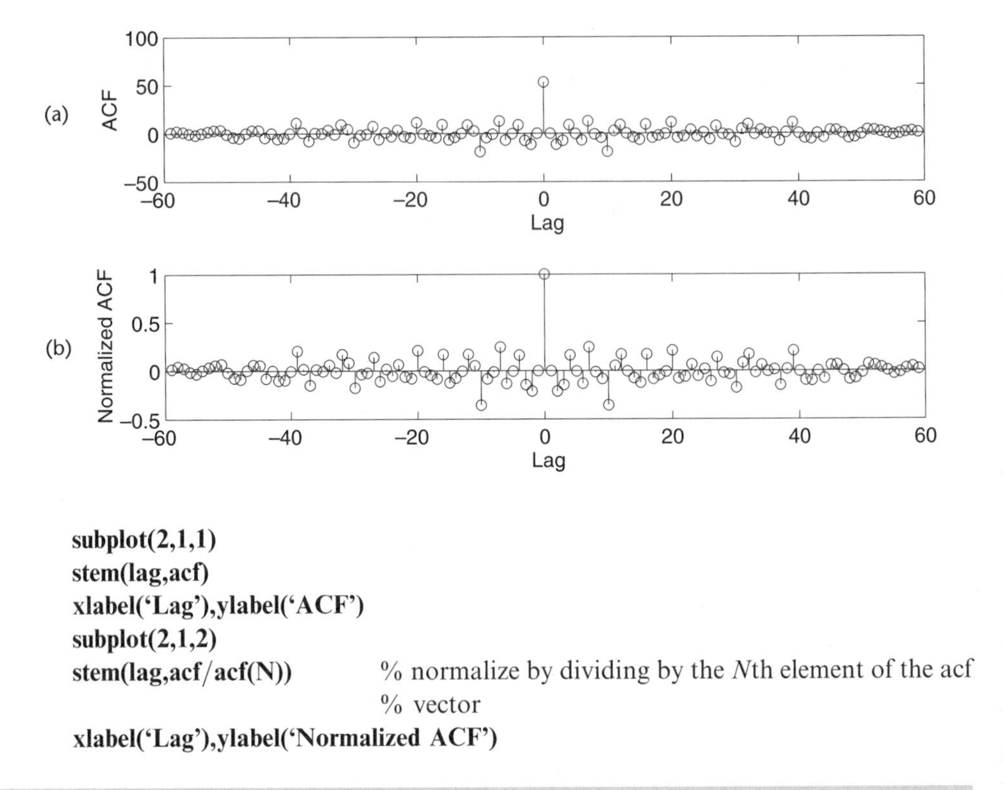

Fig. 16.9 ACF of a noise sequence: (a) unmodified; (b) normalized

```
subplot(2,1,1)
stem(lag,acf)
xlabel('Lag'),ylabel('ACF')
subplot(2,1,2)
stem(lag,acf/acf(N))          % normalize by dividing by the Nth element of the acf
                              % vector
xlabel('Lag'),ylabel('Normalized ACF')
```

As must always happen, the signal in this last example is totally correlated (or matched) with itself when there is no lag. In contrast, since successive samples of this particular sequence are random and independent, no meaningful correlation can be expected between it and a *shifted* version of itself. The only reason why the normalized ACF of this noise-like signal is not zero for non-zero lags is that the number of terms in the sequence is so low that the result is not very significant statistically. Indeed, the normalized ACF of a different set of 60 random numbers would give the same unity value for zero lag, but would have a totally different set of values from those in the above example elsewhere. Alternatively, if we had a much longer sequence of independent random numbers, the normalized ACF would remain at unity for zero lag, but would become much smaller elsewhere. In the limit, as the number of independent samples in the sequence tends to infinity, the normalized ACF tends to a discrete-time unit impulse.

Noise in finite bandwidth analogue systems has much in common with an infinite set of independent random numbers. The main differences are:

- The noise is continuous rather than discrete.
- The amplitude of continuous noise at any instant is dependent on the past history of the noise, and therefore the ACF has a peak whose width is non-zero. The width of the peak depends on the time over which the dependence lasts and this is inversely proportional to the noise bandwidth.
- By analogy with discrete noise, the contrast between the peak and the background of the ACF of continuous noise can be thought of as depending on the total

number of independent noise values involved in its evaluation. This number depends on the duration of the noise segment used for evaluating the ACF. It also depends on the rate at which the noise changes, which in turn depends on the noise bandwidth. Thus the criterion for a distinct peak in the autocorrelation function of noise is a large time bandwidth product ($BT \ll 1$).

Example 16.8

Simulate samples of band-limited continuous noise by passing 1000 random numbers having a normal distribution and a standard deviation of 1 through a lowpass digital filter whose difference equation is $y[n] = x[n] + 0.9y[n-1]$. In order to avoid transient effects from the filter select only the last 900 output samples. Plot the noise waveform and its normalized ACF.

Solution The following MATLAB program was used and produced Figure 16.10.

```
x=randn(1,1000);             % 1000 random numbers
y=zeros(size(x));            % Initialize output to zero
for n=(2:1000)
y(n)=x(n)+0.9*y(n-1);        % The difference equation
end
N=900;                       % The number of samples finally used
z=y(101:1000);               % The last 900 samples leaving the digital filter
subplot(2,1,1)
plot(z)                      % This plots the noise sequence
set(gca,'XLim',[1 900],'YTick',[0])
xlabel('Time (sample spacings)'),ylabel('Noise' )
c=conv(z,fliplr(z));         % Convolve z with a time reversed z. The indices of
                             % the acf vector extend from 1 to (2N-1) but are
                             % different from the lags which extend between
                             % -(N-1) and (N-1).
lag=(-(N-1):(N-1));          % The lags resulting from correlation
```

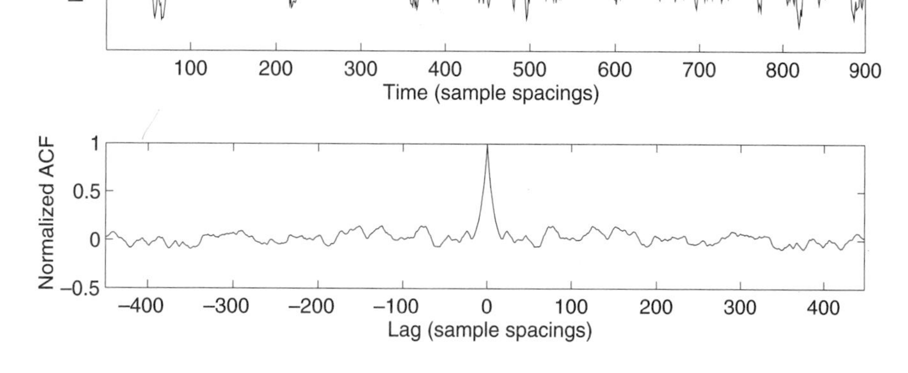

Fig. 16.10 Effect of lowpass filtering on a sequence of random numbers: (a) resulting noise signal; (b) its normalized ACF

```
subplot(2,1,2)
plot(lag,c/c(N))
set(gca,'XLim',[-450 450])  % Limit the display such that the horizontal scale
                            % matches that of the noise waveform.
xlabel('Lag (sample spacings)'),ylabel('Normalized ACF')
```

The finite width of the autocorrelation peak in this last example shows very clearly that the finite bandwidth of a noise waveform means that closely spaced samples are not inde-pendent of each other but depend on the values of the preceding samples. The fact that closely spaced samples are not independent can also be observed from the waveform directly.

16.7 Some applications of cross-correlation and autocorrelation

Applications of correlation are based upon the following:

- Cross-correlation and autocorrelation enable time delays to be determined. (In applications such as image processing *place* shifts can also be determined.)
- Cross-correlation enables the impulse response of a system to be determined without the difficulties of using an impulsive input. This process is known as system identification.
- Cross-correlation enables specific patterns to be recognized by comparing them with a set of templates. Among other possibilities these patterns might be communication signals or images.
- Cross-correlation enables signals to be detected in the presence of noise. To a lesser extent autocorrelation does the same but the main value of autocorrelation is more concerned with predicting the outcome of noise-free cross-correlation.

The following are a few examples of these principles.

16.7.1 Determination of time delays

As a first example, consider two sonobuoys using their hydrophones to listen to the sound from a distant submarine. By cross-correlating the two received signals, the time difference in the arrival time of the sound at the two hydrophones may be determined. Using a knowledge of the speed of sound in the sea, this difference in arrival time provides a difference in the path length from the submarine to the two sonobuoys. Geometry then tells us that the submarine must lie on a particular hyperbola having the two sonobuoys as their foci. A third sonobuoy enables a second pair of sonobuoys to locate the submarine on a second hyperbola, and the intersection of the two hyperbolae gives the position of the target. The technique is illustrated in Figure 16.11.

As a second but related example, consider the need to determine the range of the submarine using a single sonobuoy. There may be circumstances where there is a reflection of the submarine noise from the sea bed, thus providing a two-path system such as shown in Figure 16.12.

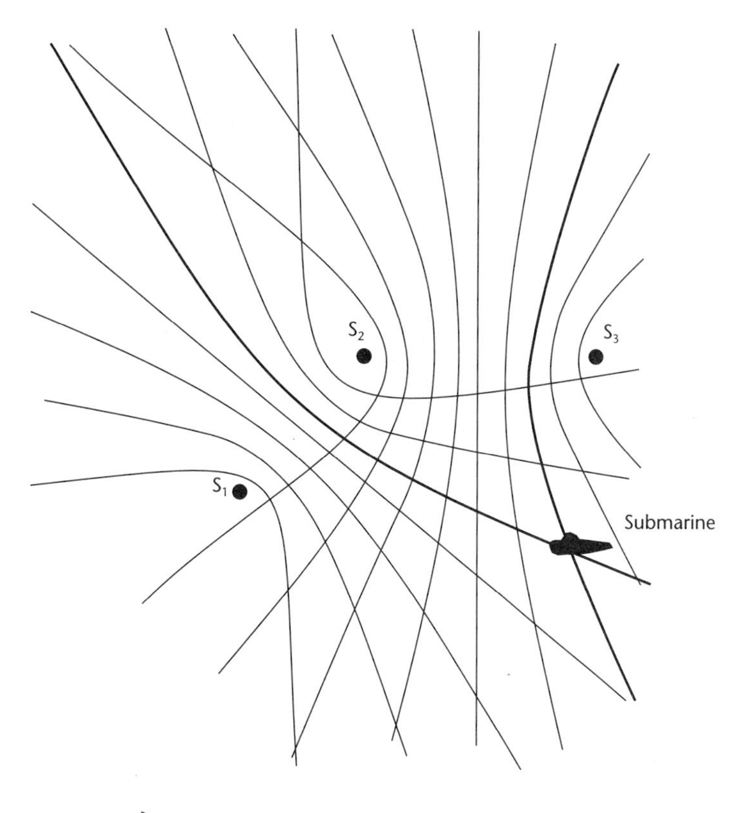

Fig. 16.11 Location of submarine by intersection of curves of constant delay, obtained by cross-correlating signals from sonobuoys

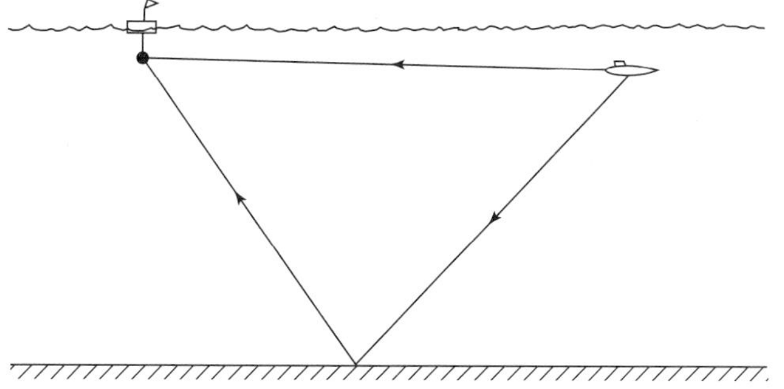

Fig. 16.12 Direct and reflected path between submarine and sonobuoy

When performing the autocorrelation of the received signal, the direct signal will correlate with itself, and the reflected signal will correlate with itself, both phenomena causing there to be a peak at zero lag. The ACF will, however, also show a peak where the direct signal correlates with the reflected signal. If the depth of the sea bed is known and the submarine can be assumed to be at a very much lesser depth, the difference in path length corresponding to the delay of this peak can be used to deduce the target range.

16.7.2 System identification

A common requirement is to characterize an unknown system by determining its impulse response. However, true impulse functions have an infinite amplitude and zero duration and are not therefore physically achievable. A short finite amplitude pulse whose bandwidth greatly exceeds that of the system can be an adequate approximation, but it contains little energy, meaning that the system output will have a very small amplitude which may be difficult to detect. If the energy of the pulse is raised by increasing its amplitude the system is likely to overload and become non-linear.

An effective and common alternative to using an input pulse is to apply noise to the system and to cross-correlate the input with the output. Consider a noise signal $n(t)$ applied to a system with impulse response $h(t)$ as shown in Figure 16.13. The output $y(t)$ results from the convolution of $n(t)$ with $h(t)$; that is

$$y(t) = n(t) * h(t) \tag{16.33}$$

Let us cross-correlate $y(t)$ with $n(t)$. As in Equation (16.14) this is seen to be equivalent to convolving $y(t)$ with $n(-t)$. Hence

$$r_{yn}(t) = y(t) * n(-t) = [n(t) * h(t)] * n(-t) \tag{16.34}$$

The order in which signals are convolved does not effect the result (see Section 2.10).

$$\therefore\ r_{yn}(t) = [n(t) * n(-t)] * h(t) \tag{16.35}$$

But, since $n(t) * n(-t) = r_{nn}(t)$, this becomes

$$r_{yn}(t) = r_{nn}(t) * h(t) \tag{16.36}$$

If the input noise has a bandwidth greatly exceeding that of the system, its autocorrelation function $r_{nn}(t)$ will have a peak that is much more narrow than the system impulse response and thus Equation (16.36) becomes

$$r_{yn}(t) \approx h(t) \tag{16.37}$$

This means that, if white noise is applied to an unknown system, the impulse response of that system can be obtained by cross-correlating the output with the input noise.

16.7.3 Pattern recognition

Consider a communication system where there are only M possible messages and where each one is conveyed by a codeword which is a sequence of 1s and 0s. For the very simple case of $M = 3$ let the three codewords be given by $x_1[n] = \{1, 1, 0, 1, 0, 0\}$, $x_2[n] = \{1, 0, 0, 1, 1, 0\}$ and $x_3[n] = \{0, 1, 0, 1, 0, 1\}$. Let us consider the simple non-recursive filter shown in Figure 16.14(a). Replacing the set of individual delays by a tapped delay line the same system can be redrawn as shown in Figure 16.14(b).

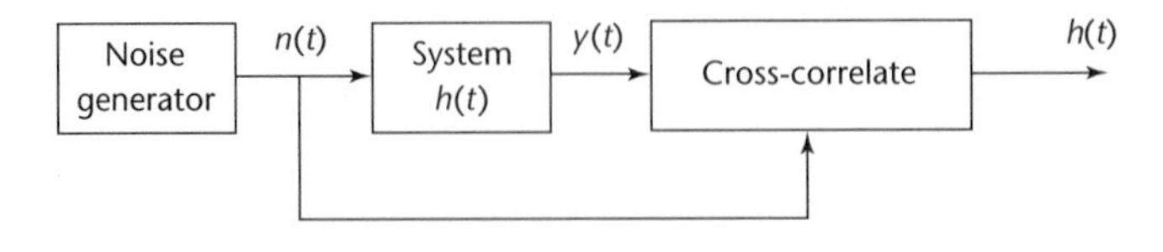

Fig. 16.13 Application of a noise signal to a system for determining impulse response

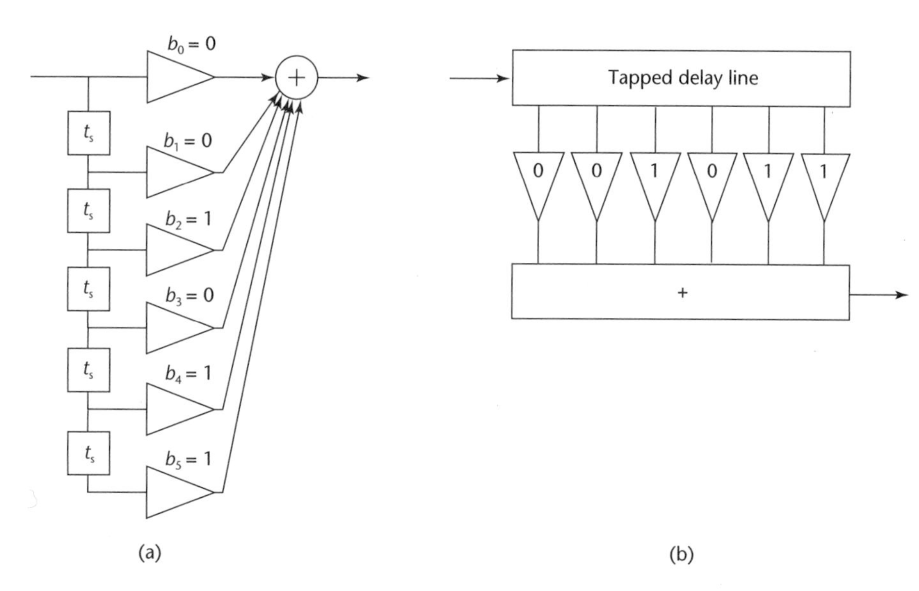

Fig. 16.14 Non-recursive filter for detecting codewords: (a) using discrete delays; (b) using a tapped delay line

If we imagine $x_1[n]$ to enter this filter it will at some time be aligned with the tapped delay line as shown in Figure 16.15. (Note that the sequence travels from left to right such that the last sample values are the last to enter, and the position of sample values in the delay line is therefore in reverse order to the way in which $x_1[n]$ is written.)

The sequence matches the set of filter coefficients and the output will be $(1 \times 1 + 1 \times 1 + 0 \times 0 + 1 \times 1 + 0 \times 0 + 0 \times 0 = 3)$. The output is greater than ever obtained from the alternative sequences of $x_2[n] = \{1, 0, 0, 1, 1, 0\}$ or $x_3[n] = \{0, 1, 0, 1, 0, 1\}$, whatever their alignments with the filter. Thus we have a filter that is capable of recognizing $x_1[n]$ by matching the pattern of $x_1[n]$ with the pattern (or *template*) stored in the filter by the choice of its coefficients. In a similar manner there can be a second filter that matches with $x_2[n]$, and a third one that matches with $x_3[n]$. The filters can be connected in parallel, each followed by a device that decides whether the relevant codeword has been received on the basis of whether a prescribed threshold is exceeded. The system is illustrated in Figure 16.16 and comprises a simple correlation receiver. The art of designing an effective system of this type lies in choosing sequences which correlate poorly except with their own filter.

16.7.4 Detection of signals in noise by cross-correlation

Consider a radar transmitting a short tone burst of electromagnetic energy, or a sonar transmitting a short tone burst of acoustic energy. In either case the

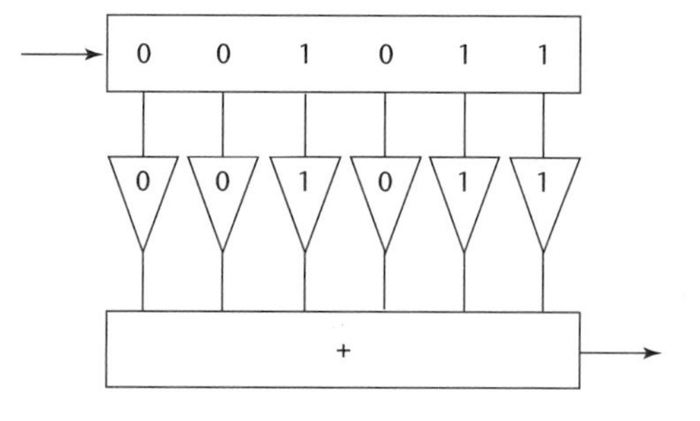

Fig. 16.15 Codeword given by $x_1[n] = \{1,1,0,1,0,0\}$ passing through filter

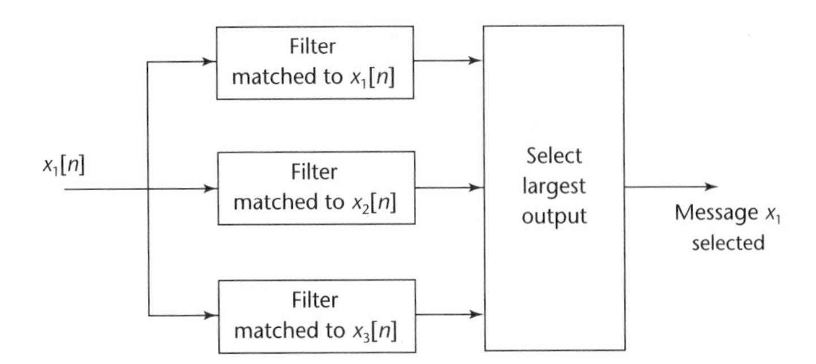

Fig. 16.16 Correlation receiver

transmitted signal would be of the form shown in Figure 16.17(a) and the weak echo from a distant target would be as shown in Figure 16.17(b). In the absence of noise the weak echo can be amplified and there is no problem with detecting it. If, however, there is a background of noise whose amplitude exceeds that of the echo, the echo will be masked and not detectable. The received signal might be as shown in Figure 16.17(c) where the echo of Figure 16.17(b) is present but not visually detectable. Under such circumstances there are several lines of action.

1. The amplitude of the transmit pulse can be increased to cause a stronger echo. However, there is a limit to this. In the case of a radar the possibility of electrical breakdown in the transmission lines feeding the antenna causes there to be a maximum amplitude for the transmit pulse. In the case of a sonar the possibility of the water breaking apart due to a negative pressure (cavitation) causes there to be a maximum amplitude for the transmit pulse. It will be assumed in what follows that a pulse of maximum amplitude is already being transmitted and that other steps are now needed to improve the detection range.
2. The duration of the transmit tone burst can be increased. Since the bandwidth of a pulse is approximately equal to the reciprocal of its duration this allows the receiver bandwidth to be reduced, thus diminishing the noise that corrupts the echo. However, the problem with this is that a long pulse signifies a loss of range resolution since echoes from closely spaced targets would overlap and not be separable. In general, therefore, this is not a good solution.

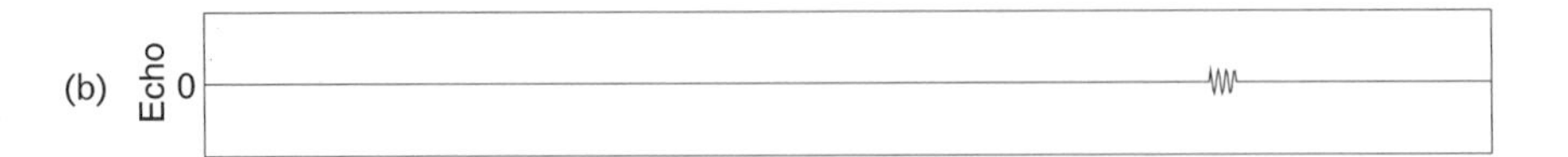

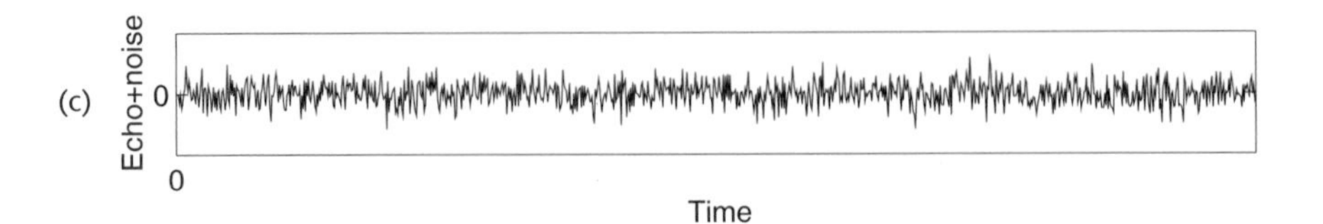

Fig. 16.17 Radar signals: (a) transmit pulse; (b) echo; (c) echo plus noise

3. The transmit pulse can be changed from a short tone burst into a long pulse of *changing* frequency. If the receiver knows the 'code' of what has been transmitted it can detect echoes having this code within a noisy receive signal. Furthermore, because the frequency is changing, the code of the echo will be very precisely indentifiable in time and overlapping echoes can be distinguished as separate entities. The principle is similar to that of the correlation receiver of Section 16.6.3 except that it is now extracting a known signal from noise rather than from other known signals. The technique is very powerful and is widely used.

A transmit pulse of particular importance in radar and sonar is the 'chirp' pulse. Described in detail in Section 16.8 this has a constant amplitude (thereby making it well suited to transmission through efficient saturating power amplifiers), but changes its frequency linearly with time. Figure 16.18(a) shows a chirp pulse and Figure 16.18(b) a weak echo. Figure 16.18(c) shows the echo buried in noise, while Figure 16.18(d) shows the cross-correlation between the noisy receive signal and the transmit pulse. The presence and location of the echo is readily apparent.

Note: The fractional change of frequency in a radar chirp pulse is very small in radar but substantial in a sonar chirp pulse. Hence the pulse shown in Figure 16.18(a) is pertinent to a sonar rather than to a radar. However, the improvement in detection that is possible with chirp pulses is applicable to both.

16.7.5 Detection of signals in noise by autocorrelation

In Example 16.5 it was shown that the ACF of a sine wave is another periodic signal, namely a cosine wave. The reason for the periodicity is that the sine wave matches up

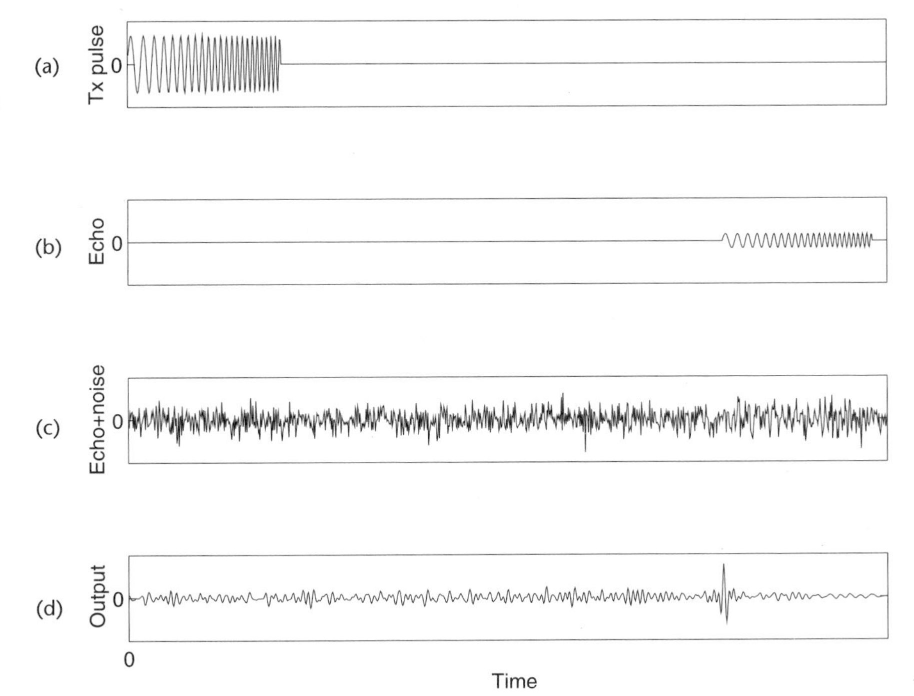

Fig. 16.18 Correlation detection in radar: (a) transmit pulse; (b) echo (c) echo plus noise; (d) cross-correlation between transmit pulse and echo pulse noise

with itself if it is delayed or advanced by an integer multiple of its period. In contrast, a noise-like signal matches up with a delayed version of itself if this delay is zero or small, but not otherwise.

If a sine wave is buried deep in noise the ACF of the combined signal has three components. There is the correlation function between the sine wave and itself, and this is the same cosine wave as before. There is the correlation function between the noise and itself, which is a peak at zero lag and little elsewhere. Finally, there is the correlation function between the sine wave and the noise. Since the sine wave and the noise are unrelated this is very small everywhere. Thus the ACF will produce a cosine wave and thus reveal the presence of the sine wave.

As described, the ACF produces a time domain signal that reveals a wanted periodic signal. In the simple example of a sinusoid, however, the frequency domain approach of spectral analysis is a better alternative. In contrast, the ACF may be more effective for revealing the presence of wideband periodic signals that have less distinctive waveforms.

16.8 The chirp pulse

Section 16.7.4 has discussed the possibility of detecting weak echoes in the presence of noise by means of correlation, and the remainder of this chapter will elaborate on this theme. Transmit pulses used in correlation detection usually have a *constant amplitude*, as this makes them well suited to efficient amplification within a saturating power amplifier at the transmitter and to providing the maximum possible signal compatible with voltage breakdown or cavitation. Chirp pulses are very common and change their frequencies linearly with time. They are also known as linear FM slides. Chirp pulses can have increasing or decreasing frequencies and an example of a chirp pulse extending over $-T/2 \leqslant t \leqslant T/2$ with a linearly *increasing* frequency is shown in Figure 16.19. Note though that practical chirp pulses would contain very many more cycles than shown here.

The law giving its instantaneous frequency as a function of time is illustrated in Figure 16.20, where it is assumed that the centre frequency is f_0 and the frequency changes by B Hz in T seconds. Mathematically, this law is given by

$$f = f_0 + (B/T)t \qquad -T/2 \leqslant t \leqslant T/2 \tag{16.38}$$

One could just as easily have a chirp pulse whose instantaneous frequency decreases linearly with time. The law of instantaneous frequency would then be given by

$$f = f_0 - (B/T)t \qquad -T/2 \leqslant t \leqslant T/2 \tag{16.39}$$

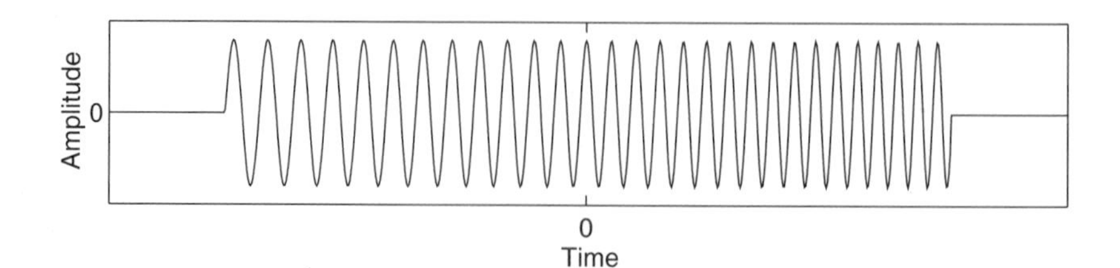

Fig. 16.19 Chirp pulse centred about origin

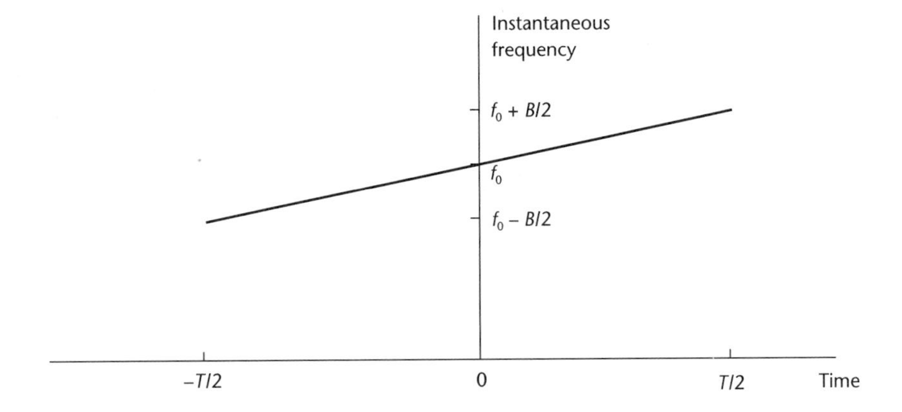

Fig. 16.20 Instantaneous frequency versus time for a Chirp pulse

Working in terms of radian frequency an alternative to Equations (16.38) and (16.39) is

$$\omega = \omega_0 + kt \qquad -T/2 \leqslant t \leqslant T/2 \tag{16.40}$$

where $k = \pm 2\pi B/T$ and is positive for an 'up chirp' and negative for a 'down chirp'.

Consider the completely general expression for a signal having any amplitude variation and any phase variation whatsoever as

$$s(t) = a(t)\cos\theta(t) \tag{16.41}$$

The instantaneous radian frequency is $\mathrm{d}\theta(t)/\mathrm{d}t$. For the chirp pulse this instantaneous radian frequency is $(\omega_0 + kt)$, thus making the instantaneous phase $\theta(t)$ be given by

$$\theta(t) = \omega_0 t + \tfrac{1}{2}kt^2 \tag{16.42}$$

Thus the equation of the chirp pulse is

$$s(t) = \mathrm{rect}(t/T)\cos(\omega_0 t + \tfrac{1}{2}kt^2) \tag{16.43}$$

or

$$s(t) = \mathrm{rect}(t/T)\cos(2\pi f_0 t + \pi B t^2/T) \tag{16.44}$$

It will be shown later that the energy spectrum of a pulse has considerable significance in determining the advantage gained by using correlation processing and, in order to determine the energy spectrum of the chirp waveform it is necessary

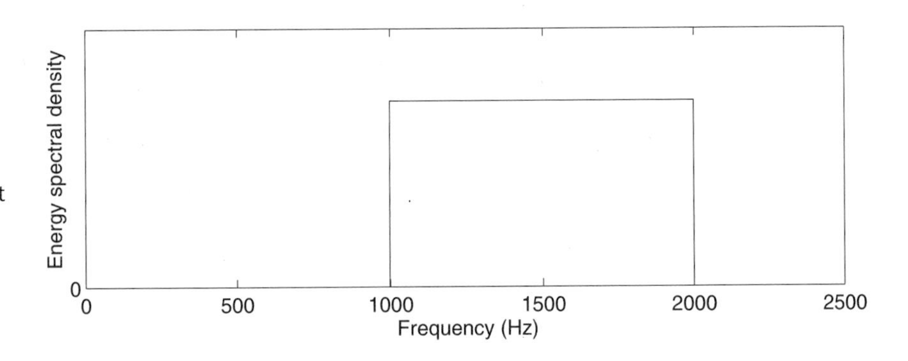

Fig. 16.21 Approximate energy spectrum of a Chirp pulse obtained by a physical argument

to take the waveform's Fourier transform. Unfortunately, it is far from easy to obtain an analytical result for the signal given by Equation (16.43). However, for pulses with a large BT product, a good approximation arises from the intuitive argument that the pulse spends an equal amount of time at all frequencies within the bandwidth B and therefore has an energy spectrum that is approximately uniform over the bandwidth B. Figure 16.21 illustrates this approximate spectrum for the case of a sonar chirp pulse whose instantaneous frequency sweeps linearly from 1000 Hz to 2000 Hz.

An accurate spectrum is most easily obtained by computational means as shown by the next example.

Example 16.9

Use MATLAB to determine the energy spectrum of a sonar chirp pulse that has a linear FM sweep from 1000 Hz to 2000 Hz in 200 ms.

Solution Equation (16.44) gives an expression for a chirp pulse centred about $t = 0$. To generate samples of this pulse in MATLAB, with some zero-padding added, we can have

```
fs=10000;                       % The sampling frequency
t=(-2048:2047)/fs;              % 4096 sampling instants -0.2048< t <0.2047
fd=fs/4096;                     % The spacing of FFT frequency coefficients
s=cos(2*pi*1500*t+(pi*1000*t.^2)/.2); % A chirp extending -0.2048< t <0.2047
s(1:1024)=zeros(1,1024);        % The first part of the overlong chirp is set to zero
s(3073:4096)=zeros(1,1024);     % The last part of the overlong chirp is set to zero
```

Next, to perform a spectral analysis on this and plot the result we use

```
spectrum=abs(fft(s)).^2;
f=(0:1023)*fd;                  % The frequencies of the first 1024 FFT cofficients.
plot(f,spectrum(1:1024))
set(gca,'Ytick',[0])
xlabel('Frequency (Hz)'),ylabel('Energy spectral density')
```

The result is shown in Figure 16.22. It is noted that the energy spectrum is

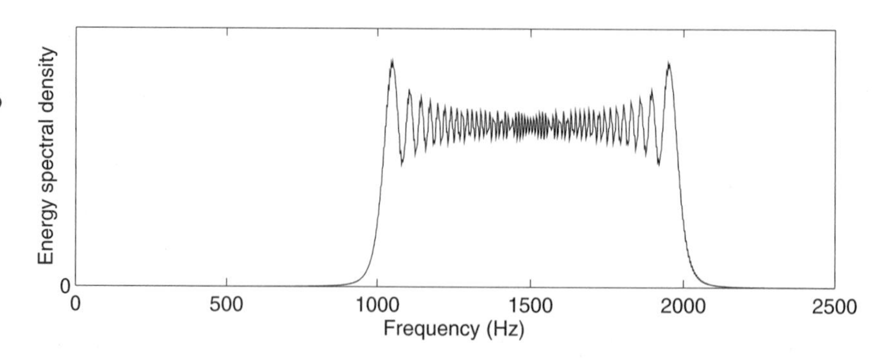

Fig. 16.22 Accurate energy spectrum of a chirp pulse

approximately uniform over the 1000 Hz bandwidth of the pulse, but has some 'ringing' effects caused by its sharp beginning and end.

16.9 Time domain correlators and matched filters

In Section 16.7.4 it was argued that, for coded transmissions, the best way of detecting an echo in noise was to cross-correlate the receive signal with the transmit pulse. It can be proved that this is strictly true only if the input noise has a uniform spectral density ('white' noise) but, in most practical circumstances, the loss in performance due to any invalidity in this assumption is negligible.

Let the transmit pulse be $s(t)$, and let the receive signal consisting of an echo plus noise be $s'(t)$. The cross-correlation function, obtained by multiplying $s'(t)$ with delayed versions of $s(t)$ and integrating the product, is given mathematically by

$$r(\tau) = \int_{-\infty}^{\infty} s'(t)s(t-\tau)\mathrm{d}t \qquad (16.45)$$

As derived in Equation (16.14) this output can also be expressed as the *convolution* of $s'(t)$ with a time-reversed version of the transmit pulse. Thus the output of a correlator can also be written as

$$r(t) = s'(t) * s(-t) \qquad (16.46)$$

Suppose, however, that the received signal $s'(t)$ is applied to a linear filter having an impulse response $h(t)$. The output now is

$$f(t) = s'(t) * h(t) \qquad (16.47)$$

and a comparison with Equation (16.46) now shows that this is the same output as from the correlator so long as the impulse response of the filter is given by

$$h(t) = s(-t) \qquad (16.48)$$

Thus a linear filter gives an identical output to that of a correlator if it has an impulse response equal to a time-reversed version of the transmit pulse. If the filter is made to have this property it is considered to be 'matched' to the transmit pulse and will correlate the received signal with the transmit pulse by virtue of its special in-built impulse response. Such a filter is a 'matched filter'. The terms 'matched filter' and 'correlator' may be regarded as synonymous, although it may be possible to make some distinction regarding the method of physical implementation. For example, the term 'cross-correlation' implies that the transmit signal is stored in memory and brought out and multipled with the receive signal. The term 'matched filtering' implies that a filter is built whose properties are matched to the transmit pulse. However, this distinction can become extremely blurred, as is now shown.

The most powerful technique of realizing a matched filter in hardware is probably that which uses a non-recursive digital filter. Such a filter has an impulse response that is the same as the distribution of the filter coefficients and, within the limitations imposed by a finite number of coefficients, can thus achieve *any* impulse response that is required. The coefficients can be obtained with multipliers and, as an example, the configuration of Figure 16.23 can be made to give the impulse response of Figure 16.24(a). This is a sampled version of the impulse response of Figure 16.24(b) which, being a time-reversed version of the chirp pulse of Figure 16.19, makes Figure 16.23 a matched filter for this chirp pulse.

It could be said that the transmit signal is stored in the filter coefficients of Figure 16.23 and that this is multiplied with the incoming signal as it passes through. On this basis it could be argued that the term 'correlator' would also be appropriate. It can be seen that the terms 'matched filter' and 'correlator' are virtually indistinguishable.

When used for correlation applications the non-recursive digital filter is often termed a 'transversal filter'. A particularly flexible version of it is the programmable transversal filter shown in Figure 16.25. Here, samples of the transmit signal are stored in memory and used to provide the filter coefficients by means of digital multipliers. A great advantage of such a filter is that the transmit pulse waveform can be changed and yet the impulse response can be adapted instantly to match it. The configuration again shows very clearly the blurred distinction between a matched filter and correlator. Should one think of Figure 16.25 as a device that provides a required impulse response, or as a device that correlates the stored signal with the received signal? The answer of course is that the two concepts are inextricably linked.

If one imagines a chirp signal passing through a transversal filter to which it is matched two points are of particular note.

- It will only match the coefficients of the filter when it is very precisely centred over those coefficients. This suggests that the output will be large for only a very short time.

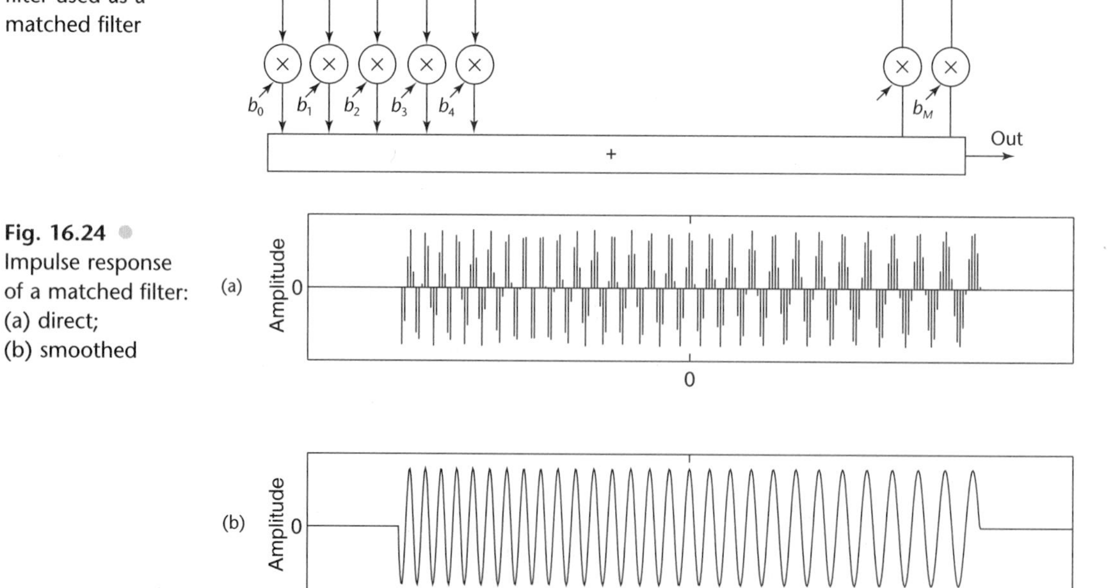

Fig. 16.23 FIR filter used as a matched filter

Fig. 16.24 Impulse response of a matched filter: (a) direct; (b) smoothed

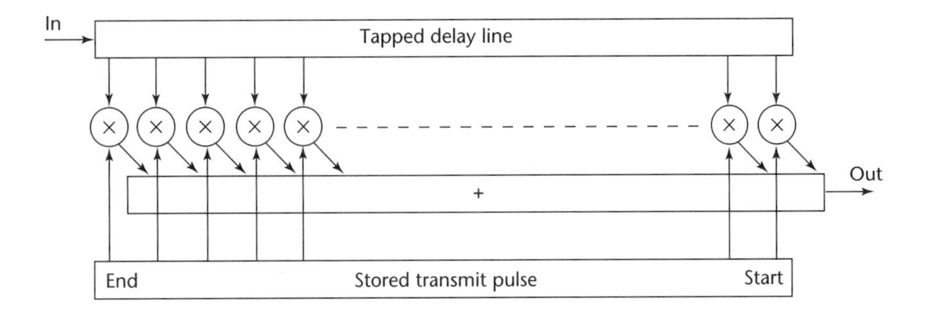

Fig. 16.25 Programmable transversal filter

- Any other signal entering the filter will not be matched to the coefficients and will give only a small output.

It is for these reasons that the correlator restores the time resolution and enhances the peak signal to noise ratio.

16.10 Frequency domain correlation

Another important property of the matched filter concerns its frequency response. This is obtained directly by taking the Fourier transform of the impulse response. Using the relationship of Equation (16.48), that the impulse response of a matched filter is the time reverse of the signal $s(t)$ to which it is matched, we obtain

$$H(f) = \int_{-\infty}^{\infty} h(t)e^{-j2\pi ft}dt = \int_{-\infty}^{\infty} s(-t)e^{-j2\pi ft}\,dt \tag{16.49}$$

Substituting the variable x in place of $-t$ gives $dx = -dt$ and hence

$$\begin{aligned} H(f) &= \int_{\infty}^{-\infty} -s(x)e^{j\pi fx}dx = \int_{-\infty}^{\infty} s(x)e^{j\pi fx}dx \\ &= \int_{-\infty}^{\infty} s(x)\cos 2\pi fx\,dx + j\int_{-\infty}^{\infty} s(x)\sin 2\pi fx\,dx \end{aligned} \tag{16.50}$$

However, the Fourier transform of the transmit pulse is

$$S(f) = \int_{-\infty}^{\infty} s(t)e^{-j2\pi ft}dt$$

which, changing the variable from t to x, gives

$$S(f) = \int_{-\infty}^{\infty} s(x)\cos 2\pi fx\,dx - j\int_{-\infty}^{\infty} s(x)\sin 2\pi fx\,dx \tag{16.51}$$

A comparison of Equations (16.50) and (16.51) shows that the transfer function of the matched filter is given by

$$H(f) = S^*(f) \tag{16.52}$$

A constant delay is likely to be acceptable. The condition of Equation (16.52) may therefore be relaxed and the transfer function of a matched filter is given by

$$H(f) = S^*(f)e^{-j2\pi ft_0} \tag{16.53}$$

Expressed in words, this says that the amplitude response must be identical to the amplitude spectrum of the transmit pulse but that the phase response should be the negative of the phase spectrum of the transmit pulse.

If one attempts to realize a matched filter using analogue circuitry the difficult part of the transfer function specification is the phase response. This is directly related to the delay response. Considering the example of an up chirp of duration T, a useful physical picture of the analogue filter is that it must have a frequency-dependent delay which causes all the different frequencies of the pulse to emerge together. In other words, the low frequencies which occur at the beginning of the chirp pulse should undergo a delay that is T greater than that of the high frequencies which occur at the end of the pulse. It will be apparent that the variation of delay of the filter with frequency is likely to be considerable and, in practice, it is not possible to achieve useful matched filters using analogue circuits.

In contrast, digital signal processing provides a very powerful and widely used technique of frequency domain correlation. From Equation (16.52) it is apparent that, if a signal $s'(t)$ with Fourier transform $S'(f)$ enters a filter that is matched to $s(t)$, the Fourier transform of the output will be $S'(f)S^*(f)$. The output waveform from the matched filter is therefore given by

$$y(t) = \mathscr{F}^{-1}\{S'(f)S^*(f)\} \tag{16.54}$$

Using the FFT the cross-correlation operation becomes

$$y(t) = IFFT\{FFT[s'(t)]\{FFT[s(t)]\}^*\} \tag{16.55}$$

Because of the efficiency of the FFT algorithm this method of cross-correlation is an important alternative to that of the transversal filter. It should be noted, however, that, just as discussed in Section 9.6 for convolution, application of Equation (16.55) results in *circular* correlation. As in Section 9.6 the data sets may need to be padded with zeros to avoid the wrap-around error and thereby to achieve linear correlation.

Example 16.10

Use MATLAB to generate a sampled approximation to a chirp pulse $x(t) = \sin(6000t + 84000t^2)$ that extends over the region $0 \leqslant t \leqslant 0.02$ s. Perform a frequency domain cross-correlation of this pulse with an identical noise-free pulse delayed by 0.015 s.

Solution A sampling period of 0.00005 s correponds to a sampling frequency of 20 kHz and is suitable for producing a good display of the Chirp pulse with its starting frequency of $6000/2\pi = 955$ Hz. The program will generate 1024 points of the waveforms covering the range from $t = 0$ up to $t = (0.00005 \times 1024) = 0.0512$ s. The necessary code is given by

```
% fig16_26.m
N=1024;
```

```
ts=.00005;
t=(0:N-1)*ts;
% We commence by generating two sequences of zeros
x=zeros(size(t));
y=zeros(size(t));
% We now superpose an appropriate chirp signal on each of these
x(1:400)=sin(6000*t(1:400)+84000*t(1:400).^2);
y(301:700)=sin(6000*t(1:400)+84000*t(1:400).^2);
% We can now plot these two waveforms
subplot(3,1,1)
plot(t,x)
axis([0 .05 -1.2 1.2])
xlabel('Time (seconds)');ylabel('x(t)')
subplot(3,1,2)
plot(t,y)
axis([0 .05 -1.2 1.2])
xlabel('Time (seconds)');ylabel('y(t)')
% Next comes the frequency domain correlation.
z=ifft(conj(fft(x)).*(fft(y)));
% We now want to plot the result. Since FFTs and inverse FFTs are complex
% quantities they cannot be plotted directly. However, since we have
% cross-correlated two real signals the result must be real and we can thus display
% the real part of the z vector.
subplot(3,1,3)
plot(t,real(z))
```

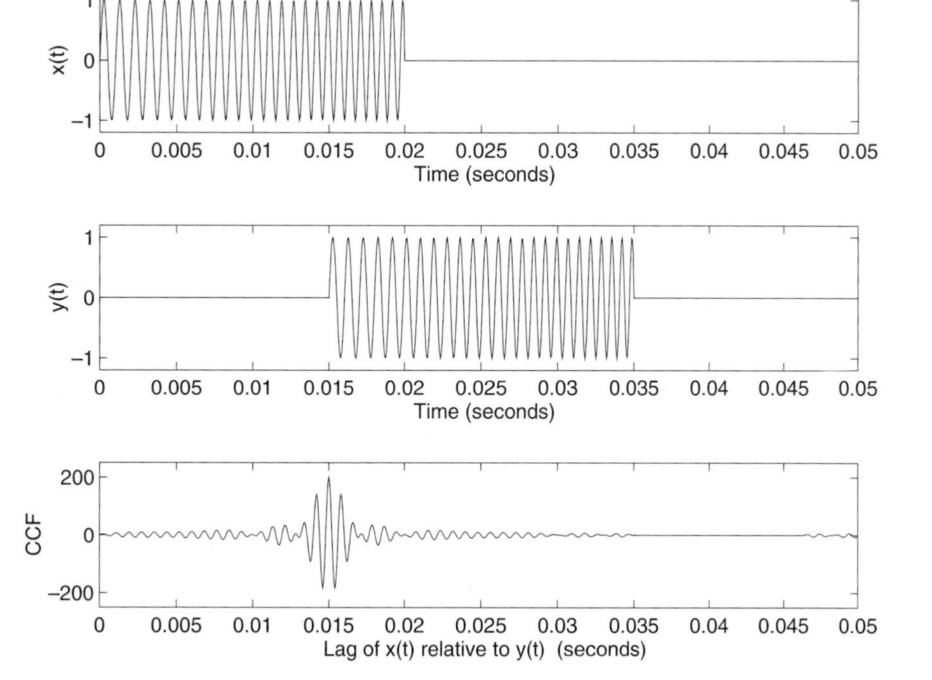

Fig. 16.26 Frequency domain cross-correlation of a chirp pulse with a delayed replica of itself

```
axis([0 .05 -250 250])
xlabel('Lag of x(t) relative to y(t) (seconds)');ylabel('CCF')
```

The outcome is shown in Figure 16.26.

16.11 The output waveform from a matched filter, and the processing gain

The output of a matched filter is obtained by correlating the receive signal with the transmit signal. In the absence of noise, and neglecting any echo delay, the receive signal would be identical to the transmit signal such that, disregarding any delay in the filter, the output of a matched filter would be the autocorrelation function of the transmit pulse. Figure 16.27 shows a transmit pulse entering a filter that is matched to it. Working in the frequency domain, the Fourier transform of the output is seen to be $S(f)S^*(f)$.

In the time domain the output signal $y(t)$ is obtained from the inverse Fourier transform of $S(f)S^*(f)$ that is;

$$y(t) = ACF\{s(t)\} = \int_{-\infty}^{\infty} S(f)S^*(f)e^{j2\pi ft}df$$
$$= \int_{-\infty}^{\infty} |S(f)|^2 e^{j2\pi ft}df \qquad (16.56)$$

This important result states that, when a signal is presented to a filter to which it is matched, the output is determined entirely by the *energy spectrum* of the signal. As an example of the application of this result, consider a single radar echo due to a chirp signal that changes its frequency from $f_0 - B/2$ to $f_0 + B/2$ over its duration T, and where $BT \ll 1$. In order to find the shape of the output from a matched filter we need only to determine the energy spectrum of this chirp signal and to then take its inverse Fourier transform.

As argued in Section 16.8, a reasonable approximation to the one-sided energy spectrum of a chirp pulse is the rectangular spectrum shown in Figure 16.21. The two-sided energy spectrum can be regarded as a rectangular spectrum between $-B/2$ and $B/2$ that is shifted up in frequency by f_0, plus a rectangular spectrum between $-B/2$ and $B/2$ that is shifted down in frequency by f_0. Using the duality properties of the Fourier transform, Example 6.10 told us that the inverse Fourier transform of the rectangle amplitude spectrum $\text{rect}(f/B)$ is the pulse $B[\sin(\pi Bt)/\pi Bt]$. Using the frequency shift property of Section 6.18.5 we thus find that the inverse Fourier transform of the chirp pulse's energy spectrum is a pulse of frequency f_0 having a sinc shaped envelope with a 4 dB width equal to $1/B$. This is shown in Figure 16.28.

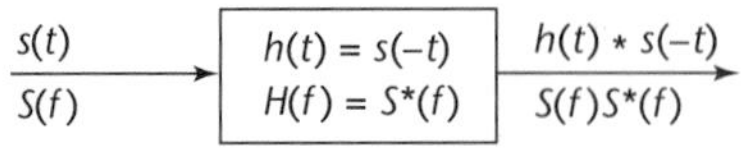

$s(t)$, $S(f)$ → [$h(t) = s(-t)$, $H(f) = S^*(f)$] → $h(t) * s(-t)$, $S(f)S^*(f)$

Fig. 16.27 Frequency domain response of a matched filter

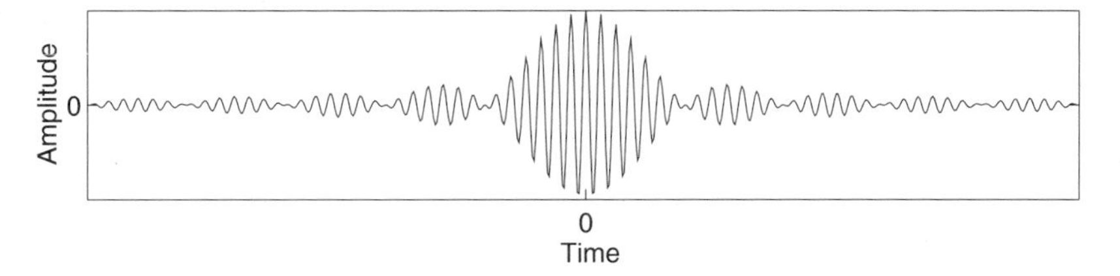

Fig. 16.28 Approximate output from a filter matched to a Chirp pulse of bandwidth B

From this it follows that, when the radar echo from a chirp transmission enters a matched filter, it is *compressed* from its original duration T to a new duration $1/B$. The pulse compression is BT. However, Equation (16.52) tells us that the amplitude response of the matched filter equals the amplitude spectrum of the chirp pulse, which, extrapolating from Figure 16.22, is approximately constant over the pulse bandwidth. Thus the amplitude response of the matched filter is approximately constant over the bandwidth of the pulse and, considering this response to be unity, it is apparent that none of the frequency components of the echo encounter any attenuation in passing through the matched filter. It follows that the *energy* of the echo is unchanged by the matched filter. Since the energy of a pulse of amplitude A and duration τ is proportional to $A^2\tau$ it follows that a reduction of BT in echo duration must be accompanied by an increase in echo amplitude of $\sqrt{BT}$ in order to maintain the same energy.

If a matched filter whose amplitude response is unity over the pulse bandwidth B is compared with the alternative of a simple bandpass filter whose amplitude response is unity over the pulse bandwidth B it will be apparent that both have the same effect on any input noise. *Therefore the matched filter does not reduce the noise.* Instead, it compresses the echo in time such that *its peak amplitude increases* and it thus *rises up out of the noise*. The ratio of peak signal amplitude to r.m.s. noise amplitude is increased by a factor of $\sqrt{BT}$ due to the matched filter. Thus, compared with what is possible using a simple bandpass filter, the r.m.s. noise amplitude entering the matched filter can be increased by $\sqrt{BT}$ and yet give the same ratio of output peak signal amplitude to r.m.s. noise amplitude. An increase of $\sqrt{BT}$ in amplitude corresponds to an increase in power of BT. Thus the matched filter provides a processing gain compared with that possible with a simple bandpass filter that is given by

$$PG = 10 \log_{10}(BT)\,\text{dB} \tag{16.57}$$

As an example, a modern radar might typically use a chirp pulse 1 ms long occupying a bandwidth of 100 MHz centred at 9 GHz. The BT product is 100 000, thus providing a processing gain of 50 dB. The range resolution corresponding to the transmit pulse (given by $cT/2$ where $c = 3 \times 10^8$ m/s) is 150 km, while that of the processed signal is 1.5 m. A modern sonar might typically use a coded pulse 2 s long occupying a bandwidth of 500 Hz centred at 2 kHz. Here the BT product is 1000, thus providing a processing gain of 30 dB. In this case the range resolution corresponding to the unprocessed echo (given by $cT/2$ where $c = 1500$ m/s) is 1.5 km while that of the processed signal is 1.5 m. In general, less processing gain is feasible with sonar than with radar. This is not because of any technological difficulty, but because of echo distortion caused by complex propagation phenomena in the ocean.

One problem encountered with chirp and similar pulses is that the echo from a moving target can be different from the original pulse due to Doppler effects. Such effects are discussed next.

16.12 Doppler effects in correlators

In radar and sonar the echo from a moving target (or from a moving transmitter) will differ from the transmit pulse because of Doppler effects. If, for example, the target is approaching, the two-way propagation time for the leading edge of the transmit pulse will be greater than the two-way propagation time for the trailing edge of the pulse. This is because the target has moved closer to the transmitter at the later time when the trailing edge strikes it. The effect is that the echo is a compressed version of the transmit pulse. If the transmit pulse is given by the equation

$$s(t) = a(t)\cos[\theta(t)] \tag{16.58}$$

then, neglecting any delay in the echo, the echo from an approaching target is given by

$$s_D(t) = a(\alpha t)\cos[\theta(\alpha t)] \tag{16.59}$$

where α is less than one and is the factor by which the echo is compressed relative to the transmit pulse.

The same formula applies to a receding target except that the echo is then an expanded version of the transmit pulse and α is greater than one. The value of α is easily shown to be related to the target velocity and the velocity of propagation by the equation

$$\alpha = \frac{c - v_{\mathrm{r}}}{c + v_{\mathrm{r}}} \tag{16.60}$$

where v_{r} is the radial velocity of the target towards the radar or sonar, and c is the velocity of propagation (3×10^8 m/s for radar or approximately 1500 m/s for sonar).

It is usual that $v_{\mathrm{r}} \ll c$, in which case Equation (16.60) simplifies to

$$\alpha = \frac{1 - v_{\mathrm{r}}/c}{1 + v_{\mathrm{r}}/c} \approx 1 - \frac{2v_{\mathrm{r}}}{c} \tag{16.61}$$

The value of α is usually very close to unity in radar and reasonably close to unity for sonar. For example, an aircraft velocity of 330 m/s (Mach 1) causes α to be 0.999 998 in the radar case, and a submarine velocity of 22 m/s (40 knots) causes α to be 0.97 in the sonar case.

Consider next the specific case of a chirp pulse described in Equation (16.43) for which the instantaneous radian frequency is $(\omega_0 + kt)$ and the phase law is $\theta(t) = \omega_0 t + \frac{1}{2}kt^2$. From Equation (16.59) the echo from a moving target, neglecting the time delay, is given by

$$s_{\mathrm{D}}(t) = \mathrm{rect}(\alpha t/T)\cos\left[\alpha\omega_0 t + \tfrac{1}{2}k\alpha^2 t^2\right] \tag{16.62}$$

For most radar pulses (but not sonar pulses) the bandwidth of the pulse is very much less than the centre frequency of the pulse and the greatest difference between the transmit waveform $\mathrm{rect}(t/T)\cos\left[\omega_0 t + \frac{1}{2}kt^2\right]$ and the echo waveform

$\text{rect}(\alpha t/T)\cos[\alpha\omega_0 t + \frac{1}{2}k\alpha^2 t^2]$ lies in the $\omega_0 t$ and $\alpha\omega_0 t$ terms. On this basis the greatest effect of target velocity is to impose a constant frequency shift on all the intantaneous frequencies within the pulse, so that, neglecting the time delay, the echo of Equation (16.62) is approximated by

$$s_{\text{D}}(t) \approx \text{rect}(t/T)\cos[\alpha\omega_0 t + \tfrac{1}{2}kt^2] \qquad (16.63)$$

Using Equation (16.61) this can be simplified further to become

$$s_{\text{D}}(t) \approx \text{rect}(t/T)\cos[(\omega_0 - \omega_{\text{D}})t + \tfrac{1}{2}kt^2] \qquad (16.64)$$

where

$$\omega_0 - \omega_{\text{D}} = \alpha\omega_0 = \left(1 - \frac{2v_{\text{r}}}{c}\right)\omega_0$$

and therefore

$$\omega_{\text{D}} = \frac{2v_{\text{r}}\omega_0}{c} \qquad (16.65)$$

In terms of cyclic frequency, it follows that the Doppler shift is given by

$$f_{\text{D}} = \frac{2v_{\text{r}}f_0}{c} \qquad (16.66)$$

It should be noted that, although often adequate for radar, the constant frequency shift indicated by Equation (16.64) is rarely an adequate approximation for the echo arising from a sonar chirp pulse. This is because of the much greater fractional bandwidth of most sonar transmissions (sometimes an octave or more).

The effect of the Doppler shift between an echo and the transmit pulse is to cause the two to be imperfectly correlated. This may be *disvantageous*, in that detection of Doppler shifted echoes will then be degraded. However, it is possible to have a whole *bank* of correlators, each 'matched' to transmit pulses with different Doppler shifts imposed. In this case the decorrelation due to Doppler may be *advantageous* because only one of the correlators will give optimum detection. This means that

- 'Clutter' from stationary targets will be poorly correlated in that channel, thus enabling a moving target to be discriminated from clutter within that channel.
- The target velocity may be deduced by noting which channel gives the maximum correlation.

Considerable emphasis has been placed on the chirp pulse, and the effect of Doppler shift on the decorrelation between a chirp pulse and its echo has some

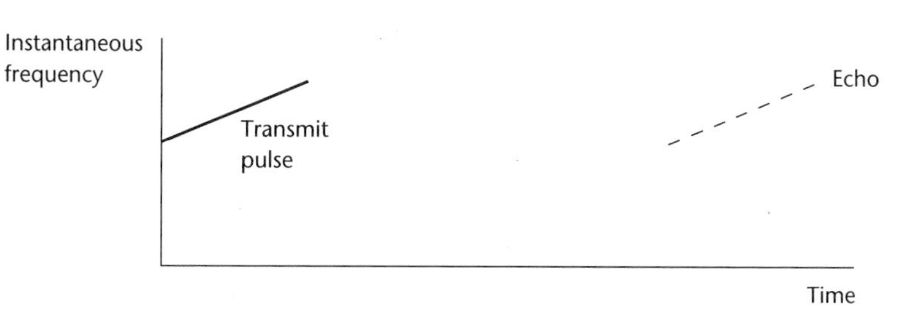

Fig. 16.29 Law of instantaneous frequency versus time for a transmit chirp pulse and an echo from a stationary target

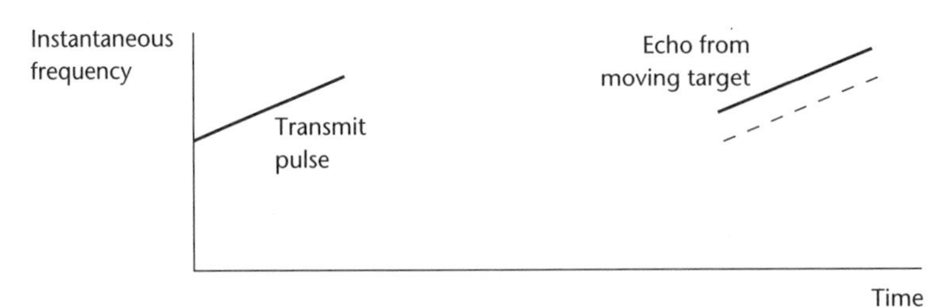

Fig. 16.30 ● Law of instantaneous frequency versus time for a transmit Chirp pulse and an echo from a moving target

interesting properties. Some physical insight to this is most easily gained from diagrams illustrating the law of instantaneous frequency versus time for the pulse. The left-hand section of Figure 16.29 shows the linear law of instantaneous frequency versus time for a chirp transmit pulse. The dotted line in the right-hand section shows the law for an echo from a stationary target after imposing a delay equal to the two way propagation time. Although Figure 16.29 only represents the laws of instantaneous frequency versus time for the transmit pulse and echo, and does not represent their waveforms, it is easy to envisage how the two laws match if the transmit pulse is delayed by the correct amount in a correlator. The two laws will only match for this one delay, thus implying a short-duration high-amplitude pulse from the correlator. This agrees with the more formalized treatment of the previous section.

Consider next what happens if the echo arises from a moving target. Figure 16.30 is a repeat of Figure 16.29 except that the solid line above the dotted line denotes the law for the delayed echo after a constant Doppler shift is imposed on it. It will be apparent that if the transmit pulse is delayed by the right amount (different from the two-way propagation time) there will still be a partial overleaf between the law of instantaneous frequency for the transmit pulse and that for the delayed Doppler-shifted echo. In other words, there is a high degree of correlation in spite of the Doppler shift. However, it occurs at the incorrect time. This property makes the output of the correlator difficult to interpret and the result may be considered 'ambiguous'.

If we consider an echo with *no* delay, a plot showing the peak output of a correlator as a function of the two parameters, lag *and* Doppler shift, is known as a diagram of the pulse's ambiguity function. A study of ambiguity functions for different types of pulse is of considerable importance in radar for deciding the best type of transmit pulse for a particular application.

16.13 ● Summary

Correlation techniques can be used for determining the delay between two signals, for system identification, for recognizing patterns, and for detecting signals buried in noise. One particularly important application is that of enhancing weak echoes in radar and sonar. Correlators are synonymous with matched filters and their implementation can be done either in the time domain or in the frequency domain by means of FFTs.

16.14 Problems

1. In an experiment the following table of input and output measurements is obtained.

Input	1.01	2.05	3.02	3.96	5.03	5.98	7.05	8.1	8.92	10.1
Output	6.2	8.7	12.8	14.5	18.2	20.5	23.9	28.2	29.9	33.2

Use MATLAB to plot these results. Calculate manually the correlation coefficient between the input and output measurements.

In MATLAB create a 10×2 matrix X of the measurements. Apply the command **corrcoef(X)** to obtain a 2×2 matrix containing the correlation coefficients between input and input (which is one), between input and output, between output and output (which is one), and between output and input. Confirm that the correlation cefficients between input and output and between output and input are identical and give the same result as obtained manually.

2. Monthly measurements over one year of the concentrations of four pollutants produce the following table:

A	18.4	22.3	25.2	31.6	28.2	28.4	24.6	22.1	19.4	17.3	17.2	17.9
B	8.4	12.3	13.1	15.2	14.7	14.2	12.3	12.6	10.9	9.3	7.4	7.7
C	30.3	24.3	22.1	20.2	21.4	21.3	23.6	24.2	27.3	29.2	28.2	29.4
D	13.5	34.2	22.5	7.4	2.4	12.6	13.4	44.2	41.7	12.6	33.2	19.7

In MATLAB create a 12×4 matrix X of the measurements. Apply the command **corrcoef(X)** to obtain a 4×4 matrix containing the correlation cofficients. Deduce the correlation coefficients between A and B, between A and D, and between B and C.

3. If two sets of measurements, $x_1, x_2, x_3, \cdots, x_N$ and $y_1, y_2, y_3, \cdots, y_N$, are related by the law $y_i = kx_i + c$, show that the correlation coefficient is +1 if k is positive and -1 if k is negative.

4. A system has the impulse response $h[n] = \{10, 9, 8, 7, 6, 5, 4, 3, 2, 1\}$. Generate a 1000 point sequence of normally distributed random numbers in MATLAB and, considering this to be an input signal to the system, determine the system output. Cross-correlate this output with the input sequence to examine whether the impulse response of a system can be learnt knowing only the output due to a known noise-like input signal. Produce a 'stem' plot of the cross-correlation function and explain why it differs from the impulse response. Show how an improved result can be obtained.

Repeat the investigation for the case where the input signal is a random *binary* sequence.

5. In a communication system a binary message signal is given by $x[n] = \{1, 1, 0, 1, 1, 0, 0, 1, 1, 0, 0, 0, 1, 0\}$. What is the impulse response of its associated matched filter? What is the output of this matched filter when the message signal is applied to it?

6. Use MATLAB to determine the cross-correlation function between the two sequences $x[n] = \{2, 5, 1, -2, 3, -1, -5, 4, 5, -1, -6, 7, 3, -5, 1, 4\}$ and $y[n] = \{7, 1, -3, 3, 6, 0, -1, 2, -2, -6, 4, 5, -1, -6, 5, 4\}$:

(a) using time domain commands;

(b) using the FFT.

Do what is necessary in both cases to obtain the correct result.

7. A radar system transmits a chirp pulse whose duration is 100 μs and whose bandwidth is 150 MHz. Its receiver requires a signal to noise ratio of around 16 dB to enter its detector in order to give a high enough probability of detecting a target coupled with a low enough probability of a false detection. If the detector is preceded by a matched filter determine a satisfactory signal to noise ratio at the input to the matched filter. What are the ratios of signal power to one-sided noise spectral density before and after the matched filter?

Chapter seventeen

Processing techniques for bandpass signals

17.1 ● Preview

It will be shown in this chapter that all the useful information of a bandpass signal can be conveyed by two lowpass signals. This means that far fewer samples are needed to convey the information of the bandpass signal than might be expected on the basis of applying the Nyquist sampling criterion to the highest signal frequency. The two lowpass signals are termed the in-phase and quadrature components of the bandpass signal, and this chapter discusses how they may be obtained and how they may be usefully processed in order to achieve envelope detection, spectral analysis, bandpass filtering and correlation detection.

17.2 Bandpass signals

A typical signal arising in communications, radar and related fields frequently has a spectrum centred about a frequency f_0, but which is confined to a bandwidth B that is much less than f_0. Such a signal is referred to as a narrowband signal. Since it is difficult to quantify precisely what is meant by a bandwidth *much less* than f_0 this chapter will consider a broader class of signals, namely that of *bandpass* signals. This incorporates signals whose half-bandwidths $B/2$ are less than f_0, but which are not

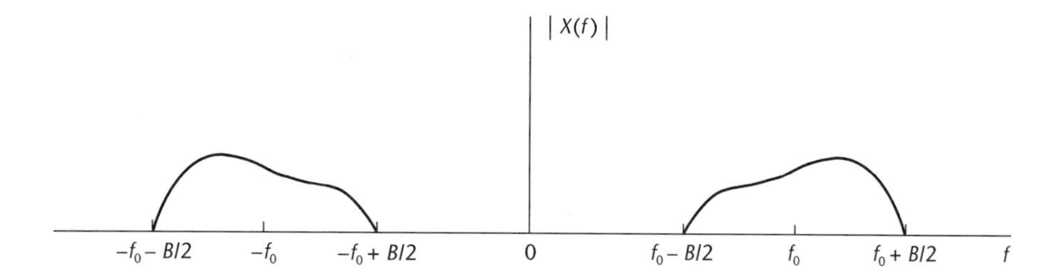

Fig. 17.1 Amplitude spectrum of a bandpass signal

necessarily *much less* than f_0. An example of the two-sided spectrum of a bandpass signal is given in Figure 17.1.

Processing operations that may need to be applied to bandpass signals include the following

- envelope detection
- bandpass filtering
- matched filtering

It will be shown in subsequent sections that a bandpass signal can be expressed in terms of lowpass signals and that this can lead to a simplification in the implementation of these processing operations. A particular example of this arises when digital processing is used. By deriving and processing the appropriate lowpass signals, a much lower sampling rate is possible than if the bandpass signal is processed directly. The reduced number of samples leads to a more efficient processor. Another important application of this principle not covered in this chapter is in data compression and communication.

17.3 Baseband representations of bandpass signals and the complex envelope

Based upon Fourier concepts we can visualize a bandpass signal $x(t)$ of bandwidth B as the sum of a very large number of very closely spaced sine waves $\Delta\omega$ apart, where

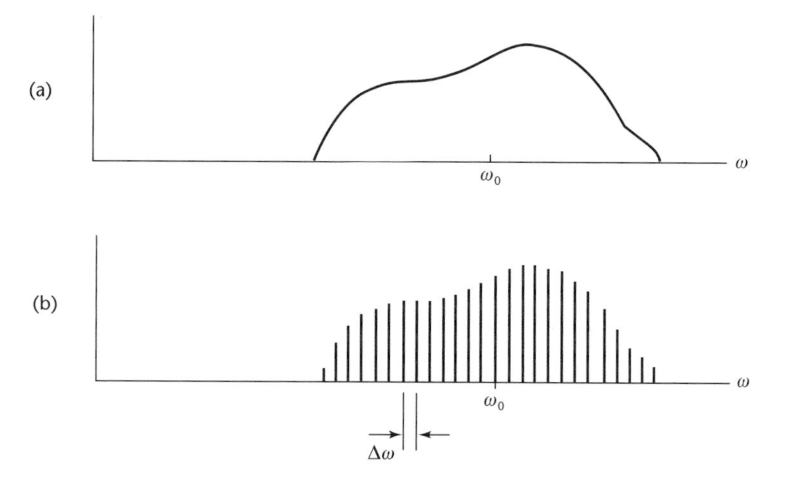

Fig. 17.2 One-sided spectral representation of a bandpass signal: (a) continuous amplitude spectrum; (b) discrete sinusoid approximation

$(\Delta\omega/2\pi) \ll B$. For example, a signal having the one-sided continuous amplitude spectrum shown in Figure 17.2(a) can be approximated by one having the discrete one-sided amplitude spectrum shown in Figure 17.2(b). Thus

$$x(t) = \lim_{\Delta\omega \to 0} \sum_{l=-\infty}^{\infty} a_l \cos[(\omega_0 + l\Delta\omega)t + \theta_l] \tag{17.1}$$

where ω_0 would normally be taken to be the frequency at the middle of the band, but need not necessarily be so, and the θ_ls respect appropriate phases.

Using the trigonometric identity

$$\cos(A + B) = \cos A \cos B - \sin A \sin B \tag{17.2}$$

this can be rewritten as

$$x(t) = \left[\lim_{\Delta\omega \to 0} \sum_{l=-\infty}^{\infty} a_l \cos(l\Delta\omega t + \theta_l)\right] \cos \omega_0 t - \left[\lim_{\Delta\omega \to 0} \sum_{l=-\infty}^{\infty} a_l \sin(l\Delta\omega t + \theta_l)\right] \sin \omega_0 t \tag{17.3}$$

Thus

$$x(t) = x_I(t) \cos \omega_0 t - x_Q(t) \sin \omega_0 t \tag{17.4}$$

where

$$x_I(t) = \lim_{\Delta\omega \to 0} \sum_{l=-\infty}^{\infty} a_l \cos(l\Delta\omega t + \theta_l) \tag{17.5}$$

and

$$x_Q(t) = \lim_{\Delta\omega \to 0} \sum_{l=-\infty}^{\infty} a_l \sin(l\Delta\omega t + \theta_l) \tag{17.6}$$

When compared with the summations representing $x(t)$ in Equation (17.1), the summations representing $x_I(t)$ and $x_Q(t)$ are seen to have components of the same amplitudes but with frequencies reduced by ω_0. Thus $x_I(t)$ and $x_Q(t)$ are low-frequency baseband signals whose *two-sided* amplitude spectra are the same as each other, and are the same as the one-sided amplitude spectrum of $x(t)$ except shifted down in frequency by ω_0. It is usual to choose ω_0 to be at the centre of the band of the bandpass signal, in which case the amplitude spectra of $x_I(t)$ and $x_Q(t)$ would be centred about zero frequency, as illustrated in Figure 17.3.

It will be noted from Equations (17.5) and (17.6) that $x_I(t)$ and $x_Q(t)$ have different though related phase spectra.

Fig. 17.3 Amplitude spectrum of in-phase and quadrature components

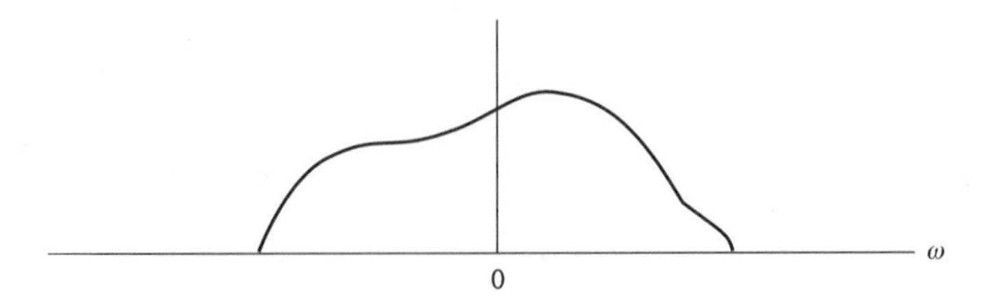

Equation (17.4) provides a representation of the bandpass signal that is in terms of two lowpass signals and two sinusoidal carrier signals. The two carrier signals $\cos(\omega_0 t)$ and $\sin(\omega_0 t)$ are of the same frequency but are 90° out of phase, such that they are termed as being 'in phase quadrature'. In turn, the two lowpass signals $x_I(t)$ and $x_Q(t)$ that multiply them are termed the *in-phase* and *quadrature* components of the bandpass signal. Often they are simply termed the I and Q components of the signal.

Another important way of writing the bandpass signal is in terms of its amplitude and phase modulation, namely

$$x(t) = a(t)\cos[\omega_0 t + \theta(t)] \tag{17.7}$$

Using the trigonometric identity of Equation (17.2), this can be expanded as

$$x(t) = [a(t)\cos\theta(t)]\cos\omega_0 t - [a(t)\sin\theta(t)]\sin\omega_0 t \tag{17.8}$$

which is of the same form as Equation (17.4) if

$$x_I(t) = a(t)\cos\theta(t) \tag{17.9}$$

and

$$x_Q(t) = a(t)\sin\theta(t) \tag{17.10}$$

We note from Equations (17.9) and (17.10) that

$$x_I^2(t) + x_Q^2(t) = a^2(t) \tag{17.11}$$

such that

$$|a(t)| = \sqrt{x_I^2(t) + x_Q^2(t)} \tag{17.12}$$

This is the envelope of the signal. It also equals the amplitude modulation $a(t)$ so long as $a(t)$ is always positive.

We also note from Equations (17.9) and (17.10) that the phase modulation $\theta(t)$ is given by

$$\theta(t) = \tan^{-1}\frac{x_Q(t)}{x_I(t)} \tag{17.13}$$

Equations (17.12) and (17.13) show that all useful information about the signal, in the form of its amplitude and phase information, is conveyed by its in-phase and quadrature components $x_I(t)$ and $x_Q(t)$. It should be noted that, in spite of the similarity between the amplitude spectrum of $x(t)$ shown in Figure 17.2(a) and the amplitude spectrum of $x_I(t)$ shown in Figure 17.3, the signal $x_I(t)$ alone does not contain the information of $x(t)$. Physically, this is best understood by comparing the *one-sided* spectra of $x(t)$ and $x_I(t)$. The one-sided spectrum of $x_I(t)$ is obtained by folding the negative frequency components of Figure 17.3 into the positive frequency region, whereupon the result bears no resemblance to the one-sided spectrum of $x(t)$. This demonstrates that information has been lost. The same applies to $x_Q(t)$ on its own.

Since $x_I(t)$ and $x_Q(t)$ together convey all useful information about $x(t)$, an alternative way of conveying all the information content of $x(t)$ is given by the complex signal $\tilde{u}(t)$, where

$$\tilde{u}(t) = x_I(t) + jx_Q(t) \tag{17.14}$$

The signal $\tilde{u}(t)$ is complex because of the j multiplier applied to the quadrature component. It is a wholly lowpass representation of the bandpass signal that is known as the *complex envelope*. As an alternative to Equation (17.12), the *real* envelope can be expressed in terms of the complex envelope as

$$|\tilde{u}(t)| = \sqrt{x_I^2(t) + x_Q^2(t)} \tag{17.15}$$

17.4 Generation of in-phase and quadrature components

Information-carrying signals are most commonly written in terms of their amplitude and phase modulation, and the bandpass signal considered in this section will therefore use the notation of Equation (17.7), that $x(t) = a(t)\cos[\omega_0 t + \theta(t)]$. The frequency ω_0 is termed the carrier frequency and may be any frequency within the signal bandwidth, but it is usually advantageous to take it as the mid-band frequency.

Let the signal $x(t)$ take two parallel paths, as shown in Figure 17.4. In the upper channel the signal is multiplied by $2\cos\omega_0 t$ and then lowpass filtered. In the lower channel the signal is multiplied by $2\sin\omega_0 t$ and then lowpass filtered. Using the trigonometric identity that

$$\cos A \cos B = \tfrac{1}{2}\cos(A+B) + \tfrac{1}{2}\cos(A-B) \tag{17.16}$$

the output of the multiplier in the top channel is given by $a(t)\cos[2\omega_0 t + \theta(t)] + a(t)\cos[\theta(t)]$. After the lowpass filter removes the $2\omega_0$ term, this becomes the signal $x_I(t)$ given by Equation (7.9).

Considering next the lower channel and using the trigonometric identity that

$$\cos A \sin B = \tfrac{1}{2}\sin(A+B) - \tfrac{1}{2}\sin(A-B) \tag{17.17}$$

the output of the multiplier in the lower channel is given by $a(t)\sin[2\omega_0 t + \theta(t)] - a(t)\sin[\theta(t)]$. After the lowpass filter removes the $2\omega_0$ term, this becomes the signal $-x_Q(t)$, where $x_Q(t)$ is given by Equation (17.11).

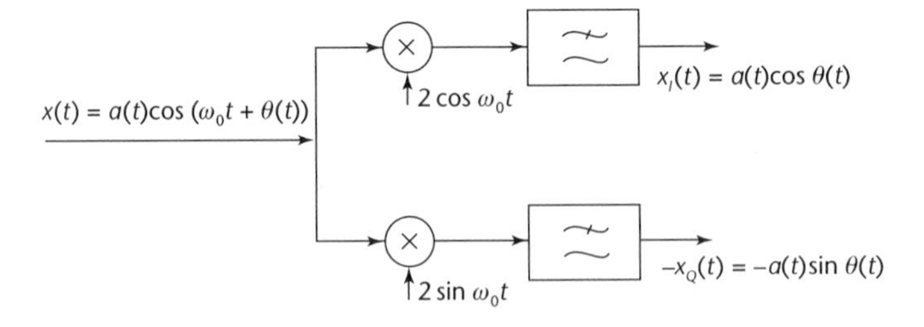

Fig. 17.4 Extraction of in-phase and quadrature components $x_I(t)$ and $x_Q(t)$

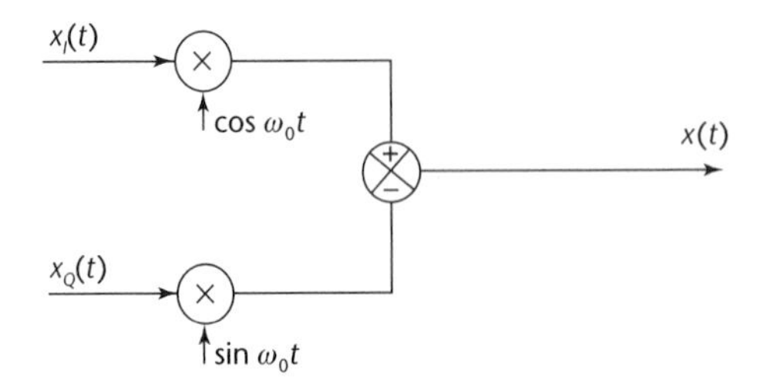

Fig. 17.5 Reconstruction of the original signal from in-phase and quadrature components

Thus, with the exception of a sign change in the lower channel, the system of Figure 17.4 produces the in-phase and quadrature components of the input signal $x(t)$.

Figure 17.5 shows the system necessary to reconstruct a bandpass signal from the quadrature components $x_I(t)$ and $x_Q(t)$. Based upon Equation (17.4), the signal $x_Q(t)\sin(\omega_0 t)$ is simply subtracted from the signal $x_I(t)\cos(\omega_0 t)$.

17.5 Envelope detection using quadrature components

Consider that the message signal $m(t)$ shown in Figure 17.6(a) is used to amplitude modulate a carrier of frequency ω_0 to produce the signal $x(t)$ shown in Figure 17.6(b), and expressed by

$$x(t) = a(t)\cos\omega_0 t = [A + m(t)]\cos\omega_0 t \tag{17.18}$$

where the constant A is sufficiently large relative to $m(t)$ to ensure that $A + m(t)$ is always positive, thus making $A + m(t)$ the envelope of the signal. The classical way of demodulating such a signal involves a circuit similar to, or based upon, that shown in Figure 17.7. When the modulated signal is applied to this circuit the diode conducts on positive half cycles such that the capacitor follows the voltage of that half cycle up to its peak value. Following the peak the capacitor then discharges slowly through the resistor, such that its voltage decays exponentially with a time constant of RC until it intercepts the rising voltage of the next half cycle. In this way the output voltage is the envelope of the waveform shown in Figure 17.8. The output ripple is at the carrier frequency and can be removed by further filtering. Similarly, the d.c. offset may be removed using a d.c. blocking capacitor. The result is the demodulated output $m(t)$.

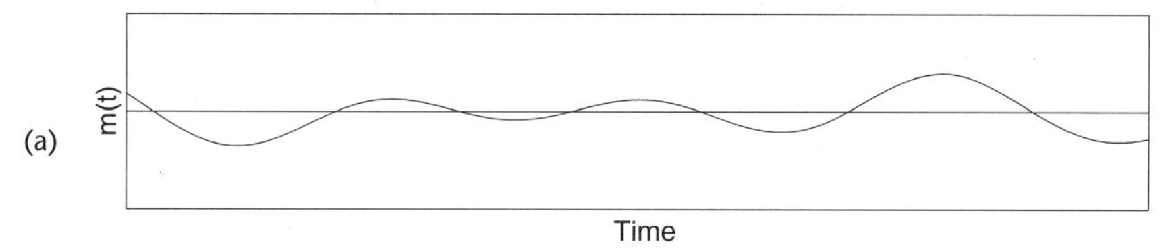

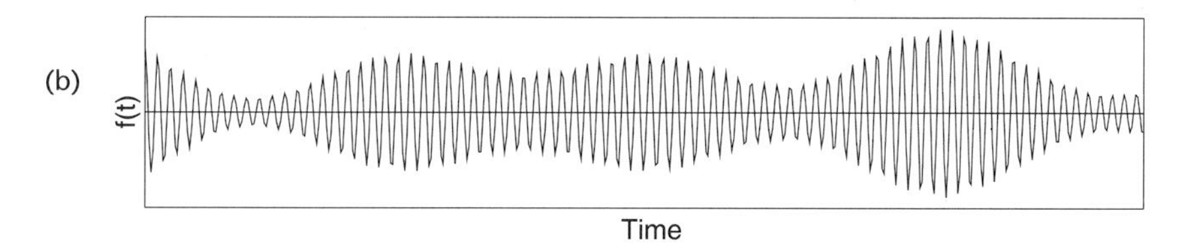

Fig. 17.6 Amplitude modulation: (a) message signal; (b) amplitude-modulated signal

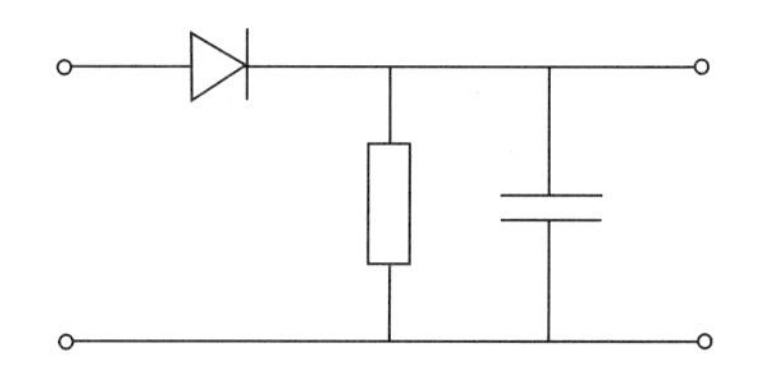

Fig. 17.7 Simple diode envelope detector

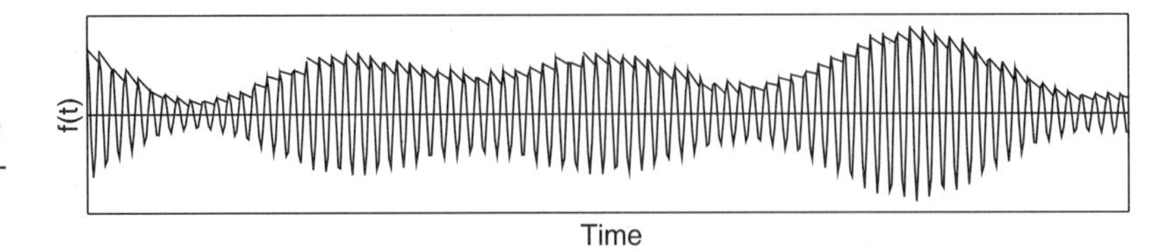

Fig. 17.8 Output of a diode detector when the input is amplitude-modulated

Although useful in many situations, there are two potential problems with this circuit.

- For the circuit to be effective the *RC* time constant must be small enough that each decaying exponential intercepts the subsequent half cycle. This can result in excessive ripple unless the carrier frequency is very much greater than the highest modulating frequency. This problem is illustrated in Figure 17.9, where we have the same waveforms $m(t)$ and $[A + m(t)]\cos(\omega_0 t)$ as in Figure 17.6 except that ω_0 is greatly reduced. It is noted that the detected waveform of Figure 17.9(b) is not very similar to the original signal and that this is likely to remain so even with further filtering.
- It is common in communication, radar and sonar applications to downconvert a high frequency signal of a frequency f_0 to a much lower 'intermediate' frequency (IF) by multiplying or 'mixing' it with a local oscillator signal of frequency $(f_0 \pm f_{\text{IF}})$. This produces a sum frequency term of frequency $(2f_0 \pm f_{\text{IF}})$ which is eliminated by lowpass filtering, plus a difference frequency term of frequency f_{IF} which is the wanted downconverted signal. If, for example, we multiply the signal $a(t)\cos(\omega_0 t + \theta)$ by a local oscillator signal we can use the trigonometric identity of Equation (17.16) to obtain

$$2a(t)\cos(\omega_0 t + \theta)\cos[(\omega_0 - \omega_{\text{IF}})t] = a(t)\cos[(2\omega_0 - \omega_{\text{IF}})t + \theta] + a(t)\cos(\omega_{\text{IF}}t + \theta) \quad (17.19)$$

  thus yielding $a(t)\cos(\omega_{\text{IF}}t + \theta)$. A problem arises however when this signal contains very few cycles, as perhaps in a radar echo, as then the waveform $a(t)\cos(\omega_{\text{IF}}t + \theta)$ can vary quite considerably depending on the phase θ of the local oscillator. This makes a diode detector even less satisfactory. This difficulty will be demonstrated in Example 17.1.

(a)

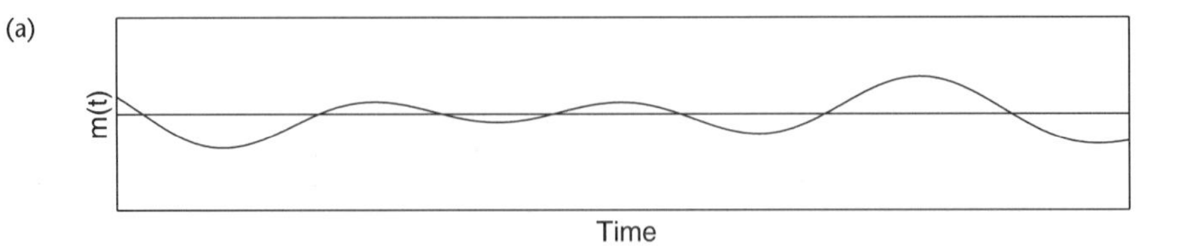

(b)

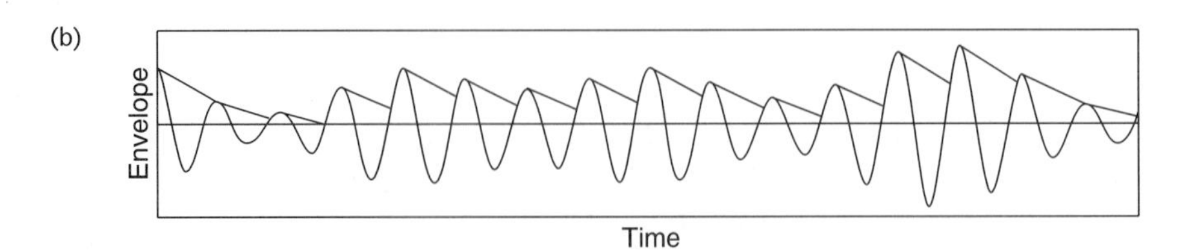

Fig. 17.9 Diode detector: (a) modulating signal; (b) output when the AM signal has low carrier frequency

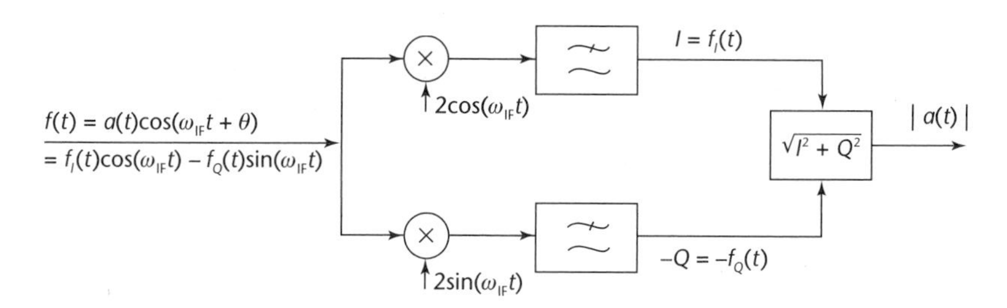

Fig. 17.10 Quadrature envelope detector

So long as the carrier frequency is known at the receiver, or can be derived there, an elegant way around both of these difficulties involves quadrature demodulation. Whether the configuration of Figure 17.4 is applied to the original signal or to the IF signal, the in-phase component $x_I(t)$ and the quadrature component $x_Q(t)$ can be obtained. As indicated by Equation (17.12), the envelope is then given by $\sqrt{x_I^2(t) + x_Q^2(t)}$.

As a reconfirmation of this, consider an IF signal given by $a(t)\cos(\omega_{IF}t + \theta)$ to enter the quadrature envelope detector of Figure 17.10, where the in-phase and quadrature local oscillator signals $2\cos(\omega_{IF}t)$ and $2\sin(\omega_{IF}t)$ are available. After the multiplier of the in-phase channel the signal is $2a(t)\cos(\omega_{IF}t + \theta)\cos(\omega_{IF}t,)$ After a suitable lowpass filter, becomes

$$x_I(t) = a(t)\cos\theta \tag{17.20}$$

Similarly, the signal after the multiplier of a quadrature channel is $2a(t)\cos(\omega_{IF}t + \theta)\sin(\omega_{IF}t)$ which, after a second lowpass filter, becomes

$$x_Q(t) = a(t)\sin\theta \tag{17.21}$$

As required, the square root of the sum of the squares produces the required envelope:

$$|a(t)| = \sqrt{x_I^2(t) + x_Q^2(t)} \tag{17.22}$$

It is interesting to note that the envelope $|a(t)|$ has lost some of the information of the modulating signal $a(t)$, since it does not convey the polarity of $a(t)$. In contrast, the complex envelope $\tilde{u}(t) = x_I(t) + jx_Q(t)$ as defined in Equation (17.14) conveys all the information about the signal.

Example 17.1

A radar transmits a 0.05 μs pulse at a frequency of 1500 MHz. After some initial amplification the received signal is mixed with a 1450 MHz signal and then passed through a frequency-selective IF amplifier centred at 50 MHz. The impulse response of this IF amplifier is $h(t) = h_{LP}(t)\cos(2\pi \times 50 \times 10^6 t)$, where $h_{LP}(t)$ is the impulse response of a second-order Butterworth lowpass filter of 15 MHz bandwidth. Use MATLAB to plot the output of this IF amplifier due to an echo from a point source: (a) when the echo and the 1450 MHz signal are in phase, and (b) when the echo and the 1450 MHz signal are 180° out of phase. In each case determine and plot the output of a quadrature envelope detector.

Solution The following program does the necessary simulation.

```
% radardet.m
t=(0:399)*0.5e-9;                              % 400 samples 0.5 ns apart
x1=1.5*sin(2*pi*50.e6*t(1:100));              % 0.05 microsecond 50 MHz echo
x1(400)=0;                                     % timespan extended
[b,a]=butter(2, 2*pi*15.e6, 's');              % Butterworth coefficients
[r,s,t]=impulse(b,a,t);                        % r is column vector of lowpass
                                               % impulse response
h=r'.*cos(2*pi*50.e6*t);                       % IF bandpass impulse response
x_IF1=conv(x1,h);                              % IF amplifier output
subplot(2,1,1)
plot(t, x_IF1(1:400));
xlabel('Time (seconds)'); ylabel('IF pulse 1')
x2=1.5*sin(2*pi*50.e6*t(1:100)+pi);           % echo given pi phase shift
x_IF2=conv(x2,h);                              % IF amplifier output
subplot(2,1,2)
plot(t, x_IF2(1:400));
xlabel('Time (seconds)'); ylabel('IF pulse 2')
keyboard                                       % this halts the program having produced
                                               % Figure 17.11. To continue we type
                                               % 'return' and enter it.
osc1=cos(2*pi*50.e6*t);                        % in-phase local oscillator
osc2=sin(2*pi*50.e6*t);                        % quadrature local oscillator
% The lowpass filters will be assumed to be the same as the Butterworth filter
% on which the IF bandpass filter was based, although they could be narrower.
% An efficient way efficient for finding their output is to use the 'lsim' command.
[xI1,k]=lsim(b,a,osc1.*x_IF1(1:400),t);
[xQ1,k]=lsim(b,a,osc2.*x_IF1(1:400),t);
envelope1=sqrt(xI1.^2+xQ1.^2);
```

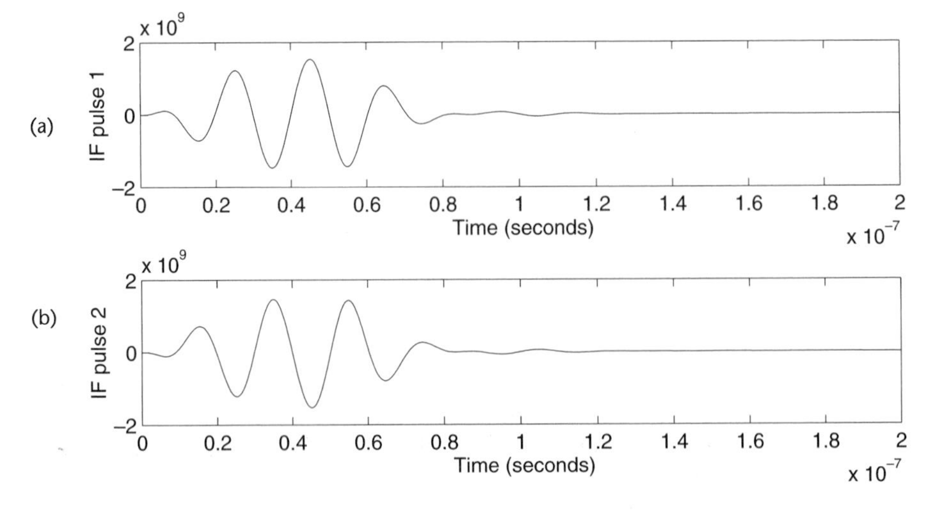

Fig. 17.11 Radar echoes after IF amplifier: (a) when received echo and downconverting local oscillator are in phase; (b) when received echo and downconverting local oscillator are 180° out of phase

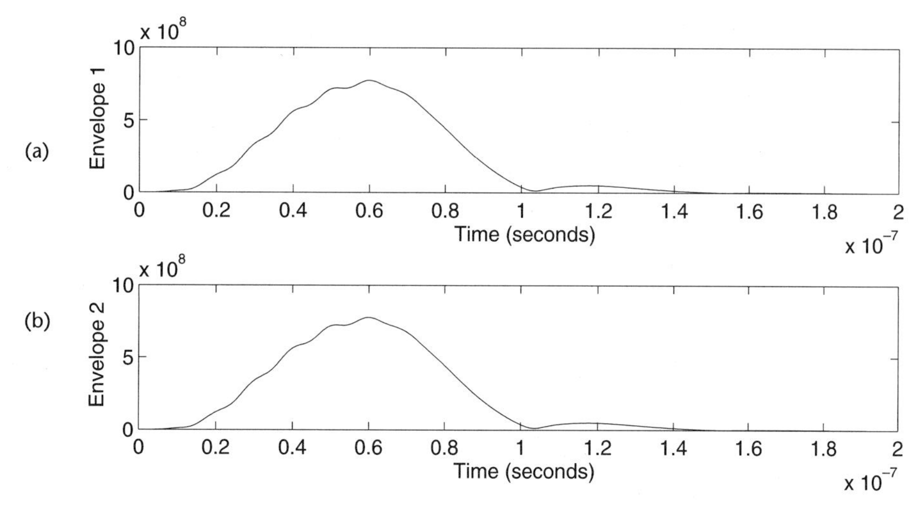

Fig. 17.12 Radar echoes after quadrature envelope detector: (a) when received echo and downconverting local oscillator are in phase; (b) when received echo and downconverting local oscillator are 180° out of phase

```
subplot(2,1,1)
plot(t, envelope1)                    % result for echo with no phase shift
xlabel('Time (seconds)'); ylabel('Envelope 1')
[xI2,k]=lsim(b, a, osc1.*x_IF2(1:400), t);
[xQ2,k]=lsim(b, a, osc2.*x_IF2(1:400), t);
envelope2=sqrt(xI2.^2+xQ2.^2);
subplot(2,1,2)
plot(t, envelope2)                    % result for echo with pi phase shift
xlabel('Time (seconds)'); ylabel('Envelope 2')
```

The first part of this program produces Figure 17.11 and demonstrates how much the processed echo can vary depending on its phase relative to the local oscillator that is used to downconvert it to IF. The second part of the program produces Figure 17.12 and demonstrates how very satisfactory and identical envelopes can be obtained from both of these two IF waveforms.

17.6 An example of quadrature channels for measuring Doppler shifts in radar

If a target has a component of velocity in the direction of a radar the radar echo experiences a Doppler shift f_D given by

$$f_D = \frac{2v_R}{c} f_0 \tag{17.23}$$

where v_R is the radial component of velocity, c is the velocity of light and f_0 is the frequency of the radar transmit pulse. By measuring the Doppler shift of the echo the radial component of velocity can be determined.

The block diagram of a suitable radar transmitter and receiver is shown in Figure 17.13.

The transmit pulse is produced by gating a continuous sinusoid on and off, but, for the practical reasons of wanting an intermediate frequency (IF) stage in the receiver to achieve amplification and frequency selectivity, this continuous sinusoid is normally generated by having a transmitter with two free-running oscillators of frequencies ω_L and ω_C.

Note: ω_C is the same as the IF frequency ω_{IF} of Section 17.5 but is now renamed in accordance with common convention, where the subscript C arises from coherent local oscillator.

These two free-running oscillators are multiplied to produce $\cos(\omega_C + \omega_L)t$. The output of this is gated on and off at a suitable pulse repetition frequency (prf) to produce transmit pulses of frequency $(\omega_C + \omega_L)$. The free-running signal $\cos[(\omega_C + \omega_L)t]$ generated within the radar, and the pulses of this leaving the radar, are shown in Figures 17.14(a) and 17.14(b) respectively.

The periodic set of echoes from a single point target is shown in Figure 17.14(c), where, because of the Doppler effect, it is assumed that the frequency is changed to $(\omega_C + \omega_L + \omega_D)$. Thus Figure 17.14(c) consists of bursts of the signal $A\cos[(\omega_C + \omega_L + \omega_D)t + \phi]$, each burst having the duration of the transmit pulse and having a delay, relative to the transmit pulse, of the two-way propagation time.

The receiver first downconverts the receive signal to an IF frequency by mixing it with the output of the free-running oscillator of frequency ω_L and thus producing bursts of the signal $A\cos[(\omega_C + \omega_D)t + \phi]$. This now undergoes quadrature

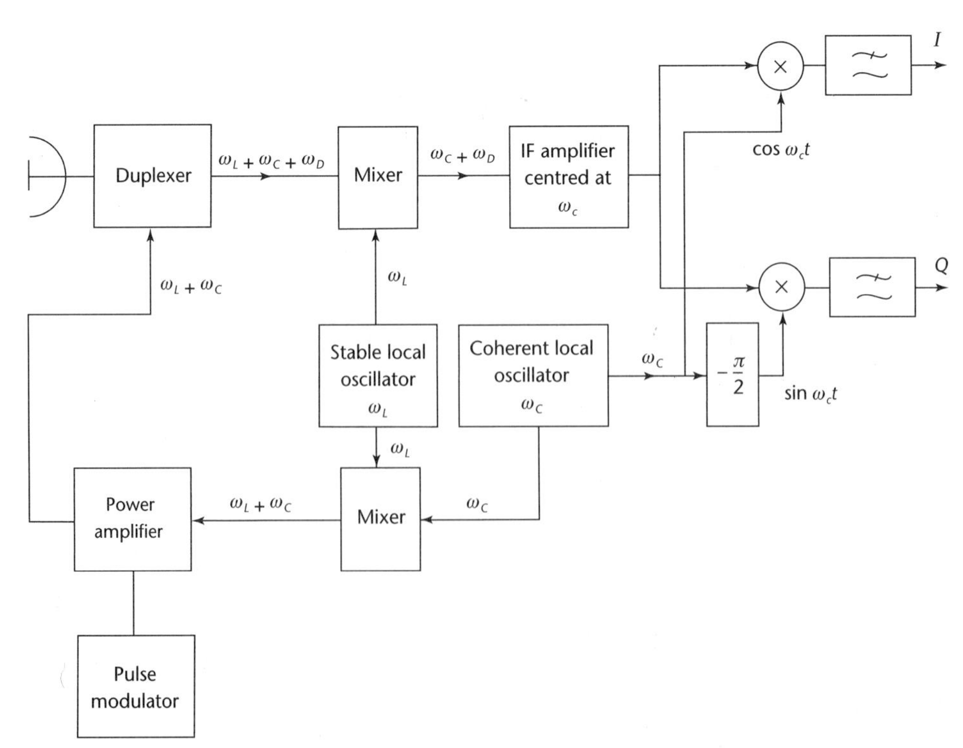

Fig. 17.13 Block diagram of a radar for measuring Doppler shift

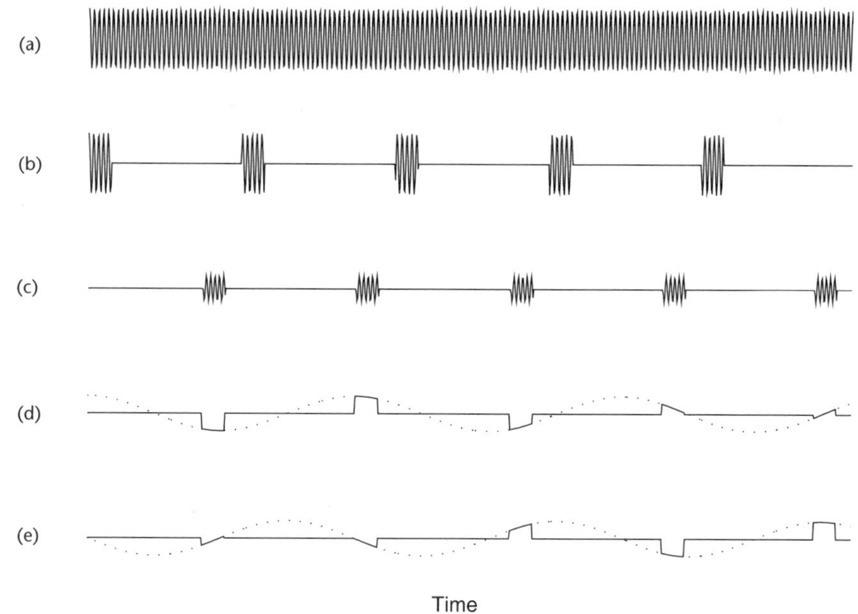

Fig. 17.14 Waveforms in Doppler measuring radar: (a) sinusoidal reference; (b) gated transmit pulses; (c) echoes from single point target; (d) in-phase component of echo; (e) quadrature component of echo

demodulation by mixing it in two parallel channels with $2\cos(\omega_C t)$ and $2\sin(\omega_C t)$ to produce bursts of $2A\cos(\omega_C t)\cos[(\omega_C + \omega_D)t + \phi]$ and of $2A\sin(\omega_C t)\cos[(\omega_C + \omega_D)t + \phi]$. Following lowpass filters the difference frequency terms correspond to bursts of $A\cos(\omega_D t + \phi)$ and $-A\sin(\omega_D t + \phi)$, as shown in Figures 17.14(d) and 17.14(e). Let us next suppose that these two waveforms are sampled and that at every sampling instant the two samples are combined to become a single sample of the *complex* envelope $\tilde{u}(t)$, where $\tilde{u}(t)$ is given by

$$\tilde{u}(t) = A\cos(\omega_D t + \phi) + jA\sin(\omega_D t + \phi) = Ae^{j(\omega_D t + \phi)} \tag{17.24}$$

Finally, we consider a set of return signals arising from a substantial number, N, of transmit pulses and, for each and every 'range bin', collect N samples of this complex envelope. In each range bin we then perform an FFT spectral analysis on these N samples to obtain an estimate of ω_D, and hence target radial velocity, in that range bin.

In practice, a radar using this 'Moving Target Detector' processing is also able to measure the azimuth of incoming echoes by virtue of its rotating antenna. This signifies that a very substantial digital memory is required to store the three-dimensional 'clutter map' of amplitude versus range, velocity and azimuth.

It is important to note that if we only had the one channel, say the in-phase channel, we would be unable to recognize the difference between an 'up-Doppler' of $+\omega_D$ and a 'down-Doppler' of $-\omega_D$. This is because the Fourier transform of $A\cos(\omega_D t + \phi)$ has frequency components at both $+\omega_D$ and $-\omega_D$, just as does the Fourier transform of $A\cos(-\omega_D t + \phi)$. The two case are indistinguishable. In contrast, the Fourier transform of $Ae^{j(\omega_D t + \phi)}$ has a frequency component solely at $+\omega_D$, and the Fourier transform of $Ae^{j(-\omega_D t + \phi)}$ has a frequency component solely at at $-\omega_D$.

17.7 An example of quadrature channels in narrowband FFT beamforming

Consider a single-frequency wave to be incident on a continuous array as shown in Figure 17.15(a). At a particular instant in time the variation of amplitude with position would be as shown in Figure 17.15(b).

The *spatial* frequency in cycles/metre or radians/metre is dependent on the direction of the source and a spectral analysis of the variation of received amplitude with position can therefore indicate that direction. This suggests we have a method of beamforming. For example let the signal leaving the centre element of the array be $A\cos(\omega_0 t)$. If the angle of incidence is $\theta = \theta_0$ it will be seen from Figure 17.16 that, at a distance x from the centre of the array, the wave still has to travel a further distance of $x \sin\theta_0$ before striking the array. This means that the wave at this point is delayed relative to that at the centre point by $(x \sin\theta_0)/\lambda$ wavelengths, or $(2\pi x \sin\theta_0)/\lambda$ radians. Defining a parameter k as $2\pi/\lambda$, we have that the signal at distance x is $\cos(\omega_0 t - kx \sin\theta_0)$. From this it can be seen that this signal varies with time at a rate of ω_0 radians/second, and with position at a rate of $-k \sin\theta_0$ radians/metre. Thus it has a spatial frequency of $-k \sin\theta_0$ radians/metre.

In practice, the array would consist of a large number of closely spaced *discrete* elements and one possibility would be to take a sample from each one at the same instant in time such that we have a *sampled* version of the signal in Figure 17.15(b). The spatial frequencies present in this could then be analysed by means of an FFT. Unfortunately, there are several snags with this.

- The system suffers from a direction ambiguity in that the sampled wavefront is identical whether the source is at $+\theta_0$ or $-\theta_0$.
- The signal on each element has the frequency of the incoming wave and this means that very accurate sampling is required.
- The received signal on each element is likely to be weak and masked by noise. A single sample is likey to be inaccurate because of this noise.

All of these difficulties can be overcome by passing the output of each array element through a quadrature demodulator and smoothing filters before the samples are taken. In this way the *complex envelope* at the output of each element of the array is sampled and an FFT of the spatial variation of this complex envelope can be

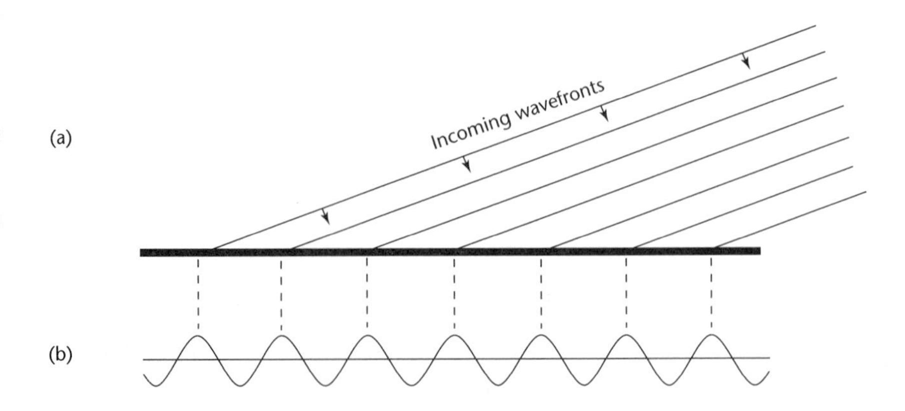

Fig. 17.15 Determination of direction from spatial frequency: (a) wavefronts striking array; (b) instantaneous amplitude across array

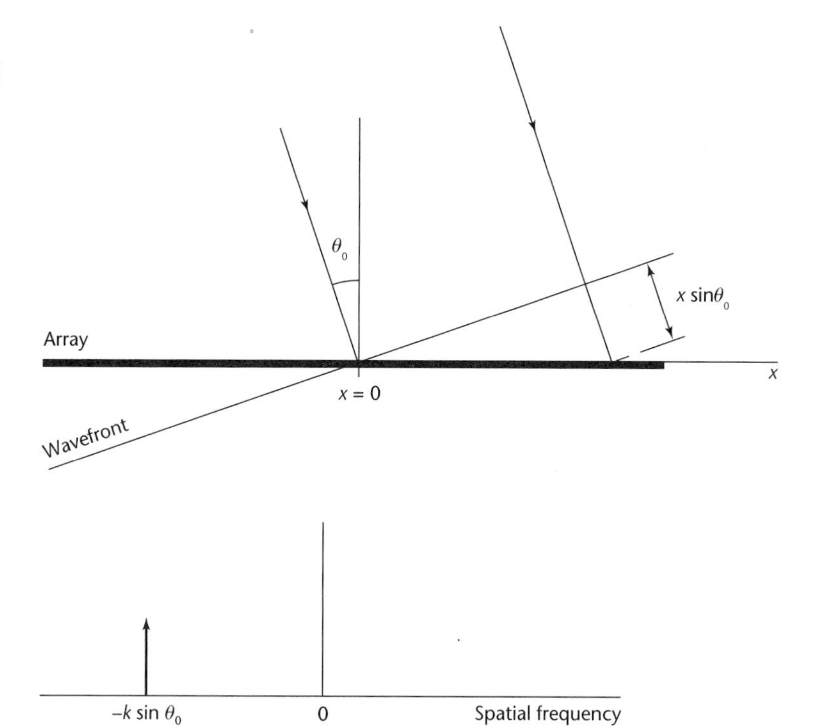

Fig. 17.16 Variation of arrival time of wavefront across array

Fig. 17.17 Amplitude spectrum of the complex signal across the array

performed to indicate the direction of the incoming signal at each instant of sampling. The in-phase and quadrature components of this incoming signal can be obtained at each element of the array by multiplying each output by $2\cos(\omega_0 t)$ and $2\sin(\omega_0 t)$ in two parallel channels to become $2\cos(\omega_0 t)\cos(\omega_0 t - kx \sin\theta_0)$ and $2\sin(\omega_0 t)\cos(\omega_0 t - kx \sin\theta_0)$, and then lowpass filtering the output of each multiplier to remove the maximum amount of noise (in practice, the incoming signal will have a finite bandwidth and the filters must be wide enough to pass this). In the absence of noise the in-phase and quadrature components for a single source at an angle θ_0 will become $\cos(kx \sin\theta_0)$ and $\sin(kx \sin\theta_0)$, respectively. If the sign of $\sin(kx \sin\theta_0)$ is changed an FFT algorithm can treat these as the real and imaginary components of the complex signal $e^{-jkx \sin\theta_0}$ and then determine its spatial spectrum. The spatial spectrum will contain an impulse function at spatial frequency $-k \sin\theta_0$ radians/metre, as shown in Figure 17.17. As required, this spectrum indicates a single source at $\theta = \theta_0$. It will be noted that there is now no ambiguity between left and right of the 'boresight'. A source to the right of the 'boresight' at an angle $-\theta_0$ would give rise to an output from the same element of $\cos(\omega_0 t + kx \sin\theta_0)$ and thus generate the complex envelope $e^{jkx \sin\theta_0}$. This gives rise to an impulse function in the spatial spectrum at the positive spatial frequency of $+k \sin\theta_0$ and this is quite distinct from an impulse function at $-k \sin\theta_0$.

17.8 Filtering bandpass signals with lowpass filters

Many signals of importance occupy a narrow bandwidth centred about some high frequency. If such a signal is to undergo narrowband analogue filtering it is often advantageous to first downconvert it to some intermediate frequency (IF). For example, a conventional FM radio obtains its frequency selectivity by mixing the incoming signal of around 100 MHz with a local oscillator offset by 10.7 MHz, and passing the difference frequency terms through a 10.7 MHz IF filter. The main concern of this section, however, concerns the processing of bandpass signals using *digital* filters.

If a narrowband high-frequency signal is to undergo digital filtering it needs first to undergo analogue to digital conversion, but the sampling of such a signal at twice its highest frequency is unnecessary and leads to an inefficient system. As in the analogue case, one possibility is to downconvert the signal to some intermediate frequency and to then digitize this. However, the number of samples needed is *minimized* by deriving the in-phase and quadrature components of the signal. The appropriate filtering operation can then be applied to these two lowpass signals before recombining them to obtain a filtered bandpass signal. It is this procedure that forms the subject of this section.

Let the unfiltered bandpass signal be $x(t)$, for which the in-phase and quadrature components are $x_{\mathrm{I}}(t)$ and $x_{\mathrm{Q}}(t)$, and let the required bandpass filter have an impulse response $h(t)$, for which the in-phase and quadrature components are $h_{\mathrm{I}}(t)$ and $h_{\mathrm{Q}}(t)$. We require to produce the signal

$$y(t) = x(t) * h(t) \tag{17.25}$$

However, we wish to use only the lowpass signals $x_{\mathrm{I}}(t)$, $x_{\mathrm{Q}}(t)$, $h_{\mathrm{I}}(t)$ and $h_{\mathrm{Q}}(t)$. We have

$$\begin{aligned} y(t) &= x(t) * h(t) \\ &= \{x_{\mathrm{I}}(t)\cos\omega_0 t - x_{\mathrm{Q}}(t)\sin\omega_0 t\} * \{h_{\mathrm{I}}(t)\cos\omega_0 t - h_{\mathrm{Q}}(t)\sin\omega_0 t\} \end{aligned} \tag{17.26}$$

Expanding this, we have four convolutions to perform. The first convolution is $\{x_{\mathrm{I}}(t)\cos\omega_0 t\} * \{h_{\mathrm{I}}(t)\cos\omega_0 t\}$. Using the definition of the convolution integral this can be written out in full as

$$\{x_{\mathrm{I}}(t)\cos\omega_0 t\} * \{h_{\mathrm{I}}(t)\cos\omega_0 t\} = \int_{-\infty}^{\infty} x_{\mathrm{I}}(\tau)\cos(\omega_0\tau)h_{\mathrm{I}}(t-\tau)\cos\omega_0(t-\tau)\,\mathrm{d}\tau \tag{17.27}$$

Using the trigonometric identity of Equation (17.16) this becomes

$$\begin{aligned} &\{x_{\mathrm{I}}(t)\cos\omega_0 t\} * \{h_{\mathrm{I}}(t)\cos\omega_0 t\} \\ &\qquad = \int_{-\infty}^{\infty} x_{\mathrm{I}}(\tau)h_{\mathrm{I}}(t-\tau)\{\tfrac{1}{2}\cos\omega_0 t + \tfrac{1}{2}\cos(2\omega_0\tau - \omega_0 t)\}\,\mathrm{d}\tau \end{aligned} \tag{17.28}$$

However, $x_{\mathrm{I}}(\tau)$ and $h_{\mathrm{I}}(t-\tau)$ are both lowpass signals and so therefore is their product. It follows that $x_{\mathrm{I}}(\tau)h_{\mathrm{I}}(t-\tau)\left\{\tfrac{1}{2}\cos(2\omega_0\tau - \omega_0 t)\right\}$ is a modulated signal centred about $2\omega_0$ that varies with time τ and that therefore

$$\int_{-\infty}^{\infty} x_{\mathrm{I}}(\tau)h_{\mathrm{I}}(t-\tau)\left\{\tfrac{1}{2}\cos(2\omega_0\tau - \omega_0 t)\right\}\mathrm{d}\tau = 0 \tag{17.29}$$

Hence Equation (17.28) simplifies to become

$$\begin{aligned} \{x_{\mathrm{I}}(t)\cos\omega_0 t\} * \{h_{\mathrm{I}}(t)\cos\omega_0 t\} &= \tfrac{1}{2}\cos\omega_0 t \int_{-\infty}^{\infty} x_{\mathrm{I}}(\tau)h_{\mathrm{I}}(t-\tau)\,\mathrm{d}\tau \\ &= \tfrac{1}{2}\cos\omega_0 t\{x_{\mathrm{I}}(t) * h_{\mathrm{I}}(t)\} \end{aligned} \tag{17.30}$$

Extending this procedure to the other three convolutions of Equation (17.26) we obtain

$$y(t) = x(t) * h(t) = \tfrac{1}{2}\{x_I(t) * h_I(t) - x_Q(t) * h_Q(t)\} \cos \omega_0 t$$
$$-\tfrac{1}{2}\{x_Q(t) * h_I(t) + x_I(t) * h_Q(t)\} \sin \omega_0 t \qquad (17.31)$$

This equation is realized by the block diagram in Figure 17.18, which therefore represents the complete bandpass filter using lowpass signals.

In some applications the requirement may be solely to extract the envelope or phase modulation of the signal at the output of the bandpass filter. For these we recognize that the output $y(t)$ can be expressed in terms of its in-phase and quadrature components as

$$y(t) = y_I(t) \cos \omega_0 t - y_Q(t) \sin \omega_0 t \qquad (17.32)$$

and that the envelope of the filtered bandpass signal is given by

$$|a(t)| = \sqrt{y_I^2(t) + y_Q^2(t)} \qquad (17.33)$$

and the phase modulation is given by

$$\theta(t) = \tan^{-1} \frac{y_Q(t)}{y_I(t)} \qquad (17.34)$$

Comparing Equation (17.32) with Equation (17.31), we see that the required waveforms for determining the amplitude and phase modulations are given by

$$y_I(t) = \tfrac{1}{2}\{x_I(t) * h_I(t) - x_Q(t) * h_Q(t)\} \qquad (17.35)$$
$$y_Q(t) = \tfrac{1}{2}\{x_Q(t) * h_I(t) + x_I(t) * h_Q(t)\} \qquad (17.36)$$

An alternative viewpoint to that based on Equation (17.32) is to write down the complex envelope of the output signal $y(t)$ as

$$\tilde{u}(t) = y_I(t) + \mathrm{j}y_Q(t) \qquad (17.37)$$

and to note that the envelope is given by

$$|a(t)| = |y_I(t) + \mathrm{j}y_Q(t)| = |\tilde{u}(t)| \qquad (17.38)$$

For many bandpass signals the configuration of Figure 17.18 can be greatly simplified. Consider a bandpass filter which has an amplitude and phase response that corresponds exactly to a frequency translated version of the amplitude and phase response of a lowpass filter; that is

$$H_{BP}(f) = \tfrac{1}{2}H_{LP}(f+f_0) + \tfrac{1}{2}H_{LP}(f-f_0) \qquad (17.39)$$

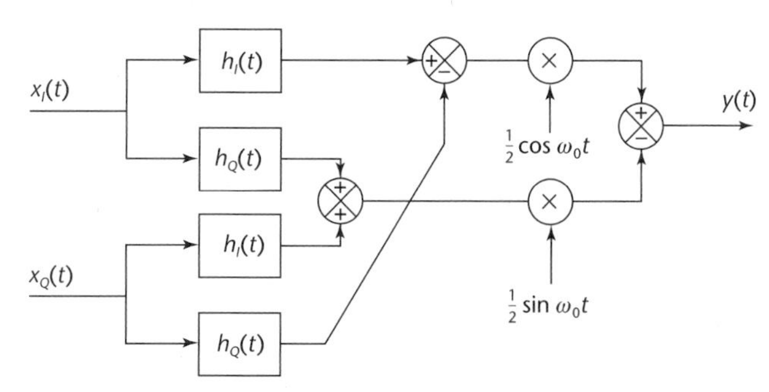

Fig. 17.18 Block diagram of bandpass filter using lowpass filters and the in-phase and quadrature components of the signal

where $H_{LP}(f)$ is the transfer function of a lowpass filter and $H_{BP}(f)$ is the transfer function of the corresponding bandpass filter centred at f_0. The impulse response of this bandpass filter, $h_{BP}(t)$, is the inverse Fourier transform of the transfer function of Equation (17.39) and the frequency shift theorem of Section 6.18.5 tells us that this is given by

$$h_{BP}(t) = \tfrac{1}{2}h_{LP}(t)e^{-j\omega_0 t} + \tfrac{1}{2}h_{LP}(t)e^{j\omega_0 t} = h_{LP}(t)\cos\omega_0 t \tag{17.40}$$

where $h_{LP}(t)$ is the inverse Fourier transform of $H_{LP}(f)$ and is the impulse response of the lowpass filter. Writing $h_{BP}(t)$ in terms of its in-phase and quadrature components, such that

$$h_{BP}(t) = h_I(t)\cos\omega_0 t - jh_Q(t)\sin\omega_0 t \tag{17.41}$$

a comparison with Equation (17.40) shows that $h_Q(t)$, the quadrature component of the bandpass impulse response, is zero. It follows that, so long as Equation (17.39) is applicable, the configuration of Figure 17.18 can be simplified to that of Figure 17.19. It is seen that all that needs to be done in the bandpass filter of Figure 17.18 is to filter the two quadrature components of the signal with separate lowpass filters, each of transfer function $H_{LP}(f)$, and then to reconstruct a bandpass signal from the resulting quadrature components using the system of Figure 17.5.

The procedure shown in Figure 17.4 for obtaining $x_I(t)$ and $-x_Q(t)$ involves two lowpass filters, and the lowpass filters of Figure 17.19 can also be used for this purpose to give the complete system shown in Figure 17.20.

Physically realizable lowpass filters will clearly have impulse responses that are real and, from Section 6.3, this means that they have amplitude responses with even symmetry and phase responses with odd symmetry. Therefore, from Equation (17.39), the condition for Figure 17.19 to be applicable is that the bandpass filter should have an amplitude response with even symmetry about its centre frequency and a phase response with odd symmetry about its centre frequency. Such a response is shown in Figures 17.21(a) and 17.21(b). An alternative and equivalent criterion for

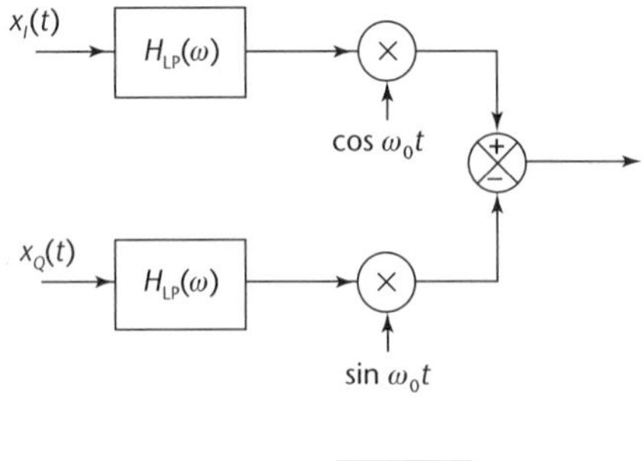

Fig. 17.19 Simplification of Figure 17.18 when the bandpass filter response has symmetry about centre frequency

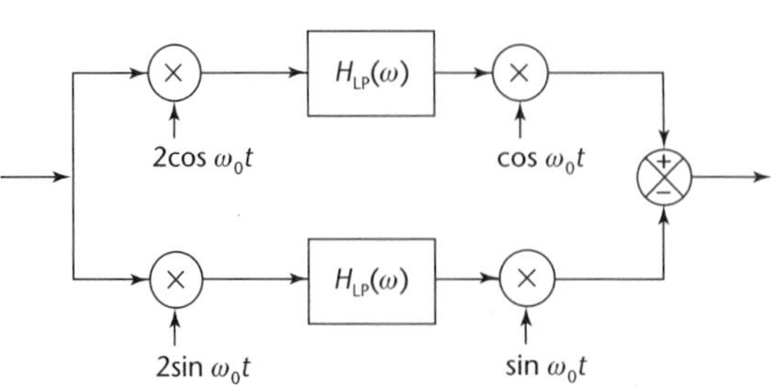

Fig. 17.20 Complete bandpass filter applicable when the response has symmetry about the centre frequency

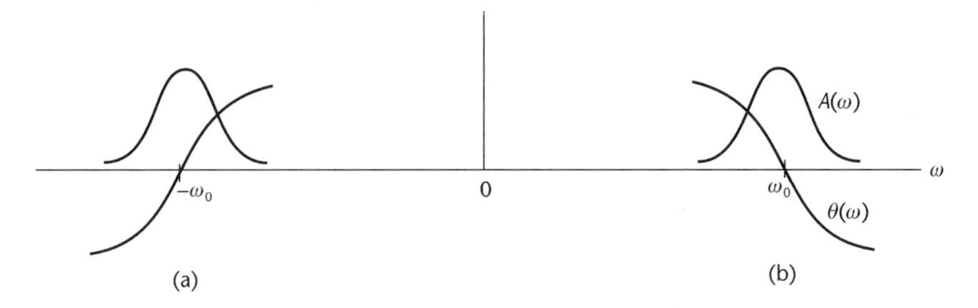

Fig. 17.21 Symmetries of frequency response needed for bandpass filter simplification: (a) amplitude response; (b) phase response

such a filter is that it should be possible to express its impulse response as $h_{\mathrm{LP}}(t)\cos\omega_0 t$, where $h_{\mathrm{LP}}(t)$ is a lowpass signal.

17.9 Correlators for bandpass signals

A particularly important example where the findings of the last section are important, but where the simplification of Figure 17.20 is not appropriate, is in the digital implementation of a matched filter (or correlator) for enhancing a radar or sonar echo.

Let us consider the requirement in a radar or sonar to derive the envelope of the output of a filter matched to a transmit pulse $x(t)$ when the input to this filter is the receive signal $x'(t)$. From Equation (16.48) we require the matched filter to have an impulse response given by

$$h(t) = x(-t) \tag{17.42}$$

Consider $x(t)$ to be given by

$$x(t) = a(t)\cos[\omega_0 t + \theta(t)] \tag{17.43}$$

This can be expressed in terms of its in-phase and quadrature components as

$$x(t) = x_{\mathrm{I}}(t)\cos\omega_0 t - x_{\mathrm{Q}}(t)\sin\omega_0 t \tag{17.44}$$

It follows from this and Equation (17.42) that the impulse response should be given by

$$h(t) = x_{\mathrm{I}}(-t)\cos(-\omega_0 t) - x_{\mathrm{Q}}(-t)\sin(-\omega_0 t) \tag{17.45}$$

However, the impulse response of a matched filter can also be written in terms of its in-phase and quadrature components as

$$h(t) = h_{\mathrm{I}}(t)\cos\omega_0 t - h_{\mathrm{Q}}(t)\sin\omega_0 t \tag{17.46}$$

Comparing this with Equation (17.45) shows that we require

$$h_{\mathrm{I}}(t) = x_{\mathrm{I}}(-t) \tag{17.47}$$

$$h_{\mathrm{Q}}(t) = -x_{\mathrm{Q}}(-t) \tag{17.48}$$

Substituting these terms into Equations (17.35) and (17.36) and determining the envelope using Equation (17.38) we find that the envelope of the cross-correlation function between the receive signal $x'(t)$ and the transmit pulse $x(t)$ is realized by the operation

$$|\tilde{v}(t)| = \tfrac{1}{2}|x'_{\mathrm{I}}(t)*x_{\mathrm{I}}(-t) + x'_{\mathrm{Q}}(t)*x_{\mathrm{Q}}(-t) + \mathrm{j}x'_{\mathrm{Q}}(t)*x_{\mathrm{I}}(-t) - \mathrm{j}x'_{\mathrm{I}}(t)*x_{\mathrm{Q}}(-t)| \tag{17.49}$$

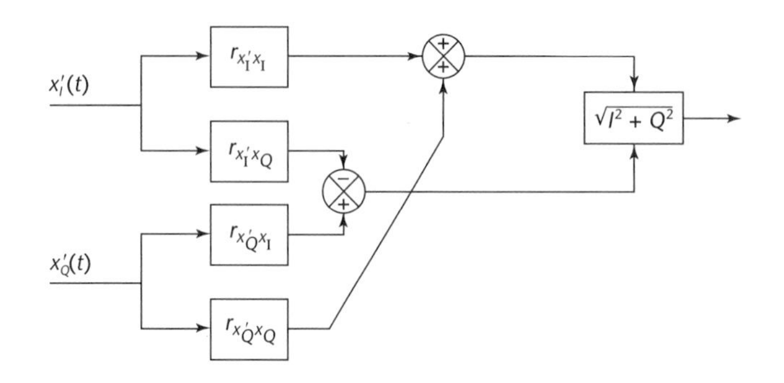

Fig. 17.22 ● Correlator using the in-phase and quadrature components of a signal

The first term, representing the convolution of $x_I'(t)$ with $x_I(-t)$, can alternatively be interpreted as the cross-correlation function between $x_I'(t)$ and $x_I(t)$. Extending this alternative interpretation to the other terms in Equation (17.49) we see that a correlator for a bandpass signal should produce an envelope given by

$$|\tilde{v}(t)| = \tfrac{1}{2}|(r_{x_I'x_I} + r_{x_Q'x_Q}) + j(r_{x_Q'x_I} - r_{x_I'x_Q})| \tag{17.50}$$

This can be realized using the configuration of four baseband correlators shown in Figure 17.22.

Assuming that the transmit pulse has a bandwidth of B Hz centred about f_0 Hz its in-phase and quadrature components will occupy the frequency band 0 to $B/2$ Hz. The Nyquist sampling rate for each of the four correlators is B Hz. If the transmit pulse is of duration T and the correlators are realized by the transversal filters of Figures 16.23 or 16.25 the theoretical minimum for the number of coefficients required by each is therefore BT. For the complete correlator the number is $4BT$. In radar and sonar applications this is far fewer than would be needed if the correlator were implemented directly on the receive signal, or even on an IF version of the receive signal.

Example 17.2

A sonar transmits a chirp pulse whose instantaneous frequency varies from 4 kHz to 5 kHz in 0.2 s. Design a digital correlator using baseband transversal filters.

Solution The general expression for a chirp pulse extending from $-T/2$ to $+T/2$ is

$$x(t) = \text{rect}(t/T)\cos(\omega_0 t + \tfrac{1}{2}kt^2) \tag{17.51}$$

The instantaneous frequency of this is given by

$$f = \frac{1}{2\pi}\frac{d\theta}{dt} = f_0 + \frac{kt}{2\pi} \qquad -\frac{T}{2} \leqslant t \leqslant \frac{T}{2} \tag{17.52}$$

In our case the instantaneous frequency in the middle of the pulse is 4.5 kHz, thus giving $f_0 = 4500$ *Hz*. In order for the instantaneous frequency to equal 5000 Hz when $t = 0.1$ s we obtain $4500 + 0.1k/2\pi = 5000$, and hence $k = 31\,416$. Therefore

$$x(t) = \text{rect}(t/0.2)\cos(28274t + 15708t^2) \tag{17.53}$$

Expansion of this into the form

$$x(t) = x_I(t)\cos\omega_0 t - x_Q(t)\sin\omega_0 t \tag{17.54}$$

shows that

$$x_{\mathrm{I}}(t) = \mathrm{rect}(t/0.2)\cos(15\,708t^2) \tag{17.55}$$

and

$$x_{\mathrm{Q}}(t) = -\mathrm{rect}(t/0.2)\sin(15\,708t^2) \tag{17.56}$$

Thus the complete correlator, or matched filter, including envelope detector, is as shown in Figure 17.22, where the in-phase and quadrature components of the input are each correlated with $x_{\mathrm{I}}(t)$ and $x_{\mathrm{Q}}(t)$.

If the correlators are realized with transversal filters the impulse responses of these filters are given, using Equations (17.47) and (17.48), as

$$h_{\mathrm{I}}(t) = x_{\mathrm{I}}(-t) = \mathrm{rect}(t/0.2)\cos(15\,708t^2) \tag{17.57}$$

and

$$h_{\mathrm{Q}}(t) = -x_{\mathrm{Q}}(-t) = \mathrm{rect}(t/0.2)\sin(15\,708t^2) \tag{17.58}$$

These two impulse responses are illustrated in Figures 17.23(a) and 17.23(b). Neglecting transient effects caused by the sudden start and end of the transmit pulse, the spectra of $x(t)$ and $h(t)$ would extend from 4000 Hz to 5000 Hz. The spectra of $h_{\mathrm{I}}(t)$ and $h_{\mathrm{Q}}(t)$ must be similar except shifted down in frequency by 4500 Hz, such that each is contained primarily within the band -500 Hz to $+500$ Hz. The same applies to the in-phase and quadrature components of the received sonar signal. The sampling frequency should be somewhat more than twice the highest frequency, and a suitable value would be 1250 Hz, corresponding to a sampling period of 0.8 ms.

Inserting the sampling period of 0.8 ms such that the nth sample occurs at a time $0.0008n$, we replace t in Equations (17.57) and (17.58) by $0.0008n$ and these impulse responses become

$$h_{\mathrm{I}}[n] = \cos(0.0016755n^2) \qquad -750 \geqslant n \geqslant 750 \tag{17.59}$$

and

$$h_{\mathrm{Q}}[n] = \sin(0.0016755n^2) \qquad -750 \geqslant n \geqslant 750 \tag{17.60}$$

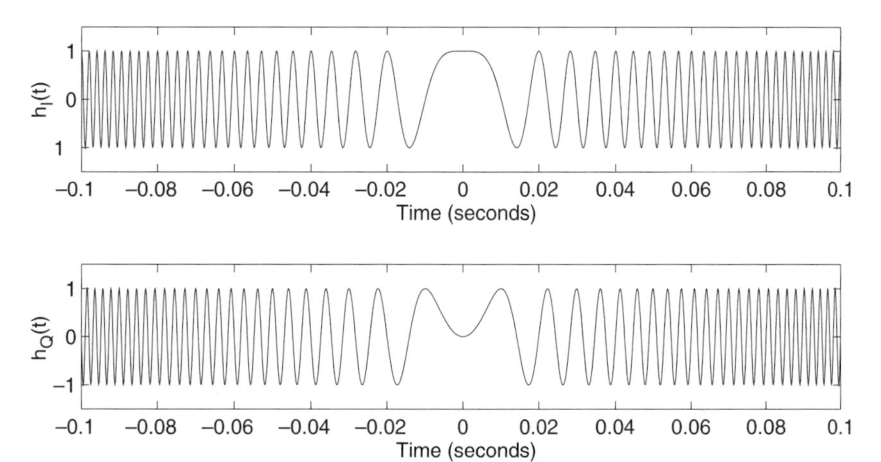

Fig. 17.23 Impulse responses of transversal filters in a correlator for a Chirp signal

These impulse responses are non-causal. They can be made causal by delaying them by half the duration of the pulse. Doing this and recognizing that the coefficients of a transversal filter are given by the impulse response we find that these coefficients are given by

$$b_i = \cos[0.0016755(i - 750)^2] \qquad 0 \leqslant i \leqslant 1500 \tag{17.61}$$

for the two in-phase filters, and by

$$b_i = \sin[0.0016755(i - 750)^2] \qquad 0 \leqslant i \leqslant 1500 \tag{17.62}$$

for the two quadrature filters. The total number of coefficients required by the four transversal filters is 6004.

A common and concise alternative way of writing Equation (17.50) is that the output of a correlator, in terms of the variable τ, should be given by

$$|r_{x'x}(\tau)| = \tfrac{1}{2}\left|\int_{-\infty}^{\infty} \tilde{u}_{\mathrm{r}}(t)\tilde{u}^*(t-\tau)\,\mathrm{d}t\right| \tag{17.63}$$

where $\tilde{u}(t)$ is the complex envelope of the transmit signal and $\tilde{u}_{\mathrm{r}}(t)$ is the complex envelope of the receive signal. We can easily check this result by substituting into Equation (17.63) the complex envelopes

$$\tilde{u}(t) = x_{\mathrm{I}}(t) + \mathrm{j}x_{\mathrm{Q}}(t) \tag{17.64}$$
$$\tilde{u}_{\mathrm{r}}^*(t) = x'_{\mathrm{I}}(t) - \mathrm{j}x'_{\mathrm{Q}}(t) \tag{17.65}$$

This gives

$$\begin{aligned} |r_{x'x}(\tau)| &= \tfrac{1}{2}\left|\int_{-\infty}^{\infty} \{x'_{\mathrm{I}}(t) + \mathrm{j}x'_{\mathrm{Q}}(t)\}\{x_{\mathrm{I}}(t-\tau) - \mathrm{j}x_{\mathrm{Q}}(t-\tau)\}\,\mathrm{d}t\right| \\ &= \tfrac{1}{2}\left|\int_{-\infty}^{\infty} x'_{\mathrm{I}}(t)x_{\mathrm{I}}(t-\tau)\,\mathrm{d}t + \int_{-\infty}^{\infty} x'_{\mathrm{Q}}(t)x_{\mathrm{Q}}(t-\tau)\,\mathrm{d}t\right. \\ &\qquad \left. +\mathrm{j}\int_{-\infty}^{\infty} x'_{\mathrm{Q}}(t)x_{\mathrm{I}}(t-\tau)\,\mathrm{d}t - \mathrm{j}\int_{-\infty}^{\infty} x'_{\mathrm{I}}(t)x_{\mathrm{Q}}(t-\tau)\,\mathrm{d}t\right| \end{aligned} \tag{17.66}$$

This is seen to be the same as Equation (17.49), thus confirming Equation (17.63).

For the case where the receive signal has a Doppler shift on it, the representation of Equation (17.63) leads to the *ambiguity function* that is important in radar for determining how the output of a correlator is affected by Doppler shift for a particular type of transmitted pulse. The ambiguity function is defined as $|\chi(\tau, f_{\mathrm{D}})|^2$ where

$$\chi(\tau, f_{\mathrm{D}}) = \int_{-\infty}^{\infty} \tilde{u}(t)\tilde{u}^*(t+\tau)\mathrm{e}^{\mathrm{j}2\pi f_{\mathrm{D}}t}\,\mathrm{d}t \tag{17.67}$$

17.10 The physical reality of complex envelopes

In the earlier parts of this book it was common to replace a real signal such as $\cos \omega t$ by the complex signal $\mathrm{e}^{\mathrm{j}\omega t}$. This was done for *mathematical* convenience because solving the differential equations describing system responses was much easier when the input was $e^{j\omega t}$ than when it was $\cos \omega t$, and the response to the real signal $\cos \omega t$ could then be determined by taking the real part of the response to the complex signal $e^{j\omega t}$.

It was never suggested in these earlier chapters that the complex signal was in any way physically realizable. It was not a signal that could be observed on an oscilloscope.

This chapter has shown that all the information of a real bandpass signal $x(t)$ can be conveyed by its real in-phase and quadrature components, $x_I(t)$ and $x_Q(t)$, and that we can mathematically combine these as $x_I(t) + jx_Q(t)$ to give us the complex envelope $u(t)$. In the sense that $x_I(t)$ and $x_Q(t)$ can be extracted from $x(t)$ and can be displayed on *two* channels of an oscilloscope to convey all the information of $x(t)$ it can almost be argued that the complex envelope has a *physical* existence (not quite so since one of its two components is multiplied by j). Certainly, however, the complex envelope has useful applications.

In contrast to the earlier use of the complex signal $e^{j\omega t} = (\cos \omega t + j \sin \omega t)$, the real and imaginary parts of a complex envelope can be totally different. The only constraint is that they relate to the same real signal.

A totally different use of the j operator that is quite separate from its application to complex envelopes is to use it as a mathematical tool for combining two *totally independent* real signals into a single complex signal. For example, all the information of the signals $a(t)$ and $b(t)$ is conveyed and can be extracted from the complex signal $c(t) = a(t) + jb(t)$. It can be useful in signal processing to exploit this possibility for speeding up the FFT evaluation of real signals.

17.11 Analytic signals

It has been shown in Equation (17.4) that a bandpass signal $x(t)$ centred around ω_0 can be expressed as

$$x(t) = x_I(t) \cos \omega_0 t - x_Q(t) \sin \omega_0 t \tag{17.68}$$

and in Equation (17.14) that a lowpass representation of the bandpass signal is given by the complex envelope

$$\tilde{u}(t) = x_I(t) + jx_Q(t) \tag{17.69}$$

The analytic signal representing $x(t)$ (also known as the pre-envelope of $x(t)$) is defined as

$$\tilde{x}(t) = \tilde{u}(t)e^{j\omega_0 t} \tag{17.70}$$

where the tilde is used to emphasize that $\tilde{x}(t)$ is complex, thus distinguishing it from the original (and real) signal $x(t)$.

Substituting Equation (17.69) into Equation (17.70) we obtain

$$\begin{aligned}\tilde{x}(t) &= \{x_I(t) + jx_Q(t)\}\{\cos \omega_0 t + j \sin \omega_0 t\} \\ &= x_I(t) \cos \omega_0 t - x_Q(t) \sin \omega_0 t + j[x_I(t) \sin \omega_0 t + x_Q(t) \cos \omega_0 t] \\ &= x(t) + j\hat{x}(t)\end{aligned} \tag{17.71}$$

where $\hat{x}(t)$ is identical to $x(t)$ except that all its frequency components are retarded by 90°. $\hat{x}(t)$ is known as the Hilbert transform of $x(t)$, and a physical system that achieves the necessary 90° retardation of all frequency components is known as a Hilbert transformer. It should be noted that achieving a 90° retardation of all frequency components is different from, and very much more difficult to achieve, than a constant time delay for all frequency components.

Equation (17.71) should not be confused with Equation (17.69). For the case of a bandpass signal the three terms of Equation (17.71), namely $\tilde{x}(t)$, $x(t)$ and $\hat{x}(t)$, are also all bandpass signals. In contrast, the three terms of Equation (17.69), namely $\tilde{u}(t)$, $x_I(t)$ and $x_Q(t)$, are all lowpass signals.

An interesting property of the analytic signal $\tilde{x}(t)$ is that its Fourier transform is entirely one-sided. This can be seen by combining Equations (17.69) and (17.70) as

$$\tilde{x}(t) = [x_I(t) + jx_Q(t)]e^{j\omega_0 t} \tag{17.72}$$

Recognizing that $x_I(t)$ and $x_Q(t)$ have low-frequency spectra and noting that multiplication of these terms by $e^{j\omega_0 t}$ represents an upconversion of these spectra such that they become centred at $+\omega_0$ it follows that the resultant spectra are thus non-zero for positive frequencies only.

In contrast to complex envelopes, analytic signals have little relevance to the signal processing objectives of this chapter. They are mentioned partly for completeness, but mostly to ensure that they are not confused with complex envelopes.

17.12 Summary

This chapter has shown that all the useful information of a bandpass signal is contained within its in-phase and quadrature components, both of which are lowpass signals that can readily be extracted. It has then been shown how these in-phase and quadrature components can be processed to achieve envelope detection, spectral analysis and bandpass filtering of the bandpass signal. The radar and sonar applications of Doppler measurement, beamforming and correlation detection are described.

17.13 Problems

1. A radar echo leaving the IF stage of a receiver is given by $x(t) = \text{rect}(t/T)(1 + \cos 2\pi t/T)\cos[2\pi(f_0 + f_D)t + \theta]$, where $T = 0.2\,\mu\text{s}$, $f_0 = 50$ MHz, and f_D is an unknown Doppler shift. Write a program in MATLAB showing that quadrature demodulation of the IF signal using 50 MHz local oscillators can retrieve the exact envelope irrespective of f_D. Base the program on Figure 17.10 and plot the input waveform, the waveforms after the two multipliers, and the output waveform. Assume the lowpass filters are third-order Butterworth filters with cutoff frequencies of 10 MHz.
2. The signal $\cos(2\pi \times 1000t)$ is contaminated by the interference signal $0.8\cos(2\pi \times 1060t)$ and it is wished to reduce the interference using a bandpass filter made up of two fourth-order 40 Hz lowpass filters in the arrangement of Figure 17.20. Assuming the oscillator frequencies are 1000 Hz, write a

program in MATLAB showing the waveforms at the input, after the first multipliers, after the two lowpass filters, and at the output. Note that there is a short 'start-up' transient before the output acquires its final form. Obtain a second set of plots assuming the oscillator frequencies are 1005 Hz.

3. It is wished to simulate a passive sonar beamformer working on the output of 16 hydrophones mounted 0.5 m apart in a line array. In MATLAB generate a random noise signal of 1591 samples 1/30 ms apart and pass this through a four-pole Butterworth bandpass filter whose passband extends from 1.5 kHz to 3 kHz. Selecting every third sample only, extract 512 samples of this filter output to represent the sampled signal on the first array element (that is, 512 samples 0.1 ms apart). Repeat for the other 15 elements but assume that, because the signal source is offset from the normal to the array, there is a delay of two samples between adjacent elements (that is, 1/15 ms). Perform an FFT on each of the element outputs. Plot the waveform and the amplitude spectrum of the signal on the first element. Next select the *n*th frequency coefficient from all 16 FFTs, apply a Hamming window across the array to reduce directional sidelobes, and then perform a *cross*-channel FFT to determine the *spatial* frequency present at this frequency. Assuming the velocity of sound in water is 1500 m/s, determine the relationship between spatial frequency and angle of incidence and thence plot signal amplitude as a function of angle of incidence. Confirm that the peak occurs at the corect angle corresponding to the delay between adjacent elements. Confirm that the angle stays the same for different frequencies between 1.5 kHz and 3 kHz, but that the beamwidth is wider as the frequency is reduced. (Note: for wideband noiselike signals, an alternative method for estimating angles of incidence is to determine the position of peaks in the cross-correlation functon between the outputs of two hydrophones.)

Appendix A

Listings of selected MATLAB programs

Figure 1.14

```
% fig1_14.m
t=(1:500);
x1=.8*sin(.5*t+.002*t.^2);                % chirp signal is generated for 0 < t < 500
x1(117:500)=zeros(size(117:500));         % long chirp signal is truncated
subplot(9,1,1)
plot(x1)
set(gca,'xtick',[],'ytick',[])
axis([0 500 -1 1])
text(-35,0,'(a)')
%
x2=.3*sin(.5*(t-300)+.002*(t-300).^2);    % a delayed chirp signal
x2(1:300)=zeros(size(1:300));             % first part of chirp signal truncated
x2(417:500)=zeros(size(417:500));         % last part of chirp signal truncated
subplot(9,1,2)
plot(x2)
set(gca,'xtick',[],'ytick',[])
axis([0 500 -1 1])
text(-35,0,'(b)')
%
n1=.12*randn(size(t));                    % this generates random numbers with a
                                          % normal distribution whose standard
                                          % deviation is 0.12.
b=1;
a=[1 -.7];
n2=filter(b,a,n1);                        % this lowpass filters the noise
subplot(9,1,3)
plot(n2)
set(gca,'xtick',[],'ytick',[])
axis([0 500 -1 1])
text(-35,0,'(c)')
```

```
%
s=x2+n2;                                % chirp added to noise
subplot(9,1,4)
plot(s)
set(gca,'xtick',[],'ytick',[])
axis([0 500 -1 1])
text(-35,0,'(d)')
%
x3=.8*sin(.5*(t-300)+.002*(t-300).^2);
x3(1:300)=zeros(size(1:300));
x3(417:500)=zeros(size(417:500));
subplot(9,1,5)
plot(x3)
set(gca,'xtick',[],'ytick',[])
axis([0 500 -1 1])
text(-35,0,'(e)')
%
x4=x2.*x3;                              % multiplication of signals
subplot(9,1,6)
plot(x4)
set(gca,'xtick',[],'ytick',[])
axis([0 500 -1 1])
text(-35,0,'(f)')
%
x5=n2.*x3;
subplot(9,1,7)
plot(x5)
set(gca,'xtick',[],'ytick',[])
axis([0 500 -1 1])
text(-35,0,'(g)')
%
x6=s.*x3;
subplot(9,1,8)
plot(x6)
set(gca,'xtick',[],'ytick',[])
axis([0 500 -1 1])
text(-35,0,'(h)')
%
x7=conv(fliplr(x1),s);                  % correlation by means of convolution
                                        % (refer to Chapter 16)
subplot(919)
plot(x7(500:999))
set(gca,'xtick',[],'ytick',[])
xlabel('Time')
text(-35,0,'(i)')
set(gcf,'Paperposition',[0.2 1 7 10])
```

```
% this last instruction does not change the image on the computer screen but
% specifies the location and size of the MATLAB figure window as it appears
% in the hardcopy from the printer (note that the figure window is larger than
% the plots within the figure window). The default units are inches. Here the
% dimensions are 7 inches by 10 inches, and the bottom left-hand corner is one
% inch up and 0.2 inches from the left. This instruction need not be inside the
% program but can, if preferred, be executed after running the program prior to
% printing. It should be noted that in the current session of MATLAB this
% instruction will remain for any subsequent prints arising from other programs
% unless modified by a new set(gcf,...) command or else deleted by the command
% reset(gcf) to return to default values.
```

Figure 3.2

```
% fig3_2.m
t=[0:500];
y=zeros(size(t));                  % it is useful to plot a zero line
subplot(3,1,1)
plot(t,0.8*cos(t/50),'-',t,.5*sin(t/50),':',t,y,'-')
set(gca,'xtick',[],'ytick',[])
axis([0 500 -1 1])
text(25,.82,'voltage')
text(130, .39,'current')
xlabel('Time')
```

Figure 3.3

```
% fig3_3.m
t=[0:500];
y=zeros(size(t));                  % it is useful to plot a zero line
subplot(3,1,1)
plot(t,cot(t/50),t,y,'-')
set(gca,'xtick',[],'ytick',[])
axis([0 500 -10 10])               % large values of cot(t/50) are excluded
                                   % from the plot by constraining the
                                   % vertical axis
xlabel('Time')
ylabel('v/i')
```

Figure 5.19

```
% fig5_19.m
wt=(0:500)/50;
a=4/pi*cos(wt);
b=-4/pi/3*cos(3*wt);
c=4/pi/5*cos(5*wt);
```

```
d=-4/pi/7*cos(7*wt);
e=4/pi/9*cos(9*wt);
x=zeros(size(wt));
% we now use the same window pane for all the plots including
% zero lines for each. We do this by applying offsets.
plot(wt,(a+12),wt,(x+12),wt,(b+10),wt,(x+10),wt,(c+8),wt,(x+8),...
wt,(d+6),wt,(x+6),wt,(e+4),wt,(x+4),wt,(a+b+c+d+e),wt,x)
axis off
text(-1.6,12,'  cos(wt)')
text(-1.6,10,'- cos(3wt)')
text(-1.6,8,'  cos(5wt)')
text(-1.6,6,'- cos(7wt)')
text(-1.6,4,'  cos(9wt)')
text(-1.6,0,'  Sum')
```

Figure 6.1

```
% fig6_1.m
t=(-200:200);
x=zeros(size(t));
x(170:230)=ones(1,61);
subplot(2,1,1)
y=x.*cos(t/1.5);
plot(t,x+2,t,y)
set(gca,'XTick',[],'YTick',[])
axis([-200 200 -2 4])
```

Figure 8.4

```
% fig8_4.m
clf                          % Used to clear any previous plots
                             % from the active figure window.

t=(0:511)/64;
a=zeros(size(t));
b=ones(size(t));
n=`(0:7);
d=ones(size(n));
subplot(8,1,1)
stem(n,d)
hold on
plot(t,b,':',t,a,'-')
axis([0 7 -1 1])
axis off
subplot(8,1,2)
```

```
stem(n,cos(pi*n/4))
hold on
plot(t,cos(pi*t/4),':',t,a,'-')
axis([0 7 -1 1])
axis off
subplot(8,1,3)
stem(n,sin(pi*n/4))
hold on
plot(t,sin(pi*t/4),':',t,a,'-')
axis([0 7 -1 1])
axis off
subplot(8,1,4)
stem(n,cos(pi*n/2))
hold on
plot(t,cos(pi*t/2),':',t,a,'-')
axis([0 7 -1 1])
axis off
subplot(8,1,5)
stem(n,sin(pi*n/2))
hold on
plot(t,sin(pi*t/2),':',t,a,'-')
axis([0 7 -1 1])
axis off
subplot(8,1,6)
stem(n,cos(pi*n*3/4))
hold on
plot(t,cos(pi*t*3/4),':',t,a,'-')
axis([0 7 -1 1])
axis off
subplot(8,1,7)
stem(n,sin(pi*n*3/4))
hold on
plot(t,sin(pi*t*3/4),':',t,a,'-')
axis([0 7 -1 1])
axis off
subplot(8,1,8)
stem(n,cos(pi*n))
hold on
plot(t,cos(pi*t),':',t,a,'-')
axis([0 7 -1 1])
axis off
```

Figure 12.31

```
% fig12_31.m
clf
wnt=(0:500)/30;
for zeta=[.1 .2 .4 .7 .99]
x=sqrt(1-zeta^2);
c=1-(1/x)*exp(-zeta*wnt).*cos(x*wnt-atan(zeta/x));
plot(wnt,c)
axis([0 16 0 2])
hold on
end
text(1.7, 1.86, 'damping')
text(1.7, 1.78, 'ratio =')
text(3.1, 1.78, '.1')
text(3.1, 1.58, '.2')
text(3.1, 1.3, '.4')
text(3.1, 1.05, '.7')
text(3.1, 0.88, '1')
ylabel('c(t)'), xlabel('wnt')
```

Figure 15.12

```
% fig15_12.m
clf
b=[.0435 .0318 0 0]; a=[1 -2.0203 1.464 -.3678];
[h,f]=freqz(b,a,200,2);
subplot(2,1,1)
plot(f,20*log10(abs(h)))
set(gca,'XLim',[0 1],'YLim',[-100 0])
xlabel('Frequency (Hz)');ylabel('Magnitude Response (dB)')
grid on
b=[1 3 3 1]; a=[105 -213 155 -39];
[h,f]=freqz(b,a,200,2);
subplot(2,1,2)
plot(f,20*log10(abs(h)))
set(gca,'XLim',[0 1],'YLim',[-100 0])
xlabel('Frequency (Hz)');ylabel('Magnitude Response (dB)')
grid on
```

Bibliography

Some of the following books have been referred to in the text. The remaining few have been selected because of their high quality and because they provide an extension of the material presented.

Ambardar, A. (1995) *Analog and Digital Signal Processing*, PWS Publishing Company, Boston.

Bode, H.W. (1945) *Network Analysis and Feedback Amplifier Design*, van Nostrand, New York.

Brigham, E.O. (1988) *The Fast Fourier Transform and its Applications*, Prentice Hall, Englewood Cliffs, NJ.

Geffe, P. (1963) *Simplified Modern Filter Design*, John F. Rider, New York.

Ifeachor, C.E. and Jervis, B.W. (1993) *Digital Signal Processing, A Practical Approach*, Addison-Wesley, Reading, MA.

MATLAB (1995) *The Student Edition of Matlab, Version 4*, Prentice Hall, Englewood Cliffs, NJ.

MATLAB (1997) *The Student Edition of Matlab, Version 5*, Prentice Hall, Englewood Cliffs, NJ.

Nise, N.S. (1992) *Control Systems Engineering*, Benjamin/Cummings, Redwood City, CA.

Proakis, J.G. and Manolakis, D.G. (1996) *Digital Signal Processing: Principles, Algorithms, and Applications*, 3rd edn, Prentice Hall, Englewood Cliffs, NJ.

Williams, A.B. and Taylor, F.J. (1988) *Electronic Filter Design Handbook*, 2nd edn, McGraw-Hill, London.

Index